Ingredients and Integration of Vision Systems

This functional flowchart illustrates the many influences impinging upon integrated vision system design.

Praise for

Machine Vision

THEORY, ALGORITHMS, PRACTICALITIES

3RD EDITION

"This book brings together the analytic aspects of image processing with the practicalities of applying the techniques in an industrial setting. It is excellent grounding for a machine vision researcher."

—*John Billingsley, University of Southern Queensland*

"The book in its previous incarnations has established its place as a unique repository of detailed analysis of important image processing and computer vision algorithms. This edition builds on these strengths and adds material to guide the reader's understanding of the latest developments in the field. The result is a comprehensive up-to-date reference text."

—*Farzin Deravi, University of Kent*

"This book is an essential reference for anyone developing techniques for machine vision analysis, including systems for industrial inspection, biomedical analysis, and much more. It comes from a long-term practitioner and is packed with the fundamental techniques required to build and prototype methods to test their applicability to the problem at hand."

—*Majid Mirmehdi, University of Bristol*

"The book contains a large number of experimental design and evaluation procedures that are of keen interest to industrial application engineers of machine vision."

—*William Wee, University of Cincinnati*

About the Author

Professor E.R. Davies graduated from Oxford University in 1963 with a First Class Honours degree in Physics. After 12 years research in solid state physics, he became interested in vision and is currently professor of machine vision at Royal Holloway, University of London. He has worked on image filtering, shape analysis, edge detection, Hough transforms, robust pattern matching, and artificial neural networks and is involved in algorithm design for inspection, surveillance, real-time vision, and a number of industrial applications. He has published over 180 papers and two books—*Electronics, Noise and Signal Recovery* (1993), and *Image Processing for the Food Industry* (2000)—as well as the present volume. Professor Davies is on the Editorial Boards of *Real-Time Imaging*, *Pattern Recognition Letters* and *Imaging Science*. He holds a DSc at the University of London and is a Fellow of the Institute of Physics (UK) and the Institution of Electrical Engineering (USA), and a Senior Member of the Institute of Electrical and Electronics Engineering (USA). He is on the Executive Committees of the British Machine Vision Association and the IEE Visual Information Engineering Professional Network.

Machine Vision

THEORY, ALGORITHMS, PRACTICALITIES

3RD EDITION

Machine Vision

THEORY, ALGORITHMS, PRACTICALITIES

3RD EDITION

E. R. DAVIES
Department of Physics
Royal Holloway
University of London

ELSEVIER

AMSTERDAM • BOSTON • HEIDELBERG • LONDON
NEW YORK • OXFORD • PARIS • SAN DIEGO
SAN FRANCISCO • SINGAPORE • SYDNEY • TOKYO

Morgan Kaufmann Publishers is an imprint of Elsevier

Publisher	Denise E.M. Penrose
Publishing Services Manager	Simon Crump
Senior Project Manager	Angela Dooley
Editorial Assistant	Valerie Witte
Cover Design	Ross Carron
Cover Image	Dynamic Abstract © Digital Vision
Text Design	Chen Design
Composition	CEPHA
Copyeditor	Betty Passagno
Proofreader	Phyllis Coyne et al.
Imdexer	Northwind Editorial
Interior printer	The Maple-Vail Book Manufacturing Group
Cover printer	Phoenix Color Corp.

Morgan Kaufmann Publishers is an imprint of Elsevier.
500 Sansome Street, Suite 400, San Francisco, CA 94111

This book is printed on acid-free paper.

Library of Congress Cataloging-in-Publication Data
Application submitted

ISBN: 0-12-206093-8

For all information on all Morgan Kauffmann publications,
visit our Web site at www.mkp.com or www.books.elsevier.com

Printed in the United States of America
04 05 06 07 08 5 4 3 2 1

Dedication

This book is dedicated to my family.

To my late mother, Mary Davies, to record her never-failing love and devotion.

To my late father, Arthur Granville Davies, who passed on to me his appreciation of the beauties of mathematics and science.

To my wife, Joan, for love, patience, support, and inspiration.

To my children, Elizabeth, Sarah, and Marion, the music in my life.

Contents

CHAPTER 3
Basic Image Filtering Operations

CHAPTER 4
Thresholding Techniques

CHAPTER 5
Edge Detection

CHAPTER 6
Binary Shape Analysis

CHAPTER 7
Boundary Pattern Analysis

CHAPTER 8
Mathematical Morphology

CHAPTER 25
Biologically Inspired Recognition Schemes

CHAPTER 26
Texture

CHAPTER 27
Image Acquisition

CHAPTER 28
Real-time Hardware and Systems Design Considerations

PART 5
PERSPECTIVES ON VISION 831

CHAPTER 29
Machine Vision: Art or Science?

APPENDIX
Robust Statistics

Foreword

An important focus of advances in mechatronics and robotics is the addition of sensory inputs to systems with increasing "intelligence." Without doubt, sight *is* the "sense of choice." In everyday life, whether driving a car or threading a needle, we depend first on sight. The addition of visual perception to machines promises the greatest improvement and at the same time presents the greatest challenge.

Until relatively recently, the volume of data in the images that make up a video stream has been a serious deterrent to progress. A single frame of very modest resolution might occupy a quarter of a megabyte, so the task of handling thirty or more such frames per second requires substantial computer resources.

Fortunately, the computer and communications industries' investment in entertainment has helped address this challenge. The transmission and processing of video signals are an easy justification for selling the consumer increased computing speed and bandwidth. A digital camera, capable of video capture, has already become a fashion accessory as part of a mobile phone. As a result, video signals have become more accessible to the serious engineer. But the task of acquiring a visual image is just the tip of the iceberg.

While generating sounds and pictures is a well-defined process (speech generation is a standard "accessibility" feature of Windows), the inverse task of recognizing connected speech is still at an unfinished state, a quarter of a century later, as any user of "dictation" software will attest. Still, analyzing sound is not even in the same league with analyzing images, particularly when they are of real-world situations rather than staged pieces with synthetic backgrounds and artificial lighting.

The task is essentially one of data reduction. From the many megabytes of the image stream, the required output might be a simple "All wheel nuts are in place" or "This tomato is ripe." But images tend to be noisy, objects that look sharp to the eye can have broken edges, boundaries can be fuzzy, and straight lines can be illusory. The task of image analysis demands a wealth of background know-how and mathematical analytic tools.

Roy Davies has been developing that rich background for well over two decades. At the time of the UK Robotics Initiative, in the 1980s, Roy had formed a relationship with the company United Biscuits. We fellow researchers might well have been amused by the task of ensuring that the blob of jam on a "Jaffacake" had been placed centrally beneath the enrobing chocolate. However, the funding of the then expensive vision acquisition and analysis equipment, together with the spur of a practical target of real economic value, gave Roy a head start that made us all envious.

Grounded in that research is the realization that human image analysis is a many-layered process. It starts with simple graphical processing, of the sort that our eyes perform without our conscious awareness. Contrasts are enhanced; changes are emphasized and brought to our attention. Next we start to code the image in terms of lines, the curves of the horizon or of a face, boundaries between one region and another that a child might draw with a crayon. Then we have to "understand" the shape – is it a broken biscuit, or is one biscuit partly hidden by another? If we are comparing a succession of images, has something moved? What action should we take?

Machine vision must follow a similar, multilayered path. Roy has captured the essentials with clear, well-illustrated examples. He demonstrates filters that by convolution can smooth or sharpen an image. He shows us how to wield the tool of the Hough transform for locating lines and boundaries that are made indistinct by noise. He throws in the third dimension with stereo analysis, structured lighting, and optical flow. At every step, however, he drags us back to the world of reality by considering some practical task. Then, with software examples, he challenges us to try it for ourselves.

The easy access to digital cameras, video cameras, and streaming video in all its forms has promoted a tidal wave of would-be applications. But an evolving and substantial methodology is still essential to underpin the "art" of image processing. With this latest edition of his book, Roy continues to surf the crest of that wave.

John Billingsley
University of Southern Queensland

Preface

Preface to the third edition

The first edition came out in 1990, and was welcomed by many researchers and practitioners. However, in the space of 14 years the subject has moved on at a seemingly accelerating rate, and topics which hardly deserved a mention in the first edition had to be considered very seriously for the second and more recently for the third edition. It seemed particularly important to bring in significant amounts of new material on mathematical morphology, 3-D vision, invariance, motion analysis, artificial neural networks, texture analysis, X-ray inspection and foreign object detection, and robust statistics. There are thus new chapters or appendices on these topics, and these have been carefully integrated with the existing material. The greater proportion of this new material has been included in Parts 3 and 4. So great has been the growth in work on 3-D vision and motion—coupled with numerous applications on (for example) surveillance and vehicle guidance—that the original single chapter on 3-D vision has had to be expanded into the set of *six* chapters on 3-D vision and motion forming Part 3. In addition, Part 4 encompasses an increased range of chapters whose aim is to cover both applications and all the components needed for design of real-time visual pattern recognition systems. In fact, it is difficult to design a logical ordering for all the chapters—particularly in Part 4—as the topics interact with each other at a variety of different levels—theory, algorithms, methodologies, practicalities, design constraints, and so on. However, this should not matter, as the reader will be exposed to the essential richness of the subject, and his/her studies should be amply rewarded by increased understanding and capability.

A typical final-year undergraduate course on vision for Electronic Engineering or Computer Science students might include much of the work of Chapters 1 to 10 and 14 to 16, and a selection of sections from other chapters, according to requirements. For MSc or PhD research students, a suitable lecture course might go on to cover Part 3 in depth, including some chapters in Part 4 and also the appendix on robust statistics,[1] with many practical exercises being undertaken on

1 The importance of this appendix should not be underestimated once one gets onto serious work, though this will probably be outside the restrictive environment of an undergraduate syllabus.

an image analysis system. Here much will depend on the research program being undertaken by each individual student. At this stage the text will have to be used more as a handbook for research, and, indeed, one of the prime aims of the volume is to act as a handbook for the researcher and practitioner in this important area.

As mentioned above, this book leans heavily on experience I have gained from working with postgraduate students: in particular I would like to express my gratitude to Barry Cook, Mark Edmonds, Simon Barker, Daniel Celano, Darrel Greenhill, and Mark Sugrue, all of whom have in their own ways helped to shape my view of the subject. In addition, it is a special pleasure to recall very many rewarding discussions with my colleagues Zahid Hussain, Ian Hannah, Dev Patel, David Mason, Mark Bateman, Tieying Lu, Adrian Johnstone, and Piers Plummer; the last two named having been particularly prolific in generating hardware systems for implementing my research group's vision algorithms.

The author owes a debt of gratitude to John Billingsley, Kevin Bowyer, Farzin Deravi, Cornelia Fermuller, Martial Hebert, Majid Mirmehdi, Qiang Ji, Stan Sclaroff, Milan Sonka, Ellen Walker, and William G. Wee, all of whom read a draft of the manuscript and made a great many insightful comments and suggestions. The author is particularly sad to record the passing of Professor Azriel Rosenfeld before he could complete his reading of the last few chapters, having been highly supportive of this volume in its various editions. Finally, I am indebted to Denise Penrose of Elsevier Science for her help and encouragement, without which this third edition might never have been completed.

Supporting Materials:

Morgan Kaufmann's website for the book contains resources to help instructors teach courses using this text. Please check the publishers' website for more information:

www.mkp.com/companions/0122060938

E.R. DAVIES
Royal Holloway,
University of London

Preface to the first edition

Over the past 30 years or so, machine vision has evolved into a mature subject embracing many topics and applications: these range from automatic (robot) assembly to automatic vehicle guidance, from automatic interpretation of documents to verification of signatures, and from analysis of remotely sensed images to checking of fingerprints and human blood cells; currently, automated visual inspection is undergoing very substantial growth, necessary improvements in quality, safety and cost-effectiveness being the stimulating factors. With so much ongoing activity, it has become a difficult business for the professional to keep up with the subject and with relevant methodologies: in particular, it is difficult for him to distinguish accidental developments from genuine advances. It is the purpose of this book to provide background in this area.

The book was shaped over a period of 10–12 years, through material I have given on undergraduate and postgraduate courses at London University and contributions to various industrial courses and seminars. At the same time, my own investigations coupled with experience gained while supervising PhD and postdoctoral researchers helped to form the state of mind and knowledge that is now set out here. Certainly it is true to say that if I had had this book 8, 6, 4, or even 2 years ago, it would have been of inestimable value to myself for solving practical problems in machine vision. It is therefore my hope that it will now be of use to others in the same way. Of course, it has tended to follow an emphasis that is my own—and in particular one view of one path towards solving automated visual inspection and other problems associated with the application of vision in industry. At the same time, although there is a specialism here, great care has been taken to bring out general principles—including many applying throughout the field of image analysis. The reader will note the universality of topics such as noise suppression, edge detection, principles of illumination, feature recognition, Bayes' theory, and (nowadays) Hough transforms. However, the generalities lie deeper than this. The book has aimed to make some general observations and messages about the limitations, constraints, and tradeoffs to which vision algorithms are subject. Thus there are themes about the effects of noise, occlusion, distortion, and the need for built-in forms of robustness (as distinct from less successful *ad hoc* varieties and those added on as an afterthought); there are also themes about accuracy, systematic design, and the matching of algorithms and architectures. Finally, there are the problems of setting up lighting schemes which must be addressed in complete systems, yet which receive scant attention in most books on image processing and analysis. These remarks will indicate that the text is intended to be read at various levels—a factor that should make it of more lasting value than might initially be supposed from a quick perusal of the Contents.

Of course, writing a text such as this presents a great difficulty in that it is necessary to be highly selective: space simply does not allow everything in a subject of this nature and maturity to be dealt with adequately between two covers. One solution might be to dash rapidly through the whole area mentioning everything that comes to mind, but leaving the reader unable to understand anything in detail or to *achieve* anything having read the book. However, in a practical subject of this nature this seemed to me a rather worthless extreme. It is just possible that the emphasis has now veered too much in the opposite direction, by coming down to practicalities (detailed algorithms, details of lighting schemes, and so on): individual readers will have to judge this for themselves. On the other hand, an author has to be true to himself and my view is that it is better for a reader or student to have mastered a coherent series of topics than to have a mish-mash of information that he is later unable to recall with any accuracy. This, then, is my justification for presenting this particular material in this particular way and for reluctantly omitting from detailed discussion such important topics as texture analysis, relaxation methods, motion, and optical flow.

As for the organization of the material, I have tried to make the early part of the book lead into the subject gently, giving enough detailed algorithms (especially in Chapters 2 and 6) to provide a sound feel for the subject—including especially vital, and in their own way quite intricate, topics such as connectedness in binary images. Hence Part 1 provides the lead-in, although it is not always trivial material and indeed some of the latest research ideas have been brought in (e.g., on thresholding techniques and edge detection). Part 2 gives much of the meat of the book. Indeed, the (book) literature of the subject currently has a significant gap in the area of intermediate-level vision; while high-level vision (AI) topics have long caught the researcher's imagination, intermediate-level vision has its own difficulties which are currently being solved with great success (note that the Hough transform, originally developed in 1962, and by many thought to be a very specialist topic of rather esoteric interest, is arguably only now coming into its own). Part 2 and the early chapters of Part 3 aim to make this clear, while Part 4 gives reasons why this particular transform has become so useful. As a whole, Part 3 aims to demonstrate some of the practical applications of the basic work covered earlier in the book, and to discuss some of the principles underlying implementation: it is here that chapters on lighting and hardware systems will be found. As there is a limit to what can be covered in the space available, there is a corresponding emphasis on the theory underpinning practicalities. Probably this is a vital feature, since there are many applications of vision both in industry and elsewhere, yet listing them and their intricacies risks dwelling on interminable detail, which some might find insipid; furthermore, detail has a tendency to date rather rapidly. Although the book could not cover 3-D vision in full (this topic would easily consume a whole volume in its own right), a careful overview of this complex mathematical and highly important subject seemed vital. It is therefore no accident that Chapter 16 is the longest in the book. Finally, Part 4 asks questions about the limitations and constraints of vision algorithms and answers them by drawing on information and experience from earlier chapters. It is tempting to call the last chapter the Conclusion. However, in such a dynamic subject area any such temptation has to be resisted, although it has still been possible to draw a good number of lessons on the nature and current state of the subject. Clearly, this chapter presents a personal view but I hope it is one that readers will find interesting and useful.

Acknowledgments

The author would like to credit the following sources for permission to reproduce tables, figures and extracts of text from earlier publications:

The Committee of the Alvey Vision Club for permission to reprint portions of the following paper as text in Chapter 15 and as Figs. 15.1, 15.2, 15.6:

Davies, E.R. (1988g).

CEP Consultants Ltd (Edinburgh) for permission to reprint portions of the following paper as text in Chapter 19:

Davies, E.R. (1987a).

Portions of the following papers from *Pattern Recognition and Signal Processing* as text in Chapters 3, 5, 9–15, 23, 24 and tables 5.1, 5.2, 5.3 and as Figs. 3.7, 3.9, 3.11, 3.12, 3.13, 3.17, 3,18, 3.19, 3.20, 3.21, 3.22, 5.2, 5.3, 5.4, 5.8, 9.1, 9.2, 9.3, 9.4, 10.4–10.14, 11.1, 11.3–11.11, 12.5–12.11, 13.1–13.5, 14.1–14.3, 14.6, 14.7, 14.8, 15.8, 23.2, 23.5:

Davies, E.R. (1986a,c, 1987c,d,g,k,l, 1988b,c,e,f, 1989 a–d).

Portions of the following papers from Image and Vision Computing as text in Chapter 5, as Tables 5.4, 5.5, 5.6 and as Figs. 3.33, 5.1, 5.5, 5.6, 5.7:

Davies, E.R. (1984b, 1987e)

Figures 5, 9; and about 10% of the text in the following paper:

Davies, E.R., Bateman, M., Mason, D.R., Chambers, J. and Ridgway, C. (2003). "Design of efficient line segment detectors for cereal grain inspection." Pattern Recogn. Lett., 24, nos. 1–3, 421–436.

Figure 4 of the following article:

Davies, E.R. (1987) "Visual inspection, automatic (robotics)", in *Encyclopedia of Physical Science and Technology*, Vol. 14, Academic Press, pp. 360–377.

About 4% of the text in the following paper:

Davies, E.R. (2003). "An analysis of the geometric distortions produced by median and related image processing filters." *Advances in Imaging and Electron Physics*, 126, Academic Press, pp. 93–193.

The above five items reprinted with permission from ***Elsevier***.

IEE for permission to reprint portions of the following papers from the IEE Proceedings and Colloquium Digests as text in Chapters 14, 22, 25, and as Figs. 14.13–14.15, 22.2–22.6, 25.13:

Davies, E.R. (1985b, 1988a)
Davies, E.R. and Johnstone, A.I.C. (1989)
Greenhill, D. and Davies, E.R. (1994b)

and

Figures 1–5; and about 44% of the text in the following paper:

Davies, E.R. (1997). Principles and design graphs for obtaining uniform illumination in automated visual inspection. Proc. 6th IEE Int. Conf. on *Image Processing and its Applications*, Dublin (14–17 July), IEE Conf. Publication no. 443, pp. 161–165.

Table 1; and about 95% of the text in the following paper:

Davies, E.R. (1997). Algorithms for inspection: constraints, tradeoffs and the design process. IEE Digest no. 1997/041, Colloquium on *Industrial Inspection*, IEE (10 Feb.), pp. 6/1–5.

Figure 1; and about 35% of the text in the following paper:

Davies, E.R., Bateman, M., Chambers, J. and Ridgway, C. (1998). Hybrid non-linear filters for locating speckled contaminants in grain. IEE Digest no. 1998/284, Colloquium on *Non-Linear Signal and Image Processing*, IEE (22 May), pp. 12/1–5.

Table 1; Figures 4–8; and about 15% of the text in the following paper:

Davies, E.R. (1999). Image distortions produced by mean, median and mode filters. IEE Proc. Vision Image Signal Processing, **146**, no. 5, 279–285.

Figure 1; and about 45% of the text in the following paper:

Davies, E.R. (2000). Obtaining optimum signal from set of directional template masks. Electronics Lett., **36**, no. 15, 1271–1272.

Figures 1–3; and about 90% of the text in the following paper:

Davies, E.R. (2000). Resolution of problem with use of closing for texture segmentation. Electronics Lett., **36**, no. 20, 1694–1696.

IEEE for permission to reprint portions of the following paper as text in Chapter 14 and as Figs. 3.5, 3.6, 3.8, 3.12, 14.4, 14.5:

Davies, E.R. (1984a, 1986d).

IFS Publications Ltd for permission to reprint portions of the following paper as text in Chapter 22 and as Figs. 10.1, 10.2, 22.7:

Davies, E.R. (1984c).

The Council of the Institution of Mechanical Engineers for permission to reprint portions of the following paper as Tables 28.1, 28.2:

Davies. E.R. and Johnstone, A.I.C. (1986).

MCB University Press (Emerald Group) for permission to reproduce Plate 1 of the following paper as Fig. 22.8:

Patel, D., Davies, E.R. and Hannah, I. (1995).

Pergamon (Elsevier Science Ltd) for permission to reprint portions of the following articles as text in Chapters 6, 7 and as Figs. 3.16, 6.7, 6.24:

Davies, E.R. and Plummer, A.P.N. (1981).
Davies, E.R. (1982).

Royal Swedish Academy of Engineering Sciences for permission to reprint portions of the following paper as text in Chapter 22 and as Fig. 7.4:

Davies, E.R. (1987f).

World Scientific for permission to reprint:

Figures 2.7(a, b), 3.4–3.6, 10.4, 12.1–12.3, 14.1; plus about 3 pp. of text extracted from p. 49 et seq; about 1 p. of text extracted from p. 132 et seq; about 2 pp. of text extracted from p. 157 et seq; about 3 pp. of text extracted from p. 187 et seq; about 7 pp. of text extracted from p. 207 et seq; about 3 pp. of text extracted from p. 219 et seq, in the following book:

Davies, E.R. (2000). *Image Processing for the Food Industry.*

in all cases excluding text which originated from earlier publications

The Royal Photographic Society for permission to reprint:

Table 1; figures 1, 2, 7, 9; and about 40% of the text in the following paper:

Davies, E.R. (2000). A generalized model of the geometric distortions produced by rank-order filters. Imaging Science Journal, **48**, no. 3, 121–130.

Figures 4(b,e,h), 5(b), 14(a,b,c) in the following paper:

Charles, D. and Davies, E.R. (2004). Mode filters and their effectiveness for processing colour images. Imaging Science Journal, **52**, no. 1, 3–25.

EURASIP for permission to reprint:

Figures 1–7; and about 65% of the text in the following paper:

Davies, E.R. (1998). Rapid location of convex objects in digital images. Proc. EUSIPCO'98, Rhodes, Greece (8–11 Sept.), pp. 589–592.

Figure 1; and about 30% of the text in the following paper:

Davies, E.R., Mason, D.R., Bateman, M., Chambers, J. and Ridgway, C. (1998). Linear feature detectors and their application to cereal inspection. Proc. EUSIPCO'98, Rhodes, Greece (8–11 Sept.), pp. 2561–2564.

CRC Press for permission to reprint:

Table 1; Figures 8, 10, 12, 15–17; and about 15% of the text in the following chapter:

Davies (2003): Chapter 7 in Cho, H. (ed.) (2003). *Opto-Mechatronic Systems Handbook*.

Springer-Verlag for permission to reprint:

about 4% of the text in the following chapter:

Davies, E.R. (2003). Design of object location algorithms and their use for food and cereals inspection. Chapter 15 in Graves, M. and Batchelor, B.G. (eds.) (2003). Machine Vision Techniques for Inspecting Natural Products, 393–420.

Unicom Seminars Ltd. for permission to reprint:

Portions of the following paper as text in Chapter 22 and as Figs. 9.2, 10.3, 12.3, 14.15, 15.9:

Davies, E.R. (1988h).

Royal Holloway, University of London for permission to reprint:

Extracts from the following examination questions, written by E.R. Davies:

EL333/01/2, 4–6; EL333/98/2; EL333/99/2, 3, 5, 6; EL385/97/2; PH4760/04/1–5; PH5330/98/3, 5; PH5330/03/1–5.

University of London for permission to reprint:

Extracts from the following examination questions, written by E.R. Davies:

PH385/92/2, 3; PH385/93/1–3; PH385/94/1–4; PH385/95/4; PH385/96/3, 6; PH433/94/3, 5; PH433/96/2, 5.

CHAPTER 1

Vision, the Challenge

1.1 Introduction—The Senses

Of the five senses—vision, hearing, smell, taste, and touch—vision is undoubtedly the one that we have come to depend upon above all others, and indeed the one that provides most of the data we receive. Not only do the input pathways from the eyes provide megabits of information at each glance, but the data rate for continuous viewing probably exceed 10 megabits per second. However, much of this information is redundant and is compressed by the various layers of the visual cortex, so that the higher centers of the brain have to interpret abstractly only a small fraction of the data. Nonetheless, the amount of information the higher centers receive from the eyes must be at least two orders of magnitude greater than all the information they obtain from the other senses.

Another feature of the human visual system is the ease with which interpretation is carried out. We see a scene as it is—trees in a landscape, books on a desk, widgets in a factory. No obvious deductions are needed, and no overt effort is required to interpret each scene. In addition, answers are effectively immediate and are normally available within a tenth of a second. Just now and again some doubt arises—for example, a wire cube might be "seen" correctly or inside out. This and a host of other optical illusions are well known, although for the most part we can regard them as curiosities—irrelevant freaks of nature. Somewhat surprisingly, it turns out that illusions are quite important, since they reflect hidden assumptions that the brain is making in its struggle with the huge amounts of complex visual data it is receiving. We have to bypass this topic here (though it surfaces now and again in various parts of this book). However, the important point is that we are for the most part unaware of the complexities of vision. Seeing is not a simple process: it is just that vision has evolved over millions of years, and there was no particular advantage in evolution giving us any indication of the difficulties of the task. (If anything, to have done so would have cluttered our minds with worthless information and quite probably slowed our reaction times in crucial situations.)

Thus, ignorance of the process of human vision abounds. However, being by nature inventive, the human species is now trying to get machines to do much of its work. For the simplest tasks there should be no particular difficulty in mechanization, but for more complex tasks the machine must be given our prime sense, that of vision. Efforts have been made to achieve this, sometimes in modest ways, for well over 30 years. At first such tasks seemed trivial, and schemes were devised for reading, for interpreting chromosome images, and so on. But when such schemes were challenged with rigorous practical tests, the problems often turned out to be more difficult. Generally, researchers react to their discovery that apparent "trivia" are getting in the way by intensifying their efforts and applying great ingenuity. This was certainly the case with some early efforts at vision algorithm design. Hence, it soon became plain that the task is a complex one, in which numerous fundamental problems confront the researcher, and the ease with which the eye can interpret scenes has turned out to be highly deceptive.

Of course, one way in which the human visual system surpasses the machine is that the brain possesses some 10^{10} cells (or neurons), some of which have well over 10,000 contacts (or synapses) with other neurons. If each neuron acts as a type of microprocessor, then we have an immense computer in which all the processing elements can operate concurrently. Probably, the largest single man-made computer still contains less than 100 million processing elements, so the majority of the visual and mental processing tasks that the eye–brain system can perform in a flash have no chance of being performed by present-day man-made systems. Added to these problems of scale is the problem of how to organize such a large processing system and how to program it. Clearly, the eye–brain system is partly hard-wired by evolution, but there is also an interesting capability to program it dynamically by training during active use. This need for a large parallel processing system with the attendant complex control problems clearly illustrates that machine vision must indeed be one of the most difficult intellectual problems to tackle.

So what are the problems involved in vision that make it apparently so easy for the eye and yet so difficult for the machine? In the next few sections we attempt to answer this question.

1.2 The Nature of Vision

1.2.1 *The Process of Recognition*

This section illustrates the intrinsic difficulties of implementing machine vision, starting with an extremely simple example—that of character recognition. Consider the set of patterns shown in Fig. 1.1*a*. Each pattern can be considered as a set of

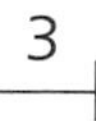

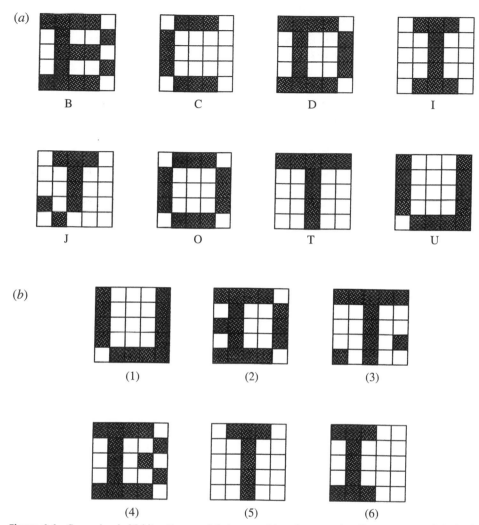

Figure 1.1 Some simple 25-bit patterns and their recognition classes used to illustrate some of the basic problems of recognition: (*a*) training set patterns (for which the known classes are indicated); (*b*) test patterns.

25 bits of information, together with an associated class indicating its interpretation. In each case, imagine a computer learning the patterns and their classes by rote. Then any new pattern may be classified (or "recognized") by comparing it with this previously learned "training set," and assigning it to the class of the nearest pattern in the training set. Clearly, test pattern (1) (Fig. 1.1*b*) will be allotted to class U on this basis. Chapter 24 shows that this method is a simple form of the nearest-neighbor approach to pattern recognition.

The scheme we have outlined in Fig. 1.1 seems straightforward and is indeed highly effective, even being able to cope with situations where distortions of the test patterns occur or where noise is present: this is illustrated by test patterns (2) and (3). However, this approach is not always foolproof. First, in some situations distortions or noise is so excessive that errors of interpretation arise. Second, patterns may not be badly distorted or subject to obvious noise and, yet, are misinterpreted: this situation seems much more serious, since it indicates an unexpected limitation of the technique rather than a reasonable result of noise or distortion. In particular, these problems arise where the test pattern is displaced or erroneously oriented relative to the appropriate training set pattern, as with test pattern (6).

As will be seen in Chapter 24, a powerful principle is operative here, indicating why the unlikely limitation we have described can arise: it is simply that there are *insufficient training set patterns* and that those that are present are *insufficiently representative* of what will arise during tests. Unfortunately, this presents a major difficulty inasmuch as providing enough training set patterns incurs a serious storage problem, and an even more serious search problem when patterns are tested. Furthermore, it is easy to see that these problems are exacerbated as patterns become larger and more real. (Obviously, the examples of Fig. 1.1 are far from having enough resolution even to display normal type fonts.) In fact, a combinatorial explosion[1] takes place. Forgetting for the moment that the patterns of Fig. 1.1 have familiar shapes, let us temporarily regard them as random bit patterns. Now the number of bits in these $N \times N$ patterns is N^2, and the number of possible patterns of this size is 2^{N^2}: even in a case where $N = 20$, remembering all these patterns and their interpretations would be impossible on any practical machine, and searching systematically through them would take impracticably long (involving times of the order of the age of the universe). Thus, it is not only impracticable to consider such brute-force means of solving the recognition problem, but also it is effectively impossible theoretically. These considerations show that other means are required to tackle the problem.

1.2.2 *Tackling the Recognition Problem*

An obvious means of tackling the recognition problem is to normalize the images in some way. Clearly, normalizing the position and orientation of any 2-D picture object would help considerably. Indeed, this would reduce the number of degrees of

1 This is normally taken to mean that one or more parameters produce fast-varying (often exponential) effects that "explode" as the parameters increase by modest amounts.

Figure 1.2 Use of thinning to regularize character shapes. Here character shapes of different limb widths—or even varying limb widths—are reduced to stick figures or skeletons. Thus, irrelevant information is removed, and at the same time recognition is facilitated.

freedom by three. Methods for achieving this involve centralizing the objects—arranging that their centroids are at the center of the normalized image—and making their major axes (deduced by moment calculations, for example) vertical or horizontal. Next, we can make use of the order that is known to be present in the image—and here it may be noted that very few patterns of real interest are indistinguishable from random dot patterns. This approach can be taken further: if patterns are to be nonrandom, isolated noise points may be eliminated. Ultimately, all these methods help by making the test pattern closer to a restricted set of training set patterns (although care has to be taken to process the training set patterns initially so that they are representative of the processed test patterns).

It is useful to consider character recognition further. Here we can make additional use of what is known about the structure of characters—namely, that they consist of limbs of roughly constant width. In that case, the width carries no useful information, so the patterns can be thinned to stick figures (called skeletons—see Chapter 6). Then, hopefully, there is an even greater chance that the test patterns will be similar to appropriate training set patterns (Fig. 1.2). This process can be regarded as another instance of reducing the number of degrees of freedom in the image, and hence of helping to minimize the combinatorial explosion—or, from a practical point of view, to minimize the size of the training set necessary for effective recognition.

Next, consider a rather different way of looking at the problem. Recognition is necessarily a problem of discrimination—that is, of discriminating between patterns of different classes. In practice, however, considering the natural variation of patterns, including the effects of noise and distortions (or even the effects of breakages or occlusions), there is also a problem of generalizing over patterns of the same class. In practical problems a tension exists between the need to discriminate and the need to generalize. Nor is this a fixed situation. Even for the character recognition task, some classes are so close to others (*n*'s and *h*'s will be similar) that less generalization is possible than in other cases. On the other hand, extreme forms of generalization arise when, for example, an *A* is to be recognized as an *A* whether it is a capital or small letter, or in italic, bold, or other form of font—even if it is handwritten. The variability is determined largely by the training set initially provided. What we emphasize here, however, is that generalization is as necessary a prerequisite to successful recognition as is discrimination.

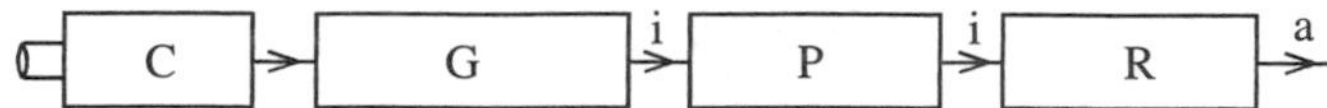

Figure 1.3 The two-stage recognition paradigm: C, input from camera; G, grab image (digitize and store); P, preprocess; R, recognize (i, image data; a, abstract data). The classical paradigm for object recognition is that of (i) preprocessing (image processing) to suppress noise or other artifacts and to regularize the image data, and (ii) applying a process of abstract (often statistical) pattern recognition to extract the very few bits required to classify the object.

At this point it is worth giving more careful consideration to the means whereby generalization was achieved in the examples cited above. First, objects were positioned and orientated appropriately; second, they were cleaned of noise spots; and third, they were thinned to skeleton figures (although the last-named process is relevant only for certain tasks such as character recognition). In the third case we are generalizing over characters drawn with all possible limb widths, width being an irrelevant degree of freedom for this type of recognition task. Note that we could have generalized the characters further by normalizing their size and saving another degree of freedom. The common feature of all these processes is that they aim to give the characters a high level of standardization against known types of variability before finally attempting to recognize them.

The standardization (or generalization) processes we have outlined are all realized by image processing, that is, the conversion of one image into another by suitable means. The result is a two-stage recognition scheme: first, images are converted into more amenable forms containing the same numbers of bits of data; and second, they are classified, with the result that their data content is reduced to very few bits (Fig. 1.3). In fact, recognition is a process of data abstraction, the final data being abstract and totally unlike the original data. Thus we must imagine a letter *A* starting as an array of perhaps 20×20 bits arranged in the form of an *A* and then ending as the 7 bits in an ASCII representation of an *A*, namely, 1000001 (which is essentially a random bit pattern bearing no resemblance to an *A*).

The last paragraph reflects to a large extent the history of image analysis. Early on, a good proportion of the image analysis problems being tackled was envisaged as consisting of an image "preprocessing" task carried out by image processing techniques, followed by a recognition task undertaken by statistical pattern recognition methods (Chapter 24). These two topics—image processing and statistical pattern recognition—consumed much research effort and effectively dominated the subject of image analysis, whereas other approaches such as the Hough transform were not given enough recognition and as a result were not researched adequately. One aim of this book is to ensure that such "intermediate-level" processing techniques are given due emphasis and attention, and that the best range of techniques is applied to any machine vision task.

1.2.3 *Object Location*

The problem tackled in the preceding section—that of character recognition—is a highly constrained one. In many practical applications, it is necessary to search pictures for objects of various types rather than just interpreting a small area of a picture.

The search task can involve prodigious amounts of computation and is also subject to a combinatorial explosion. Imagine the task of searching for a letter *E* in a page of text. An obvious way of conducting this search is to move a suitable "template" of size $n \times n$ over the whole image, of size $N \times N$, and to find where a match occurs (Fig. 1.4). A match can be defined as a position where there is exact agreement between the template and the local portion of the image, but, in keeping with the ideas of Section 1.2.1, it will evidently be more relevant to look for a best local match (i.e., a position where the match is locally better than in adjacent regions) and where the match is also good in some more absolute sense, indicating that an *E* is present.

One of the most natural ways of checking for a match is to measure the Hamming distance between the template and the local $n \times n$ region of the image, that is, to sum the number of differences between corresponding bits. This is essentially the process described in Section 1.2.1. Then places with a low Hamming distance are places where the match is good. These template matching ideas can be extended to cases where the corresponding bit positions in the template and the image do not just have binary values but may have intensity values over a range 0 to 255. In that case, the sums obtained are no longer Hamming distances but may be generalized to the form:

$$\mathscr{D} = \sum_{\mathrm{t}} |I_{\mathrm{i}} - I_{\mathrm{t}}| \tag{1.1}$$

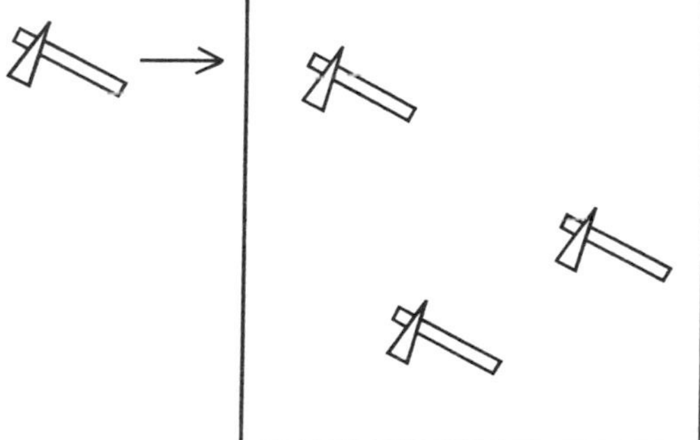

Figure 1.4 Template matching, the process of moving a suitable template over an image to determine the precise positions at which a match occurs—hence revealing the presence of objects of a particular type.

where I_t is the local template value, I_i is the local image value, and the sum is taken over the area of the template. This makes template matching practicable in many situations; the possibilities are examined in more detail in subsequent chapters.

We referred earlier to a combinatorial explosion in this search problem too. The reason this arises is as follows. First, when a 5×5 template is moved over an $N \times N$ image in order to look for a match, the number of operations required is of the order of 5^2N^2, totaling some 1 million operations for a 256×256 image. The problem is that when larger objects are being sought in an image, the number of operations increases in proportion to the square of the size of the object, the total number of operations being N^2n^2 when an $n \times n$ template is used.[2] For a 30×30 template and a 256×256 image, the number of operations required rises to some 60 million. The time it takes to achieve this on any conventional computer will be many seconds—well away from the requirements of real-time[3] operation.

Next, recall that, in general, objects may appear in many orientations in an image (*E*'s on a printed page are exceptional). If we imagine a possible 360 orientations (i.e., one per degree of rotation), then a corresponding number of templates will in principle have to be applied in order to locate the object. This additional degree of freedom pushes the search effort and time to enormous levels, so far away from the possibility of real-time implementation that new approaches must be found for tackling the task. Fortunately, many researchers have applied their minds to this problem, and as a result we now have many ideas for tackling it. Perhaps the most important general means of saving effort on this sort of scale is that of two-stage (or multistage) template matching. The principle is to search for objects via their features. For example, we might consider searching for *E*'s by looking for characters that have horizontal line segments within them. Similarly, we might search for hinges on a manufacturer's conveyor by looking first for the screw holes they possess. In general, it is useful to look for small features, since they require smaller templates and hence involve significantly less computation, as we have demonstrated. This means that it may be better to search for *E*'s by looking for corners instead of horizontal line segments.

Unfortunately, noise and distortions give rise to problems if we search for objects via small features; there is even a risk of missing the object altogether. Hence, it is necessary to collate the information from a number of such features. This is the point where the many available methods start to differ from each other. How many features should be collated? Is it better to take a few larger features

---

2 Note that, in general, a template will be larger than the object it is used to search for, because some background will have to be included to help demarcate the object.

3 A commonly used phrase meaning that the information has to be processed as it becomes available. This contrasts with the many situations (such as the processing of images from space probes) where the information may be stored and processed at leisure.

than a lot of smaller ones? And so on. Also, we have not fully answered the question of what types of features are the best to employ. These and other questions are considered in the following chapters.

Indeed, in a sense, these questions are the subject of this book. Search is one of the fundamental problems of vision, yet the details and the application of the basic idea of two-stage template matching give the subject much of its richness: to solve the recognition problem the dataset needs to be explored carefully. Clearly, any answers will tend to be data-dependent, but it is worth exploring to what extent generalized solutions to the problem are possible.

1.2.4 *Scene Analysis*

The preceding subsection considered what is involved in searching an image for objects of a certain type. The result of such a search is likely to be a list of centroid coordinates for these objects, although an accompanying list of orientations might also be obtained. The present subsection considers what is involved in scene analysis—the activity we are continually engaged in as we walk around, negotiating obstacles, finding food, and so on. Scenes contain a multitude of objects, and it is their interrelationships and relative positions that matter as much as identifying what they are. There is often no need for a search per se, and we could in principle passively take in what is in the scene. However, there is much evidence (e.g., from analysis of eye movements) that the eye–brain system interprets scenes by continually asking questions about what is there. For example, we might ask the following questions: Is this a lamppost? How far away is it? Do I know this person? Is it safe to cross the road? And so on. Our purpose here is not to dwell on these human activities or to indulge in long introspection about them but merely to observe that scene analysis involves enormous amounts of input data, complex relationships between objects within scenes, and, ultimately, descriptions of these complex relationships. The latter no longer take the form of simple classification labels, or lists of object coordinates, but have a much richer information content. Indeed, a scene will, to a first approximation, be better described in English than as a list of numbers. It seems likely that a much greater combinatorial explosion is involved in determining relationships between objects than in merely identifying and locating them. Hence all sorts of props must be used to aid visual interpretation: there is considerable evidence of this in the human visual system, where contextual information and the availability of immense databases of possibilities clearly help the eye to a very marked degree.

Note also that scene descriptions may initially be at the level of factual content but will eventually be at a deeper level—that of meaning, significance, and relevance. However, we shall not be able to delve further into these areas in this book.

1.2.5 *Vision as Inverse Graphics*

It has often been said that vision is "merely" inverse graphics. There is a certain amount of truth in this statement. Computer graphics is the generation of images by computer, starting from abstract descriptions of scenes and knowledge of the laws of image formation. Clearly, it is difficult to quarrel with the idea that vision is the process of obtaining descriptions of sets of objects, starting from sets of images and knowledge of the laws of image formation. (Indeed, it is good to see a definition that explicitly brings in the need to know the laws of image formation, since it is all too easy to forget that this is a prerequisite when building descriptions incorporating heuristics that aid interpretation.)

This similarity in formulation of the two processes, however, hides some fundamental points. First, graphics is a "feedforward" activity; that is, images can be produced in a straightforward fashion once sufficient specification about the viewpoint and the objects, and knowledge of the laws of image formation, has been obtained. True, considerable computation may be required, but the process is entirely determined and predictable. The situation is not so straightforward for vision because search is involved and there is an accompanying combinatorial explosion. Indeed, certain vision packages incorporate graphics or CAD (computer-aided design) packages (Tabandeh and Fallside, 1986) which are inserted into feedback loops for interpretation. The graphics package is then guided iteratively until it produces an acceptable approximation to the input image, when its input parameters embody the correct interpretation. (There is a close parallel here with the problem of designing analog-to-digital converters by making use of digital-to-analog converters). Hence it seems inescapable that vision is intrinsically more complex than graphics.

We can clarify the situation somewhat by noting that, as a scene is observed, a 3-D environment is compressed into a 2-D image and a considerable amount of depth and other information is lost. This can lead to ambiguity of interpretation of the image (both a helix viewed end-on and a circle project into a circle), so the 3-D to 2-D transformation is many-to-one. Conversely, the interpretation must be one-to-many, meaning that many interpretations are possible, yet we know that only one can be correct. Vision involves not merely providing a list of all possible interpretations but providing the most likely one. Some additional rules or constraints must therefore be involved in order to determine the single most likely interpretation. Graphics, in contrast, does not have these problems, for it has been shown to be a many-to-one process.

1.3 From Automated Visual Inspection to Surveillance

So far we have considered the nature of vision but not the possible uses of man-made vision systems. There is in fact a great variety of applications for artificial vision systems—including, of course, all of those for which we employ our visual senses. Of particular interest in this book are surveillance, automated inspection, robot assembly, vehicle guidance, traffic monitoring and control, biometric measurement, and analysis of remotely sensed images. By way of example, fingerprint analysis and recognition have long been important applications of computer vision, as have the counting of red blood cells, signature verification and character recognition, and aeroplane identification (both from aerial silhouettes and from ground surveillance pictures taken from satellites). Face recognition and even iris recognition have become practical possibilities, and vehicle guidance by vision will in principle soon be sufficiently reliable for urban use.[4] However, among the main applications of vision considered in this book are those of manufacturing industry—particularly, automated visual inspection and vision for automated assembly.

In the last two cases, much the same manufactured components are viewed by cameras: the difference lies in how the resulting information is used. In assembly, components must be located and oriented so that a robot can pick them up and assemble them. For example, the various parts of a motor or brake system need to be taken in turn and put into the correct positions, or a coil may have to be mounted on a TV tube, an integrated circuit placed on a printed circuit board, or a chocolate placed into a box. In inspection, objects may pass the inspection station on a moving conveyor at rates typically between 10 and 30 items per second, and it has to be ascertained whether they have any defects. If any defects are detected, the offending parts will usually have to be rejected: that is the feedforward solution. In addition, a feedback solution may be instigated—that is, some parameter may have to be adjusted to control plant further back down the production line. (This is especially true for parameters that control dimensional characteristics such as product diameter.) Inspection also has the potential for amassing a wealth of information, useful for management, on the state of the parts coming down the line: the total number of products per day, the number of defective products per day, the distribution of sizes of products, and so on. The important feature of artificial vision is that it is tireless and that *all* products can be scrutinized and measured. Thus, quality control can be maintained to a very high standard. In automated assembly, too, a considerable amount of on-the-spot inspection can be performed, which may help to avoid the problem of complex assemblies being

4 Whether the public will accept this, with all its legal implications, is another matter, although it is worth noting that radar blind-landing aids for aircraft have been in wide use for some years. As discussed further in Chapter 18, last-minute automatic action to prevent accidents is a good compromise.

rejected, or having to be subjected to expensive repairs, because (for example) a proportion of screws were threadless and could not be inserted properly.

An important feature of most industrial tasks is that they take place in real time: if applied, machine vision must be able to keep up with the manufacturing process. For assembly, this may not be too exacting a problem, since a robot may not be able to pick up and place more than one item per second—leaving the vision system a similar time to do its processing. For inspection, this supposition is rarely valid: even a single automated line (for example, one for stoppering bottles) is able to keep up a rate of 10 items per second (and, of course, parallel lines are able to keep up much higher rates). Hence, visual inspection tends to press computer hardware very hard. Note in addition that many manufacturing processes operate under severe financial constraints, so that it is not possible to employ expensive multiprocessing systems or supercomputers. Great care must therefore be taken in the design of hardware accelerators for inspection applications. Chapter 28 aims to give some insight into these hardware problems.

Finally, we return to our initial discussion about the huge variety of applications of machine vision—and it is interesting to note that, in a sense, surveillance tasks are the outdoor analogs of automated inspection. They have recently been acquiring close to exponentially increasing application; as a result, they have been included in this volume, not as an afterthought but as an expanding area wherein the techniques used for inspection have acquired a new injection of vitality. Note, however, that in the process they have taken in whole branches of new subject matter, such as motion analysis and perspective invariants (see Part 3). It is also interesting that such techniques add a new richness to such old topics as face recognition (Section 24.12.1).

1.4 What This Book Is About

The foregoing sections have examined the nature of machine vision and have briefly considered its applications and implementation. Clearly, implementing machine vision involves considerable practical difficulties, but, more important, these practical difficulties embody substantial fundamental problems, including various mechanisms giving rise to excessive processing load and time. With ingenuity and care practical problems may be overcome. However, by definition, truly fundamental problems cannot be overcome by *any* means—the best that we can hope for is that they be minimized following a complete understanding of their nature.

Understanding is thus a cornerstone for success in machine vision. It is often difficult to achieve, since the dataset (i.e., all pictures that could reasonably be expected to arise) is highly variegated. Much investigation is required to determine the nature of a given dataset, including not only the objects being observed but also the noise levels, and degrees of occlusion, breakage, defect, and distortion that are to be expected, and the quality and nature of lighting schemes. Ultimately, sufficient knowledge might be obtained in a useful set of cases so that a good understanding of the milieu can be attained. Then it remains to compare and contrast the various methods of image analysis that are available. Some methods will turn out to be quite unsatisfactory for reasons of robustness, accuracy or cost of implementation, or other relevant variables: and who is to say in advance what a relevant set of variables is? This, too, needs to be ascertained and defined. Finally, among the methods that could reasonably be used, there will be competition: tradeoffs between parameters such as accuracy, speed, robustness, and cost will have to be worked out first theoretically and then in numerical detail to find an optimal solution. This is a complex and long process in a situation where workers have usually aimed to find solutions for their own particular (often short-term) needs. Clearly, there is a need to raise practical machine vision from an art to a science. Fortunately, this process has been developing for some years, and it is one of the aims of this book to throw additional light on the problem.

Before proceeding further, we need to fit one or two more pieces into the jigsaw. First, there is an important guiding principle: *if the eye can do it, so can the machine*. Thus, if an object is fairly well hidden in an image, but the eye can still see it, then it should be possible to devise a vision algorithm that can also find it. Next, although we can expect to meet this challenge, should we set our sights even higher and aim to devise algorithms that can beat the eye? There seems no reason to suppose that the eye is the ultimate vision machine: it has been built through the vagaries of evolution, so it may be well adapted for finding berries or nuts, or for recognizing faces, but ill-suited for certain other tasks. One such task is that of measurement. The eye probably does not need to measure the sizes of objects, at a glance, to better than a few percent accuracy. However, it could be distinctly useful if the robot eye could achieve remote size measurement, at a glance, and with an accuracy of say 0.001%. Clearly, it is worth noting that the robot eye will have some capabilities superior to those of biological systems. Again, this book aims to point out such possibilities where they exist.

Machine Vision

Machine vision is the study of methods and techniques whereby artificial vision systems can be constructed and usefully employed in practical applications. As such, it embraces both the science and engineering of vision.

Finally, it is good to have a working definition of machine vision (see box).[5] Its study includes not only the software but also the hardware environment and image acquisition techniques needed to apply it. As such, it differs from computer vision, which appears from most books on the subject to be the realm of the *possible* design of the software, without too much attention on what goes into an integrated vision system (though modern books on computer vision usually say a fair amount about the "nasty realities" of vision, such as noise elimination and occlusion analysis).

1.5 The Following Chapters

On the whole the early chapters of the book (Chapters 2–4) cover rather simple concepts, such as the nature of image processing, how image processing algorithms may be devised, and the restrictions on intensity thresholding techniques. The next four chapters (Chapters 5–8) discuss edge detection and some fairly traditional binary image analysis techniques. Then, Chapters 9–15 move on to intermediate-level processing, which has developed significantly in the past two decades, particularly in the use of transform techniques to deduce the presence of objects. Intermediate-level processing is important in leading to the inference of complex objects, both in 2-D (Chapter 15) and in 3-D (Chapter 16). It also enables automated inspection to be undertaken efficiently (Chapter 22). Chapter 24 expands on the process of recognition that is fundamental to many inspection and other processes—as outlined earlier in this chapter. Chapters 27 and 28, respectively, outline the enabling technologies of image acquisition and vision hardware design, and, finally, Chapter 29 reiterates and highlights some of the lessons and topics covered earlier in the book.

To help give the reader more perspective on the 29 chapters, the main text has been divided into five parts: Part 1 (Chapters 2–8) is entitled Low-Level Vision, Part 2 (Chapters 9–15) Intermediate-Level Vision, Part 3 (Chapters 16–21) 3-D Vision and Motion, Part 4 (Chapters 22–28) Toward Real-Time Pattern Recognition Systems, and Part 5 (Chapter 29) Perspectives on Vision. The heading Toward Real-Time Pattern Recognition Systems is used to emphasize real-world applications with immutable data flow rates, and the need to integrate all the necessary recognition processes into reliable working systems. The purpose of Part 5 is to obtain a final analytic view of the topics already encountered.

---

5 Interestingly, this definition is much broader than that of Batchelor (2003) who relates the subject to manufacturing in general and inspection and assembly in particular.

Although the sequence of chapters follows the somewhat logical order just described, the ideas outlined in the previous section—understanding of the visual process, constraints imposed by realities such as noise and occlusion, tradeoffs between relevant parameters, and so on—are mixed into the text at relevant junctures, as they reflect all-pervasive issues.

Finally, many other topics would ideally have been included in the book, yet space did not permit it. The chapter bibliographies, the main list of references, and the indexes are intended to compensate in some measure for some of these deficiencies.

1.6 Bibliographical Notes

This chapter introduces the reader to some of the problems of machine vision, showing the intrinsic difficulties but not at this stage getting into details. For detailed references, the reader should consult the later chapters. Meanwhile, some background on the world of pattern recognition can be obtained from Duda et al. (2001). In addition, some insight into human vision can be obtained from the fascinating monograph by Hubel (1995).

PART 1

Low-Level Vision

This part of the book introduces images and image processing, and then proceeds to show how image processing may be developed in order to start the process of image analysis. By the end of Chapter 8 image analysis is taken, via this "traditional" route, to a reasonably useful level. Although Chapter 8 encompasses and summarizes much of the earlier work, Chapter 7 is better envisaged as forming a bridge with Part 2, as it already incorporates a greater degree of global analysis for real-world objects.

Images and Imaging Operations

Images are at the core of vision, and there are many ways—from simple to sophisticated—to process and analyze them. This chapter concentrates on simple algorithms, which nevertheless need to be treated carefully, for important subtleties need to be learned. Above all, the chapter aims to show that quite a lot can be achieved with such algorithms, which can readily be programmed and tested by the reader.

Look out for:

- the different types of images—binary, gray-scale, and color.
- a useful, compact notation for presenting image processing operations.
- basic pixel operations—clearing, copying, inverting, thresholding.
- basic window operations—shifting, shrinking, expanding.
- gray-scale brightening and contrast-stretching operations.
- binary edge location and noise removal operations.
- multi-image and convolution operations.
- the distinction between sequential and parallel operations, and complications that can arise in the sequential case.
- problems that arise around the edge of the image.

Though elementary, this chapter provides the basic methodology for the whole of Part 1 and much of Part 2 of the book, and its importance should not be underestimated; neither should the subtleties be ignored. Full understanding at this stage will save many complications later, when one is programming more sophisticated algorithms. In particular, obvious fundamental difficulties arise when applying a sequence of window operations to the same original image.

CHAPTER 2

Images and Imaging Operations

pĭx'ĕllātėd, a. picture broken into a regular tiling
pĭx'ĭlātėd, a. pixie-like, crazy, deranged

2.1 Introduction

This chapter focuses on images and simple image processing operations and is intended to lead to more advanced image analysis operations that can be used for machine vision in an industrial environment. Perhaps the main purpose of the chapter is to introduce the reader to some basic techniques and notation that will be of use throughout the book. However, the image processing algorithms introduced here are of value in their own right in disciplines ranging from remote sensing to medicine, and from forensic to military and scientific applications.

The images discussed in this chapter have already been obtained from suitable sensors: the sensors are covered in a later chapter. Typical of such images is that shown in Fig. 2.1*a*. This is a gray-tone image, which at first sight appears to be a normal "black and white" photograph. On closer inspection, however, it becomes clear that it is composed of a large number of individual picture cells, or pixels. In fact, the image is a 128×128 array of pixels. To get a better feel for the limitations of such a digitized image, Fig. 2.1*b* shows a section that has been enlarged so that the pixels can be examined individually. A threefold magnification factor has been used, and thus the visible part of Fig. 2.1*b* contains only a 42×42 array of pixels.

It is not easy to see that these gray-tone images are digitized into a gray-scale containing just 64 gray levels. To some extent, high spatial resolution compensates for the lack of gray-scale resolution, and as a result we cannot at once see the difference between an individual shade of gray and the shade it

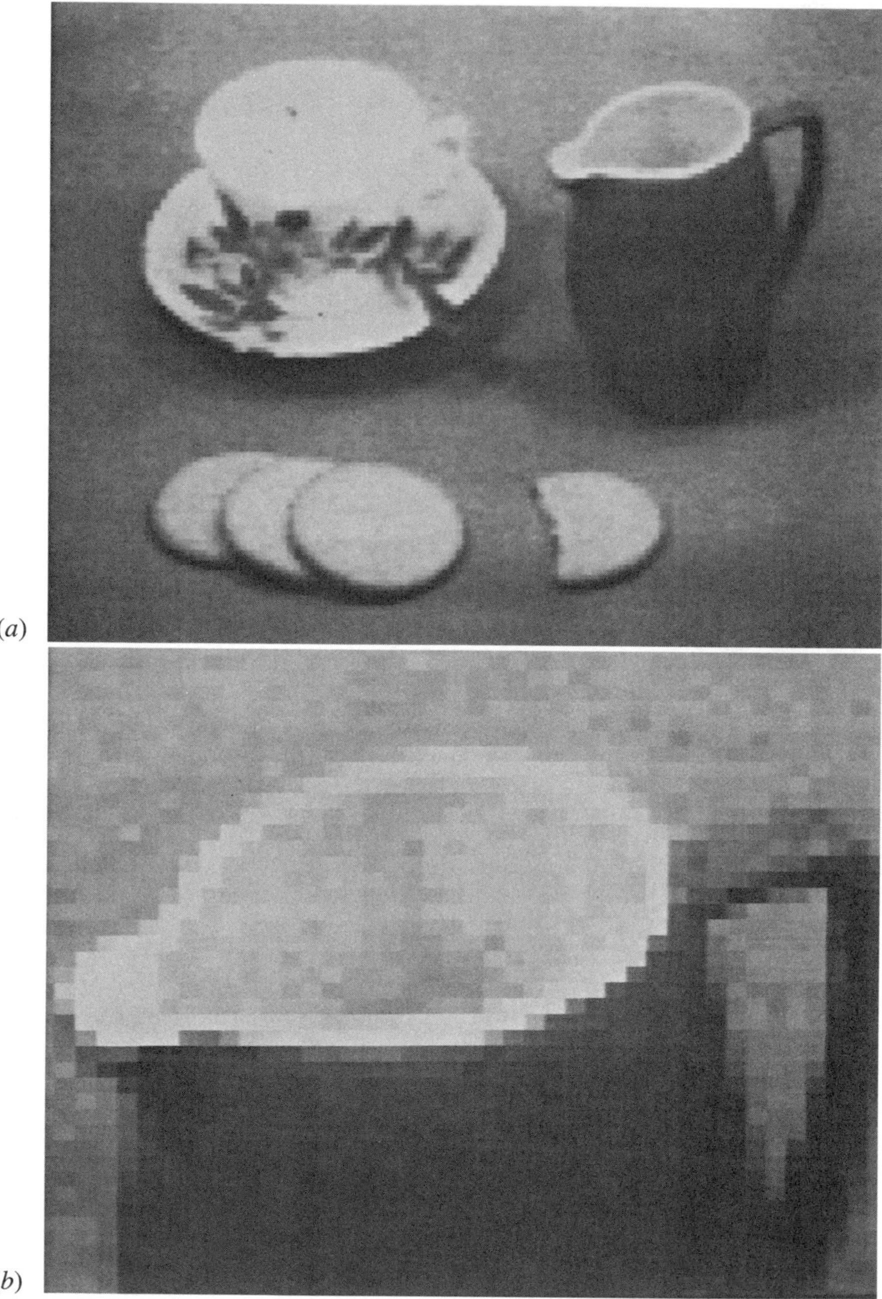

Figure 2.1 Typical gray-scale images: (*a*) gray-scale image digitized into a 128×128 array of pixels; (*b*) section of image shown in (*a*) subjected to threefold linear magnification: the individual pixels are now clearly visible.

would have had in an ideal picture. In addition, when we look at the magnified section of image in Fig. 2.1*b*, it is difficult to understand the significance of the individual pixel intensities—the whole is becoming lost in a mass of small parts. Older television cameras typically gave a gray-scale resolution that was accurate only to about one part in 50, corresponding to about 6 bits of useful information per pixel. Modern solid-state cameras commonly give less noise and may allow 8 or even 9 bits of information per pixel. However, it is often not worthwhile to aim for such high gray-scale resolutions, particularly when the result will not be visible to the human eye, and when, for example, there is an enormous amount of other data that a robot can use to locate a particular object within the field of view. Note that if the human eye can see an object in a digitized image of particular spatial and gray-scale resolution, it is in principle possible to devise a computer algorithm to do the same thing.

Nevertheless, there is a range of applications for which it is valuable to retain good gray-scale resolution, so that highly accurate measurements can be made from a digital image. This is the case in many robotic applications, where high-accuracy checking of components is critical. More will be said about this subject later. In addition, in Part 2 we will see that certain techniques for locating components efficiently require that local edge orientation be estimated to better than 1°, and this can be achieved only if at least 6 bits of gray-scale information are available per pixel.

2.1.1 Gray-Scale versus Color

Returning now to the image of Fig. 2.1*a*, we might reasonably ask whether it would be better to replace the gray-scale with color, using an RGB (red, green, blue) color camera and three digitizers for the three main colors. Two aspects of color are important for the present discussion. One is the *intrinsic value* of color in machine vision: the other is the *additional storage and processing penalty* it might bring. It is tempting to say that the second aspect is of no great importance given the cheapness of modern computers which have both high storage and high speed. On the other hand, high-resolution images can arrive from a collection of CCTV cameras at huge data rates, and it will be many years before it will be possible to analyze *all* the data arriving from such sources. Hence, if color adds substantially to the storage and processing load, its use will need to be justified.

The *potential* of color in helping with many aspects of inspection, surveillance, control, and a wide variety of other applications including medicine (color plays a crucial role in images taken during surgery) is enormous. This potential is illustrated with regard to agriculture in Fig. 2.2; for robot navigation and

Figure 2.2 Value of color in agricultural applications. In agricultural scenes such as this one, color helps with segmentation and with recognition. It may be crucial in discriminating between weeds and crops if selective robot weedkilling is to be carried out. (See color insert.)

driving in Figs. 2.3 and 2.4; for printed circuit board inspection in Fig. 2.5; for food inspection in Figs. 2.6 and 2.7; and for color filtering in Figs. 2.8 and 2.9. Notice that some of these images almost have color for color's sake (especially in Figs. 2.6–2.8), though none of them is artificially generated. In others the color is more subdued (Figs. 2.2 and 2.4), and in Fig. 2.7 (excluding the tomatoes) it is quite subtle. The point here is that for color to be useful it need not be garish; it can be subtle as long as it brings the right sort of information to bear on the task in hand. Suffice it to say that in some of the simpler inspection applications, where mechanical components are scrutinized on a conveyor or workbench, it is quite likely the shape that is in question rather than the color of the object or its parts. On the other hand, if an automatic fruit picker is to be devised, it is much more likely to be crucial to check color rather than specific shape. We leave it to the reader to imagine when and where color is particularly useful or merely an unnecessary luxury.

Next, it is useful to consider the processing aspect of color. In many cases, good color discrimination is required to separate and segment two types of objects from each other. Typically, this will mean not using one or another specific

Figure 2.3 Value of color for segmentation and recognition. In natural outdoor scenes such as this one, color helps with segmentation and with recognition. While it may have been important to the early human when discerning sources of food in the wild, robot drones may benefit by using color to aid navigation. (See color insert.)

color channel,[1] but subtracting two, or combining three in such a way as to foster discrimination. In the worst case of combining three color channels by simple arithmetic processing in which each pixel is treated identically, the processing load will be very light. In contrast, the amount of processing required to *determine* the optimal means of combining the data from the color channels and to carry out different operations dynamically on different parts of the image may be far from negligible, and some care will be needed in the analysis. These problems arise because color signals are inhomogeneous. This contrasts with the situation for gray-scale images, where the bits representing the gray-scale are all of the same type and take the form of a number representing the pixel intensity: they can thus be processed as a single entity on a digital computer.

---

1 Here we use the term *channel* not just to refer to the red, green, or blue channel, but any derived channel obtained by combining the colors in any way into a single color dimension.

Figure 2.4 Value of color in the built environment. Color plays an important role for the human in managing the built environment. In a vehicle, a plethora of bright lights, road signs, and markings (such as yellow lines) are coded to help the driver. They may also help a robot to drive more safely by providing crucial information. (See color insert.)

2.2 Image Processing Operations

The images of Figs 2.1 and 2.11*a* are considered in some detail here, including some of the many image processing operations that can be performed on them. The resolution of these images reveals a considerable amount of detail and at the same time shows how it relates to the more "meaningful" global information. This should help to show how simple imaging operations contribute to image interpretation.

When performing image processing operations, we start with an image in one storage area and generate a new processed image in another storage area. In practice these storage areas may either be in a special hardware unit called a frame store that is interfaced to the computer, or else they may be in the main memory of the computer or on one of its discs. In the past, a special frame store was required to store images because each image contains a good fraction of a megabyte of information, and this amount of space was not available for normal users in the computer main memory. Today this is less of a problem,

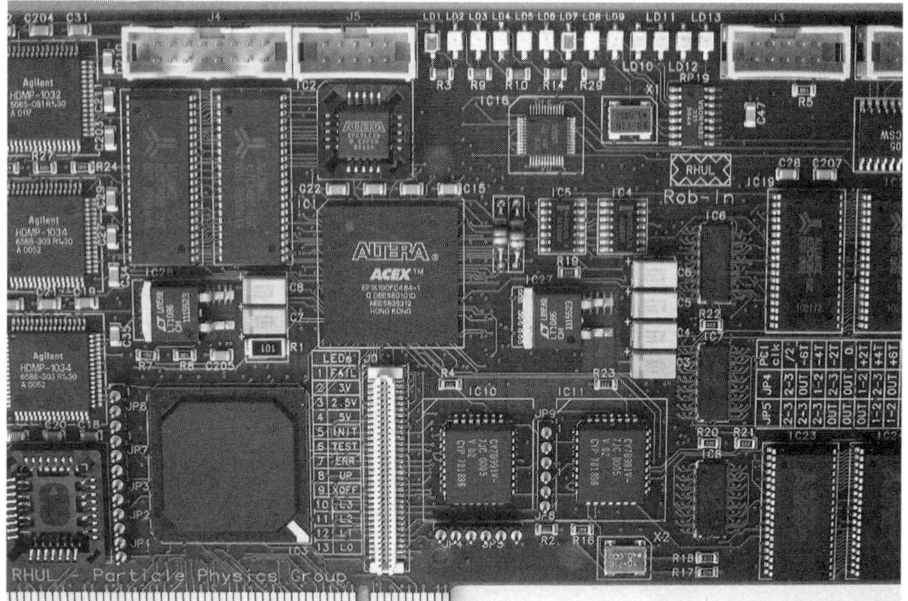

Figure 2.5 Value of color for unambiguous checking. In many inspection applications, color can be of help to unambiguously confirm the presence of certain objects, as on this printed circuit board. (See color insert.)

but for image acquisition a frame store is still required. However, we shall not worry about such details here; instead it will be assumed that all images are inherently visible and that they are stored in various image "spaces" P, Q, R, and so on. Thus, we might start with an image in space P and copy it to space Q, for example.

2.2.1 *Some Basic Operations on Gray-scale Images*

Perhaps the simplest imaging operation is that of clearing an image or setting the contents of a given image space to a constant level. We need some way of arranging this; accordingly, the following C++ routine may be written for implementing it:[2]

```
for (j = 0; j <= 127; j++)
  for (i = 0; i <= 127; i++)        (2.1)
    P[j][i] = alpha;
```

2 Readers who are unfamiliar with C++ should refer to books such as Stroustrup (1991) and Schildt (1995).

Figure 2.6 Value of color for food inspection. Much food is brightly colored, as with this Japanese meal. Although this may be attractive to the human, it could also help the robot to check quickly for foreign bodies or toxic substances. (See color insert.)

In this routine the local pixel intensity value is expressed as P[i][j], since P-space is taken to be a two-dimensional array of intensity values. In what follows it will be advantageous to rewrite such routines in the more succinct form:

$$\text{SETP: [[P0} = \text{alpha;]]} \tag{2.2}$$

since it is then much easier to understand the image processing operations. In effect, we are attempting to expose the fundamental imaging operation by removing irrelevant programming detail. The double square bracket notation is intended to show that the operation it encloses is applied at every pixel location in the image, for whatever size of image is currently defined. The reason for calling the pixel intensity P0 will become clear later.

Figure 2.7 Subtle shades of color in food inspection. Although much food is brightly colored, as for the tomatoes in this picture, green salad leaves show much more subtle combinations of color and may indeed provide the only reliable means of identification. This could be important for inspection both of the raw product and its state when it reaches the warehouse or the supermarket. (See color insert.)

Another simple imaging operation is to copy an image from one space to another. This is achieved, without changing the contents of the original space P, by the routine:

$$\text{COPY: [[Q0 = P0;]]} \tag{2.3}$$

Again, the double bracket notation indicates a double loop that copies each individual pixel of P-space to the corresponding location in Q-space.

Figure 2.8 Color filtering of brightly colored objects. (*a*) Original color image of some sweets. (*b*) Vector median filtered version. (*c*) Vector mode filtered version. (*d*) Version to which a mode filter has been applied to each color channel separately. Note that (*b*) and (*c*) show no evidence of color bleeding, though it is strongly evident in (*d*). It is most noticeable as isolated pink pixels, plus a few green pixels, around the yellow sweets. For further details on color bleeding, see Section 3.14. (See color insert.)

A more interesting operation is that of inverting the image, as in the process of converting a photographic negative to a positive. This process is represented as follows:

$$\text{INVERT: [[Q0} = 255 - \text{P0;]]} \tag{2.4}$$

In this case it is assumed that pixel intensity values lie within the range 0–255, as is commonly true for frame stores that represent each pixel as one byte of information. Note that such intensity values are commonly unsigned, and this is assumed generally in what follows.

(c)

Figure 2.9 Color filtering of images containing substantial impulse noise. (*a*) Version of the Lena image containing 70% random color impulse noise. (*b*) Effect of applying a vector median filter, and (*c*) effect of applying a vector mode filter. Although the mode filter is designed more for enhancement than for noise suppression, it has been found to perform remarkably well at this task when the noise level is very high (see Section 3.4). (See color insert.)

There are many operations of these types. Some other simple operations are those that shift the image left, right, up, down, or diagonally. They are easy to implement if the new local intensity is made identical to that at a neighboring location in the original image. It is evident how this would be expressed in the double suffix notation used in the original C++ routine. In the new shortened notation, it is necessary to name neighboring pixels in some convenient way,

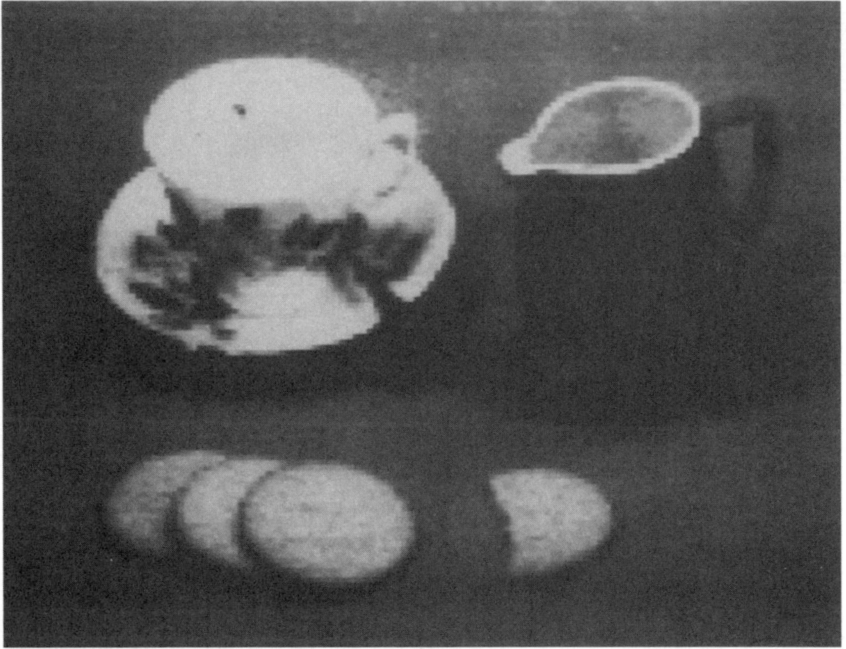

Figure 2.10 Contrast-stretching: effect of increasing the contrast in the image of Fig. 2.1*a* by a factor of two and adjusting the mean intensity level appropriately. The interior of the jug can now be seen more easily. Note, however, that there is no additional information in the new image.

and we here employ a commonly used numbering scheme:

P4	**P3**	**P2**
P5	**P0**	**P1**
P6	**P7**	**P8**

with a similar scheme for other image spaces. With this notation, it is easy to express a left shift of an image as follows:

$$\text{LEFT: [[Q0 = P1;]]} \tag{2.5}$$

Similarly, a shift down to the bottom right is expressed as:

$$\text{DOWNR: [[Q0 = P4;]]} \tag{2.6}$$

Figure 2.2 Value of color in agricultural applications. In agricultural scenes such as this one, color helps with segmentation and with recognition. It may be crucial in discriminating between weeds and crops if selective robot weedkilling is to be carried out.

Figure 2.3 Value of color for segmentation and recognition. In natural outdoor scenes such as this one, color helps with segmentation and with recognition. While it may have been important to the early human when discerning sources of food in the wild, robot drones may benefit by using color to aid navigation.

Figure 2.4 Value of color in the built environment. Color plays an important role for the human in managing the built environment. In a vehicle, a plethora of bright lights, road signs, and markings (such as yellow lines) are coded to help the driver. They may also help a robot to drive more safely by providing crucial information.

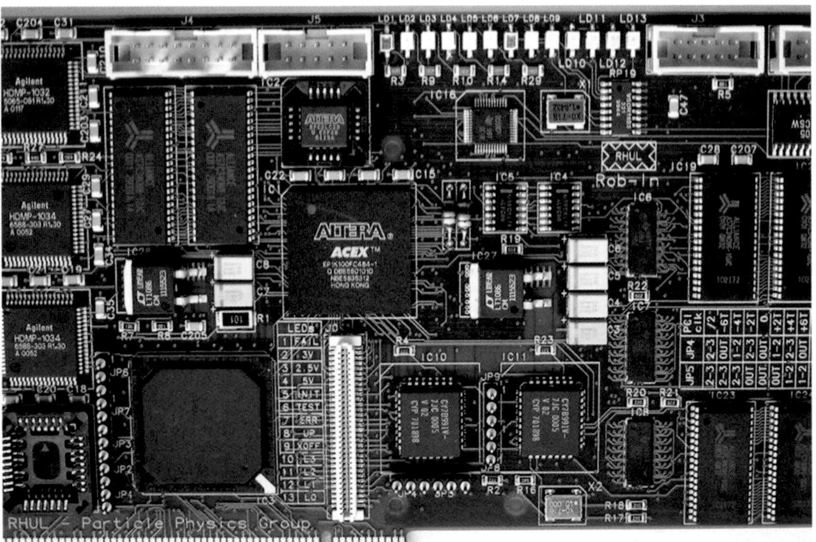

Figure 2.5 Value of color for unambiguous checking. In many inspection applications, color can be of help to unambiguously confirm the presence of certain objects, as on this printed circuit board.

Figure 2.6 Value of color for food inspection. Much food is brightly colored, as with this Japanese meal. Although this may be attractive to the human, it could also help the robot to check quickly for foreign bodies or toxic substances.

Figure 2.7 Subtle shades of color in food inspection. Although much food is brightly colored, as for the tomatoes in this picture, green salad leaves show much more subtle combinations of color and may indeed provide the only reliable means of identification. This could be important for inspection both of the raw product and its state when it reaches the warehouse or the supermarket.

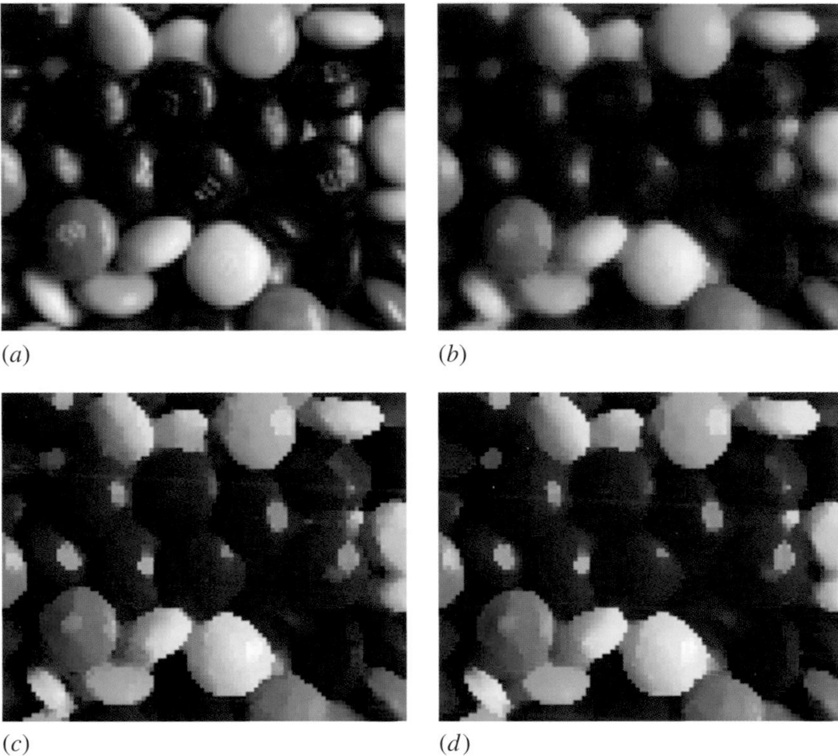

(*a*) (*b*) (*c*) (*d*)

Figure 2.8 Color filtering of brightly colored objects. (*a*) Original color image of some sweets. (*b*) Vector median filtered version. (*c*) Vector mode filtered version. (*d*) Version to which a mode filter has been applied to each color channel separately. Note that (*b*) and (*c*) show no evidence of color bleeding, though it is strongly evident in (*d*). It is most noticeable as isolated pink pixels, plus a few green pixels, around the yellow sweets. For further details on color bleeding, see Section 3.14.

(*a*) (*b*) (*c*)

Figure 2.9 Color filtering of images containing substantial impulse noise. (*a*) Version of the Lena image containing 70% random color impulse noise. (*b*) Effect of applying a vector median filter, and (*c*) effect of applying a vector mode filter. Although the mode filter is designed more for enhancement than for noise suppression, it has been found to perform remarkably well at this task when the noise level is very high (see Section 3.4).

(Note that the shift appears in a direction opposite to that which one might at first have suspected.)

It will now be clear why P0 and Q0 were chosen for the basic notation of pixel intensity: the "0" denotes the central pixel in the "neighborhood" or "window," and corresponds to zero shift when copying from one space to another.

A whole range of possible operations is associated with modifying images in such a way as to make them more satisfactory for a human viewer. For example, adding a constant intensity makes the image brighter:

$$\text{BRIGHTEN: [[Q0} = \text{P0} + \text{beta;]]} \tag{2.7}$$

and the image can be made darker in the same way. A more interesting operation is to stretch the contrast of a dull image:

$$\text{STRETCH: [[Q0} = \text{P0} * \text{gamma} + \text{beta;]]} \tag{2.8}$$

where gamma > 1. In practice (as for Fig. 2.10) it is necessary to ensure that intensities do not result that are outside the normal range, for example, by using an operation of the form:

$$\begin{aligned}\text{STRETCH: [[QQ} &= \text{P0} * \text{gamma} + \text{beta;}\\ &\text{if (QQ} < 0\text{) Q0} = 0;\\ &\text{else if (QQ} > 255\text{) Q0} = 255;\\ &\text{else Q0} = \text{QQ;]]}\end{aligned} \tag{2.9}$$

Most practical situations demand more sophisticated transfer functions—either nonlinear or piecewise linear—but such complexities are ignored here.

A further simple operation that is often applied to gray-scale images is that of thresholding to convert to a binary image. This topic is covered in more detail later in this chapter since it is widely used to detect objects in images. However, our purpose here is to look on it as another basic imaging operation. It can be implemented using the routine[3]

$$\text{THRESH: [[if (P0} > \text{thresh) A0} = 1\text{; else A0} = 0\text{;]]} \tag{2.10}$$

3 The first few letters of the alphabet (A, B, C,...) are used consistently to denote binary image spaces, and later letters (P, Q, R,...) to denote gray-scale images. In software, these variables are assumed to be predeclared, and in hardware (e.g., frame store) terms they are taken to refer to dedicated memory spaces containing only the necessary 1 or 8 bits per pixel. The intricacies of data transfer between variables of different types are important considerations which are not addressed in detail here: it is sufficient to assume that both A0 = P0 and P0 = A0 correspond to a single-bit transfer, except that in the latter case the top 7 bits are assigned the value 0.

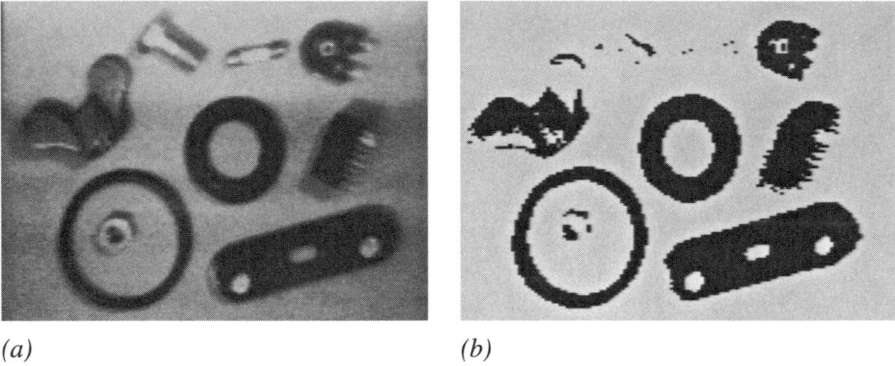

(*a*) (*b*)

Figure 2.11 Thresholding of gray-scale images: (*a*) 128 × 128 pixel gray-scale image of a collection of parts; (*b*) effect of thresholding the image.

If, as often happens, objects appear as dark objects on a light background, it is easier to visualize the subsequent binary processing operations by inverting the thresholded image using a routine such as

$$\text{INV: [[A0} = 1 - \text{A0;]]} \tag{2.11}$$

However, it would be more usual to combine the two operations into a single routine of the form:

$$\text{INVTHRESH: [[if (P0} > \text{thresh) A0} = 0\text{; else A0} = 1\text{;]]} \tag{2.12}$$

To display the resulting image in a form as close as possible to the original, it can be reinverted and given the full range of intensity values (intensity values 0 and 1 being scarcely visible):

$$\text{INVDISPLAY: [[R0} = 255 * (1 - \text{A0);]]} \tag{2.13}$$

Figure 2.11 shows the effect of these two operations.

2.2.2 *Basic Operations on Binary Images*

Once the image has been thresholded, a wide range of binary imaging operations becomes possible. Only a few such operations are covered here, with the aim of being instructive rather than comprehensive. With this in mind, a routine may be

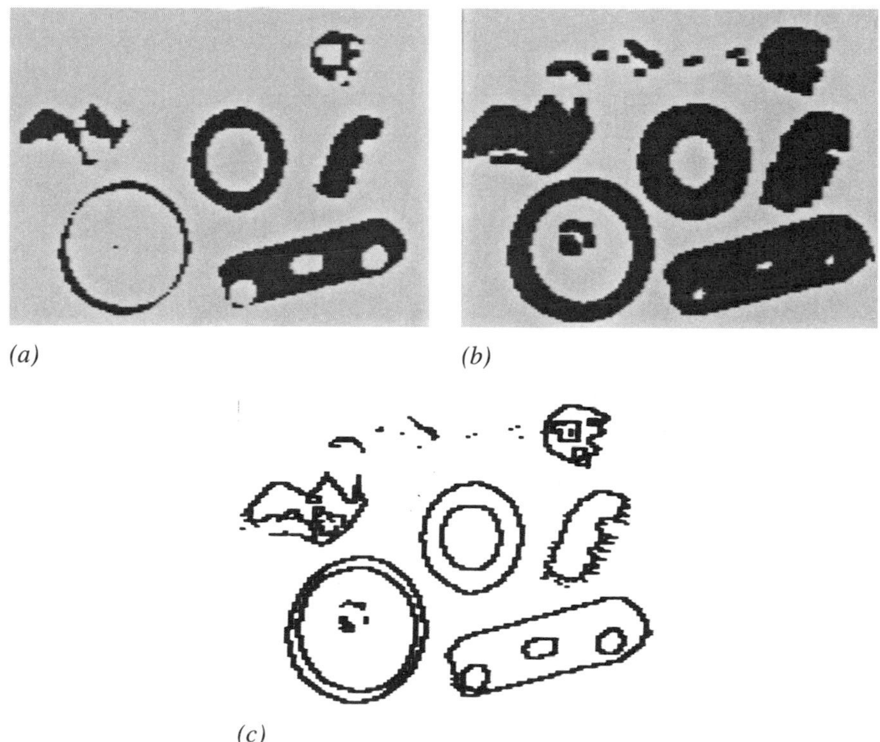

Figure 2.12 Simple operations applied to binary images: (*a*) effect of shrinking the dark-thresholded objects appearing in Fig. 2.11*b*; (*b*) effect of expanding these dark objects; (*c*) result of applying an edge location routine. Note that the SHRINK, EXPAND, and EDGE routines are applied to the *dark* objects. This implies that the intensities are initially inverted as part of the thresholding operation and then reinverted as part of the display operation (see text).

written for shrinking dark thresholded objects (Fig. 2.12*a*), which are represented here by a set of 1's in a background of 0's:

```
SHRINK: [[ sigma = A1 + A2 + A3 + A4 + A5 + A6 + A7 + A8;
           if (A0 == 0) B0 = 0;
           else if (sigma < 8) B0 = 0;
           else B0 = 1; ]]                                  (2.14)
```

In fact, the logic of this routine can be simplified to give the following more compact version:

```
SHRINK: [[ sigma = A1 + A2 + A3 + A4 + A5 + A6 + A7 + A8;
           if (sigma < 8) B0 = 0; else B0 = A0; ]]          (2.15)
```

Note that the process of shrinking[4] dark objects also expands light objects, including the light background. It also expands holes in dark objects. The opposite process, that of expanding dark objects (or shrinking light ones), is achieved (Fig. 2.12*b*) with the routine:

$$\begin{aligned}\text{EXPAND: [[sigma} &= \text{A1} + \text{A2} + \text{A3} + \text{A4} + \text{A5} + \text{A6} + \text{A7} + \text{A8;} \\ &\text{if (sigma} > 0\text{) B0} = 1\text{; else B0} = \text{A0;]]}\end{aligned} \tag{2.16}$$

Each of these routines employs the same technique for interrogating neighboring pixels in the original image: as will be apparent on numerous occasions in this book, the sigma value is a useful and powerful descriptor for 3×3 pixel neighborhoods. Thus, "if (sigma > 0)" can be taken to mean "if next to a dark object," and the consequence can be read as "then expand it." Similarly, "if (sigma< 8)" can be taken to mean "if next to a light object" or "if next to light background," and the consequence can be read as "then expand the light background into the dark object."

The process of finding the edge of a binary object has several possible interpretations. Clearly, it can be assumed that an edge point has a sigma value in the range 1–7 inclusive. However, it may be defined as being within the object, within the background, or in either position. Taking the definition that the edge of an object has to lie within the object (Fig. 2.12*c*), the following edge-finding routine for binary images results:

$$\begin{aligned}\text{EDGE: [[sigma} &= \text{A1} + \text{A2} + \text{A3} + \text{A4} + \text{A5} + \text{A6} + \text{A7} + \text{A8;} \\ &\text{if (sigma} == 8\text{) B0} = 0\text{; else B0} = \text{A0;]]}\end{aligned} \tag{2.17}$$

This strategy amounts to canceling out object pixels that are not on the edge. For this and a number of other algorithms (including the SHRINK and EXPAND algorithms already encountered), a thorough analysis of exactly which pixels should be set to 1 and 0 (or which should be retained and which eliminated), involves drawing up tables of the form:

		sigma	
EDGE		**0 to 7**	**8**
A0	0	0	0
	1	1	0

4 The processes of shrinking and expanding are also widely known by the respective terms *erosion* and *dilation*. (See also Chapter 8.)

This reflects the fact that algorithm specification includes a recognition phase and an action phase. That is, it is necessary first to locate situations within an image where (for example) edges are to be marked or noise eliminated, and then action must be taken to implement the change.

Another function that can usefully be performed on binary images is the removal of "salt and pepper" noise, that is, noise that appears as a light spot on a dark background or a dark spot on a light background. The first problem to be solved is that of recognizing such noise spots; the second is the simpler one of correcting the intensity value. For the first of these tasks, the sigma value is again useful. To remove salt noise (which has binary value 0 in our convention), we arrive at the following routine:

```
SALT: [[ sigma = A1 + A2 + A3 + A4 + A5 + A6 + A7 + A8;
        if (sigma == 8) B0 = 1; else B0 = A0; ]]                    (2.18)
```

which can be read as leaving the pixel intensity unchanged unless it is proven to be a salt noise spot. The corresponding routine for removing pepper noise (binary value 1) is:

```
PEPPER: [[ sigma = A1 + A2 + A3 + A4 + A5 + A6 + A7 + A8;
          if (sigma == 0) B0 = 0; else B0 = A0; ]]                  (2.19)
```

Combining these two routines into one operation (Fig. 2.13*a*) gives:

```
NOISE: [[ sigma = A1 + A2 + A3 + A4 + A5 + A6 + A7 + A8;
         if (sigma == 0) B0 = 0;
         else if (sigma == 8) B0 = 1;
         else B0 = A0; ]]                                           (2.20)
```

The routine can be made less stringent in its specification of noise pixels, so that it removes spurs on objects and background. This is achieved (Fig. 2.13*b*) by a variant such as

```
NOIZE: [[ sigma = A1 + A2 + A3 + A4 + A5 + A6 + A7 + A8;
         if (sigma < 2) B0 = 0;
         else if (sigma > 6) B0 = 1;
         else B0 = A0; ]]                                           (2.21)
```

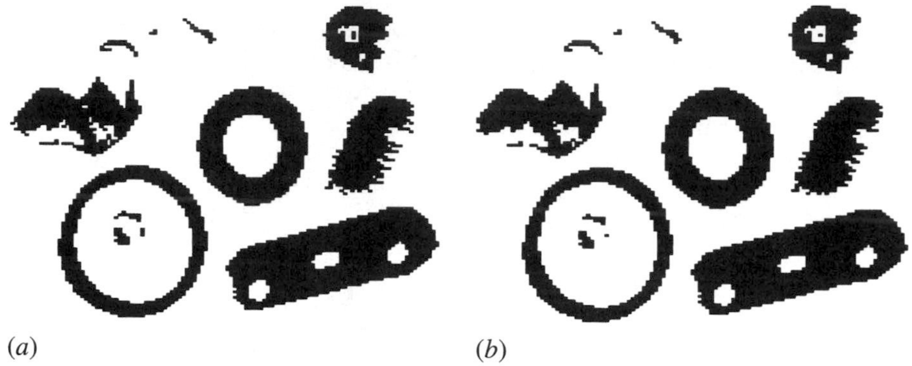

(*a*) (*b*)

Figure 2.13 Simple binary noise removal operations: (*a*) result of applying a "salt and pepper" noise removal operation to the thresholded image in Fig. 2.11*b*; (*b*) result of applying a less stringent noise removal routine: this is effective in cutting down the jagged spurs that appear on some of the objects.

As before, if any doubt about the algorithm exists, its specification should be set up rigorously—in this case with the following table:

		sigma		
NOIZE		**0 or 1**	**2 to 6**	**7 or 8**
A0	0	0	0	1
	1	0	1	1

Many other simple operations can usefully be applied to binary images, some of which are dealt with in Chapter 6. Meanwhile, we should consider the use of common logical operations such as NOT, AND, OR, EXOR. The NOT function is related to the inversion operation described earlier, in that it produces the complement of a binary image:

$$\text{NOT: [[B0} = \text{! A0;]]} \tag{2.22}$$

This has the same action as

$$\text{INV [[B0} = 1 - \text{A0;]]} \tag{2.23}$$

The other logical operations operate on two or more images to produce a third image. This may be illustrated for the AND operation:

$$\text{AND: [[C0} = \text{A0 \&\& B0;]]} \tag{2.24}$$

This has the effect of masking off those parts of A-space that are not within a certain region defined in B-space (or locating those objects that are visible in both A-space and B-space).

2.2.3 *Noise Suppression by Image Accumulation*

The AND routine combines information from two or more images to generate a third image. It is frequently useful to average gray-scale images in a similar way in order to suppress noise. A routine for achieving this is:

$$\text{AVERAGE: [[R0} = (\text{P0} + \text{Q0})/2\text{;]]} \tag{2.25}$$

This idea can be developed so that noisy images from a TV camera or other input device are progressively averaged until the noise level is low enough for useful application. It is, of course, well known that it is necessary to average 100 measurements to improve the signal-to-noise ratio by a factor of 10: that is, the signal-to-noise ratio is multiplied only by the square root of the number of measurements averaged. Since whole-image operations are computationally costly, it is usually practicable to average only up to 16 images. To achieve this the routine shown in Fig. 2.14 is applied. Note that overflow from the usual single byte per pixel storage limit will prevent this routine from working properly if the input images do not have low enough intensities. Sometimes, it may be permissible to reserve more than one byte of storage per pixel. If this is not possible, a useful technique is to employ the "learning" routine of Fig. 2.15. The action of this routine is somewhat unusual, for it retains one-sixteenth of the last image obtained from the camera and progressively lower portions of all earlier images. The effectiveness of this technique can be seen in Fig. 2.16.

```
ADDAVG: {
   [[ P0 = 0; ]]
   for (i = 1; i <= 16; i++) {
      input(Q); // obtain new image in Q-space
      [[ P0 = P0 + Q0; ]]
   }
   [[ P0 = P0 / 16; ]]
}
```

Figure 2.14 Routine for averaging a set of input images.

```
LEARNAVG: {
    input(P); // obtain new image in P-space
    for (i = 1; i <= 15; i++) {
        input(Q); // obtain new image in Q-space
        [[ P0 = (P0 * 15 + Q0 + 8) / 16; ]]
    }
}
```

Figure 2.15 Alternative "learning" routine for averaging images.

Figure 2.16 Noise suppression by averaging a number of images using the LEARNAVG routine of Fig. 2.15. Note that this is highly effective at removing much of the noise and does not introduce any blurring. (Clearly this is true only if, as here, there is no camera or object motion.) Compare this image with that of Fig. 2.1*a* for which no averaging was used.

Note that in this last routine 8 is added before dividing by 16 in order to offset rounding errors. If accuracy is not to suffer, this is a vital feature of most imaging routines that involve small-integer arithmetic. Surprisingly, this problem is seldom referred to in the literature, although it must account for many attempts at perfectly good algorithm strategies that mysteriously turn out not to work well in practice!

2.3 Convolutions and Point Spread Functions

Convolution is a powerful and widely used technique in image processing and other areas of science. Because it appears in many applications throughout this book, it is useful to introduce it at an early stage. We start by defining the convolution of two functions $f(x)$ and $g(x)$ as the integral:

$$f(x) * g(x) = \int_{-\infty}^{\infty} f(u)g(x-u)\mathrm{d}u \tag{2.26}$$

The action of this integral is normally described as the result of applying a point spread function $g(x)$ to all points of a function $f(x)$ and accumulating the contributions at every point. It is significant that if the point spread function (PSF) is very narrow,[5] then the convolution is identical to the original function $f(x)$. This makes it natural to think of the function $f(x)$ as having been spread out under the influence of $g(x)$. This argument may give the impression that convolution necessarily blurs the original function, but this is not always so if, for example, the PSF has a distribution of positive and negative values.

When convolution is applied to digital images, the above formulation changes in two ways: (1) a double integral must be used in respect of the two dimensions; and (2) integration must be changed into discrete summation. The new form of the convolution is:

$$F(x, y) = f(x, y) * g(x, y) = \sum_i \sum_j f(i, j)g(x-i, y-j) \tag{2.27}$$

where g is now referred to as a spatial convolution mask. When convolutions are performed on whole images, it is usual to restrict the sizes of masks as far as possible in order to save computation. Thus, convolution masks are seldom larger than 15 pixels square, and 3×3 masks are typical. The fact that the mask has to be inverted before it is applied is inconvenient for visualizing the process of convolution. In this book we therefore present only preinverted masks of the form:

$$h(x, y) = g(-x, -y) \tag{2.28}$$

5 Formally, it can be a delta function, which is infinite at one point and zero elsewhere while having an integral of unity.

Convolution can then be calculated using the more intuitive formula:

$$F(x, y) = \sum_i \sum_j f(x + i, y + j)h(i, j) \tag{2.29}$$

Clearly, this involves multiplying corresponding values in the modified mask and the neighborhood under consideration. Reexpressing this result for a 3×3 neighborhood and writing the mask coefficients in the form:

$$\begin{bmatrix} \mathrm{h4} & \mathrm{h3} & \mathrm{h2} \\ \mathrm{h5} & \mathrm{h0} & \mathrm{h1} \\ \mathrm{h6} & \mathrm{h7} & \mathrm{h8} \end{bmatrix}$$

the algorithm can be obtained in terms of our earlier notation:

$$\begin{aligned} \text{CONVOLV: [[} \mathrm{Q0} = {} & \mathrm{P0} * \mathrm{h0} + \mathrm{P1} * \mathrm{h1} + \mathrm{P2} * \mathrm{h2} + \mathrm{P3} * \mathrm{h3} + \mathrm{P4} * \mathrm{h4} \\ & + \mathrm{P5} * \mathrm{h5} + \mathrm{P6} * \mathrm{h6} + \mathrm{P7} * \mathrm{h7} + \mathrm{P8} * \mathrm{h8}; \text{]]} \end{aligned} \tag{2.30}$$

We are now in a position to apply convolution to a real situation. At this stage we merely extend the ideas of the previous section, attempting to suppress noise by averaging not over corresponding pixels of different images, but over nearby pixels in the same image. A simple way of achieving this is to use the convolution mask:

$$\frac{1}{9}\begin{bmatrix} 1 & 1 & 1 \\ 1 & 1 & 1 \\ 1 & 1 & 1 \end{bmatrix}$$

where the number in front of the mask weights all the coefficients in the mask and is inserted to ensure that applying the convolution does not alter the mean intensity in the image. As hinted earlier, this particular convolution has the effect of blurring the image as well as reducing the noise level (Fig. 2.17). More will be said about this in the next chapter.

Convolutions are linear operators and are the most general spatially invariant linear operators that can be applied to a signal such as an image. Note that linearity is often of interest in that it permits mathematical analysis to be performed that would otherwise be intractable.

Figure 2.17 Noise suppression by neighborhood averaging achieved by convolving the original image of Fig. 2.1*a* with a uniform mask within a 3 × 3 neighborhood. Note that noise is suppressed only at the expense of introducing significant blurring.

2.4 Sequential versus Parallel Operations

Most of the operations defined so far have started with an image in one space and finished with an image in a different space. Usually, P-space was used for an initial gray-scale image and Q-space for the processed image, or A-space and B-space for the corresponding binary images. This may seem somewhat inconvenient, since if routines are to act as interchangeable modules that can be applied in any sequence, it will be necessary to interpose restoring routines[6] between modules:

$$\text{RESTORE: [[P0 = Q0;]]} \tag{2.31}$$

6 Restoring is not actually necessary within a sequence of routines if the computer is able to keep track of where the currently useful images are stored and to modify the routines accordingly. Nonetheless, it can be a particular problem when it is wished to see a sequence of operations as they occur, using a frame store with only one display space.

Unfortunately, many of the operations cannot work satisfactorily if we do not use separate input and output spaces in this way. This is because they are inherently "parallel processing" routines. This term is used in as much as these are the types of processes that would be performed by a parallel computer possessing a number of processing elements equal to the number of pixels in the image, so that all the pixels are processed simultaneously. If a serial computer is to *simulate* the operation of a parallel computer, then it must have separate input and output image spaces and rigorously work in such a way that it uses the original image values to compute the output pixel values. This means that an operation such as the following cannot be an ideal parallel process:

$$\begin{aligned}\text{BADSHRINK: [[sigma} &= \text{A1} + \text{A2} + \text{A3} + \text{A4} + \text{A5} + \text{A6} + \text{A7} + \text{A8;} \\ &\text{if (sigma} < 8\text{) A0} = 0\text{; else A0} = \text{A0;]]}\end{aligned} \tag{2.32}$$

This is so because, when the operation is half completed, the output pixel intensity will depend not only on some of the unprocessed pixel values but also on some values that have already been processed. For example, if the computer makes a normal (forward) TV raster scan through the image, the situation at a general point in the scan will be

$$\begin{array}{|ccc|}\hline \surd & \surd & \surd \\ \surd & \times & \times \\ \times & \times & \times \\ \hline\end{array}$$

where the ticked pixels have already been processed and the others have not. As a result, the BADSHRINK routine will shrink all objects to nothing!

A much simpler illustration is obtained by attempting to shift an image to the right using the following routine:

$$\text{BADSHIFT: [[P0} = \text{P5;]]} \tag{2.33}$$

All this achieves is to fill up the image with values corresponding to those off its left edge,[7] whatever they are assumed to be. Thus, we have shown that the BADSHIFT process is inherently parallel.

---

7 Note that when the computer is performing a 3×3 (or larger) window operation, it has to assume some value for off-image pixel intensities. Usually, whatever value is selected will be inaccurate, and so the final processed image will contain a border that is also inaccurate. This will be so whether the off-image pixel addresses are trapped in software or in specially designed circuitry in the frame store.

As will be seen later, some processes are inherently sequential—that is, the processed pixel has to be returned immediately to the *original* image space. Meanwhile, note that not all of the routines described so far need to be restricted rigorously to parallel processing. In particular, all single-pixel routines (essentially, those that only refer to the single pixel in a 1×1 neighborhood) can validly be performed as if they were sequential in nature. Such routines include the following inverting, intensity adjustment, contrast stretching, thresholding, and logical operations:

$$\text{INVERT: [[P0} = 255 - \text{P0;]]} \tag{2.34}$$

$$\text{BRIGHTEN: [[P0} = \text{P0} + \text{beta;]]} \tag{2.35}$$

$$\text{STRETCH: [[P0} = \text{P0} * \text{gamma} + \text{beta;]]} \tag{2.36}$$

$$\text{THRESH: [[if (P0} > \text{thresh) P0} = 1; \text{else P0} = 0; \text{]]} \tag{2.37}$$

$$\text{NOT: [[A0} = \text{! A0;]]} \tag{2.38}$$

$$\text{AND: [[A0} = \text{A0 \&\& B0;]]} \tag{2.39}$$

These remarks are intended to act as a warning. It may well be safest to design algorithms that are exclusively parallel processes unless there is a definite need to make them sequential. Later we will see how this need can arise.

2.5 Concluding Remarks

This chapter has introduced a compact notation for representing imaging operations and has demonstrated some basic parallel processing routines. The following chapter extends this work to see how noise suppression can be achieved in gray-scale images. This leads to more advanced image analysis work that is directly relevant to machine vision applications. In particular, Chapter 4 studies in more detail the thresholding of gray-scale images, building on the work of Section 2.2.1, while Chapter 6 studies object shape analysis in binary images.

Pixel–pixel operations can be used to make radical changes in digital images. However, this chapter has shown that window–pixel operations are far more powerful and capable of performing all manner of size-and-shape changing operations, as well as eliminating noise. But *caveat emptor*—sequential operations can have some odd effects if adventitiously applied.

2.6 Bibliographical and Historical Notes

Since this chapter seeks to provide a succinct overview of basic techniques rather than cover the most recent material, it will not be surprising that most of the topics discussed were discovered at least 20 years ago and have been used by a large number of workers in many areas. For example, thresholding of gray-scale images was first reported at least as long ago as 1960, while shrinking and expanding of binary picture objects date from a similar period. Averaging of a number of images to suppress noise has been reported many times. Although the learning routine described here has not been found in the literature, it has long been used in the author's laboratory, and it was probably developed independently elsewhere.

Discussion of the origins of other techniques is curtailed: for further detail the reader is referred to the texts by (for example) Gonzalez and Woods (1992), Hall (1979), Jähne and Haussecker (2000), Nixon and Aguado (2002), Petrou and Bosdogianni (1999), Pratt (2001), Rosenfeld and Kak (1981), Russ (1998), and Sonka et al. (1999). More specialized texts will be referred to in the following chapters. (Note that as a matter of policy this book refers to a good many other specialized texts that have the space to delve more deeply into specific issues.) Here we refer to three texts that cover programming aspects of image processing in some depth: Parker (1994), which covers C programming; Whelan and Molloy (2001), which covers Java programming; and Batchelor (1991), which covers Prolog programming.

2.7 Problems

1. Derive an algorithm for finding the edges of binary picture objects by applying a SHRINK operation and combining the result with the original image. Is the result

the same as that obtained using the edge-finding routine (2.17)? Prove your statement rigorously by drawing up suitable algorithm tables as in Section 2.2.2.

2. In a certain frame store, each off-image pixel can be taken to have either the value 0 or the intensity for the nearest image pixel. Which will give the more meaningful results for (a) shrinking, (b) expanding, and (c) blurring convolution?

3. Suppose the NOISE and NOIZE routines of Section 2.2.2 were reimplemented as sequential algorithms. Show that the action of NOISE would be unchanged, whereas NOIZE would produce very odd effects on some binary images.

Basic Image Filtering Operations

Image filtering involves the application of window operations that achieve useful effects, such as noise removal or image enhancement. This chapter deals particularly with what can be achieved with quite basic approaches, such as application of local mean, median, or mode filters to digital images. The focus is on gray-scale images, though some aspects of color processing are also covered.

Look out for:

- what can be achieved by low-pass filtering in the spatial frequency domain.
- how the same process can be carried out by convolution in the spatial domain.
- the problem of impulse noise and what can be achieved with a limiting filter.
- the value of median, mode, and rank order filters.
- how computational load can be reduced.
- what can be achieved by sharp–unsharp masking.
- the distinction between image enhancement and image restoration.
- the distortions produced by standard filters—mean, Gaussian, median, mode, and rank order filters.

This chapter delves into the properties of a variety of standard types of filters in order to show what they can achieve and what their limitations are. The edge shifts produced by most of these filters are small but predictable, and therefore correctable in principle. The exception is the rank order filter, for which the shifts can be large—but then this is the *advantage* of this type of filter and is at the core of mathematical morphology (see Chapter 8).

CHAPTER 3

Basic Image Filtering Operations

3.1 Introduction

In this chapter the discussion is extended to noise suppression and enhancement in gray-scale images. Although these types of operations can for the most part be avoided in industrial applications of vision, it is useful to examine them in some depth because of their wide use in a variety of other image processing applications and because they set the scene for much of what follows. In addition, some fundamental issues come to light which are of vital importance.

It has already been seen that noise can arise in real images; hence it is necessary to have sound techniques for suppressing it. Commonly, in electrical engineering applications, noise is removed by means of low-pass or other filters that operate in the frequency domain (Rosie, 1966). Applying these filters to 1-D time-varying analog signals is straightforward, since it is necessary only to place them at suitable stages in the sequence of black boxes through which the signals pass. For digital signals, the situation is more complicated, since first the frequency transform of the signal must be computed, then the low-pass filter applied, and finally the signal obtained from the modified transform by converting back to the time domain. Thus, two Fourier transforms have to be computed, although modifying the signal while it is in the frequency domain is a straightforward task (Fig. 3.1). It is by now well known that the amount of processing involved in computing a discrete Fourier transform of a signal represented by N samples is of order N^2 (we shall write this as $O(N^2)$), and that the amount of computation can be cut down to $O(N \log_2 N)$ by employing the fast Fourier transform (FFT) (Gonzalez and Woods, 1992). This then becomes a practical approach for the elimination of noise.

When applying these ideas to images, we must first note that the signal is a spatial rather than a time-varying quantity and must be filtered in the spatial frequency domain. Mathematically, this makes no real difference, but there are nevertheless significant problems. First, there is no satisfactory analog shortcut,

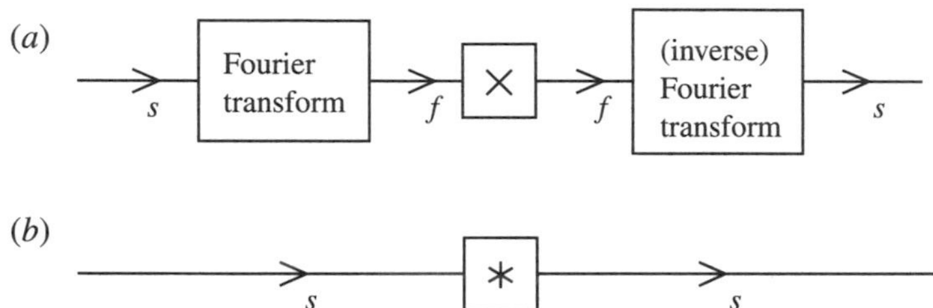

Figure 3.1 Low-pass filtering for noise suppression: *s*, spatial domain; *f*, spatial frequency domain; ×, multiplication by low-pass characteristic; ∗, convolution with Fourier transform of low-pass characteristics. (*a*) Low-pass filtering achieved most simply, by a process of multiplication in the (spatial) frequency domain; (*b*) low-pass filtering achieved by a process of convolution. Note that (*a*) may require more computation overall because of the two Fourier transforms that have to be performed.

and the whole process has to be carried out digitally. (Here we ignore optical processing methods despite their obvious power, speed, and high resolution, because they are by no means trivial to marry with digital computer technology; also, while there is much ongoing research in this area, it is difficult to anticipate the future position with any accuracy.) Second, for an $N \times N$ pixel image, the number of operations required to compute a Fourier transform is $O(N^3)$ and the FFT only reduces this to $O(N^2 \log_2 N)$, so the amount of computation is quite considerable. (Here it is assumed that the 2-D transforms are implemented by successive passes of 1-D transforms: see Gonzalez and Woods, 1992.) Note also that two Fourier transforms are required for the purpose of noise suppression (Fig. 3.1). Nevertheless, in many imaging applications it is worth proceeding in this way, not only so that noise can be removed but also so that TV scan lines and other artifacts can be filtered out. This situation particularly applies in remote sensing and space technology. However, in industrial applications the emphasis is always on real-time processing; in many cases, therefore, it is not practicable to remove noise by spatial frequency domain operations. A secondary problem is that low-pass filtering is suited to removing Gaussian noise but distorts the image if it is used to remove impulse noise.

Section 3.2 discusses Gaussian smoothing in both the spatial frequency and the spatial domains. The subsequent three sections introduce median filters, mode filters, and then general rank order filters, and contrast their main properties and uses. In Section 3.6 consideration is given to reducing computational load—with particular reference to the median filter. Section 3.7 introduces the sharp–unsharp masking technique, which provides a rather simple route to image enhancement. Then follow a number of sections that concentrate on the edge shifts produced by the various filters. In the case of the median filter, the discrete theory (Section 3.9) is much more exact than the continuum model (Section 3.8). All edge shifts are quite small, except for rank order filters

(Section 3.12); these are treated fairly fully because of their relevance to widely used morphological operators (Chapter 8) where the shifts are turned to advantage.

3.2 Noise Suppression by Gaussian Smoothing

Low-pass filtering is normally thought of as the elimination of signal components with high spatial frequencies; it is therefore natural to carry it out in the spatial frequency domain. Nevertheless, it is possible to implement it directly in the spatial domain. That this is possible is due to the well-known fact (Rosie, 1966) that multiplying a signal by a function in the spatial frequency domain is equivalent to convolving it with the Fourier transform of the function in the spatial domain (Fig. 3.1). If the final convolving function in the spatial domain is sufficiently narrow, then the amount of computation involved will not be excessive. In this way, a satisfactory implementation of the low-pass filter can be sought. It now remains to find a suitable convolving function.

If the low-pass filter is to have a sharp cutoff, then its transform in image space will be oscillatory. An extreme example is the sinc ($\sin x/x$) function, which is the spatial transform of a low-pass filter of rectangular profile (Rosie, 1966). It turns out that oscillatory convolving functions are unsatisfactory since they can introduce halos around objects, hence distorting the image quite grossly. Marr and Hildreth (1980) suggested that the right types of filters to apply to images are those that are well behaved (nonoscillatory) both in the frequency and in the spatial domain. Gaussian filters are able to fulfill this criterion optimally: they have identical forms in the spatial and spatial frequency domains. In 1-D these forms are

$$f(x) = \left[1/(2\pi\sigma^2)^{1/2}\right] \exp\left[-x^2/(2\sigma^2)\right] \tag{3.1}$$

$$F(\omega) = \exp(-\sigma^2\omega^2/2) \tag{3.2}$$

Thus, the type of spatial convolving operator required for the purpose of noise suppression by low-pass filtering is one that approximates to a Gaussian profile. Many such approximations appear in the literature; these vary with the size of the neighborhood chosen and in the precise values of the convolution mask coefficients.

One of the most common masks is the following one, first introduced in Chapter 2, which is used more for simplicity of computation than for its fidelity to a

Gaussian profile:

$$\frac{1}{9}\begin{bmatrix} 1 & 1 & 1 \\ 1 & 1 & 1 \\ 1 & 1 & 1 \end{bmatrix}$$

Another commonly used mask, which approximates more closely to a Gaussian profile, is the following:

$$\frac{1}{16}\begin{bmatrix} 1 & 2 & 1 \\ 2 & 4 & 2 \\ 1 & 2 & 1 \end{bmatrix}$$

In both cases, the coefficients that precede the mask are used to weight all the mask coefficients. As mentioned in Section 2.2.3, these weights are chosen so that applying the convolution to an image does not affect the average image intensity. These two convolution masks probably account for over 80% of all discrete approximations to a Gaussian. Notice that as they operate within a 3×3 neighborhood, they are reasonably narrow and hence incur a relatively small computational load.

Let us next study the properties of this type of operator, deferring for now consideration of Gaussian operators in larger neighborhoods. First, imagine such an operator applied to a noisy image whose intensity is inherently uniform. Then clearly noise is suppressed, as it is now averaged over 9 pixels. This averaging model is obvious for the first of the two masks above but in fact applies equally to the second mask, once it is accepted that the averaging effect is differently distributed in accordance with the improved approximation to a Gaussian profile.

Although this example shows that noise is suppressed, it will be plain that the signal is also affected. This problem arises only where the signal is initially nonuniform. Indeed, if the image intensity is constant, or if the intensity map approximates to a plane, there is again no problem. However, if the signal is uniform over one part of a neighborhood and rises in another part of it, as is bound to occur adjacent to the edge of an object, then the object will make itself felt at the center of the neighborhood in the filtered image (see Fig. 3.2). As a result, the edges of objects become somewhat blurred. Looking at the operator as a "mixing operator" that forms a new picture by mixing together the intensities of pixels fairly close to each other, it is intuitively obvious why blurring occurs.

It is also apparent, from a spatial frequency viewpoint, why blurring should occur. Basically, we are aiming to give the signal a sharp cutoff in the spatial frequency domain, and as a result it will become slightly blurred in the spatial domain. Clearly, the blurring effect can be reduced by using the narrowest

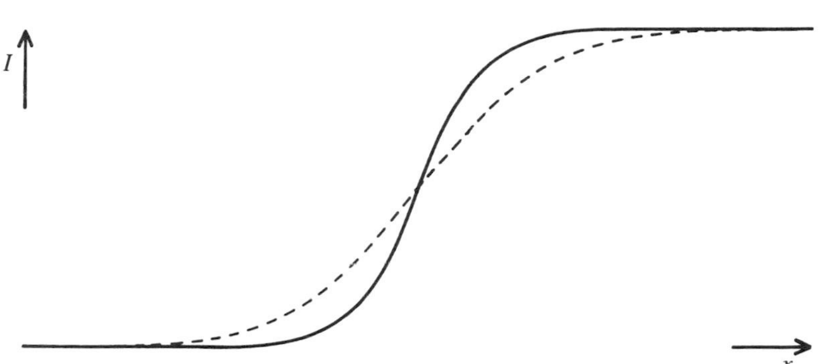

Figure 3.2 Blurring of object edges by simple Gaussian convolutions. The simple Gaussian convolution can be regarded as a gray-scale neighborhood "mixing" operator, hence explaining why blurring arises.

possible approximation to a Gaussian convolution filter, but at the same time the noise suppression properties of the filter are lessened. Assuming that the image was initially digitized at roughly the correct spatial resolution, it will not normally be appropriate to smooth it using convolution masks larger than 3×3 or at most 5×5 pixels. (Here we ignore methods of analyzing images that use a number of versions of the image with different spatial resolutions: see, for example, Babaud et al., 1986.)

Overall, low-pass filtering and Gaussian smoothing are not appropriate for the applications considered here because they introduce blurring effects. Notice also that where interference occurs, which can give rise to impulse or "spike" noise (corresponding to a number of individual pixels having totally the wrong intensities), merely averaging this noise over a larger neighborhood can make the situation worse, since the spikes will be smeared over a sizable number of pixels and will distort the intensity values of all of them. This consideration is important, leading naturally to the concepts of limit and median filtering.

3.3 Median Filters

The idea explored here is to locate those pixels in the image which have extreme and therefore highly improbable intensities and to ignore their actual intensities, replacing them with more suitable values. This is akin to drawing a graph through a set of plots and ignoring those plots that are evidently a long way from the best fit curve. An obvious way of using this technique is to apply a "limit" filter that prevents any pixel from having an intensity outside the intensity range of its neighbors:

$$
\begin{aligned}
\text{LIMIT: [[} & \min \mathrm{P} = \min(\mathrm{P1}, \mathrm{P2}, \mathrm{P3}, \mathrm{P4}, \mathrm{P5}, \mathrm{P6}, \mathrm{P7}, \mathrm{P8}); \\
& \max \mathrm{P} = \max(\mathrm{P1}, \mathrm{P2}, \mathrm{P3}, \mathrm{P4}, \mathrm{P5}, \mathrm{P6}, \mathrm{P7}, \mathrm{P8}); \\
& \text{if} (\mathrm{P0} < \min \mathrm{P})\ \mathrm{Q0} = \min \mathrm{P}; \\
& \text{else if } (\mathrm{P0} > \max \mathrm{P})\ \mathrm{Q0} = \max \mathrm{P}; \\
& \text{else } \mathrm{Q0} = \mathrm{P0}; \\
]] &
\end{aligned}
\tag{3.3}
$$

To develop this technique, it is necessary to examine the local intensity distribution within a particular neighborhood. Points at the extremes of the distribution are quite likely to have arisen from impulse noise. Not only is it sensible to eliminate these points, as in the limit filter, but it is reasonable to try taking the process further, removing equal areas at either end of the distribution and ending with the median. Thus, we arrive at the median filter, which takes all the local intensity distributions and generates a new image corresponding to the set of median values. As the preceding argument indicates, the median filter is excellent at impulse noise suppression; this is amply confirmed in practice (see Fig. 3.3).

Figure 3.3 Effect of applying a 3 × 3 median filter to the image of Fig. 2.1*a*. Note the slight loss of fine detail and the rather "softened" appearance of the whole image.

In view of the blurring caused by Gaussian smoothing operators, it is pertinent to ask whether the median filter also induces blurring. Figure 3.3 shows that any blurring is only marginal, although some slight loss of fine detail takes place, which can give the resulting pictures a "softened" appearance. Theoretical discussion of this point is deferred for now. The lack of blur makes good the main deficiency of the Gaussian smoothing filter and results in the median filter being almost certainly the single most widely used filter in general image processing applications.

The median filter may be implemented in many ways: Figure 3.4 reproduces only an obvious algorithm that essentially implements the above description. The notation of Chapter 2 is used but is augmented suitably to permit the nine pixels in a 3×3 neighborhood to be accessed in turn with a running suffix (specifically, P0 to P8 are written as P[m], where m runs from 0 to 8).

The operation of the algorithm is as follows: first, the histogram array is cleared and the image is scanned, generating a new image in Q-space; next, for each neighborhood, the histogram of intensity values is constructed; then the median is found; and, finally, points in the histogram array that have been incremented are cleared. This last feature eliminates the need to clear the whole histogram and hence saves computation. Further savings could be made by finding the extreme intensity values and limiting the search for the median to this range. Unlike the general situation when the median of a distribution is being located, only one (half-) scan through the distribution is required, since the total area is known in advance (in this case it is 9).

As is clear from the above, methods of computing the median involve pixel intensity sorting operations. If a bubble sort (Gonnet, 1984) were used for this purpose, then up to $O(n^4)$ operations would be required for an $n \times n$ neighborhood, compared with some 256 operations for the histogram method we have described. Thus, sorting methods such as the bubble sort are faster for

```
MEDIAN: {
    for (i = 0; i <= 255; i++) hist[i] = 0;
    [[ for (m = 0; m <= 8; m++) hist[ P[m] ]++;
        i = 0; sum = 0;
        while (sum < 5) {
            sum = sum + hist[i];
            i = i + 1;
        }
        Q0 = i – 1;
        for (m = 0; m <= 8; m++) hist[ P[m] ] = 0;
    ]]
}
```

Figure 3.4 An implementation of the median filter.

small neighborhoods where n is 3 or 4 but not for neighborhoods where n is greater than about 5, or where pixel intensity values are more restricted.

Much of the discussion of the median filter in the literature is concerned with saving computation (Narendra, 1978; Huang et al., 1979; Danielsson, 1981). In particular, it has been observed that, on proceeding from one neighborhood to the next, relatively few new pixels are encountered. Thus, the new median value can be found by updating the old value rather than starting from scratch (Huang et al., 1979).

3.4 Mode Filters

Having considered the mean and the median of the local intensity distribution as candidate intensity values for noise smoothing filters, we also find it relevant to consider the mode of the distribution. Indeed, we might imagine that this is, if anything, more important than the mean or the median, since the mode represents the most probable value of any distribution.

A tedious problem arises, however, as soon as we attempt to apply this idea. The local intensity distribution is calculated from relatively few pixel intensity values (Fig. 3.5). This means that instead of a smooth intensity distribution whose mode is easily located, we are almost certain to have a multimodal distribution whose highest point does not indicate the position of the *underlying* mode. Clearly, the distribution needs to be smoothed out considerably before the mode is computed. Another tedious problem is that the width of the distribution varies widely from neighborhood to neighborhood (e.g., from close to zero to close to 256), so that it is difficult to know quite how much to smooth the distribution in any instance. For these reasons, it is likely to be better to choose an indirect measure of the position of the mode rather than to attempt to measure it directly.

The position of the mode can be estimated with reasonable accuracy once the median has been located (Davies, 1984a, 1988c). To understand the technique, it is necessary to consider how local intensity distributions of various sorts arise in practical situations. At most positions in an image, variations in pixel intensity are generated by steady changes in background illumination, or by steady variations in surface orientation, or else by noise. Thus, a symmetrical unimodal local intensity distribution is to be expected. It is well known that the mean, median, and mode are coincident in such cases. More problematic is what happens to the intensity variation near the edge of an object in the image. Here the local intensity distribution is unlikely to be symmetrical, and, more important, it may not even be unimodal. In fact, near an edge the distribution is in general inherently *bimodal*, since the neighborhood contains pixels with intensities

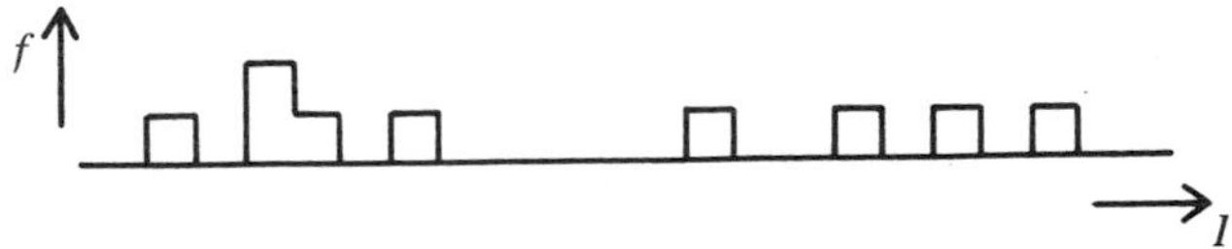

Figure 3.5 The sparse nature of the local intensity histogram for a small neighborhood. This situation clearly causes significant problems for estimation of the mode. It also has definite implications for rigorous estimation of the underlying median, assuming that the observed intensities are only noisy samples of the ideal intensity pattern (see Section 3.8.4).

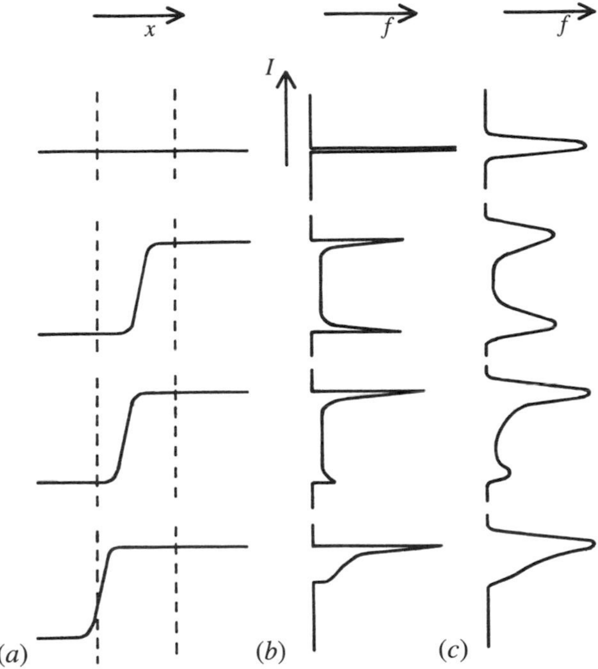

Figure 3.6 Local models of image data near the edge of an object: (*a*) cross sections of an edge falling in the vicinity of a filter neighborhood; (*b*) corresponding local intensity distributions when very little image noise is present; (*c*) situation when the noise level is increased.

corresponding to the values they would have on either side of the edge (Fig. 3.6). Considering the image as a whole, this will be the most likely alternative to a symmetrical unimodal distribution; any further possibilities such as trimodal distributions are rare and of varied causes (e.g., odd glints on the edges of metal objects) and are outside the scope of the present discussion.[1]

1 Here we are ignoring the effects of noise and just considering the underlying image signal.

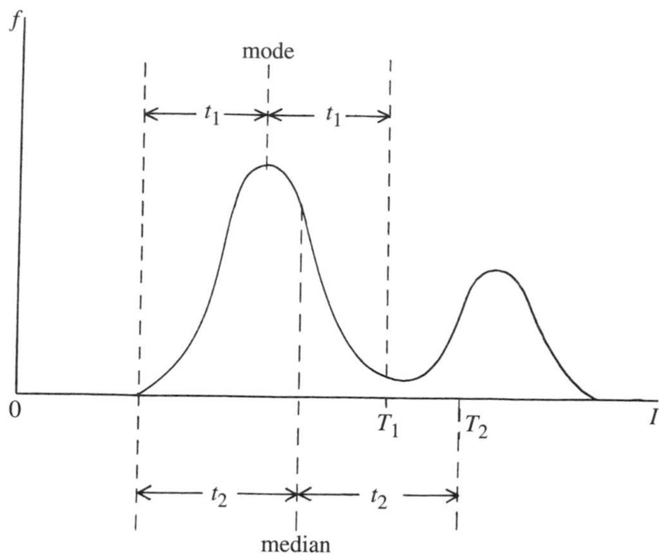

Figure 3.7 Rationale for the method of truncation. The obvious position at which to truncate the distribution is T_1. Since the position of the mode is not initially known, it is suboptimal but safe to truncate instead at T_2.

If the neighborhood straddles an edge and the local intensity distribution is bimodal, the larger peak position should clearly be selected as the most probable intensity value. A good strategy for finding the larger peak is to eliminate the smaller peak. If we knew the position of the mode, we could find where to truncate the smaller peak by first finding which extreme of the distribution was closer to the mode, and then moving an equal distance to the opposite side of the mode (Fig. 3.7). Since we start off *not* knowing the position of the mode, one option is to use the position of the median as an estimator of the position of the mode, and then to use that position to find where to truncate the distribution. Since the three means invariably take the order mean, median, and mode (see Fig. 3.8), except when distributions are badly behaved or multimodal, it turns out that this method is cautious in the sense that it truncates less of the distribution than the required amount: this makes it a safe method to use. When we now find the median of the truncated distribution, the position is much closer to the mode than the original median was, a good proportion of the second peak having been removed (Fig. 3.9). Iteration could be used to find an even closer approximation to the position of the mode. However, the results show that the method gives a marked enhancement in the image even when this is not done (Fig. 3.10).

What has been achieved by the above procedure? When we compare the results of applying a median filter and the new filter (which we call a truncated

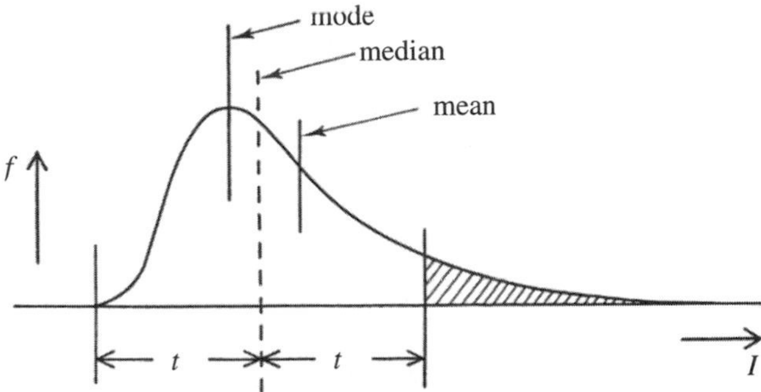

Figure 3.8 Relative positions of the mode, median, and mean for a typical unimodal distribution. This ordering is unchanged for a bimodal distribution, as long as it can be approximated by two Gaussian distributions of similar width.

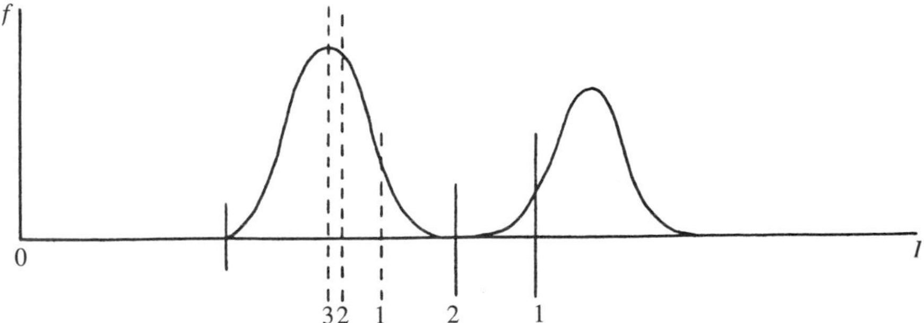

Figure 3.9 Iterative truncation of the local intensity distribution. Here the median converges on the mode within three iterations of the truncation procedure. This is possible since at each stage the mode of the new truncated distribution remains the same as that of the previous distribution.

median filter, for it is not a true mode filter—see below), we find that whereas the median filter is highly successful at removing noise, the new filter not only removes noise but also enhances the image so that edges are made sharper. By reference to Fig. 3.11 it is quite easy to see why this should happen. Basically, at a location even very slightly to one side of an edge, a majority of the pixel intensities contribute to the larger peak and the filter ignores the pixel intensities contributing to the smaller peak. Thus, the filter makes an informed binary choice about which side of the edge it is on. At first this seems to mean that it pushes a nearby edge further away. However, it must be remembered that it actually "pushes the edge away" from both sides, and the result is that its sides are made sharper and object outlines are crispened up. Particularly striking is the effect of applying this filter to an image a number of times, when objects start to

Figure 3.10 Effect of a single application of 3×3 truncated median filter to the image of Fig. 2.1*a*.

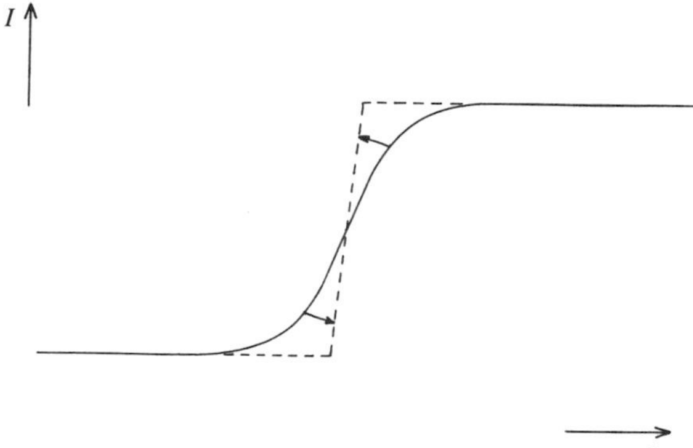

Figure 3.11 Image enhancement performed by the mode filter. Here the onset of the edge is pushed laterally by the action of the mode filter within one neighborhood. Since the same happens from the other side within an adjacent neighborhood, the actual position of the edge is unchanged in first order. The overall effect is to sharpen the edge.

become segmented into regions of fairly uniform intensity (Fig. 3.12). The complete algorithm for achieving this is outlined in Fig. 3.13.

We have dealt with this problem at some length for several reasons. First, it seems that the mode filter has not hitherto received the attention it deserves. Second, the median filter seems to be used fairly universally, often without very

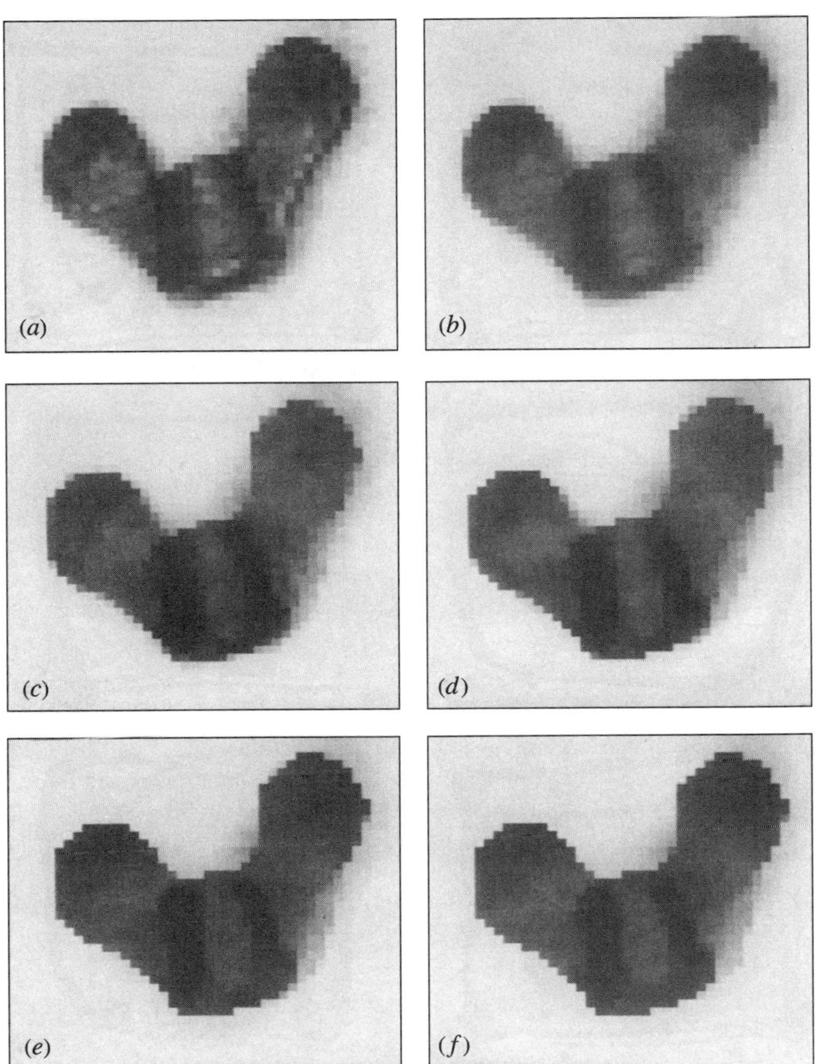

Figure 3.12 Results of repeated action of the truncated median filter: (*a*) the original, moderately noisy picture; (*b*) effect of a 3×3 median filter; (*c*)–(*f*) effect of 1–4 passes of the basic truncated median filter, respectively.

much justification or thought. Third, all these filters show what markedly different characteristics are available merely by analyzing the contents of the local intensity distribution and ignoring totally where in the neighborhood the different intensities appear. It is perhaps remarkable that there is sufficient information in the local intensity distribution for this to be possible. All this shows the danger of applying operators that have been derived in an ad hoc manner without first

```
do { // as many passes over image as necessary
   do { // for each pixel
      compute local intensity distribution;
      do { // iterate to improve estimate of mode
         find minimum, median, and maximum intensity values;
         decide from which end local intensity distribution should be truncated;
         deduce where local intensity distribution should be truncated;
         truncate local intensity distribution;
         find median of truncated local intensity distribution;
      } until median sufficiently close to mode of local distribution;
      transfer estimate of mode to output image space;
   } until all pixels processed;
} until sufficient enhancement of image;
```

Figure 3.13 Outline of algorithm for implementing the truncated median filter. In this algorithm (i) The outermost and innnermost loops can normally be omitted (i.e., they need to be executed once only). (ii) The *final* estimate of the position of the mode can be performed by simple averaging instead of computing the median: this has been found to save computation with negligible loss of accuracy. (iii) Instead of the minimum and maximum intensity values, the positions of the outermost octiles (for example) may be used to give more stable estimates of the extremes of the local intensity distribution.

making a specification of what is required and then designing an operator with the required characteristics. In fact, it appears that if we want a filter that has maximum impulse noise suppression capability, then we should use a median filter; and if we want a filter that enhances images by sharpening edges, then we should use a mode or truncated median filter. (Note that the truncated median filter should in some ways be an improvement on the mode filter in that it is more cautious when very close to an edge transition, where noise prevents an exact judgement from being made as to which side of the edge a pixel is on: see Davies, 1984a, 1988c.)

While considering enhancement, attention has been restricted to filters based on the local intensity distribution; many filters enhance images without the aid of the local intensity distribution (Lev et al., 1977; Nagao and Matsuyama, 1979), but they are not within the scope of this chapter. Note that the method of "sharp–unsharp masking" (Section 3.7) performs an enhancement function, although its main purpose is to restore images that have inadvertently become blurred, for example, by a hazy atmosphere or defocussed camera.

Finally, although this section has concentrated on the gray-scale properties of mode filters, Charles and Davies (2003a, 2004) have recently shown how to devise versions of the mode filter that operate on color images. Typical results are shown in Fig. 2.8. In addition, Fig. 2.9 shows that the mode filter has the hitherto unexpected property of being able to eliminate very large amounts of impulse noise from images—significantly more than a median filter—in spite of being designated as an image enhancement filter.

3.5 Rank Order Filters

The principle employed in rank order filters is to take all the intensity values in a given neighborhood; place these in order of increasing value; and finally select the rth of the n values and return this value as the filter local output value. Clearly, n rank order filters can be specified in terms of the value r that is used, but these filters are all intrinsically nonlinear; that is, the output intensity cannot be expressed as a linear sum of the component intensities within the neighborhood. In particular, the median filter (for which $r=(n+1)/2$, and which is only defined if n is odd[2]) does not normally give the same output image as a mean filter. Indeed, it is well known that the mean and median of a distribution are in general only coincident for symmetrical distributions. Note that minimum and maximum filters (corresponding to $r=1$ and $r=n$, respectively) are also often classed as morphological filters (see Chapter 8).

3.6 Reducing Computational Load

Significant efforts have been made to speed the operation of the Gaussian filter since implementations in large neighborhoods require considerable amounts of computation (Wiejak et al., 1985). For example, smoothing images using a 30 × 30 Gaussian convolution mask in an image of 256 × 256 pixels involves 64 million basic operations. For such a basic operation as smoothing, this is unacceptable. However, the amount of computation can be cut down drastically, since a 2-D Gaussian convolution can be factorized into two 1-D Gaussian convolutions which can be applied in turn:

$$\exp(-r^2/2\sigma^2) = \exp(-x^2/2\sigma^2)\exp(-y^2/2\sigma^2) \tag{3.4}$$

The decomposition is rigorously provable and is not an approximation; we shall refer to this below in the context of the median filter. Meanwhile, the decompositions for the two 3 × 3 Gaussian filters we met earlier are:

$$\frac{1}{9}\begin{bmatrix} 1 & 1 & 1 \\ 1 & 1 & 1 \\ 1 & 1 & 1 \end{bmatrix} = \frac{1}{3}\begin{bmatrix} 1 \\ 1 \\ 1 \end{bmatrix} \frac{1}{3}\begin{bmatrix} 1 & 1 & 1 \end{bmatrix} \tag{3.5}$$

2 If n is even, the mean of the central two values in the distribution is usually taken as representing the median.

and

$$\frac{1}{16}\begin{bmatrix} 1 & 2 & 1 \\ 2 & 4 & 2 \\ 1 & 2 & 1 \end{bmatrix} = \frac{1}{4}\begin{bmatrix} 1 \\ 2 \\ 1 \end{bmatrix}\frac{1}{4}\begin{bmatrix} 1 & 2 & 1 \end{bmatrix} \tag{3.6}$$

Overall, this approach replaces a single $n \times n$ operator whose load is $O(n^2)$ with two operators of load $O(n)$, and for $n > 3$ there are always worthwhile savings. Ignoring scanning and other overheads we are summarize the situation in Table 3.1, the final column of which represents the saving factor.

It turns out that it is not possible to decompose the median filter in the same way without making approximations. However, it is quite common to try to perform a similar function by applying two 1-D median filters in turn (Narendra, 1978). Although the effect is similar in its outlier rejection properties to the corresponding 2-D median filter, by appealing to specific examples of image data it is simple to show that the two are not formally equivalent. An instance in which the same effect is obtained by a 3×3 median filter or by 3×1 and 1×3 median filters applied in either order is the following:

$$\text{original image segment:} \quad \begin{matrix} 0 & 0 & 0 & 0 & 0 & 0 \\ 0 & 0 & 1 & 0 & 0 & 0 \\ 0 & 0 & 0 & 1 & 0 & 0 \\ 0 & 1 & 0 & 0 & 1 & 0 \\ 0 & 0 & 0 & 0 & 0 & 0 \end{matrix}$$

In this case, all the 1's immediately disappear after applying a 3×3 or a 3×1 or a 1×3 median filter. In the following instance, the situation is slightly more complex:

$$\text{original image segment:} \quad \begin{matrix} 0 & 0 & 0 & 0 & 0 & 0 \\ 0 & 0 & 1 & 1 & 2 & 0 \\ 0 & 0 & 2 & 2 & 0 & 0 \\ 0 & 0 & 0 & 0 & 0 & 0 \end{matrix}$$

Table 3.1 Savings achieved by factorizing an $n \times n$ Gaussian operator

n	*2n*	n^2	*n/2*
1	2	1	0.5
3	6	9	1.5
5	10	25	2.5
7	14	49	3.5
9	18	81	4.5
11	22	121	5.5
13	26	169	6.5
15	30	225	7.5

$$\text{after applying a } 3 \times 3 \text{ filter: } \begin{matrix} 0 & 0 & 0 & 0 & 0 & 0 \\ 0 & 0 & 0 & 1 & 0 & 0 \\ 0 & 0 & 0 & 1 & 0 & 0 \\ 0 & 0 & 0 & 0 & 0 & 0 \end{matrix}$$

after applying a 3 × 1 and then a 1 × 3 filter:

$$\begin{matrix} 0 & 0 & 0 & 0 & 0 & 0 \\ 0 & 0 & 1 & 1 & 1 & 0 \\ 0 & 0 & 2 & 2 & 0 & 0 \\ 0 & 0 & 0 & 0 & 0 & 0 \end{matrix} \qquad \begin{matrix} 0 & 0 & 0 & 0 & 0 & 0 \\ 0 & 0 & 1 & 1 & 0 & 0 \\ 0 & 0 & 1 & 1 & 0 & 0 \\ 0 & 0 & 0 & 0 & 0 & 0 \end{matrix}$$

after applying a 3 × 1 and then a 1 × 3 filter:

$$\begin{matrix} 0 & 0 & 0 & 0 & 0 & 0 \\ 0 & 0 & 1 & 1 & 0 & 0 \\ 0 & 0 & 1 & 1 & 0 & 0 \\ 0 & 0 & 0 & 0 & 0 & 0 \end{matrix} \qquad \begin{matrix} 0 & 0 & 0 & 0 & 0 & 0 \\ 0 & 0 & 1 & 1 & 0 & 0 \\ 0 & 0 & 1 & 1 & 0 & 0 \\ 0 & 0 & 0 & 0 & 0 & 0 \end{matrix}$$

The following case tests the situation more thoroughly and is one of the simplest for which the three final image segments are all different:

$$\text{original image segment: } \begin{matrix} 0 & 0 & 0 & 0 & 0 & 0 \\ 0 & 0 & 0 & 1 & 0 & 0 \\ 0 & 0 & 1 & 1 & 1 & 0 \\ 0 & 0 & 1 & 0 & 0 & 0 \\ 0 & 0 & 0 & 0 & 0 & 0 \end{matrix}$$

$$\text{after applying a } 3 \times 3 \text{ filter: } \begin{matrix} 0 & 0 & 0 & 0 & 0 & 0 \\ 0 & 0 & 0 & 0 & 0 & 0 \\ 0 & 0 & 0 & 1 & 0 & 0 \\ 0 & 0 & 0 & 0 & 0 & 0 \\ 0 & 0 & 0 & 0 & 0 & 0 \end{matrix}$$

after applying a 3 × 1 and then a 1 × 3 filter:

$$\begin{matrix} 0 & 0 & 0 & 0 & 0 & 0 \\ 0 & 0 & 0 & 0 & 0 & 0 \\ 0 & 0 & 1 & 1 & 1 & 0 \\ 0 & 0 & 0 & 0 & 0 & 0 \\ 0 & 0 & 0 & 0 & 0 & 0 \end{matrix} \qquad \begin{matrix} 0 & 0 & 0 & 0 & 0 & 0 \\ 0 & 0 & 0 & 0 & 0 & 0 \\ 0 & 0 & 0 & 0 & 0 & 0 \\ 0 & 0 & 0 & 0 & 0 & 0 \\ 0 & 0 & 0 & 0 & 0 & 0 \end{matrix}$$

after applying a 1×3 and then a 3×1 filter:

$$\begin{matrix} 0 & 0 & 0 & 0 & 0 & 0 \\ 0 & 0 & 0 & 1 & 0 & 0 \\ 0 & 0 & 1 & 1 & 0 & 0 \\ 0 & 0 & 1 & 0 & 0 & 0 \\ 0 & 0 & 0 & 0 & 0 & 0 \end{matrix} \qquad \begin{matrix} 0 & 0 & 0 & 0 & 0 & 0 \\ 0 & 0 & 0 & 0 & 0 & 0 \\ 0 & 0 & 1 & 1 & 0 & 0 \\ 0 & 0 & 0 & 0 & 0 & 0 \\ 0 & 0 & 0 & 0 & 0 & 0 \end{matrix}$$

In general, spots and streaks not more than one pixel wide are eliminated quite effectively by the original or by the separated forms of the filter. Larger filters should effectively eliminate wider spots and streaks, although exact functional equivalence between the original and its separated forms is not to be expected, as has been indicated.

Finally, the problem of inexact decomposition is not an exclusive property of nonlinear filters: many linear filters cannot be decomposed exactly either, because the number of independent coefficients in an $n \times n$ mask is n^2 which is much greater than the number in an $n \times 1$ or $1 \times n$ component mask.

3.6.1 *A Bit-based Method for Fast Median Filtering*

A number of other means have been employed to speed the operation of the median filter. Of especial interest is the bit-based method of Ataman et al. (1980), since it is surprisingly simple, yet highly effective. Space does not permit a full account of the method to be given here, so a simplified version retaining the same main underlying principle is given.

The basic idea derives from the observation that only the most significant bit (MSB) of each of the original numbers is relevant for computing the MSB of the median. This suggests that it ought to be possible to determine the median in a number of operations of the order of the number of bits required to represent the original numbers. Since pixel intensities are frequently represented by just 6 to 8 bits of data, this makes the idea a particularly attractive one for further development in image processing applications.

To proceed, the discussion is specialized to the case where the numbers involved are the N pixel intensities within a given neighborhood, each represented by L bits, and t is the number of the median in the ordered sequence of the pixel intensities (so that $t = N/2 + 1$). Bits are here labeled from the MSB to the LSB in the order $1, 2, \ldots, L$ in which they will be examined. Finally, $p(r)$ denotes the number of pixels for which the rth bit is 1, while the first $r - 1$ bits are identical to those of the median (as so far determined during the processing).

In this notation, the MSB of the median will be 1 if $p(1)$ is at least t and 0 otherwise:

$$\text{if (p(1) >= t) median[1] = 1; else median[1] = 0;} \tag{3.7}$$

It is relevant to what follows to note that if $p(1) < t$, this identifies $p(1)$ pixel intensities as being greater than the median: otherwise none is so identified.

This simple result is somewhat tricky to generalize; in fact, it is now necessary to restate the problem. At any stage, let us denote by the variable G the number of pixel intensities that have already been shown to be greater than the median. Let us also imagine that we are making a trial to determine whether, after considering the rth bit, this number should be increased. If $G + p(r)$ is at least t, then the additional pixels which contribute to the value of $p(r)$ (having rth bit equal to 1) ensure that the rth bit of the median is 1, but we are prevented from knowing any more about the number of pixels that have intensities greater than the median. However, if $G + p(r)$ is less than t, then (a) we know that median[r] is 0, and (b) we have identified additional $p(r)$ pixel intensities that are larger than the median. This can be written algorithmically in the form:

$$\begin{aligned} &\text{if (G + p(r) >= t) median[r] = 1;} \\ &\text{else } \{\text{median[r] = 0; G = G + p(r);}\}; \end{aligned} \tag{3.8}$$

Clearly, the processing terminates when all bits have been examined.

The overall algorithm now appears as in Fig. 3.14. Note that $p(r)$ will have to be a (C++) function requiring separate declaration, in accordance with the definition given earlier. If the whole algorithm including $p(r)$ is expressed in C++, then it is unlikely to be significantly faster than other relevant methods. However, if it is written in machine code or implemented partly in suitable hardware, then it should achieve its potential speed. This is ultimately due to the problem of bit packing and unpacking, which certain high-level languages are unable to handle efficiently.

3.7 Sharp–Unsharp Masking

When images are blurred either before or as part of the process of acquisition, it is frequently possible to restore them substantially to their ideal state. Properly, this is achieved by making a model of the blurring process and applying an inverse transformation that is intended exactly to cancel the blurring. This is a complex task to carry out rigorously, but in some cases a rather simple method called sharp–unsharp masking is able to produce significant improvement (Gonzalez and Woods, 1992). This technique involves first obtaining an even

```
t = N / 2 + 1;
G = 0;
r = 0;
do {
   r = r + 1;
   if (G + p(r) >= t) median[r] = 1;
   else { median[r] = 0; G = G + p(r) }
} until r = L;
/* It is assumed that p(r) has already been implemented as a software or hardware
function */
```

Figure 3.14 The complete bit-based median algorithm.

more blurred version of the image (e.g., with the aid of a Gaussian filter) and then subtracting this image from the original:

```
UNSHARP: {
      [[ Q0 = (P0 * 4 + (P1 + P3 + P5 + P7) * 2 + P2 + P4 + P6 + P8)/16; ]]
      [[ P0 = P0 * (1 + alpha) − Q0 * alpha; ]] }                    (3.9)
```

The principle of the technique is illustrated in Fig. 3.15 and its effect is seen in Fig. 3.16. Unfortunately, this particular technique is difficult to set up rigorously, since the amount of artificial blurring to apply and the proportion of the blurred image to subtract are rather arbitrary quantities and are usually adjusted by eye. (The value of alpha is typically 1/4.) Thus, the method is in practice better categorized under the heading "enhancement" than as "restoration," for it is not the precise mathematical technique normally understood by the latter term. With regard to such enhancement techniques, Hall (1979) states: "Much of the art of enhancement is knowing when to stop."

3.8 Shifts Introduced by Median Filters

Despite knowledge of the main characteristics of the different types of filters, some factors remain unknown. In particular, it is often important (especially when making precision measurements on manufactured components) to ensure that noise is removed in such a way that object locations and sizes are unchanged. It is necessary to examine how the various filters satisfy this criterion. Actually, we seem to have answered this question already when we pointed out that the mode filter operates by pushing back edges equally from both sides. Not only does this

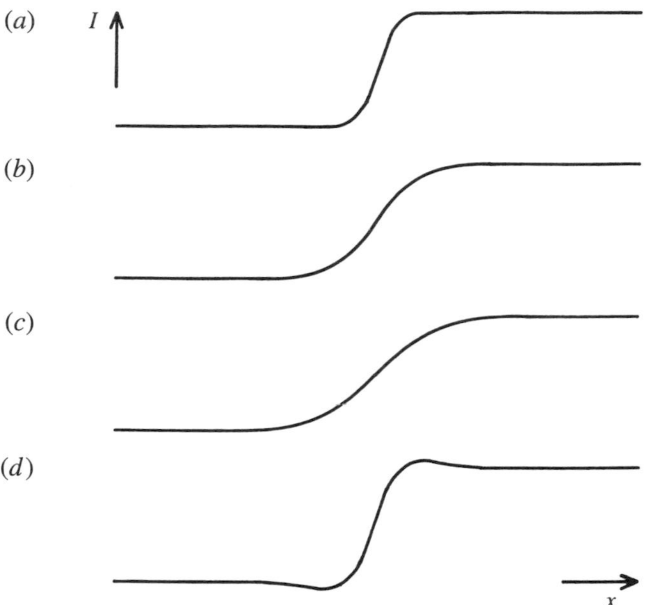

Figure 3.15 The principle of sharp–unsharp masking: (*a*) cross section of an idealized edge; (*b*) observed edge; (*c*) artificially blurred version of (*b*); (*d*) result of subtracting a proportion of (*c*) from (*b*).

filter not move the center of an edge, but also it sharpens the edge so that it is actually easier to measure its position accurately. However, this simple answer presents two problems.

First, it has been assumed that the intensity profile of an edge is symmetrical. If this is so, then the mean, median, and mode of the local intensity distribution will be coincident and there will be no overall bias for any of them. However, when the edge profile is asymmetrical, in the absence of a detailed model of the situation the result for any of the filters will not be obvious. The situation is even more involved when significant noise is present (Yang and Huang, 1981; Bovik et al., 1987). Since the problem is extremely data-dependent, it is not profitable to consider it further here.

The other problem concerns the situation for a curved edge. In this case, there is again a variety of possibilities, and filters employing the different means will modify the edge position in ways that depend markedly on its shape. In robot vision applications, the median filter is the one we are most likely to use because its main purpose is to suppress noise without introducing blurring. Hence, it is important for us to consider the bias produced by this type of filter in some detail. This is done in the next subsection.

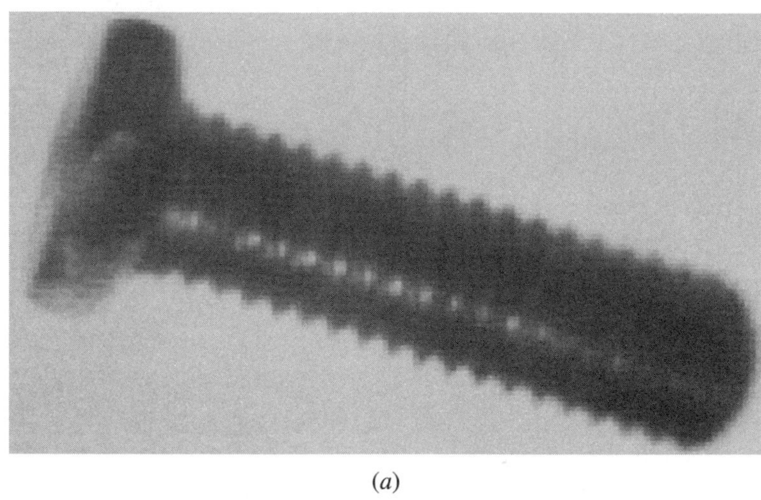

(*a*)

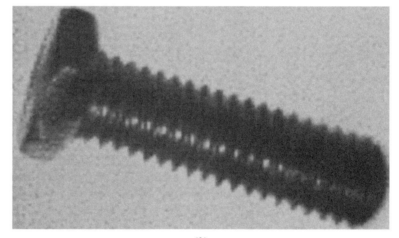

(*b*)

Figure 3.16 Sharp–unsharp masking: (*a*) original, slightly blurred image; (*b*) improvement after sharp–unsharp masking.

3.8.1 *Continuum Model of Median Shifts*

This section focuses on the case of a continuous image (i.e., a nondiscrete lattice), assuming (1) that the image is binary, (2) that neighborhoods are exactly circular, and (3) that images are noise-free. To proceed, we notice that binary edges have symmetrical cross sections, wherever straight edges extend this symmetry into 2-D. Hence, applying a median filter in a (symmetrical) circular neighborhood cannot pull a straight edge to one side or the other.

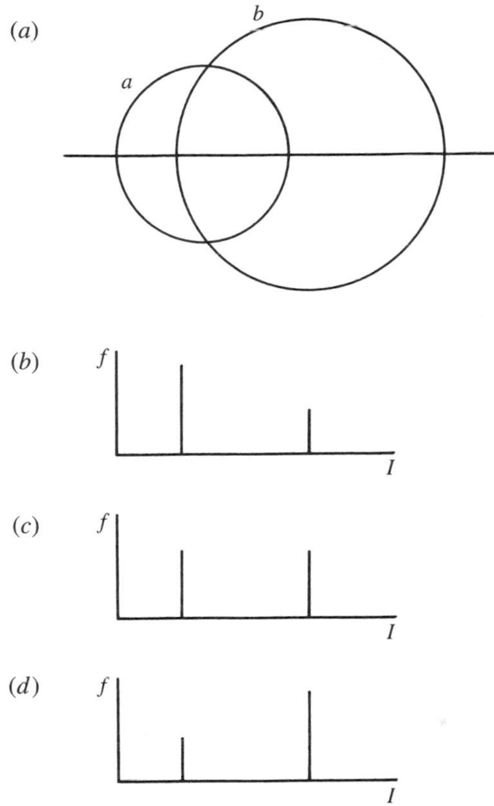

Figure 3.17 Variation in local intensity distribution with position of neighborhood: (*a*) neighborhood of radius *a* overlapping a dark circular object of radius *b*; (*b*)–(*d*) intensity distributions *I* when the separations of the centers are, respectively, less than, equal to, or greater than the distance *d* for which the object bisects the area of the neighborhood.

Now consider what happens when the filter is applied to an edge that is not straight. If, for example, the edge is circular, the local intensity distribution will contain two peaks whose relative sizes will vary with the precise position of the neighborhood (Fig. 3.17). At some position, the sizes of the two peaks will be identical. This happens when the center of the neighborhood is at a unique distance from the center of a circular object; this is the position at which the output of the median filter changes from dark to light (or vice versa). Thus, the median filter produces an inward shift toward the center of a circular object (or the center of curvature), whether the object is dark on a light background or light on a dark background.

Next suppose that the edge is irregular with several "bumps" (i.e., prominences or indentations) within the filter neighborhood. The filter will now tend to average out the bumps and straighten the edge, because it acts in such a way as to form a boundary on which the amounts of dark and light within the neighborhood are equalized (Fig. 3.18). This means that the edge will be locally biased but only by a reduced amount, since the various bumps will tend to pull the final edge in opposite directions. On the other hand, if there is one gross bump within the neighborhood—that is if the curvature has the same sign and is roughly constant at all points on the edge within the filter neighborhood—then all these parts of the edge will act in consort and it will be pulled sideways a significant amount by the filter. Thus, a circular section of the boundary constitutes a worst-case situation, for which the filter produces the largest bias in the position of the edge. It is clearly worth finding the size of the worst-case shift, and for this reason we concentrate our attention on circular objects, in the knowledge that all other shapes will give less serious shifts and distortions.

The worst-case calculation is a matter of elementary geometry: we need to find at what distance d from the center of a circular object (of radius b) the area of a circular neighborhood (of radius a) is bisected by the object boundary.

From Fig. 3.19 the area of the sector of angle 2β is βb^2 whereas the area of the triangle of angle 2β is $b^2 \sin\beta \cos\beta$. Hence, the area of the segment shown shaded is:

$$B = b^2(\beta - \sin\beta \cos\beta) \tag{3.10}$$

Making a similar calculation of the area A of a circular segment of radius a and angle 2α, we may deduce the area of overlap (Fig. 3.19) between the circular neighborhood of radius a and the circular object of radius b as:

$$C = A + B \tag{3.11}$$

For a median filter this is equal to $\pi a^2/2$. Hence:

$$F = a^2(\alpha - \sin\alpha \cos\alpha) + b^2(\beta - \sin\beta \cos\beta) - \pi a^2/2 = 0 \tag{3.12}$$

where

$$a^2 = b^2 + d^2 - 2bd \cos\beta \tag{3.13}$$

and

$$b^2 = a^2 + d^2 - 2ad \cos\alpha \tag{3.14}$$

To solve this set of equations, we take a given value of d, deduce values of α and β, calculate the value of F, and then adjust the value of d until $F = 0$. Since d is the modified value of b obtained after filtering, the shift produced by the filtering process is:

$$D = b - d \tag{3.15}$$

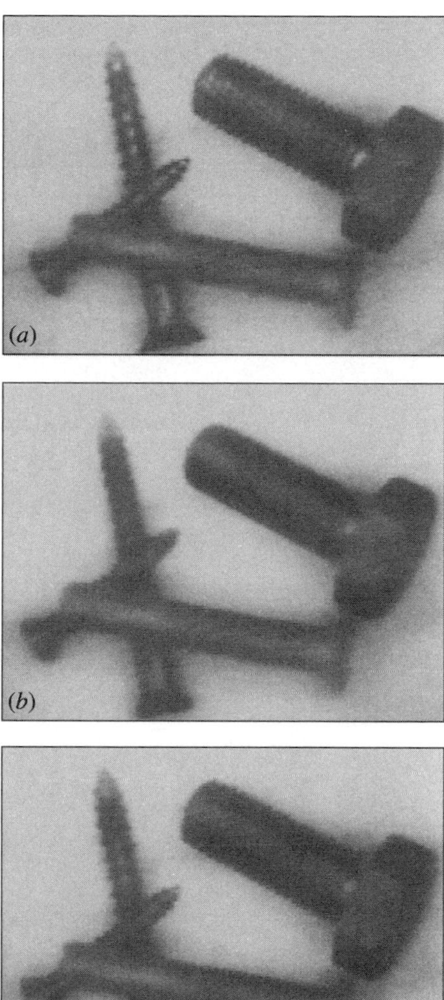

Figure 3.18 Edge smoothing property of the median filter. (*a*) Original image, (*b*) median filter smoothing of irregularities, in particular those around the boundaries (notice how the threads on the screws are virtually eliminated, although detail larger in scale than half the filter area is preserved), using a 21-element filter operating within a 5 × 5 neighborhood on a 128 × 128 pixel image of 6-bit gray-scale; (*c*) effect of a new type of "detail-preserving" filter (see Section 3.8.3).

Davies (1989b) has found the results of doing this computation numerically. As expected, $D \to 0$ as $b \to \infty$ or as $a \to 0$. Conversely, the shift becomes very large as a first approaches and then exceeds b. Note, however, that when $a > \sqrt{2}b$ the object is ignored, being small enough to be regarded as irrelevant noise by the

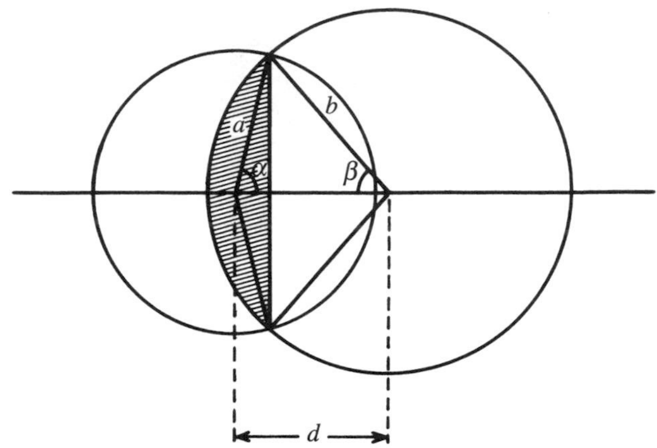

Figure 3.19 Geometry for calculating neighborhood and object overlap.

filter: beyond this point it has no effect at all on the final image. The maximum edge shift before the object finally disappears is $(2-\sqrt{2})b \approx 0.586b$.

It is instructive to approximate the above equations for the case when edge curvature is small; that is, $a \ll b$. Under these conditions, β is small, $\alpha \approx \pi/2$, and $d \approx b$. Hence we find:

$$\beta \approx a/b \tag{3.16}$$

After some manipulation, the edge shift D is obtained in the form:

$$D \approx a^2/6b = \kappa a^2/6 \tag{3.17}$$

$\kappa = 1/b$ being the local curvature. In Chapter 14 this equation is found to be useful for estimating the signals from a median-based corner detector.

3.8.2 *Generalization to Gray-scale Images*

To extend these results to gray-scale images, first consider the effect of applying a median filter near a smooth step edge in 1-D. Here the median filter gives zero shift, since for equal distances from the center to either end of the neighborhood there are equal numbers of higher and lower intensity values and hence equal areas under the corresponding portions of the intensity histogram. This is always valid where the intensity increases monotonically from one end of the neighborhood to the other—a property first pointed out by Gallagher and Wise (1981). (For more recent discussions on related "root" [invariance] properties of signals under median filtering, see Fitch et al., 1985 and Heinonen and Neuvo, 1987.)

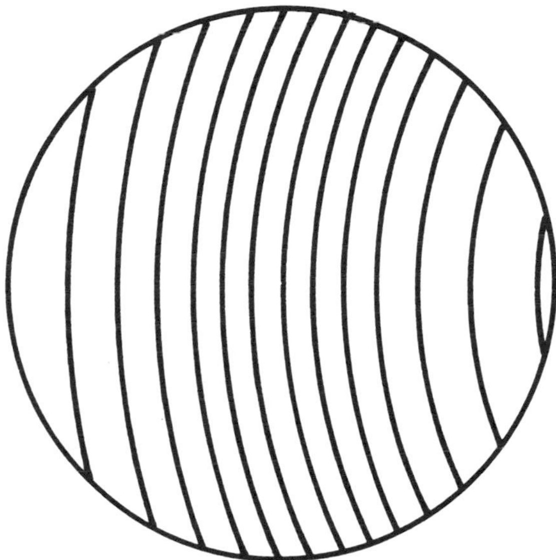

Figure 3.20 Contours of constant intensity on the edge of a large circular object, as seen within a small circular neighborhood.

Next, it is clear that for 2-D images, the situation is again unchanged in the vicinity of a straight edge, since the situation remains highly symmetrical. Hence, the median filter gives zero shift, as in the binary case.

For curved edges, it again turns out that circular boundaries constitute a worst case that should be considered carefully. However, gray-scale edges are unlike binary edges in that they have finite slope. This means that it is necessary to take account of the exact form of the intensity function within the neighborhood.

When boundaries are roughly circular, contours of constant intensity often appear as in Fig. 3.20. To find how a median filter acts, we merely need to identify the contour of median intensity (in 2-D the median intensity value labels a whole contour), which divides the area of the neighborhood into two equal parts. The geometry of the situation is identical to that already examined in Section 3.8.1. The main difference here is that for every position of the neighborhood, there is a corresponding median contour with its own particular value of shift depending on the curvature. Intriguingly, the formulas already deduced may immediately be applied for calculating the shift for each contour. Figure 3.20 shows an idealized case in which the contours of constant intensity have similar curvature, so that they are all moved inward by similar amounts. This means that, to a first approximation, the edges of the object retain their cross-sectional profile as it becomes smaller.

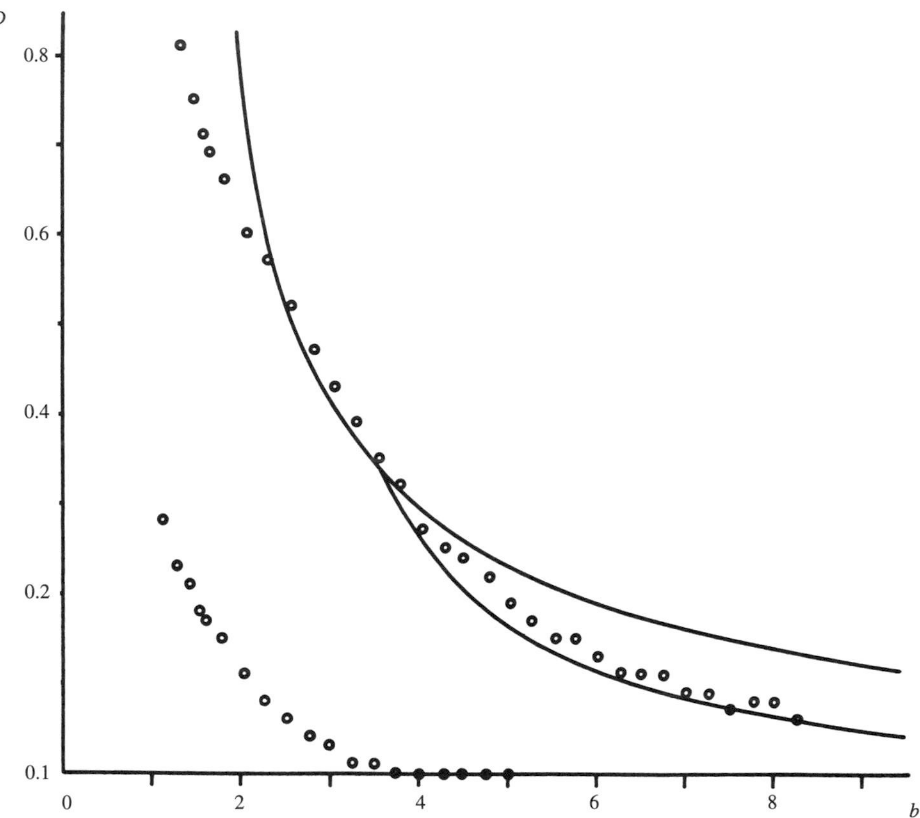

Figure 3.21 Edge shifts for 5 × 5 median filter applied to a gray-scale image. The upper set of plots represents the experimental results, and the upper continuous curve is derived from the theory of Section 3.8.1. The lower continuous curve is derived from a more accurate model (Davies, 1989b). The lower set of plots represents the much reduced shifts obtained with the detail-preserving type of filter (see Section 3.8.3).

For gray-scale images the shifts predicted by this theory (with certain additional corrections: see Davies, 1989b) agree with experimental shifts within approximately 10% for a large range of circle sizes in a discrete lattice (see Fig. 3.21). Paradoxically, the agreement is less perfect for binary images, since circles of certain sizes show stability effects (akin to median root behavior). These effects tend to average out for gray-scale images, owing to the presence of many contours of different sizes at different gray levels. Overall, it appears that the edge shifts obtained with median filters are now quite well understood. Figures 3.18 and 3.22 give some indication of the magnitudes of these shifts in practical situations. Note that once image detail such as a small hole or screw thread has been eliminated by a filter, it is not possible to apply any edge shift correction formula to recover it. For larger features, however, such formulas are useful for deducing true edge positions.

3.8.3 Shifts Arising with Hybrid Median Filters

Although median filters preserve edges in digital images, they are also known to remove fine image detail such as lines. For example, 3 × 3 median filters remove lines 1 pixel wide, and 5 × 5 median filters remove lines 2 pixels wide. In many applications such as remote sensing and X-ray imaging, this is exceedingly important and efforts have been made to develop filters that overcome the problem. In 1987, Nieminen et al. reported a new class of "detail preserving" filters. These employ linear subfilters whose outputs are combined by median operations. There are a great variety of such filters, employing different subfilter shapes and having the possibility of several layers of median operations. Hence we cannot describe them fully here in the space available. Although these filters are aimed particularly at retention of line detail and are readily understood in this context, they turn out to have some corner preserving properties and to be resistant to the edge shifts that arise when there is a nonzero curvature.

Perhaps the best of the filters in the new class, from the point of view of preserving edge position, is the two-level "bi-directional" linear-median hybrid filter termed 2LH+ (Nieminen et al., 1987). Its operation in a 5 × 5 neighborhood may be illustrated as follows. It employs the subfilters A to I in the 5 × 5 region:

E	–	D	–	C
–	E	D	C	–
F	F	A	B	B
–	G	H	I	–
G	–	H	–	I

pixels marked as being in the same subfilter having their intensities averaged, and dashed pixels being ignored. Nonlinear filtering then proceeds using two levels of median filtering, with the final center-pixel intensity being taken as:

$$A' = \mathrm{med}(A, \mathrm{med}(A, B, D, F, H), \mathrm{med}(A, C, E, G, I)) \tag{3.18}$$

Here we ignore the line-preserving properties of this filter and concentrate on its corner-preserving, low-edge-shift characteristics. It is quite easy to see that the 5×5 regions

$$\begin{matrix} 0 & 0 & 0 & 0 & 0 \\ 0 & 0 & 0 & 0 & 0 \\ 0 & 0 & 1 & 1 & 1 \\ 0 & 0 & 1 & 1 & 1 \\ 0 & 0 & 1 & 1 & 1 \end{matrix} \qquad \begin{matrix} 0 & 0 & 0 & 0 & 0 \\ 0 & 0 & 0 & 0 & 0 \\ 0 & 0 & 1 & 0 & 0 \\ 0 & 1 & 1 & 1 & 0 \\ 1 & 1 & 1 & 1 & 1 \end{matrix}$$

are preserved by this filter, although these examples represent limiting cases that could be disrupted by minor amounts of noise or slight changes of orientation. Thus, the filter seems *guaranteed* to preserve corners only if the internal angle is greater than 135°. This figure should be compared with the 180° obtained using similar arguments for the normal median filter in 5×5 regions such as

$$\begin{matrix} 0 & 0 & 0 & 1 & 1 \\ 0 & 0 & 0 & 1 & 1 \\ 0 & 0 & 1 & 1 & 1 \\ 0 & 0 & 1 & 1 & 1 \\ 0 & 0 & 1 & 1 & 1 \end{matrix}$$

Figure 3.21 shows plots obtained with this filter under the same conditions as for the 5×5 median filters. It always gives at least a fourfold improvement in edge shift over that for the median filter, and this performance improves with increasing radius of curvature b until there is zero shift for $b > 4$. (Note that $b = 4$ is approximately the figure that would be expected from the corner angle of 135° noted above, within a 5×5 neighborhood.) Hence, such detail-preserving filters improve the situation dramatically but do not completely overcome the underlying problem described earlier. In addition, this improvement may not have been obtained without cost, since in some cases the filter seems to *insert* structure where none exists (Davies, 1989b). The result is to cast some doubt on the usefulness of this type of filter in all possible situations. Nevertheless, its effect on real images appears to be generally very good (see Figs 3.18*c* and 3.22*c*).

3.8.4 ***Problems with Statistics***

Thus far it has been seen that computations of the position of the mode are made more difficult because of the sparse statistics of the local intensity distribution. This also affects the median calculation. Suppose the median value

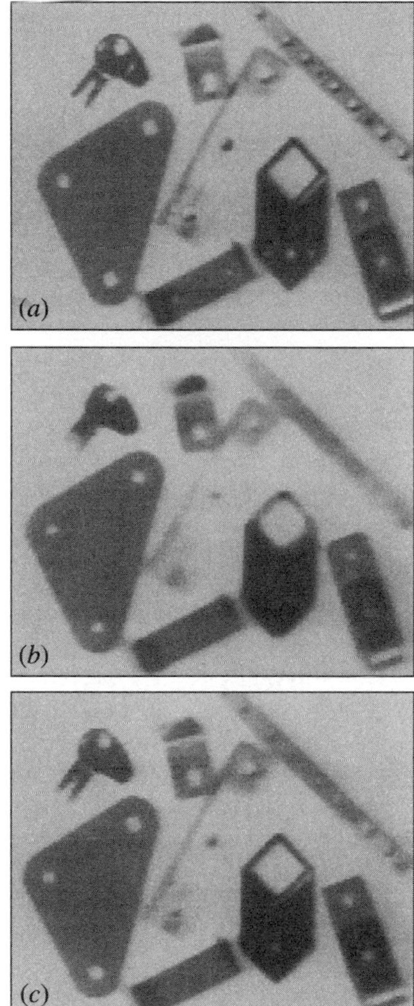

Figure 3.22 Circular holes in metal objects before and after filtering: (*a*) original 128 × 128 pixel image with 6-bit gray-scale; (*b*) 5 × 5 median-filtered image: the diminution in size of the hole is clearly visible and such distortions would have to be corrected for when taking measurement from real filtered images of this type; (*c*) result using a detail-preserving filter: some distortions are present although the overall result is much better than in (*b*).

happens to be well spaced near the center of the distribution (Fig. 3.5). Then a small error in this one intensity value is immediately reflected in full when calculating the median: i.e., the poor statistics have biased the median in a particular way. Ideally, what is required is a stable estimator of the median of the underlying distribution. Thus, the distribution should be made smoother before arriving at a specific value for the median. In practice, this procedure adds significant computational load to the filter calculation and is commonly not carried out. As a consequence, the median filter tends to result in runs of constant intensity, thereby giving images the "softened" appearance noted earlier. This is apparent on studying the following 1-D example:

orginal:	0	0	1	0	0	1	1	2	1	2	2	1	2	3	3	4	3	2	2	3
filtered:	?	0	0	0	0	1	1	1	2	2	2	2	2	3	3	3	3	2	2	?

Although histogram smoothing is not commonly carried out, some workers have felt it necessary to adjust the relative weights of the various pixels in the neighborhood according to their distance from the central pixel (Akey and Mitchell, 1984). This mimics what happens for a Gaussian filter and is theoretically necessary, although it is not generally implemented. However, recently further developments have been recorded on this front (e.g., see Charles and Davies, 2003b).

3.9 Discrete Model of Median Shifts

To produce a discrete model we need to recognize explicitly the positions of the pixels within the chosen neighborhood. We approximate by assuming that the intensity of any pixel is the mean intensity over the whole pixel and is represented by a sample positioned at the center of the pixel. We start by examining the case of a 3×3 neighborhood, and proceed by taking the underlying analog intensity variation to have contours of curvature κ, as shown in Fig. 3.20. Following what happened in the continuum case, it will not matter whether the contours of constant intensity are those of a step edge or those of a slowly varying slant edge; it is what happens at the median contour that determines the shift that arises.

The starting point is that zero shift occurs for $\kappa = 0$. Next, if κ is even minutely greater than zero, the center pixel will not necessarily be the median pixel. Consider the case when the circular median intensity contour passes close to the center of the neighborhood at a small angle θ to the positive x-axis (Figs. 3.23 and 3.24*a*). In that case, the filter will produce a definite shift, whose value is:

$$D_\theta \approx \frac{1}{2}\kappa a_0^2 - a_0\theta = \frac{1}{2}\kappa - \theta \tag{3.19}$$

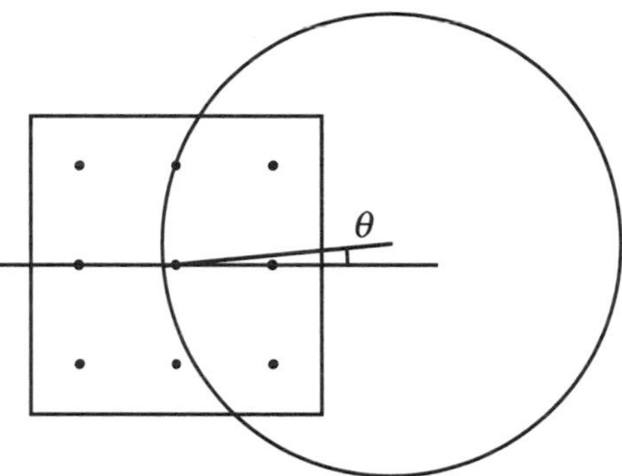

Figure 3.23 Geometry for calculation of median shifts on the discrete model.

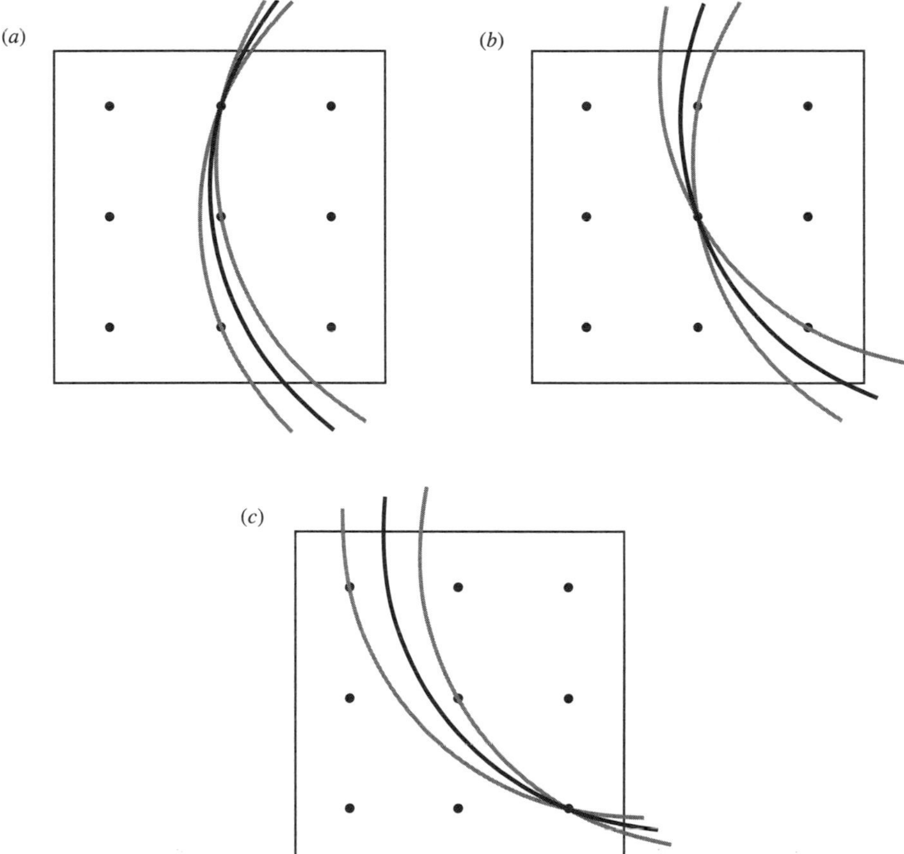

Figure 3.24 Geometry for calculation of median shifts at low κ. These three diagrams show the positions of the median pixels and the ranges of orientations of circular intensity contours for which they apply, (*a*) for low θ, (*b*) for intermediate θ, and (*c*) for high θ.

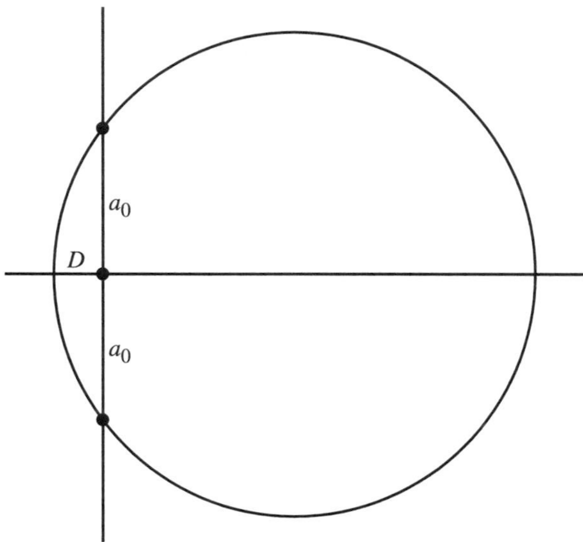

Figure 3.25 Geometry for calculation of shift when the median contour passes through the centers of two pixels.

where a_0 is the inter-pixel separation—here taken as unity. The first term arises from the following simple result for the geometry of the circle of radius $b = 1/\kappa$ in Fig. 3.25:

$$a_0^2 = D \times (2b - D) \approx 2Db \tag{3.20}$$

If θ is close to 45°, the shift will have the value:

$$D_\varphi \approx \frac{1}{2}\kappa\left(\sqrt{2}a_0\right)^2 - \left(\sqrt{2}a_0\right)\varphi = \kappa - \sqrt{2}\varphi \tag{3.21}$$

where

$$\varphi = \frac{\pi}{4} - \theta \tag{3.22}$$

and the relevant pixel separation is $\sqrt{2}a_0$ rather than a_0 (see Fig. 3.24*c*).

Equations (3.19) and (3.21) show that at the ends of the range $0 \leq \theta \leq \pi/4$, D_θ varies in proportion to $\Delta\theta$. The next problem is understanding what happens when D_θ falls to zero at intermediate values of θ. D_θ remains at zero in this range because the median contour reverts to passing through the central pixel center in the neighborhood (Fig. 3.24*b*). The resulting approximately piecewise-linear variation in D_θ (Fig. 3.26*a*) is far from what would be expected on the

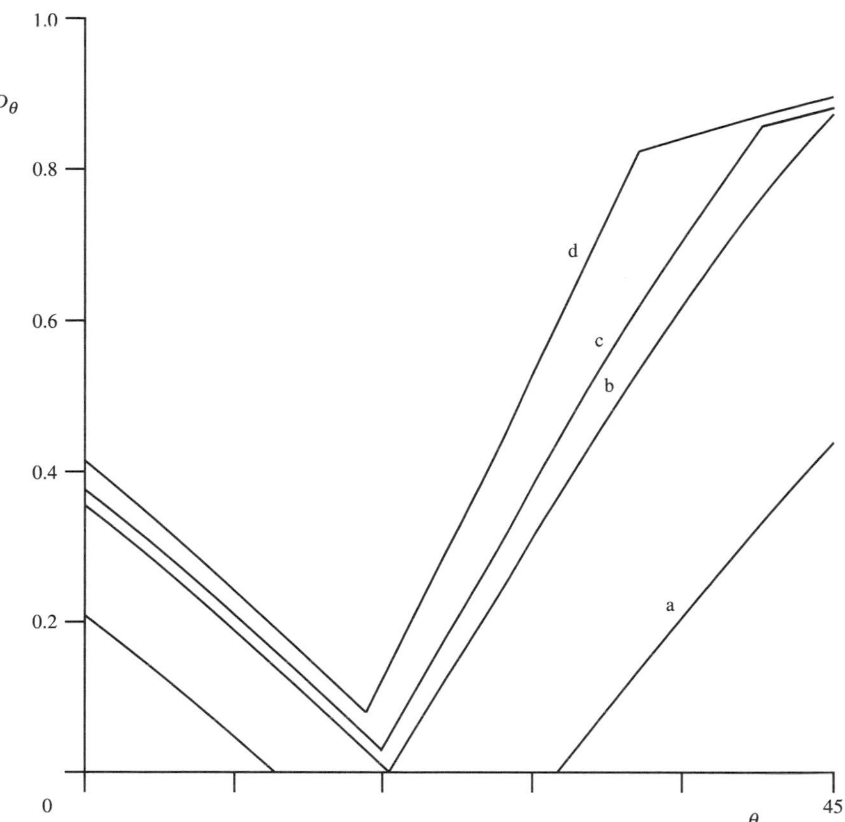

Figure 3.26 Angular variations of 3 × 3 median shifts. (*a*) Graph showing the situation for $\kappa = 0.4$: note that the θ-axis constitutes the middle part of the variation. (*b*) Graph for higher κ (0.632) when the middle part of the variation just vanishes. (*c*) Graph for a slightly higher value of κ (0.66). (*d*) Graph for the highest value of κ (0.7071) for which a valid value of D_θ exists for all θ. Note that (*b*) is the highest graph for which no change of gradient occurs at high θ. All the graphs presented in this figure are calculated from nonapproximated formulas (Davies, 1999g).

continuum model. To make a realistic comparison, we must average over all θ. In that case we obtain the result:

$$\begin{aligned} D &\approx \left\{ \int_0^{\kappa/2} (\kappa/2 - \theta)\mathrm{d}\theta + \int_0^{\kappa/\sqrt{2}} \left(\kappa - \sqrt{2}\varphi\right)\mathrm{d}\varphi \right\} / (\pi/4) \\ &= \frac{4}{\pi}\left\{ \left[-\frac{1}{2}(\kappa/2 - \theta)^2 \right]_0^{\kappa/2} + \left[-\frac{1}{2\sqrt{2}}\left(\kappa - \sqrt{2}\varphi\right)^2 \right]_0^{\kappa/\sqrt{2}} \right\} \\ &= \left(\frac{1 + 2\sqrt{2}}{2\pi} \right)\kappa^2 \approx 0.609\kappa^2 \end{aligned} \tag{3.23}$$

This shows that D follows a quadratic rather than a linear law at low values of κ, unlike the situation for the continuum model. However, for high values of κ, the variation would be expected to revert to a linear model. This should occur when κ reaches such high values that the range of values of θ for which $D_\theta = 0$ falls to zero. At that stage the whole variation of D_θ with θ should rise bodily as θ increases further (Fig. 3.26). Equating to zero the values of D_θ and D_φ given by the approximate equations (3.19) and (3.21), we deduce that this should happen for values of κ above about $(\sqrt{2}-1)\pi/2 \approx 0.8$. (The accurate value is 0.632—see below.)

Perhaps the most surprising thing is that this hardly happens for a 3×3 neighborhood (Figs. 3.26 and 3.27). For the necessary high values of κ, the median contour fails to reach all the outermost pixels in the neighborhood and there are orientations for which the contour represents an object that is eliminated entirely by the median filter. Averaging over all θ is then not meaningful: here we do not consider such cases further.

There is one problem with the above interpretation—that for quite high values of κ one other pixel separation than a_0 and $\sqrt{2}a_0$ becomes important. This value is $a_0/\sqrt{2}$ (Fig. 3.28*c*). This leads to the following equation taking over from equation (3.21) at high values of θ:

$$D_\varphi \approx \left(a_0/\sqrt{2}\right) + \frac{1}{2}\kappa\left(a_0/\sqrt{2}\right)^2 - \left(a_0/\sqrt{2}\right)\varphi$$

$$= \frac{1}{\sqrt{2}} + \frac{1}{4}\kappa - \frac{1}{\sqrt{2}}\varphi \tag{3.24}$$

To understand in detail when this happens, we should compare Figs. 3.24*c* and 3.28*c*. The changeover from one situation to the other occurs when an extreme intensity contour passes through the following three pixel centers: $(-1, 0)$, $(0, -1)$, $(1, -1)$. Such a contour will have a radius $b = [(3/2)^2 + (1/2)^2]^{1/2} = \sqrt{2.5} \approx 1.581$, leading to a curvature $\kappa \approx 0.632$. Curiously, this is the same value as that (noted above) for which the value zero for D_θ drops out of consideration—as is seen by considering Fig. 3.24*b*, when we find that both extreme intensity contours pass through $(1, 0)$, $(0, 0)$, and $(1, -1)$.

Finally, exact formulas (Davies, 1999g) are used in place of the approximate results of equations (3.19)–(3.23) to produce the D_θ graphs in Fig. 3.26 and the average graph in Fig. 3.27.

3.9.1 *Generalization to Gray-scale Images*

In this section, we consider the results of experiments carried out to check the predictions of the discrete theory. It turned out that the discrete theory was

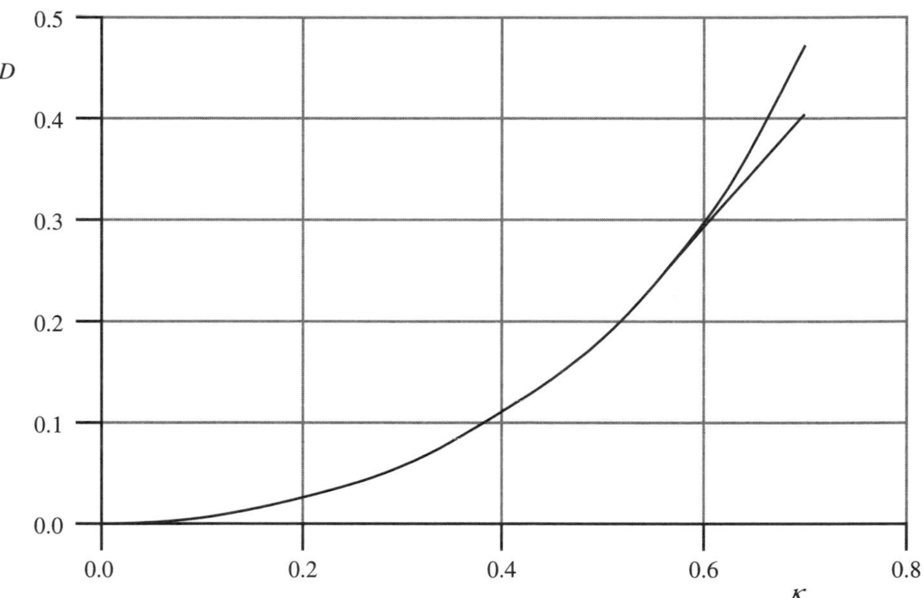

Figure 3.27 Comparisons of 3×3 median shifts. The lower solid curve shows the nonapproximated results of the discrete model (Davies, 1999g): the upper solid curve shows the results of experiments on gray-scale circles.

much more accurate than the continuum theory. However, a crucial step in demonstrating this was the realization that the gray-scale circles used in the tests had to be placed at, and the results averaged over, all possible sub-pixel positions. Once this was carried out, the results obtained showed precise agreement over the entire range of κ up to $\kappa \approx 0.6$ (Fig. 3.27).

The divergence above $\kappa \approx 0.6$ is readily explained, as the theory is based on edge shifts, where the edges are taken to correspond simply to *single* edges within a 3×3 neighborhood, whereas the experimental results corresponded to gray-scale circles with dome-shaped intensity profiles. Thus, for high values of κ, even if a circle was not located entirely within the neighborhood, some of its gray-levels might appear entirely within the neighborhood. A number of these would then be eliminated (a process that can be visualized as "dome-slicing"), and the remaining gray levels would be subject to different shifts that would be combined in a nonlinear manner by the measurement process. This meant that agreement would only be expected where κ was sufficiently small that all shifts produced within a *given* neighborhood would be in very much the same direction, as indicated by the intensity paradigm of Fig. 3.20.

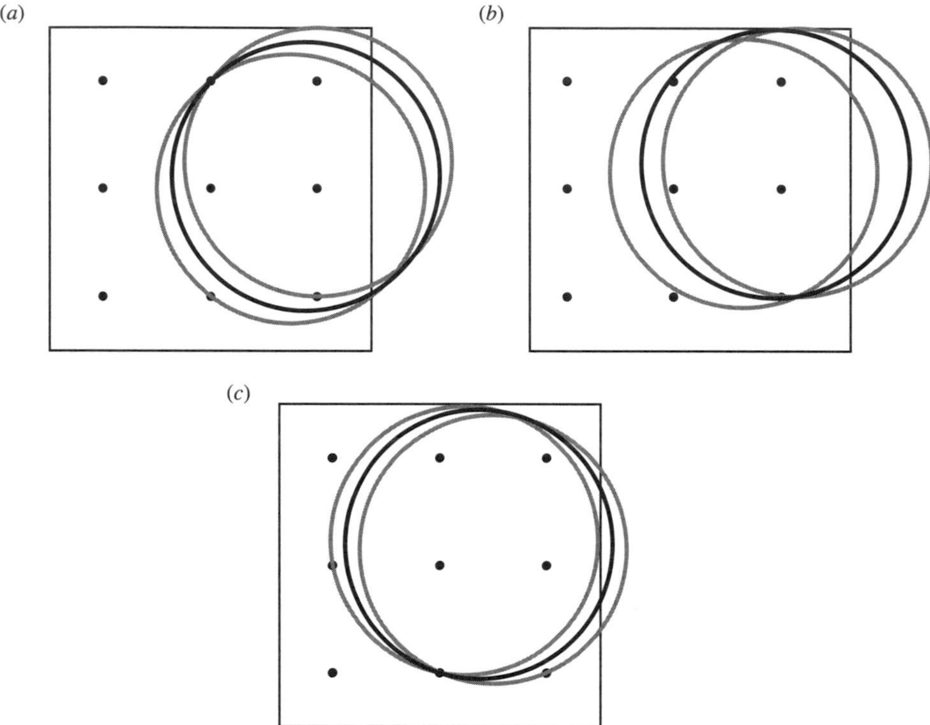

Figure 3.28 Geometry for calculation of median shifts at high κ. These three diagrams show the positions of the median pixels and the ranges of orientations of circular intensity contours for which they apply, (*a*) for low θ, (*b*) for intermediate θ, and (*c*) for high θ.

3.10 Shifts Introduced by Mode Filters

In this section we consider the shifts produced by mode filters in continuous images. As in the cases of median filters, straight edges with symmetrical profiles cannot be shifted by mode filters because of symmetry. We proceed to the two paradigm cases—step edges and slant edges with circular boundaries. Again, the effects of noise will be ignored as we are considering the intrinsic rather than the noise-induced behavior of the mode filtering operation.

The situation for a curved step edge can again be understood by appealing to Fig. 3.17. The result for the mode also has to be identical to that for the median, because the local intensity distribution is exactly symmetrical and bimodal at the point where the median filter switches from a left-hand to a right-hand decision. At that point the mode must give the same result, since the median and the mode are coincident for a symmetrical distribution. Hence, we conclude that the mode also gives a shift of $1/6\ \kappa a^2$ for a curved step edge.

Next, we calculate edge shifts in the case where smoothly varying intensity functions exist or—within the confines of a small neighborhood—*appear* to be smoothly varying. In this case the calculation is especially simple (Davies, 1997c). Using the geometry of Fig. 3.20, we consider the intensity pattern within a circular neighborhood C. Of all the circular intensity contours appearing within C, the one possessing the most frequently occurring intensity, as selected by a mode filter, is the longest. This is the one (M) whose ends are at opposite ends of a diameter of C. To estimate the shift in this case, all we need to do is to calculate the position of M and determine its distance from the center of C. To proceed, we use the well-known formula relating the lengths of parts of intersecting chords of a circle, which gives:[3]

$$a^2 = D(2b - D) \approx 2Db \tag{3.25}$$

Hence:

$$D = \frac{1}{2}\kappa a^2 \tag{3.26}$$

That is there is a *right* shift of the contour, toward the local center of curvature, of $1/2\ \kappa a^2$. If we regard this set of contours as forming part of a gray-scale edge profile, then the mode filter shifts the edge through $1/2\ \kappa a^2$ toward the center of curvature.

A comment is needed on the marked difference between the cases of step edges and linear intensity profiles. This is all the more interesting as the median filter produces identical shifts of $1/6\ \kappa a^2$, for the two profiles (see Table 3.2). Of all the cases listed in Table 3.2, the outstanding one is the large shift for a mode filter operating on a linear intensity profile. Special in this case is that the result relies on a single extreme contour length rather than an average of lengths

Table 3.2 Summary of edge shifts for neighborhood averaging filters

Edge Type	Filter		
	Mean	**Median**	**Mode**
Step	$1/6\ \kappa a^2$	$1/6\ \kappa a^2$	$1/6\ \kappa a^2$
Intermediate	$\sim 1/7\ \kappa a^2$	$1/6\ \kappa a^2$	$1/2\ \kappa a^2$
Linear	$1/8\ \kappa a^2$	$1/6\ \kappa a^2$	$1/2\ \kappa a^2$

3 This equation is a more general form of equation (3.20), as a can have any value: the proof follows similarly, on replacing a_0 by a in Fig. 3.22.

amounting to an area measure. Hence, it is not surprising that the mode filter gives an exceptionally large shift in this case.

Finally, we note that edge shifts are not avoided merely by choosing an alternative method of neighborhood averaging, but rather that they are intrinsic to the averaging process and can be avoided only by specially designed operators (e.g., see Greenhill and Davies, 1994a).

3.11 Shifts Introduced by Mean and Gaussian Filters

In this section, we consider the shifts produced by mean and Gaussian filters in continuous images. As in the cases of median and mode filters, straight edges with symmetrical profiles cannot be shifted by mean and Gaussian filters, because of symmetry. We again consider the two paradigm cases—step edges and slant edges with circular boundaries. Again, we will ignore the effects of noise because we are considering the intrinsic rather than the noise-induced behavior of the filters.

The situation for a curved step edge can again be understood by appealing to Fig. 3.17. It is identical to that for the median and mode filters, and it follows because of the symmetry of the local intensity distribution at the point where the filter switches from a left-hand to a right-hand decision. At that point the mean filter must give the same result, since all three statistics coincide for a symmetrical distribution. Hence, it also gives a shift of $1/6\ \kappa a^2$ for a curved step edge.

In the case of a smoothly varying slant edge, the result for the mean filter has to be calculated by integrating over the area of the neighborhood. The results cannot be obtained by intuitive or simple geometric or intuitive arguments, and here we merely quote the shift for the mean filter as being $1/8\ \kappa a^2$.

These considerations complete the entries in Table 3.2. The results for Gaussian filters do not differ substantially from those for mean filters but have to be obtained by integration, taking account of the Gaussian weighting function (Davies, 1991b). All such filters have similar shifting effects because they all incorporate a measure of signal averaging. The shifting effect is not avoided simply by employing a different central-seeking statistic to perform the averaging.

3.12 Shifts Introduced by Rank Order Filters

This section focuses on rank order filters (Bovik et al., 1983), which form a whole family of filters that can be applied to digital images—often in combination with other filters of the family—in order to give a variety of effects (Goetcherian, 1980; Hodgson et al., 1985). Other notable members of the family are max and min filters. Because rank order filters generalize the concept of the median filter, it is relevant to study the types of distortions they produce on straight and curved intensity contours. These filters are of central importance in the design of filters for morphological image analysis and measurement. In addition, they have some advantages when used for this purpose in that they help to suppress noise (Harvey and Marshall, 1995) (though the effect vanishes in the special cases of max and min filters).

Section 3.12.1 examines the reasons underlying the shifts produced by rank order filters and makes calculations of their extent for rectangular neighborhoods. Section 3.12.2 generalizes these results to circular neighborhoods. Section 3.12.3 examines the extent to which the theoretical predictions of the previous sections are borne out in practice by measurements of the shifts produced by 5×5 rank order filters on circular discs of varying sizes. It will be taken as axiomatic that the application of rank order filters produces edge shifts on real images (they are well attested in the case of max, min, and median filters). The main question to be answered here is the exact numerical extent of these shifts and how they may be modeled for general rank order filters.

3.12.1 *Shifts in Rectangular Neighborhoods*

In common with previous work in this area we concentrate here on the ideal noiseless case, in which the filter operates within a small neighborhood, over which the signal is basically a monotonically increasing intensity function in some direction. The most complex intensity variation that will be considered is that in which the intensity contours are curved with curvature κ. In spite of this simplified configuration, it will be found that valuable statements can be made about the level of distortion likely to be produced in practice by rank order filters.

Because of the complexity of the calculations that arise in the case of rank order filters, which involve an additional parameter vis-à-vis the median filter, it is worth studying their properties first for the simple case of rectangular neighborhoods (Davies, 2000f). Let us presume that a rank order filter is being applied in a situation in which straight intensity contours are aligned parallel to the short sides of a rectangular neighborhood, which we initially take to be a $1 \times n$ array of pixels (Fig. 3.29). In this case, we can assume without loss of generality that the successive pixels within the neighborhood will have *increasing* values of intensity. We next take the basic property of the rank order filter as being (effectively or in fact) to construct an intensity histogram of the local intensity

(*a*)

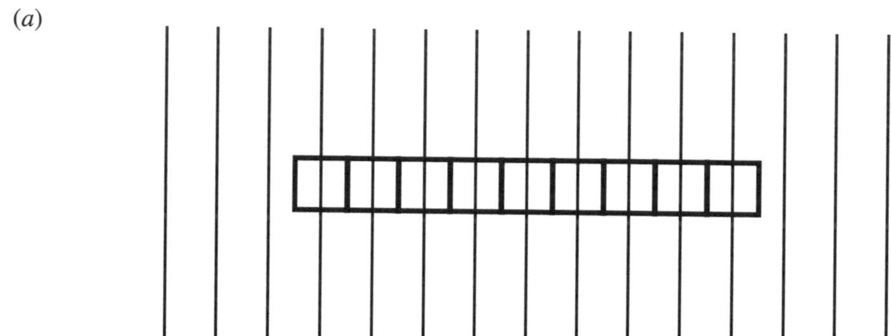

(*b*)

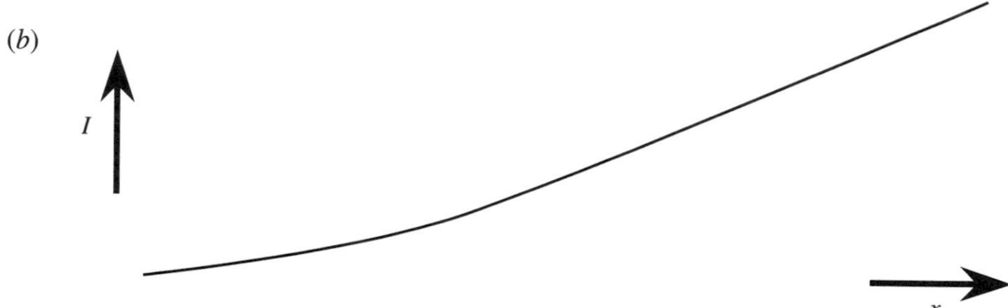

Figure 3.29 Basic situation for a rank-order filter in a rectangular neighborhood. This figure illustrates the problem of applying a rank-order filter within a rectangular neighborhood consisting of a $1 \times n$ array of pixels. The intensity is taken to increase monotonically from left to right, as in (*b*); the intensity contours in (*a*) are taken to be parallel to the short sides of the neighborhood.

distribution and return the value of the *r*th of the *n* intensity values within the neighborhood. Thus, the rank order filter selects an intensity that has physical separation *B* from the lowest intensity pixel of the neighborhood and *C* from the highest intensity pixel, where:

$$B = r - 1 \tag{3.27}$$

$$C = n - r \tag{3.28}$$

$$A = B + C = n - 1 \tag{3.29}$$

These definitions emphasize that a rank order filter will in general produce a *D*-pixel shift, whose value is:

$$D = \frac{1}{2}(n + 1) - r \tag{3.30}$$

Table 3.3 Properties of the three paradigm filters

Filter	r	η	B	C	D
Median	$1/2\ (n+1)$	0	$1/2\ A$	$1/2\ A$	0
Max	n	-1	A	0	$-1/2\ (n-1)$
Min	1	1	0	A	$1/2\ (n-1)$

Before proceeding further, it will be useful to introduce a parameter η that is more symmetrical than r and has value $+1$ at $r=1$ and -1 at $r=n$:

$$\eta = (n - 2r + 1)/(n - 1) \tag{3.31}$$

With this notation, which we will use in preference to r throughout the remainder of the chapter, we can write down new formulas for B, C, and D:

$$B = \frac{1}{2}A(1 - \eta) \tag{3.32}$$

$$C = \frac{1}{2}A(1 + \eta) \tag{3.33}$$

$$D = \frac{1}{2}\eta(n - 1) \tag{3.34}$$

The properties of the three paradigm filters are summarized in Table 3.3 in terms of these parameters.

We now proceed to a continuum model, assuming a large number of pixels in any neighborhood (i.e., $n \to \infty$). The main difference will be that we shall specify distance in terms of the half-length a of the neighborhood rather than in terms of numbers of pixels:

$$D = \eta a \tag{3.35}$$

Next note that this formulation is independent of the width of the neighborhood, as long as the neighborhood is rectangular. We now generalize the situation by taking the neighborhood to be rectangular and of dimensions $2a$ by $2\tilde{a}$ (Fig. 3.30).

The next task is to determine the result of a curvature $\kappa = 1/b$ in the intensity contours. Here we approximate the equation of a circle of radius b, with its diameter on the positive x-axis and passing through the origin, as:

$$x = y^2/2b \tag{3.36}$$

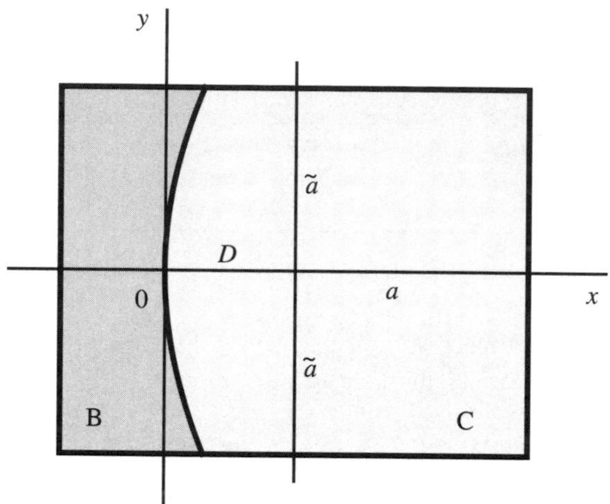

Figure 3.30 Geometry of a rectangular neighborhood with curved intensity contours. Here the neighborhood is a general rectangular neighborhood of dimensions $2a \times 2\tilde{a}$. Again, the intensity is taken to increase monotonically from left to right; the intensity contours are taken to be parallel and in this case are curved with identical curvature κ. x- and y-axes needed for area calculations are also shown. B and C represent the areas of the two shaded regions on either side of the thick intensity contour.

We can integrate the area under an intensity contour (see Fig. 3.30) as follows:

$$\begin{aligned} K &= \int_{-\tilde{a}}^{\tilde{a}} x\,\mathrm{d}y = (1/2b)\int_{-\tilde{a}}^{\tilde{a}} y^2\,\mathrm{d}y = (1/2b)[y^3/3]_{-\tilde{a}}^{\tilde{a}} \\ &= (1/6b)\,2\tilde{a}3 = \frac{1}{3}\tilde{a}^3/b = \frac{1}{3}\kappa\tilde{a}^3 \end{aligned} \tag{3.37}$$

We deduce that the shift D is given by

$$B = 2\tilde{a}(a - D) + \frac{1}{3}\kappa\tilde{a}^3 \tag{3.38}$$

$$C = 2\tilde{a}(a + D) - \frac{1}{3}\kappa\tilde{a}^3 \tag{3.39}$$

$$\eta = (C - B)/A = 4\tilde{a}D/A - \frac{2}{3}\kappa\tilde{a}^3/A \tag{3.40}$$

where

$$A = 4a\tilde{a} \tag{3.41}$$

Hence:

$$D = \eta A/4\tilde{a} + \frac{1}{6}\kappa\tilde{a}^2 = \eta a + \frac{1}{6}\kappa\tilde{a}^2 \tag{3.42}$$

Equation (3.42) shows that the effects of rank order and of curvature can be calculated and summed separately, the first term being that obtained above for the case of zero curvature and the second term being exactly that calculated for a median filter when the intensity contour is of length $2\tilde{a}$ (the earlier calculation (Davies, 1989b) related to a circular neighborhood). Thus, in principle we merely need to recompute the first term for any appropriate shape of neighborhood. However, there is a complication in that the value of $\tilde{a}$ depends on the value of η for any neighborhood other than a rectangle. Davies (2000f) has shown how to allow for this complication.

3.12.2 *Case of High Curvature*

When the curvatures are very high, they may arise from spots that are entirely within the neighborhood, and then there is the possibility that they will be completely eliminated by the rank order filter. (Note that noise points are entirely eliminated by a median filter, which indeed is the prime practical use of that type of filter.) More important, the assumptions of both our model and the exact numerical solution break down when there is no intersection of the circular neighborhood and the intensity contour of radius $b = 1/\kappa$. The limiting situation has been studied by Davies (2000f).

Some of the conclusions of this work are quite important. In particular, the result for a median filter is the special case that arises when $\eta = 0$, and is in agreement with the calculations of Section 3.8. Next, the max and min filters are also special cases and occur for $\eta = -1$ and 1, respectively. In these limiting cases, the shifts are $D = -a$ and a, respectively, the results being independent of κ. This is as might be expected since the value of $\tilde{a}$ is zero in each case. Between the max and min filters and the median filter, there is a continuous gradation of performance, with very significant but opposite shifts for the max and min filters, and the two basic effects canceling out for median filters—though the cancellation is only exact for straight contours. The full situation is summarized in Fig. 3.31.

3.12.3 *Test of the Model in a Discrete Case*

Tests of the model were carried out using a truncated 5×5 neighborhood with the four corner pixels excluded, in an effort to make the shape a somewhat closer approximation to circular. The resulting rank order filters (with $n = 21$) were applied to circular discs of various radii. These tests led to the shift variations

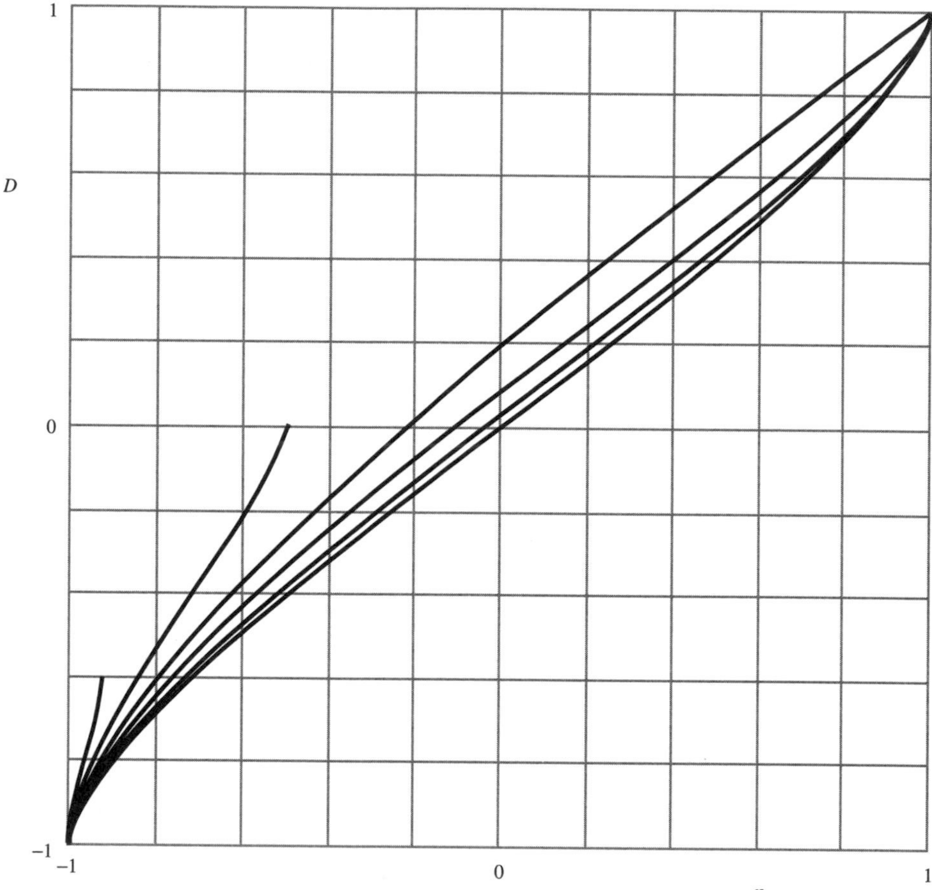

Figure 3.31 Graphs of shift D against rank-order parameter η for various κ. This diagram summarizes the operation of rank-order filters, with graphs, bottom to top, respectively, for $\kappa = 0$, $0.2/a$, $0.5/a$, $1/a$, $2a$, $5a$. Note that graphs for which $b < a$ ($\kappa > 1/a$) apply for restricted ranges of η and D (see Section 3.12.2). A multiplier of a must be included in the D-values.

shown in Fig. 3.32. The uppermost curve represents the theoretical limiting value (Davies, 2000f).

The variations shown in Fig. 3.32 are very close to what would be expected from Fig. 3.31 (but notice that in Fig. 3.31, a multiplier of a must be included in the D-values, whereas in Fig. 3.32 the D-values are absolute figures for the particular 5×5 neighborhood). The upward and downward curl at the ends of the curves—especially that for $\kappa = 0$—is not as pronounced in Fig. 3.32 as it is in Fig. 3.31, and the overall shape, while similar, is by no means identical. On the other hand, it is extremely close considering that Fig. 3.31 results from the continuum model, whereas Fig. 3.32 results from a discrete test employing

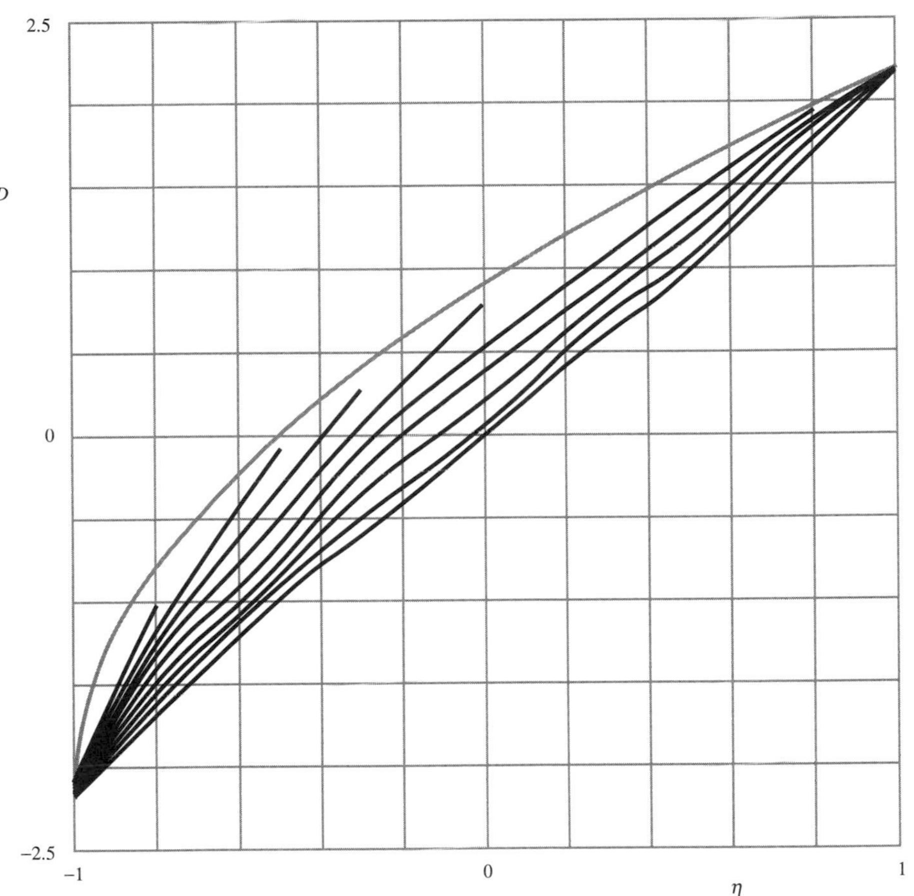

Figure 3.32 Shifts obtained for a typical discrete neighborhood. These shifts were obtained for rank-order filters operating within a truncated 5×5 neighborhood when applied to eight discrete circular discs with radii ranging from 10.0 down to 1.25 pixels, the mean curvatures being 0.1 to 0.8 in steps of 0.1. The lowest curve was obtained by averaging the responses from circular discs of radius ± 20.0 pixels, with curvatures ± 0.05, and to the given scale is indistinguishable from the result that would be obtained with zero curvature. The uppermost curve represents the theoretical limiting value. However, because of the directional effects that occur in the discrete case, the upper limit is actually lower than indicated by this curve (see text).

a small neighborhood. It is doubtful whether a more detailed correspondence could be produced without attempting a full discrete model of the shifts. Nevertheless, the limiting values at $\eta = \pm 1$ are observed to be 2.173—well within 1% of the theoretical value 2.174 obtained by calculating the mean distance to the outermost pixels in the 5×5 neighborhood (Davies, 2000f).

3.13 The Role of Filters in Industrial Applications of Vision

We have shown how the median filter can successfully remove noise and artifacts such as spots and streaks from images. Unfortunately, many useful features such as fine lines and important points and holes are effectively indistinguishable from spots and streaks. In addition, it has been seen that the median filter "softens" pictures by removing fine detail. It has also been found to clip corners of objects—another generally undesirable trait (but see Chapter 14). Finally, although it does not blur edges, it can still shift them slightly. Shifting of curved edges seems likely to be a general characteristic of noise suppression filters (but see Section 3.8.3).

Such distortions are quite alarming and mitigate against the indiscriminate use of filters. If applied in situations where accurate measurements are to be made on images, particular care must be taken to test whether the data are being biased in any way. Although it is possible to make suitable corrections to the data, it seems a good general policy to employ noise removal filters only where they are absolutely essential for object visibility. The alternative is to employ edge detection and other operators that automatically suppress noise as an integral part of their function. This is the general approach taken in subsequent chapters. Indeed, one of the principles underlined in this book is that algorithms should be "robust" against noise or other artifacts that might upset measurements. There is quite significant scope for the design of robust algorithms, since images contain so much information that it is usually possible to arrange for erroneous information to be ignored.

3.14 Color in Image Filtering

In Chapter 2 we indicated that color often adds to the complexity of image analysis algorithms and could also add to the associated computational costs. From these points of view color might, except for applications such as assessing the ripeness of fruit, be regarded as an irrelevant luxury. Nevertheless, in the field of image processing and image filtering, where good quality images have to be presented to human operators, it is a vital concern. In recent years, much effort has been devoted to the development of effective color filtering algorithms. Here we shall consider mainly median and related impulse noise filtering procedures.

Perhaps the first point to note is that median filtering is defined in terms of sorting operations and is thus undefined in the color domain, which normally contains three dimensions. However, a simple solution is to apply a standard median filter to each of the color channels and then to reassemble the color image. Unfortunately, this approach leads to certain problems, the most obvious one being that of color "bleeding" (Fig. 2.8). This occurs when an impulse noise point appears in just one of the channels and is situated near an edge or other image feature. The case of an impulse noise point near an edge is illustrated in simplified form

original: 0 0 0 0 1 0 1 1 1 1 1 1
filtered: ? 0 0 0 0 1 1 1 1 1 1 ?

We see that a three-element median filter eliminates the impulse noise point but at the same time moves the edge toward it. The end result for a color image is that the edge will be tinted with the color of the impulse noise point.

Fortunately, there is a standard solution to this problem. First, note that it is possible to express single-channel median filtering as the minimization of a distance metric, and this metric is trivially extendible to three color channels (or indeed any number of channels). The relevant single channel metric is:

$$\text{median} = \min_i \sum\nolimits_j |d_{ij}| \tag{3.43}$$

where d_{ij} is the distance between sample points i and j in the single-channel (grayscale) space. In the three-color domain the metric is readily extended to:

$$\text{median} = \min_i \sum\nolimits_j |\tilde{d}_{ij}| \tag{3.44}$$

where $\tilde{d}_{ij}$ is the generalized distance between sample points i and j, and we typically take the L_2 norm to define the distance measure for three colors:

$$\tilde{d}_{ij} = \left[\sum_{k=1}^{3} (I_{i,k} - I_{j,k})^2 \right]^{1/2} \tag{3.45}$$

Here $\mathbf{I}_j$, $\mathbf{I}_j$ are RGB vectors and $I_{i,k}$, $I_{j,k}$ ($k = 1, 2, 3$) are their color components.

Although the resulting vector median filter no longer treats the individual color components separately, it is by no means guaranteed to eliminate color bleeding completely. Like the standard median, it replaces any noisy intensity $\mathbf{I}_\mathrm{n}$, (including color) by the intensity $\mathbf{I}_j$ of another pixel that exists in the same window—rather than by an ideal intensity **I**. Hence, color bleeding is reduced but not eliminated. If indeed there is a confluence of colors at any one point in an image, even in the absence of any impulse noise these sorts of algorithms might become confused and inadvertently introduce small amounts of color bleeding: ultimately, the effect is due to the increased dimensionality of the data, which means that the algorithm has to contend with a greatly increased number of possible outcomes, in spite of being an ad hoc procedure that does not embody specific understanding of images.

Figure 2.8 demonstrates the nature of color bleeding, albeit in the case of mode filtering. This figure shows vector median and vector mode filters to be remarkably free from color bleeding, but the same does not apply to scalar mode filters—for similar reasons to those indicated above for median filters.

3.15 Concluding Remarks

Although this chapter has concentrated on the implementation of noise suppression and image enhancement operators based on the local intensity distribution, it has made certain other points as well. In particular, it has shown the need to make a specification of the required imaging process and only then to work out the algorithm design strategy. Not only does this ensure that the algorithm will perform its function effectively, but also it should make it possible to optimize the algorithm for various practical criteria including speed, storage, and other parameters of interest. In addition, this chapter has demonstrated that any undesirable properties of the particular design strategy chosen (such as the inadvertent shifting of edges) should be sought out. Next, it has demonstrated a number of fundamental problems to do with imaging in discrete lattices—not least being problems of statistics that arise with small pixel neighborhoods. Finally, the edge shift work is particularly important in the case of the rather large shifts produced by rank order filters because of their relevance to widely used morphological operators (Chapter 8) where the shifts are turned to advantage.

The next chapter moves on to a particularly vital problem in machine vision—that of segmenting images in order to find where objects are situated. This work builds on what has been learned in the present chapter about edge profiles and how they are "seen" by neighborhood operators.

Median filters have long been used to eliminate impulse noise without blurring edges. However, this chapter has shown that significant shifting of edges can result from use of median filters, and this property extends to mode filters and *a fortiori* to rank order filters—so much so that the latter form the basis for morphological processing.

3.16 Bibliographical and Historical Notes

Much of this chapter has built on a paper by the author (Davies, 1988c) which itself rests on considerable earlier work on Gaussian, median, and other rank order filters (Hodgson et al., 1985; Duin et al., 1986). The edge shifts that occur for median filters are not limited to this type of filter but apply almost equally to mean filters (Davies, 1991b). In addition, other inaccuracies have been found with median filters, and methods have been found to correct them (Davies, 1992f).

The early literature hardly mentions mode filters, presumably because of the difficulty of finding simple mode estimators that are not unduly muddled by noise and that still operate rapidly. Indeed, only one overt reference has been found (Coleman and Andrews, 1979). Other work referred to here is that on decomposing Gaussian and median filters (Narendra, 1978; Wiejak et al., 1985), and the many papers on fast implementation of median filters, both in software and in hardware (e.g., Narendra, 1978; Huang et al., 1979; Ataman et al., 1980; Danielsson, 1981; Davies, 1992a). In particular, there was a strong feeling that VLSI (very large-scale integration) microcircuits would provide the best route for development in this area (Oflazer, 1983; Offen, 1985).

Considerable effort has been devoted to studying the "root" behavior of the median filter, that is, the result of applying median filtering operations until no further change occurs. In fact, most of this work has been carried out on 1-D signals, including cardiac and speech waveforms, rather than on images (Gallagher and Wise, 1981; Fitch et al., 1985; Heinonen and Neuvo, 1987). Root behavior is of interest as it relates to the underlying structure of signals, although its realization involves considerable amounts of processing.

Some of the work on filtering aims to improve on rather than to emulate the median filter. Work of this type includes that of Heinonen and others (Nieminen et al., 1987). See also the neural network approach to this topic discussed in Chapter 25.

Davies (1987e) has reported methods of optimizing linear smoothing filters in small neighborhoods by minimizing the total error in fitting them to a continuous Gaussian function. It turns out that a balance has to be struck between subpixel errors within the neighborhood and errors that arise from the proportion of the distribution that lies outside the neighborhood (Fig. 3.33).

More recent work on nonlinear filtering appears in Marshall et al. (1998): see Marshall (2004) for a new design method for weighted order statistic filters.

With the advent of extremely low-cost color frame grabbers on PCs, and the widespread use of digital cameras, digital color images have become ubiquitous, and this has extended to (or even necessitated) much research on color filtering. A convenient summary of work in this area up to 1998 appears in Sangwine and Horne (1998). More recent work on vector (color) filtering includes that of Lukac (2003). Charles and Davies (2003b) describe new distance-weighted median filters and their application to color images. They also extend the author's earlier mode filter work to color images (Charles and

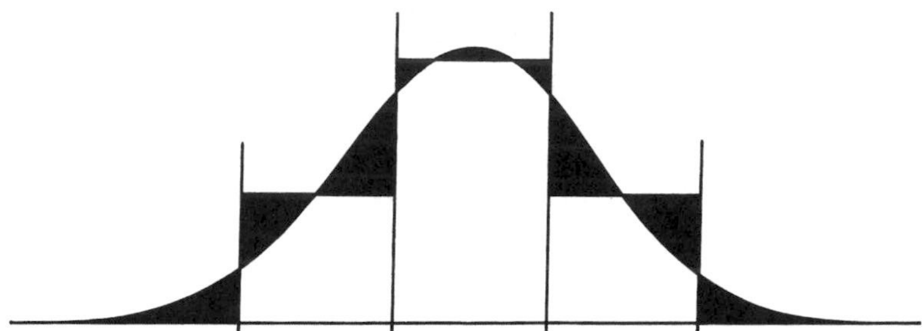

Figure 3.33 Approximating a discrete to a continuous Gaussian. This diagram shows how a balance needs to be struck between subpixel errors and those arising from the truncated part of the function.

Davies, 2003a, 2004). Davies (2000e) has published a theorem proving that restricting a multichannel (color) filter output to the vector value of one of the input sample points (i.e., from the current window in the image) will increase the inaccuracy present in the final image, for a large proportion of pixels. Since this represents the usual vector median strategy that is employed to minimize color bleeding, the effectiveness of the current generation of color filter algorithms needs to be looked at further.

Davies (1997c, 1998a, 1999g, 2000f) has further analyzed the distortions and edge shifts produced by a range of rank order, mean, and mode filters, and has produced a unified review of the subject (Davies, 2003e). In the case of median filters, it proved possible, and necessary for high accuracy, to produce a discrete model of the situation (Davies, 1998c, 1999d, 1999g, 2003c), rather than extending the continuum model described much earlier (Davies, 1989b).

3.17 Problems

1. Draw up a table showing the number of operations required to implement a median filter in various sizes of the neighborhood. Include in your table (a) results for a straight bubble sort of all n^2 pixels, (b) results for bubble sorts in separated $1 \times n$ and $n \times 1$ neighborhoods, and (c) results for the histogram method of Section 3.3. Discuss the results, taking account of possible computational overheads.

2. Show how to perform a median filtering operation on a binary image. Also show that if a set of binary images is formed by thresholding a gray-scale image at various levels, and each of these binary images is median filtered, then a gray-scale image can be reconstructed that is a median filtered version of the original gray-scale image. Consider to what extent the reduced amount of computation in

filtering a binary image compensates for the number of separate thresholded images to be filtered.

3. An "extremum" filter is an image-parallel operation that assigns to every pixel the intensity value of the closer of the two extreme values in its local intensity distribution. Show that it should be possible to use such a filter to enhance images. What would be the *disadvantage* of such a filter?

4. Under what conditions is a 1-D signal that has been filtered once by a median filter a root signal? What truth is there in the statement that a straight edge in an image is neither shifted nor blurred by a median filter, whatever its cross section?

5. (a) Explain the action of the following median filtering algorithm:

```
[[ for (i = 0; i <= 255; i++) hist[i] = 0;
   for (m = 0; m <= 8; m++) hist[ P[m] ]++;
   i = 0;  sum = 0;
   while (sum < 5){
         sum = sum + hist[i];
         i = i + 1;
   }
   Q0 = i - 1;
]]
```

(b) Show how this algorithm can be speeded up (i) by a more efficient histogram clearing technique and (ii) by calculating the minimum intensity in each 3×3 window. In each case estimate approximately how much the algorithm will be speeded up.

(c) Explain why a median filter is able to smooth images without introducing blurring.

(d) A 1-D cross section of an image has the following intensity profile:

1 2 1 1 2 3 0 2 2 3 1 1 2 2 9 2 2 8 8 8 7 8 8 7 9 9 9

Apply (i) a three-element median filter and (ii) a five-element median filter to this profile. With the aid of these examples, show that median filters tend to produce "runs" of constant values in 1-D profiles. Show also that under some circumstances an edge in the profile can be shifted by a nearby spike. Give a rule showing when this is likely to occur for an n-element median filter in one dimension.

6. (a) A *mode* filter is defined as one in which the new pixel intensity at any pixel takes the most probable value in the local intensity distribution of a window placed around that pixel in the original image space. Show for a gray-scale image that a

mode filter will, if anything, sharpen the image, while a *mean* filter will tend to blur the image.

(b) A *max* filter is one that takes the maximum value of the local intensity distribution in a window around each pixel. Explain what will be seen when a max filter is applied to an image. Consider whether any similar effects are liable to happen when a mode filter is applied to an image.

(c) Explain the purpose of a *median* filter. Why are 2-D median filters sometimes implemented as two 1-D median filters applied in sequence?

(d) Contrast the behavior of five-element 1-D mean, max, and median filters as applied to the following waveform:†

0 1 1 2 3 2 2 0 2 3 9 3 2 4 4 6 5 6 7 0 8 8 9 1 1 8 9

(e) Work out what would happen if the 1-D median filter were applied many times, starting with this waveform.

7. (a) Determine the effect of applying (i) a 3×3 median filter and (ii) a 5×5 median filter to the portion of an image shown in Fig. 3.P1.

```
0 0 0 0 0 0 0 0 0 0
0 0 0 0 0 0 0 2 0 0
0 1 0 0 0 0 0 0 0 0
0 0 1 0 0 0 0 0 0 0
0 0 0 0 9 9 9 9 9 9
0 0 0 0 9 9 8 9 9 9
0 0 1 0 9 8 9 9 7 9
0 0 0 0 7 9 9 8 9 9
0 1 0 8 9 9 9 9 9 9
0 0 0 0 9 9 9 9 9 9
0 0 0 0 8 9 9 9 9 9
```

Figure 3.P1

(b) Show that it should be possible to develop a corner detector based on the properties of these median filters. What advantages or disadvantages might result from employing this design strategy?

8. (a) Distinguish between *mean* and *median filtering*. Explain why a mean filter would be expected to blur an image, while a median filter would not have this effect. Illustrate your answer by showing what happens in the following 1-D case with a

† For the mean filter, give the nearest integer value in each case.

window of size 1×3:

1 1 1 1 2 1 1 2 3 4 4 0 4 4 4 5 6 7 6 5 4 3 3

(b) Give a complete median filter algorithm based on histograms and operating within a 3×3 window. Explain why it operates relatively slowly.

(c) A computer language has the *max*(a, b) operation as standard. Show how it may be used to find the maximum intensity within a 3×3 window. Show also how it may be used to find the median by successively replacing the maximum values by zeros. If the *max*(a, b) operation is about the same speed as the $a+b$ operation, determine whether the median can be found any faster by this method.

(d) Discuss whether splitting a 3×3 median operation into 1×3 and 3×1 median operations is likely to be effective in eliminating impulse noise in images. How would the speed of this approach be affected by use of the *max*(a, b) operation?

9. (a) Determine the result of applying a three-element median filter to the following 1-D signals:

(i)	0 0 0 0 0 1 0 1 1 1 1 1 1 1
(ii)	2 1 2 3 2 1 2 2 3 2 4 3 3 4
(iii)	1 1 2 3 3 4 5 8 6 6 7 8 9 9

(b) What general lessons can be learned from the results? In the first case, consider also the corresponding situation for a gray-scale edge in a 2-D image.

(c) 2-D median filters are sometimes implemented as two 1-D median filters applied in sequence in order to improve the speed of processing. Estimate the gain in speed that could be achieved in this way for (i) a 3×3 median filter, (ii) a 7×7 median filter, and (iii) in the general case.

Thresholding Techniques

An important practical aim of image processing is the demarcation of objects appearing in digital images. This process is called segmentation, and a good approximation to it can be achieved by thresholding. Broadly, this process involves separating the dark and light regions of the image, and thus identifying dark objects on a light background (or the inverse). This chapter discusses the effectiveness of this idea and the means for achieving it.

Look out for:

- the segmentation, region-growing, and thresholding concepts.
- the problem of threshold selection.
- the limitations of global thresholding.
- problems in the form of shadows or glints (highlights).
- the possibility of modeling the image background.
- the idea of adaptive thresholding.
- the rigorous Chow and Kaneko approach.
- what can be achieved with simple local adaptive thresholding algorithms.
- more thoroughgoing variance, entropy-based, and maximum likelihood methods.

Thresholding is limited in what it can achieve, and there are severe difficulties in automatically estimating the optimum threshold—as evidenced by the many available techniques that have been devised for the purpose. Segmentation is an ill-posed problem, and it is misleading that the human eye appears to perform thresholding reliably. Nevertheless, the task can sometimes be simplified, for example, by suitable lighting schemes, so that thresholding becomes effective. Hence, it is a useful technique that needs to be included in the toolbox of available algorithms for use when appropriate. However, edge detection (Chapter 5) provides a more general means to key into complex image data.

CHAPTER 4

Thresholding Techniques

4.1 Introduction

One of the first tasks to be undertaken in vision applications is to segment objects from their background. When objects are large and do not possess very much surface detail, segmentation is often imagined as splitting the image into a number of regions, each having a high level of uniformity in some parameter such as brightness, color, texture, or even motion. Hence, it should be straightforward to separate objects from one another and from their background, as well as to discern the different facets of solid objects such as cubes.

Unfortunately, this concept of segmentation is an idealization that is sometimes reasonably accurate, but more often in the real world it is an invention of the human mind, generalized inaccurately from certain simple cases. This problem arises because of the ability of the eye to understand real scenes at a glance, and hence to segment and perceive objects within images in the form they are known to have. Introspection is not a good way of devising vision algorithms, and it must not be overlooked that segmentation is actually one of the central and most difficult practical problems of machine vision.

Thus, the common view of segmentation as looking for regions possessing some degree of uniformity is to a large extent invalid. There are many examples of this in the world of 3-D objects. One example is a sphere lit from one direction, the brightness in this case changing continuously over the surface so that there is no distinct region of uniformity. Another is a cube where the direction of the lighting may lead to several of the facets having equal brightness values, so that it is impossible from intensity data alone to segment the image completely as desired.

Nevertheless, there is sufficient correctness in the concept of segmentation by uniformity measures for it to be worth pursuing for practical applications. The reason is that in many (especially industrial) applications only a very restricted range and number of objects are involved. In addition, it is possible to have almost complete control over the lighting and the general environment. The fact that a particular method may not be completely general need not be problematic, since by employing tools that are appropriate for the task at hand, a cost-effective solution will have been achieved in that case at least. However, in practical situations a tension exists between the simple cost-effective solution and the general-purpose but more computationally expensive solution. This tension must always be kept in mind in severely practical subjects such as machine vision.

4.2 Region-growing Methods

The segmentation idea as we have outlined it leads naturally to the region-growing technique (Zucker, 1976b). Here pixels of like intensity (or other suitable property) are successively grouped together to form larger and larger regions until the whole image has been segmented. Clearly, there have to be rules about not combining adjacent pixels that differ too much in intensity, while permitting combinations for which intensity changes gradually because of variations in background illumination over the field of view. However, this is not enough to make a viable strategy, and in practice the technique has to include the facility not only to merge regions together but also to split them if they become too large and inhomogeneous (Horowitz and Pavlidis, 1974). Particular problems are noise and sharp edges and lines that form disconnected boundaries, and for which it is difficult to formulate simple criteria to decide whether they form true region boundaries. In remote sensing applications, for example, it is often difficult to separate fields rigorously when edges are broken and do not give continuous lines. In such applications segmentation may have to be performed interactively, with a human operator helping the computer. Hall (1979) found that, in practice, regions tend to grow too far,[1] so that to make the technique work well it is necessary to limit their growth with the aid of edge detection schemes.

Thus, the region-growing approach to segmentation turns out to be quite complex to apply in practice. In addition, region-growing schemes usually operate iteratively, gradually refining hypotheses about which pixels belong to which regions. The technique is complicated because, carried out properly, it involves global as well as local image operations. Thus, each pixel intensity will in principle have to be examined many times, and as a result the process tends to be quite computationally intensive. For this reason it is not considered further here, since we are particularly interested in methods involving low computational load which are amenable to real-time implementation.

1 The danger is clearly that even one small break will in principle join two regions into a single larger one.

4.3 Thresholding

If background lighting is arranged so as to be fairly uniform, and we are looking for rather flat objects that can be silhouetted against a contrasting background, segmentation can be achieved simply by thresholding the image at a particular intensity level. This possibility was apparent from Fig. 2.3. In such cases, the complexities of the region-growing approach are bypassed. The process of thresholding has already been covered in Chapter 2, the basic result being that the initial gray-scale image is converted into a binary image in which objects appear as black figures on a white background or as white figures on a black background. Further analysis of the image then devolves into analysis of the shapes and dimensions of the figures. At this stage, object identification should be straightforward. Chapter 6 concentrates on such tasks. Meanwhile there is one outstanding problem—how to devise an automatic procedure for determining the optimum thresholding level.

4.3.1 *Finding a Suitable Threshold*

One simple technique for finding a suitable threshold arises in situations such as optical character recognition (OCR) where the proportion of the background that is occupied by objects (i.e., print) is substantially constant in a variety of conditions. A preliminary analysis of relevant picture statistics then permits subsequent thresholds to be set by insisting on a fixed proportion of dark and light in a sequence of images (Doyle, 1962). In practice, a series of experiments is performed in which the thresholded image is examined as the threshold is adjusted, and the best result is ascertained by eye. At that stage, the proportions of dark and light in the image are measured. Unfortunately, any changes in noise level following the original measurement will upset such a scheme, since they will affect the relative amounts of dark and light in the image. However, this technique is frequently useful in industrial applications, especially when particular details within an object are to be examined. Typical examples are holes in mechanical components such as brackets. (Notice that the mark–space ratio for objects may well vary substantially on a production line, but the proportion of hole area *within* the object outline would not be expected to vary.)

The technique that is most frequently employed for determining thresholds involves analyzing the histogram of intensity levels in the digitized image (see Fig. 4.1). If a significant minimum is found, it is interpreted as the required threshold value (Weska, 1978). The assumption being made here is that the peak on the left of the histogram corresponds to dark objects, and the peak on the

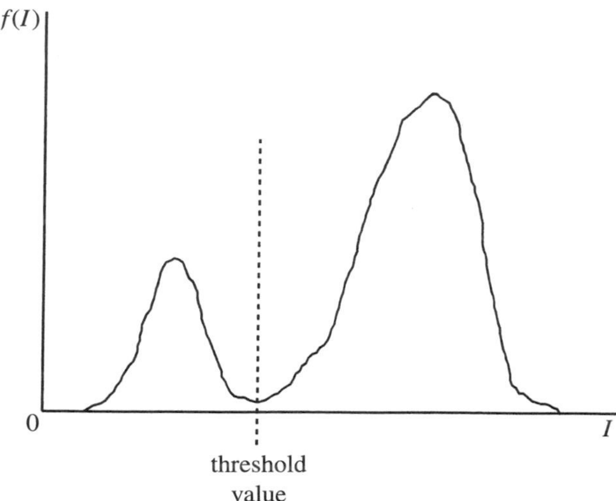

Figure 4.1 Idealized histogram of pixel intensity levels in an image. The large peak on the right results from the light background; the smaller peak on the left is due to dark foreground objects. The minimum of the distribution provides a convenient intensity value to use as a threshold.

right corresponds to light background (i.e., it is assumed that, as in many industrial applications, objects appear dark on a light background).

This method is subject to the following major difficulties:

1. The valley may be so broad that it is difficult to locate a significant minimum.
2. There may be a number of minima because of the type of detail in the image, and selecting the most significant one will be difficult.
3. Noise within the valley may inhibit location of the optimum position.
4. There may be no clearly visible valley in the distribution because noise may be excessive or because the background lighting may vary appreciably over the image.
5. Either of the major peaks in the histogram (usually that due to the background) may be much larger than the other, and this will then bias the position of the minimum.
6. The histogram may be inherently multimodal, making it difficult to determine which is the relevant thresholding level.

Perhaps the worst of these problems is the last: if the histogram is inherently multimodal, and we are trying to employ a single threshold, we are applying what is essentially an ad hoc technique to obtain a meaningful result. In general, such efforts are unlikely to succeed, and this is clearly a case where full image interpretation must be performed before we can be satisfied that the results are valid. Ideally, thresholding rules have to be formed after many images have been analyzed. In what follows, such problems of meaningfulness are eschewed. Instead attention is concentrated on how best to find a genuine single threshold when its position is obscured, as suggested by problems (1)–(6) above (which can be ascribed to image "clutter," noise, and lighting variations).

4.3.2 *Tackling the Problem of Bias in Threshold Selection*

This subsection considers problem (5) above—that of eliminating the bias in the selection of thresholds that arises when one peak in the histogram is larger than the other. First, note that if the relative heights of the peaks are known, this effectively eliminates the problem, since the "fixed proportion" method of threshold selection outlined above can be used. However, this is not normally possible. A more useful approach is to prevent bias by weighting down the extreme values of the intensity distribution and weighting up the intermediate values in some way. To achieve this effect, note that the intermediate values are special in that they correspond to object edges. Hence, a good basic strategy is to find positions in the image where there are significant intensity gradients—corresponding to pixels in the regions of edges—and to analyze the intensity values of these locations while ignoring other points in the image.

To develop this strategy, consider the "scattergram" of Fig. 4.2. Here pixel properties are plotted on a 2-D map with intensity variation along one axis and intensity gradient magnitude variation along the other. Broadly speaking, three main populated regions are seen on the map: (1) a low-intensity, low-gradient region corresponding to the dark objects; (2) a high-intensity, low-gradient region corresponding to the background; and (3) a medium-intensity, high-gradient region corresponding to object edges (Panda and Rosenfeld, 1978). These three regions merge into each other, with the high-gradient region that corresponds to edges forming a path from the low-intensity to the high-intensity region. The fact that the three regions merge in this way gives a choice of methods for selecting intensity thresholds. The first choice is to consider the intensity distribution along the zero gradient axis, but this alternative has the same disadvantage as the original histogram method—namely, introducing a substantial bias. The second choice is that at high gradient, which means looking for a peak rather than a valley in the intensity distribution. The final choice is that at moderate values of gradient, which was suggested earlier as a possible strategy for avoiding bias.

It is now apparent that there is likely to be quite a narrow range of values of gradient for which the third method will be a genuine improvement. In addition, it may be necessary to construct a scattergram rather than a simple intensity

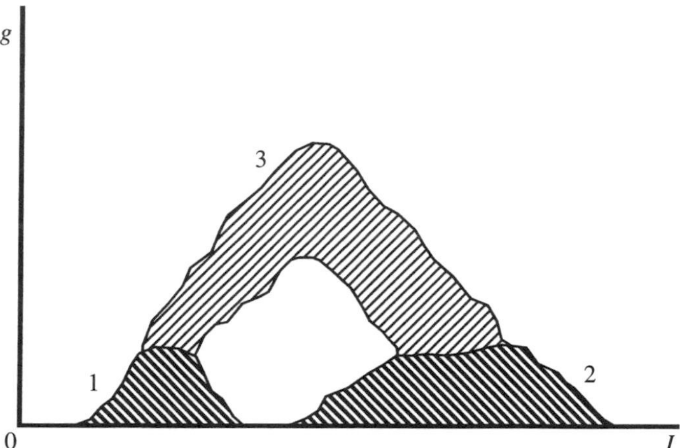

Figure 4.2 Scattergram showing the frequency of occurrence of various combinations of pixel intensity I and intensity gradient magnitude g in an idealized image. There are three main populated regions of interest: (1) a low I, low g region; (2) a high I, low g region; and (3) a medium I, high g region. Analysis of the scattergram provides useful information on how to segment the image.

histogram in order to locate it. If, instead, we try to weight the plots on a histogram with the magnitudes of the intensity gradient values, we are likely to end with a peaked intensity distribution (i.e., method 2) rather than one with a valley (Weska and Rosenfeld, 1979). A mathematical model of the situation is presented in Section 4.3.3 in order to explore the situation more fully.

4.3.2.1 *Methods Based on Finding a Valley in the Intensity Distribution*

This section considers how to weight the intensity distribution using a parameter other than the intensity gradient, in order to locate accurately the valley in the intensity distribution. A simple strategy is first to locate all pixels that have a significant intensity gradient and then to find the intensity histogram not only of these pixels but also of nearby pixels. This means that the two main modes in the intensity distribution are still attenuated very markedly and hence the bias in the valley position is significantly reduced. Indeed, the numbers of background and foreground pixels that are now being examined are very similar, so the bias from the relatively large numbers of background pixels is virtually eliminated. (Note that if the modes are modeled as two Gaussian distributions of equal widths and they also have equal heights, then the minimum lies exactly halfway between them.)

Though obvious, this approach includes the edge pixels themselves, and from Fig. 4.2 it can be seen that they will tend to fill the valley between the two modes. For the best results, the points of highest gradient must actually be removed from the intensity histogram. A well-attested way of achieving this is to weight pixels in the intensity histogram according to their response to a Laplacian filter (Weska et al., 1974). Since such a filter gives an isotropic estimate of the second derivative of the image intensity (i.e., the magnitude of the first derivative of the intensity gradient), it is zero where intensity gradient magnitude is high. Hence, it gives such locations zero weight, but it nevertheless weights up those locations on the shoulders of edges. It has been found that this approach is excellent at estimating where to place a threshold within a wide valley in the intensity histogram (Weska et al., 1974). Section 4.3.3 shows that this is far from being an ad hoc solution to a tedious problem. Rather, it is in many ways ideal in giving a very exact indication of the optimum threshold value.

4.3.2.2 Methods That Concentrate on the Peaked Intensity Distribution at High Gradient

The other approach indicated earlier is to concentrate on the peaked intensity distribution at high-gradient values. It seems possible that location of a peak will be less accurate (for the purpose of finding an effective intensity threshold value) than locating a valley between two distributions, especially when edges do not have symmetrical intensity profiles. However, an interesting result was obtained by Kittler et al. (1984, 1985), who suggested locating the threshold directly (i.e., without constructing an intensity histogram or a scattergram) by calculating the statistic:

$$T = \frac{\sum_i I_i g_i}{\sum_i g_i} \tag{4.1}$$

where I_i is the local intensity and g_i is the local intensity gradient magnitude. This statistic was found to be biased by noise, resulting in noisy background and foreground pixels that contributed to the average value of intensity in the threshold calculation. This effect occurs only if the image contains different numbers of background and foreground pixels. It is a rather different manifestation of the bias between the two histogram modes that was noted earlier. In this case, a formula *including* bias was obtained in the form

$$T = T_0 + C(0.5 - b) \tag{4.2}$$

where T_0 is the inherent (i.e., unbiased) threshold, C is the image contrast, and b is the proportion of background pixels in the image.

To understand the reasoning behind this formula, denote the background and foreground intensities as B and F, respectively, and the *proportions* of background and foreground pixels as b and f, where $f + b = 1$. (Here we are effectively assuming that a negligible proportion of pixels are situated on object edges. This is a reasonable assumption as the number of foreground and background pixels will be of order N^2 for an $N \times N$ pixel image, whereas the number of edge pixels will be of order N.) Assuming a simple model of intensity variation, we find that

$$T_0 = (B + F)/2 \tag{4.3}$$

and

$$C = F - B \tag{4.4}$$

Hence:

$$\begin{aligned} T &= (B + F)/2 + (F - B)(f - b)/2 \\ &= B(1 - f + b)/2 + F(1 + f - b)/2 \\ &= Bb + Ff \end{aligned} \tag{4.5}$$

Equation (4.5) can be interpreted as follows. The threshold T is here calculated by averaging the intensity over image pixels, after weighting them in proportion to the local intensity gradient. Within the background, the pixels with appreciable intensity gradient are noise pixels, and these are proportional in number to b and have mean intensity B. Similarly, within the foreground, pixels with appreciable intensity gradient are proportional in number to f and have mean intensity F. As a result, the mean intensity of all these noise pixels is $Bb + Ff$, which is in agreement with the above equation for T. On the other hand, if pixels with low-intensity gradient are excluded, this progressively eliminates normal noise pixels and weights up the edge pixels (not represented in the above equation), so that the threshold reverts to the unbiased value T_0. By applying the right corrections, Kittler et al. were able to eliminate the bias and to obtain a corrected threshold value T_0 in practical situations, even when pixels with low-intensity gradient values were not excluded.

This method is useful because it results in unbiased estimates of the threshold T_0 without the need to construct histograms or scattergrams. However, it is probably somewhat risky not to obtain histograms inasmuch as they give so much information—for example, on the existence of multiple modes. Thus, the statistic T may in complex cases amount to a bland average, which means very little. Ultimately, it seems that the safest threshold determination procedure must involve seeking and identifying *specific* peaks on the intensity histogram for relatively high values of intensity gradient, or *specific* troughs for relatively low values of intensity gradient.

4.3.3 *A Convenient Mathematical Model*

To get a clearer understanding of these problems, it is useful to construct a theoretical model of the situation. It is assumed that a number of dark, fairly flat objects are lying on a light worktable and are subject to soft, reasonably uniform lighting. In the absence of noise and other perturbations, the only variations in intensity occur near object boundaries (Fig. 2.3). With little loss of generality, the model can be simplified to a single dimension. It is easy to see that the distribution of intensities in the image is then as shown in Fig. 4.3 and is closely related to the shape of the edge profile. The intensity distribution is effectively the derivative of the edge profile:

$$f(I) = c|\mathrm{d}x/\mathrm{d}I| \tag{4.6}$$

since for I in a given range ΔI, $f(I)\Delta I$ is proportional to the range of distances Δx that contribute to it. Taking account now of the fact that the entire image may contain many edges at specific locations, we can calculate the distribution $f(I)$ in principle for all values of I, including the two very large peaks at the maximum and minimum values of intensity.

So far the model is incomplete in that it takes no account of noise or other variations in intensity. Essentially, these act in such a way as to broaden the

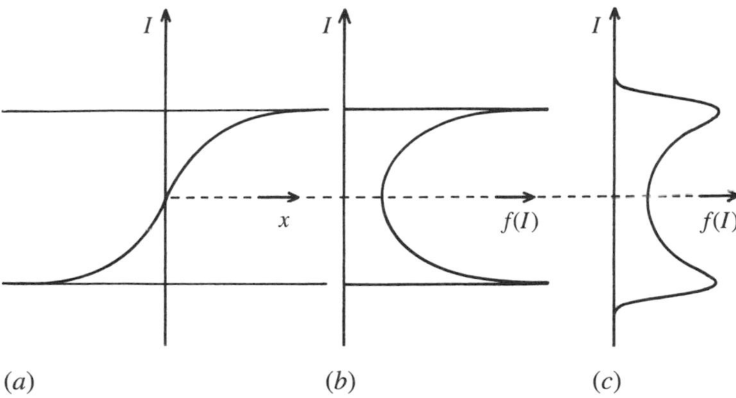

Figure 4.3 Relation of edge profile to intensity distribution: (*a*) shape of a typical edge profile; (*b*) resulting distribution of intensities; (*c*) effect of applying a broadening convolution (e.g., resulting from noise) to (*b*).

intensity values from those predicted from edge profiles alone. This means that they can be allowed for by applying a broadening convolution $b(I)$ to $f(I)$:

$$F(I) = \int_{-\infty}^{\infty} b(u - I)f(u)\mathrm{d}u \tag{4.7}$$

This gives the more realistic distribution shown in Fig. 4.3c. These ideas confirm that a high histogram peak at the highest or lowest value of intensity can pull the minimum one way or the other and result in the original automatic threshold selection criterion being inaccurate.

In what follows, we simplify the analysis by returning to the unbroadened distribution $f(I)$. We make the following definitions of intensity gradient g and rate of change of intensity gradient g':

$$g = \mathrm{d}I/\mathrm{d}x \tag{4.8}$$

$$g' = \mathrm{d}g/\mathrm{d}x = (\mathrm{d}g/\mathrm{d}I)(\mathrm{d}I/\mathrm{d}x) = g(\mathrm{d}g/\mathrm{d}I) \tag{4.9}$$

Applying equation (4.6), we find that

$$f(I) = c|g|^{-1} \tag{4.10}$$

If we wish to weight the contributions to $f(I)$ using some power p of $|g|$, the new distribution becomes:

$$h(I) = f(I)|g|^p = c|g|^{p-1} \tag{4.11}$$

This is not a sharp unimodal distribution as required by the earlier discussion, unless $p > 1$. We could conveniently set $p = 2$, although in practice workers have sometimes opted for a nonlinear, sharply varying weighting which is unity if $|g|$ exceeds a given threshold and zero otherwise (Weska and Rosenfeld, 1979). This matter is not pursued further here. Instead, we consider the effect of weighting the contributions to $f(I)$ using $|g'|$ and obtain the new distribution:

$$k(I) = f(I)|g'| \tag{4.12}$$

Applying equations (4.6) and (4.9) now gives

$$k(I) = c|\mathrm{d}g/\mathrm{d}I| \tag{4.13}$$

To proceed further, an exact mathematical model is needed. A convenient one is obtained using the tanh function:

$$I = I_0 \tanh x \tag{4.14}$$

with I being restricted to the range $-I_0 \leq I \leq I_0$. (Note that we have simplified the model by making I zero at the center of the range.) This leads to:

$$\begin{aligned} \mathrm{d}I/\mathrm{d}x &= I_0 \operatorname{sech}^2 x \\ &= I_0(1 - \tanh^2 x) \\ &= I_0(1 - I^2/I_0^2) \end{aligned} \tag{4.15}$$

Applying equations (4.8) and (4.9) now gives

$$g = I_0(1 - I^2/I_0^2) \tag{4.16}$$

and

$$g' = -2I(1 - I^2/I_0^2) \tag{4.17}$$

Note that the form of g given by the model, as a function of x, is:

$$g = I_0 \operatorname{sech}^2 x = 4I_0/(\mathrm{e}^x + \mathrm{e}^{-x})^2 \tag{4.18}$$

and for high values of $|x|$

$$g \approx 4I_0 \,\mathrm{e}^{-2|x|} \tag{4.19}$$

Although a Gaussian variation for g might fit real data better, the present model has the advantage of simplicity and is able to provide useful insight into the problem of thresholding.

Applying equations (4.10)–(4.13) and (4.16), we may deduce that

$$f(I) = c|I_0(1 - I^2/I_0^2)|^{-1} \tag{4.20}$$

$$h(I) = c|I_0(1 - I^2/I_0^2)|^{p-1} \tag{4.21}$$

$$k(I) = c|2I/I_0| \tag{4.22}$$

We have now arrived at an interesting point: starting with a particular edge profile, we have calculated the shape of the basic intensity distribution $f(I)$, the shape of the intensity distribution $h(I)$ resulting from weighting pixel contributions according to a power of the intensity gradient value, and the distribution $k(I)$ resulting from weighting them according to the spatial derivative of the intensity gradient, that is, according to the output of a Laplacian operator. The $k(I)$ distribution has a very sharp valley with linear variation near the minimum.

By contrast, $h(I)$ varies rather slowly and smoothly near its maximum. This shows that the Laplacian weighting method should provide the most sensitive indicator of the optimum threshold value and thereby provides strong justification for the use of this method. However, it must be made absolutely clear that the advantages of this approach may well be washed away in practice if the assumptions made in the model are invalid. In particular, broadening of the distribution by noise, variations in background lighting, image clutter, and other artifacts will affect the situation adversely for all of the above methods. Further detailed analysis seems pointless, in part because these interfering factors make choice of the threshold more difficult in a very data-dependent way, but more importantly because they ultimately make the concept of thresholding at a single level invalid (see Figs 4.4–4.6).

4.3.4 *Summary*

It has been shown that available techniques can provide values at which intensity thresholding can be applied, but they do not themselves solve the problems caused by uneven lighting. They are even less capable of coping with glints, shadows, and image clutter. Unfortunately, these artifacts are common in most real situations and are only eliminated with difficulty. Indeed, in industrial applications where shiny metal components are involved, glints are the rule rather than the exception, whereas shadows can seldom be avoided with any sort of object. Even flat objects are liable to have quite a strong shadow contour around them because of the particular placing of lights. Lighting problems are studied in detail in Chapter 27. Meanwhile, it may be noted that glints and shadows can only be allowed for properly in a two-stage image analysis system, where tentative assignments are made first and these are firmed up by exact explanation of all pixel intensities. We now return to the problem of making the most of the thresholding technique—by finding how we can allow for variations in background lighting.

4.4 Adaptive Thresholding

The problem that arises when illumination is not sufficiently uniform may be tackled by permitting the threshold to vary adaptively (or "dynamically") over the whole image. In principle, there are several ways of achieving this. One involves modeling the background within the image. Another is to work out a local threshold value for each pixel by examining the range of intensities in its

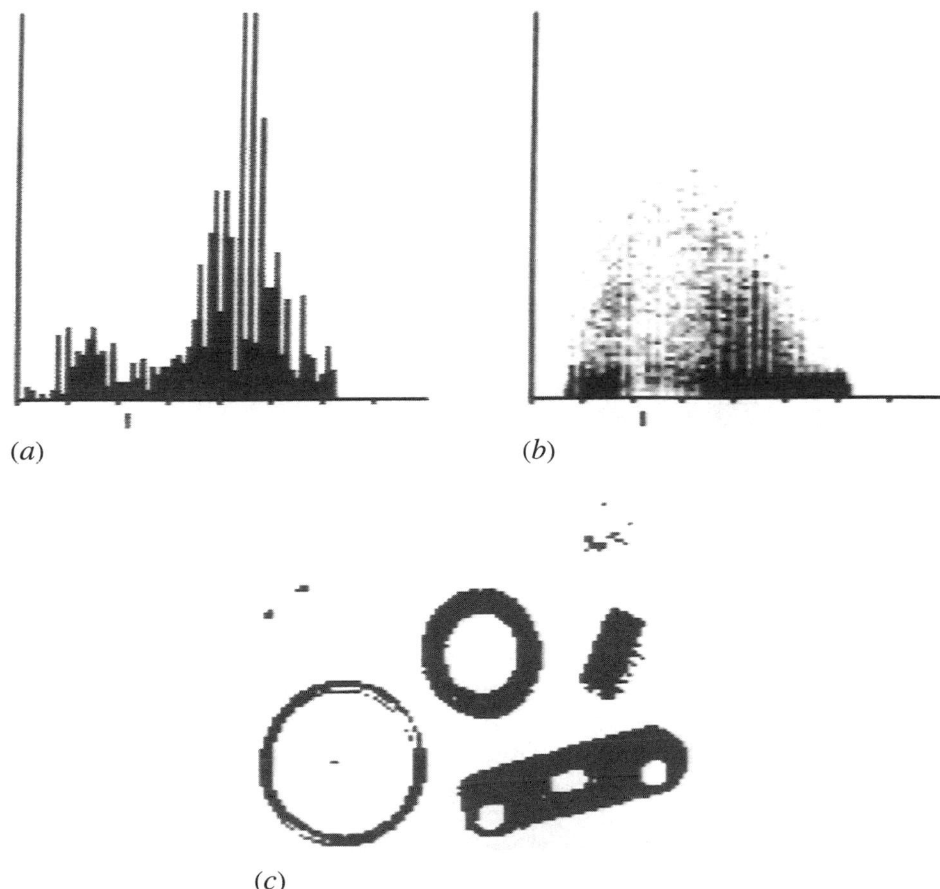

Figure 4.4 (*a*) Histogram and (*b*) scattergram for the image shown in Fig. 2.11*a*. In neither case is the diagram particularly close to the ideal form of Figs. 4.1 and 4.2. Hence, the threshold obtained from (*a*) (indicated by the short lines beneath the two scales) does not give ideal results with all the objects in the binarized image (*c*). Nevertheless, the results are better than for the arbitrarily thresholded image of Fig. 2.11*b*.

neighborhood. A third approach is to split the image into subimages and deal with them independently. Though "obvious," this last method will clearly run into problems at the boundaries between subimages, and by the time these problems have been solved it will look more like one of the other two methods. Ultimately, all such methods must operate on identical principles. The differences arise in the rigor with which the threshold is calculated at different locations and in the amount of computation required in each case. In real-time applications the problem amounts to finding how to estimate a set of thresholds with a minimum amount of computation.

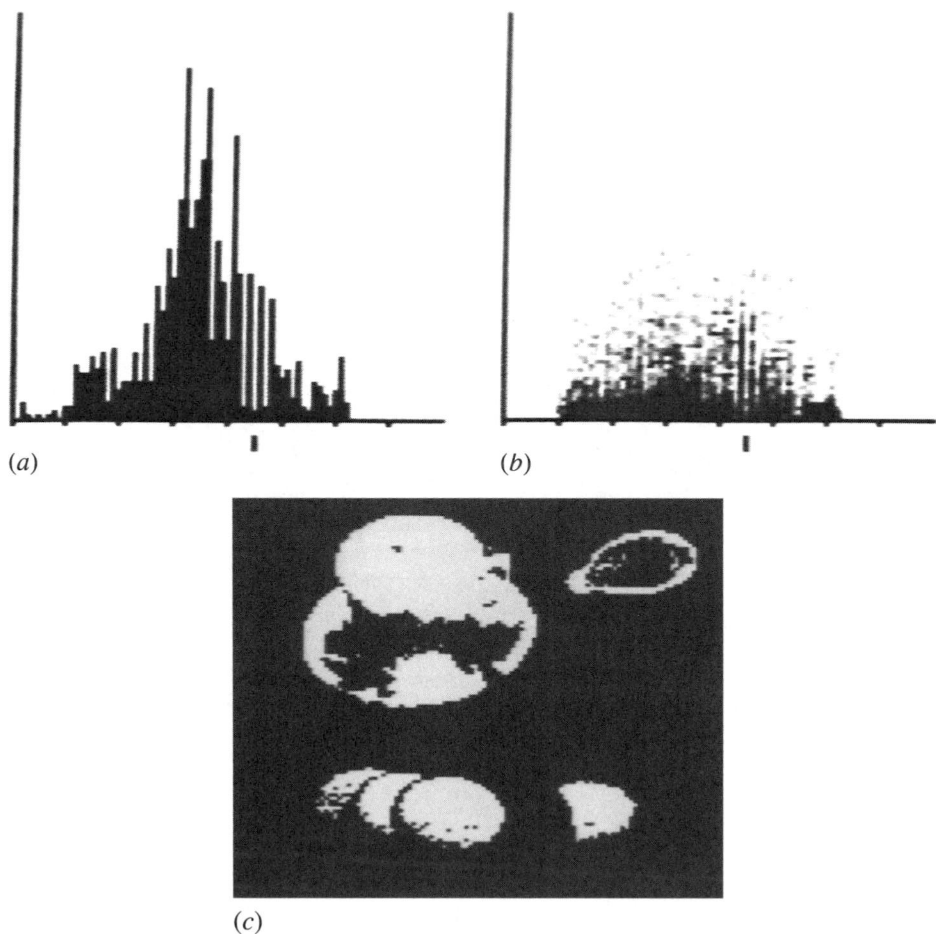

Figure 4.5 (*a*) Histogram and (*b*) scattergram for the image shown in Fig. 2.1*a*. In neither case is the diagram at all close to the idealized form, and the results of thresholding (*c*) are not a particularly useful aid to interpretation.

The problem can sometimes be solved rather neatly in the following way. On some occasions—such as in automated assembly applications—it is possible to obtain an image of the background in the absence of any objects. This appears to solve the problem of adaptive thresholding in a rigorous manner, since the tedious task of modeling the background has already been carried out. However, some caution is needed with this approach. Objects bring with them not only shadows (which can in some sense be regarded as part of the objects), but also an additional effect due to the reflections they cast over the background and other objects. This additional effect is nonlinear, in the sense that it is necessary to add not only the difference between the object and the background intensity in each case but also an intensity that depends on the products of the reflectances of

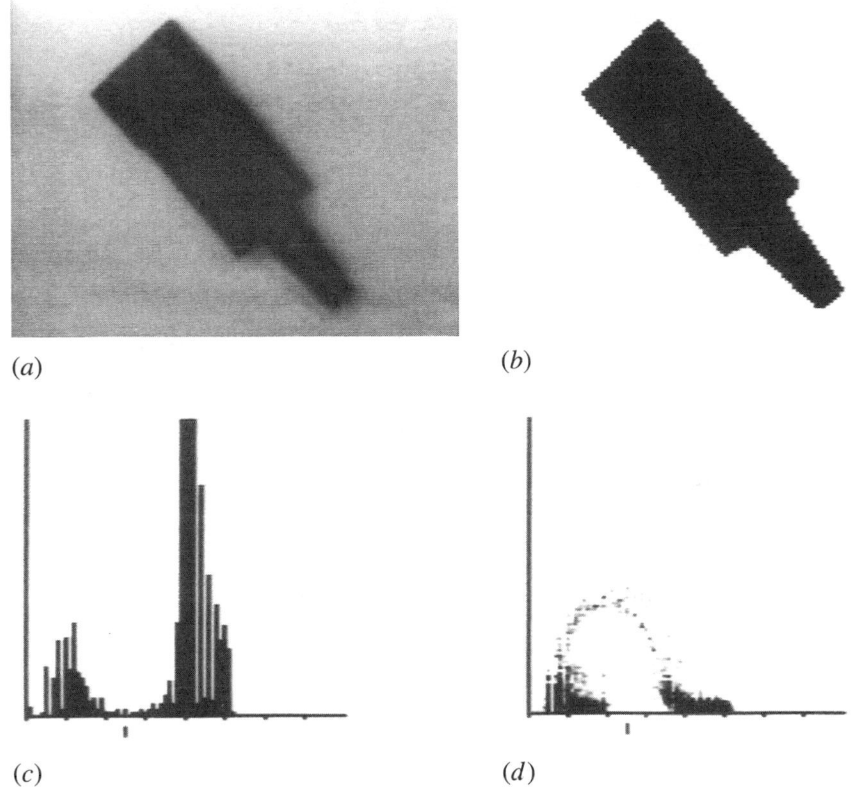

Figure 4.6 A picture with more ideal properties. (*a*) Image of a plug that has been lit fairly uniformly. The histogram (*c*) and scattergram (*d*) approximate to the ideal forms, and the result of thresholding (*b*) is acceptable. However, much of the structure of the plug is lost during binarization.

pairs of objects. These considerations mean that using the no-object background as the equivalent background when several objects are present is ultimately invalid. However, as a first approximation it is frequently possible to assume an equivalence. If this proves impracticable, the only option is to model the background from the actual image to be segmented.

On other occasions, the variation in background intensity may be rather slowly varying, in which case it may be possible to model it by the following technique. (this is a form of Hough transform—see Chapter 9.) First, an equation is selected which can act as a reasonable approximation to the intensity function, for example, a linear or quadratic variation:

$$I = a + bx + cy + dx^2 + exy + fy^2 \tag{4.23}$$

Next, a parameter space for the six variables a, b, c, d, e, f is constructed. Then each pixel in the image is taken in turn, and all sets of values of the parameters that could have given rise to the pixel intensity value are accumulated in parameter space. Finally, a peak is sought in parameter space which represents an optimal fit to the background model. So far it appears that this has been carried out only for a linear variation, the analysis being simplified initially by considering only the differences in intensities of pairs of points in image space (Nixon, 1985). Note that a sufficient number of pairs of points must be considered so that the peak in parameter space resulting from background pairs is sufficiently well populated. This implies quite a large computational load for quadratic or higher variations in intensity. Hence, it is unlikely that this method will be generally useful if the background intensity is not approximated well by a linear variation.

4.4.1 *The Chow and Kaneko Approach*

As early as 1972, Chow and Kaneko introduced what is widely recognized as the standard technique for dynamic thresholding. It performs a thoroughgoing analysis of the background intensity variation, making few compromises to save computation (Chow and Kaneko, 1972). In this method, the image is divided into a regular array of overlapping subimages, and individual intensity histograms are constructed for each one. Those that are unimodal are ignored since they are assumed not to provide any useful information that can help in modeling the background intensity variation. However, the bimodal distributions are well suited to this task. These are individually fitted to pairs of Gaussian distributions of adjustable height and width, and the threshold values are located. Thresholds are then found, by interpolation, for the unimodal distributions. Finally, a second stage of interpolation is necessary to find the correct thresholding value at each pixel.

One problem with this approach is that if the individual subimages are made very small in an effort to model the background illumination more exactly, the statistics of the individual distributions become worse, their minima become less well defined, and the thresholds deduced from them are no longer statistically significant. This means that it does not pay to make subimages too small and that ultimately only a certain level of accuracy can be achieved in modeling the background in this way. The situation is highly data dependent, but little can be expected to be gained by reducing the subimage size below 32×32 pixels. Chow and Kaneko employed 256×256 pixel images and divided these into a 7×7 array of 64×64 pixel subimages with 50% overlap.

Overall, this approach involves considerable computation, and in real-time applications it may well not be viable for this reason. However, this type of approach is of considerable value in certain medical, remote sensing, and space applications. In addition, note that the method of Kittler et al. (1985) referred to earlier may be used to obtain threshold values for the subimages with rather less computation.

4.4.2 *Local Thresholding Methods*

In real-time applications the alternative approach mentioned earlier is often more useful for finding local thresholds. It involves analyzing intensities in the neighborhood of each pixel to determine the optimum local thresholding level. Ideally, the Chow and Kaneko histogramming technique would be repeated at each pixel, but this would significantly increase the computational load of this already computationally intensive technique. Thus, it is necessary to obtain the vital information by an efficient sampling procedure. One simple means of achieving this is to take a suitably computed function of nearby intensity values as the threshold. Often the mean of the local intensity distribution is taken, since this is a simple statistic and gives good results in some cases. For example, in astronomical images stars have been thresholded in this way. Niblack (1985) reported a case in which a proportion of the local standard deviation was added to the mean to give a more suitable threshold value, the reason (presumably) being to help suppress noise. (Addition is appropriate where bright objects such as stars are to be located, whereas subtraction will be more appropriate in the case of dark objects.)

Another frequently used statistic is the mean of the maximum and minimum values in the local intensity distribution. Whatever the sizes of the two main peaks of the distribution, this statistic often gives a reasonable estimate of the position of the histogram minimum. The theory presented earlier shows that this method will only be accurate if (1) the intensity profiles of object edges are symmetrical, (2) noise acts uniformly everywhere in the image so that the widths of the two peaks of the distribution are similar, and (3) the heights of the two distributions do not differ markedly. Sometimes these assumptions are definitely invalid—for example, when looking for (dark) cracks in eggs or other products. In such cases the mean and maximum of the local intensity distribution can be found, and a threshold can be deduced using the statistic

$$T = \text{mean} - (\text{maximum} - \text{mean}) \tag{4.24}$$

where the strategy is to estimate the lowest intensity in the bright background assuming the distribution of noise to be symmetrical (Fig. 4.7). Use of the mean here is realistic only if the crack is narrow and does not affect the value of the mean significantly. If it does, then the statistic can be adjusted by use of an ad hoc parameter:

$$T = \text{mean} - k(\text{maximum} - \text{mean}) \tag{4.25}$$

where k may be as low as 0.5 (Plummer and Dale, 1984).

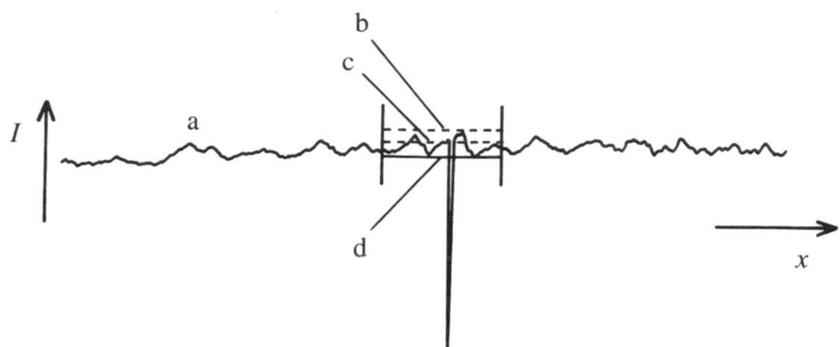

Figure 4.7 Method for thresholding the crack in an egg. (*a*) Intensity profile of an egg in the vicinity of a crack: the crack is assumed to appear dark (e.g., under oblique lighting); (*b*) local maximum of intensity on the surface of the egg; (*c*) local mean intensity. Equation (4.24) gives a useful estimator T of the thresholding level (*d*).

This method is essentially the same as that of Niblack (1985), but the computational load in estimating the standard deviation is minimized. Each of the last two techniques relies on finding local extrema of intensity. Using these measures helps save computation, but they are somewhat unreliable because of the effects of noise. If this is a serious problem, quartiles or other statistics of the distribution may be used. The alternative of prefiltering the image to remove noise is unlikely to work for crack thresholding because cracks will almost certainly be removed at the same time as the noise. A better strategy is to form an image of T-values obtained using equation (4.24) or (4.25). Smoothing this image should then permit the initial image to be thresholded effectively.

Unfortunately, all these methods work well only if the size of the neighborhood selected for estimating the required threshold is sufficiently large to span a significant amount of foreground and background. In many practical cases, this is not possible and the method then adjusts itself erroneously, for example, so that it finds darker spots within dark objects as well as segmenting the dark objects themselves. However, in certain applications there is little risk of this occurring. One notable case is that of OCR. Here the widths of character limbs are likely to be known in advance and should not vary substantially. If this is so, then a neighborhood size can be chosen to span or at least sample both character and background. It is thus possible to threshold the characters highly efficiently using a simple functional test of the type described above. The effectiveness of this procedure (Fig. 4.8) is demonstrated in Fig. 4.9.

Finally, before leaving this topic, it should be noted that hysteresis thresholding is a type of adaptive thresholding—effectively permitting the threshold value to vary locally. This topic will be investigated in Section 7.1.1.

```
minrange = 255 / 5;
/* minimum likely difference in intensity between print and background: this
parameter can be preset manually or "learned" by a previous routine */
do { // for each pixel
   find minimum and maximum of local intensity distribution;
   range = maximum – minimum;
   if (range > minrange)
      T = (minimum + maximum)/2; // print is visible in neighborhood
   else T = maximum – minrange/2; // neighborhood is all white
   if (P0 > T) Q0 = 255; else Q0 = 0; // now binarize print
} until all pixels processed;
```

Figure 4.8 A simple algorithm for adaptively thresholding print.

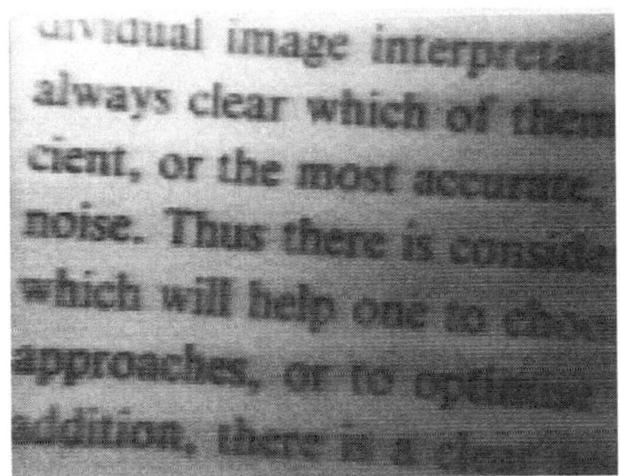

(*a*) (*b*)

(*c*)

Figure 4.9 Effectiveness of local thresholding on printed text. Here a simple local thresholding procedure (Fig. 4.8), operating within a 3×3 neighborhood, is used to binarize the image of a piece of printed text (*a*). Despite the poor illumination, binarization is performed quite effectively (*b*). Note the complete absence of isolated noise points in (*b*), while by contrast the dots on all the *i*'s are accurately reproduced. The best that could be achieved by uniform thresholding is shown in (*c*).

4.5 More Thoroughgoing Approaches to Threshold Selection

At this point, we return to global threshold selection and describe some especially important approaches that have a rigorous mathematical basis. The first of these approaches is variance-based thresholding; the second is entropy-based thresholding; and the third is maximum likelihood thresholding. All three are widely used, the second having achieved an increasingly wide following over the past 20 to 30 years. The third is a more broadly based technique that has its roots in statistical pattern recognition—a subject that is dealt with in depth in Chapter 24.

4.5.1 *Variance-based Thresholding*

The standard approach to thresholding outlined earlier involved finding the neck of the global image intensity histogram. However, this is impracticable when the dark peak of the histogram is minuscule in size, for it will then be hidden among the noise in the histogram and it will not be possible to extract it with the usual algorithms.

Many investigators have studied this problem (e.g., Otsu, 1979; Kittler et al., 1985; Sahoo et al., 1988; Abutaleb, 1989). Among the most well-known approaches are the variance-based methods. In these methods, the image intensity histogram is analyzed to find where it can best be partitioned to optimize criteria based on ratios of the within-class, between-class, and total variance. The simplest approach (Otsu, 1979) is to calculate the between-class variance, as will now be described.

First, we assume that the image has a gray-scale resolution of L gray levels. The number of pixels with gray level i is written as n_i, so the total number of pixels in the image is $N = n_1 + n_2 + \cdots + n_L$. Thus, the probability of a pixel having gray level i is:

$$p_i = n_i / N \tag{4.26}$$

where

$$p_i \geq 0 \qquad \sum_{i=1}^{L} p_i = 1 \tag{4.27}$$

For ranges of intensities up to and above the threshold value k, we can now calculate the between-class variance σ_B^2 and the total variance σ_T^2:

$$\sigma_B^2 = \pi_0(\mu_0 - \mu_T)^2 + \pi_1(\mu_1 - \mu_T)^2 \tag{4.28}$$

$$\sigma_T^2 = \sum_{i=1}^{L} (i - \mu_T)^2 p_i \tag{4.29}$$

where

$$\pi_0 = \sum_{i=1}^{k} p_i \qquad \pi_1 = \sum_{i=k+1}^{L} p_i = 1 - \pi_0 \tag{4.30}$$

$$\mu_0 = \sum_{i=1}^{k} ip_i/\pi_0 \qquad \mu_1 = \sum_{i=k+1}^{L} ip_i/\pi_1 \qquad \mu_T = \sum_{i=1}^{L} ip_i \tag{4.31}$$

Making use of the latter definitions, the formula for the between-class variance can be simplified to:

$$\sigma_B^2 = \pi_0\pi_1(\mu_1 - \mu_0)^2 \tag{4.32}$$

For a single threshold the criterion to be maximized is the ratio of the between-class variance to the total variance:

$$\eta = \sigma_B^2/\sigma_T^2 \tag{4.33}$$

However, the total variance is constant for a given image histogram; thus, maximizing η simplifies to maximizing the between-class variance.

The method can readily be extended to the dual threshold case $1 \leq k_1 \leq k_2 \leq L$, where the resultant classes, C_0, C_1, and C_2, have respective gray-level ranges of $[1, \ldots, k_1]$, $[k_1 + 1, \ldots, k_2]$, and $[k_2 + 1, \ldots, L]$.

In some situations (e.g., Hannah et al., 1995), this approach is still not sensitive enough to cope with histogram noise, and more sophisticated methods must be used. One such technique is that of entropy-based thresholding, which has become firmly embedded in the subject (Pun, 1980; Kapur et al., 1985; Abutaleb, 1989; Brink, 1992).

4.5.2 *Entropy-based Thresholding*

Entropy measures of thresholding are based on the concept of entropy. The entropy statistic is high if a variable is well distributed over the available range, and low if it is well ordered and narrowly distributed. Specifically, entropy is a measure of disorder and is zero for a perfectly ordered system. The concept of entropy thresholding is to threshold at an intensity for which the sum of the entropies of the two intensity probability distributions thereby separated is maximized. The reason for this is to obtain the greatest reduction in entropy—that is, the greatest increase in order—by applying the threshold. In other words, the most appropriate threshold level is the one that imposes the greatest order on the system, and thus leads to the most meaningful result.

To proceed, the intensity probability distribution is again divided into two classes—those with gray levels up to the threshold value k, and those with gray levels above k (Kapur et al., 1985). This leads to two probability distributions A and B:

$$\text{A:}\quad \frac{p_1}{P_k}, \frac{p_2}{P_k}, \ldots, \frac{p_k}{P_k} \tag{4.34}$$

$$\text{B:}\quad \frac{p_{k+1}}{1-P_k}, \frac{p_{k+2}}{1-P_k}, \ldots, \frac{p_L}{1-P_k} \tag{4.35}$$

where

$$P_k = \sum_{i=1}^{k} p_i \qquad 1 - P_k = \sum_{i=k+1}^{L} p_i \tag{4.36}$$

The entropies for each class are given by

$$H(A) = -\sum_{i=1}^{k} \frac{p_i}{P_k} \ln \frac{p_i}{P_k} \tag{4.37}$$

$$H(B) = -\sum_{i=k+1}^{L} \frac{p_i}{1-P_k} \ln \frac{p_i}{1-P_k} \tag{4.38}$$

and the total entropy is

$$\mathscr{H}(k) = H(A) + H(B) \tag{4.39}$$

Substitution leads to the final formula:

$$\mathscr{H}(k) = \ln\left(\sum_{i=1}^{k} p_i\right) + \ln\left(\sum_{i=k+1}^{L} p_i\right) - \frac{\sum_{i=1}^{k} p_i \ln p_i}{\sum_{i=1}^{k} p_i} - \frac{\sum_{i=k+1}^{L} p_i \ln p_i}{\sum_{i=k+1}^{L} p_i} \tag{4.40}$$

and it is this parameter that has to be maximized.

This approach can give very good results—see, for example, Hannah et al. (1995). Again, it is straightforwardly extended to dual thresholds, but we shall not go into the details here (Kapur et al., 1985). Probabilistic analysis to find mathematically ideal dual thresholds may not be the best approach in practical situations. An alternative technique for determining dual thresholds sequentially has been devised by Hannah et al. (1995) and applied to an X-ray inspection task—as described in Chapter 22.

4.5.3 *Maximum Likelihood Thresholding*

When dealing with distributions such as intensity histograms, it is important to compare the actual data with the data that might be expected from a previously constructed model based on a training set. This is in keeping with the methods of statistical pattern recognition (see Chapter 24), which takes full account of prior probabilities. For this purpose, one option is to model the training set data using a known distribution function such as a Gaussian. The Gaussian model has many advantages, including its accessibility to relatively straightforward mathematical analysis. In addition, it is specifiable in terms of two well-known parameters—the mean and standard deviation—which are easily measured in practical situations. Indeed, for any Gaussian distribution we have:

$$p_i(x) = \frac{1}{(2\pi\sigma_i^2)^{1/2}} \exp\left[-\frac{(x-\mu_i)^2}{2\sigma_i^2}\right] \tag{4.41}$$

where the suffix i refers to a specific distribution, and of course when thresholding is being carried out two such distributions are assumed to be involved. Applying the respective *a priori* class probabilities P_1, P_2 (Chapter 24), careful analysis (Gonzalez and Woods, 1992) shows that the condition $p_1(x) = p_2(x)$ reduces to the form:

$$x^2\left(\frac{1}{\sigma_1^2} - \frac{1}{\sigma_2^2}\right) - 2x\left(\frac{\mu_1}{\sigma_1^2} - \frac{\mu_2}{\sigma_2^2}\right) + \left(\frac{\mu_1^2}{\sigma_1^2} - \frac{\mu_2^2}{\sigma_2^2}\right) + 2\log\left(\frac{P_2\sigma_1}{P_1\sigma_2}\right) = 0 \tag{4.42}$$

Note that, in general, this equation has two solutions,[2] implying the need for two thresholds, though when $\sigma_1 = \sigma_2$ there is a single solution:

$$x = \frac{1}{2}(\mu_1 + \mu_2) + \frac{\sigma^2}{\mu_1 - \mu_2}\ln\left(\frac{P_2}{P_1}\right) \tag{4.43}$$

In addition, when the prior probabilities for the two classes are equal, the equation reduces to the altogether simpler and more obvious form:

$$x = \frac{1}{2}(\mu_1 + \mu_2) \tag{4.44}$$

Of all the methods described in this chapter, only the maximum likelihood method makes use of *a priori* probabilities. Although this makes it appear to be

2 Two solutions exist because one solution represents a threshold in the area of overlap between the two Gaussians; the other solution is necessary mathematically and lies either at very high or very low intensities. It is the latter solution that disappears when the two Gaussians have equal variance, as the distributions clearly never cross again. In any case, it seems unlikely that the distributions being modeled would in practice approximate so well to Gaussians that the noncentral solution could ever be important—that is, it is essentially a mathematical fiction that needs to be eliminated from consideration.

the only rigorous method, and indeed that all other methods are automatically erroneous and biased in their estimations, this need not be the actual position. The reason lies in the fact that the other methods incorporate actual frequencies of sample data, which should embody the *a priori* probabilities (see Section 24.4). Hence, the other methods could well give correct results. Nevertheless, it is refreshing to see *a priori* probabilies brought in explicitly, for this gives a greater confidence of getting unbiased results in any doubtful situations.

4.6 Concluding Remarks

A number of factors are crucial to the process of thresholding: first, the need to avoid bias in threshold selection by arranging roughly equal populations in the dark and light regions of the intensity histogram; second, the need to work with the smallest possible subimage (or neighborhood) size so that the intensity histogram has a well-defined valley despite variations in illumination; and third, the need for subimages to be sufficiently large so that statistics are good, permitting the valley to be located with precision.

A moment's thought will confirm that these conditions are not compatible and that a definite compromise will be needed in practical situations. In particular, it is generally not possible to find a neighborhood size that can be applied everywhere in an image, on all occasions yielding roughly equal populations of dark and light pixels. Indeed, if the chosen size is small enough to span edges ideally, hence yielding unbiased local thresholds, it will be totally at a loss inside large objects. Attempting to avoid this situation by resorting to alternative methods of threshold calculation—as for the statistic of equation (4.1)—does not solve the problem since inherent in such methods is a built-in region size. (In equation (4.1) the size is given by the summation limits.) It is therefore not surprising that a number of workers have opted for variable resolution and hierarchical techniques in an attempt to make thresholding more effective (Wu et al., 1982; Wermser et al., 1984; Kittler et al., 1985).

At this stage we call to question the complications involved in such thresholding procedures—which become even worse when intensity distributions start to become multimodal. Note that the overall procedure is to find local intensity gradients in order to obtain accurate, unbiased estimates of thresholds, so that it then becomes possible to take a horizontal slice through a grey-scale image and hence, ultimately, find "vertical" (i.e., spatial) boundaries within the image. Why not use the gradients *directly* to estimate the boundary positions? Such an approach, for example, leads to no problems from large regions where intensity histograms are essentially unimodal, although it would be foolish to pretend that no other problems exist (see Chapters 5 and 7).

On the whole, many approaches (region-growing, thresholding, edge detection, etc.), *taken to the limits of approximation*, will give equally good results. After all, they are all limited by the same physical effects—image noise, variability of lighting, the presence of shadows, and so on. However, some methods are easier to coax into working well, or need minimal computation, or have other useful properties such as robustness. Thus, thresholding can be a highly efficient means of aiding the interpretation of certain types of images, but as soon as image complexity rises above a certain critical level, it suddenly becomes more effective and considerably less complicated to employ overt edge detection techniques. These techniques are studied in depth in the next chapter. Meanwhile, we must not overlook the possibility of easing the thresholding process by going to some trouble to optimize the lighting system and to ensure that any worktable or conveyor is kept clean and white. This turns out to be a viable approach in a surprisingly large number of industrial applications.

The end result of thresholding is a set of silhouettes representing the shapes of objects, these constitute a "binarized" version of the original image. Many techniques exist for performing binary shape analysis, and some of these are described in Chapter 6. Meanwhile, it should be noted that many features of the original scene—for example, texture, grooves, or other surface structure—will not be present in the binarized image. Although the use of multiple thresholds to generate a number of binarized versions of the original image can preserve relevant information present in the original image, this approach tends to be clumsy and impracticable, and sooner or later one is likely to be forced to return to the original gray-scale image for the required data.

Thresholding is among the simplest of image processing operations and is an intrinsically appealing way of performing segmentation. This chapter has shown that even with the most scientific means of determining thresholds, it is a limited technique and can only form one of many tools in the programmer's toolkit.

4.7 Bibliographical and Historical Notes

Segmentation by thresholding started many years ago from simple beginnings and in recent years has been refined into a set of mature procedures. In some ways these may have been pushed too far, or at least further than the basic concepts—which were after all intended to save computation—can justify.

For example, Milgram's (1979) method of "convergent evidence" works by trying many possible threshold values and finding iteratively which of these is best at making thresholding agree with the results of edge detection—a method involving considerable computation. Pal et al. (1983) devised a technique based on fuzzy set theory, which analyzes the consequences of applying all possible threshold levels and minimizes an "index of fuzziness." This technique also has the disadvantage of large computational cost. The gray-level transition matrix approach of Deravi and Pal (1983) involves constructing a matrix of adjacent pairs of image intensity levels and then seeking a threshold within this matrix. This achieves the capability for segmentation even in cases where the intensity distribution is unimodal but with the disadvantage of requiring large storage space.

The Chow and Kaneko method (1972) has been seen to involve considerable amounts of processing (though Greenhill and Davies, 1995, have sought to reduce it with the aid of neural networks). Nakagawa and Rosenfeld (1979) studied the method and developed it for cases of trimodal distributions but without improving computational load. For bimodal distributions, however, the approach of Kittler et al. (1985) should require far less processing because it does not need to find a best fit to pairs of Gaussian functions.

Several workers have investigated the use of relaxation methods for threshold selection. In particular, Bhanu and Faugeras (1982) devised a gradient relaxation scheme that is useful for unimodal distributions. However, relaxation methods as a class are well known to involve considerable amounts of computation even if they converge rapidly, since they involve several iterations of whole-image operations (see also Section 15.6). Indeed, in relation to the related region-growing approach, Fu and Mui (1981) stated explicitly that: "All the region extraction techniques process the pictures in an iterative manner and usually involve a great expenditure in computation time and memory." Finally, Wang and Haralick (1984) developed a recursive scheme for automatic multithreshold selection, which involves successive analysis of the histograms of relatively dark and relatively light edge pixels.

Fu and Mui (1981) provided a useful general survey on image segmentation, which has been updated by Haralick and Shapiro (1985). These papers review many topics that could not be covered in the present chapter for reasons of space and emphasis. A similar comment applies to Sahoo et al.'s (1988) survey of thresholding techniques.

As hinted in Section 4.4, thresholding (particularly local adaptive thresholding) has had many applications in optical character recognition. Among the earliest were the algorithms described by Bartz (1968) and Ullmann (1974): two highly effective algorithms have been described by White and Rohrer (1983).

During the 1980s, the entropy approach to automatic thresholding evolved (e.g., Pun, 1981; Kapur et al., 1985; Abutaleb, 1989; Pal and Pal, 1989). This approach (Section 4.5.2) proved highly effective, and its development continued during the 1990s (e.g., Hannah et al., 1995). Indeed, it may surprise the reader that thresholding and adaptive thresholding are still hot subjects (see for example Yang et al., 1994; Ramesh et al., 1995).

Most recently, the entropy approach to threshold selection has remained important, in respect to both conventional region location and ascertaining the transition region between objects and background to make the segmentation process more reliable (Yan et al., 2003). In one instance it was found useful to employ fuzzy entropy and genetic algorithms (Tao et al., 2003). Wang and Bai (2003) have shown how threshold selection may be made more reliable by clustering the intensities of boundary pixels, while ensuring that a continuous rather than a discrete boundary is considered (the problem being that in images that approximate to binary images over restricted regions, the edge points will lie preferentially in the object or the background, not neatly between both). However, it has to be said that in complex outdoor scenes and for many medical images such as brain scans, thresholding alone will not be sufficient, and resort may even have to be made to graph matching (Chapter 15) to produce the best results—reflecting the important fact that segmentation is necessarily a high-level rather than a low-level process (Wang and Siskind, 2003). In rather less demanding cases, deformable model-guided split-and-merge techniques may, on the other hand, still be sufficient (Liu and Sclaroff, 2004).

4.8 Problems

1. Using the methods of Section 4.3.3, model the intensity distribution obtained by finding all the edge pixels in an image and including also all pixels adjacent to these pixels. Show that while this gives a sharper valley than for the original intensity distribution, it is not as sharp as for pixels located by the Laplacian operator.

2. Because of the nature of the illumination, certain fairly flat objects are found to have a fine shadow around a portion of their boundary. Show that this could confuse an adaptive thresholder that bases its threshold value on the statistic of equation (4.1).

3. Consider whether it is more accurate to estimate a suitable threshold for a bimodal, dual-Gaussian distribution by (a) finding the position of the minimum or (b) finding the mean of the two peak positions. What corrections could be made by taking account of the magnitudes of the peaks?

4. Obtain a complete derivation of equation (4.42). Show that, in general (as stated in Section 4.5.3), it has two solutions. What is the physical reason for this? How can it have only one solution when $\sigma_1 = \sigma_2$?

Edge Detection

Edge detection provides a more rigorous means than thresholding for initiating image segmentation. There is a large history of ad hoc edge detection algorithms, and this chapter aims to distinguish what is principled from what is ad hoc, and to provide theory and practical knowledge underpinning available techniques.

Look out for:

- the variety of template matching operators that have been used for edge detection—for example, the Prewitt, Kirsch, and Robinson operators.
- the differential gradient approach to edge detection—exemplified by the Roberts, Sobel, and Frei-Chen operators.
- theory explaining the performance of the template matching operators.
- methods for the optimal design of differential gradient operators, and the value of "circular" operators.
- tradeoffs between resolution, noise suppression capability, location accuracy, and orientation accuracy.
- outlines of more modern operators.
- the distinction between edge enhancement and edge detection.

In discussing the process of edge detection, this chapter shows that it is possible to estimate edge orientation with surprising accuracy within a small window—the secret being the considerable information residing in the gray-scale values. High orientation accuracy is of particular value when using the Hough transform to locate extended objects in digital images—as will be seen in several chapters in Part 2 of this volume.

CHAPTER 5

Edge Detection

5.1 Introduction

In Chapter 4 segmentation was tackled by the general approach of finding regions of uniformity in images—on the basis that the areas found in this way would have a fair likelihood of coinciding with the surfaces and facets of objects. The most computationally efficient means of following this approach was that of thresholding, but this turns out for real images to be failure-prone or else quite difficult to implement satisfactorily. Indeed, making it work well seems to require a multiresolution or hierarchical approach, coupled with sensitive measures for obtaining suitable local thresholds. Such measures have to take account of local intensity gradients as well as pixel intensities, and the possibility of proceeding more simply—by taking account of intensity gradients alone—was suggested.

Edge detection has long been an alternative path to image segmentation, and it is the method pursued in this chapter. Whichever way is inherently the better approach, edge detection has a considerable additional advantage in that it immediately reduces by a large factor (typically around 100) the redundancy inherent in the image data. This is useful because it immensely reduces both the space needed to store the information and the amount of processing subsequently required to analyze it.

Edge detection has gone through an evolution spanning more than 30 years. Two main methods of edge detection have been apparent over this period, the first of these being template matching (TM) and the second being the differential gradient (DG) approach. In either case, the aim is to find where the intensity gradient magnitude g is sufficiently large to be taken as a reliable indicator of the edge of an object. Then g can be thresholded in a similar way to that in which intensity was thresholded in Chapter 4. (It is possible to look for local maxima of g instead of thresholding it, but this possibility and its particular complexities are ignored until a later section.) The TM and DG methods differ mainly in how they proceed to estimate g locally. However, important differences exist in how they determine local edge orientation, which is an important variable in certain object detection schemes.

5.2 Basic Theory of Edge Detection

Both DG and TM operators estimate local intensity gradients with the aid of suitable convolution masks. In the case of the DG type of operator, only two such masks are required—for the x and y directions. In the TM case, it is usual to employ up to 12 convolution masks capable of estimating local components of gradient in the different directions (Prewitt, 1970; Kirsch, 1971; Robinson, 1977; Abdou and Pratt, 1979).

In the TM approach, the local edge gradient magnitude (for short, the edge "magnitude") is approximated by taking the maximum of the responses for the component masks:

$$g = \max(g_i\colon\ i = 1, \ldots, n) \tag{5.1}$$

where n is usually 8 or 12.

In the DG approach, the local edge magnitude may be computed vectorially using the nonlinear transformation:

$$g = \left(g_x^2 + g_y^2\right)^{1/2} \tag{5.2}$$

In order to save computational effort, it is common practice (Abdou and Pratt, 1979) to approximate this formula by one of the simpler forms:

$$g = |g_x| + |g_y| \tag{5.3}$$

or

$$g = \max(|g_x|, |g_y|) \tag{5.4}$$

which are, on average, equally accurate (Föglein, 1983).

In the TM approach, edge orientation is estimated simply as that of the mask giving rise to the largest value of gradient in equation (5.1). In the DG approach, it is estimated vectorially by the more complex equation:

$$\theta = \arctan(g_y/g_x) \tag{5.5}$$

(a) Masks for the Roberts 2×2 operator:

$$\mathrm{R}_{x'} = \begin{bmatrix} 0 & 1 \\ -1 & 0 \end{bmatrix} \qquad \mathrm{R}_{y'} = \begin{bmatrix} 1 & 0 \\ 0 & -1 \end{bmatrix}$$

(b) Masks for the Sobel 3×3 operator:

$$\mathrm{S}_x = \begin{bmatrix} -1 & 0 & 1 \\ -2 & 0 & 2 \\ -1 & 0 & 1 \end{bmatrix} \qquad \mathrm{S}_y = \begin{bmatrix} 1 & 2 & 1 \\ 0 & 0 & 0 \\ -1 & -2 & -1 \end{bmatrix}$$

(c) Masks for the Prewitt 3×3 "smoothed gradient" operator:

$$\mathrm{P}_x = \begin{bmatrix} -1 & 0 & 1 \\ -1 & 0 & 1 \\ -1 & 0 & 1 \end{bmatrix} \qquad \mathrm{P}_y = \begin{bmatrix} 1 & 1 & 1 \\ 0 & 0 & 0 \\ -1 & -1 & -1 \end{bmatrix}$$

Figure 5.1 Masks of well-known differential edge operators. In this figure masks are presented in an intuitive format (viz. coefficients increasing in the positive x and y directions) by rotating the normal convolution format through 180°. This convention is employed throughout this chapter. The Roberts 2×2 operator masks (a) can be taken as being referred to axes x', y' at 45° to the usual x, y.

DG equations (5.2) and (5.5) are considerably more computationally intensive than TM equation (5.1), although they are also more accurate. In some situations, however, orientation information is not required. In addition, image contrast may vary widely, so there may be little to be gained from thresholding a more accurate estimate of g. This may explain why so many workers have employed the TM instead of the DG approach. Since both approaches essentially involve estimation of local intensity gradients, it is not surprising that TM masks often turn out to be identical to DG masks—see Figs. 5.1 and 5.2.

5.3 The Template Matching Approach

Figure 5.2 shows four sets of well-known TM masks for edge detection. These masks were originally (Prewitt, 1970; Kirsch, 1971; Robinson, 1977) introduced on an intuitive basis, starting in two cases from the DG masks shown in Fig. 5.1. In all cases, the eight masks of each set are obtained from a given mask by permuting the mask coefficients cyclically. By symmetry, this is a good strategy for even permutations, but symmetry alone does not justify it for odd permutations. The situation is explored in more detail below.

	0°	45°
(*a*) Prewitt masks:	$\begin{bmatrix} -1 & 1 & 1 \\ -1 & -2 & 1 \\ -1 & 1 & 1 \end{bmatrix}$	$\begin{bmatrix} 1 & 1 & 1 \\ -1 & -2 & 1 \\ -1 & -1 & 1 \end{bmatrix}$
(*b*) Kirsch masks:	$\begin{bmatrix} -3 & -3 & 5 \\ -3 & 0 & 5 \\ -3 & -3 & 5 \end{bmatrix}$	$\begin{bmatrix} -3 & 5 & 5 \\ -3 & 0 & 5 \\ -3 & -3 & -3 \end{bmatrix}$
(*c*) Robinson "3-level" masks:	$\begin{bmatrix} -1 & 0 & 1 \\ -1 & 0 & 1 \\ -1 & 0 & 1 \end{bmatrix}$	$\begin{bmatrix} 0 & 1 & 1 \\ -1 & 0 & 1 \\ -1 & -1 & 0 \end{bmatrix}$
(*d*) Robinson "5-level" masks:	$\begin{bmatrix} -1 & 0 & 1 \\ -2 & 0 & 2 \\ -1 & 0 & 1 \end{bmatrix}$	$\begin{bmatrix} 0 & 1 & 2 \\ -1 & 0 & 1 \\ -2 & -1 & 0 \end{bmatrix}$

Figure 5.2 Masks of well-known 3×3 template matching edge operators. The figure illustrates only two of the eight masks in each set: the remaining masks can in each case be generated by symmetry operations. For the 3-level and 5-level operators, four of the eight available masks are inverted versions of the other four (see text).

Note first that four of the "3-level" and four of the "5-level" masks can be generated from the other four of their set by sign inversion. This means that in either case only four convolutions need to be performed at each pixel neighborhood, thereby saving computational effort. This is an obvious procedure if the basic idea of the TM approach is regarded as one of comparing intensity gradients in the eight directions. The two operators that do not employ this strategy were developed much earlier on some unknown intuitive basis.

Before proceeding, we note the rationale behind the Robinson "5-level" masks. These were intended (Robinson, 1977) to emphasize the weights of diagonal edges in order to compensate for the characteristics of the human eye, which tends to enhance vertical and horizontal lines in images. Normally, image analysis is concerned with computer interpretation of images, and an isotropic response is required. Thus, the "5-level" operator is a special-purpose one that need not be discussed further here.

Table 5.1 Angular variations for the Robinson 3-level TM operator

Sudden Step-edge Response			Planar Edge Response		
Edge Angle	**0° Mask**	**45° Mask**	**Edge Angle**	**0° Mask**	**45° Mask**
0.00	3.00	2.00	0.00	3.00	2.00
5.00	3.00	2.17	5.00	2.99	2.17
10.00	3.00	2.35	10.00	2.95	2.32
15.00	3.00	2.54	15.00	2.90	2.45
20.00	2.99	2.72	20.00	2.82	2.56
25.00	2.91	2.85	25.00	2.72	2.66
30.00	2.77	2.92	30.00	2.60	2.73
35.00	2.57	2.97	35.00	2.46	2.79
40.00	2.32	2.99	40.00	2.30	2.82
45.00	2.00	3.00	45.00	2.12	2.83
Changeover angle = 26.5°			Changeover angle = 26.5°		

These considerations show that the four template operators mentioned earlier have limited theoretical justification. It is therefore worth studying the situation in more depth, and this is done in the next section.

5.4 Theory of 3 × 3 Template Operators

Before analyzing the performance of TM operators, we note that they are likely to be used with a variety of types of edges, including in particular the "sudden step" edge, the "slanted step" edge, the "planar" edge, and various intermediate edge profiles (see Fig. 5.3). The sudden step edge has frequently been used as a paradigm with which to investigate and test the performance of edge template masks. This is highly convenient because its use eases much of the mathematical analysis, as will be seen below.

It is assumed here that eight masks are to be used, for angles differing by 45°. In addition, four of the masks differ from the others only in sign, since this seems unlikely to result in any loss of performance. Symmetry requirements then lead to the following masks for 0° and 45°, respectively.

$$\begin{bmatrix} -A & 0 & A \\ -B & 0 & B \\ -A & 0 & A \end{bmatrix} \qquad \begin{bmatrix} 0 & C & D \\ -C & 0 & C \\ -D & -C & 0 \end{bmatrix}$$

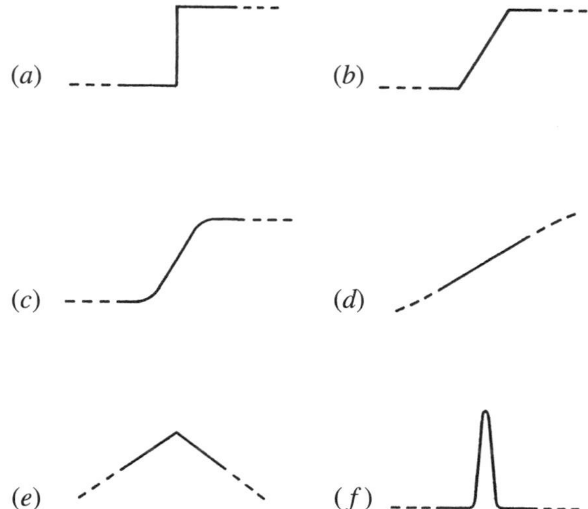

Figure 5.3 Edge models: (*a*) sudden step-edge; (*b*) slanted step edge; (*c*) smooth step edge; (*d*) planar edge; (*e*) roof edge; (*f*) line edge. The effective profiles of edge models are nonzero only within the stated neighborhood. The slanted step and the smooth step are approximations to realistic edge profiles: the sudden step and the planar edge are extreme forms that are useful for comparisons (see text). The roof and line edge models are shown for completeness only and are not considered further in this chapter.

For a step edge, the computation devolves into a determination of the areas of various triangles and other shapes shown in Fig. 5.4. As a result, the step edge responses are computed for the 0° mask as:

$$0^\circ\text{response} = 2A + B \tag{5.6}$$

$$45^\circ\text{response} = A + B \tag{5.7}$$

$$\begin{aligned}\alpha\text{-axis response} &= (1 + X + Y - W)A + B \\ &= 2(1 - W)A + B\end{aligned} \tag{5.8}$$

and for the 45° mask as:

$$0^\circ\text{response} = C + D \tag{5.9}$$

$$45^\circ\text{response} = 2C + D \tag{5.10}$$

$$\begin{aligned}\alpha\text{-axis response} &= (1 + T + U - V)C + D \\ &= 2(1 - V)C + D\end{aligned} \tag{5.11}$$

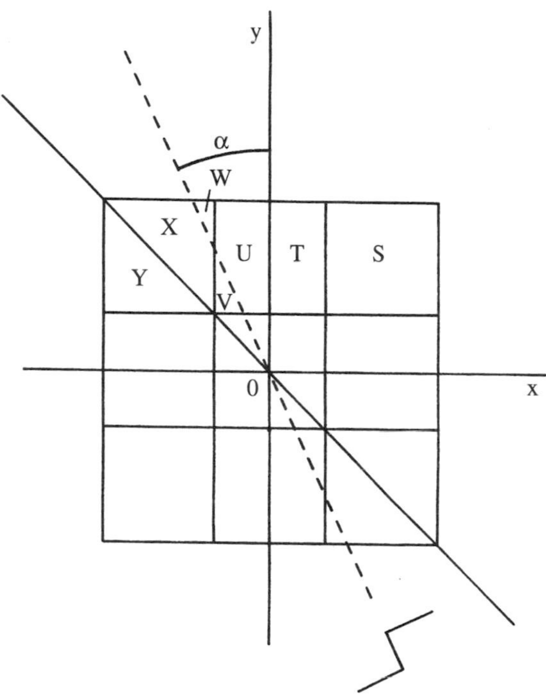

Figure 5.4 Geometry for computation of step-edge responses in a 3 × 3 neighborhood: α, angle of the step-edge; T, U, V, W, X, Y, areas of various triangles and other shapes; S, area of a single pixel (taken as unity).

Note that the α-axis response expressions given above are correct only for α in the range $\arctan(1/3) \leq \alpha \leq \arctan(1)$. Values for the areas V and W are given in terms of α by the expressions:

$$V = (1 - \tan\alpha)^2 / 8\tan\alpha \tag{5.12}$$

and

$$W = (3\tan\alpha - 1)^2 / 8\tan\alpha \tag{5.13}$$

Applying these formulas to the 3-level and 5-level operators (see Fig. 5.2) immediately leads to the additional conditions $A = C$, $B = D$. Thus, the operator responses for 0° and 45° step edges are exactly equal. (It is easy to see that this is also true for the Prewitt and Kirsch operators.) This was clearly one of the original design ideas in each case. It is also possible to determine the changeover angle α at which the 0° and 45° masks give equal responses. This occurs for

$$2(1 - W)A + B = 2(1 - V)C + D \tag{5.14}$$

Table 5.2 Angular variations for the "vector addition" operator*

Sudden Step-edge Response			Planar Edge Response		
Edge Angle	**0° Mask**	**45° Mask**	**Edge Angle**	**0° Mask**	**45° Mask**
0.00	4.06	2.87	0.00	4.06	2.87
5.00	4.06	3.12	5.00	4.04	3.11
10.00	4.06	3.38	10.00	3.99	3.32
15.00	4.06	3.65	15.00	3.92	3.51
20.00	4.05	3.92	20.00	3.81	3.67
25.00	3.97	4.10	25.00	3.68	3.81
30.00	3.82	4.21	30.00	3.51	3.92
35.00	3.62	4.27	35.00	3.32	3.99
40.00	3.37	4.31	40.00	3.11	4.04
45.00	3.06	4.32	45.00	2.87	4.06
Changeover angle = 22.5°			Changeover angle = 22.5°		

$$\text{*0°mask:} \begin{bmatrix} -1.000 & 0.000 & 1.000 \\ -2.055 & 0.000 & 2.055 \\ -1.000 & 0.000 & 1.000 \end{bmatrix} \qquad \text{45°mask:} \begin{bmatrix} 0.000 & 1.453 & 1.414 \\ -1.453 & 0.000 & 1.453 \\ -1.414 & -1.453 & 0.000 \end{bmatrix}$$

which leads (for operators with $A = C$ and $B = D$) to

$$W = V \tag{5.15}$$

This means that $\alpha = \arctan(1/2)$, or 26.5°. Thus, it has been shown that 26.5° is the changeover angle for *any* operator of the above type in which 45° masks are generated from 0° masks merely by permuting coefficients cyclically. A more complex calculation shows that 26.5° is the changeover angle for any set of masks in which the 0° mask has reflection symmetry in the x-axis and in which the 45° mask is obtained by permuting the coefficients cyclically. This explains the fact that all four masks in Fig. 5.2 have a changeover angle of 26.5°—that is, in no case is the changeover angle equal to the ideal value of 22.5° (see, for example, Table 5.1).

As noted earlier, equalization of the respective mask response in the 0° and 45° directions was essentially one of the main specifications underlying the masks of Fig. 5.2. However, another relevant specification is that of making estimates of local edge orientation as accurate as possible: in practice, this means making the changeover angle equal to 22.5°.

To find how this affects the mask coefficients, we employ the strategy of ensuring that intensity gradients follow the rules of vector addition. If the pixel intensity values within a 3×3 neighborhood are

$$\boxed{\begin{matrix} a & b & c \\ d & e & f \\ g & h & i \end{matrix}}$$

then estimation of the 0°, 90°, and 45° components of gradient by the earlier general masks gives:

$$g_0 = A(c + i - a - g) + B(f - d) \tag{5.16}$$

$$g_{90} = A(a + c - g - i) + B(b - h) \tag{5.17}$$

$$g_{45} = C(b + f - d - h) + D(c - g) \tag{5.18}$$

If vector addition is to be valid, then:

$$g_{45} = (g_0 + g_{90})/\sqrt{2} \tag{5.19}$$

Equating coefficients of a, b, c, d, e, f, g, h, i leads to the self-consistent pair of conditions:

$$C = \mathrm{B}/\sqrt{2} \tag{5.20}$$

$$D = A\sqrt{2} \tag{5.21}$$

A further requirement is for the masks to give equal responses at 22.5°. This leads to the formula

$$B/A = \sqrt{2}\frac{9t^2 - (14 - 4\sqrt{2})t + 1}{t^2 - (10 - 4\sqrt{2})t + 1} \tag{5.22}$$

where $t = \tan 22.5°$, so that

$$B/A = \left(13\sqrt{2} - 4\right)/7 = 2.055 \tag{5.23}$$

This gives a value for B/A in line with that for an optimized Sobel operator (see below).

The angular variations obtained with these masks are shown in Table 5.2. They are peculiar in giving optimal estimates of edge orientation for both step and planar edges and in giving relatively accurate estimates of edge magnitude for planar edges. However, their magnitude response for step edges is not particularly accurate. In this respect, it is interesting to compare the results of Table 5.1 with those of Table 5.2.

Table 5.3 Angular variations for a simple 3 × 3 TM operator*

Sudden Step-edge Response			Planar Edge Response		
Edge Angle	**0° Mask**	**45° Mask**	**Edge Angle**	**0° Mask**	**45° Mask**
0.00	1.00	1.00	0.00	1.00	1.00
5.00	1.00	1.00	5.00	1.00	1.08
10.00	1.00	1.00	10.00	0.98	1.16
15.00	1.00	1.00	15.00	0.97	1.22
20.00	1.00	1.00	20.00	0.94	1.28
25.00	1.00	1.00	25.00	0.91	1.33
30.00	1.00	1.00	30.00	0.87	1.37
35.00	1.00	1.00	35.00	0.82	1.39
40.00	1.00	1.00	40.00	0.77	1.41
45.00	1.00	1.00	45.00	0.71	1.41

Changeover angle = 22.5° | Changeover angle = 0°

$$\text{*}0^\circ\text{mask}: \begin{bmatrix} 0 & 0 & 0 \\ -1 & 0 & 1 \\ 0 & 0 & 0 \end{bmatrix} \qquad 45^\circ\text{mask}: \begin{bmatrix} 0 & 0 & 1 \\ 0 & 0 & 0 \\ -1 & 0 & 0 \end{bmatrix}$$

Finally, careful examination of equations (5.6)–(5.11) shows that coefficients B and D produce an isotropic contribution to the step-edge response for the 0° and 45° masks over a wide range of angles. This suggests that for greatest angular sensitivity these coefficients should be kept small, although the situation is inevitably more complicated for a planar edge. On the other hand, for *least* angular sensitivity, and hence greater constancy and accuracy in the estimation of edge magnitude, we may set B = D and A = C = 0. As shown in Table 5.3, this strategy works well for step edges but is not good for planar edges. Note that in the masks of Table 5.3, only two of the coefficients are nonzero, so there is very little averaging effect which can help to suppress pixel intensity noise. These considerations show clearly that a number of competing factors are involved in the design of effective edge detection operators and that it will be impossible to meet all of them with a single set of masks: hence the desired application must be specified clearly in advance.

5.5 Summary—Design Constraints and Conclusions

So far we have studied existing template masks for edge detection and examined the underlying theory. Detailed calculations have led to the design of TM operators having a variety of characteristics. At the same time, a number of design aims have become evident:

1. The need to optimize accuracy in the estimation of edge magnitude
2. The need to optimize accuracy in the estimation of edge orientation
3. The need to suppress noise during edge detection
4. The need to optimize operators for different edge profiles

These design aims have been found to conflict with each other; thus, tradeoffs between them exist and compromises have to be made. In general, it is clear (Davies, 1986a) that:

1. Different masks are needed for the accurate estimation of edge magnitude and orientation.
2. Optimization of noise suppression imposes further conditions on TM masks.
3. Obtaining sets of masks by permuting coefficients "cyclically" in a square neighborhood is ad hoc and cannot be relied upon to produce useful results.
4. The vector addition approach to the design of template masks is optimal for planar edge profiles but also gives acceptable results for step edges.
5. There is some danger in tailoring masks specifically for step-edge profiles, since they may then have poor responses for planar and other edge profiles.

Having obtained some insight into the process of designing TM masks for edge detection, we next move on to study the design of DG masks.

5.6 The Design of Differential Gradient Operators

This section studies the design of DG operators, including the Roberts 2×2 pixel operator and the Sobel and Prewitt 3×3 pixel operators (Roberts, 1965; Prewitt, 1970; for the Sobel operator see Pringle, 1969; Duda and Hart, 1973, p. 271) (see Fig. 5.1). The Prewitt or "gradient smoothing" type of operator has been extended to larger pixel neighborhoods by Prewitt (1970) and others (Brooks, 1978; Haralick, 1980) (see Fig. 5.5). In these instances the basic rationale is to model local edges by the best fitting plane over a convenient size of neighborhood. Mathematically, this amounts to obtaining suitably weighted averages to estimate slope in the x and y directions. As Haralick (1980) has pointed out, the use of

	M_x	M_y
(*a*) 2 × 2 neighborhood	$\begin{bmatrix} -1 & 1 \\ -1 & 1 \end{bmatrix}$	$\begin{bmatrix} 1 & 1 \\ -1 & -1 \end{bmatrix}$
(*b*) 3 × 3 neighborhood	$\begin{bmatrix} -1 & 0 & 1 \\ -1 & 0 & 1 \\ -1 & 0 & 1 \end{bmatrix}$	$\begin{bmatrix} 1 & 1 & 1 \\ 0 & 0 & 0 \\ -1 & -1 & -1 \end{bmatrix}$
(*c*) 4 × 4 neighborhood	$\begin{bmatrix} -3 & -1 & 1 & 3 \\ -3 & -1 & 1 & 3 \\ -3 & -1 & 1 & 3 \\ -3 & -1 & 1 & 3 \end{bmatrix}$	$\begin{bmatrix} 3 & 3 & 3 & 3 \\ 1 & 1 & 1 & 1 \\ -1 & -1 & -1 & -1 \\ -3 & -3 & -3 & -3 \end{bmatrix}$
(*d*) 5 × 5 neighborhood	$\begin{bmatrix} -2 & -1 & 0 & 1 & 2 \\ -2 & -1 & 0 & 1 & 2 \\ -2 & -1 & 0 & 1 & 2 \\ -2 & -1 & 0 & 1 & 2 \\ -2 & -1 & 0 & 1 & 2 \end{bmatrix}$	$\begin{bmatrix} 2 & 2 & 2 & 2 & 2 \\ 1 & 1 & 1 & 1 & 1 \\ 0 & 0 & 0 & 0 & 0 \\ -1 & -1 & -1 & -1 & -1 \\ -2 & -2 & -2 & -2 & -2 \end{bmatrix}$

Figure 5.5 Masks for estimating components of gradient in square neighborhoods. The above masks can be regarded as extended Prewitt masks. The 3 × 3 masks are Prewitt masks, included in this table for completeness. The 2 × 2 masks can be shown to be identical to Roberts operator masks rotated through 45°. In all cases weighting factors have been omitted in the interests of simplicity, as they are throughout this chapter.

equally weighted averages to measure slope in a given direction is incorrect; the proper weightings to use are given by the masks listed in Fig. 5.5. Thus, the Roberts and Prewitt operators are apparently optimal, whereas the Sobel operator is not. More will be said about this below.

A full discussion of the edge detection problem involves consideration of the accuracy with which edge magnitude and orientation can be estimated when the local intensity pattern cannot be assumed to be planar. To analyze the situation, we again take the "worst case," that is, that of a sudden step-edge profile, which is as distinct as possible from the planar edge approximation.

A number of analyses have been done on the angular dependencies of edge detection operators for a step-edge approximation. In particular, O'Gorman (1978) considered the variation of estimated versus actual angle resulting from a step edge observed within a square neighborhood (see also Brooks, 1978). Note that the case considered was that of a *continuous* rather than a discrete lattice of pixels. This was found to lead to a smooth variation, with angular error varying from zero at 0° and 45° to a maximum of 6.63° at 28.37° (where the estimated orientation was 21.74°), the variation for angles outside this range being replicated by symmetry. Abdou and Pratt (1979) obtained similar variations for the Sobel and Prewitt operators (in a *discrete* lattice), the respective maximum angular errors being 1.36° and 7.38° (Davies, 1984b). The Sobel operator apparently has angular accuracy that is close to optimal because it is close to being a "truly circular" operator. This point is discussed in more detail below.

5.7 The Concept of a Circular Operator

As stated earlier, when step-edge orientation is estimated in a square neighborhood, an error of up to 6.6° can result. Such an error does not arise with a planar edge approximation, since fitting of a plane to a planar edge profile within a square window can be carried out exactly. Errors appear only when the edge profile differs from the ideal planar form, within the square neighborhood—with the step-edge probably being a "worst case."

One way to limit errors in the estimation of edge orientation might be to restrict observation of the edge to a circular neighborhood. In the continuous case, this is sufficient to reduce the error to zero for all orientations, since symmetry dictates that there is only one way of fitting a plane to a step edge within a circular neighborhood, assuming that all planes pass through the same central point. The estimated orientation θ is then equal to the actual angle φ. A rigorous calculation along the lines indicated by Brooks (1976), which results in the following formula for a square neighborhood (O'Gorman, 1978):

$$\tan\theta = 2\tan\varphi/(3 - \tan^2\varphi) \quad 0^\circ \leq \varphi \leq 45^\circ \tag{5.24}$$

leads to the formula

$$\tan\theta = \tan\varphi, \quad \text{that is,} \ \theta = \varphi \tag{5.25}$$

for a circular neighborhood (Davies, 1984b). Similarly, zero angular error results from fitting a plane to an edge of *any* profile within a circular neighborhood, in the continuous approximation. Indeed, for an edge surface of arbitrary shape, the only problem is whether the mathematical best fit plane coincides with one that is subjectively desirable (and, if not, a *fixed* angular correction will be required). Ignoring such cases, the basic problem is how to approximate a circular neighborhood in a digitized image of small dimensions, containing typically 3×3 or 5×5 pixels.

A simple way to attempt this in a 3×3 window might be to approximate the neighborhood by an octagon, giving the corner pixels a weight reflecting the active area of the octagon within them. A calculation of this type results in a relative corner weighting of 0.614—a value that already gives the Sobel operator significantly greater credibility. However, this approach is somewhat ad hoc and misleading. To proceed more systematically, we recall a fundamental principle stated by Haralick (1980): "[T]he fact that the slopes in two orthogonal directions determine the slope in any direction is well known in vector calculus. However, it seems not to be so well known in the image processing community."

Essentially, appropriate estimates of slopes in two orthogonal directions permit the slope in any direction to be computed. For this principle to apply, appropriate estimates of the slopes have to be made first. If the components of slope are inappropriate, they will not act as components of true vectors, and the resulting estimates of edge orientation will be in error. This appears to be the main source of error with the Prewitt and other operators. It is not so much that the components of slope are in any instance incorrect, but rather that they are inappropriate for the purpose of vector computation since *they do not match one another adequately in the required way* (Davies, 1984b).

Following the arguments for the continuous case discussed earlier, slopes must be rigorously estimated within a circular neighborhood. Then the operator design problem devolves into determining how best to simulate a circular neighborhood on a discrete lattice so that errors are minimized. In order to carry this out, it is insufficient simply to weight the corner pixels after computing the masks, as implied by the octagon idea. Instead, it is necessary to apply the circular (or octagonal) weighting while computing the masks, so that correlations between the gradient weighting and circular weighting factors are taken properly into account.

5.8 Detailed Implementation of Circular Operators

In practice, the task of computing angular variations and error curves has to be tackled numerically, dividing each pixel in the neighborhood into arrays of suitably small subpixels. Each subpixel is then assigned a gradient weighting

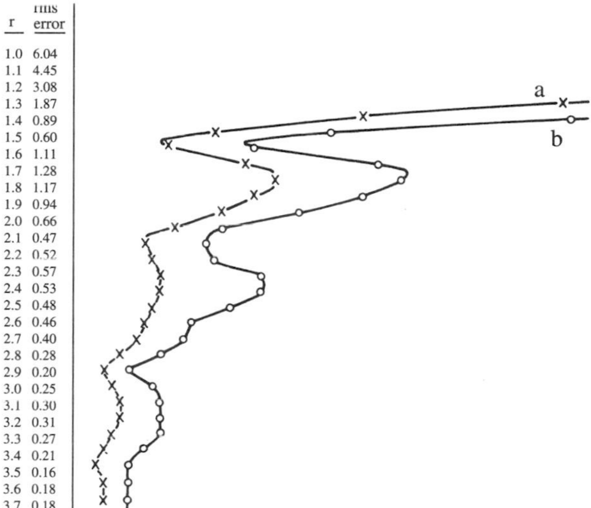

r
rms error
1.0 6.04
1.1 4.45
1.2 3.08
1.3 1.87
1.4 0.89
1.5 0.60
1.6 1.11
1.7 1.28
1.8 1.17
1.9 0.94
2.0 0.66
2.1 0.47
2.2 0.52
2.3 0.57
2.4 0.53
2.5 0.48
2.6 0.46
2.7 0.40
2.8 0.28
2.9 0.20
3.0 0.25
3.1 0.30
3.2 0.31
3.3 0.27
3.4 0.21
3.5 0.16
3.6 0.18
3.7 0.18
a
b

Figure 5.6 Variations in angular error as a function of radius r: (*a*) RMS angular error; (*b*) maximum angular error.

(equal to the x or y displacement) and a neighborhood weighting (equal to 1 for inside and 0 for outside a circle of radius r). The angular accuracy of "circular" differential gradient edge detection operators must depend on the radius of the circular neighborhood. In particular, poor accuracy would be expected for small values of r and reasonable accuracy for large values of r, as the discrete neighborhood approaches a continuum.

The results of this study are presented in Fig. 5.6. The variations depicted represent (1) RMS angular errors and (2) maximum angular errors in the estimation of edge orientation. The structures on each variation are surprisingly smooth. They are so closely related and systematic that they can only represent statistics of the arrangement of pixels in neighborhoods of various sizes. Details of these statistics are discussed in the next section.

Before proceeding with a detailed interpretation of the effects depicted in Fig. 5.6, three features will be apparent. First, as expected, there is a general trend to zero angular error as r tends to infinity. This is understandable, since the discrete neighborhood is then tending toward a continuum. Second, there is a very marked periodic variation, with particularly good accuracy resulting where the circular operators best match the tessellation of the digital lattice. The third feature of interest is the fact that errors do not vanish for any finite value of r—clearly, the constraints of the problem do not permit more than the minimization of errors. Overall, these curves show that it is possible to generate a family of optimal operators (at the minima of the error curves), the first of which corresponds closely to an operator (the Sobel operator) that is known to be nearly optimal.

5.9 Structured Bands of Pixels in Neighborhoods of Various Sizes

In order to explain the variations shown in Fig. 5.6, a study was made (Davies, 1984b) of the number of pixels at various distances from a given pixel, in a square lattice. Since symmetry demands that in most cases 4 or 8 pixels lie at each valid distance from the central pixel (Table 5.4), the most useful measure of the

Table 5.4 Numbers of pixels at different distances from central pixel

Shell	Radius	No. of Pixels
a	0.000	1
b	1.000	4
c	1.414	4
d	2.000	4
e	2.236	8
f	2.828	4
g	3.000	4
h	3.162	8
i	3.606	8
j	4.000	4
k	4.123	8
l	4.243	4
m	4.472	8
n	5.000	12
o	5.099	8
p	5.385	8
q	5.657	4
r	5.831	8
s	6.000	4
t	6.083	8
u	6.325	8
v	6.403	8
w	6.708	8
x	7.000	4
y	7.071	12
z	7.211	8
A	7.280	8
B	7.616	8
C	7.810	8
D	8.000	4
E	8.062	16
F	8.246	8

Table 5.5 Differences in radii between shells of pixels

Difference between Radii			
Shells		**Value**	**Histogram**
a	b	1.00	*
b	c	0.41	*
c	d	0.59	*
d	e	0.24	*
e	f	0.59	*
f	g	0.17	*
g	h	0.16	*
h	i	0.44	*
i	j	0.39	*
j	k	0.12	*
k	l	0.12	*
l	m	0.23	*
m	n	0.53	*
n	o	0.10	*
o	p	0.29	*
p	q	0.27	*
q	r	0.17	*
r	s	0.17	*
s	t	0.08	*
t	u	0.24	*
u	v	0.08	*
v	w	0.31	*
w	x	0.29	*
x	y	0.07	*
y	z	0.14	*
z	A	0.07	*
A	B	0.34	*
B	C	0.19	*
C	D	0.19	*
D	E	0.06	*
E	F	0.18	*

relevant statistics seems to be the difference between consecutive pixel distances. These are listed out to quite a large distance (representing neighborhoods of the order of 13×13) in Table 5.5. The histogram displayed in Table 5.5 indicates that a few radial differences are substantially larger than the rest. It is possible to envisage these larger values as demarcating bands of pixels which are internally well packed and approximate continua as closely as is possible. Given such continua, circular neighborhood approximations within them should be relatively accurate, whereas neighborhoods that straddle bands would be expected to be poor approximations to continua. This substantially explains

Table 5.6 Angular variations for the best operators tested*

Actual Angle (degrees)	Estimated Angle (degrees)†						
	Prew	Sob	a–c	circ	a–e	a–h	a–i
0	0.00	0.00	0.00	0.00	0.00	0.00	0.00
5	3.32	4.97	5.05	5.14	5.42	5.22	5.01
10	6.67	9.95	10.11	10.30	10.81	10.28	9.94
15	10.13	15.00	15.24	15.52	15.83	14.81	14.74
20	13.69	19.99	20.29	20.64	20.07	19.73	19.90
25	17.72	24.42	24.73	25.10	24.62	25.00	25.34
30	22.62	28.86	29.14	29.48	29.89	30.02	30.16
35	28.69	33.64	33.86	34.13	35.43	34.86	34.91
40	35.94	38.87	39.00	39.15	40.30	39.71	40.01
45	45.00	45.00	45.00	45.00	45.00	45.00	45.00
RMS error	5.18	0.73	0.60	0.53	0.47	0.19	0.16

**Key*: Prew, Prewitt; Sob, Sobel; a–c, theoretical optimum: closed band containing shells a–c; circ, actual optimum circular operator; a–e, theoretical optimum: closed band containing shells a–e; a–h, theoretical optimum: closed band containing shells a–h; a–i, theoretical optimum: closed band containing shells a–i.
†Values should be accurate to within 0.02° in each case.

the results of Fig. 5.6 and even suggests that these variations are not entirely specific to edge detection but reflect more fundamental limitations of digital image processing methodology. Thus, the design of any image processing operator should take into account banding of pixels and look for solutions that suppress pixel position statistics as far as possible.

The "closed-band" operators predicted above are listed in Fig. 5.7; their angular variations appear in Table 5.6. The Sobel operator, which is already the most accurate of the 3×3 edge gradient operators suggested previously, can be made some 30% more accurate by adjusting its coefficients to make it more circular. In addition, the closed-bands idea indicates that the corner pixels of 5×5 or larger operators are best removed altogether. Not only does this require less computation but also (e.g., compared with the extended Prewitt masks of Fig. 5.5) it actually improves performance!

Many sets of mask coefficients within a 3×3 neighborhood correspond to circular operators—that is, some value of r can be found which generates the masks in the manner outlined above. However, the Sobel operator has been shown to be more ideal in that it also corresponds approximately to a closed band situation. Since it is then closer to the ideal analog circular mask case, we may say that it is close to being a "truly circular" operator, thus justifying the statement at the end of Section 5.6. For neighborhoods larger than 3×3, there are so many mask coefficients that arbitrarily designed operators will not in general be "circular" or (*a fortiori*) "truly circular."

(*a*) Band containing shells a–c (effective radius = 1.500)

$$\begin{bmatrix} -0.464 & 0.000 & 0.464 \\ -0.959 & 0.000 & 0.959 \\ -0.464 & 0.000 & 0.464 \end{bmatrix}$$

(*b*) Band containing shells a–e (effective radius = 2.121)

$$\begin{bmatrix} 0.000 & -0.294 & 0.000 & 0.294 & 0.000 \\ -0.582 & -1.000 & 0.000 & 1.000 & 0.582 \\ -1.085 & -1.000 & 0.000 & 1.000 & 1.085 \\ -0.582 & -1.000 & 0.000 & 1.000 & 0.582 \\ 0.000 & -0.294 & 0.000 & 0.294 & 0.000 \end{bmatrix}$$

(*c*) Band containing shells a–h (effective radius = 2.915)

$$\begin{bmatrix} 0.000 & 0.000 & -0.191 & 0.000 & 0.191 & 0.000 & 0.000 \\ 0.000 & -1.085 & -1.000 & 0.000 & 1.000 & 1.085 & 0.000 \\ -0.585 & -2.000 & -1.000 & 0.000 & 1.000 & 2.000 & 0.585 \\ -1.083 & -2.000 & -1.000 & 0.000 & 1.000 & 2.000 & 1.083 \\ -0.585 & -2.000 & -1.000 & 0.000 & 1.000 & 2.000 & 0.585 \\ 0.000 & -1.085 & -1.000 & 0.000 & 1.000 & 1.085 & 0.000 \\ 0.000 & 0.000 & -0.191 & 0.000 & 0.191 & 0.000 & 0.000 \end{bmatrix}$$

(*d*) Band containing shells a–i (effective radius = 3.500)

$$\begin{bmatrix} 0.000 & -0.646 & -0.815 & 0.000 & 0.815 & 0.646 & 0.000 \\ -0.963 & -1.997 & -1.000 & 0.000 & 1.000 & 1.997 & 0.963 \\ -2.458 & -2.000 & -1.000 & 0.000 & 1.000 & 2.000 & 2.458 \\ -2.962 & -2.000 & -1.000 & 0.000 & 1.000 & 2.000 & 2.962 \\ -2.458 & -2.000 & -1.000 & 0.000 & 1.000 & 2.000 & 2.458 \\ -0.963 & -1.997 & -1.000 & 0.000 & 1.000 & 1.997 & 0.963 \\ 0.000 & -0.646 & -0.815 & 0.000 & 0.815 & 0.646 & 0.000 \end{bmatrix}$$

Figure 5.7 Masks of "closed band" differential gradient edge operators. In all cases only x-mask is shown: the y-mask may be obtained by a trivial symmetry operation. Mask coefficients are accurate to ~0.003 but would in normal practical applications be rounded to 1- or 2-figure accuracy.

The optimal 3×3 masks obtained above numerically by consideration of circular operators are very close indeed to those obtained purely analytically in Section 5.4, for TM masks, following the rules of vector addition. In the latter case, a value of 2.055 was obtained for the ratio of the two mask coefficients, whereas for circular operators the value $0.959/0.464 = 2.067 \pm 0.015$ is obtained.

This is no accident and it is very satisfying that a coefficient that was formerly regarded as ad hoc (Kittler, 1983) is in fact optimizable and can be obtained in closed form (see Section 5.4).

5.10 The Systematic Design of Differential Edge Operators

The family of "circular" differential gradient edge operators studied in Sections 5.6–5.9 incorporates only one design parameter—the radius r. Only a limited number of values of this parameter permit optimum accuracy for estimation of edge orientation to be attained.

It is worth considering what additional properties this one parameter can control and how it should be adjusted during operator design. It affects the signal-to-noise ratio of the final edges, as well as the resolution. (These effects are important in the design of any operator, for example, in the design of an averaging filter.) The signal-to-noise ratio varies linearly with the radius of the circular neighborhood, since signal is proportional to area and Gaussian noise is proportional to $\sqrt{\text{area}}$, that is, to radius. This type of averaging process for improving signal-to-noise ratio is, of course, well known. Similarly, the accuracy of linear measurement is determined by the number of pixels over which averaging occurs and hence is proportional to operator radius. In contrast, resolution and the resulting "scale" of the edge image vary inversely with radius, since relevant linear properties of the image are averaged over the active area of the neighborhood. Note that the accuracy with which the positions of objects may be measured also depends on resolution. However, achievable resolution depends on available signal-to-noise ratio, and, as has been seen, this itself varies with the radius of the neighborhood. Here the final tradeoff depends on the scale of the noise one is attempting to suppress. Finally, computational load, and the associated cost of hardware for speeding up the processing, is generally at least in proportion to the number of pixels in the neighborhood, and hence in proportion to r^2.

These facts explain various points that appear in the literature, such as the better fidelity of the Roberts' operator (see, for example, Bryant and Bouldin, 1979) and its higher noise levels (Abdou and Pratt, 1979). In addition, it is now seen that an exact tradeoff is operative, in that operator radius carries with it four immediate properties that have not always been seen to be intimately related: viz. signal-to-noise ratio, resolution, accuracy, and hardware/computational cost. Since none of these can be altered without the others being modified as well, one is always in an engineering situation of having to make a compromise to suit circumstances. On the other hand, the single-design parameter concept is complicated by the discreteness of the digital image processing milieu. Hence, in practice, it is necessary to fall back on one or other of the "closed-band" radii discussed above for the best combination of properties.

5.11 Problems with the Above Approach—Some Alternative Schemes

Interesting though these ideas may be, they have their own inherent problems. In particular, they take no account of the displacement E of the edge from the center of the neighborhood, or of the effects of noise in biasing the estimates of edge magnitude and orientation. Now it is possible to show that a Sobel operator gives *zero* error in the estimation of step-edge orientation when $|\theta| \le \arctan(1/3)$ and $|E| \le (\cos\theta - 3\sin|\theta|)/2$. For a 3×3 operator of the form

$$\begin{bmatrix} -1 & 0 & 1 \\ -\mathrm{B} & 0 & \mathrm{B} \\ -1 & 0 & 1 \end{bmatrix} \quad \begin{bmatrix} 1 & \mathrm{B} & 1 \\ 0 & 0 & 0 \\ -1 & -\mathrm{B} & -1 \end{bmatrix}$$

applied to the edge

$$\begin{array}{|lll|} a & a+h(0.5 - E\sec\theta + \tan\theta) & a+h \\ a & a+h(0.5 - E\sec\theta) & a+h \\ a & a+h(0.5 - E\sec\theta - \tan\theta) & a+h \end{array}$$

Lyvers and Mitchell (1988) found that the estimated orientation is:

$$\varphi = \arctan[2B\tan\theta/(B+2)] \tag{5.26}$$

which immediately shows why the Sobel operator should give zero error for a specific range of θ and E. However, this is somewhat misleading, since considerable errors arise outside this region. Not only do they arise when $E=0$, as assumed in the foregoing sections, but also they vary strongly with E. Indeed, the maximum errors for the Sobel and Prewitt operators rise to 2.90° and 7.43°, respectively, in this more general case (the corresponding RMS errors being 1.20° and 4.50°). Hence, a full analysis should be performed to determine how to reduce the maximum and average errors. Lyvers and Mitchell (1988) carried out an empirical analysis and constructed a lookup table with which to correct the orientations estimated by the Sobel operator, the maximum error being reduced to 2.06°. Their approach was motivated by the "iterated Sobel" of Iannino and Shapiro (1979).

Another scheme that reduces the error is the moment-based operator of Reeves et al. (1983). This leads to Sobel-like 3×3 masks that are essentially identical to the 3×3 masks of Davies (1984b), both having $\mathrm{B} = 2.067$ (for $\mathrm{A} = 1$). However, the moment method can also be used to estimate the edge position E if additional masks are used to compute second-order moments of intensity.

Hence, it is possible to make a very significant improvement in performance by using a 2-D lookup table to estimate orientation: the result is that the maximum error is reduced from 2.83° to 0.135° for 3×3 masks and from 0.996° to 0.0042° for 5×5 masks.

Lyvers and Mitchell (1988) found, however, that much of this additional accuracy is lost in the presence of noise, and root mean square (RMS) standard deviations of edge orientation estimates are already around 0.5° for all 3×3 operators at 40 db signal-to-noise ratios. The reasons for this are quite simple. Each pixel intensity has a noise component that induces errors in its weighted mask components. The combined effects of these errors can be estimated assuming that they arise independently, so that their variances add (Davies, 1987k). Thus, noise contributions to the x and y components of gradient can be computed. These provide estimates for the components of noise along and perpendicular to the edge gradient vector (Fig. 5.8). The edge orientation for a Sobel operator is affected by an amount $\sqrt{12}\sigma/4h$ radians, where σ is the standard deviation on the pixel intensity values and h is the edge contrast. This explains the angular errors given by Lyvers and Mitchell, if Pratt's (2001) definition of signal-to-noise ratio (in db) is used:

$$\mathrm{S/N} = 20 \log_{10}(h/\sigma) \tag{5.27}$$

An alternative approach, that of Haralick (1984), uses a "facet model" for edge detection and computes edge parameters from a polynomial surface fit at every pixel location. To estimate edge orientation accurately, a third-order model

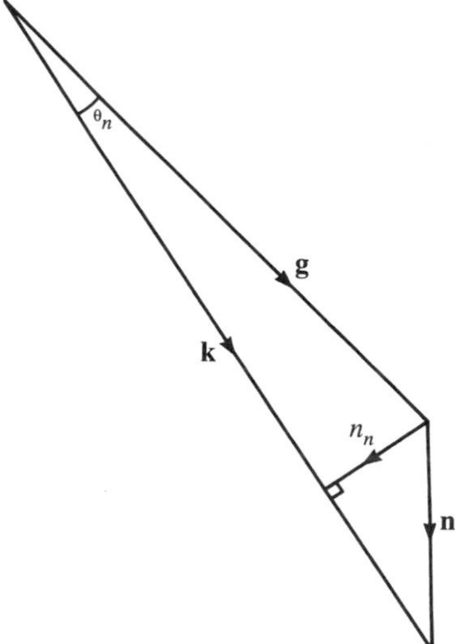

Figure 5.8 Calculating angular errors arising from noise: **g**, intensity gradient vector; **n**, noise vector; **k**, resultant of intensity gradient and noise vector; n_n, normal component of noise; θ_n, noise-induced orientation error.

had to be used. This work was extended by Zuniga and Haralick (1987) using a special averaging technique called the integrated directional derivative (IDD). This involves integrating the directional derivative of the intensity over rectangular regions of length L and width W centered over each pixel and aligned in all possible directions θ. At any location, the edge magnitude is taken as the maximum value of the IDD, and the edge orientation as the direction that gives the maximum value. In practice, equal values of L and W have been employed—1.8 for a 5×5 neighborhood, 2.5 for 7×7. The theory permits new masks to be computed, and, indeed, application of the operator then reduces merely to application of these masks with the usual formulas for edge direction and magnitude. Under zero-noise conditions, the 5×5 and 7×7 operators give worst bias values of 1.05° and 0.65°, respectively. These accuracies are not quite as good as for the moment-based detectors and certainly not as good as for the moment detectors with 2-D lookup tables. However, performance in the presence of noise (S/N ratio of 30 db or worse) is virtually identical. That it is possible to obtain such improvements in edge orientation accuracy is ultimately due to the availability of many adjustable mask coefficients, though clearly suitable strategies have to be found for deducing them.

Canny (1986) developed a totally different approach. He used functional analysis to derive an optimal function for edge detection, starting with three optimization criteria—good detection, good localization, and only one response per edge under white noise conditions. The analysis is too technical to be discussed in detail here. However, the 1-D function found by Canny is accurately approximated by the derivative of a Gaussian. This is then combined with a Gaussian of identical σ in the perpendicular direction, truncated at 0.001 of its peak value, and split into suitable masks. Underlying this method is the idea of locating edges at local maxima of gradient magnitude of a Gaussian-smoothed image. In addition, the Canny implementation employs a hysteresis operation (Section 7.1.1) on edge magnitude in order to make edges reasonably connected. Finally, a multiple-scale method is employed to analyze the output of the edge detector. More will be said about these points later in this chapter. Lyvers and Mitchell (1988) tested the Canny operator and found it to be significantly less accurate for orientation estimation than the moment and IDD operators. In addition, it needed to be implemented using 180 masks and hence took enormous computation time, though many practical implementations of this operator are much faster than this early paper indicates.[1]

Other operators have been designed for detecting edges. Of particular interest is the Hueckel approach developed in the early 1970s. This approach involved expanding intensity variations over (ideally) rather large circular regions in terms of complete sets of Fourier-like orthonormal basis functions (Hueckel, 1971, 1973).

1 It is nowadays necessary to ask "Which Canny?", as there are a great many implementations of it: this leads to problems for any realistic comparison between operators.

Because it is quite computationally intensive, it is not described in detail here. Suffice it to state that the tests of Lyvers and Mitchell (1988) showed that it has certain problems of convergence, especially in the presence of noise, while it involves typically 10 times the computation of other operators of comparable mask size.

An operator that has been of great historical importance is that of Marr and Hildreth (1980). The motivation for the design of this operator was the modeling of certain psychophysical processes in mammalian vision. The basic rationale is to find the Laplacian of the Gaussian-smoothed $(\nabla^2 G)$ image and then to obtain a "raw primal sketch" as a set of zero-crossing lines. The Marr–Hildreth operator does not use any form of threshold since it merely assesses where the $\nabla^2 G$ image passes through zero. This feature is attractive, for working out threshold values is a difficult and unreliable task. However, the Gaussian smoothing procedure can be applied at a variety of scales, and in one sense the scale is a new parameter that substitutes for the threshold. In fact, a major feature of the Marr–Hildreth approach that has been very influential in later work (Witkin, 1983; Bergholm, 1986) is the fact that zero-crossings can be obtained at several scales, giving the potential for more powerful semantic processing. This necessitates finding the means to combine all the information in a systematic and meaningful way. This may be carried out by bottom-up or top-down approaches, and there is much discussion in the literature about methods for carrying out these processes. However, in many (especially industrial inspection) applications, one is interested in working at a particular resolution, and considerable savings in computation can then be made. It is also noteworthy that the Marr–Hildreth operator is reputed to require neighborhoods of at least 35×35 for proper implementation (Brady, 1982). Nevertheless, other workers have implemented the operator in much smaller neighborhoods, down to 5×5. Wiejak et al. (1985) showed how to implement the operator using linear smoothing operations to save computation. Lyvers and Mitchell (1988) reported orientation accuracies using the Marr–Hildreth operator that are not especially high (2.47° for a 5×5 operator and 0.912° for a 7×7 operator, in the absence of noise).

As noted earlier, those edge detection operators that are applied at different scales lead to different edge maps at different scales. In such cases, certain edges that are present at lower scales disappear at larger scales; in addition, edges that are present at both low and high scales may be shifted or merged at higher scales. Bergholm (1986) demonstrated the possible occurrence of elimination, shifting, and merging, while Yuille and Poggio (1986) showed that edges that are present at high resolution should not disappear at some lower resolution. These aspects of edge location are now becoming well understood.

Finally, Overington and Greenway (1987) sound a cautionary note about the accuracy of the zero-crossing approach to edge detection. They found that it is significantly more accurate to estimate edge position and edge orientation by interpolating to local maximum values of edge gradient than by locating zero-crossings of the ∇^2G image. Their rationale was that accurate estimation of these zero-crossing positions relies on *third* derivatives of the image intensity function, which tend to be affected seriously by noise.

5.12 Concluding Remarks

The design of edge detection operators has by now been taken to quite an advanced stage, so that edges can be located to subpixel accuracy and orientated to fractions of a degree. In addition, edge maps may be made at several scales and the results correlated as an aid to image interpretation. Unfortunately, some of the schemes that have been devised to achieve these things (and especially those outlined in the previous section) are fairly complex and tend to consume considerable computation. In many applications, this complexity may not be justified because the application requirements are, or can reasonably be made, quite restricted. Furthermore, there is often the need to save computation for real-time implementation. For these reasons, the remainder of this book reverts to employing edges found by a single low-resolution detector. In practice, this will normally be a Sobel operator, which provides a good balance between computational load and orientation accuracy. Indeed, virtually all the examples in Part 2 of the book have been implemented using a Sobel operator. (In addition, the RMS error in estimating edge orientation using this operator was found by Lyvers and Mitchell to be 1.20°, and in later chapters this quantity will generally be taken as a round 1°.) This does not in any way invalidate the latest methods, particularly those involving studies of edges at various scales. Such methods come into their own in applications such as general scene analysis, where vision systems are required to cope with largely unconstrained image data.

This chapter has largely completed the task of segmentation of images at low level. The following chapter moves on to consider the shapes of objects that have been found by the thresholding and edge detection schemes discussed in the last two chapters. In fact, Chapter 6 studies shapes by analysis of the regions over which objects extend, while Chapter 7 studies shapes by considering their boundary patterns.

Edge detection is perhaps the most widely used means of locating and identifying objects in digital images. While different edge detection strategies vie with each other for acceptance, this chapter has shown that all obey fundamental laws—such as sensitivity, noise suppression capability, and computation cost *all* increasing with footprint size.

5.13 Bibliographical and Historical Notes

Early attempts at edge detection tended to employ numbers of template masks that could locate edges at various orientations. Often these masks were ad hoc in nature, and after 1980 this approach finally gave way to the differential gradient approach that had already existed in various forms for a considerable period (see the influential paper by Haralick, 1980).

The Frei–Chen approach is of particular interest in that it takes a set of nine 3×3 masks forming a complete set within this size of neighborhood—of which one tests for brightness, four test for edges, and four test for lines (Frei and Chen, 1977). Though interesting, the Frei–Chen edge masks do not correspond to those devised for optimal edge detection, and it appears possible that edge detection in the set as a whole may have been compromised in some way in order to make a complete "toolbox." Indeed, Lacroix (1988) found errors in the Frei–Chen approach which may ultimately show how to rationalize the situation.

The Hueckel scheme (1971, 1973) provided an early alternative to these approaches, although it was rather computationally intensive and therefore perhaps not so widely used. To tackle this problem, Mérõ and Vassy (1975) developed a modified form of the scheme, but it does not appear to have been used particularly widely.

Meanwhile, psychophysical work by Marr (1976), Wilson and Giese (1977), and others provided another line of development for edge detection. This led to the well-known paper by Marr and Hildreth (1980), which was highly influential in the following few years. This spurred others to think of alternative schemes, and the Canny (1986) operator emerged from this vigorous milieu. The Marr–Hildreth operator seems to have been the first to preprocess images in order to study them at different scales—a technique that has survived and expanded considerably (see for example, Witkin, 1983; Bergholm, 1986; Yuille and Poggio, 1986). The computational problems of the Marr–Hildreth operator have kept others thinking along more traditional lines, and the work by Reeves et al. (1983), Haralick (1984), and Zuniga and Haralick (1987) falls into this category. Lyvers and Mitchell (1988) reviewed many of these papers and made their own suggestions. Other studies (McIvor, 1988; Petrou and Kittler, 1988) show continuing development in the subject, especially regarding operator optimization, although it is not improbable that diminishing returns have now set in and that from now on the main advances will be at a higher level. In this context, the work of Sjöberg and Bergholm (1988), which finds rules for discerning shadow edges from object edges, is of great interest. Nevertheless, other nuances are worth considering, such as the accuracy of edge orientation for partly saturated gray-scale images (Davies and Celano, 1993a), and the development of edge detectors that are suited to particular purposes, such as finding high accuracy subedges for use in conjunction with a Hough transform (Davies, 1992c).

Most recently, there has been a move to achieve greater robustness and confidence in edge detection by careful elimination of local outliers. In Meer and Georgescu's (2001) method, this is achieved by estimating the gradient vector, suppressing nonmaxima, performing hysteresis thresholding (Section 7.1.1), and integrating with a confidence measure to produce a more general robust result. Each pixel is assigned a confidence value *before* the final two steps of the algorithm. Kim et al. (2004) have taken this technique a step further and eliminated the need for setting a threshold by using a fuzzy reasoning approach. Yitzhaky and Peli (2003) have expressed similar sentiments and they aim to find an optimal parameter set for edge detectors by ROC and chi-square measures, which actually give very similar results. Prieto and Allen (2003) designed a similarity metric for edge images, which could be used to test the effectiveness of a variety of edge detectors. They pointed out that metrics need to allow slight latitude in the positions of edges, in order to compare the similarity of edges reliably. They report a new approach that takes into account not only displacement of edge positions but also edge strengths in determining the similarity between edge images.

Not content with handcrafted algorithms, Suzuki et al. (2003) devised a backpropagation neural edge enhancer that undergoes supervised learning on model data to permit it to cope well (in the sense of giving clear, continuous edges) with noisy images. It was found to give results superior to those of conventional algorithms (including Canny, Heuckel, Sobel, and Marr–Hildreth) in similarity tests relative to the desired edges. The disadvantage was a long learning time, though the final execution time was short.

Finally, Cheng (2003) produced a rapidly operating moment-preserving edge detector working in image blocks in order to facilitate fast image database searches for likely image structures without the need for special hardware. This is a use for which increasing speed will be demanded for the foreseeable future, regardless of computer speeds, as more and more images and image sequences have to be searched. However, the method presented cannot be described as a temporary expedient "fix," as its results seem impressive and yield exactly the type of data needed in such applications.

5.14 **Problems**

1. Prove equations (5.22) and (5.23).
2. Check the results quoted in Section 5.11 giving the conditions under which the Sobel operator leads to zero error in the estimation of edge orientation. Proceed to prove equation (5.26).

Binary Shape Analysis

Although binary images contain much less information than their gray-scale counterparts, they embody shape and size information that is highly relevant for object recognition. However, this information resides in a digital lattice of pixels, and this results in intricacies appearing in the geometry. This chapter resolves these problems and explores a number of important algorithms for processing shapes.

Look out for:

- the connectedness paradox and how it is resolved.
- object labeling and how labeling conflicts are resolved.
- problems related to measurement in binary images.
- size filtering techniques.
- the convex hull as a means of characterizing shape, and methods for determining it.
- distance functions and how they are obtained using parallel and sequential algorithms.
- the skeleton and how it is found by thinning: the crucial role played by the crossing number, both in determining the skeleton and in analyzing it.
- simple measures for shape recognition, including circularity and aspect ratio.
- more rigorous measures of shape, including moments and boundary descriptors.

In reality, this chapter almost exclusively covers area-based methods of shape analysis, leaving boundary-based procedures to Chapter 7—though circularity measures and boundary tracking are both covered. However, chapter boundaries cannot be completely exclusive, as any new concept requires "hooks" that have been laid down in a variety of places. Indeed, it is often valuable to meet an idea before finding out in detail how to put flesh on it.

Returning briefly to the present chapter, it is interesting to note how intricate some of the algorithmic processes are: connectedness, in particular, pervades the whole subject of digital shape analysis and comes with a serious health warning.

CHAPTER 6

Binary Shape Analysis

6.1 Introduction

Over the past few decades, 2-D shape analysis has provided the main means by which objects are recognized and located in digital images. Fundamentally, 2-D shape has been important because it uniquely characterizes many types of object, from keys to characters, from gaskets to spanners, and from fingerprints to chromosomes, while in addition it can be represented by patterns in simple binary images. Chapter 1 showed how the template matching approach leads to a combinatorial explosion even when fairly basic patterns are to be found, so preliminary analysis of images to find features constitutes a crucial stage in the process of efficient recognition and location. Thus, the capability for binary shape analysis is a very basic requirement for practical visual recognition systems.

Forty years of progress have provided an enormous range of shape analysis techniques and a correspondingly large range of applications. It will be impossible to cover the whole field within the confines of a single chapter, so completeness will not even be attempted. (The alternative of a catalog of algorithms and methods, all of which are covered only in brief outline, is eschewed.) At one level, the main topics covered are examples with their own intrinsic interest and practical application, and at another level they introduce matters of fundamental principle. Recurring themes are the central importance of connectedness for binary images, the contrasts between local and global operations on images and between different representations of image data, and the need to optimize accuracy and computational efficiency, and the compatibility of algorithms and hardware. The chapter starts with a discussion of how connectedness is measured in binary images.

6.2 Connectedness in Binary Images

This section begins with the assumption that objects have been segmented, by thresholding or other procedures, into sets of 1's in a background of 0's (see Chapters 2 to 4). At this stage it is important to realize that a second assumption is already being made implicitly—that it is easy to demarcate the boundaries between objects in binary images. However, in an image that is represented digitally in rectangular tessellation, a problem arises with the definition of connectedness. Consider a dumbell-shaped object:[1]

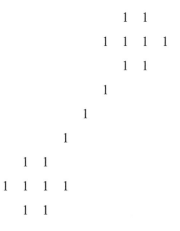

At its center, this object has a segment of the form

0 1
1 0

which separates two regions of background. At this point, diagonally adjacent 1's are regarded as being connected, whereas diagonally adjacent 0's are regarded as disconnected—a set of allocations that seems inconsistent. However, we can hardly accept a situation where a connected diagonal pair of 0's crosses a connected diagonal pair of 1's without causing a break in either case. Similarly, we cannot accept a situation in which a disconnected diagonal pair of 0's crosses a disconnected diagonal pair of 1's without there being a join in either case. Hence, a symmetrical definition of connectedness is not possible, and it is conventional to regard diagonal neighbors as connected only if they are foreground; that is, the foreground is "8-connected" and the background is "4-connected." This convention is followed in the subsequent discussion.

1 All unmarked image points are taken to have the binary value 0.

6.3 Object Labeling and Counting

Now that we have a consistent definition of connectedness, we can unambiguously demarcate all objects in binary images and we should be able to devise algorithms for labeling them uniquely and counting them. Labeling may be achieved by scanning the image sequentially until a 1 is encountered on the first object. A note is then made of the scanning position, and a "propagation" routine is initiated to label the whole of the object with a 1. Since the original image space is already in use, a separate image space has to be allocated for labeling. Next, the scan is resumed, ignoring all points already labeled, until another object is found. This is labeled with a 2 in the separate image space. This procedure is continued until the whole image has been scanned and all the objects have been labeled (Fig. 6.1). Implicit in this procedure is the possibility of propagating through a connected object. Suppose at this stage that no method is available for limiting the field of the propagation routine, so that it has to scan the whole image space. Then the propagation routine takes the form:

```
do
  for all points in image
    if point is in object
      and next to a propagating region labeled N
    label this point N also
until no further change;                                    (6.1)
```

the kernel of the do–until loop being expressed more explicitly as:

```
PROP:
/*original image in A-space; labels inserted in P-space */
[[ if ((A0 == 1) && ((P1 == N) || (P2 ==N) || (P3 ==N) || (P4 ==N) ||
   (P5 ==N) || (P6 ==N) || (P7 ==N) || (P8 ==N)))
   P0 = N;
]]                                                          (6.2)
```

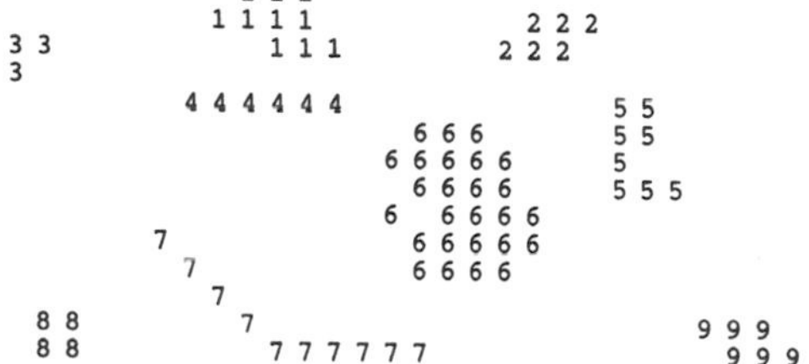

Figure 6.1 A process in which all binary objects are labeled.

```
// start with binary image containing objects in A-space
// clear label space
[[ P0 = 0; ]]
// start with no objects
N = 0;
/* look for objects using a sequential scan and propagate labels through them */
do { // search for an unlabeled object
    found = false;
    [[+    if ((A0 == 1) && (P0 == 0) && not found) {
                N = N + 1;
                P0 = N;
                found = true;
            }
    +]]
    if (found) // label the object just found
        do {
            finished = true;
            [[ if ((A0 == 1) && (P0 == 0) &&
                   ((P1 == N) || (P2 == N) || (P3 == N) || (P4 == N) ||
                   (P5 == N) || (P6 == N) || (P7 == N) || (P8 == N))) {
                       P0 = N ;
                       finished = false;
                }
            ]]
        } until finished;
} until not found; // i.e. no (more) objects
// N is the number of objects found and labeled
```

Figure 6.2 A simple algorithm for object labeling.

At this stage a fairly simple type of algorithm for object labeling is obtained, as shown in Fig. 6.2; the [[+ ··· +]] notation denotes a *sequential* scan over the image.

The above object counting and labeling routine requires a *minimum* of $2N+1$ passes over the image space, and in practice the number will be closer to $NW/2$ where W is the average width of the objects. Hence, the algorithm is inherently inefficient. This prompts us to consider how the number of passes over the image can be reduced to save computation. One possibility would be to scan forward through the image, propagating new labels through objects as they are discovered. Although this would work mostly straightforwardly with convex objects, problems would be encountered with objects possessing concavities—for example, "U" shapes—since different parts of the same object would end with

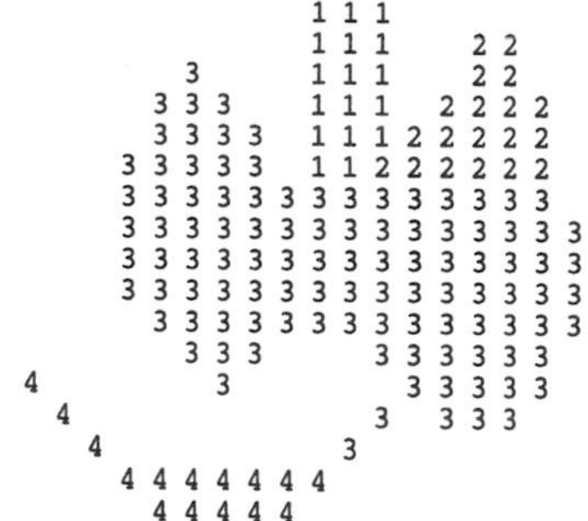

Figure 6.3 Labeling U-shaped objects: a problem that arises in labeling certain types of objects if a simple propagation algorithm is used. Some provision has to be made to accept "collisions" of labels, although the confusion can be removed by a subsequent stage of processing.

different labels. Also, a means would have to be devised for coping with "collisions" of labels (e.g., the largest local label could be propagated through the remainder of the object: see Fig. 6.3). Then inconsistencies could be resolved by a reverse scan through the image. However, this procedure will not resolve all problems that can arise, as in the case of more complex (e.g., spiral) objects. In such cases, a general parallel propagation, repeatedly applied until no further labeling occurs, might be preferable—though as we have seen, such a process is inherently rather computationally intensive. However, it is implemented very conveniently on certain types of parallel SIMD (single instruction stream, multiple data stream) processors (see Chapter 28).

Ultimately, the least computationally intensive procedures for propagation involve a different approach: objects and parts of objects are labeled on a *single* sequential pass through the image, at the same time noting which labels coexist on objects. Then the labels are sorted separately, in a stage of abstract information processing, to determine how the initially rather ad hoc labels should be interpreted. Finally, the objects are relabeled appropriately in a second pass over the image. (This latter pass is sometimes unnecessary, since the image data are merely labeled in an overcomplex manner and what is needed is simply a key to interpret them.) The improved labeling algorithm now takes the form shown in Fig. 6.4. This algorithm with its single sequential scan is intrinsically far more efficient than the previous one, although the presence of particular dedicated hardware or a suitable SIMD processor might alter the situation and justify the use of alternative procedures.

Minor amendments to the above algorithms permit the areas and perimeters of objects to be determined. Thus, objects may be labeled by their areas or perimeters instead of by numbers representing their order of appearance in the image. More important, the availability of propagation routines means that objects can be considered in turn in their entirety—if necessary by transferring

```
// clear label space
[[ P0 = 0; ]]
// start with no objects
N = 0;
// clear the table that is to hold the label coexistence data
for (i = 1; i <= Nmax; i++)
    for (j = 1; j <= Nmax; j++)
        coexist[i][j] = false;
// label objects in a single sequential scan
[[+ if (A0 == 1)
        if ((P2 == 0) && (P3 == 0) && (P4 == 0) && (P5 == 0)) {
            N = N + 1;
            P0 = N;
        }
        else {
            P0 = max(P2, P3, P4, P5);
            // now note which labels coexist in objects
            coexist[P0][P2] = true;
            coexist[P0][P3] = true;
            coexist[P0][P4] = true;
            coexist[P0][P5] = true;
        }
+]]
analyze the coexist table and decide ideal labeling scheme;
relabel image if necessary;
```

Figure 6.4 The improved algorithm for object labeling.

them individually to separate image spaces or storage areas ready for unencumbered independent analysis. Evidently, if objects appear in individual binary spaces, maximum and minimum spatial coordinates are trivially measurable, centroids can readily be found, and more detailed calculations of moments (see below) and other parameters can easily be undertaken.

6.3.1 *Solving the Labeling Problem in a More Complex Case*

In this section we add substance to the all too facile statement at the end of Fig. 6.4: "analyze coexist table and decide ideal labeling scheme." First, we have to make the task nontrivial by providing a suitable example. Figure 6.5 shows an example image, in which sequential labeling has been carried out in line with the algorithm of Fig. 6.4. However, one variation has been adopted—that of

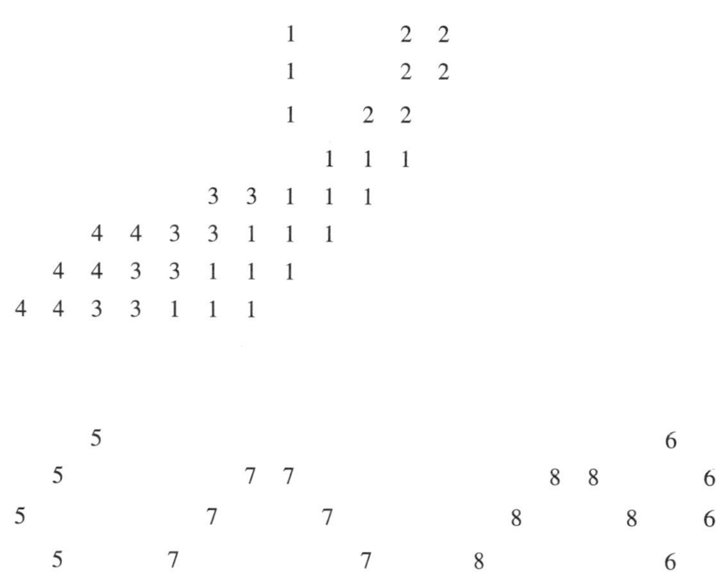

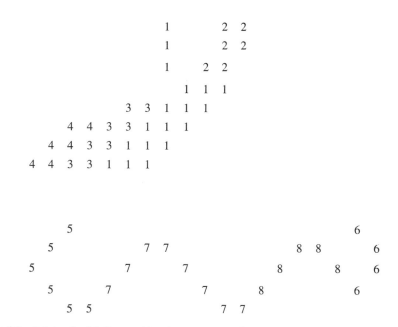

Figure 6.5 Solving the labeling problem in a more complex case.

using a minimum rather than a maximum labeling convention, so that the values are in general slightly closer to the eventual ideal labels. (This also serves to demonstrate that there is not just one way of designing a suitable labeling algorithm.) The algorithm itself indicates that the coexist table should now appear as in Table 6.1. However, the whole process of calculating ideal labels can be made more efficient by inserting numbers instead of ticks, and also adding the right numbers along the leading diagonal, as in Table 6.2. For the same reason, the numbers below the leading diagonal, which are technically redundant, are retained here.

The next step is to minimize the entries along the individual rows of the table, as in Table 6.3; next we minimize along the individual columns (Table 6.4); and then we minimize along rows again (Table 6.5). This process is iterated to completion, which has already happened here after three stages of minimization. We can now read off the final result from the leading diagonal. Note that a further stage of computation is needed to make the resulting labels consecutive integers, starting with unity. However, the procedure needed to achieve this is much more basic and does not need to manipulate a 2-D table of data. This will be left as a simple programming task for the reader.

At this point, some comment on the nature of the process described above will be appropriate. What has happened is that the original image data have effectively been condensed into the minimum space required to express the

Table 6.1 Coexist table for the image of Fig. 6.4*

	1	2	3	4	5	6	7	8
1		√	√					
2	√							
3	√			√				
4			√					
5							√	
6								√
7					√			√
8						√	√	

*The ticks correspond to clashes of labels.

Table 6.2 Coexist table with additional numerical information*

	1	2	3	4	5	6	7	8
1	1	1	1					
2	1	2						
3	1		3	3				
4			3	4				
5					5		5	
6						6		6
7					5		7	7
8						6	7	8

*This coexist table is an enhanced version of Table 6.3. Technically, the numbers along, and below, the leading diagonal are redundant, but nevertheless they speed up the subsequent computation.

Table 6.3 Coexist table redrawn with minimized rows. At this stage the table is no longer symmetrical

	1	2	3	4	5	6	7	8
1	1	1	1					
2	1	1						
3	1		1	1				
4			3	3				
5					5		5	
6						6		6
7					5		5	5
8						6	6	6

Table 6.4 Coexist table redrawn again with minimized columns

	1	**2**	**3**	**4**	**5**	**6**	**7**	**8**
1	1	1	1					
2	1	1						
3	1		1	1				
4			1	1				
5					5		5	
6						6		5
7					5		5	5
8						6	5	5

Table 6.5 Coexist table redrawn yet again with minimized rows. At this stage the table is in its final form and is once again symmetrical

	1	**2**	**3**	**4**	**5**	**6**	**7**	**8**
1	1	1	1					
2	1	1						
3	1		1	1				
4			1	1				
5					5		5	
6						5		5
7					5		5	5
8						5	5	5

labels—namely, just one entry per original clash. This explains why the table retains the 2-D format of the original image: lower dimensionality would not permit the image topology to be represented properly. It also explains why minimization has to be carried out, to completion, in two orthogonal directions. On the other hand, the particular implementation, including both above- and below-diagonal elements, is able to minimize computational overheads and finalize the operation in remarkably few iterations.

Finally, it might be felt that too much attention has been devoted to finding connected components of binary images. This is a highly important topic in practical applications such as industrial inspection, where it is crucial to locate all the objects unambiguously before they can individually be identified and scrutinized. In addition, Fig. 6.5 makes clear that it is not only U-shaped objects that give problems, but also those that have shape subtleties—as happens at the left of the upper object in this figure.

6.4 Metric Properties in Digital Images

Measurement of distance in a digital image is complicated by what is and what is not easy to calculate, given the types of neighborhood operations that are common in image processing. However, it must be noted that rigorous measures of distance depend on the notion of a *metric*. A metric is defined as a measure $d(r_i, r_j)$ that fulfills three conditions:

1. It must always be greater than or equal to zero, equality occurring only when points r_i and r_j are identical:

$$d(r_i, r_j) > 0, \quad i \neq j; \quad d(r_i, r_i) = 0 \tag{6.3}$$

2. It must be symmetrical:

$$d(r_i, r_j) = d(r_j, r_i) \tag{6.4}$$

3. It must obey the triangle rule:

$$d(r_i, r_j) + d(r_j, r_k) \geq d(r_i, r_k) \tag{6.5}$$

Distance can be estimated accurately in a digital image using the Euclidean metric:

$$d_{\mathrm{E}}=\left[(x_i - x_j)^2 + (y_i - y_j)^2\right]^{1/2} \tag{6.6}$$

However, it is interesting that the 8-connected and 4-connected definitions of connectedness given above also lead to measures that are metrics:

$$d_8 = \max\left(|x_i - x_j|, |y_i - y_j|\right) \tag{6.7}$$

and

$$d_4 = |x_i - x_j| + |y_i - y_j| \tag{6.8}$$

On the other hand, the latter two metrics are far from ideal in that they lead to constant distance loci that appear as squares and diamonds rather than as circles. In addition, it is difficult to devise other more accurate measures that fulfill the requirements of a metric. Clearly, this means that if we restrict ourselves to local neighborhood operations, we will not be able to obtain very high accuracy of measurement. Hence, if we require the maximum accuracy possible, it will be necessary to revert to using the Euclidean metric. For example, if we wish to draw an accurate ellipse, then we should employ some scheme such as that in Fig. 6.6 rather than attempting to employ neighborhood operations.

```
// local integer coordinates are (x, y)
[[ A0 = 0; ]]
for (i = 0; i <= 359; i++) {
    theta = i * 2 * pi/360;
    // now write coordinates to framestore registers
    x = (int)(a * sin(theta));
    y = (int)(b * cos(theta));
    A0 = 1;
}
```

Figure 6.6 Routine for drawing an ellipse.

Similarly, if we find two features in an image and wish to measure their distance apart, it is trivial to do this by "abstract" processing using the two pairs of coordinates but difficult to do it efficiently by operations taking place entirely within image space. Nevertheless, it is useful to see how much can be achieved in image space. This aim has given rise to such approximations as the following widely used scheme for rapid estimation of the perimeter of an object: treat horizontal and vertical edges normally (i.e., as unit steps), but treat diagonal edges as $\sqrt{2}$ times the step length. This rather limited approach (analyzed more fully in Chapter 7) is acceptable for certain simple recognition tasks.

6.5 Size Filtering

Although use of the d_4 and d_8 metrics is bound to lead to certain inaccuracies, it is useful to see what can be achieved with the use of local operations in binary images. This section studies how simple size filtering operations can be carried out, using merely local (3×3) operations. The basic idea is that small objects may be eliminated by applying a series of SHRINK operations. N SHRINK operations will eliminate an object (or those parts of an object) that are $2N$ or fewer pixels wide (i.e., across their narrower dimension). Of course this process shrinks *all* the objects in the image, but in principle a subsequent N EXPAND operation will restore the larger objects to their former size.

If complete elimination of small objects but perfect retention of larger objects is required, this will, however, not be achieved by the above procedure, since in many cases the larger objects will be distorted or fragmented by these operations (Fig. 6.7). To recover the larger objects in their *original* form, the proper approach is to use the shrunken versions as "seeds" from which to grow the originals via a propagation process. The algorithm of Fig. 6.8 is able to achieve this.

Having seen how to remove whole (connected) objects that are everywhere narrower than a specified width, it is possible to devise algorithms for removing

Figure 6.7 The effect of a simple size filtering procedure. When size filtering is attempted by a set of N SHRINK and N EXPAND operations, larger objects are restored aproximately to their original sizes, but their shapes frequently become distorted or even fragmented. In this example, (*b*) shows the effect of applying two SHRINK and then two EXPAND operations to the image in (*a*).

```
// save original image
[[ C0 = A0; ]]
// now SHRINK the original objects N times
for (i = 1; i <= N; i++) {
   [[ sigma = A1 + A2 + A3 + A4 + A5 + A6 + A7 + A8;
      if (sigma < 8) B0 = 0; else B0 = A0;
   ]]
   [[ A0 = B0; ]]
}
// next propagate the shrunken objects using the original image
do {
   finished = true;
   [[ sigma = A1 + A2 + A3 + A4 + A5 + A6 + A7 + A8;
      if ((A0 == 0) && (sigma > 0) && (C0 == 1)) {
         A0 = 1;
         finished = false;
      }
   ]]
} until finished;
```

Figure 6.8 Algorithm for recovering original forms of shrunken objects.

any subset of objects that are characterized by a given range of widths. Large objects may be filtered out by first removing lesser-sized objects and then performing a logical masking operation with the original image, while intermediate-sized objects may be filtered out by removing a larger subset and then restoring small objects that have previously been stored in a separate image space. Ultimately, all these schemes depend on the availability of the propagation technique, which in turn depends on the internal connectedness of individual objects.

Finally, note that EXPAND operations followed by SHRINK (or thinning—see below) operations may be useful for joining nearby objects, filling in holes, and so on. Numerous refinements and additions to these simple techniques are possible. A particularly interesting one is to separate the silhouettes of touching objects such as chocolates by a shrinking operation. This then permits them to be counted reliably (Fig. 6.9).

6.6 The Convex Hull and Its Computation

This section moves on to consider potentially more precise descriptions of shape, and in particular the *convex hull*. This is basically an analog descriptor of

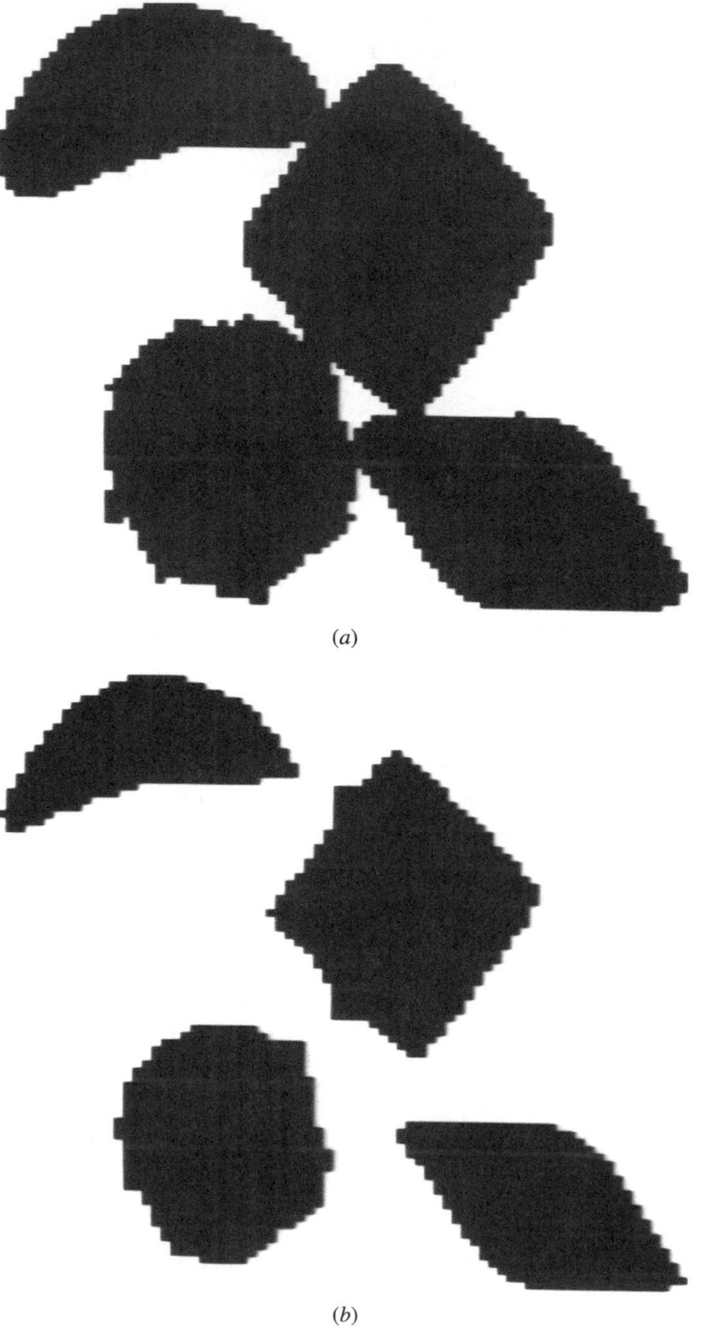

(*a*)

(*b*)

Figure 6.9 Separation of touching objects by SHRINK operations. Here objects (chocolates) in (*a*) are shrunk (*b*) in order to separate them so that they may be counted reliably.

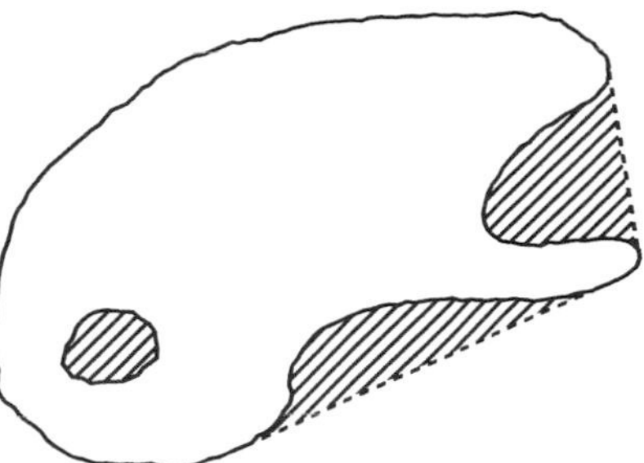

Figure 6.10 Convex hull and convex deficiency. The convex hull is the shape enclosed on placing an elastic band around an object. The shaded portion is the convex deficiency that is added to the shape to create the convex hull.

shape, being defined (in two dimensions) as the shape enclosed by an elastic band placed around the object in question. The *convex deficiency* is defined as the shape that has to be added to a given shape to create the convex hull (Fig. 6.10).

The convex hull may be used to simplify complex shapes to provide a rapid indication of the extent of an object. A particular application might be locating the hole in a washer or nut, since the hole is that part of the convex deficiency that is *not* adjacent to the background. In addition, the convexity or otherwise of a figure is determined by finding whether the convex deficiency is (to the required precision) an empty set. A fuller description of the shape of an object may be obtained by means of *concavity trees*. Here the convex hull of an object is first obtained with its convex deficiencies, next the convex hulls and deficiencies of the convex deficiencies are found, then the convex hulls and deficiencies of these convex deficiencies—and so on until all the derived shapes are convex. Thus, a tree is formed that can be used for systematic shape analysis and recognition (Fig. 6.11). We shall not dwell on this approach beyond noting its inherent utility and observing that at its core is the need for a reliable means of determining the convex hull of a shape.

A simple means of obtaining the convex hull (but not an accurate Euclidean one) is to repeatedly fill in the center pixel of all neighborhoods that exhibit a concavity, including each of the following:

$$\begin{matrix} 1 & 1 & 1 \\ 1 & 0 & 0 \\ 0 & 0 & 0 \end{matrix} \qquad \begin{matrix} 0 & 1 & 1 \\ 1 & 0 & 0 \\ 0 & 0 & 0 \end{matrix}$$

until no further change occurs. If, as in the second example, a corner pixel in the neighborhood is 0 and is adjacent to two 1's, it makes no difference to the 8-connectedness condition whether the corner is 0 or 1. Hence, we can pretend that it has been replaced by a 1. We then get a simple rule for determining whether to fill

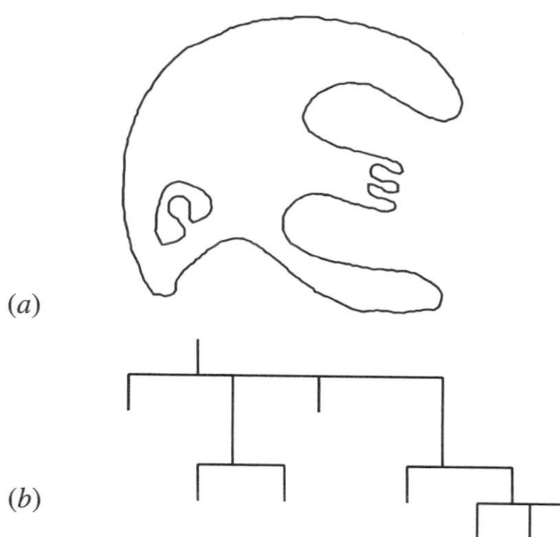

Figure 6.11 A simple shape and its concavity tree. The shape in (*a*) has been analyzed by repeated formation of convex hulls and convex deficiencies until all the constituent regions are convex (see text). The tree representing the entire process is shown in (*b*): at each node, the branch on the left is the convex hull, and the branches on the right are convex deficiencies.

in the center pixel. If there are four or more 1's around the boundary of the neighborhood, then there is a concavity that must be filled in. This can be expressed as follows:

$$\begin{aligned}\text{BASICFILL: [[sigma} &= \text{A1} + \text{A2} + \text{A3} + \text{A4} + \text{A5} + \text{A6} + \text{A7} + \text{A8};\\ &\text{if (sigma} > 3)\ \text{B0} = 1;\ \text{else B0} = \text{A0;]]}\end{aligned} \tag{6.9}$$

However, the corner pixels may not in fact be present. It is therefore necessary to correct the routine as follows:[2]

$$\begin{aligned}\text{FILL: [[sigma} = {} & \text{A1} + \text{A2} + \text{A3} + \text{A4} + \text{A5} + \text{A6} + \text{A7} + \text{A8}\\ & +(\text{A1}\ \&\&\ !\,\text{A2}\ \&\&\ \text{A3}) + (\text{A3}\ \&\&\ !\,\text{A4}\ \&\&\ \text{A5})\\ & +(\text{A5}\ \&\&\ !\,\text{A6}\ \&\&\ \text{A7}) + (\text{A7}\ \&\&\ !\,\text{A8}\ \&\&\ \text{A1});\\ \text{if (sigma} > 3)\ \text{B0} = 1;\ & \text{else B0} = \text{A0;]]}\end{aligned} \tag{6.10}$$

Finally, a complex convex hull algorithm (Fig. 6.12) requires the FILL routine to be included in a loop, with a suitable flag to test for completion. Note that the algorithm is forced to perform a final pass in which nothing happens. Although this seems to waste computation, it is necessary, for otherwise it will

2 For a more detailed explanation of this notation, see Section 6.12.

```
do {
  finished = true;
  [[ sigma = Al + A2 + A3 + A4 + A5 + A6 + A7 + A8
              + (Al && ! A2 && A3) + (A3 && ! A4 && A5)
              + (A5 && ! A6 && A7) + (A7 && ! A8 && Al);
    if ((A0 == 0) && (sigma > 3)) {
      B0 = 1;
      finished = false;
    }
    else B0 = A0;
  ]]
  [[ A0 = B0; ]]
} until finished;
```

Figure 6.12 The complete convex hull algorithm.

not be known that the process has run to completion. This situation arises repeatedly in this sort of work—there are seven examples in this chapter alone.

The shapes obtained by the above algorithm are actually overconvex and approximate to octagonal (or degenerate octagonal) shapes such as the following:

```
    . 1 1 1 1 1 . . . . . . . . . .
  1 1 1 1 1 1 1 1 1 1 . . . . 1 1 1 1
1 1 1 1 . . . . 1 1 1 . . . 1 . . .
1 1 1 . . . . . . 1 1 1 1 1 . . .
. 1 1 1 . . . . . . . 1 1 . . .
  1 1 1 1 . 1 . . . . . . . .
    . 1 1 1 . . . . . . . .
```

The method may be refined so that an approximately 16-sided polygon results (Hussain, 1988). This is achieved by forming the above "overconvex hull"; then forming an "underconvex hull" by finding the convex hull of the background with the constraint that it is permitted to eat into the convex deficiency but must not eat into the original object; then finding a median line (Fig. 6.13) between the over- and underconvex hulls by a suitable technique such as thinning (see Section 6.8).[3]

3 Although the effect of this algorithm is as described above, it is formulated in rather different terms which cannot be gone into here.

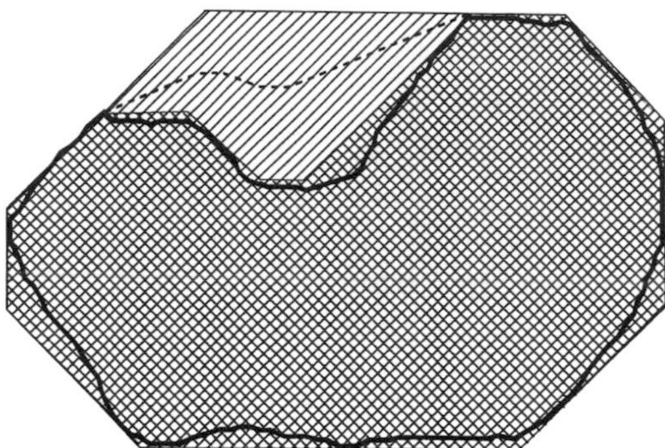

Figure 6.13 Finding an approximate convex hull using over- and underconvex hulls. The shaded region is the overconvex hull of the original shape, and the doubly shaded region is the underconvex hull. A (potentially) 16-sided approximation to the ideal convex hull is found by thinning the singly shaded region, as indicated by the dotted line.

Although such techniques making using of solely local operations are possible, and are ideally implemented on certain types of parallel processor (Hussain, 1988), they have only limited accuracy and can only be improved with considerable ingenuity. Ultimately, for connected objects, the alternative option of tracking around the perimeter of the object and deducing the position of the convex hull has to be adopted. The basic strategy for achieving this is simple—search for positions on the boundary that have common tangent lines. Note that changing from a body-oriented to a boundary-oriented approach can in principle save considerable computation. However, the details of this strategy are not trivial to implement because of the effects of spatial quantization which complicate these inherently simple analog ideas. Overall, there is no escape from the necessity to take connectedness fully into account. Hence, algorithms are required which can reliably track around object boundaries. This aspect of the problem is dealt with in a later section. Further details of convex hull algorithms based on this approach are not given here: the reader is referred to the literature for additional information (see, for example, Rutovitz, 1975).

Before leaving the topic of convex hull determination, a possible complication should be noted. This problem arises when the convex hulls of the individual objects overlap (though not the original objects). In this case, the final convex hull is undefined. Should it be the union of the separate convex hulls or the convex hull of their union? The local operator algorithm considered earlier will find the latter, whereas algorithms based on boundary tracking will find the former. In general, obtaining a convex hull from an original shape is not trivial, as tests must be incorporated to ensure that separate limbs are in all cases connected, and to ensure that the convex hulls of other objects are not inadvertently combined. Ultimately, such problems are probably best dealt with by transferring each object to its own storage area before starting convex hull computations (see also Section 6.3).

6.7 Distance Functions and Their Uses

The distance function of an object is a very simple and useful concept in shape analysis. Essentially, each pixel in the object is numbered according to its distance from the background. As usual, background pixels are taken as 0's; then edge pixels are counted as 1's; object pixels next to 1's become 2's; next to those are the 3's; and so on throughout all the object pixels (Fig. 6.14).

The parallel algorithm of Fig. 6.15 finds the distance function of binary objects by propagation. Note that, as in the computation of convex hulls, this algorithm performs a final pass in which nothing happens. Again, this is inevitable if it is to be certain that the process runs to completion. An alternative way of propagating the distance function is shown in Fig. 6.16. Note the complication in ensuring completion of the procedure. If it could be assumed that N passes are needed for objects of known maximum width, then the algorithm would be simply:

```
PPROP3:
  for (i = 1; i <= N; i++)
    [[ Q0 = min(Q0 - 1, Q1, Q2, Q3, Q4, Q5, Q6, Q7, Q8) + 1;]]     (6.11)
```

This routine operates entirely within a single image space and appears to be simpler and more elegant than PPROP1. Note, however, that in every pass

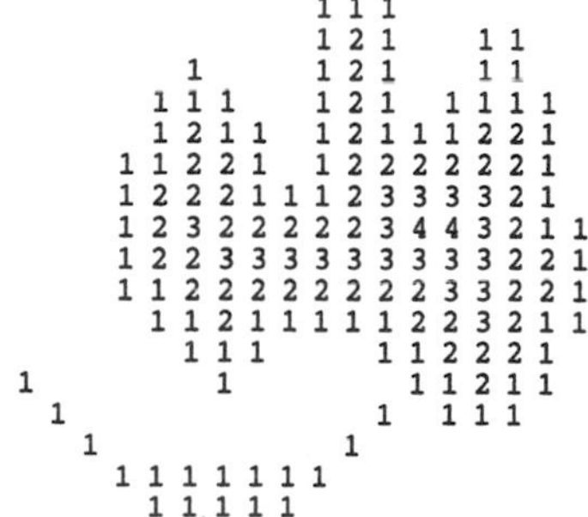

Figure 6.14 The distance function of a binary shape: the value at every pixel is the distance (in the d_8 metric) from the background.

```
PPROP1: { // Start with binary image containing objects in A-space
   [[ Q0 = A0 * 255; ]]
   N = 0;
   do {
      finished = true;
      [[ if ((Q0 == 255) // in object and no answer yet
            && ((Q1 == N) || (Q2 == N) || (Q3 == N) || (Q4 == N) ||
                  (Q5 == N) || (Q6 == N) || (Q7 == N) || (Q8 == N)) {
            // next to an N
            Q0 = N + 1;
            finished = false; // some action has been taken
         }
      ]]
      N = N + 1;
   } until finished;
}
```

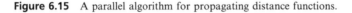

Figure 6.15 A parallel algorithm for propagating distance functions.

```
PPROP2: {
   [[ Q0 = A0 * 255; ]]
   do {
      finished = true;
      [[ minplusone = min(Q0 – 1,Q1,Q2,Q3,Q4,Q5,Q6,Q7,Q8) + 1;
         if (Q0 > minplusone) {
            Q0 = minplusone;
            finished = false; // some action has been taken
         }
      ]]
   } until finished;
}
```

Figure 6.16 An alternative parallel algorithm for propagating distance functions.

over the image it writes a value at every location, which under some circumstances could require more computation. (This is not true for PPROP1, which writes a value only at those object locations that are next to previously assigned values.) For SIMD parallel processing machines, which have a processing element at every pixel location, this is no problem, and PPROP3 may be more useful.

It is possible to perform the propagation of a distance function with far fewer operations if the process is expressed as two sequential operations, one being a normal forward raster scan and the other being a reverse raster scan:

$$
\begin{aligned}
&\text{SPROP1: } \{\\
&\qquad [[\ Q0 = A0 * 255;\ ]]\\
&\qquad [[+\ \text{minplusone} = \min(Q2, Q3, Q4, Q5) + 1;\\
&\qquad\quad \text{if } (Q0 > \text{minplusone})\ Q0 = \text{minplusone}; +]]\\
&\qquad [[-\ \text{minplusone} = \min(Q6, Q7, Q8, Q1) + 1;\\
&\qquad\quad \text{if } (Q0 > \text{minplusone})\ Q0 = \text{minplusone}; -]]\\
&\}
\end{aligned}
\tag{6.12}
$$

Note the compact notation being used to distinguish between forward and reverse raster scans over the image: $[[+\cdots+]]$ denotes a forward raster scan, while $[[-\cdots-]]$ denotes a reverse raster scan. A more succinct version of this algorithm is the following:

$$
\begin{aligned}
&\text{SPROP2: } \{\\
&\qquad [[Q0 = A0 * 255;]]\\
&\qquad [[+\ Q0 = \min(Q0 - 1, Q2, Q3, Q4, Q5) + 1; +]]\\
&\qquad [[-\ Q0 = \min(Q0 - 1, Q6, Q7, Q8, Q1) + 1; -]]\\
&\}
\end{aligned}
\tag{6.13}
$$

An interesting application of distance functions is that of data compression. To achieve it, operations are carried out to locate those pixels that are local maxima of the distance function (Fig. 6.17), since storing these pixel values and positions permits the original image to be regenerated by a process of downward propagation (see below). Note that although finding the local maxima of the distance function provides the basic information for data compression, the actual compression occurs only when the data are stored as a list of points rather than in the original picture format. In order to locate the local maxima, the following parallel routine may be employed:

$$
\begin{aligned}
&\text{LMAX1: } [[\ \text{maximum} = \max(Q1, Q2, Q3, Q4, Q5, Q6, Q7, Q8);\\
&\qquad \text{if } ((Q0 > 0)\ \&\&\ (Q0 >= \text{maximum}))\ B0 = 1;\ \text{else } B0 = 0;]]
\end{aligned}
\tag{6.14}
$$

Alternatively, the compressed data can be transferred to a single image space:

$$
\begin{aligned}
&\text{LMAX2: } [[\ \text{maximum} = \max(Q1, Q2, Q3, Q4, Q5, Q6, Q7, Q8);\\
&\qquad \text{if } ((Q0 > 0)\ \&\&\ (Q0 >= \text{maximum}))\ P0 = Q0; \text{else } P0 = 0;]]
\end{aligned}
\tag{6.15}
$$

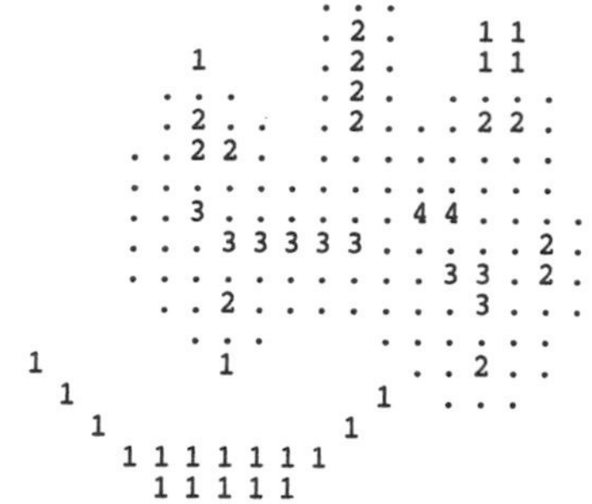

Figure 6.17 Local maxima of the distance function of the shape shown in Fig. 6.14, the remainder of the shape being indicated by dots and the background being blank. Notice that the local maxima group themselves into clusters, each containing points of equal distance function value, while clusters of different values are clearly separated.

Note that the local maxima that are retained for the purpose of data compression are not absolute maxima but are maximal in the sense of not being adjacent to larger values. If this were not so, insufficient numbers of points would be retained for completely regenerating the original object. As a result, it is found that the absolute maxima group themselves into clusters of connected points, each cluster having a common distance value and being separated from points of different distance values (Fig. 6.17). Thus, the set of local maxima of an object is not a connected subset. This fact has an important bearing on skeleton formation (see below).

Having seen how data compression may be performed by finding local maxima of the distance function, it is relevant to consider a parallel downward propagation algorithm (Fig. 6.18) for recovering the shapes of objects from an image into which the values of the local maxima have been inserted. Note again that if it can be assumed that at most N passes are needed to propagate through objects of known maximum width, then the algorithm becomes simply:

DPROP2:

for (i = 1; i <= N;i++)

$$[[\ Q0 = \max(Q0 + 1, Q1, Q2, Q3, Q4, Q5, Q6, Q7, Q8) - 1;\ ]] \tag{6.16}$$

Finally, although the local maxima of the distance function form a subset of points that may be used to reconstruct the object shape exactly, it is frequently possible to find a subset M of this subset that has this same property and thereby corresponds to a greater degree of data compression. This can be seen by appealing to Fig. 6.19. This is possible because although downward propagation from M permits the object shape to be recovered immediately, it does not also provide a correct set of distance function values. If required, these must be

```
DPROP1: {
    /* assume that input image is in Q-space, and that non maximum values have
    value 0, as at output of LMAX2 routine */
    do {
        finished = true;
        [[ maxminusone = max(Q0 + 1,Q1,Q2,Q3,Q4,Q5,Q6,Q7,Q8) – 1;
            if (Q0 < maxminusone)
                {
                    Q0 = maxminusone;
                    finished = false; // some action has been taken
                }
        ]]
    } until finished;
}
```

Figure 6.18 A parallel algorithm for recovering objects from the local maxima of the distance functions.

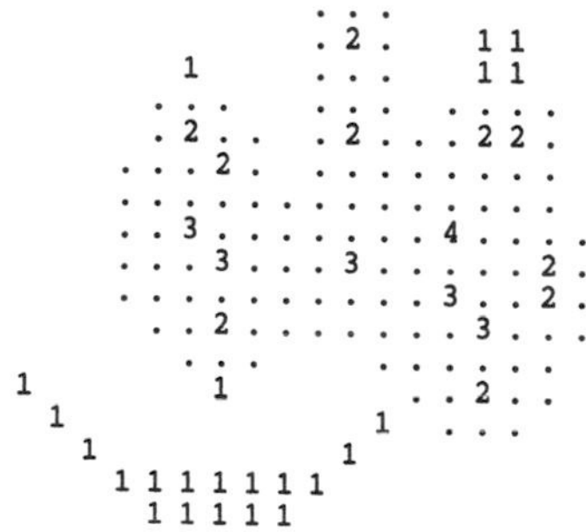

Figure 6.19 Minimal subset of points for object reconstruction. The numbered pixels may be used to reconstruct the entire shape shown in Fig. 6.14 and form a more economical subset than the set of local maxima of Fig. 6.17.

obtained by subsequent upward propagation. It turns out that the process of finding a minimal set of points M involves quite a complex and computationally intensive algorithm and is too specialized to describe in detail here (see Davies and Plummer, 1980).

6.8 Skeletons and Thinning

Like the convex hull, the skeleton is a powerful analog concept that may be employed for the analysis and description of shapes in binary images. A skeleton may be defined as a connected set of medial lines along the limbs of a figure. For example, in the case of thick hand-drawn characters, the skeleton may be the path actually traveled by the pen. The basic idea of the skeleton is that of eliminating redundant information while retaining only the topological information concerning the shape and structure of the object that can help with recognition. In the case of hand-drawn characters, the thickness of the limbs is taken to be irrelevant. It may be constant and therefore carry no useful information, or it may vary randomly and again be of no value for recognition (Fig. 1.2).

The skeleton definition presented earlier leads to the idea of finding the loci of the centers of maximal discs inserted within the object boundary. First, suppose the image space is a continuum. Then the discs are circles, and their centers form loci that may be modeled conveniently when object boundaries are approximated by linear segments. Sections of the loci fall into three categories:

1. They may be angle bisectors, that is, lines that bisect corner angles and reach right up to the apexes of corners.
2. They may be lines that lie halfway between boundary lines.
3. They may be parabolas that are equidistant from lines and from the nearest points of other lines—namely, corners where two lines join.

Clearly, (1) and (2) are special forms of a more general case.

These ideas lead to unique skeletons for objects with linear boundaries, and the concepts are easily generalizable to curved shapes. This approach tends to give rather more detail than is commonly required, even the most obtuse corner having a skeleton line going into its apex (Fig. 6.20). Hence, a thresholding scheme is often employed, so that skeleton lines only reach into corners having a specified minimum degree of sharpness.

We now have to see how the skeleton concept will work in a digital lattice. Here we are presented with an immediate choice: which metric should we employ? If we select the Euclidean metric, we are liable to engender a considerable computational load (although many workers do choose this option). If we select the d_8 metric we will immediately lose accuracy but the computational requirements should be more modest. (Here we do not consider the d_4 metric, since we are dealing with foreground objects.) In the following discussion we concentrate on the d_8 metric.

At this stage, some thought shows that the application of maximal discs in order to locate skeleton lines amounts essentially to finding the positions of local maxima of the distance function. Unfortunately, as seen in the previous section, the set of local maxima does not form a connected graph within a given object. Nor is it necessarily composed of thin lines, and indeed in places it may be 2 pixels wide. Thus, problems arise in trying to use this approach to obtain a

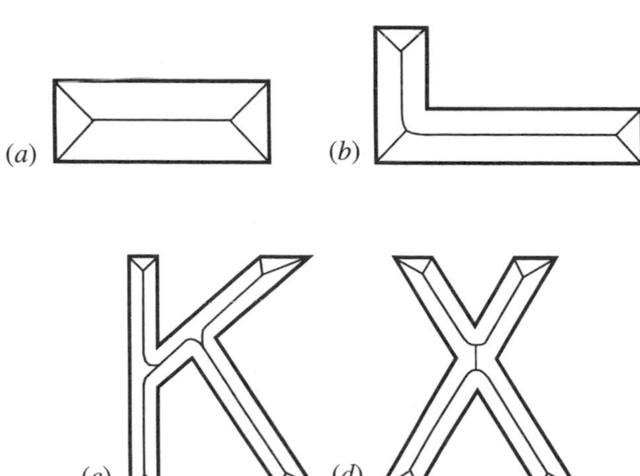

Figure 6.20 Four shapes whose boundaries consist entirely of straight-line segments. The idealized skeletons go right to the apex of each corner, however obtuse. In certain parts of shapes (*b*), (*c*), and (*d*), the skeleton segments are parts of parabolas rather than straight lines. As a result, the detailed shape of the skeleton (or the approximations produced by most algorithms operating in discrete images) is not exactly what might initially be expected or what would be preferred in certain applications.

connected unit-width skeleton that can conveniently be used to represent the shape of the object. We shall return to this approach again below. Meanwhile, we pursue an alternative idea—that of thinning.

Thinning is perhaps the simplest approach to skeletonization. It may be defined as the process of systematically stripping away the outermost layers of a figure until only a connected unit-width skeleton remains (see, for example, Fig. 6.21). A number of algorithms are available to implement this process, with varying degrees of accuracy, and we will discuss how a specified level of precision can be achieved and tested for. First, however, it is necessary to discuss the mechanism by which points on the boundary of a figure may validly be removed in thinning algorithms.

6.8.1 *Crossing Number*

The exact mechanism for examining points to determine whether they can be removed in a thinning algorithm must now be considered. This may be decided by reference to the *crossing number* χ (chi) for the 8 pixels around the outside of a particular 3×3 neighborhood. χ is defined as the total number of 0-to-1 and 1-to-0

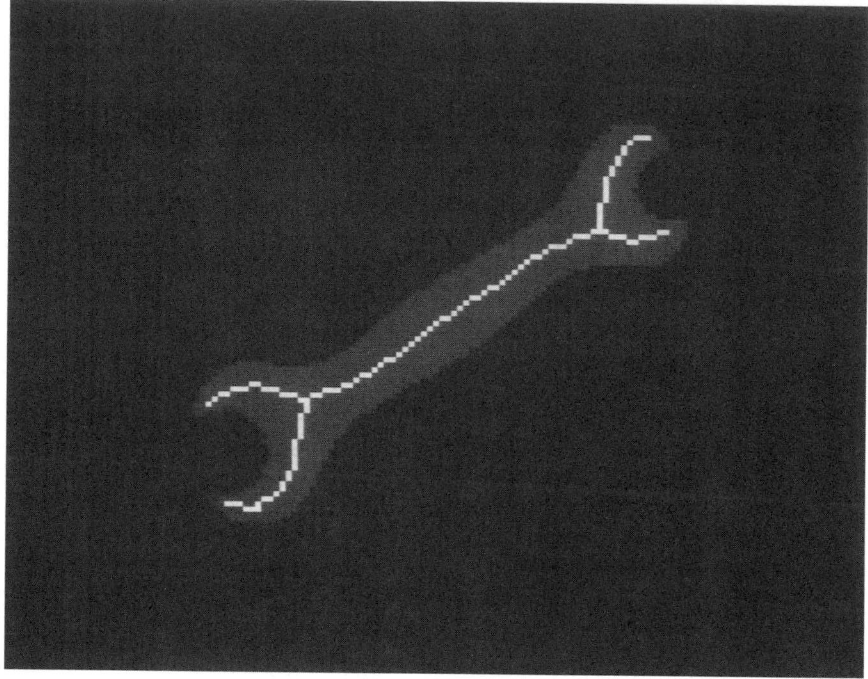

Figure 6.21 Typical result of a thinning algorithm operating in a discrete lattice.

0 0 0	0 0 0	1 0 0	1 0 0	1 0 1	1 0 0	1 0 1
0 1 0	0 1 1	0 1 1	0 1 0	0 1 0	0 1 0	0 1 0
0 0 0	1 1 1	1 1 1	0 1 0	1 1 1	1 0 1	1 0 1
0	2	4	4	6	6	8

Figure 6.22 Some examples of the crossing number values associated with given pixel neighborhood configurations (0, background; 1, foreground).

transitions on going once round the outside of the neighborhood. This number is in fact twice the number of potential connections joining the remainder of the object to the neighborhood (Fig. 6.22). Unfortunately, the formula for χ is made more complex by the 8-connectedness criterion. Basically, we would expect:

$$\begin{aligned} \text{badchi} = {} & (\text{A1} \mathrel{!=} \text{A2}) + (\text{A2} \mathrel{!=} \text{A3}) + (\text{A3} \mathrel{!=} \text{A4}) + (\text{A4} \mathrel{!=} \text{A5}) \\ & +(\text{A5} \mathrel{!=} \text{A6}) + (\text{A6} \mathrel{!=} \text{A7}) + (\text{A7} \mathrel{!=} \text{A8}) + (\text{A8} \mathrel{!=} \text{A1}); \end{aligned} \tag{6.17}$$

However, this is incorrect because of the 8-connectedness criterion. For example, in the case

$$\begin{matrix} 0 & 1 & 0 \\ 0 & 1 & 1 \\ 1 & 1 & 1 \end{matrix}$$

the formula gives the value 4 for χ instead of 2. The reason is that the isolated 0 in the top right-hand corner does not prevent the adjacent 1's from being joined. It is therefore tempting to use the modified formula:

$$\text{wrongchi} = (\text{A1} \mathrel{!=} \text{A3}) + (\text{A3} \mathrel{!=} \text{A5}) + (\text{A5} \mathrel{!=} \text{A7}) + (\text{A7} \mathrel{!=} \text{A1}); \tag{6.18}$$

This too is wrong, however, since in the case

$$\begin{matrix} 0 & 0 & 1 \\ 0 & 1 & 0 \\ 1 & 1 & 1 \end{matrix}$$

it gives the answer 2 instead of 4. It is therefore necessary to add four extra terms to deal with isolated 1's in the corners:[4]

$$\begin{aligned} \text{chi} = {} & (\text{A1} \mathrel{!=} \text{A3}) + (\text{A3} \mathrel{!=} \text{A5}) + (\text{A5} \mathrel{!=} \text{A7}) + (\text{A7} \mathrel{!=} \text{A1}) \\ & + 2 * ((!\,\text{A1} \,\&\&\, \text{A2} \,\&\&\, !\,\text{A3}) + (!\,\text{A3} \,\&\&\, \text{A4} \,\&\&\, !\,\text{A5}) \\ & + (!\,\text{A5} \,\&\&\, \text{A6} \,\&\&\, !\,\text{A7}) + (!\,\text{A7} \,\&\&\, \text{A8} \,\&\&\, !\,\text{A1})); \end{aligned} \tag{6.19}$$

This (now correct) formula for crossing number gives values 0, 2, 4, 6, or 8 in different cases (Fig. 6.22). The rule for removing points during thinning is that points may only be removed if they are at those positions on the boundary of an object where χ is 2: when χ is greater than 2, the point *must* be retained, for it forms a vital connecting point between two parts of the object. In addition, when χ is 0, it must be retained since removing it would create a hole.

Finally, one more condition must be fulfilled before a point can be removed during thinning—that the sum σ (sigma) of the eight pixel values around the outside of the 3×3 neighborhood (see Chapter 2) must not be equal to 1. The reason for this is to preserve line ends, as in the following cases:

$$\begin{matrix} 0 & 0 & 0 \\ 0 & 1 & 0 \\ 0 & 1 & 0 \end{matrix} \qquad \begin{matrix} 0 & 0 & 0 \\ 0 & 1 & 0 \\ 0 & 0 & 1 \end{matrix}$$

4 For further explanation of this notation, see Section 6.12.

If line ends are eroded as thinning proceeds, the final skeleton will not represent the shape of an object (including the relative dimensions of its limbs) at all accurately. (However, we might sometimes wish to shrink an object while preserving connectedness, in which case this extra condition need not be implemented.) Having covered these basics, we are now in a position to devise complete thinning algorithms.

6.8.2 *Parallel and Sequential Implementations of Thinning*

Thinning is "essentially sequential" in that it is easiest to ensure that connectedness is maintained by arranging that only one point may be removed at a time. As indicated earlier, this is achieved by checking before removing a point that has a crossing number of 2. Now imagine applying the "obvious" sequential algorithm of Fig. 6.23 to a binary image. Assuming a normal forward raster scan, the result of this process is to produce a highly distorted skeleton, consisting of lines along the right-hand and bottom edges of objects. It may now be seen that the $\chi = 2$ condition is necessary but not sufficient, since it says nothing about the order in which points are removed. To produce a skeleton that is unbiased, giving a set of truly medial lines, it is necessary to remove points as evenly as possible around the object boundary. A scheme that helps with this involves a novel processing sequence: mark edge points on the first pass over an image; on the second pass, strip points sequentially as in the above algorithm, *but only where they have already been marked*; then mark a new set of edge points; then perform another stripping

```
STHIN: {
  do {
    finished = true;
    [[+   sigma = A1 + A2 + A3 + A4 + A5 + A6 + A7 + A8;
          chi = (A1 != A3) + (A3 != A5) + (A5 != A7) + (A7 != A1)
                + 2 * ((! A1 && A2 && ! A3) + (! A3 && A4 && ! A5)
                + (! A5 && A6 && ! A7) + (! A7 && A8 && ! A1));
          if ((A0 == 1) && (chi == 2) && (sigma != 1) {
             A0 = 0;
             finished = false; // some action has been taken
          }
    +]]
  } until finished;
}
```

Figure 6.23 An "obvious" sequential thinning algorithm.

pass; then finally repeat this marking and stripping sequence until no further change occurs. An early algorithm working on this principle is that of Beun (1973).

Although the novel sequential thinning algorithm described above can be used to produce a reasonable skeleton, it would be far better if the stripping action could be performed symmetrically around the object, thereby removing any possible skeletal bias. In this respect, a parallel algorithm should have a distinct advantage. However, parallel algorithms result in several points being removed at once. This means that lines 2 pixels wide will disappear (since masks operating in a 3×3 neighborhood cannot "see" enough of the object to judge whether or not a point may validly be removed), and as a result shapes can become disconnected. The general principle for avoiding this problem is to strip points lying on different parts of the boundary in different passes, so that there is no risk of causing breaks. This can be achieved in a large number of ways by applying different masks and conditions to characterize different parts of the boundary. If boundaries were always convex, the problem would no doubt be reduced; however, boundaries can be very convoluted and are subject to quantization noise, so the problem is a complex one. With so many potential solutions to the problem, we concentrate here on one that can conveniently be analyzed and that gives acceptable results.

The method discussed is that of removing north, south, east, and west points cyclically until thinning is complete. North points are defined as the following:

$$\begin{matrix} \times & 0 & \times \\ \times & 1 & \times \\ \times & 1 & \times \end{matrix}$$

where $\times$ means either a 0 or a 1: south, east, and west points are defined similarly. It is easy to show that all north points for which $\chi = 2$ and $\sigma \neq 1$ may be removed in parallel without any risk of causing a break in the skeleton—and similarly for south, east, and west points. Thus, a possible format for a parallel thinning algorithm in rectangular tessellation is the following:

```
do
   strip appropriate north points;
   strip appropriate south points;
   strip appropriate east points;
   strip appropriate west points;
until no further change;                                    (6.20)
```

where the basic parallel routine for stripping "appropriate" north points is:

```
NSTRIP:
  [[ sigma = A1 + A2 + A3 + A4 + A5 + A6 + A7 + A8;
     chi = (A1 != A3) + (A3 != A5) + (A5 != A7) + (A7 != A1)
           +2 * ((A1 && A2 && ! A3) + (! A3 && A4 && ! A5)
           +(! A5 && A6 && ! A7) + (! A7 && A8 && ! A1));
     if ((! A3 && A0 && A7 )//north point
           && (chi == 2) && (sigma != 1))
             B0 = 0;
     else    B0 = A0;
  ]]                                                          (6.21)
```

(But extra code needs to be inserted to detect whether any changes have been made in a given pass over the image.)

Algorithms of the above type can be highly effective, although their design tends to be rather intuitive and ad hoc. In Davies and Plummer (1981), a great many such algorithms exhibited problems. Ignoring cases where the algorithm design was insufficiently rigorous to maintain connectedness, four other problems were evident:

1. The problem of skeletal bias.
2. The problem of eliminating skeletal lines along certain limbs.
3. The problem of introducing "noise spurs."
4. The problem of slow speed of operation.

Problems (2) and (3) are opposites in many ways: if an algorithm is designed to suppress noise spurs, it is liable to eliminate skeletal lines in some circumstances. Contrariwise, if an algorithm is designed never to eliminate skeletal lines, it is unlikely to be able to suppress noise spurs. This situation arises because the masks and conditions for performing thinning are intuitive and ad hoc, and therefore have no basis for discriminating between valid and invalid skeletal lines. Ultimately this is because it is difficult to build overt global models of reality into purely local operators. In a similar way, algorithms that proceed with caution, that is, that do not remove object points in the fear of making an error or causing bias, tend to be slower in operation than they might otherwise be. Again, it is difficult to design algorithms that can make correct global decisions rapidly via intuitively designed local operators. Hence, a totally different approach is needed if solving one of the above problems is not to cause difficulties with the others. Such an alternative approach is discussed in the next section.

6.8.3 Guided Thinning

This section returns to the ideas discussed in the early part of Section 6.8, where it was found that the local maxima of the distance function do not form an ideal skeleton because they appear in clusters and are not connected. In addition, the clusters are often 2 pixels wide. On the plus side, the clusters are accurately in the correct positions and should therefore not be subject to skeletal bias. Hence, an ideal skeleton should result if (1) the clusters could be reconnected appropriately and (2) the resulting structure could be reduced to unit width. In this respect it should be noted that a unit width skeleton can only be perfectly unbiased where the object is an odd number of pixels wide: where the object is an even number of pixels wide, the unit-width skeleton is bound to be biased by 1/2 pixel, and the above strategy can give a skeleton that is as unbiased as is in principle possible.

A simple means of reconnecting the clusters is to use them to guide a conventional thinning algorithm (see Section 6.8.2). As a first stage, thinning is allowed to proceed normally but with the proviso that no cluster points may be removed. This gives a connected graph that is in certain places 2 pixels wide. Then a routine is applied to strip the graph down to unit width. At this stage, an unbiased skeleton (within 1/2 pixel) should result. The main problem here is the presence of noise spurs. The opportunity now exists to eliminate these systematically by applying suitable global rules. A simple rule is that of eliminating lines on the skeletal graph that terminate in a local maximum of value (say) 1 (or, better, stripping them back to the next local maximum), since such a local maximum corresponds to fairly trivial detail on the boundary of the object. Thus the level of detail that is ignored can be programmed into the system (Davies and Plummer, 1981). The whole guided thinning process is shown in Fig. 6.24.

6.8.4 A Comment on the Nature of the Skeleton

At the beginning of Section 6.8, the case of character recognition was taken as an example, and it was stated that the skeleton was supposed to be the path traveled by the pen in drawing out the character. However, in one important respect this is not valid. The reason is seen both in the analog reasoning and from the results of thinning algorithms. (Ultimately, these reflect the same phenomena.) Take the case of a letter *K*. The vertical limb on the left of the skeleton is most unlikely to turn out to be straight, because theoretically it will consist of two linear segments joined by two parabolic segments leading into the junction (Fig. 6.20). It will therefore be very difficult to devise a skeletonizing

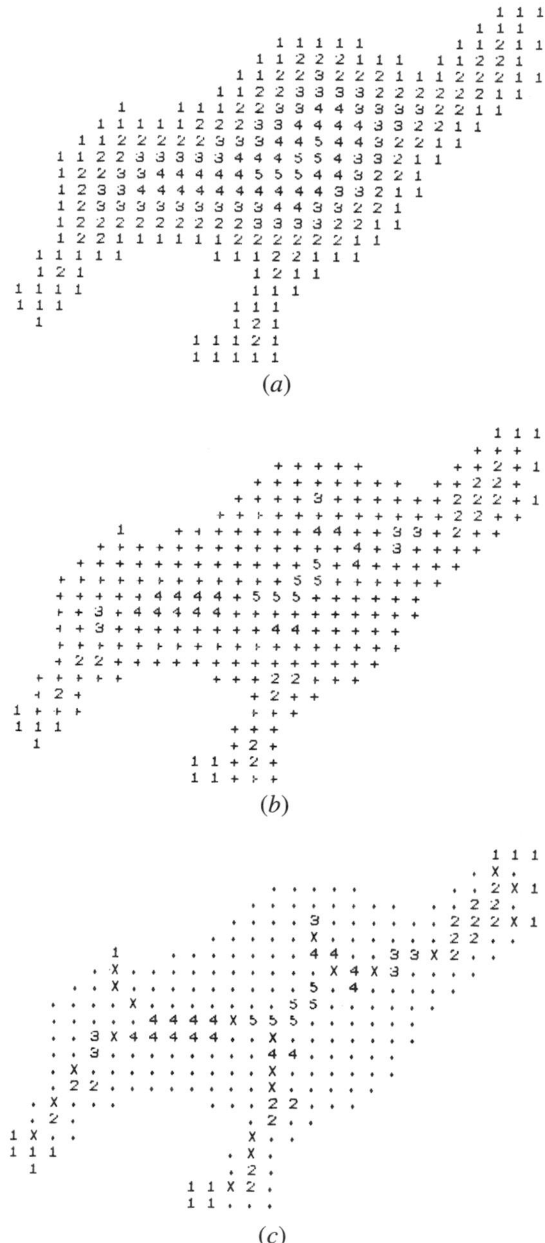

Figure 6.24 Results of a guided thinning algorithm: (*a*) distance function on the original shape; (*b*) set of local maxima; (*c*) set of local maxima now connected by a simple thinning algorithm; (*d*) final thinned skeleton. The effect of removing noise spurs systematically, by cutting limbs terminating in a 1 back to the next local maximum, is easily discernible from the result in (*d*): the general shape of the object is not perturbed by this process.

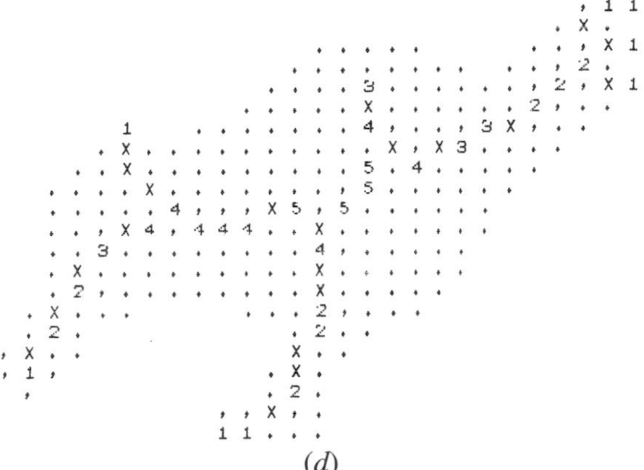

(d)

Figure 6.24 Continued.

algorithm that will give a single straight line in such cases, for all possible orientations of the object. The most reliable means of achieving this would probably be via some high-level interpretation scheme that analyzes the skeleton shape and deduces the ideal shape, for example, by constrained least-squares line fitting.

6.8.5 *Skeleton Node Analysis*

Skeleton node analysis may be carried out very simply with the aid of the crossing number concept. Points in the middle of a skeletal line have a crossing number of 4; points at the end of a line have a crossing number 2; points at skeletal "T" junctions have a crossing number of 6; and points at skeletal "X" junctions have a crossing number of 8. However, there is a situation to beware of—places that look like a "+" junction:

$$\begin{matrix} 0 & 1 & 0 \\ 1 & 1 & 1 \\ 0 & 1 & 0 \end{matrix}$$

In such places, the crossing number is actually 0 (see formula), although the pattern is definitely that of a cross. At first, the situation seems to be that there is insufficient resolution in a 3×3 neighborhood to identify a "+" cross. The best option is to look for this particular pattern of 0's and 1's and use a more sophisticated construct than the 3×3 crossing number to check whether or not a cross is present. The problem is that of distinguishing between two situations such as:

$$\begin{matrix} 0 & 0 & 0 & 0 & 0 \\ 0 & 0 & 1 & 0 & 0 \\ 0 & 1 & 1 & 1 & 0 \\ 0 & 0 & 1 & 0 & 0 \\ 0 & 0 & 0 & 0 & 0 \end{matrix} \qquad \text{and} \qquad \begin{matrix} 0 & 0 & 1 & 0 & 0 \\ 1 & 0 & 1 & 0 & 0 \\ 1 & 1 & 1 & 1 & 1 \\ 0 & 0 & 1 & 0 & 0 \\ 0 & 0 & 0 & 1 & 0 \end{matrix}$$

However, further analysis shows that the first of these two cases would be thinned down to a dot (or a short bar), so that if a "+" node appears on the final skeleton (as in the second case) it actually signifies that a cross is present despite the contrary value of χ. Davies and Celano (1993b) have shown that the proper measure to use in such cases is the *modified* crossing number $\chi_{skel} = 2\sigma$, this crossing number being different from χ because it is required not to test whether points can be eliminated from the skeleton, but to ascertain the meaning of points that are at that stage *known* to lie on the final skeleton. Note that χ_{skel} can have values as high as 16—it is not restricted to the range 0 to 8!

Finally, note that sometimes insufficient resolution really is a problem, in that a cross with a shallow crossing angle appears as two "T" junctions:

$$\begin{matrix} 0 & 0 & 0 & 0 & 0 & 0 & 0 & 1 \\ 1 & 1 & 1 & 0 & 0 & 1 & 1 & 0 \\ 0 & 0 & 0 & 1 & 1 & 0 & 0 & 0 \\ 0 & 1 & 1 & 0 & 0 & 1 & 1 & 1 \\ 1 & 0 & 0 & 0 & 0 & 0 & 0 & 0 \end{matrix}$$

Resolution makes it impossible to recognize an asterisk or more complex figure from its crossing number, within a 3×3 neighborhood. Probably, the best solution is to label junctions tentatively, then to consider all the junction labels in the image, and to analyze whether a particular local combination of junctions should be reinterpreted—for example, two "T" junctions may be deduced to form a cross. This is especially important in view of the distortions that appear on a skeleton in the region of a junction (see Section 6.8.4).

6.8.6 *Application of Skeletons for Shape Recognition*

Shape analysis may be carried out simply and conveniently by analysis of skeleton shapes and dimensions. Clearly, study of the nodes of a skeleton (points for which there are other than two skeletal neighbors) can lead to the classification of simple shapes but not, for example, discrimination of all block capitals from each other. Many classification schemes exist which can complete the analysis, in terms of limb lengths and positions. Methods for achieving this are touched on in later chapters.

A similar situation exists for analysis of the shapes of chromosomes, which take the form of a cross or a "V." For small industrial components, more detailed shape analysis is necessary. This can still be approached with the skeleton technique, by examining distance function values along the lines of the skeleton. In general, shape analysis using the skeleton proceeds by examination in turn of nodes, limb lengths and orientations, and distance function values, until the required level of characterization is obtained.

The particular importance of the skeleton as an aid in the analysis of connected shapes is not only that it is invariant under translations and rotations but also that it embodies what is for many purposes a highly convenient representation of the figure which (with the distance function values) essentially carries all the original information. If the original shape of an object can be deduced exactly from a representation, this is generally a good sign because it means that it is not merely an ad hoc descriptor of shape but that considerable reliance may be placed on it (compare other methods such as the circularity measure—see Section 6.9).

6.9 Some Simple Measures for Shape Recognition

Many simple tests of shape can be made to confirm the identity of products or to check for defects. These include measurements of product area and perimeter, length of maximum dimension, moments relative to the centroid, number and area of holes, area and dimensions of the convex hull and enclosing rectangle, number of sharp corners, number of intersections with a check circle and angles between intersections (Fig. 6.25), and numbers and types of skeleton nodes.

The list would not be complete without mention of the widely used shape measure $C=\text{area}/(\text{perimeter})^2$. This quantity is often called *circularity* or *compactness*, because it has a maximum value of $1/4\pi$ for a circle, decreases as shapes become more irregular, and approaches zero for long narrow objects. Alternatively, its reciprocal is sometimes used, being called *complexity* since it increases in size as shapes become more complex. Note that both measures are dimensionless, so that they are independent of size and are therefore sensitive only to the shape of an object. Other dimensionless measures of this type include rectangularity and aspect ratio.

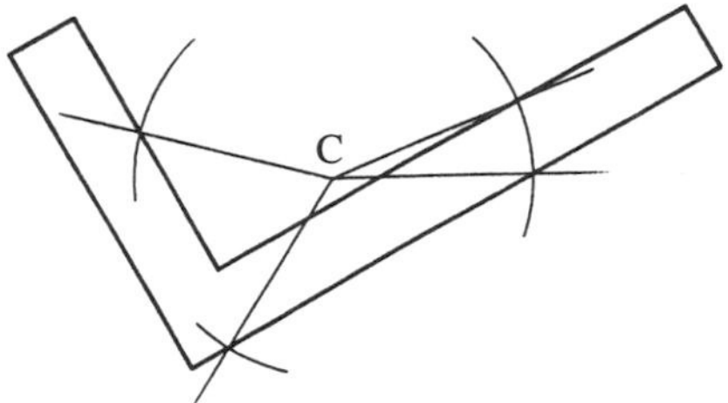

Figure 6.25 Rapid product inspection by polar checking.

All these measures have the property of characterizing a shape but not of describing it uniquely. Thus, it is easy to see that there are in general many different shapes having the same value of circularity. Hence, these rather ad hoc measures are on the whole less valuable than approaches such as skeletonization (Section 6.8) or moments (Section 6.10) that can be used to represent and reproduce a shape to any required degree of precision. Nevertheless, it should be noted that rigorous checking of even one measured number to high precision often permits a machined part to be identified positively.

6.10 Shape Description by Moments

Moment approximations provide one of the more obvious means of describing 2-D shapes and take the form of series expansions of the type:

$$M_{pq} = \sum_{x}\sum_{y} x^p y^q f(x, y) \tag{6.22}$$

for a picture function $f(x, y)$. Such a series may be curtailed when the approximation is sufficiently accurate. By referring axes to the centroid of the shape, moments can be constructed that are position-invariant. They can also be renormalized so that they are invariant under rotation and change of scale (Hu, 1962; see also Wong and Hall, 1978). The main value of using moment descriptors is that in certain applications the number of parameters may be made small without significant loss of precision—although the number required may not be clear without considerable experimentation. Moments can prove particularly valuable in describing shapes such as cams and other fairly round objects, although they have also been used in a variety of other applications including airplane silhouette recognition (Dudani et al., 1977).

6.11 Boundary Tracking Procedures

The preceding sections have described methods of analyzing shape on the basis of body representations of one sort or another—convex hulls, skeletons, moments, and so on. However, an important approach has so far been omitted—the use of boundary pattern analysis. This approach has the potential advantage of requiring considerably reduced computation, since the number of pixels to be examined is equal to the number of pixels on the boundary of any object rather than the much larger number of pixels within the boundary. Before proper use can be made of boundary pattern analysis techniques, the means must be found for tracking systematically around the boundaries of all the objects in an image. In addition, care must be taken not to ignore any holes that are present or any objects within holes.

In one sense, the problem has been analyzed already, in that the object labeling algorithm of Section 6.3 systematically visits and propagates through all objects in the image. All that is required now is some means of tracking around object boundaries once they have been encountered. Quite clearly, it will be useful to mark in a separate image space all points that have been tracked. Alternatively, an object boundary image may be constructed and the tracking performed in this space, all tracked points being eliminated as they are passed.

In the latter procedure, objects having unit width in certain places may become disconnected. Hence, we ignore this approach and adopt the previous one. There is still a problem when objects have unit-width sections, since these can cause badly designed tracking algorithms to choose a wrong path, going back around the previous section instead of on to the next (Fig. 6.26). To avoid this circumstance, it is best to adopt the following strategy:

1. Track around each boundary, keeping to the left path consistently.
2. Stop the tracking procedure only when passing through the starting point in the original direction (or passing through the first two points in the same order).

Apart from necessary initialization at the start, a suitable tracking procedure is given in Fig. 6.27. Boundary tracking procedures such as this require careful coding. In addition, they have to operate rapidly despite the considerable amount of computation required for every boundary pixel. (See the amount of code within the main "do" loop of the procedure of Fig. 6.27.) For this reason it is normally best to implement the main direction-finding routine using lookup tables, or else using special hardware: details are, however, beyond the scope of this chapter.

Having seen how to track around the boundaries of objects in binary images, we are now in a position to embark on boundary pattern analysis. (See Chapter 7).

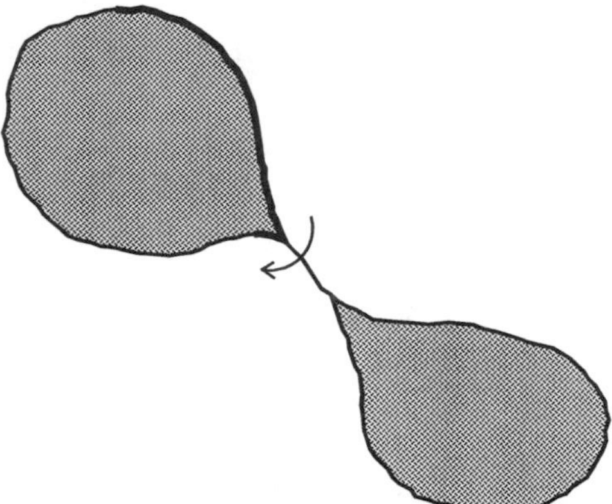

Figure 6.26 A problem with an oversimple boundary tracking algorithm. The boundary tracking procedure takes a shortcut across a unit-width boundary segment instead of continuing and keeping to the left path at all times.

```
do {
   // find direction to move next
   start with current tracking direction;
   reverse it;
   do {
      rotate tracking direction clockwise
   } until the next 1 is met on outer pixels of 3 × 3 neighborhood;
   record this as new current direction;
   move one pixel along this direction;
   increment boundary index;
   store current position in boundary list;
} until (position = original position) && (direction = original direction)
```

Figure 6.27 Basic procedure for tracking around a single object.

6.12 More Detail on the Sigma and Chi Functions

In this section, we present more explanation on the notation used in Sections 6.6 and 6.8.1. Essentially, this combines logical and unsigned integer arithmetic in a single expression relating to binary pixel values.

$$
\begin{aligned}
\text{sigma} = {} & \text{A1} + \text{A2} + \text{A3} + \text{A4} + \text{A5} + \text{A6} + \text{A7} + \text{A8} \\
& +(\text{A1}\ \&\&\ !\,\text{A2}\ \&\&\ \text{A3}) + (\text{A3}\ \&\&\ !\,\text{A4}\ \&\&\ \text{A5}); \\
& +(\text{A5}\ \&\&\ !\,\text{A6}\ \&\&\ \text{A7}) + (\text{A7}\ \&\&\ !\,\text{A8}\ \&\&\ \text{A1});
\end{aligned} \tag{6.23}
$$

In case of confusion, this statement may be written more conventionally in the form:

$$\begin{aligned}&\text{sigma} = \text{A1} + \text{A2} + \text{A3} + \text{A4} + \text{A5} + \text{A6} + \text{A7} + \text{A8};\\ &\text{if (A1 \&\& ! A2 \&\& A3) sigma} ++;\\ &\text{if (A3 \&\& ! A4 \&\& A5) sigma} ++;\\ &\text{if (A5 \&\& ! A6 \&\& A7) sigma} ++;\\ &\text{if (A7 \&\& ! A8 \&\& A1) sigma} ++;\end{aligned} \tag{6.24}$$

Similarly, the statement:

$$\begin{aligned}\text{chi} = {} & (\text{A1 != A3}) + (\text{A3 != A5}) + (\text{A5 != A7}) + (\text{A7 != A1})\\ & +2 * ((\text{! A1 \&\& A2 \&\& ! A3}) + (\text{! A3 \&\& A4 \&\&! A5});\\ & +(\text{! A5 \&\& A6 \&\& ! A7}) + (\text{! A7 \&\& A8 \&\& ! A1}));\end{aligned} \tag{6.25}$$

may be rewritten in the form:

$$\begin{aligned}&\text{chi} = (\text{A1 != A3}) + (\text{A3 != A5}) + (\text{A5 != A7}) + (\text{A7 != A1});\\ &\text{if (! A1 \&\& A2 \&\& ! A3) chi} = \text{chi} + 2;\\ &\text{if (! A3 \&\& A4 \&\& ! A5) chi} = \text{chi} + 2;\\ &\text{if (! A5 \&\& A6 \&\& ! A7) chi} = \text{chi} + 2;\\ &\text{if (! A7 \&\& A8 \&\& ! A1) chi} = \text{chi} + 2;\end{aligned} \tag{6.26}$$

Note that, in general, declarations of variables are not given in the text but should be clear from the context. However, image variables are taken to be predeclared as bit or byte unsigned integers (see Fig. 6.28), bit variables being needed to hold shape information in binary images, and byte variables to hold information on up to 256 gray levels in gray-scale images (see Chapter 2).

6.13 Concluding Remarks

This chapter has concentrated on rather traditional methods of performing image analysis—using image processing techniques. This has led naturally to area representations of objects, including, for example, moment and convex hull-based schemes, although the skeleton approach appeared as a rather special case in that it converts objects into graphical structures. An alternative schema is to represent shapes by their boundary patterns, after applying suitable tracking algorithms; this approach is considered in Chapter 7. Meanwhile, it should be noted that connectedness has been an underlying theme in the

A0 . . . A8 B0 . . . B8 C0 . . . C8	bit image variables
P0 . . . P8 Q0 . . . Q8 R0 . . . R8	byte image variables
P[0], . . .	equivalent to P0, . . .
[[. . .]]	double for loop over image (parallel processing)
[[+ . . . +]]	double for loop for forward sequential raster scan
[[− . . . −]]	double for loop for reverse sequential raster scan
x, y	local coordinates for random access to image
input(P)	grab a new image into P-space

Figure 6.28 Special syntax.

present chapter, objects being separated from each other by regions of background, thereby permitting objects to be considered and characterized individually. Connectedness has been seen to involve rather more intricate computations than might have been expected, necessitating great care in the design of algorithms. This must partly explain why after so many years, new thinning algorithms are still being developed (e.g., Kwok, 1989; Choy et al., 1995). (Ultimately, these complexities arise because global properties of images are being computed by purely local means.)

Although boundary pattern analysis is in certain ways more attractive than region pattern analysis, this comparison cannot be divorced from considerations of the hardware the algorithms have to run on. In this respect, it should be noted that many of the algorithms of this chapter can be performed efficiently on SIMD processors, which have one processing element per pixel (see Chapter 28), whereas boundary pattern analysis will be seen to better match the capabilities of more conventional serial computers.

Shape analysis can be attempted by boundary or region representations. Both are deeply affected by connectedness and related metric issues for a digital lattice of pixels. This chapter has shown that these issues are only solved by carefully incorporating global knowledge alongside local information—for example, by use of distance transforms.

6.14 Bibliographical and Historical Notes

The development of shape analysis techniques has been extensive: hence, only a brief perusal of the history is attempted here. The all-important theory of connectedness and the related concept of adjacency in digital images were developed largely by Rosenfeld (see, e.g., Rosenfeld, 1970); for a review of this work, see Rosenfeld (1979). The connectedness concept led to the idea of a distance function in a digital picture (Rosenfeld and Pfaltz, 1966, 1968) and the related skeleton concept (Pfaltz and Rosenfeld, 1967). However, the basic idea of a skeleton dates from the classic work by Blum (1967)—see also Blum and Nagel (1978). Important work on thinning has been carried out by Arcelli et al. (1975, 1981; Arcelli and di Baja, 1985) and parallels work by Davies and Plummer (1981). Note that the Davies–Plummer paper demonstrates a rigorous method for testing the results of *any* thinning algorithm, however generated, and in particular for detecting skeletal bias. More recently, Arcelli and Ramella (1995) have reconsidered the problem of skeletons in gray-scale images.

Returning momentarily to the distance function concept, which is of some importance for measurement applications, there have been important developments to generalize this concept and to make distance functions uniform and isotropic. See, for example, Huttenlocher et al. (1993).

Sklansky has carried out much work on convexity and convex hull determination (see, for example, Sklansky, 1970; Sklansky et al., 1976), while Batchelor (1979) developed concavity trees for shape description. Early work by Moore (1968) using shrinking and expanding techniques for shape analysis has been developed considerably by Serra and his co-workers (see, for example, Serra, 1986). Haralick et al. (1987) have generalized the underlying mathematical (morphological) concepts, including the case of gray-scale analysis. For more details, see Chapter 8. Use of invariant moments for pattern recognition dates from the two seminal papers by Hu (1961, 1962) and has continued unabated ever since. The field of shape analysis has been reviewed a number of times by Pavlidis (1977, 1978), and he has also drawn attention to the importance of unambiguous ("asymptotic") shape representation schemes (Pavlidis, 1980).

Finally, the design of a modified crossing number χ_{skel} for the analysis of skeletal shape has not been considered until relatively recently. As pointed out in Section 6.8.5, χ_{skel} is different from χ because it judges the *remaining* (i.e., skeletal) points rather than points that *might* be eliminated from the skeleton (Davies and Celano, 1993b).

Most recently, skeletons have maintained their interest and utility, becoming if anything more precise by reference to exact analog shapes (Kégl and Krzyżak, 2002), and giving rise to the concept of a shock graph, which characterizes the result of the much earlier grass-fire transformation (Blum, 1967) more rigorously

(Giblin and Kimia, 2003). Wavelet transforms have also been used to implement skeletons more accurately in the discrete domain (Tang and You, 2003). In contrast, shape matching has been carried out using self-similarity analysis coupled with tree representation—an approach that has been especially valuable for tracking articulated object shapes, including particularly human profiles and hand motions (Geiger et al., 2003). Finally, it is of interest to see graphical analysis of skeletonized hand-written character shapes performed taking account of catastrophe theory (Chakravarty and Kompella, 2003). This is relevant because (1) critical points—where points of inflection exist—can be deformed into pairs of points, each corresponding to a curvature maximum plus a minimum; (2) crossing of t's can be actual or noncrossing; and (3) loops can turn into cusps or corners (many other possibilities also exist). The point is that methods are needed for mapping between *variations* of shapes rather than making snap judgments as to classification. (This corresponds to the difference between scientific understanding of process and ad hoc engineering.)

6.15 Problems

1. Write a routine for removing inconsistencies in labeling U-shaped objects by implementing a reverse scan through the image (see Section 6.3).
2. Write the full C++ routine required to sort the lists of labels, to be inserted at the end of the algorithm of Fig. 6.4.
3. Devise an algorithm for determining the "underconvex" hull described in Section 6.6.
4. Show that, as stated in Section 6.8.2 for a parallel thinning algorithm, all north points may be removed in parallel without any risk of causing a break in the skeleton.
5. Show that the convex-hull algorithm of Fig. 6.12 enlarges objects that may then coalesce. Find a simple means of locally inhibiting the algorithm to prevent objects from becoming connected in this way. Take care to consider whether a parallel or sequential process is to be used.
6. Describe methods for locating, labeling, and counting objects in binary images. You should consider whether methods based on propagation from a "seed" pixel, or those based on progressively shrinking a skeleton to a point, would provide the more efficient means for achieving the stated aims. Give examples for objects of various shapes.

7. (a) Give a simple one-pass algorithm for labeling the objects appearing in a binary image, making clear the role played by connectedness. Give examples showing how this basic algorithm goes wrong with real objects: illustrate your answer with clear pixel diagrams, that show the numbers of labels that can appear on objects of different shapes.

   (b) Show how a table-oriented approach can be used to eliminate multiple labels in objects. Make clear how the table is set up and what numbers have to be inserted into it. Are the number of iterations needed to analyze the table similar to the number that would be needed in a multipass labeling algorithm taking place entirely within the original image? Consider how the *real* gain in using a table to analyze the labels arises.

8. (a) Using the normal notation for a 3×3 window:

A4	A3	A2
A5	A0	A1
A6	A7	A8

   work out the effect of the following algorithm on a binary image containing small foreground objects of various shapes:

```
do
  [[ sum = (A1 && A3) + (A3 && A5) + (A5 && A7) + (A7 && A1);
     if(sum > 0) B0 = 1; else B0 = A0 ]];
  [[ A0 = B0 ]];
until no further change;
```

   (b) Show in detail how to implement the *do ... until no further change* function in this algorithm.

9. (a) Give a simple algorithm operating in a 3×3 window for generating a *rectangular* convex hull around each object in a binary image. Include in your algorithm any necessary code for implementing the required *do ... until no further change* function.

   (b) A more sophisticated algorithm for finding accurate convex hulls is to be designed. Explain why this would employ a boundary tracking procedure. State the general strategy of an algorithm for tracking around the boundaries of objects in binary images and write a program for implementing it.

   (c) Suggest a strategy for designing the complete convex hull algorithm and indicate how rapidly you would expect it to operate, in terms of the size of the image.

10. What are the problems in producing ideal object boundaries from the output of an edge detection operator? Show how disconnected edges might, in favorable conditions, be reconnected by a simple procedure such as extending lines in the same direction, until they meet other lines. Show that a better strategy is to join line ends only if the joining line would lie within a small angle of both line ends. What other factors should be taken into account in devising edge linking strategies? To what extent can any problems be overcome by any following algorithms, for example, Hough transforms?

11. (a) Explain the meaning of the term *distance function*. Give examples of the distance functions of simple shapes, including that shown in Fig. 6.P1.

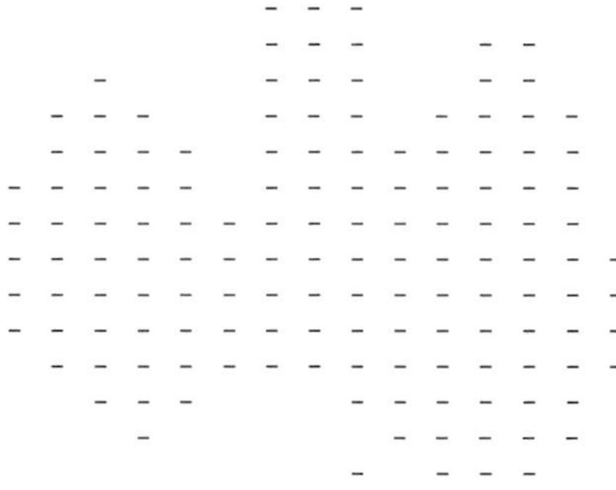

Figure 6.P1

(b) Rapid image transmission is to be performed by sending only the coordinates and values of the local maxima of the distance functions. Give a complete algorithm for finding the local maxima of the distance functions in an image, and devise a further algorithm for reconstructing the original binary image.

(c) Discuss which of the following sets of data would give more compressed versions of the binary picture object shown in Fig. 6.P1:

(i) the list of local maxima coordinates and values;

(ii) a list of the coordinates of the boundary points of the object;

(iii) a list consisting of one point on the boundary and the relative directions (each expressed as a 3-bit code) between each *pair* of boundary points on tracking round the boundary

12. (a) What is the *distance function* of a binary image? Illustrate your answer for the case where a 128×128 image P contains just the object shown in Fig. 6.P2. How many passes of (i) a parallel algorithm and (ii) a sequential algorithm would be required to find the distance function of the image?

```
1 1 1 1 1 1 1 1 1 1 1 1 1 1 1 1 1
1 1 1 1 1 1 1 1 1 1 1 1 1 1 1 1 1 1
1 1 1 1 1 1 1 1 1 1 1 1 1 1 1 1 1 1 1
1 1 1 1 1 1 1 1 1 1 1 1 1 1 1 1 1 1 1 1
1 1 1 1 1 1 1 1 1 1 1 1 1 1 1 1 1 1 1 1
  1 1 1 1 1 1 1 1 1 1 1 1 1 1 1 1 1 1 1
                        1 1 1 1 1 1 1
                        1 1 1 1 1 1 1
                        1 1 1 1 1 1 1
                        1 1 1 1 1 1 1
                        1 1 1 1 1 1 1
                        1 1 1 1 1 1 1 1 1
```

Figure 6.P2

(b) Give a complete parallel *or* sequential algorithm for finding the distance function, and explain how it operates.

(c) Image P is to be transmitted rapidly by determining the coordinates of locations where the distance function is locally maximum. Indicate the positions of the local maxima, and explain how the image could be reconstituted at the receiver.

(d) Determine the compression factor η if image P is transmitted in this way. Show that η can be increased by eliminating some of the local maxima before transmission, and estimate the resulting increase in η.

13. (a) The local maxima of the distance function can be defined in the following two ways:

    (i) those pixels whose values are greater than those of all the neighboring pixels;

    (ii) those pixels whose values are greater than *or equal to* the values of all the neighboring pixels. Which definition of the local maxima would be more useful for reproducing the original object shapes? Why is this?

    (b) Give an algorithm that is capable of reproducing the original object shapes from the local maxima of the distance function, and explain how it operates.

    (c) Explain the *run-length encoding* approach to image compression. Compare the run-length encoding and local maxima methods for compressing binary images. Explain why the one method would be expected to lead to a greater degree of compression with some types of images, while the other method would be expected to be better with other types of images.

14. (a) Explain how propagation of a distance function may be carried out using a parallel algorithm. Give in full a simpler algorithm that operates using two sequential passes over the image.

(b) It has been suggested that a four-pass sequential algorithm will be even faster than the two-pass algorithm, as each pass can use just a 1-D window involving at most three pixels. Write down the code for *one* typical pass of the algorithm.

(c) Estimate the approximate speeds of these three algorithms for computing the distance function, in the case of an $N \times N$ pixel image. Assume a conventional serial computer is used to perform the computation.

15. (a) A new wavefront method is devised for obtaining unbroken edges in gray-scale images. It involves (i) detecting edges and thinning them to unit width, (ii) placing a 1 on the high-intensity side and a -1 on the other side of each edge point (in neither case affecting the edge points themselves, which are counted as having value 0), (iii) propagating distance function values from the 1's and negative distance function values from the -1's, and (iv) assigning 0 to any points that are adjacent to equal positive and negative values. Decide whether this specification is sufficient and whether any ambiguities remain. In the latter case, suggest suitable modifications to the procedure.

    (b) Determine the effect of applying the complete algorithm. In your assessment of the algorithm, show what happens to edges that are initially (i) only slightly broken and (ii) widely spaced laterally. How could this algorithm be extended for use in practical situations?

16. Small dark insects are to be located among cereal grains. The insects approximate to rectangular bars of dimensions 20 by 7 pixels, and the cereal grains are approximately elliptical with dimensions 40 by 24 pixels. The proposed algorithm design strategy is: (i) apply an edge detector that will mark all the edge points in the image as 0's in a 1's background, (ii) propagate a distance function in the *background* region, (iii) locate the local maxima of the distance function, (iv) analyze the values of the local maxima, and (v) carry out necessary further processing to identify the nearly parallel sides of the insects. Explain how to design stages (iv) and (v) of the algorithm in order to identify the insects, ignoring the cereal grains. Assume that the image is not so large that the distance function will overflow the byte limit. Determine how robust the method will be if the edge is broken in a few places.

17. Give the general strategy of an algorithm for tracking around the boundaries of objects in binary images. If the tracker has reached a particular boundary point with crossing number $\chi = 2$, such as

$$\begin{array}{|ccc|} 0 & 0 & 1 \\ 0 & 1 & 1 \\ 1 & 1 & 1 \end{array},$$

decide in which direction it should now proceed. Hence, give a complete procedure for determining the direction code of the next position on the boundary for cases where $\chi = 2$. Take the direction codes starting *from* the current pixel (*) as being specified by the following diagram:

4	3	2
5	*	1
6	7	8

How should the procedure be modified for cases where $\chi \neq 2$?

18. (a) Explain the principles involved in tracking around the boundaries of objects in binary images to produce reliable outlines. Outline an algorithm that can be used for this purpose, assuming it is to get its information from a normal 3 × 3 window.

(b) A binary image is obtained and the data in it are to be compressed as much as possible. The following range of modified images and datasets is to be tested for this purpose:

(i) the boundary image;

(ii) the skeleton image;

(iii) the image of the local maxima of the distance function;

(iv) the image of a suitably chosen subset of the local maxima of the distance function;

(v) a set of run-length data, that is, a series of numbers obtained by counting runs of 0's, then runs of 1's, then runs of 0's, and so on, in a continuous line-by-line scan over the image.

In (i) and (ii) the lines may be encoded using chain code, that is, giving the *coordinates* of the first point met, and the *direction* of each subsequent point using the direction codes 1 to 8 defined relative to the current position C by:

4	3	2
5	C	1
6	7	8

With the aid of suitably chosen examples, discuss which of the methods of data compression should be most suitable for different types of data. Wherever possible, give numerical estimates to support your arguments.

(c) Indicate what you would expect to happen if noise were added to initially noise-free input images.

Boundary Pattern Analysis

Recognition of binary objects by boundary pattern analysis should be a straightforward process, but as this chapter shows, a number of problems need to be overcome. In particular, any boundary distortions such as those due to breakage or several objects being in contact may result in total failure of the matching process. This chapter discusses the problems and analyzes possible solutions.

Look out for:

- an analysis of hysteresis thresholding procedures.
- the centroidal profile approach and its limitations.
- how the method may be speeded up.
- how recognition based on the (s, ψ) boundary plot is significantly more robust.
- how the (s, ψ) plot leads to the more convenient (s, κ) plot.
- the exact relation between κ and ψ.
- more rigorous ways of dealing with the occlusion problem.
- chain coding.
- the (r, s) plot and its advantages.
- discussion of the accuracy of boundary length measures.

Disparate ways are available for representing object boundaries and for measuring and recognizing objects using them. All the methods are subject to the same ultimate difficulties—particularly that of managing occlusion (which necessarily removes relevant data) and that of inaccuracy in the pixel-based description. Sound ways of managing occlusion are suggested in Section 7.7. This work presages Part 2 of this volume where the Hough transform is employed widely for object recognition—albeit not by overt boundary tracking, but by an altogether more robust parallel processing approach.

CHAPTER 7

Boundary Pattern Analysis

7.1 Introduction

An earlier chapter showed how thresholding may be used to binarize gray-scale images and hence to present objects as 2-D shapes. This method of segmenting objects is successful only when considerable care is taken with lighting and when the object can be presented conveniently, for example, as dark patches on a light background. In other cases, adaptive thresholding may help to achieve the same end: as an alternative, edge detection schemes can be applied, these schemes generally being rather more resistant to problems of lighting. Nevertheless, thresholding of edge-enhanced images still poses certain problems; in particular, edges may peter out in some places and may be thick in others (Fig. 7.1). For many purposes, the output of an edge detector is ideally a connected unit-width line around the periphery of an object, and steps will need to be taken to convert edges to this form.

Thinning algorithms can be used to reduce edges to unit thickness, while maintaining connectedness (Fig. 7.1*d*). Many algorithms have been developed for this purpose, and the main problems here are: (1) slight bias and inaccuracy due to uneven stripping of pixels, especially in view of the fact that even the best algorithm can only produce a line that is locally within 1/2 pixel of the ideal position; and (2) introduction of a certain number of noise spurs. The first of these problems can be minimized by using gray-scale edge thinning algorithms which act directly on the original gray-scale edge-enhanced image (e.g., Paler and Kittler, 1983). Noise spurs around object boundaries can be eliminated quite efficiently by removing lines that are shorter than (say) 3 pixels. Overall, the major problem to be dealt with is that of reconnecting boundaries that have become fragmented during thresholding.

A number of rather ad hoc schemes are available for relinking broken boundaries. For example, line ends may be extended along their existing directions—a very limited procedure since there are (at least for binary edges)

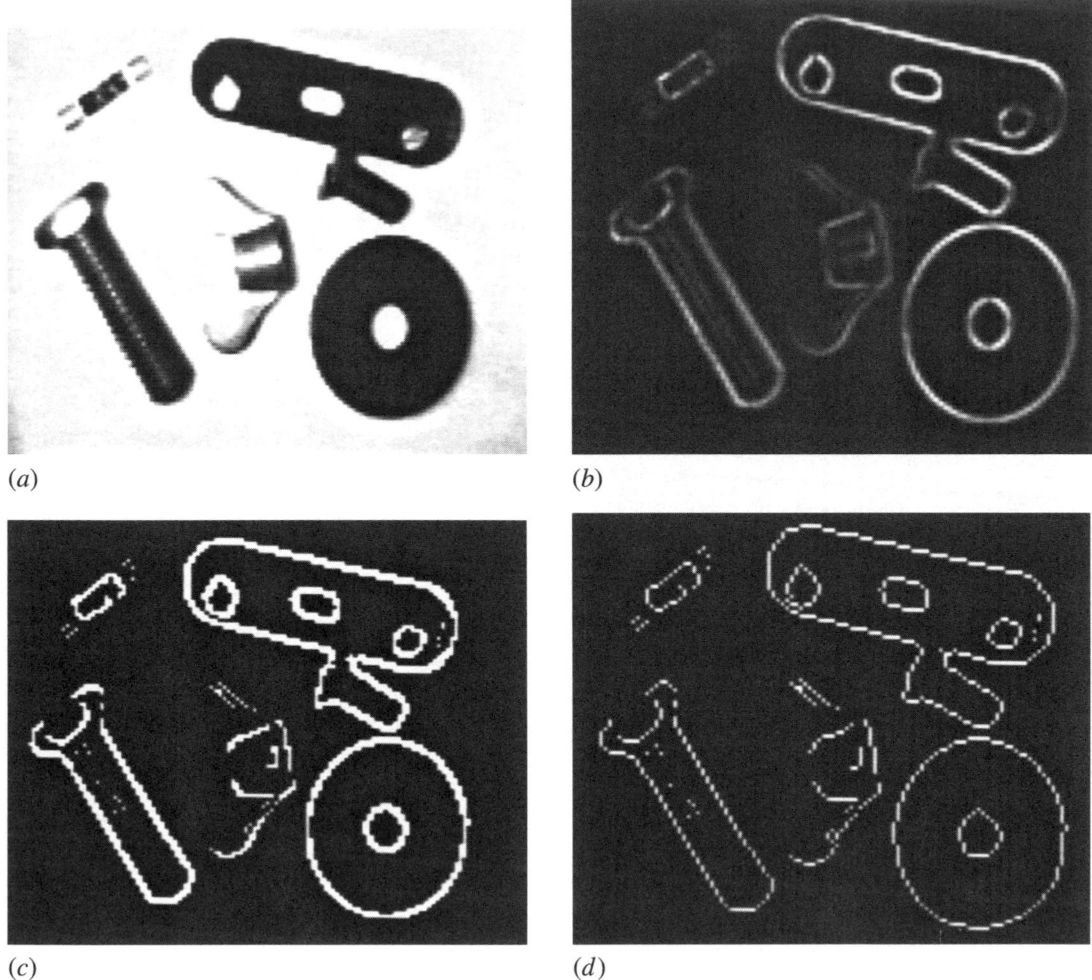

Figure 7.1 Some problems with edges. The edge-enhanced image (*b*) from an original image (*a*) is thresholded as in (*c*). The edges so detected are found to peter out in some places and to be thick in other places. A thinning algorithm is able to reduce the edges to unit thickness (*d*), but ad hoc (i.e., not model-based) linking algorithms are liable to produce erroneous results (not shown).

only eight possible directions, and it is quite possible for the extended line ends not to meet. Another approach is to join line ends which are (1) sufficiently close together and (2) pointing in similar directions to each other and to the direction of the vector between the two ends. This approach can be made quite credible in principle, but in practice it can lead to all sorts of problems as it is still ad hoc and not model driven. Hence, adjacent lines that arise from genuine surface markings and from shadows may be arbitrarily linked together by such algorithms. In many situations it is therefore best if the process is model driven—for example, by finding the best fit to some appropriate idealized boundary such as an ellipse. Yet another approach is that of relaxation labeling, which iteratively enhances the original image, progressively making decisions as to where the original gray levels reinforce each other. Thus, edge linking is permitted only where evidence is available in the original image that this is permissible. A similar but computationally more efficient line of attack is the hysteresis thresholding method of Canny (1986). Here intensity gradients above a certain upper threshold are taken to give definite indication of edge positions, whereas those above a second, lower threshold are taken to indicate edges only if they are adjacent to positions that have already been accepted as edges (for a detailed analysis, see Section 7.1.1).

It may be thought that the Marr–Hildreth and related edge detectors do not run into these problems because they give edge contours that are necessarily connected. However, the result of using methods that force connectedness is that sometimes (e.g., when edges are diffuse, or of low contrast, so that image noise is an important factor) parts of a contour will lack meaning. Indeed, a contour may meander over such regions following noise rather than useful object boundaries. Furthermore, it is as well to note that the problem is not merely one of pulling low-level signals out of noise, but rather that sometimes no signal at all is present that could be enhanced to a meaningful level. One reason may be that the lighting is such as to give zero contrast (as, for example, when a cube is lit in such a way that two faces have equal brightness) and occlusion. Lack of spatial resolution can also cause problems by merging together several lines on an object.

It is now assumed that all of these problems have been overcome by sufficient care with the lighting scheme, appropriate digitization, and other means. It is also assumed that suitable thinning and linking algorithms have been applied so that all objects are outlined by connected unit-width boundary lines. At this stage, it should be possible (1) to *locate* the objects from the boundary image, (2) to *identify* and *orient* them accurately, and (3) to *scrutinize* them for shape and size defects.

7.1.1 *Hysteresis Thresholding*

The concept of hysteresis thresholding is a general one that can be applied in a range of applications, including both image and signal processing. The Schmitt trigger is a very widely used electronic circuit for converting a varying voltage into a pulsed (binary) waveform. In this latter case, there are two thresholds, and the input has to rise above the upper threshold before the output is allowed to switch on and has to fall below the lower threshold before the output is allowed to switch off. This gives considerable immunity against noise in the input waveform—far more than where the difference between the upper and lower switching thresholds is zero (the case of zero hysteresis), since then a small amount of noise can cause an undue amount of switching between the upper and lower output levels.

When the concept is applied in image processing, it is usually with regard to edge detection, in which case there is an analogous 1-D waveform to be negotiated around the boundary of an object—though as we shall see, some specifically 2-D complications arise. The basic rule is to threshold the edge at a high level and then to allow extension of the edge down to a lower level threshold, but only adjacent to points that have already been assigned edge status.

Figure 7.2 shows the results of making tests on the edge gradient image in Fig. 8.4*b*. Figures 7.2*a* and *b* show the result of thresholding at the upper and lower hysteresis levels, respectively, and Fig. 7.2*c* shows the result of hysteresis thresholding using these two levels. For comparison, Fig. 7.2*d* shows the effect of thresholding at a suitably chosen intermediate level. Note that isolated edge points within the object boundaries are ignored by hysteresis thresholding, though noise spurs can occur and are retained. We can envision the process of hysteresis thresholding in an edge image as the location of points that

1. form a superset of the upper threshold edge image.
2. form a subset of the lower threshold edge image.
3. form that subset of the lower threshold image which is connected to points in the upper threshold image via the usual rules of connectedness (Chapter 6).

Edge points survive only if they are seeded by points in the upper threshold image.

Although the result in Fig. 7.2*c* is better than in Fig. 7.2*d*, in that gaps in the boundaries are eliminated or reduced in length, in a few cases noise spurs are introduced. Nevertheless, the aim of hysteresis thresholding is to obtain a better balance between false positives and false negatives by exploiting connectedness in the object boundaries. Indeed, if managed correctly, the additional parameter will normally lead to a net (average) reduction in boundary pixel classification error. However, guidelines for selecting hysteresis thresholds are primarily the following:

1. Use a pair of hysteresis thresholds that provide immunity against the known range of noise levels.
2. Choose the lower threshold to limit the possible extent of noise spurs (in principle, the highest lower threshold whose subset contains *all* true boundary points).

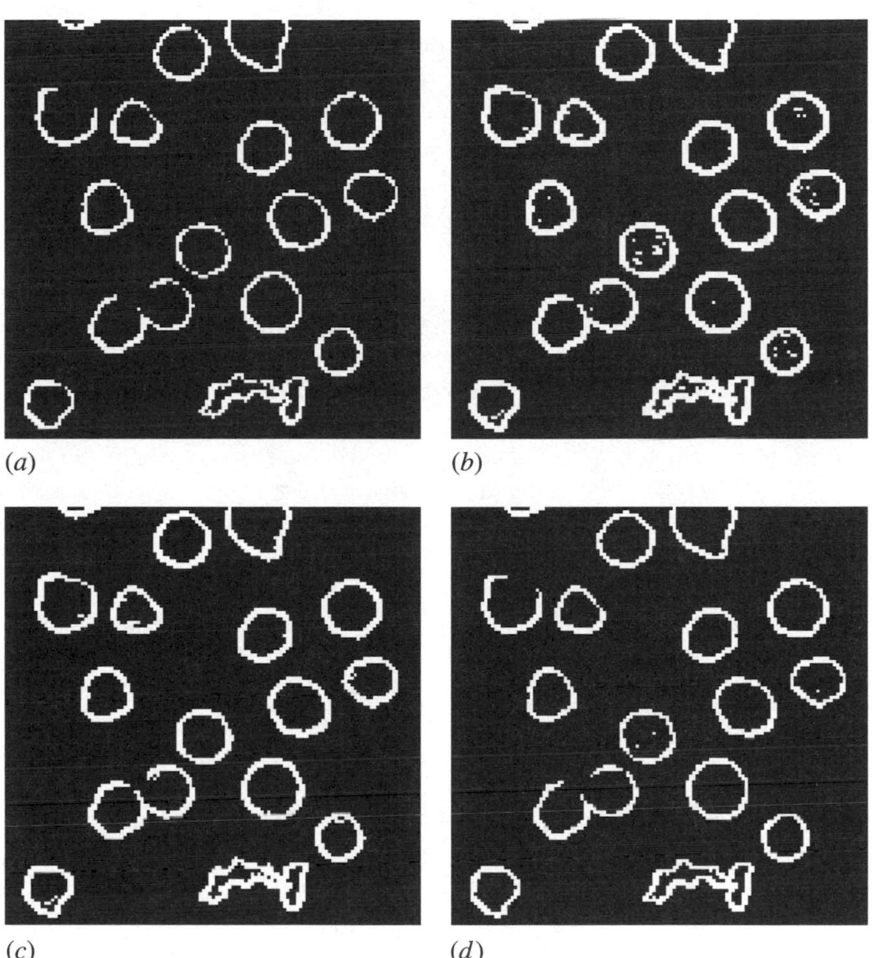

Figure 7.2 Effectiveness of hysteresis thresholding. This figure shows tests made on the edge gradient image of Fig. 8.4*b*. (*a*) Effect of thresholding at the upper hysteresis level. (*b*) Effect of thresholding at the lower hysteresis level. (*c*) Effect of hysteresis thresholding. (*d*) Effect of thresholding at an intermediate level.

3. Select the upper threshold to guarantee as far as possible the seeding of important boundary points (in principle, the highest upper threshold whose subset is connected to *all* true boundary points).

Unfortunately, in the limit of high signal variability, rules 2 and 3 appear to suggest eliminating hysteresis altogether! Ultimately, this means that the only rigorous way of treating the problem is to perform a complete statistical analysis of false positives and false negatives for a large number of images in any new application.

7.2 Boundary Tracking Procedures

Before objects can be matched from their boundary patterns, means must be found for tracking systematically around the boundaries of all the objects in an image. Means have already been demonstrated for achieving this in the case of regions such as those that result from intensity thresholding routines (Chapter 6). However, when a connected unit-width boundary exists, the problem of tracking is much simpler, since it is necessary only to move repeatedly to the next edge pixel storing the boundary information, for example, as a suitable pair of 1-D periodic coordinate arrays $x[i]$, $y[i]$. Note that each edge pixel must be marked appropriately as we proceed around the boundary, to ensure (1) that we never reverse direction, (2) that we know when we have been around the whole boundary once, and (3) that we record which object boundaries have been encountered. As when tracking around regions, we must ensure that in each case we end by passing through the starting point in the same direction.

7.3 Template Matching—A Reminder

Before considering boundary pattern analysis, recall the object location problem discussed in Chapter 1. With 2-D images there is a significant matching problem because of the number of degrees of freedom involved. Usually, there will be two degrees of freedom for position and one for orientation. For an object of size 30×30 pixels in a 256×256 image, matching the position in principle requires some 256×256 trials, each one involving 30×30 operations. Since the orientation is generally also unknown, another factor of 360 is required to cope with varying orientation (i.e., one trial per degree of rotation), hence making a total of $256^2 \times 30^2 \times 360$, that is, some 20,000 million operations. If more than one type of object is present, or if the objects have various possible scales, or if color variations have to be coped with,[1] the combinatorial explosion will be even more serious.

1 Although gray-scale and color variations are not considered in detail in this chapter, they do form part of the overall picture, and it will be clear that systematic means are needed to overcome the combinatorial explosion arising from the various possible degrees of freedom.

This is clearly an unacceptable situation, since even a modern computer could take several hours to locate objects within the image.

7.4 Centroidal Profiles

The matching problems outlined here make it attractive to attempt to locate objects in a smaller search space. This can be done simply by matching the boundary of each object in a single dimension. Perhaps the most obvious such scheme uses an (r, θ) plot. Here the centroid of the object is first located.[2] Then a polar coordinate system is set up relative to this point, and the object boundary is plotted as an (r, θ) graph—often called a centroidal profile (Fig. 7.3). Next, the 1-D graph so obtained is matched against the corresponding graph for an idealized object of the same type. Since the object in general has an arbitrary orientation, it is necessary to "slide" the idealized graph along that obtained from the image data until the best match is obtained. The match for each possible orientation α_j of the object is commonly tested by measuring the differences in radial distance between the boundary graph B and the template

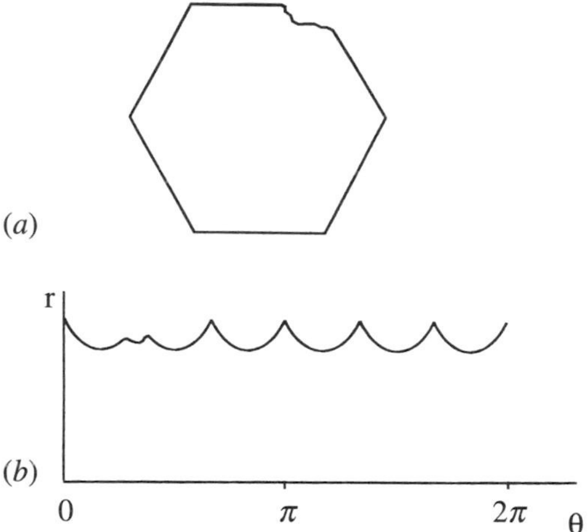

Figure 7.3 Centroidal profiles for object recognition and scrutiny: (*a*) hexagonal nut shape in which one corner has been damaged; (*b*) centroidal profile, which permits both straightforward identification of the object and detailed scrutiny of its shape.

2 Note that the position of the centroid is deducible directly from the list of boundary pixel coordinates: there is no need to start with a region-based description of the object for this purpose.

graph T for various values of θ and summing their squares to give a difference measure D_j for the quality of the fit:

$$D_j = \sum_i \left[r_B(\theta_i) - r_T(\theta_i + \alpha_j)\right]^2 \tag{7.1}$$

Alternatively, the absolute magnitudes of the differences are used:

$$D_j = \sum_i \left|r_B(\theta_i) - r_T(\theta_i + \alpha_j)\right| \tag{7.2}$$

The latter measure has the advantage of being easier to compute and of being less biased by extreme or erroneous difference values. This factor of 360 referred to above still affects the amount of computation, but at least the basic 2-D matching operation is reduced to 1-D. Interestingly, the scale of the problem is reduced in two ways, since both the idealized template and the image data are now 1-D. In the above example, the result is that the number of operations required to test each object drops to around 360^2 (which is only about 100,000), so there is a *very* substantial saving in computation.

The 1-D boundary pattern matching approach is able to identify objects as well as to find their orientations. Initial location of the centroid of the object also solves one other part of the problem as specified at the end of Section 7.1. At this stage it may be noted that the matching process also leads to the possibility of inspecting the object's shape as an inherent part of the process (Fig. 7.3). In principle, this combination of capabilities makes the centroidal profile technique quite powerful.

Finally, note that the method is able to cope with objects of identical shapes but different sizes. This is achieved by using the maximum value of r to normalize the profile, giving a variation (ρ, θ) where $\rho = r/r_{\max}$.

7.5 Problems with the Centroidal Profile Approach

In practice, the procedure we have outlined poses several problems. First, any major defect or occlusion of the object boundary can cause the centroid to be moved away from its true position, and the matching process will be largely spoiled (Fig. 7.4). Thus, instead of concluding that this is an object of type X with a specific part of its boundary damaged, the algorithm will most probably not recognize it at all. Such behavior would be inadequate in many automated inspection applications, where positive identification and fault-finding are required, and the object would have to be rejected without a satisfactory diagnosis being made.

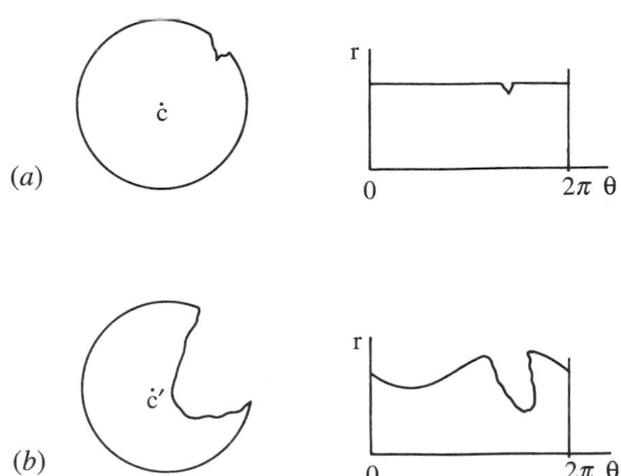

Figure 7.4 Effect of gross defects on the centroidal profile: (*a*) a chipped circular object and its centroidal profile; (*b*) a severely damaged circular object and its centroidal profile. The latter is distorted not only by the break in the circle but also by the shift in its centroid to C′.

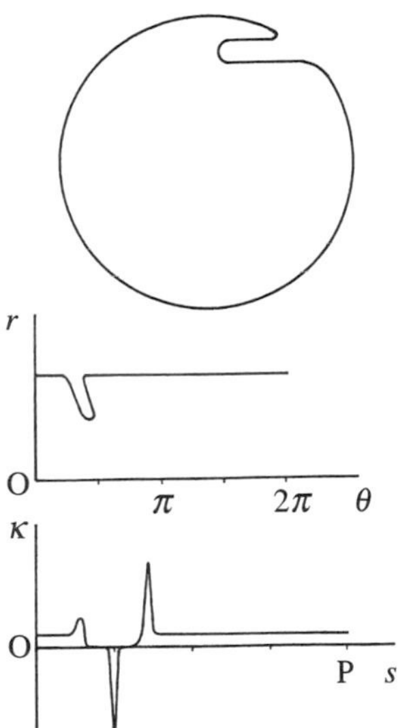

Figure 7.5 Boundary pattern analysis via (r, θ) and (s, κ) plots.

Second, the (r, θ) plot will be multivalued for a certain class of object (Fig. 7.5). This has the effect of making the matching process partly 2-D and leads to complication and excessive computation.

Third, the very variable spacing of the pixels when plotted in (r, θ) space is a source of complication. It leads to the requirement for considerable smoothing

Figure 7.6 A problem in obtaining a centroidal profile for elongated objects. This figure highlights the pixels around the boundary of an elongated object—a spanner—showing that it will be difficult to obtain an accurate centroidal profile for the region near the centroid.

of the 1-D plots, especially in regions where the boundary comes close to the centroid—as for elongated objects such as spanners or screwdrivers (Fig. 7.6). However, in other places accuracy will be greater than necessary, and the overall process will be wasteful. The problem arises because quantization should ideally be uniform along the θ-axis so that the two templates can be moved conveniently relative to one another to find the orientation of best match.

Finally, computation times can still be quite significant, so some timesaving procedure is required.

7.5.1 *Some Solutions*

All four of the above problems can be tackled in one way or another, with varying degrees of success. The first problem, that of coping with occlusions and gross defects, is probably the most fundamental and the most resistant to satisfactory solution. For the method to work successfully, a stable reference point must be found within the object. The centroid is naturally a good candidate for this since the averaging inherent in its location tends to eliminate most forms of noise or minor defect. However, major distortions such as those arising from breakages or occlusions are bound to affect it adversely. The centroid of the boundary is no better and may also be less successful at suppressing noise. Other possible candidates are the positions of prominent features such as corners, holes, and centers of arcs. In general, the smaller such a feature is the more likely it is to be missed altogether in the event of a breakage or occlusion, although the larger such a feature is the more likely it is to be affected by the defect.

Circular arcs can be located accurately (at their centers) even if they are partly occluded (see Chapter 10), so these features are very useful for leading to suitable reference points. A set of symmetrically placed holes may sometimes be suitable, since even if one of them is obscured, one of the others is likely to be visible and can act as a reference point.

Such features can help the method to work adequately, but their presence also calls into question the value of the 1-D boundary pattern matching procedure, since they make it likely that superior methods can be used for object recognition (see Parts 2 and 3). For the present we therefore accept (1) that severe complications arise when part of an object is missing or occluded, and (2) that it may be possible to provide some degree of resistance to such problems by using a prominent feature as a reference point instead of the centroid. Indeed, the only significant *further* change that is required to cope with occlusions is that difference $(r_B - r_T)$ values of greater than (say) 3 pixels should be ignored, and the best match then becomes one for which the greatest number of values of θ give good agreement between B and T.

The second problem, of multivalued (r, θ) plots, is solved very simply by employing the heuristic of taking the smallest value of r for any given θ and then proceeding with matching as normal. (Here it is assumed that the boundaries of any holes present within the boundary of the object are dealt with separately, information about any object and its holes being collated at the end of the recognition process.) This ad hoc procedure should in fact be acceptable when making a preliminary match of the object to its 1-D template and may be discarded at a later stage when the orientation of the object is known accurately.

The third problem described above arises because of uneven spacing of the pixel boundaries along the θ dimension of the (r, θ) graph. To some extent, this problem can be avoided by deciding in advance on the permissible values of θ and querying a list of boundary points to find which has the closest θ to each permissible value. Some local smoothing of the ordered set of boundary points can be undertaken, but this is in principle unnecessary, since for a connected boundary, there will always be one pixel that is closest to a line from the centroid at a given value of θ.

The two-stage approach to matching can also be used to help with the last of the problems mentioned in Section 7.5—the need to speed up the processing. First, a coarse match is obtained between the object and its 1-D template by taking θ in relatively large steps of (say) 5° and ignoring intermediate angles in both the image data and the template. Then a better match is obtained by making fine adjustments to the orientations, obtaining a match to within 1°. In this way, the coarse match is obtained perhaps 20 times faster than the previous full match, whereas the final fine match takes a relatively short time, since very few distinct orientations have to be tested.

This two-stage process can be optimized by making a few simple calculations. The coarse match is given by steps $\delta\theta$, so the computational load is proportional to $(360/\delta\theta)^2$, whereas the load for the fine match is proportional to $360\,\delta\theta$, giving a total load of:

$$\lambda = (360/\delta\theta)^2 + 360\,\delta\theta \tag{7.3}$$

This should be compared with the original load of:

$$\lambda_0 = 360^2 \tag{7.4}$$

Hence, the load is reduced (and the algorithm speeded up) by the factor:

$$\eta = \lambda_0/\lambda = 1/\left[(1/\delta\theta)^2 + \delta\theta/360\right] \tag{7.5}$$

This is a maximum for $\mathrm{d}\eta/\mathrm{d}\delta\theta = 0$, giving:

$$\delta\theta = \sqrt[3]{2 \times 360} \approx 9^\circ \tag{7.6}$$

In practice, this value of $\delta\theta$ is rather large, and there is a risk that the coarse match will give such a poor fit that the object will not be identified. Hence values of $\delta\theta$ in the range 2–5° are more usual (see, for example, Berman et al., 1985). Note that the optimum value of η is 26.8 and that this reduces only to 18.6 for $\delta\theta = 5^\circ$, although it goes down to 3.9 for $\delta\theta = 2^\circ$.

Another way of approaching the problem is to search the (r, θ) graph for some characteristic feature such as a sharp corner (this step constituting the coarse match) and then to perform a fine match around the object orientation so deduced. In situations where objects have several similar features—as in the case of a rectangle—error may occur. However, the individual trials are relatively inexpensive, and so it is worth invoking this procedure if the object possesses appropriate well-defined features. Note that it is possible to use the position of the maximum value, $r_{\max}$, as an orientating feature, but this is frequently inappropriate because a smooth maximum gives a relatively large angular error.

7.6 The (s, ψ) Plot

As can be seen from the above considerations, boundary pattern analysis should usually be practicable except when problems from occlusions and gross defects can be expected. These latter problems do give motivation, however, for employing alternative methods if these can be found. The (s, ψ) graph has proved especially popular since it is inherently better suited than the (r, θ) graph to situations where defects and occlusions may occur. In addition, it does not suffer from the multiple values encountered by the (r, θ) method.

The (s, ψ) graph does not require prior estimation of the centroid or some other reference point since it is computed directly from the boundary, in the form of a plot of the tangential orientation ψ as a function of boundary distance s. The method is not without its problems, and, in particular, distance along the boundary needs to be measured accurately. The commonly used approach is to count horizontal and vertical steps as unit distance, and to take diagonal steps as distance $\sqrt{2}$. This idea must be regarded as a rather ad hoc solution; the situation is discussed further in Section 7.10.

When considering application of the (s, ψ) graph for object recognition, it will immediately be noticed that the graph has a ψ value that increases by 2π for each circuit of the boundary; that is, $\psi(s)$ is not periodic in s. The result is that the graph becomes essentially 2-D; that is, the shape has to be matched by moving the ideal object template both along the s-axis and along the ψ-axis directions. Ideally, the template could be moved diagonally along the direction of the graph. However, noise and other deviations of the actual shape relative to the ideal shape mean that in practice the match must be at least partly 2-D, hence adding to the computational load.

One way of tackling this problem is to make a comparison with the shape of a circle of the same boundary length P. Thus, an $(s, \Delta\psi)$ graph is plotted, which reflects the difference $\Delta\psi$ between the ψ expected for the shape and that expected for a circle of the same perimeter:

$$\Delta\psi = \psi - 2\pi s/P \tag{7.7}$$

This expression helps to keep the graph 1-D, since $\Delta\psi$ automatically resets itself to its initial value after one circuit of the boundary (i.e., $\Delta\psi$ is periodic in s).

Next it should be noticed that the $\Delta\psi(s)$ variation depends on the starting position where $s = 0$ and that this is randomly sited on the boundary. It is useful to eliminate this dependence, which may be done by subtracting from $\Delta\psi$ its mean value μ. This gives the new variable:

$$\tilde{\psi} = \psi - 2\pi s/P - \mu \tag{7.8}$$

At this stage, the graph is completely 1-D and is also periodic, being similar in these respects to an (r, θ) graph. Matching should now reduce to the straightforward task of sliding the template along the $\tilde{\psi}(s)$ graph until a good fit is achieved.

At this point, there should be no problems as long as (1) the scale of the object is known and (2) occlusions or other disturbances cannot occur. Suppose next that the scale is unknown; then the perimeter P may be used to normalize the value of s. If, however, occlusions *can* occur, then no reliance can be placed on P for normalizing s, and hence the method cannot be guaranteed to work. This problem does not arise if the scale of the object is known, since a standard perimeter P_{T} can be assumed. However, the possibility of occlusion gives further problems that are discussed in the next section.

Another way in which the problem of nonperiodic $\psi(s)$ can be solved is by replacing ψ by its derivative $\mathrm{d}\psi/\mathrm{d}s$. Then the problem of constantly expanding ψ (which results in its increasing by 2π after each circuit of the boundary) is eliminated—the addition of 2π to ψ does not affect $\mathrm{d}\psi/\mathrm{d}s$ locally, since $\mathrm{d}(\psi+2\pi)/\mathrm{d}s = \mathrm{d}\psi/\mathrm{d}s$. It should be noticed that $\mathrm{d}\psi/\mathrm{d}s$ is actually the local curvature function $\kappa(s)$ (see Fig. 7.5), so the resulting graph has a simple physical interpretation. Unfortunately, this version of the method has its own problems in that κ approaches infinity at any sharp corner. For industrial components, which frequently have sharp corners, this is a genuine practical difficulty and may be tackled by approximating adjacent gradients and ensuring that κ integrates to the correct value in the region of a corner (Hall, 1979).

Many workers take the (s, κ) graph idea further and expand $\kappa(s)$ as a Fourier series:

$$\kappa(s) = \sum_{n=-\infty}^{\infty} c_n \exp(2\pi ins/P) \tag{7.9}$$

This results in the well-known Fourier descriptor method. In this method, shapes are analyzed in terms of a series of Fourier descriptor components that are truncated to zero after a sufficient number of terms. Unfortunately, the amount of computation involved in this approach is considerable, and the tendency is to approximate curves with relatively few terms. In industrial applications where computations have to be performed in real time, this can give problems, so it is often more appropriate to match to the basic (s, κ) graph. In this way, critical measurements between object features can be made with adequate accuracy in real time.

7.7 Tackling the Problems of Occlusion

Whatever means are used for tackling the problem of continuously increasing ψ, problems still arise when occlusions occur. However, the whole method is not immediately invalidated by missing sections of boundary as it is for the basic (r, θ) method. As noted above, the first effect of occlusions is that the perimeter of the object is altered, so P can no longer be used to indicate its scale. Hence, the scale has to be known in advance: this is assumed in what follows. Another practical result of occlusions is that certain sections correspond correctly to parts of the object, whereas other parts correspond to parts of occluding objects. Alternatively, they may correspond to unpredictable boundary segments where damage has occurred. Note that if the overall boundary is that of two overlapping objects, the observed perimeter P_B will be greater than the ideal perimeter P_T.

Segmenting the boundary between relevant and irrelevant sections is, *a priori*, a difficult task. However, a useful strategy is to start by making positive matches wherever possible and to ignore irrelevant sections—that is, try to match as usual, ignoring any section of the boundary that is a bad fit. We can imagine achieving a match by sliding the template T along the boundary B. However, a problem arises since T is periodic in s and should not be cut off at the ends of the range $0 \leq s \leq P_T$. As a result, it is necessary to attempt to match over a length $2P_T$. At first sight, it might be thought that the situation ought to be symmetrical between B and T. However, T is known in advance, whereas B is partly unknown, there being the possibility of one or more breaks in the ideal boundary into which foreign boundary segments have been included. Indeed, the positions of the breaks are unknown, so it is necessary to try matching the whole of T at all positions on B. Taking a length $2P_T$ in testing for a match effectively permits the required break to arise in T at any relevant position (see Fig. 7.7).

When carrying out the match, we basically use the difference measure:

$$D_{jk} = \sum_i \left[\psi_B(s_i) - \left(\psi_T(s_i + s_k) + \alpha_j\right)\right]^2 \tag{7.10}$$

where j and k are the match parameters for orientation and boundary displacement, respectively. Notice that the resulting D_{jk} is roughly proportional to the length L of the boundary over which the fit is reasonable. Unfortunately, this means that the measure D_{jk} appears to *improve* as L decreases. Hence, when variable occlusions can occur, the best match must be taken as the one for which the greatest length L gives good agreement between B and T. (This may be measured as the greatest number of values of s in the sum of equation (7.10) which give good agreement between B and T, that is, the sum over all i such that the difference in square brackets is numerically less than, say, 5°.)

If the boundary is occluded in more than one place, then L is at most the largest single length of unoccluded boundary (not the total length of unoccluded boundary), since the separate segments will in general be "out of phase" with the template. This is a disadvantage when trying to obtain an accurate result, since extraneous matches add noise that degrades the fit that is obtainable—hence adding to the risk that the object will not be identified and reducing accuracy of registration. This suggests that it might be better to use only short sections of the boundary template for matching. Indeed, this

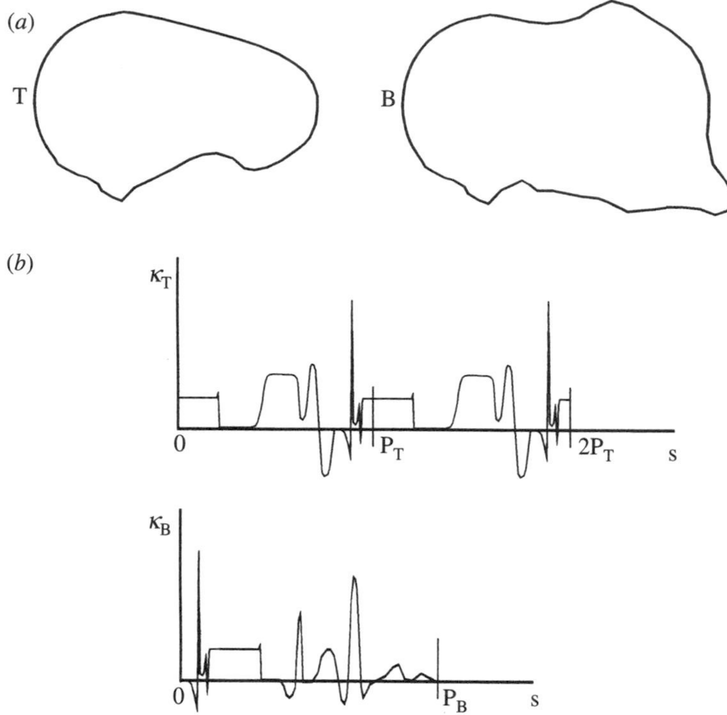

Figure 7.7 Matching a template against a distorted boundary. When a boundary B is broken (or partly occluded) but continuous, it is necessary to attempt to match between B and a template T that is doubled to length $2P_T$, to allow for T being severed at any point: (*a*) the basic problem; (*b*) matching in (s, κ)-space.

strategy can be advantageous since speed is enhanced and registration accuracy can be retained or even improved by careful selection of salient features. (Note that nonsalient features such as smooth curved segments could have originated from many places on object boundaries and are not very helpful for identifying and accurately locating objects; hence it is reasonable to ignore them.) In this version of the method, we now have $P_T < P_B$, and it is necessary to match over a length P_T rather than $2P_T$, since T is no longer periodic (Fig. 7.8). Once various segments have been located, the boundary can be reassembled as closely as possible into its full form, and at that stage defects, occlusions, and other distortions can be recognized overtly and recorded. Reassembly of the object boundaries can be performed by techniques such as the Hough transform and relational pattern matching techniques (see Chapters 11 and 15). Work of this type has been carried out by Turney et al. (1985), who found that the salient features should be short boundary segments where corners and other distinctive "kinks" occur.

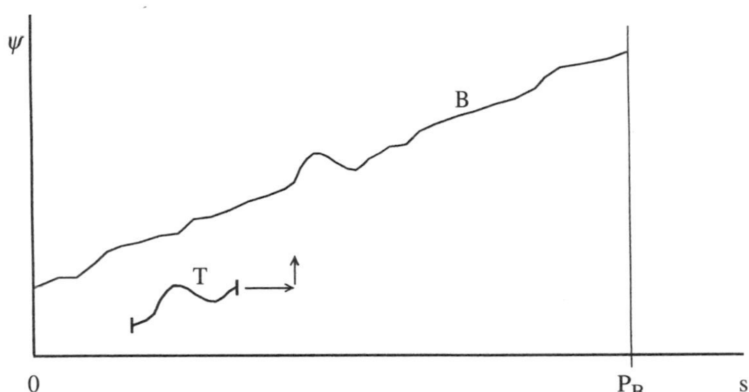

Figure 7.8 Matching a short template to part of a boundary. A short template T, corresponding to part of an idealized boundary, is matched against the observed boundary B. Strictly speaking, matching in (s, ψ)-space is 2-D, although there is very little uncertainty in the vertical (orientation) dimension.

It should be noted that $\tilde{\psi}$ can no longer be used when occlusions are present, since although the perimeter can be assumed to be known, the mean value of $\Delta\psi$ (equation (7.8)) cannot be deduced. Hence, the matching task reverts to a 2-D search (though, as stated earlier, very little unrestrained search in the ψ direction need be made). However, when small salient features are being sought, it is a reasonable working assumption that no occlusion occurs in any of the individual cases—a feature is either entirely present or entirely absent. Hence, the average slope $\bar{\psi}$ over T can validly be computed (Fig. 7.8), and this again reduces the search to 1-D (Turney et al., 1985).

Overall, missing sections of object boundaries necessitate a fundamental rethink as to how boundary pattern analysis should be carried out. For quite small defects, the (r, θ) method is sufficiently robust, but in less trivial cases it is vital to use some form of the (s, ψ) approach, while for really gross occlusions it is not particularly useful to try to match for the full boundary. Rather, it is better to attempt to match small salient features. This sets the scene for the Hough transform and relational pattern matching techniques of later chapters (see Parts 2 and 3).

7.8 Chain Code

This section briefly describes the chain code devised in 1961 by Herbert Freeman. This method recodes the boundary of a region into a sequence of octal digits, each one representing a boundary step in one of the eight basic directions relative to the current pixel position, denoted here by an asterisk:

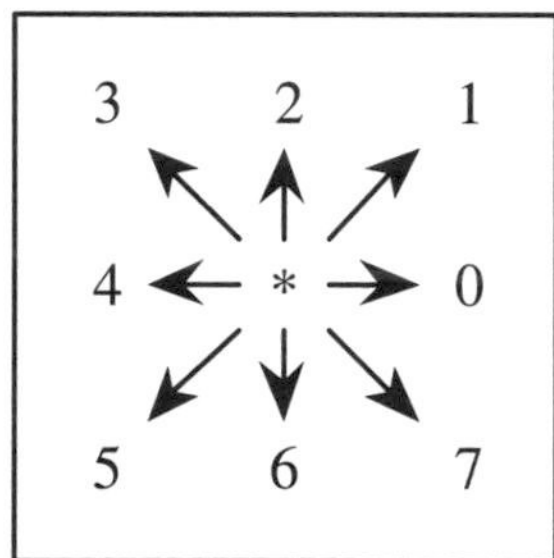

This code is important for presenting shape information in such a way as to save computer storage. In particular, note that the code ignores (and therefore does not store) the absolute position of the boundary pixels, so that only three bits are required for each boundary step. However, in order to regenerate a digital image, additional information is required, including the starting point of the chain and the length.

Chain code is useful in cutting down storage relative to the number of bits required to hold a list of the coordinates of all the boundary pixels (and *a fortiori* relative to the number of bits required to hold a whole binary image). In addition, chain code is able to provide data in a format that is suitable for certain recognition processes. However, chain coding may sometimes be too compressed a representation for convenient computer manipulation of the data. Hence, in many cases direct (x,y) storage may be more suitable, or otherwise direct (s, ψ) coding. Note also that with modern computers, the advantage of reduced storage requirements is becoming less relevant than in the early days of pattern recognition.

7.9 The (*r*, *s*) Plot

Before leaving the topic of 1-D plots for analysis and recognition of 2-D shapes, one further method should be mentioned—the (r, s) plot. Freeman (1978) introduced this method as a natural follow-up from the use of chain code (which necessarily involves the parameter s). Since quite a large family of possible 1-D plot representations exist, it is useful to demonstrate the particular advantages of any new one. The (r, s) plot is in many ways similar to the (r, θ) plot (particularly in using the centroid as a reference point) but it overcomes the difficulty of the (r, θ) plot sometimes being a multivalued function: for every value of s there is a unique value of r. However, against this advantage must be weighed the fact that it is not in general possible to recover the original shape from the 1-D plot, since an ambiguity is introduced whenever the tangent to the boundary passes through the centroid. To overcome this problem, a phase function $\Phi(s)$ (whose value is always either $+1$ or -1, the minus value corresponding to the curve turning back on itself) must be stored in addition to the function $r(s)$. Freeman made a slight variation on this scheme by relying on the fact that r is always positive, so that all the information can be stored in a single function that is the product of r and Φ. Note, however, that the derived function is no longer continuous, so minor changes in shape could lead to misleadingly large changes in cross-correlation coefficient during comparison with a template.

7.10 Accuracy of Boundary Length Measures

Next we examine the accuracy of the idea expressed earlier—that adjacent pixels on an 8-connected curve should be regarded as separated by 1 pixel if the vector joining them is aligned along the major axes, and by $\sqrt{2}$ pixels if the vector is in a diagonal orientation. It turns out that this estimator in general overestimates the distance along the boundary. The reason is quite simply seen by appealing to the following pair of situations:

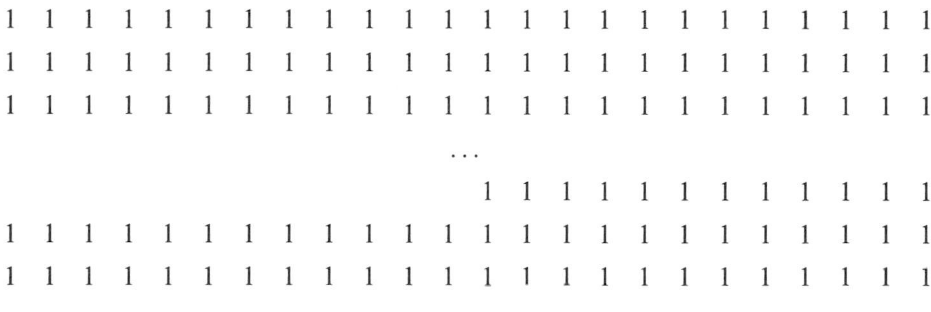

In either case, we are considering only the top of the object. In the first example, the boundary length along the top of the object is exactly that given by the rule.

However, in the second case the estimated length is increased by amount $\sqrt{2}-1$ because of the step. Now as the length of the top of the object tends to large values, say p pixels, the actual length approximates to p while the estimated length is $p+\sqrt{2}-1$. Thus, a definite error exists. Indeed, this error initially increases in importance as p decreases, since the actual length of the top of the object (when there is one step) is still

$$L = \left(1+p^2\right)^{1/2} \approx p \tag{7.11}$$

so the fractional error is

$$\xi \approx \left(\sqrt{2}-1\right)/p \tag{7.12}$$

which increases as p becomes smaller.

This result can be construed as meaning that the fractional error ξ in estimating boundary length increases initially as the boundary orientation ψ increases from zero. A similar effect occurs as the orientation decreases from 45°. Thus, the ξ variation possesses a maximum between 0° and 45°. This systematic overestimation of boundary length may be eliminated by employing an improved model in which the length per pixel is s_m along the major axes directions and s_d in diagonal directions. A complete calculation (Kulpa, 1977; see also Dorst and Smeulders, 1987) shows that:

$$s_m = 0.948 \tag{7.13}$$

and

$$s_d = 1.343 \tag{7.14}$$

It is perhaps surprising that this solution corresponds to a ratio s_d/s_m that is still equal to $\sqrt{2}$, although the arguments given above make it obvious that s_m should be less than unity.

Unfortunately, an estimator that has just two free parameters can still permit quite large errors in estimating the perimeters of individual objects. To reduce this problem, it is necessary to perform more detailed modeling of the step pattern around the boundary (Koplowitz and Bruckstein, 1989), which seems certain to increase the computational load significantly.

It is important to underline that the basis of this work is to estimate the length of the original continuous boundary rather than that of the digitized boundary. Furthermore, it must be noted that the digitization process loses information, so the best that can be done is to obtain a best estimate of the original boundary length. Thus, employing the values 0.948, 1.343 given above, rather than the values 1, $\sqrt{2}$, reduces the estimated errors in boundary length measurement from 6.6% to 2.3%—but then only under certain assumptions about correlations between orientations at neighboring boundary pixels (Dorst and Smeulders, 1987). For a more detailed insight into the situation, see Davies (1991c).

7.11 Concluding Remarks

The boundary patterns analyzed in this chapter were imagined to arise from edge detection operations, which have been processed to make them connected and of unit width. However, if intensity thresholding methods were employed for segmenting images, boundary tracking procedures would also permit the boundary pattern analysis methods of this chapter to be used. Conversely, if edge detection operations led to the production of connected boundaries, these could be filled in by suitable algorithms (which are more tricky to devise than might at first be imagined!) (Ali and Burge, 1988) and converted to regions to which the binary shape analysis methods of Chapter 6 could be applied. Hence, shapes are representable in region or boundary form. If they initially arise in one representation, they may be converted to the alternative representation. This means that boundary or regional means may be employed for shape analysis, as appropriate.

It might be thought that if shape information is available initially in one representation, the method of shape analysis that is employed should be one that is expressed in the same representation. However, this conclusion is dubious, since the routines required for conversion between representations (in either direction) require relatively modest amounts of computation. This means that it is important to cast the net quite widely in selecting the best available means for shape analysis and object recognition.

An important factor here is that a positive advantage is often gained by employing boundary pattern analysis, since computation should be inherently lower (in proportion to the number of pixels required to describe the shapes in the two representations). Another important determining factor seen in the present chapter is that of occlusion. If occlusions are present, then many of the methods described in Chapter 6 operate incorrectly—as also happens for the basic centroidal profile method described early in this chapter. The (s, ψ) method then provides a good starting point. As has been seen, this is best applied to detect small salient boundary features, which can then be reassembled into whole objects by relational pattern matching techniques (see especially Chapter 15).

A variety of boundary representations are available for shape analysis. However, this chapter has shown that intuitive schemes raise fundamental robustness issues—issues indeed that will only be resolved later on by forgoing deduction in favor of inference. *Underlying* analog shape estimation in a digital lattice is also an issue.

7.12 Bibliographical and Historical Notes

Many of the techniques described in this chapter have been known since the early days of image analysis. Boundary tracking has been known since 1961 when Freeman introduced his chain code. Indeed, Freeman is responsible for much subsequent work in this area (see, for example, Freeman, 1974). Freeman (1978) introduced the notion of segmenting boundaries at "critical points" in order to facilitate matching. Suitable critical points were corners (discontinuities in curvature), points of inflection, and curvature maxima. This work is clearly strongly related to that of Turney et al. (1985). Other workers have segmented boundaries into piecewise linear sections as a preliminary to more detailed pattern analysis (Dhome et al., 1983; Rives et al., 1985), since this procedure reduces information and hence saves computation (albeit at the expense of accuracy). Early work on Fourier boundary descriptors using the (r, θ) and (s, ψ) approaches was carried out by Rutovitz (1970), Barrow and Popplestone (1971), and Zahn and Roskies (1972). Another notable paper in this area is that by Persoon and Fu (1977). In an interesting development, Lin and Chellappa (1987) were able to classify partial (i.e., nonclosed) 2-D curves using Fourier descriptors.

As noted, there are significant problems in obtaining a thin connected boundary for every object in an image. Since 1988, the concept of active contour models (or "snakes") has acquired a large following and appears to solve many of these problems. It operates by growing an active contour that slithers around the search area until it ends in a minimum energy situation (Kass et al., 1988). See Section 18.12 for a brief introduction to snakes and their application to vehicle location, together with further references.

There has been increased attention to accuracy over the past 20 years or so. This is seen, for example, in the length estimators for digitized boundaries discussed in Section 7.10 (see Kulpa, 1977; Dorst and Smeulders, 1987; Beckers and Smeulders, 1989; Koplowitz and Bruckstein, 1989; Davies, 1991c). For a recent update on the topic, see Coeurjolly and Klette (2004).

In recent times, there has been an emphasis on characterizing and classifying families of shapes rather than just individual isolated shapes. See in particular Cootes et al. (1992), Amit (2002), and Jacinto et al. (2003). Klassen et al. (2004) provide a further example of this in their analysis of planar (boundary) shapes using geodesic paths between the various shapes of the family. In their work they employ the Surrey fish database (Mokhtarian et al., 1996). The same general idea is also manifest in the self-similarity analysis and matching approach of Geiger et al. (2003), which they used for human profile and hand motion analysis. Horng (2003) describes an adaptive smoothing approach for fitting digital planar curves with line segments and circular arcs. The motivation for this approach is to obtain significantly greater accuracy than can be achieved with the widely used polygonal approximation, yet with lower computational load than the spline-fitting type of approach. It can also be imagined that any fine accuracy restriction imposed by a line plus circular arc model will have little relevance in a discrete lattice of pixels. da Gama Leitão and Stolfi (2002) have developed a multiscale contour-based method for matching and reassembling 2-D fragmented objects. Although this method is targeted at reassembly of pottery fragments in archaeology, da Gama Leitão and Stolfi imply that it is also likely to be of value in forensic science, in art conservation, and in assessment of the causes of failure of mechanical parts following fatigue and the like.

Finally, two recent books have appeared which cover the subject of shape and shape analysis in rather different ways. One is by Costa and Cesar (2000), and the other is by Mokhtarian and Bober (2003). The Costa–Cesar work is fairly general in coverage, but emphasizes Fourier methods, wavelets, and multiscale methods. The Mokhtarian-Bober work sets up a scale-space (especially curvature scale space) representation (which is multiscale in nature), and develops the subject quite widely from there. With regard to representations, which abound in the area of shape analysis, note that they can be very powerful indeed, though they may also embody restrictions that do not apply for other representations.

7.13 Problems

1. Devise a program for finding the centroid of a region, starting from an ordered list of the coordinates of its boundary points.

2. Devise a program for finding a thinned (8-connected) boundary of an object in a binary image.

3. Determine the saving in storage that results when the boundaries of objects that are held as lists of x and y coordinates are reexpressed in chain code. Assume that the

image size is 256×256 and that a typical boundary length is 256. How would the result vary (i) with boundary length and (ii) with image size?

4. (a) Describe the centroidal profile approach to shape analysis. Illustrate your answer for a circle, a square, a triangle, and defective versions of these shapes.

   (b) Obtain a general formula expressing the shape a straight line presents in the centroidal profile.

   (c) Show that there are two means of recognizing objects from their centroidal profiles, one involving analysis of the profile and the other involving comparison with a template.

   (d) Show how the latter approach can be speeded up by implementing it in two stages, first at low resolution and then at full resolution. If the low resolution has $1/n$ of the detail of the full resolution, obtain a formula for the total computational load. Estimate from the formula for what value of n the load will be minimized. Assume that the full angular resolution involves 360 one-degree steps.

5. (a) Give a simple algorithm for eliminating salt and pepper noise in binary images, and show how it can be extended to eliminate short spurs on objects.

   (b) Show that a similar effect can be achieved by a "shrink" + "expand" type of procedure. Discuss how much such procedures affect the shapes of objects. Give examples illustrating your arguments, and try to quantify exactly what sizes and shapes of object would be completely eliminated by such procedures.

   (c) Describe the (r, θ) graph method for describing the shapes of objects. Show that applying 1-D median filtering operations to such graphs can be used to smooth the described object shapes. Would you expect this approach to be more or less effective at smoothing object boundaries than methods based on shrinking and expanding?

6. (a) Outline the (r, θ) graph method for recognizing objects in two dimensions, and state its main advantages and limitations. Describe the shape of the (r, θ) graph for an equilateral triangle.

   (b) Write down a complete algorithm, operating in a 3×3 window, for producing an approximation to the convex hull of a 2-D object. Show that a more accurate approximation to the convex hull can be obtained by joining humps with straight lines in an (r, θ) graph of the object. Give reasons why the result for the latter case will only be an approximation, and suggest how an exact convex hull might be obtained.

7. An alternative approach to shape analysis involves measuring distance around the boundary of any object and estimating increments of distance as 1 unit when progressing to the next pixel in a horizontal or vertical direction and $\sqrt{2}$ units when progressing in a diagonal direction. Taking a square of side 20 pixels, which is aligned parallel to the image axes, and rotating it through small angles, show that distance around the boundary of the square is not estimated accurately by the $1{:}\sqrt{2}$ model. Show that a similar effect occurs when the square is orientated at about 45° to the image axes. Suggest ways in which this problem might be tackled.

Mathematical Morphology

Historically, the study of shape took place over a long period of time and resulted in a highly variegated set of algorithms and methods. Over the past 20 years the formalism of mathematical morphology was set up and provided a background theory into which many of the individual advances could be slotted. This chapter takes a journey through this interesting subject but aims to steer an intuitive path between the many mathematical theorems, concentrating particularly on finding practically useful results.

Look out for:

- how the concepts of expanding and shrinking are transformed into the more general concepts of dilation and erosion.
- how dilation and erosion operations may be combined to form more complex operations whose properties may be predicted mathematically.
- how the concepts of closing and opening are defined, and how they are used to find defects in binary object shapes, via residue (or "top-hat") operations.
- the hit-and-miss transform and its applications.
- how thinning is brought into the formalism.
- how mathematical morphology is generalized to cover gray-scale processing.
- how noise affects morphological grouping operations.

This chapter on morphology follows last in Part 1 of this volume because, being rather mathematical, it could demotivate inexperienced readers if placed earlier. In any case, the theory is largely valuable by the way in which it integrates the other topics. Nevertheless, once the methods have been learned, morphology should be of distinct value in taking the earlier ideas forward and optimizing any algorithms that use them.

CHAPTER 8

Mathematical Morphology

8.1 Introduction

In Chapter 2 we met the operations of erosion and dilation, and in Chapter 6 we applied them to the filtering of binary images, showing that with suitable combinations of these operators it is possible to eliminate certain types of objects from images and to locate other objects. These possibilities are not fortuitous, but on the contrary reflect important mathematical properties of shape that are dealt with in the subject known as mathematical morphology. This subject has grown up over the past two or three decades, and over the past few years knowledge in this area has become consolidated and is now understood in considerable depth. This chapter gives some insight into this important area of study. The mathematical nature of the topic mitigated against including the material in the earlier shape analysis chapters, where it might have impeded the flow of the text and at the same time put off less mathematically minded readers. However, these considerations aside, mathematical morphology is not a mere sideline. On the contrary, it is highly important, for it provides a backbone for the whole study of shape, that is capable of unifying techniques as disparate as noise suppression, shape analysis, feature recognition, skeletonization, convex hull formation, and a host of other topics.

Section 8.2 starts the discussion by extending the concepts of expanding and shrinking first encountered in Section 2.2. Section 8.3 then develops the theory of mathematical morphology, arriving at many important results—with emphasis deliberately being placed on understanding of concepts rather than on mathematical rigor. Section 8.4 shows how mathematical morphology is able to cover connectedness aspects of shape. Section 8.5 goes on to show how morphology can be generalized to cope with gray-scale images. Also included in the chapter is a discussion (Section 8.6) of the noise behavior of morphological grouping operations, arriving at a formula explaining the shifts introduced by noise.

8.2 Dilation and Erosion in Binary Images

8.2.1 *Dilation and Erosion*

As we have seen in Chapter 2, dilation expands objects into the background and is able to eliminate "salt" noise within an object. It can also be used to remove cracks in objects which are less than three pixels in width.

In contrast, erosion shrinks binary picture objects and has the effect of removing "pepper" noise. It also removes thin object "hairs" whose widths are less than 3 pixels.

As we shall see in more detail, erosion is strongly related to dilation in that a dilation acting on the inverted input image acts as an erosion, and vice versa.

8.2.2 *Cancellation Effects*

An obvious question is whether erosions cancel out dilations, or vice versa. We can easily answer this question. If a dilation has been carried out, salt noise and cracks will have been removed, and once they are gone, erosion cannot bring them back; hence, exact cancellation will not in general occur. Thus, for the set S of object pixels in a general image I, we may write:

$$\text{erode}(\text{dilate}(S)) \neq S \tag{8.1}$$

equality only occurring for certain specific types of images. (These will lack salt noise, cracks, and fine boundary detail.) Similarly, pepper noise or hairs that are eliminated by erosion will not in general be restored by dilation:

$$\text{dilate}(\text{erode}(S)) \neq S \tag{8.2}$$

Overall, the most general statements that can be made are:

$$\text{erode}(\text{dilate}(S)) \supseteq S \tag{8.3}$$

$$\text{dilate}(\text{erode}(S)) \subseteq S \tag{8.4}$$

We may note, however, that large objects will be made 1 pixel larger all around by dilation and will be reduced by 1 pixel all around by erosion. Therefore, a considerable amount of cancellation will normally take place when the two operations are applied in sequence. This means that sequences of erosions and dilations provide a good basis for filtering noise and unwanted detail from images.

8.2.3 Modified Dilation and Erosion Operators

Images sometimes contain structures that are aligned more or less along the image axes, directions; in such cases it is useful to be able to process these structures differently. For example, it might be useful to eliminate fine vertical lines, without altering broad horizontal strips. In that case the following "vertical erosion" operator will be useful:

$$\text{ERODE: [[sigma} = \text{A1} + \text{A5;}$$
$$\text{if(sigma} < 2)\ \text{B0} = 0;\ \text{else B0} = \text{A0;]];} \tag{8.5}$$

though it will be necessary to follow it with a compensating dilation operator[1] so that horizontal strips are not shortened:

$$\text{DILATE: [[sigma} = \text{A1} + \text{A5;}$$
$$\text{if(sigma} > 0)\ \text{B0} = 1;\ \text{else B0} = \text{A0;]];} \tag{8.6}$$

This example demonstrates some of the potential for constructing more powerful types of image filters. To realize these possibilities, we next develop a more general mathematical morphology formalism.

8.3 Mathematical Morphology

8.3.1 Generalized Morphological Dilation

The basis of mathematical morphology is the application of set operations to images and their operators. We start by defining a generalized dilation mask as a set of locations within a 3×3 neighborhood. When referred to the center of the neighborhood as origin, each of these locations causes a shift of the image in the direction defined by the vector from the origin to the location. When several shifts are prescribed by a mask, the 1 locations in the various shifted images are combined by a set union operation.

1 Here and elsewhere in this chapter, any operations required to restore the image to the original image space are not considered or included (see Section 2.4).

The simplest example of this type is the identity operation which leaves the image unchanged:

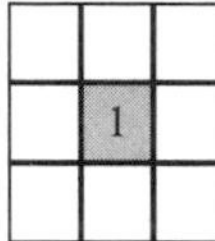

(Notice that we leave the 0's out of this mask, as we are now focusing on the set of elements at the various locations, and set elements are either present or absent.)

The next operation to consider is:

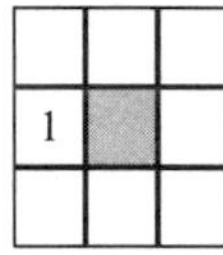

which is a left shift, equivalent to the one discussed in Section 2.2. Combining the two operations into a single mask:

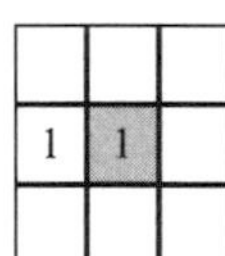

leads to a horizontal thickening of all objects in the image, by combining it with a left-shifted version of itself. An isotropic thickening of all objects is achieved by the operator:

1	1	1
1	1	1
1	1	1

(this is equivalent to the dilation operator discussed in Sections 2.2 and 8.2), while a symmetrical horizontal thickening operation (see Section 8.2.3) is achieved by the mask:

1	1	1

A rule of such operations is that if we want to guarantee that all the original object pixels are included in the output image, then we must include a 1 at the center (origin) of the mask.

Finally, there is no compulsion for all masks to be 3×3. Indeed, all but one of those listed above are effectively smaller than 3×3, and in more complex cases larger masks could be used. To emphasize this point, and to allow for asymmetrical masks in which the full 3×3 neighborhood is not given, we shall shade the origin—as shown in the above cases.

8.3.2 *Generalized Morphological Erosion*

We now move on to describe erosion in terms of set operations. The definition is somewhat peculiar in that it involves reverse shifts, but the reason for this will become clear as we proceed. Here the masks define directions as before, but in this case we shift the image in the reverse of each of these directions and perform intersection operations to combine the resulting images. For masks with a single element (as for the identity and shift left operators of Section 8.3.1), the intersection operation is improper, and the final result is as for the corresponding dilation operator but with a reverse shift. For more complex cases, the intersection operation results in objects being reduced in size. Thus the mask:

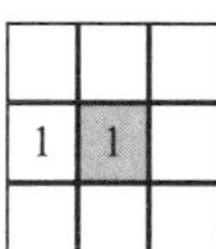

has the effect of stripping away the left sides of objects (the object is moved right and *and*ed with itself). Similarly, the mask:

1	1	1
1	1	1
1	1	1

results in an isotropic stripping operation, and is hence identical to the erosion operation described in Section 8.2.1.

8.3.3 *Duality between Dilation and Erosion*

We shall write the dilation and erosion operations formally as $A \oplus B$ and $A \ominus B$, respectively, where A is an image and B is the mask of the relevant operation:

$$A \oplus B = \cup_{b \in B} A_b \tag{8.7}$$

$$A \ominus B = \cap_{b \in B} A_{-b} \tag{8.8}$$

In these equations, A_b indicates a basic shift operation in the direction of element b of B and A_{-b} indicates the reverse shift operation.

We next prove an important theorem relating the dilation and erosion operations:

$$(A \ominus B)^{\mathrm{c}} = A^{\mathrm{c}} \oplus B^{\mathrm{r}} \tag{8.9}$$

where A^{c} represents the complement of A, and B^{r} represents the reflection of B in its origin. We first note that:[2]

$$x \in A^{\mathrm{c}} \Leftrightarrow x \notin A \tag{8.10}$$

and

$$b \in B^{\mathrm{r}} \Leftrightarrow -b \in B \tag{8.11}$$

We now have:[3]

$$\begin{aligned} x \in (A \ominus B)^{\mathrm{c}} &\Leftrightarrow x \notin A \ominus B \\ &\Leftrightarrow \exists\, b \text{ such that } x \notin A_{-b} \\ &\Leftrightarrow \exists\, b \text{ such that } x + b \notin A \\ &\Leftrightarrow \exists\, b \text{ such that } x + b \in A^{\mathrm{c}} \\ &\Leftrightarrow \exists\, b \text{ such that } x \in (A^{\mathrm{c}})_{-b} \\ &\Leftrightarrow x \in \cup_{b \in B} (A^{\mathrm{c}})_{-b} \\ &\Leftrightarrow x \in \cup_{b \in B^{\mathrm{r}}} (A^{\mathrm{c}})_{b} \\ &\Leftrightarrow x \in A^{\mathrm{c}} \oplus B^{\mathrm{r}} \end{aligned} \tag{8.12}$$

2 The sign "⇔" means "if and only if," that is, the statements so connected are equivalent.

3 This proof is based on that of Haralick et al. (1987). The sign "∃" means "there exists," and in this context it should be interpreted as "there is a value of." The symbol "∈" means "is a member of the following set": the symbol "∉" means "is *not* a member of the following set."

This completes the proof. The related theorem:

$$(A \oplus B)^c = A^c \ominus B^r \tag{8.13}$$

is proved similarly.

The fact that there are two such closely related theorems, following the related union and intersection definitions of dilation and erosion given above, indicates an important duality between the two operations. Indeed, as stated earlier, erosion of the objects in an image corresponds to dilation of the background, and vice versa. However, the two theorems indicate that this relation is not absolutely trivial, on account of the reflections of the masks required in the two cases. It is perhaps curious that, in contrast with the case of the de Morgan rule for complementation of an intersection:

$$(P \cap Q)^c = P^c \cup Q^c \tag{8.14}$$

the effective complementation of the dilating or eroding mask is its reflection rather than its complement per se, while that for the operator is the alternate operator.

8.3.4 *Properties of Dilation and Erosion Operators*

Dilation and erosion operators have some very important and useful properties. First, note that successive dilations are associative:

$$(A \oplus B) \oplus C = A \oplus (B \oplus C) \tag{8.15}$$

whereas successive erosions are not. The corresponding relation for erosions is:

$$(A \ominus B) \ominus C = A \ominus (B \oplus C) \tag{8.16}$$

The apparent symmetry between the two operators is more subtle than their simple origins in expanding and shrinking might indicate.

Next, the property:

$$X \oplus Y = Y \oplus X \tag{8.17}$$

means that the order in which dilations of an image are carried out does not matter, and the same applies to the order in which erosions are carried out:

$$(A \oplus B) \oplus C = (A \oplus C) \oplus B \tag{8.18}$$

$$(A \ominus B) \ominus C = (A \ominus C) \ominus B \tag{8.19}$$

In addition to the above relations, which use only the morphological operators $\oplus$ and $\ominus$, many more relations involve set operations. In the examples that follow, great care must be exercised to note which particular distributive operations are actually valid:

$$A \oplus (B \cup C) = (A \oplus B) \cup (A \oplus C) \tag{8.20}$$

$$A \ominus (B \cup C) = (A \ominus B) \cap (A \ominus C) \tag{8.21}$$

$$(A \cap B) \ominus C = (A \ominus C) \cap (B \ominus C) \tag{8.22}$$

In certain other cases, where equality might *a priori* have been expected, the strongest statements that can be made are typified by the following:

$$A \ominus (B \cap C) \supseteq (A \ominus B) \cup (A \ominus C) \tag{8.23}$$

Note that the associative relations show how large dilations and erosions might be factorized so that they can be implemented more efficiently as two smaller dilations and erosions applied in sequence. Similarly, the distributive relations show that a large mask may be split into two separate masks that may then be applied separately and the resulting images *or*ed together to create the same final image. These approaches can be useful for providing efficient implementations, especially in cases where very large masks are involved. For example, we could dilate an image horizontally and vertically by two separate operations, which would then be merged together—as in the following instance:

1	1	1

followed by

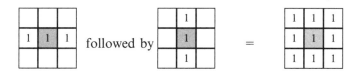

	1	
	1	
	1	

=

1	1	1
1	1	1
1	1	1

Next, let us consider the importance of the identity operation I, which corresponds to a mask with a single 1 at the central (A0) position:

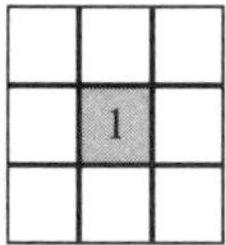

By way of example, we take equations (8.20) and (8.21) and replace C by I in each of them. If we write the union of B and I as D, so that mask D is bound to contain a central 1 (i.e., $D \supseteq I$), we have:

$$A \oplus D = A \oplus (B \cup I) = (A \oplus B) \cup (A \oplus I) = (A \oplus B) \cup A \tag{8.24}$$

which always contains A:

$$A \oplus D \supseteq A \tag{8.25}$$

Similarly:

$$A \ominus D = A \ominus (B \cup I) = (A \ominus B) \cap (A \ominus I) = (A \ominus B) \cap A \tag{8.26}$$

which is always contained within A:

$$A \ominus D \subseteq A \tag{8.27}$$

Operations (such as dilation by a mask containing a central 1) that give outputs guaranteed to contain the inputs are termed *extensive*, while those (such as erosion by a mask containing a central 1) for which the outputs are guaranteed to be contained by the inputs are termed *antiextensive*. Extensive operations extend objects, and antiextensive operations contract them, or in either case, leave them unchanged.

Another important type of operation is the *increasing* type of operation—that is an operation such as union which preserves order in the size of the objects on which it operates. If object F is small enough to be contained within object G, then applying erosions or dilations will not affect the situation, even though the objects change their sizes and shapes considerably. We can write these conditions in the form:

$$\text{if} \quad F \subseteq G \tag{8.28}$$

$$\text{then} \quad F \oplus B \subseteq G \oplus B \tag{8.29}$$

$$\text{and} \quad F \ominus B \subseteq G \ominus B \tag{8.30}$$

Next, we note that erosion can be used for locating the boundaries of objects in binary images:[4]

$$P = A - (A \ominus B) \tag{8.31}$$

4 Technically, we are dealing here with sets, and the appropriate set operation is the *andnot* function rather than minus. However, the latter admirably conveys the required meaning without ambiguity.

Many practical applications of dilation and erosion follow particularly from using them together, as we shall see later in this chapter.

Finally, we explore why the morphological definition of erosion involves a reflection. The idea is that dilation and erosion are able, under the right circumstances, to cancel each other out. Take the left shift dilation operation and the right shift erosion operation. These are both achieved via the mask:

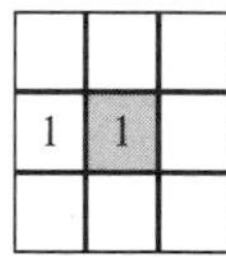

but in the erosion operation it is applied in its reflected form, thereby producing the right shift required to erode the left edge of any objects. This makes it clear why an operation of the type $(A \oplus B) \ominus B$ has a chance of canceling to give A. More specifically, there must be shifts in opposite directions as well as appropriate subtractions produced by *and*ing instead or *or*ing in order for cancellation to be possible. Of course, in many cases the dilation mask will have 180° rotation symmetry, and then the distinction between B^{r} and B will be purely academic.

8.3.5 *Closing and Opening*

Dilation and erosion are basic operators from which many others can be derived. Earlier, we were interested in the possibility of an erosion canceling a dilation and vice versa. Hence, it is an obvious step to define two new operators that express the degree of cancellation: the first is called *closing* since it often has the effect of closing gaps between objects; the other is called *opening* because it often has the effect of opening gaps (Fig. 8.1). Closing and opening are formally defined by the formulas:

$$A \bullet B = (A \oplus B) \ominus B \tag{8.32}$$

$$A \circ B = (A \ominus B) \oplus B \tag{8.33}$$

Closing is able to eliminate salt noise, narrow cracks or channels, and small holes or concavities.[5] Opening is able to eliminate pepper noise, fine hairs, and small protrusions. Thus, these operators are extremely important for practical

5 Here we continue to take the convention that dark objects have become 1's in binary images, and light background or other features have become 0's.

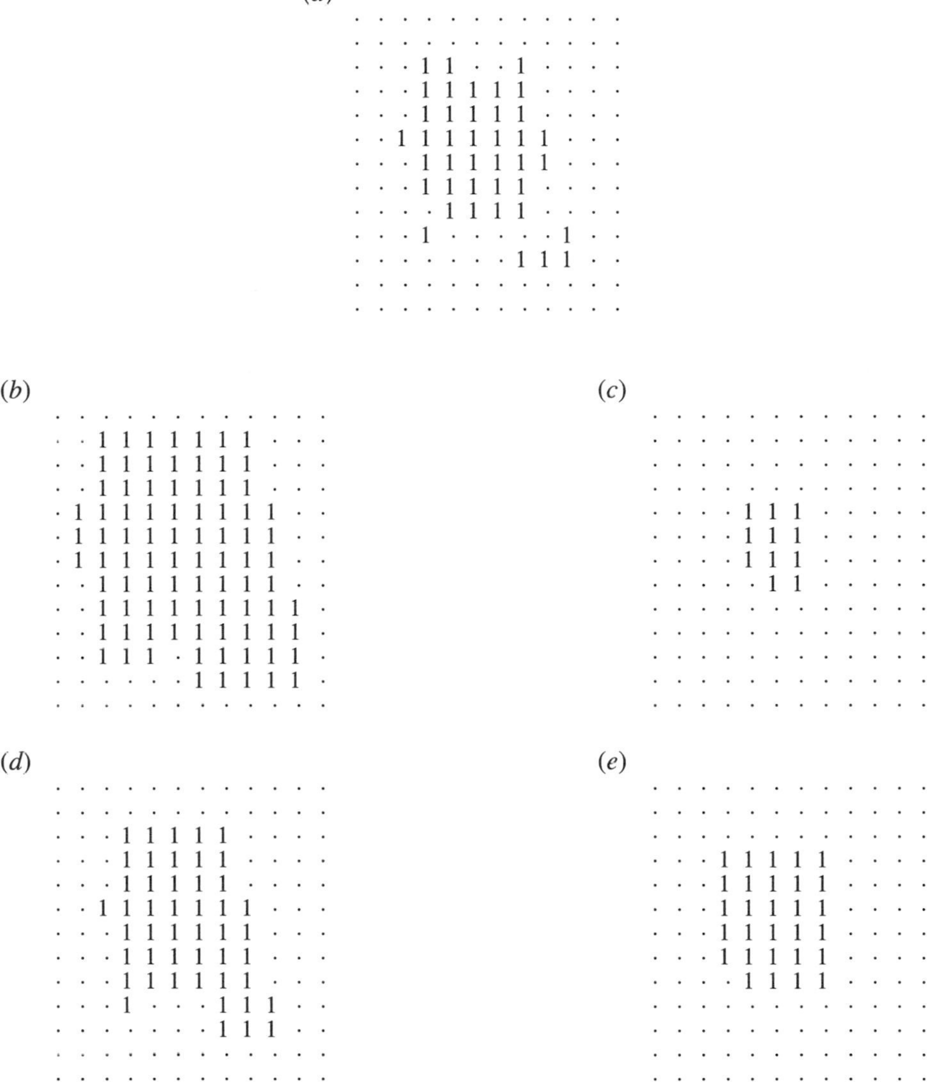

Figure 8.1 Results of morphological operations. (*a*) shows the original image, (*b*) the dilated image, (*c*) the eroded image, (*d*) the closed image, and (*e*) the opened image.

applications. Furthermore, by subtracting the derived image from the original image, it is possible to locate many sorts of defects, including those cited above as being eliminated by opening and closing: this possibility makes the two operations even more important. For example, we might use the following operation to locate all the fine hairs in an image:

$$Q = A - A \circ B \tag{8.34}$$

This operator and its dual using opening instead of closing:

$$R = A \bullet B - A \tag{8.35}$$

are extremely important for defect detection tasks. They are often, respectively, called the white and black top-hat operators.[6] (Practical applications of these two operators include location of solder bridges and cracks in printed circuit board tracks.)

Closing and opening have the interesting property that they are idempotent—that is, repeated application of either operation has no further effect. (This property contrasts strongly with what happens when dilation and erosion are applied a number of times.) We can write these results formally as follows:

$$(A \bullet B) \bullet B = A \bullet B \tag{8.36}$$

$$(A \circ B) \circ B = A \circ B \tag{8.37}$$

From a practical point of view, these properties are to be expected, since any hole or crack that has been filled in remains filled in, and there is no point in repeating the operation. Similarly, once a hair or protrusion has been removed, it cannot again be removed without first recreating it. Not quite so obvious is the fact that the combined closing and opening operation is idempotent:

$$\{[(A \bullet B) \circ C] \bullet B\} \circ C = (A \bullet B) \circ C \tag{8.38}$$

The same applies to the combined opening and closing operation. A simpler result is the following:

$$(A \oplus B) \circ B = (A \oplus B) \tag{8.39}$$

which shows that there is no point in opening with the same mask that has already been used for dilation. Essentially, the first dilation produces some effects that are not reversed by the erosion (in the opening operation), and the second dilation then merely reverses the effects of the erosion. The dual of this result is also valid:

$$(A \ominus B) \bullet B = (A \ominus B) \tag{8.40}$$

6 It is dubious whether "top-hat" is a very appropriate name for this type of operator: *a priori*, the term *residue function* (or simply *residue*) would appear to be better, as it conjures up the right functional connotations.

Among the most important other properties of closing and opening are the following set containment properties which apply when $D \supseteq I$:

$$A \oplus D \supseteq A \bullet D \supseteq A \tag{8.41}$$

$$A \ominus D \subseteq A \circ D \subseteq A \tag{8.42}$$

Thus, closing an image will, if anything, increase the sizes of objects, while opening an image will, if anything, make objects smaller, though there are clear limits on how much change closing and opening operations can induce.

Finally, it should be noted that closing and opening are subject to the same duality as for dilation and erosion:

$$(A \bullet B)^{\mathrm{c}} = A^{\mathrm{c}} \circ B^{\mathrm{r}} \tag{8.43}$$

$$(A \circ B)^{\mathrm{c}} = A^{\mathrm{c}} \bullet B^{\mathrm{r}} \tag{8.44}$$

8.3.6 *Summary of Basic Morphological Operations*

The past few sections have by no means exhausted the properties of the morphological operations *dilate*, *erode*, *close*, and *open*. However, they have outlined some of their properties and have demonstrated some of the practical results obtained using them. Perhaps the main aim of including the mathematical analysis has been to show that these operations are not ad hoc and that their properties are mathematically provable. Furthermore, the analysis has also indicated (1) how sequences of operations can be devised for a number of eventualities, and (2) how sequences of operations can be analyzed to save computation (for instance) by taking care not to use idempotent operations repeatedly and by breaking masks down into smaller more efficient ones.

Overall, the operations devised here can help to eliminate noise and irrelevant artifacts from images, so as to obtain more accurate recognition of shapes. They can also help to identify defects on objects by locating specific features of interest. In addition, they can perform grouping functions such as locating regions of images where small objects such as seeds may reside (Section 8.6). In general, elimination of artifacts is carried out by operations such as closing and opening, while location of such features is carried out by finding how the results of these operations differ from the original image (cf. equations (8.20) and (8.21)); and locating regions where clusters of small objects occur may be achieved by larger scale closing operations. Care in the choice of scales and mask sizes is of vital importance in the design of complete algorithms for all these tasks. Figures 8.2 and 8.3 illustrate some of these possibilities in the case of a peppercorn image: some of the interest in this image relates to the presence of a twiglet and how it is eliminated from consideration and/or identified.

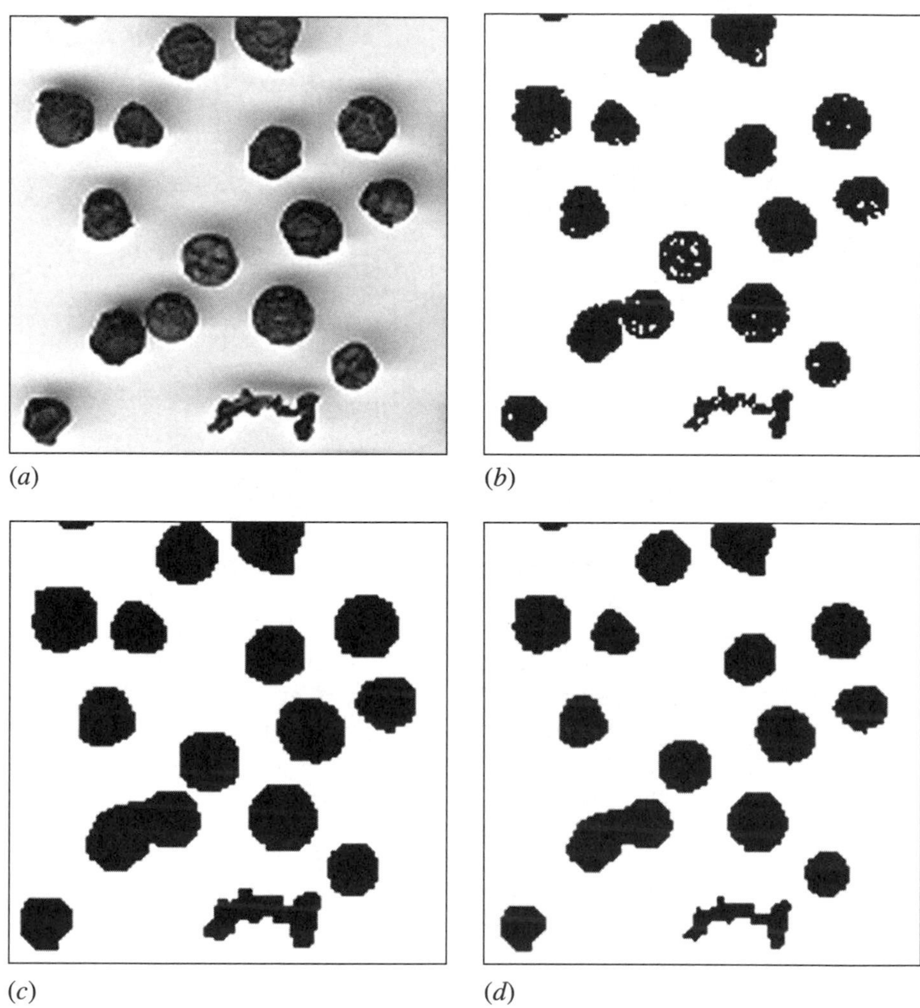

Figure 8.2 Use of the closing operation. (*a*) shows a peppercorn image, and (*b*) shows the result of thresholding. (*c*) shows the result of applying a 3×3 dilation operation to the object shapes, and (*d*) shows the effect of subsequently applying a 3×3 erosion operation. The overall effect of the two operations is a "closing" operation. In this case, closing is useful for eliminating the small holes in the objects. This would, for example, be useful for helping to prevent misleading loops from appearing in skeletons. For this picture, extremely large window operations would be required to group peppercorns into regions.

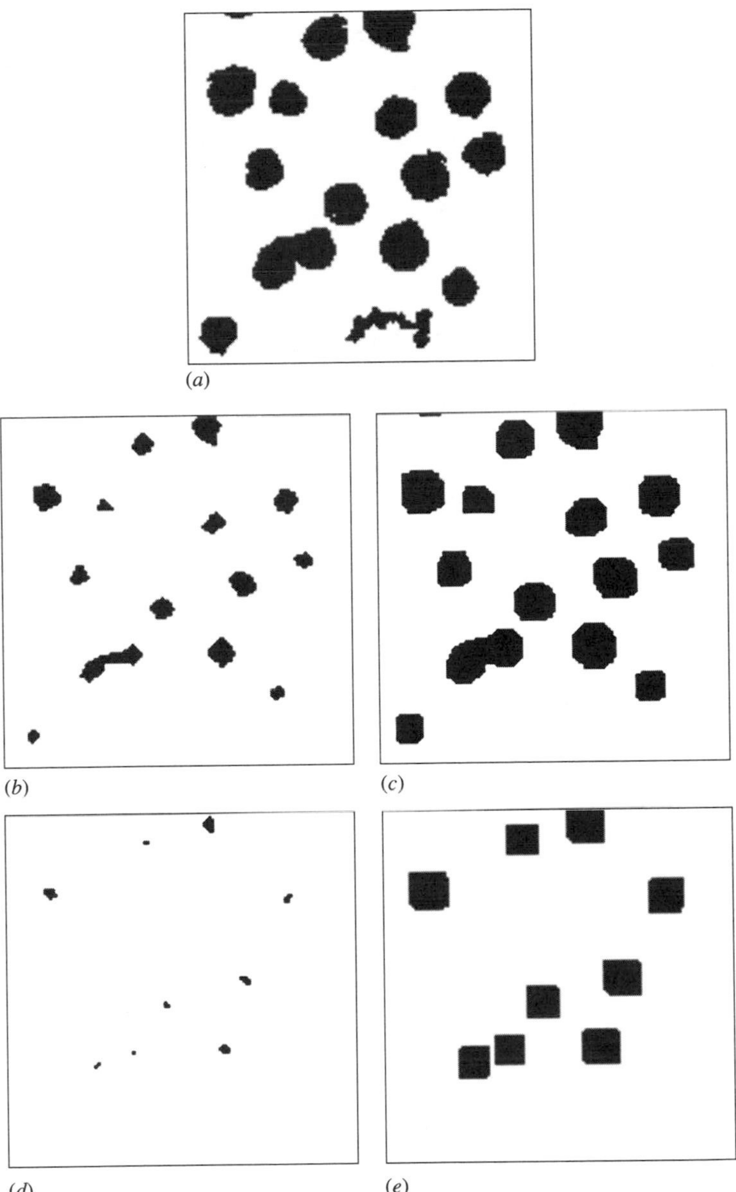

Figure 8.3 Use of the opening operation. (*a*) shows a thresholded peppercorn image. (*b*) shows the result of applying a 7×7 erosion operation to the object shapes, and (*c*) shows the effect of subsequently applying a 7×7 dilation operation. The overall effect of the two operations is an "opening" operation. In this case, opening is useful for eliminating the twiglet. (*d*) and (*e*) show the same respective operations when applied within an 11×11 window. Here some size filtering of the peppercorns has been achieved, and all the peppercorns have been separated—thereby helping with subsequent counting and labeling operations.

8.3.7 *Hit-and-Miss Transform*

We now move on to a transform that can be used to search an image for particular instances of a shape or other characteristic image feature. It is defined by the expression:

$$A \otimes (B, C) = (A \ominus B) \cap (A^{c} \ominus C) \tag{8.45}$$

The fact that this expression contains an *and* (intersection) sign would appear to make the description "hit-and-miss" transform more appropriate than the name "hit-or-miss" transform, which is sometimes used.

First, we consider the image analyzing properties of the $A \ominus B$ bracket. Applying B in this way (i.e., using erosion) leads to the location of a set of pixels where the 1's in the image match the 1's in the mask. Now notice that applying C to image A^{c} reveals a set of pixels where the 1's in C match the 1's in A^{c} or the 0's in A. However, it is usual to take a C mask that does not intersect with B (i.e., $B \cap C = \emptyset$). This indicates that we are matching a set of 0's in a more general mask D against the 0's in A. Thus, D is an augmented form of B containing a specified set of 0's (corresponding to the 1's in C) and a specified set of 1's (corresponding to the 1's in B). Certain other locations in D contain neither 0's nor 1's and can be regarded as "don't care" elements (which we will mark with an $\times$). As an example, take the following lower left concave corner locating masks:

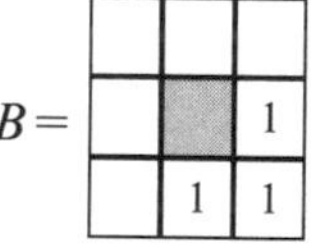

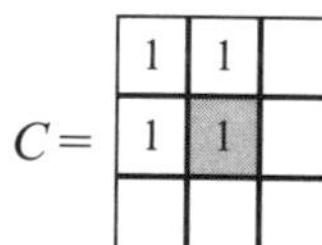

$D =$

0	0	×
0	0	1
×	1	1

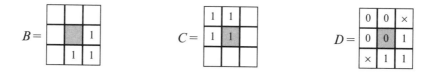

In reality, this transform is equivalent to a code coincidence detector, with the power to ignore irrelevant pixels. Thus, it searches for instances where specified pixels have the required 0 values and other specified pixels have the required 1 values. When it encounters such a pixel, it marks it with a 1 in the new image space; otherwise it returns a 0. This approach is quite general and can be used to detect a wide variety of local image conditions, including particular features, structures, shapes, or objects. It is by no means restricted to 3×3 masks as the above example might indicate. In short, it is a general binary template matching operator. Its power might serve to indicate that erosion is ultimately a more important concept than dilation—at least in its discriminatory, pattern recognition properties.

8.3.8 Template Matching

As indicated in the previous subsection, the hit-and-miss transform is essentially a template matching operator and can be used to search images for characteristic features or structures. However, it needs to be generalized significantly before it can be used for practical feature detection tasks because features may appear in a number of different orientations or guises, and any one of them will have to trigger the detector. To achieve this, we merely need to take the union of the results obtained from the various pairs of masks:

$$\cup_i A \otimes (B_i, C_i) = \cup_i (A \ominus B_i) \cap (A^c \ominus C_i) \tag{8.46}$$

or, equivalently, to take the union of the results obtained from the various D-type masks (see previous subsection).

As an example, we consider how the ends of thin lines may be located. To achieve this, we need eight masks of the following types:

$$D_1 = \begin{array}{|c|c|c|}\hline 0 & 0 & 0 \\ \hline 0 & 1 & 1 \\ \hline 0 & 0 & 0 \\ \hline\end{array} \qquad D_2 = \begin{array}{|c|c|c|}\hline 0 & 0 & 1 \\ \hline 0 & 1 & 0 \\ \hline 0 & 0 & 0 \\ \hline\end{array}$$

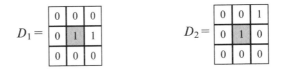

A much simpler algorithm could be used in this case to count the number of neighbors of each pixel and check whether the number is unity. (Alternative approaches can sometimes be highly beneficial in cutting down on excessive computational loads. On the other hand, morphological processing is of value in this general context because of the extreme simplicity of the masks being used for analyzing images.) Further examples of template matching will appear in the following section.

8.4 Connectivity-based Analysis of Images

This chapter has space for just one illustration of the application of morphology to connectedness-based analysis. We have chosen the case of skeletonization and thinning.

8.4.1 *Skeletons and Thinning*

A widely used thinning algorithm uses eight *D*-type hit-and-miss transform masks (see Section 8.3.5):

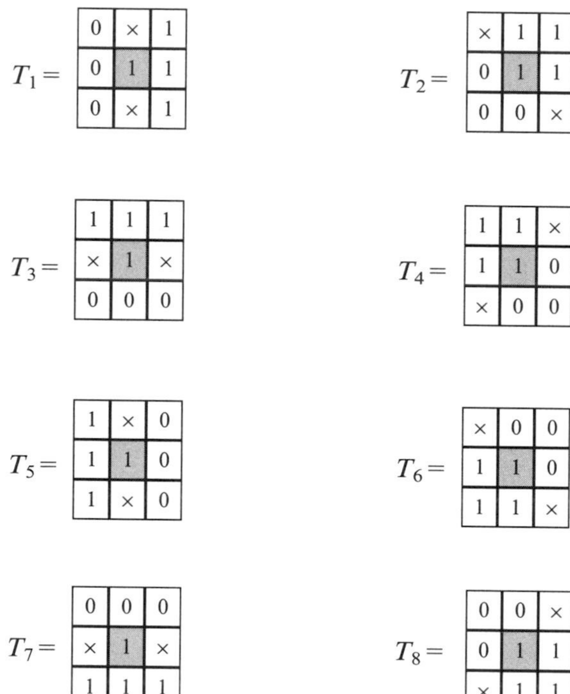

Each of these masks recognizes a particular situation where the central pixel is unnecessary for maintaining connectedness and can be eliminated. (This assumes that the foreground is regarded as 8-connected.) Thus, all pixels can be subjected to the T_1 mask and eliminated if the mask matches the neighborhood. Then all pixels can be subjected to the T_2 mask and eliminated if appropriate; and so on until upon repeating with all the masks there is no change in the output image (see Chapter 6 for a full description of the process).

Although this particular procedure is guaranteed not to disconnect objects, it does not terminate the skeleton location procedure, as there remain a number of locations where pixels can be removed without disconnecting the skeleton. The advantage of the eight masks given here is that they are very simple and produce fast convergence toward the skeleton. Their disadvantage is that a final stage of processing is required to complete the process. (For example, one of the crossing-number-based methods of Chapter 6 may be used for this purpose, but we do not pursue the matter further here.)

8.5 Gray-scale Processing

The generalization of morphology to gray-scale images can be achieved in a number of ways. A particularly simple approach is to employ "flat" structuring elements. These perform morphological processing in the same way for each of the gray levels, acting as if the shapes at each level were separate, independent binary images. If dilation is carried out in this way, the result turns out to be identical to the effect of applying an intensity maximum operation of the same shape. That is, we replace set inclusion by a magnitude comparison. Needless to say, this is mathematically identical in action for a normal binary image, but when applied to a gray-scale image it neatly generalizes the dilation concept. Similarly, erosion can be carried out by applying an intensity minimum structuring element of the same shape as the original binary structuring element. This discussion assumes that we focus on light objects against dark backgrounds.[7] These will be dilated when the maximum intensity operation is applied, and they will be eroded when the minimum intensity operation is applied. We could of course reverse the convention, depending on what type of objects we are concentrating on at any moment, or in any application. We can summarize the situation as follows:

$$A \oplus B = \max_{b \in B} A_b \tag{8.47}$$

$$A \ominus B = \min_{b \in B} A_{-b} \tag{8.48}$$

Other more complex gray-scale analogs of dilation and erosion take the form of 3-D structuring elements whose output at any gray level depends not just on the shape of the image at that gray level but also on the shapes at a number of nearly gray levels. Although such "nonflat" structuring elements are useful, for a good many applications they are not necessary, for flat structuring elements already embody considerable generalization relative to the binary case.

Next we consider how edge detection is carried out using grey-scale morphology.

7 This is the opposite convention to that employed in Chapter 2, but, as we shall see, in gray-scale processing, it is probably more general to focus on intensities than on specific objects.

8.5.1 Morphological Edge Enhancement

In Section 2.2.2 we showed how edge detection could be carried out in binary images. We defined edge detection asymmetrically, in the sense that the edge was the part of the object that was next to the background. This definition was useful because including the part of the background next to the object would merely have served to make the boundary wider and less precise. However, edge detection in gray-scale images does not need to embody such an asymmetry[8] because it starts by performing edge enhancement and then carrying out a thresholding type of operation—the width being controlled largely by the manner of thresholding and by whether nonmaximum suppression or other factors are brought to bear. Here we start by formulating the original binary edge detector in morphological form; then we make it more symmetrical; finally, we generalize it to gray-scale operation.

The original binary edge detector may be written in the form (cf. equation (8.31)):

$$E = A - (A \ominus B) \tag{8.49}$$

Making it symmetrical merely involves adding the background edge $((A \oplus B) - A)$:

$$G = (A \oplus B) - (A \ominus B) \tag{8.50}$$

To convert to gray-scale operation involves employing maximum and minimum operations in place of dilation and erosion. In this case we are concentrating on intensity per se, and so these respective assignments of dilation and erosion are used. (The alternate arrangement would result in negative edge contrast.) Thus, here we use equations (8.47) and (8.48) to *define* dilation and erosion for gray-scale processing, so equation (8.50) already represents morphological edge enhancement for grey-scale images. The argument G is often called the morphological gradient of an image (Fig. 8.4). Notice that it is not accompanied by an accurate edge orientation value, though approximate orientations can, of course, be computed by determining which part of the structuring element gives rise to the maximum signal.

8.5.2 Further Remarks on the Generalization to Gray-scale Processing

In the previous subsection, we found that reinterpreting equation (8.50) permitted edge detection to be generalized immediately from binary to gray-scale. That this

8 Indeed, any asymmetry would lead to an unnecessary bias and hence inaccuracy in the location of edges.

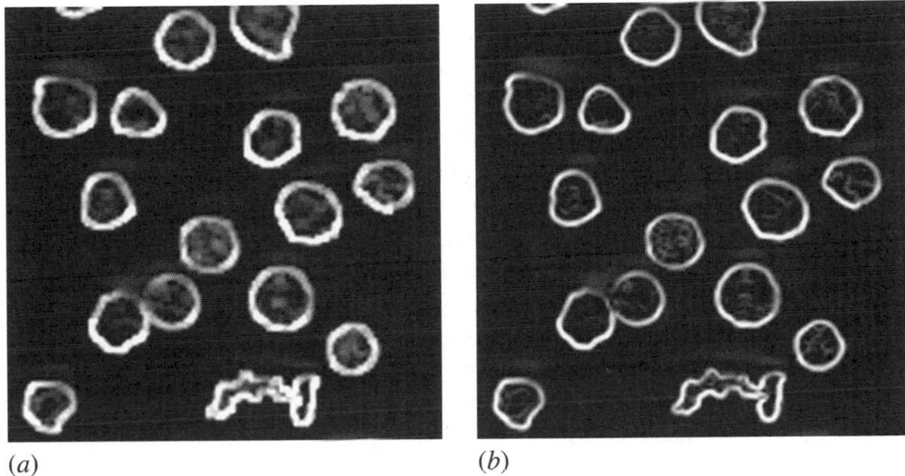

(*a*) (*b*)

Figure 8.4 Determination of the morphological gradient of an image. The original image is that of Fig. 8.2*a*. (*a*) shows the morphological gradient, obtained using 3 × 3 window operations. (*b*) shows the result for a Sobel operator. Notice that the latter gives less diffuse responses.

is so is a consequence of the extremely powerful *umbra homomorphism theorem*. This starts with the knowledge that intensity I is a single-valued function of position within the image. This means that it represents a surface in a 3-D (gray-scale) space. However, as we have seen, it is useful to take into account the individual gray levels. In particular, we note that the set of pixels of gray level g_i is a subset of the set of pixels of gray level g_{i-1}, where $g_i \geq g_{i-1}$. The important step forward is to interpret the 3-D volume containing all these gray levels under the intensity surface as constituting an umbra—a 3-D shadow region for the relevant part of the surface. We write the umbra volume of I as $U(I)$, and clearly we also have $I = T(U(I))$, where the operator $T(\cdot)$ recalculates the top surface.

The umbra homomorphism theorem then states that a dilation has to be defined and interpreted as an operation on the umbras:

$$U(I \oplus K) = U(I) \oplus U(K) \tag{8.51}$$

To find the relevant intensity function, we merely need to apply the top-surface operator to the umbra:

$$I \oplus K = T(U(I \oplus K)) = T(U(I) \oplus U(K)) \tag{8.52}$$

A similar statement applies for erosion.

The next step is to note that the generalization from binary to gray-scale dilation using flat structuring elements involved applying a maximum operation in place of a set union operation. The operation can, for the simple case of a 1-D image, be rewritten in the form:

$$(I \oplus K)(x) = \max(I(x - z) + K(z)) \tag{8.53}$$

where $K(z)$ takes values 0 or 1 only, and $x - z$, z must lie within the domains of I and K. However, another vital step is now to notice that this form generalizes to nonbinary $K(z)$, whose values can, for example, be the integer gray-level values. This makes the dilation operation considerably more powerful, yielding the nonflat structuring element concept—which will also work for 2-D images with full grey-scale.

Working out the responses of this sort of operation can be tedious, but it is susceptible to a neat geometric interpretation. If the function $K(z)$ is inverted and turned into a template, this may be run over the image $I(x)$ in such a way as to remain just in contact with it. Thus, the origin of the inverted template will trace out the top surface of the dilated image. The process is depicted in Fig. 8.5 for the case of a triangular structuring element in a 1-D image.

Similar relations apply for erosion, closing, opening, and a variety of set functions. This means that the standard binary morphological relations, equations (8.15)–(8.23), apply for gray-scale images as well as for binary images. Furthermore, the dilation–erosion and closing–opening dualities (equations (8.9), (8.13), (8.43), (8.44)) also apply for gray-scale images. These are extremely powerful results and allow one to apply morphological concepts in an intuitive

(*a*)

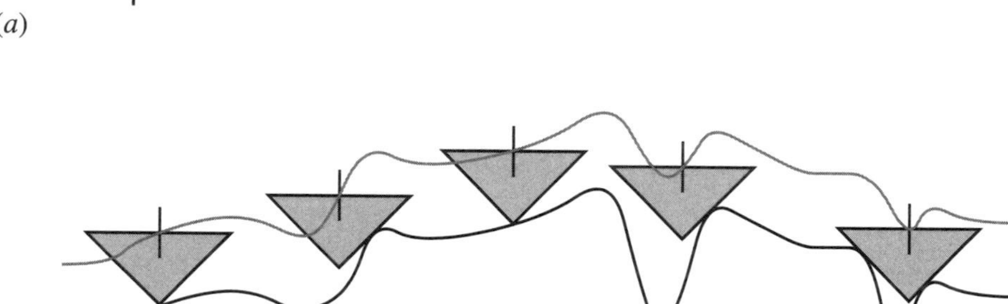

(*b*)

Figure 8.5 Dilation of 1-D grey-scale image by triangular structuring element. (*a*) shows the structuring element, with the vertical line at the bottom indicating the origin of coordinates. (*b*) shows the original image (continuous black line), several instances of the inverted structuring element being applied, and the output image (continuous gray line). This geometric construction automatically takes account of the maximum operation in equation (8.53). Note that as no part of the structuring element is below the origin in (*a*), the output intensity is increased at every point in the image.

manner. In that case, the practically important factor devolves into choosing the right gray-scale structuring element for the application.

Another interesting factor is the possibility of using morphological operations instead of convolutions in many places where convolution is employed throughout image analysis. We have already seen how edge enhancement and detection can be performed using morphology in place of convolution. In addition, noise suppression by Gaussian smoothing can be replaced by opening and closing operations. However, it must always be borne in mind that convolutions are linear operations and are thereby grossly restricted, whereas morphological operations are highly nonlinear, their very structure embodying multiple *if* statements, so that outward appearances of similarity are bound to hide deep differences of operation, effectiveness, and applicability. Consider, for example, the optimality of the mean filter for suppressing Gaussian noise, and the optimality of the median filter (a morphological operator) for eliminating impulse noise.

One example is worth including here. When edge enhancement is performed, prior to edge detection, it is possible to show that when both a differential gradient (e.g., Sobel) operator and a morphological gradient operator are applied to a noise-free image with a steady intensity gradient, the results will be identical, within a constant factor. However, if one impulse noise pixel arises, the maximum or minimum operations of the morphological gradient operator will select this value in calculating the gradient, whereas the differential gradient operator will average its effect over the window, giving significantly less error.

Space prevents gray-scale morphological processing from being considered in more detail here; see, for example, Haralick and Shapiro (1992) and Soille (2003).

8.6 Effect of Noise on Morphological Grouping Operations

Texture analysis is an important area of machine vision and is relevant not only for segmenting one region of an image from another (as in many remote sensing applications), but also for characterizing regions absolutely—as is necessary when performing surface inspection (for example, when assessing the paint finish on automobiles). Many methods have been employed for texture analysis. These range from the widely used gray-level co-occurrence matrix approach to Law's texture energy approach, and from use of Markov random fields to fractal modeling (Chapter 26).

Some approaches involve even less computation and are applicable when the textures are particularly simple and the shapes of the basic texture elements are not especially critical. For example, if it is required to locate regions

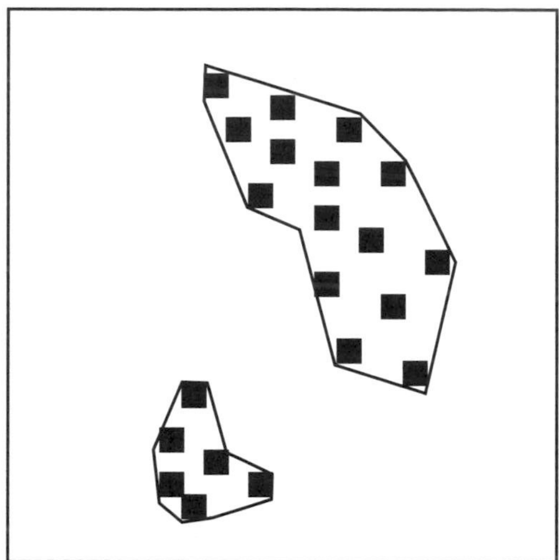

Figure 8.6 Idealized grouping of small objects into regions, such as might be attempted using closing operations.

containing small objects, simple morphological operations applied to thresholded versions of the image are often appropriate (Fig. 8.6) (Haralick and Shapiro, 1992; Bangham and Marshall, 1998). Such approaches can be used for locating regions containing seeds, grains, nails, sand, or other materials, either for assessing the overall quantity or spread or for determining whether some regions have not yet been covered. The basic operation to be applied is the dilation operation, which combines the individual particles into fully connected regions. This method is suitable not only for connecting individual particles but also for separating regions containing high and low densities of such particles. The expansion characteristic of the dilation operation can be largely canceled by a subsequent erosion operation, using the same morphological kernel. Indeed, if the particles are always convex and well separated, the erosion should exactly cancel the dilation, though in general the combined closing operation is not a null operation, and this is relied upon in the above connecting operation.

Closing operations have been applied to images of cereal grains containing dark rodent droppings in order to consolidate the droppings (which contain significant speckle—and therefore holes when the images are thresholded) and thus to make them more readily recognizable from their shapes (Davies et al., 1998a). However, the result was rather unsatisfactory as dark patches on the grains tend to combine with the dark droppings; this has the effect of distorting the shapes and also makes the objects larger. This problem was partially overcome by a subsequent erosion operation, so that the overall procedure is dilate + erode + erode.

Originally, adding this final operation seemed to be an ad hoc procedure, but on analysis it was found that the size increase actually applies quite generally when segmentation of textures containing different densities of particles is carried out. It is this general effect that we now consider.

8.6.1 Detailed Analysis

Let us take two regions containing small particles with occurrence densities ρ_1, ρ_2, where $\rho_1 > \rho_2$. In region 1 the mean distance between particles will be d_1, and in region 2 the mean distance will be d_2, where $d_1 < d_2$. If we dilate using a kernel of radius a, where $d_1 < 2a < d_2$, this will tend to connect the particles in region 1 but should leave the particles in region 2 separate. To *ensure* connecting the particles in region 1, we can make $2a$ larger than $(1/2)(d_1 + d_2)$, but this may risk connecting the particles in region 2. (The risk will be reduced when the subsequent erosion operation is taken into account.) Selecting an optimum value of a depends not only on the mean distances d_1, d_2 but also on their distributions. Space prevents us from entering into a detailed discussion of this subject; we merely assume that a suitable selection of a is made and that it is effective. The problem that is tackled here is whether the size of the final regions matches the *a priori* desired segmentation, that is, whether any size distortion takes place. We start by taking this to be an essentially 1-D problem, which can be modeled as in Fig. 8.7. (The 1-D particle densities will now be given an x suffix.)

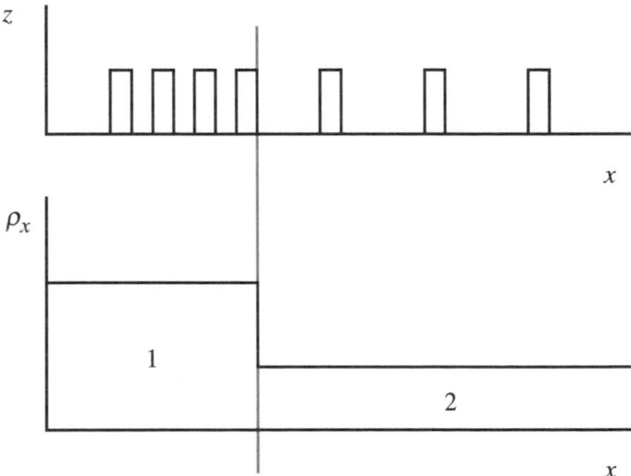

Figure 8.7 1-D particle distribution. z indicates the presence of a particle, and ρ_x shows the densities in the two regions.

Suppose first that $\rho_{2x}=0$. Then in region 2 the initial dilation will be counteracted exactly (in 1-D) by the subsequent erosion. Next take $\rho_{2x} > 0$: when dilation occurs, a number of particles in region 2 will be enveloped, and the erosion process will not exactly reverse the dilation. If a particle in region 2 is within $2a$ of an outermost particle in region 1, they will merge and will remain merged when erosion occurs. The probability P that this will happen is the integral over a distance $2a$ of the probability density for particles in region 2. In addition, when the particles are well separated, we can take the probability density as being equal to the mean particle density ρ_{2x}. Hence:

$$P = \int_0^{2a} \rho_{2x}\, dx = 2a\rho_{2x} \tag{8.54}$$

If such an event occurs, then region 1 will be expanded by amounts ranging from a to $3a$, or 0 to $2a$ after erosion, though these figures must be increased by b for particles of width b. Thus, the *mean* increase in size of region 1 after dilation + erosion is $2a\rho_{2x} \times (a+b)$, where we have assumed that the particle density in region 2 remains uniform right up to region 1.

Next we consider what additional erosion operation will be necessary to cancel this increase in size. We just make the radius $\tilde{a}_{1\text{-D}}$ of the erosion kernel equal to the increase in size:

$$\tilde{a}_{1\text{-D}} = 2a\rho_{2x}(a + b) \tag{8.55}$$

Finally, we must recognize that the required process is 2-D rather than 1-D, and we take y to be the lateral axis, normal to the original (1-D) x-axis. For simplicity we assume that the dilated particles in region 2 are separated laterally and are not touching or overlapping (Fig. 8.8). As a result, the change of size of region 1 given by equation (8.55) will be diluted relative to the 1-D case by the reduced density along the direction (y) of the border between the two regions; that is, we must multiply the right-hand side of equation (8.55) by $b\rho_{2y}$. We now obtain the relevant 2-D equation:

$$\tilde{a}_{2\text{-D}} = 2ab\rho_{2x}\rho_{2y}(a + b) = 2ab\rho_2(a + b) \tag{8.56}$$

where we have finally reverted to the appropriate 2-D *area* particle density ρ_2.

For low values of ρ_2 an additional erosion will not be required, whereas for high values of ρ_2 substantial erosion will be necessary, particularly if b is comparable to or larger than a. If $\tilde{a}_{2\text{-D}} < 1$, it will be difficult to provide an accurate correction by applying an erosion operation, and all that can be done is to bear in mind that any measurements made from the image will require correction. (Note that if, as often happens, $a \gg 1$, $\tilde{a}_{2\text{-D}}$ could well be at least 1.)

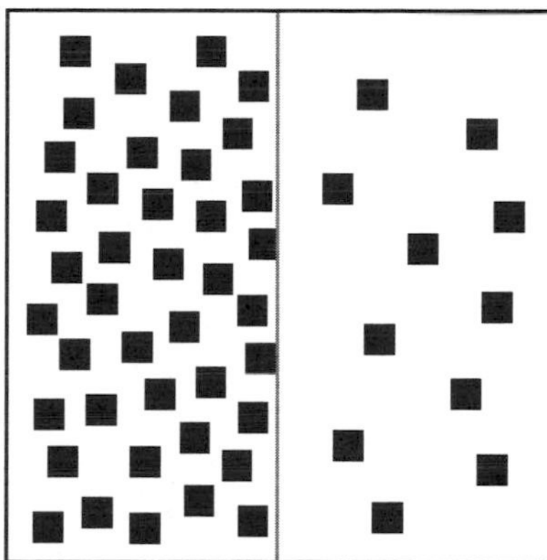

Figure 8.8 Model of the incidence of particles in two regions. Region 2 (on the right) has sufficiently low density that the dilated particles will not touch or overlap.

8.6.2 *Discussion*

The work described earlier (Davies, 2000c) was motivated by analysis of cereal grain images containing rodent droppings, which had to be consolidated by dilation operations to eliminate speckle, followed by erosion operations to restore size.[9] It has been found that if the background contains a low density of small particles that tend, upon dilation, to increase the sizes of the foreground objects, additional erosion operations will in general be required to accurately represent the sizes of the regions. The effect would be similar if impulse noise were present, though theory shows what is observed in practice—that the effect is enhanced if the particles in the background are not negligible in size. The increases in size are proportional to the occurrence density of the particles in the background, and the kernel for the final erosion operation is calculable, the overall process being a necessary measure rather than an ad hoc technique.

8.7 Concluding Remarks

Binary images contain all the data needed to analyze the shapes, sizes, positions, and orientations of objects in two dimensions, and thereby to recognize them and even to inspect them for defects. Many simple small neighborhood operations exist for processing binary images and moving toward the goals stated above. At first sight, these may appear as a somewhat random set, reflecting historical

9 For further background on this application, see Chapter 23.

development rather than systematic analytic tools. However, in the past few years, mathematical morphology has emerged as a unifying theory of shape analysis; we have aimed to give the flavor of the subject in this chapter. Mathematical morphology, as its name suggests, is mathematical in nature, and this can be a source of difficulty,[10] but a number of key theorems and results are worth remembering. A few of these have been considered here and placed in context. For example, generalized dilation and erosion have acquired a central importance, since further vital concepts and constructs are based on them—closing, opening, template matching (via the hit-and-miss transform), to name but a few. And when one moves on to connectedness properties and concepts such as skeletonization, the relevance of mathematical morphology has already been proven (though space has prevented a detailed discussion of its application to these latter topics). For further information on gray-scale morphological processing, see Haralick and Shapiro (1992) and Soille (2003).

Mathematical morphology is one of the standard methodologies that has evolved over the past few decades. This chapter has demonstrated how the mathematical aspects make the subject of shape analysis rigorous and less ad hoc. Its extensions to gray-scale image processing are interesting but not necessarily superior to those for more traditional approaches.

8.8 Bibliographical and Historical Notes

The book by Serra (1982) is an important early landmark in the development of morphology. Many subsequent papers helped to lay the mathematical foundations, perhaps the most important and influential being that by Haralick et al. (1987); see also Zhuang and Haralick (1986) for methods for decomposing morphological operators, and Crimmins and Brown (1985) for more practical aspects of shape recognition. The paper by Dougherty and Giardina (1988) was important in the development of methods for gray-scale morphological processing; Heijmans (1991) and Dougherty and Sinha (1995a,b) have extended

10 The rigor of mathematics is a cause for celebration, but at the same time it can make the arguments and the results obtained from them less intuitive. On the other hand, the real benefit of mathematics is to leapfrog what is possible by intuition alone and to arrive at results that are new and unexpected—the very antithesis of intuition.

this earlier work, while Haralick and Shapiro (1992) provide a useful general introduction to the topic. The work of Huang and Mitchell (1994) on gray-scale morphology decomposition and that of Jackway and Deriche (1996) on multiscale morphological operators show that the theory of the subject is by no means fully developed, though more applied developments such as morphological template matching (Jones and Svalbe, 1994), edge detection in medical images (Godbole and Amin, 1995) and boundary detection of moving objects (Yang and Li, 1995) demonstrate the potential value of morphology. The reader is referred to the volumes by Dougherty (1992) and Heijmans (1994) for further information on the subject.

An interesting point is that it is by no means obvious how to decide on the sequence of morphological operations that is required in any application. This is an area where genetic algorithms might be expected to contribute to the systematic generation of complete systems and seems likely to be a focus for much future research (see, e.g., Harvey and Marshall, 1994).

Further work on mathematical morphology appears in Maragos et al. (1996). Work up to 1998 is reviewed in a useful tutorial paper by Bangham and Marshall (1998). More recently, Soille (2003) has produced a thoroughgoing volume on the subject. Gil and Kimmel (2002) address the problem of rapid implementation of dilation, erosion, and opening and closing algorithms and arrive at a new approach based on deterministic calculations. These give low computational complexity for calculating max and min functions, and similar complexities for the four cited filters and other derived filters. The paper goes on to state some open problems and to suggest how they might be tackled. It is clear that some apparently simple tasks that ought perhaps to have been dealt with before the 2000s are still unsolved—such as how to compute the median in better than $O(\log^2 p)$ time per filtered point (in a $p \times p$ window); or how to *optimally* extend 1-D morphological operations to circular rather than square window (2-D) operations. Note that the new implementation is immediately applicable to determining the morphological gradient of an image.

8.9 Problem

1. A morphological gradient binary edge enhancement operator is defined by the formula:

$$G = (A \oplus B) - (A \ominus B)$$

Using a 1-D model of an edge, or otherwise, show that this will give wide edges in binary images. If grey-scale dilation ($\oplus$) is equated to taking a local *maximum* of the intensity function within a 3×3 window, and gray-scale erosion ($\ominus$) is equated to taking a local *minimum* within a 3×3 window, sketch the result of applying the operator G. Show that it is similar in effect to a Sobel edge enhancement operator, if edge orientation effects are ignored by taking the Sobel magnitude:

$$g = (g_x^2 + g_y^2)^{1/2}$$

PART 2

Intermediate-Level Vision

Intermediate-level image analysis is concerned with obtaining abstract information about images, starting with the images themselves. At this stage we are no longer interested in converting one image into another, as in the subject of image processing. In fact, intermediate-level processing was already being developed in parts of Chapter 7. However, what is special about Part 2 is that intermediate-level analysis is investigated for its own sake, in particular with the aid of transform methods that have been designed systematically for the purpose.

The chapters in Part 2 largely result in abstract information on the positions and orientations of various image features: they do not aim to provide real-world data, a function that is left to Parts 3 and 4. (Thus, Part 2 may indicate that a circle exists in one part of an image; it is left to later parts to interpret this as a wheel and to find any defects it may have.) However, an exception to this general strategy is the relational descriptor analysis of scenes that appears toward the end of Chapter 15: this approach may be regarded as constituting the foothills of "high-level" vision.

Line Detection

Detection of macroscopic features is an important part of image analysis and visual pattern recognition. Of particular interest is the identification of straight lines in images, as these are ubiquitous—appearing both on manufactured parts and in the built environment. The Hough transform provides the means for locating these features highly robustly in digital images. This chapter describes the processes and principles needed to locate the features in this way.

Look out for:

- the basic Hough transform technique for locating straight lines in images.
- an alternative approach called the foot-of-normal method.
- how lines can be localized along their length.
- how final line fitting can be made more accurate.
- why the Hough transform is robust against noise and background clutter.
- how speed of processing can be improved.

Note that while this chapter seems to provide an interesting means for detecting line features in images, it may appear somewhat specialized. However, subsequent chapters in Part 2 of this volume will demonstrate that the power of the Hough transform is far greater than such arguments might suggest.

CHAPTER 9

Line Detection

9.1 Introduction

Straight edges are among the most common features of the modern world, arising in perhaps the majority of manufactured objects and components, not least in the very buildings in which we live. Yet it is arguable whether true straight lines ever arise in the natural state. Possibly the only example of their appearance in virgin outdoor scenes is the horizon—although even this is clearly seen from space as a circular boundary! The surface of water is essentially planar, although it is important to realize that this is a deduction. The fact remains that straight lines seldom appear in completely natural scenes. Be all this as it may, it is vital both in city pictures and in the factory to have effective means of detecting straight edges. This chapter studies available methods for locating these important features.

Historically, the Hough transform (HT) has been the main means of detecting straight edges, and over the 40 years since the method was originally invented (Hough, 1962) it has been developed and refined for this purpose. Hence, this chapter concentrates on this particular technique; it also prepares the ground for applying the HT to the detection of circles, ellipses, corners, and so on, in the next few chapters. We start by examining the original Hough scheme, even though it is now seen to be wasteful in computation, since important principles are involved.

9.2 Application of the Hough Transform to Line Detection

The basic concept involved in locating lines by the HT is point-line duality. A point P can be defined either as a pair of coordinates or in terms of the set of lines passing through it. The concept starts to make sense if we consider a set of collinear points P_i, then list the sets of lines passing through each of them, and finally note that just one line is common to all these sets. Thus, it is possible to find the line containing all the points P_i merely by eliminating those that are not multiple hits. Indeed, it is easy to see that if a number of noise points Q_j are intermingled with the signal points P_i, the method will be able to discriminate the collinear points from among the noise points at the same time as finding the line containing them, merely by searching for multiple hits. Thus, the method is inherently robust against noise, as indeed it is in discriminating against currently unwanted signals such as circles.

The duality goes further. For just as a point can define (or be defined by) a set of lines, so a line can define (or be defined by) a set of points, as is obvious from the above argument. This makes the above approach to line detection a mathematically elegant one, and it is perhaps surprising that the Hough detection scheme was first published as a patent (Hough, 1962) of an electronic apparatus for detecting the tracks of high-energy particles rather than as a paper in a learned journal.

The form in which the method was originally applied involves parametrizing lines by the slope–intercept equation

$$y = mx + c \tag{9.1}$$

Every point on a straight edge is then plotted as a line in (m, c) space corresponding to all the (m, c) values consistent with its coordinates, and lines are detected in this space. The embarrassment of unlimited ranges of the (m, c) values (near-vertical lines require near-infinite values of these parameters) is overcome by using two sets of plots, the first corresponding to slopes of less than 1.0 and the second to slopes of 1.0 or more. In the latter case, equation (9.1) is replaced by the form:

$$x = \tilde{m}y + \tilde{c} \tag{9.2}$$

where

$$\tilde{m} = 1/m \tag{9.3}$$

The need for this rather wasteful device was removed by the Duda and Hart (1972) approach, which replaces the slope–intercept formulation with the so-called normal (θ, ρ) form for the straight line (see Fig. 9.1):

$$\rho = x \cos\theta + y \sin\theta \tag{9.4}$$

To apply the method using this form, the set of lines passing through each point P_i is represented as a set of sine curves in (θ, ρ) space. For example, for point $P_1(x_1, y_1)$

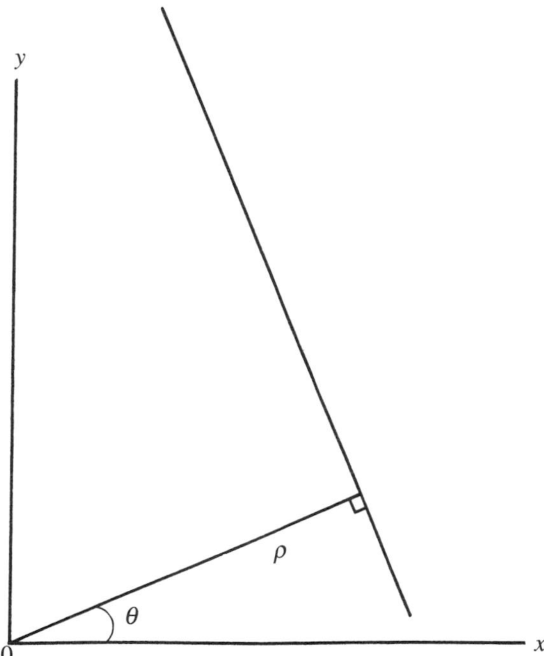

Figure 9.1 Normal (θ, ρ) parametrization of a straight line.

the sine curve has the equation

$$\rho = x_1 \cos\theta + y_1 \sin\theta \tag{9.5}$$

Then multiple hits in (θ, ρ) space indicate, via their θ, ρ values, the presence of lines in the original image.

Each of the methods described above employs an "abstract" parameter space in which multiple hits are sought. Earlier we talked about "plotting" points in parameter space, but in fact the means of looking for hits is to seek peaks that have been built by *accumulation* of data from various sources. Although it might be possible to search for hits by logical operations such as use of the AND function, the Hough method gains considerably by *accumulating evidence* for events by a *voting scheme*. It will be seen below that this is the source of the method's high degree of robustness.

Although the methods described above are mathematically elegant and are capable of detecting lines (or sets of collinear points—which may be completely isolated from each other) amid considerable interfering signals and noise, they are subject to severe computational problems. The reason for this is that every prominent point[1] in the original image gives rise to a great many votes in parameter space. Thus, for a 256×256 image the (m, c) parametrization requires 256 votes to be accumulated, while the (θ, ρ) parametrization requires a similar number—say 360 if the θ quantization is to be fine enough to resolve 1° changes in line orientation.

A vital key to overcoming this problem was discovered by O'Gorman and Clowes (1976), who noted that points on lines are usually not isolated but instead are joined in fragments that are sufficiently long that their directions can be measured. Supposing that direction is known with high accuracy, it is then sufficient to accumulate just one vote in parameter space for every potential line point. (If the local gradient is known with lesser accuracy, then parameter space can be quantized more coarsely—say in 10° steps (O'Gorman and Clowes, 1976)—and again a single vote per point can be accumulated.) This method is much more modest in its computational requirements and was soon adopted by other workers.

The computational load is still quite substantial, however. Not only is a large 2-D storage area needed, but this must be searched carefully for significant peaks—a tedious task if short line segments are being sought. Various means have been tried for cutting down computation further. Dudani and Luk (1978) tackled the problem by trying to separate out the θ and ρ estimations. They accumulated votes first in a 1-D parameter space for θ—that is, a histogram of θ values. (It must not be forgotten that such a histogram is itself a simple form of HT.)[2] Having found suitable peaks in the θ histogram, they then built a ρ histogram for all the points that contributed votes to a given θ peak, and they repeated this for all θ peaks. Thus, two 1-D spaces replace the original 2-D parameter space, with very significant savings in storage and load. Dudani and Luk applied their method to images of outdoor scenes containing houses. These images have the following characteristics: (1) many parallel lines, (2) many lines that are naturally broken into smaller segments (e.g., parts of a roof line broken by windows), and (3) many short-line segments. It seems that the method is particularly well adapted to such scenes, although it may not be so suitable for more general images. For example (as will also be seen with circle and ellipse detection), two-stage methods tend to be less accurate since the first stage is less selective. Biased θ values may result from pairs of lines that would be well separated in a 2-D space. For the same reason, problems of noise tend to be more acute. In addition, any error in estimating θ values is propagated to the ρ determination stage, making values of ρ even less accurate. (The latter problem may be reduced by searching the full (θ, ρ) space locally around the important θ-values, although this procedure increases computation again.)

---

1 For the present purpose it does not matter in what way these points are prominent. They may in fact be edge points, dark specks, centers of holes, and so on. Later we shall consistently take them to be edge points.

2 It is now common for any process to be called an HT if it involves accumulating votes in a parameter space, with the intention of searching for significant peaks to find properties of the original data.

The more efficient methods outlined above require θ to be estimated for each line fragment. Before showing how this is carried out, it is worth noting that straight lines and straight edges are different and should ultimately be detected by different means. Straight edges are probably more common than lines, true lines (such as telephone wires in outdoor scenes) being picked up by Laplacian-type operators and straight edges by edge detectors (such as the Sobel) described in Chapter 5. The two are different in that edges have a unique direction defined by that of the edge normal, whereas lines have 180° rotation symmetry and are not vectors. Hence, application of the HT to the two cases will be subtly different. For concreteness, we here concentrate on straight *edge* detection, although minor modification to the techniques (changing the range of θ from 360° to 180°) permits this method to be adapted to lines.

To proceed with line detection, we first obtain the local components of intensity gradient, then deduce the magnitude g, and threshold it to locate each edge pixel in the image. The gradient magnitude g may readily be computed from the components of intensity gradient:

$$g = \left(g_x^2 + g_y^2\right)^{1/2} \tag{9.6}$$

or using a suitable approximation (Chapter 5). θ may be estimated using the arctan function in conjunction with the local edge gradient components g_x, g_y:

$$\theta = \arctan\left(g_y/g_x\right) \tag{9.7}$$

Since the arctan function has period π, an additional π may have to be added. This can be decided from the signs of g_x and g_y. Once θ is known, ρ can be found from equation (9.4).

9.3 The Foot-of-Normal Method

Davies (1986c) has worked on an alternative means of saving computation that eliminates the use of trigonometric functions such as arctan by employing a different parametrization scheme. As noted earlier, the methods so far described all employ an abstract parameter space in which points bear no immediate relation to image space. In the author's scheme, the parameter space is a second image space, which may be said to be congruent[3] to image space. In some ways, this is

3 That is, parameter space is like image space, *and* each point in parameter space holds information that is immediately relevant to the corresponding point in image space.

highly convenient—for example, while setting up the algorithm—since working parameters and errors can be visualized and estimated more easily. Note also that it is common for there to be an additional space in the framestore which can be used for this purpose (see Chapter 2).

This type of parameter space is obtained in the following way. First, each edge fragment in the image is produced much as required previously so that ρ can be measured, but this time the foot of the normal from the origin is itself taken as a voting position in parameter space (Fig. 9.1). The foot-of-normal position embodies all the information previously carried by the ρ and θ values, and the methods are mathematically essentially equivalent. However, the details differ, as will be seen.

The detection of straight edges starts with the analysis of (1) local pixel coordinates (x, y) and (2) the corresponding local components of intensity gradient (g_x, g_y) for each edge pixel. Taking (x_0, y_0) as the foot of the normal

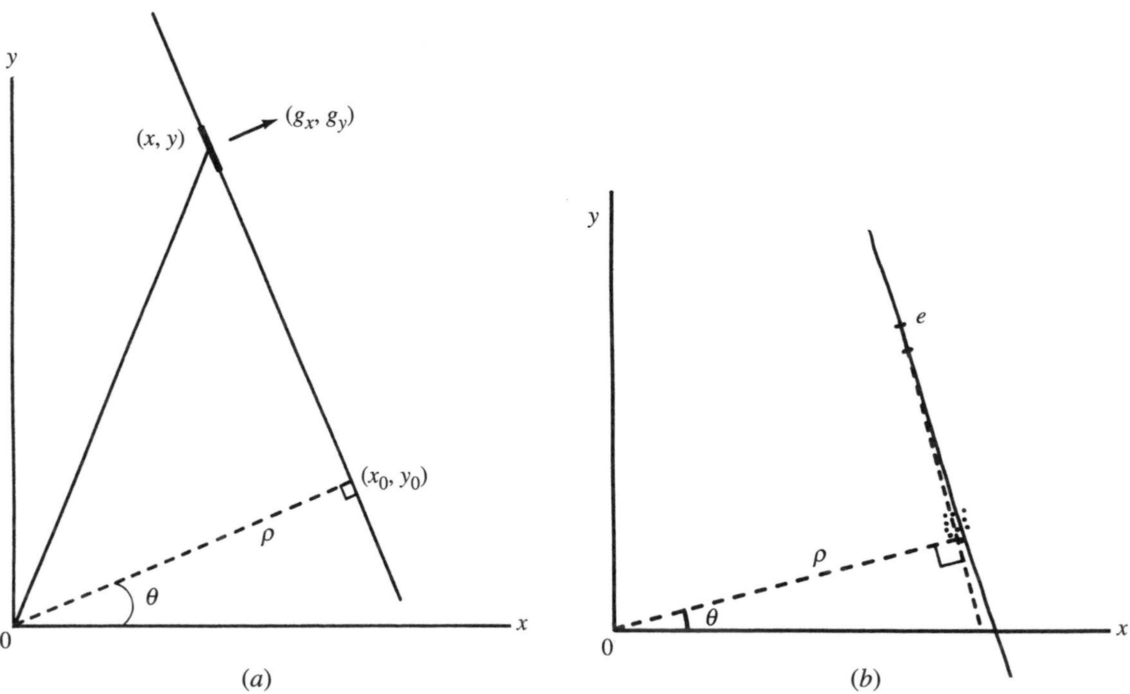

Figure 9.2 Image space parametrization of a straight line: (*a*) parameters involved in the calculation (see text); (*b*) buildup of foot-of-normal positions in parameter space for a more practical situation, where the line is not exactly straight: *e* is a typical edge fragment leading to a single vote in parameter space.

from the origin to the relevant line (produced if necessary—see Fig. 9.2), it is found that

$$g_y/g_x = y_0/x_0 \tag{9.8}$$

$$(x - x_0)x_0 + (y - y_0)y_0 = 0 \tag{9.9}$$

These two equations are sufficient to compute the two coordinates (x_0, y_0). (This must be so since the point (x_0, y_0) is uniquely determined as the join of the line to be detected and the normal through the origin.) Solving for x_0 and y_0 gives

$$x_0 = vg_x \tag{9.10}$$

$$y_0 = vg_y \tag{9.11}$$

where

$$v = \frac{xg_x + yg_y}{g_x^2 + g_y^2} \tag{9.12}$$

Notice that these expressions involve only additions, multiplications, and just one division.

For completeness, we also compute the distance of the line from the origin:

$$\rho = \frac{xg_x + yg_y}{\left(g_x^2 + g_y^2\right)^{1/2}} \tag{9.13}$$

and the vector from (x_0, y_0) to the current pixel (x, y). This has components

$$x - x_0 = wg_y \tag{9.14}$$

$$y - y_0 = -wg_x \tag{9.15}$$

where

$$w = \frac{xg_y - yg_x}{g_x^2 + g_y^2} \tag{9.16}$$

The magnitude of this vector is

$$s = \frac{|xg_y - yg_x|}{\left(g_x^2 + g_y^2\right)^{1/2}} \tag{9.17}$$

The last equation also turns out to be useful for image interpretation, as will be seen below. The next section considers the errors involved in estimating (x_0, y_0) from the raw data and shows how they may be minimized.

9.3.1 *Error Analysis*

The most serious cause of error in estimating (x_0, y_0) is that involved in estimating edge orientation. Pixel location is generally accurate to within 1 pixel, and even when this is not quite valid, errors are largely averaged out during peak location in parameter space. In the following we ignore errors in pixel location and concentrate on edge orientation errors. Figure 9.3 shows the geometry

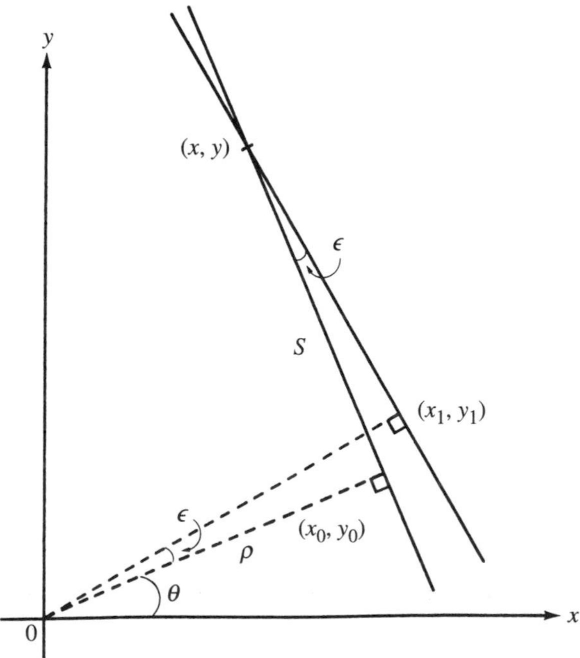

Figure 9.3 Notation for analysis of errors with line parametrization.

of the situation. A small angular error ε causes the estimate of (x_0, y_0) to shift to (x_1, y_1).

To a first approximation, the errors in estimating ρ and s are:

$$\delta\rho = \varepsilon s \tag{9.18}$$

$$\delta s = -\varepsilon\rho \tag{9.19}$$

Thus, errors in both ρ and s due to angular error ε depend on distance from the origin. These errors are minimized overall if the origin is chosen at the center of the image.

The error ε arises from two main sources—intrinsic edge detector errors and noise. The intrinsic error of the Sobel operator is about 1°; thus, image noise will generally be the main cause of inaccuracy. In this respect, it should be noted that many mechanical components possess edges that are straight to a very high precision and that if each pixel of the image is quantized into at least 8 bits (256 gray levels), then aliasing[4] problems are minimized and the ground is laid for accurate line transforms.

Since errors in the foot-of-normal position are proportional to s and ρ, and these in turn are limited by the image size, errors may be reduced overall by dividing the image into a number of separate subimages, each with an origin located at its center. In principle, by this means errors may be reduced to arbitrarily small values. However, decomposing the picture into subimages may reduce the number of (useful) points constituting a peak in parameter space, thereby decreasing the accuracy with which that peak can be located. Sometimes the number of useful points is limited by the available data, as when lines are short in length. In such cases, the subimage method helps by reducing the clutter of noise points in parameter space, thereby improving the detection of such small segments.

The subimage method should be used to increase accuracy and reduce noise when line segments are significantly shorter than the image size. The optimum subimage size depends on the lengths of the lines being detected and the lengths of the short line segments to be detected. If a long line is broken into two or more parts by analysis in subimages, this may not be too much of a disadvantage, since all the information in each line is still available at the end of the computation, although the data from the various subimage peaks will need to be correlated in order to get a complete picture. Indeed, it can be *advantageous* to break down long lines into shorter ones in this way, since this procedure gives crisp output data even if the line is slightly curved and slowly changes its

4 Aliasing is the process whereby straight lines in digital images take a stepped form following sampling decisions that they pass through one pixel but not the next one. The process is often marked in graphics (binary) images and is less marked, though never absent, in gray-scale images. See also Chapter 27.

orientation along its length. Thus, the subimage approach permits an operator to be designed to locate lines of different lengths or moderately curved lines of different shapes.

9.3.2 *Quality of the Resulting Data*

Although the foot-of-normal method is mathematically similar to the (θ, ρ) method, it is unable to determine line orientation to quite the same degree of accuracy. This is because it is mapped onto a discrete image-like parameter space so that the θ accuracy depends on the fractional accuracy in determining ρ—which in turn depends on the absolute magnitude of ρ. Hence, for small ρ the orientation of a line that is predicted from the position of the peak in parameter space will be relatively inaccurate. Indeed, if ρ happens to be zero, then no information on the orientation of the line is available in parameter space (although, as has been seen, the *position* of the foot-of-normal is known accurately).

The difficulty can be overcome in two ways. The first is by referring to a different origin if ρ is too small. If for each edge point three noncollinear origins are available, then it will always be possible to select a value of ρ that is greater than a lower bound. Errors are easy to compute if the subimage method is employed and the origins that are available are those of adjacent subimages. Basically, the situation is that errors in estimating line orientation should not be worse than $2/S$ radians (for subimages of size S), although for large S they would be limited by those (around 1°) introduced by the gradient operator.

The second and more rigorous way of obtaining accurate values of line orientation is by a two-stage method. In this case, of the original set of prominent points, the subset that contributed to a given peak in the first parameter space is isolated and made to contribute to a θ histogram, from which line orientation may be determined accurately.

It is often possible to ignore cases of small ρ or simply to acknowledge that cases of small ρ provide little line orientation information. Although this is inelegant theoretically, the amount of redundancy of many images sometimes makes this practicable, thereby saving computation. For example, if a rectangular object is to be located, there is little chance that more than one side will have ρ close to zero (see Section 9.4). Hence, the object can be located tentatively first of all from the peaks in parameter space for which ρ is large, and the interpretation confirmed with the aid of peaks for which ρ is small or zero. Figure 9.4 confirms that this can be a useful approach. Clearly, any outstanding instances of ambiguity can still be resolved by one of the methods outlined here.

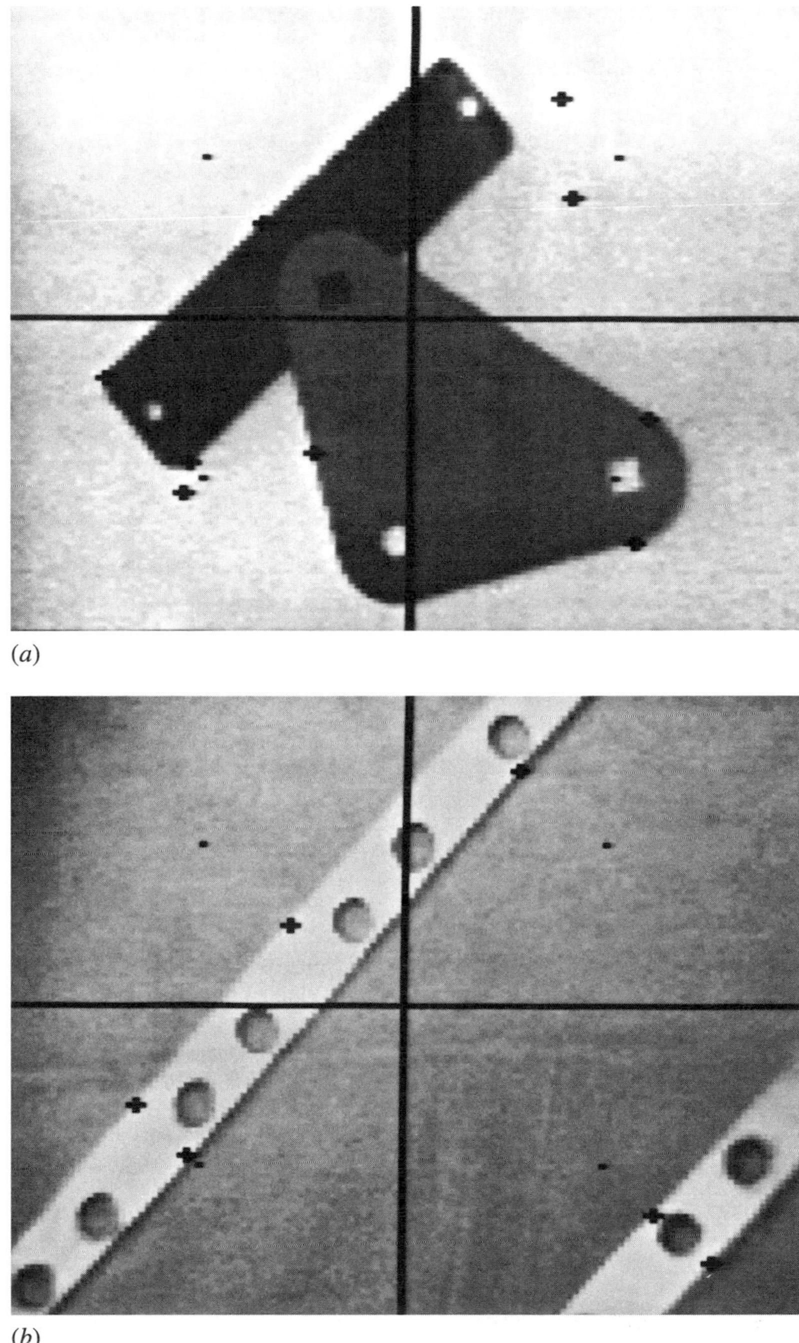

(*a*)

(*b*)

Figure 9.4 Results of image space parametrization of mechanical parts. The dot at the center of each quadrant is the origin used for computing the image space transform. The crosses are the positions of peaks in parameter space, which mark the individual straight edges (produced if necessary). For further explanation, see text.

9.3.3 *Application of the Foot-of-Normal Method*

The methods outlined here have been tested on real images containing various mechanical components. For objects with straight edges varying between about 20 and 80 pixels in length, it was found convenient to work with subimages of size 64×64.

Typical results (in this case for 128×128 images with 64 gray levels) are shown in Fig. 9.4. Some of the objects in these pictures are grossly overdetermined by their straight edges, so low ρ values are not a major problem. For those peaks where $\rho > 10$, line orientation is estimated within approximately 2°. As a result, these objects are located within 1 pixel and oriented within 1° by this technique, without the necessity for θ histograms. Figures 9.4*a* and *b* contain some line segments that are not detected. This is due partly to their relatively low contrast, higher noise levels, above-average fuzziness, or short length. It is also due to the thresholds set on the initial edge detector and on the final peak detector. When these were set at lower values, additional lines were detected, but other noise peaks also became prominent in parameter space, and each of these needed to be checked in detail to confirm the presence of the corresponding line in the image. This is one aspect of a general problem that arises right through the field of image analysis.

9.4 Longitudinal Line Localization

The preceding sections have provided a variety of means for locating lines in digital images and finding their orientations. However, these methods are insensitive to where along an infinite idealized line an observed segment appears. The reason is that the fit includes only two parameters. Some advantage can be gained in that partial occlusion of a line does not prevent its detection. Indeed, if several segments of a line are visible, they can all contribute to the peak in parameter space, thereby improving sensitivity. On the other hand, for full image interpretation, it is useful to have information about the longitudinal placement of line segments.

This is achieved by a further stage of processing (which may make line finding a two- or three-stage algorithm overall). The additional stage involves finding which points contributed to each peak in the main parameter space and carrying out connectivity analysis in each case. Dudani and Luk (1978) called this process "*xy*-grouping." It is not vital that the line segments should be 4- or 8-connected—just that there should be sufficient points on them so that adjacent points are within a threshold distance apart; that is, groups of points are merged if they are within the prespecified threshold distance. (Dudani and Luk took this distance as being 5 pixels.) Finally, segments shorter than a certain minimum length can be ignored as too insignificant to help with image interpretation.

The distance s in the foot-of-normal method can be employed as a suitable grouping parameter, although an alternative is to use either the x or y projection, depending on the orientation of the line (i.e., use x if $|m| < 1$, and y otherwise).

9.5 Final Line Fitting

Dudani and Luk (1978) found it useful to include a least-squares fitting stage to complete the analysis of the straight edges present in an image. This is carried out by taking the x, y coordinates of points on the line as input data and minimizing the sum of squares of distances from the data points to the best fit line. They added this stage in order to eliminate local edge orientation accuracy as a source of error. This motivation is generally reasonable if the highest possible accuracy is required (e.g., obtaining line orientation to significantly better than 1°). However, many workers have criticized the use of the least-squares technique, since it tends to weight up the contribution of less accurate points—including those points that have nothing to do with the line in question but that arise from image "clutter." This criticism is probably justified, although the least-squares technique is convenient in yielding to mathematical analysis and in this case in giving explicit equations for θ and ρ in terms of the input data (Dudani and Luk, 1978). Dudani and Luk obtained the endpoints by reference to the best fit line thus obtained.

9.6 Concluding Remarks

All the techniques for finding lines and straight edges in digital images described here are based on the HT, since the HT is virtually universally used for this purpose. The HT is very important throughout the field of image analysis, permitting certain types of global data to be extracted systematically from images and ignoring "local" problems due, for example, to occlusions and noise. The HT is therefore an example of intermediate-level processing, which may be defined as the conversion of crucial features in the image to a more abstract form that is no longer restricted to the image representation.

The specific techniques covered in this chapter have involved a particular parametrization of a straight line, as well as the means of improving efficiency and accuracy. Speed is improved particularly by resorting to a two-stage line-finding procedure—a method that is useful in other applications of the HT, as will be seen in later chapters. Accuracy tends to be reduced by such two-stage processing because of error propagation *and* because the first stage is subject to too many interfering signals. However, it is possible to take an approximate solution (in this case an approximation to a set of lines in an image) and to improve accuracy by including yet another stage of processing. Two examples in the chapter were adding a line orientation stage to the foot-of-normal method and adding a line-fitting procedure to refine the output of an HT procedure (although there are problems with using the least-squares technique for this purpose). Again, later chapters illustrate these types of accuracy-enhancing procedures applied in other tasks such as circle center location.

Overall, use of the HT is central to image analysis and will be encountered many times in the remainder of the book. However, it is always important to remember its limitations, especially its considerable storage and computational requirements. Further discussion of HT line finding schemes, including those aimed at saving computation, is found in Chapter 11.

The Hough transform is one way of inferring the presence of objects from their feature points. This chapter has shown how line detection may be achieved by a simple voting scheme in a carefully selected parameter space representation, starting with edge features. The transform is highly robust as it focuses only on positive evidence for the existence of objects.

9.7 Bibliographical and Historical Notes

The Hough transform (HT) was developed in 1962 (Hough, 1962) with the aim of finding (straight) particle tracks in high-energy nuclear physics, and it was brought into the mainstream image analysis literature much later by Rosenfeld (1969). Duda and Hart (1972) developed the method further and applied it to the detection of lines and curves in digital pictures. O'Gorman and Clowes (1976) soon developed a Hough-based scheme for finding lines efficiently, by making use of edge orientation information, at much the same time that Kimme et al. (1975) applied the same method (apparently independently) to the efficient location of circles. Many of the ideas for fast effective line finding described in this chapter emerged in a paper by Dudani and Luk (1978). Davies' foot-of-normal method (1986c) was developed much later. During the 1990s, work in this area progressed further—see, for example Atiquzzaman and Akhtar's (1994) method for the efficient determination of lines together with their end coordinates and lengths; Lutton et al.'s (1994) application of the transform to the determination of vanishing points; Marchant and Brivot's (1995) application to the location of plant rows for tractor guidance (see also Chapter 20); and Kamat-Sadekar and Ganesan's (1998) extensions of the technique to cover the reliable detection of multiple line segments, particularly in relation to road scene analysis. Other applications of the HT are covered in the following few chapters, while further methods for locating lines are described in Chapter 11 on the generalized HT.

Some mention should be made of the related Radon transform. It is formed by integrating the picture function $I(x, y)$ along infinitely thin straight strips of the image, with normal coordinate parameters (θ, ρ), and recording the results in a (θ, ρ) parameter space. The Radon transform is a generalization of the Hough transform for line detection (Deans, 1981). For straight lines, the Radon transform reduces to the Duda and Hart (1972) form of the HT, which, as remarked earlier, involves considerable computation. For this reason the Radon transform is not covered in depth in this book. The transforms of real lines have a characteristic "butterfly" shape (a pencil of segments passing through the corresponding peak) in parameter space. This phenomenon has been investigated by Leavers and Boyce (1987), who have devised special 3×3 convolution filters for sensitively detecting these peaks.

A few years ago there was a body of opinion that the HT is completely worked over and no longer a worthwhile topic for research. However, this view seems misguided, and there is continuing strong interest in the subject—as the following chapters will show. After all, the computational difficulties of the method reflect a problem that is inescapable in image analysis as a whole, so development of methods must continue. In this context, note especially Sections 11.10 and 11.11.

Most recently, Schaffalitsky and Zisserman (2000) carried out an interesting extension of earlier ideas on vanishing lines and points by considering the case of repeated lines such as those that occur on certain types of fences and brick buildings. Song et al. (2002) developed HT methods for coping with the problems of fuzzy edges and noise in large-sized images; Davies (2003b) explored how the size of the parameter space could usefully be minimized; and

Guru et al. (2004) demonstrated that there are viable alternatives to the Hough transform, based, for example, on heuristic search achieved by small eigenvalue analysis.

9.8 Problems

1. (a) In the foot-of-normal Hough transform, straight edges are found by locating the foot-of-the-normal F (x_f, y_f) from an origin O (0, 0) at the center of the image to an extended line containing each edge fragment E (x, y), and placing a vote at F in a separate image space.

   (b) By examining the possible positions of lines within a square image and the resulting foot-of-normal positions, determine the exact extent of the parameter space that is theoretically required to form the Hough transform.

   (c) Would this form of the Hough transform be expected to be (i) more or less robust and (ii) more or less computation intensive than the (ρ, θ) Hough transform for line location?

2. (a) Why is it sometimes stated that a Hough transform generates *hypotheses* rather than actual solutions for object location? Is this statement justified?

   (b) A new type of Hough transform is to be devised for detecting straight lines. It will take every edge fragment in the image and extend it in either direction until it meets the boundary of the image, and then it will accumulate a vote at each position. Thus, *two* peaks should result for every line. Explain why finding these peaks will require *less* computation than for the standard Hough transform, but that deducing the presence of lines will then require *extra* computation. How will these amounts of computation be likely to vary with (i) the size of the image, and (ii) the number of lines in the image?

   (c) Discuss whether this approach would be an improvement on the standard approach for straight-line location and whether it would have any disadvantages.

3. (a) Describe the *Hough transform* approach to object location. Explain its advantages relative to the *centroidal* (r, θ) *plot* approach: illustrate your answer with reference to the location of circles of known radius R.

   (b) Describe how the Hough transform may be used to locate straight edges. Explain what is seen in the parameter space if many curved edges also appear in the original image.

(c) Explain what happens if the image contains a square object of arbitrary size and *nothing* else. How would you deduce from the information in parameter space that a square object is present in the image? Give the main features of an algorithm to decide that a square object is present and to locate it.

(d) Examine in detail whether an algorithm using the strategy described in (c) would become confused if (i) parts of some sides of the square were occluded; (ii) one or more sides of the square were missing; (iii) several squares appeared in the image; (iv) several of these complications occurred together.

(e) How important is it to this type of algorithm to have edge detectors that are capable of accurately determining edge orientation? Describe a type of edge detector that is capable of determining orientation accurately.

Circle Detection

As in the case of straight lines, circle features occur very widely in digital images taken from manufactured parts (including many food products). Again they can conveniently be located by using the Hough transform approach. This chapter describes the processes and principles that are involved and shows the means by which speed and accuracy can be improved.

Look out for:

- the basic Hough transform technique for locating circular objects in images.
- how the method has to be adapted if the radius is unknown.
- how accuracy of the center location can be improved.
- how speed of processing can be improved.
- how speed can be improved further by a line sampling procedure.
- how robustness only falls gradually as available signal decreases.

This chapter shows how the Hough transform is again able to provide a useful means for detecting a selected type of macroscopic feature in digital images. Again the method is seen to rely on the accumulation of votes in a parameter space, and the robustness of the technique to result from concentration on *positive* evidence for the objects.

CHAPTER 10

Circle Detection

10.1 Introduction

The location of round objects is important in many areas of image analysis, but it is especially important in industrial applications such as automatic inspection and assembly. In the food industry alone, a sizable proportion of products are round—biscuits, cakes, pizzas, pies, oranges, and so on (Davies, 1984c). In the automotive industry many circular components are used—washers, wheels, pistons, heads of bolts, and so on—while round holes of various sizes appear in such items as casings and cylinder blocks. In addition, buttons and many other everyday objects are round. All these manufactured parts need to be checked, assembled, inspected, and measured with high precision. Finally, objects can frequently be located by their holes, so finding round holes or features is part of a larger problem. Since round picture objects form a special category of their own, efficient algorithms should be available for analyzing digital images containing them. This chapter addresses some aspects of this problem.

An earlier chapter showed how objects may be found and identified using algorithms that involve boundary tracking (Table 10.1). Unfortunately, such algorithms are insufficiently robust in many applications; for example, they tend to be muddled by artifacts such as shadows and noise. This chapter shows that the Hough transform (HT) technique is able to overcome many of these problems. It is found to be particularly good at dealing with all sorts of difficulties, including quite severe occlusions. It achieves this capability not by adding robustness but by having robustness built in as an integral part of the technique.

The application of the HT to circle detection is perhaps the most straightforward use of the technique. However, several enhancements and adaptations can be applied in order to improve the accuracy and speed of operation, and in addition to make the method work efficiently when detecting circles with a range of sizes. We will examine these modifications following a description of the basic HT techniques.

Table 10.1 Procedure for finding objects using (r, θ) boundary graphs

1.	Locate edges within the image.
2.	Link broken edges.
3.	Thin thick edges.
4.	Track around object outlines.
5.	Generate a set of (r, θ) plots.
6.	Match (r, θ) plots to standard templates.

This procedure is not sufficiently robust with many types of real data (e.g., in the presence of noise, distortions in product shape, etc). In fact, it is quite common to find the tracking procedure veering off and tracking around shadows or other artifacts.

10.2 Hough-based Schemes for Circular Object Detection

In the original HT method for finding circles (Duda and Hart, 1972), the intensity gradient is first estimated at all locations in the image and is then thresholded to give the positions of significant edges. Then the positions of all possible center locations—namely, all points a distance R away from every edge pixel—are accumulated in parameter space, R being the anticipated circle radius. Parameter space can be a general storage area, but when looking for circles it is convenient to make it congruent to image space; in that case, possible circle centers are accumulated in a new plane of image space. Finally, parameter space is searched for peaks that correspond to the centers of circular objects. Since edges have nonzero width and noise will always interfere with the process of peak location, accurate center location requires the use of suitable averaging procedures (Davies, 1984c; Brown, 1984).

This approach requires that a large number of points be accumulated in parameter space, and so a revised form of the method has now become standard. In this approach, locally available edge orientation information at each edge pixel is used to enable the exact positions of circle centers to be estimated (Kimme et al., 1975). This is achieved by moving a distance R along the edge normal at each edge location. Thus, the number of points accumulated is equal to the number of edge pixels in the image.[1] This represents a significant saving in computational load. For this saving to be possible, the edge detection operator that is employed must be highly accurate. Fortunately, the Sobel operator is able to estimate edge

1 We assume that objects are known to be *either* lighter *or* darker than the background, so that it is only necessary to move along the edge normal in one direction.

orientation to 1° and is very simple to apply (Chapter 5). Thus, the revised form of the transform is viable in practice.

As was seen in Chapter 5, once the Sobel convolution masks have been applied, the local components of intensity gradient g_x and g_y are available, and the magnitude and orientation of the local intensity gradient vector can be computed using the formulas:

$$g = \left(g_x^2 + g_y^2\right)^{1/2} \tag{10.1}$$

and

$$\theta = \arctan\left(g_y/g_x\right) \tag{10.2}$$

Use of the arctan operation is not necessary when estimating center location coordinates (x_c, y_c) since the trigonometric functions can be made to cancel out:

$$x_c = x - R(g_x/g) \tag{10.3}$$

$$y_c = y - R\left(g_y/g\right) \tag{10.4}$$

the values of cos θ and sin θ being given by:

$$\cos\theta = g_x/g \tag{10.5}$$

$$\sin\theta = g_y/g \tag{10.6}$$

In addition, the usual edge thinning and edge linking operations—which normally require considerable amounts of processing—can be avoided if a little extra smoothing of the cluster of candidate center points can be tolerated (Davies, 1984c) (Table 10.2). Thus, this Hough-based approach can be a very efficient one for

Table 10.2 A Hough-based procedure for locating circular objects

1.	Locate edges within the image.
2.	Link broken edges.
3.	Thin thick edges.
4.	For every edge pixel, find a candidate center point.
5.	Locate all clusters of candidate centers.
6.	Average each cluster to find accurate center locations.

This procedure is particularly robust. It is largely unaffected by shadows, image noise, shape distortions, and product defects. Note that stages 1–3 of the procedure are identical to stages 1–3 in Table 10.1. However, in the Hough-based method, computation can be saved, and accuracy actually increased, by omitting stages 2 and 3.

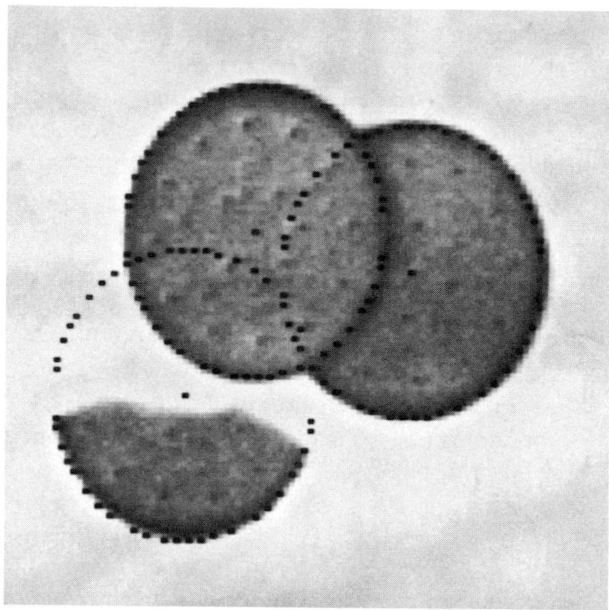

Figure 10.1 Location of broken and overlapping biscuits, showing the robustness of the center location technique. Accuracy is indicated by the black dots, which are each within 1/2 pixel of the radial distance from the center.

locating the centers of circular objects, virtually all the superfluous operations having been eliminated, leaving only edge detection, location of candidate center points, and center point averaging to be carried out. It is also relevant that the method is highly robust in that if part of the boundary of an object is obscured or distorted, the object center is still located accurately. Some remarkable instances of this can be seen in Figs. 10.1 and 10.2. The reason for this useful property is clear from Fig. 10.3.

This technique is so efficient that it takes slightly less time to perform the actual Hough transform part of the calculation than to evaluate and threshold the intensity gradient over the whole image. Part of the reason for these relative timings is that the edge detector operates within a 3×3 neighborhood and necessitates some 12 pixel accesses, four multiplications, eight additions, two subtractions, and an operation for evaluating the square root of sum of squares (equation (10.1)). As seen in Chapter 5, the latter type of operation is commonly approximated by taking a sum or maximum of absolute magnitudes of the component intensity gradients, in order to estimate the magnitude of the local intensity gradient vector:

$$g \approx |g_x| + |g_y| \tag{10.7}$$

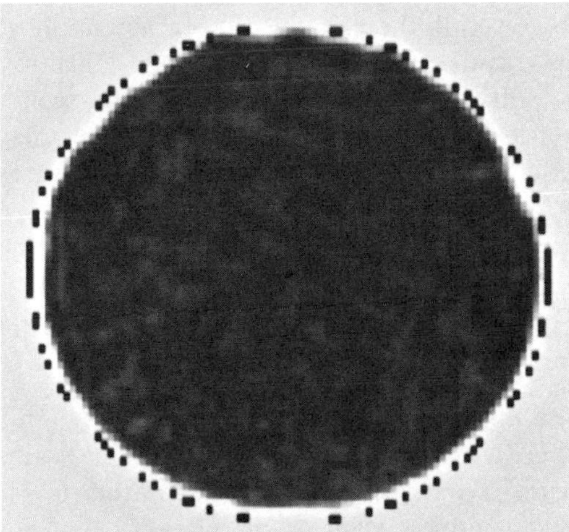

Figure 10.2 Location of a biscuit with a distortion, showing a chocolate-coated biscuit with excess chocolate on one edge. Note that the computed center has not been "pulled" sideways by the protuberances. For clarity, the black dots are marked 2 pixels outside the normal radial distance.

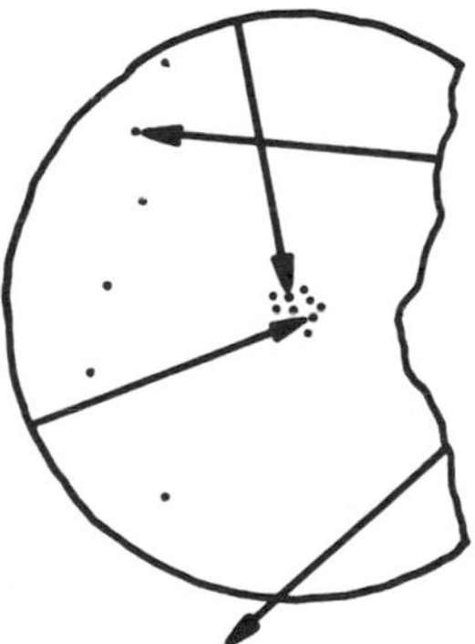

Figure 10.3 Robustness of the Hough transform when locating the center of a circular object. The circular part of the boundary gives candidate center points that focus on the true center, whereas the irregular broken boundary gives candidate center points at random positions.

or

$$g \approx \max(|g_x|, |g_y|) \tag{10.8}$$

However, this type of shortcut is not advisable in the present context, since an accurate value for the magnitude of this vector is required in order to compute the position of the corresponding candidate center location with sufficient precision. However, lookup tables provide a viable way around this problem.

Overall, the dictates of accuracy imply that candidate center location requires significant computation. However, substantial increases in speed are still possible by software means alone, as will be seen later in the chapter. Meanwhile, we consider the problems that arise when images contain circles of many different radii, or for one reason or another radii are not known in advance.

10.3 The Problem of Unknown Circle Radius

In a number of situations circle radius is initially unknown. One such situation is where a number of circles of various sizes are being sought—as in the case of coins, or different types of washers. Another is where the circle size is variable—as for food products such as biscuits—so that some tolerance must be built into the system. In general, all circular objects have to be found and their radii measured. In such cases, the standard technique is to accumulate candidate center points simultaneously in a number of parameter planes in a suitably augmented parameter space, each plane corresponding to one possible radius value. The centers of the peaks detected in parameter space give not only the location of each circle in two dimensions but also its radius. Although this scheme is entirely viable in principle, there are several problems in practice:

1. Many more points have to be accumulated in parameter space.
2. Parameter space requires much more storage.
3. Significantly greater computational effort is involved in searching parameter space for peaks.

To some extent this is to be expected, since the augmented scheme enables more objects to be detected directly in the original image.

We will shortly show that problems (2) and (3) may largely be eliminated by using just one parameter plane to store all the information for locating circles of different radii, that is, accumulating not just one point per edge pixel but a whole line of points along the direction of the edge normal in this one plane. In practice, the line need not be extended indefinitely in either direction but only over the restricted range of radii over which circular objects or holes might be expected.

Even with this restriction, a large number of points are being accumulated in a single parameter plane, and it might initially be thought that this would lead to such a proliferation of points that almost any "blob" shape would lead to a peak in parameter space might be interpreted as a circle center. However, this is not so, and significant peaks normally result only from genuine circles and some closely related shapes.

To understand the situation, consider how a sizable peak can arise at a particular position in parameter space. This can happen only when a large number of radial vectors from this position meet the boundary of the object normally. In the absence of discontinuities in the boundary, a contiguous set of boundary points can be normal to radius vectors only if they lie on the arc of a circle. (Indeed, a circle could be *defined* as a locus of points that are normal to the radius vector and that form a thin connected line.) If a limited number of discontinuities are permitted in the boundary, it may be deduced that shapes like that of a fan[2] will also be detected using this scheme. Since it is in any case useful to have a scheme that can detect such shapes, the main problem is that there will be some ambiguity in interpretation—that is, does peak P in parameter space arise from a circle or a fan? In practice, it is quite straightforward to distinguish between such shapes with relatively little additional computation; the really important problem is to cut down the amount of computation needed to key into the initially unstructured image data. It is often a good strategy to prescreen the image to eliminate most of it from further detailed consideration and then to analyze the remaining data with tools having much greater discrimination. This two-stage template matching procedure frequently leads to significant savings in computation (VanderBrug and Rosenfeld, 1977; Davies, 1988i).

A more significant problem arises because of errors in the measurement of local edge orientation. As stated earlier, edge detection operators such as the Sobel introduce an inherent inaccuracy of about 1°. Image noise typically adds a further 1° error, and for objects of radius 60 pixels the result is an overall uncertainty of about 2 pixels in the estimation of center location. This makes the scheme slightly less specific in its capability for detecting objects. In particular, edge pixels can now be accepted when their orientations are not exactly normal to the radius vector, and an inaccurate position will be accumulated in parameter space as a possible center point. Thus, the method may occasionally accept cams or other objects with a spiral boundary as potential circles.

Overall, the scheme is likely to accept the following object shapes during its prescreening stage, and the need for a further discrimination stage will be highly application-dependent:

1. Circles of various sizes
2. Shapes such as fans which contain arcs of circles
3. Spirals with almost normal radius vectors
4. Partly occluded or broken versions of these shapes

2 In this chapter all references to "fan" mean a four-blade fan shape bounded by two concentric circles with eight equally spaced radial lines joining them.

If occlusions and breakages can be assumed to be small, then it will be safe to apply a threshold on peak size at, say, 70% of that expected from the smallest circle.[3] In that case, it will be necessary to consider only whether the peaks arise from objects of types (1), (2), or (3). If, as usually happens, objects of types (2) or (3) are known *not* to be present, the method is immediately a very reliable detector of circles of arbitrary size. Sometimes specific rigorous tests for objects of types (2) and (3) will still be needed. They may readily be distinguished, for example, with the aid of corner detectors or from their radial intensity histograms (see Chapters 14 and 22).

Next, note that the information on radial distance has been lost by accumulating all votes in a single parameter plane. Hence, a further stage of analysis is needed to measure (or classify) object radius. This is so even if a stage of disambiguation is no longer required to confirm that the object is of the desired type. This extra stage of analysis normally involves negligible additional computation because the search space has been narrowed down so severely by the initial circle location procedure (see Figs. 10.4–10.6). The radial histogram technique already referred to can be used to measure the radius: in addition it can be used to perform a final object inspection function (see Chapter 22).

10.3.1 *Experimental Results*

The method described above works much as expected, the main problems arising with small circular objects (of radii less than about 20 pixels) at low resolution (Davies, 1988b). Essentially, the problems are those of lack of discrimination in the precise shapes of small objects (see Figs. 10.4–10.6), as anticipated earlier. More specifically, if a particular shape has angular deviations $\delta\psi$ relative to the ideal circular form, then the lateral displacement at the center of the object is $r\delta\psi$. If this lateral displacement is 1 pixel or less, then the peak in parameter space is broadened only slightly and its height is substantially unchanged. Hence, detection sensitivity is similar to that for a circle. This model is broadly confirmed by the results of Fig. 10.6.

Theoretical analysis of the response can be carried out for any object by obtaining the shape of the evolute (envelope of the normals—see, for example, Tuckey and Armistead, 1953)—and determining the extent to which it spreads the peak in parameter space. However, performance is much as expected intuitively on the basis of the ideas already described, except that for small objects shape

3 Note, however, that size-dependent effects can be eliminated by appropriate weighting of the votes cast at different radial distances.

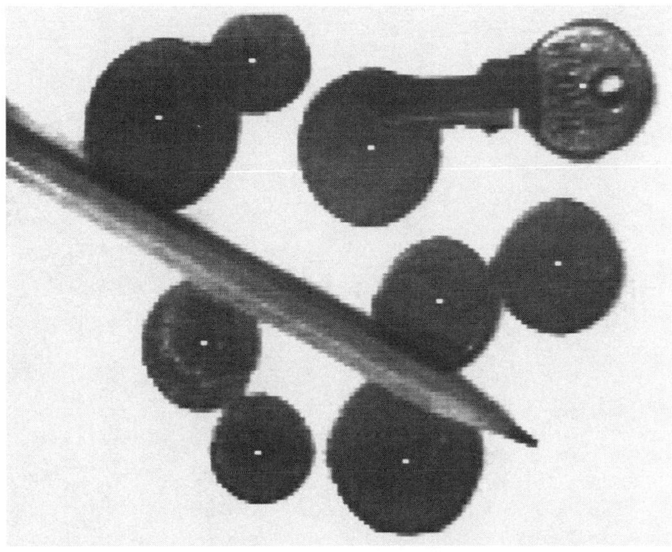

(*a*)

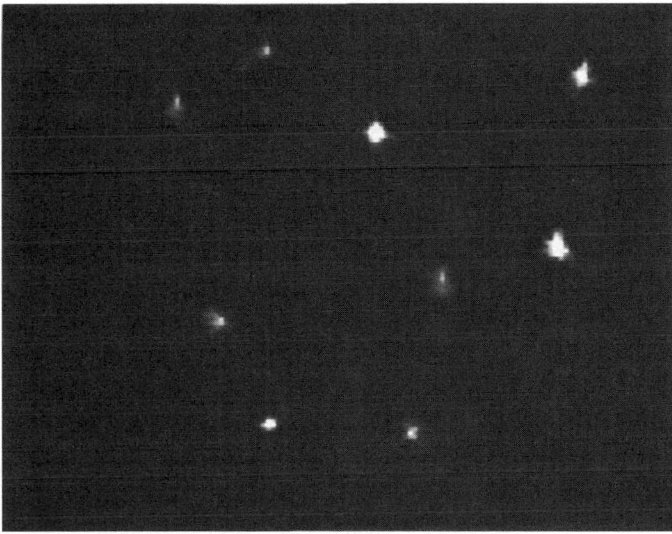

(*b*)

Figure 10.4 (*a*) Simultaneous location of coins and a key with the modified Hough scheme: the various radii range from 10 to 17 pixels; (*b*) transform used to compute the centers indicated in (*a*). Detection efficiency is unimpaired by partial occlusions, shape distortions, and glints. However, displacements of some centers are apparent. In particular, one coin (top left) has only two arcs showing. The fact that one of these is distorted, giving a lower curvature, leads to a displacement of the computed center. The shape distortions are due to rather uneven illumination.

(*a*)

(*b*)

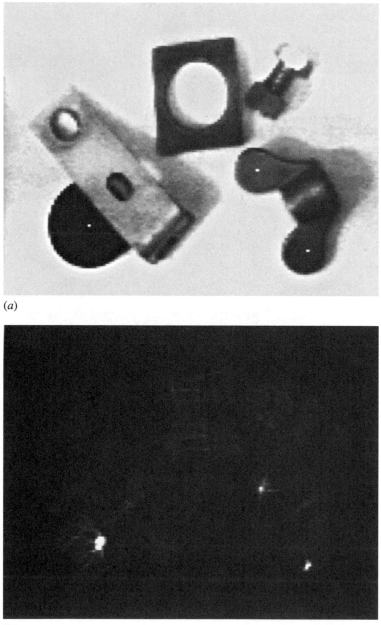

Figure 10.5 (*a*) Accurate simultaneous detection of a lens cap and a wing nut when radii are assumed to range from 4 to 17 pixels; (*b*) response in parameter space that arises with such a range of radii: note the overlap of the transforms from the lens cap and the bracket; (*c*) hole detection in the image of (*a*) when radii are assumed to fall in the range –26 to –9 pixels (negative radii are used since holes are taken to be objects of negative contrast). Clearly, in *this* image a smaller range of negative radii could have been employed.

(c)

Figure 10.5 Continued.

discrimination is relatively poor. As suggested earlier, this can frequently be turned to advantage in that the method becomes a circular feature detector for small radii (see Fig. 10.5, where a wing nut is located).

Finally, objects are detected reliably even when they are partly occluded. However, occlusions can result in the centers of small objects being "pulled" laterally (Fig. 10.4)—a situation that seldom occurs with the original Hough scheme. When this is a problem in practical situations, it may be reduced by (1) cutting down the range of values of r that are employed in any one parameter plane (and hence using more than one plane), and (2) increasing resolution. Generally, subpixel accuracy of center location cannot be expected when a single-parameter plane is used to detect objects over a range of sizes exceeding a factor of about two. This is because of the amount of irrelevant "clutter" that appears in parameter space, which has the effect of reducing the signal-to-noise ratio (see Fig. 10.5). Similarly, attempting to find circular objects *and* holes simultaneously in the same parameter plane mitigates against good performance unless the sum of the ranges of radius values is restricted.

Thus, the approach outlined here involves a tradeoff between speed and accuracy. For high accuracy a relatively small range of values of r should be

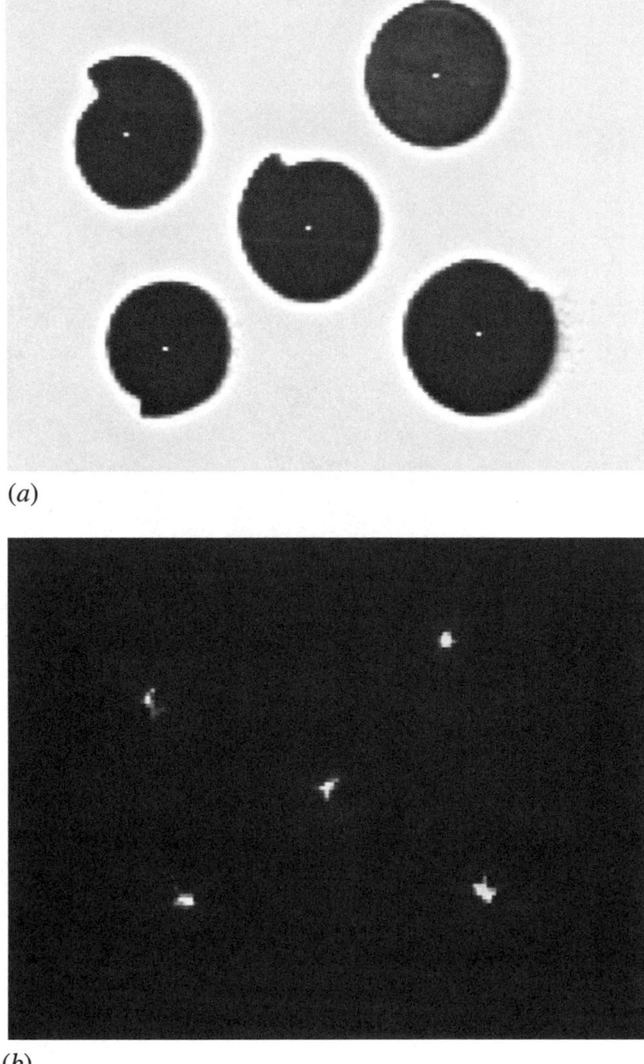

(*a*)

(*b*)

Figure 10.6 (*a*) Simultaneous detection of four (approximately) equiangular spirals and a circle of similar size (16 pixels radius), for assumed range of radii 9 to 22; (*b*) transform of (*a*). In the spiral with the largest step, the step size is about 6 pixels, and detection is possible only if some displacement of the center is acceptable. In such cases, the center is effectively the point on the evolute of the shape in parameter space that happens to have the maximum value. Note that cams frequently have shapes similar to those of the spirals shown here.

employed in any one parameter plane, whereas for high speed one parameter plane will often cover the detection of all sizes of circles with sufficient accuracy. There is also the possibility of two-stage recognition (essentially an extension of the approach already outlined in Section 10.3) in order to find circles first with moderate and then with optimal accuracy.

These results using the modified Hough scheme confirm that we can locate objects of various radii within a parameter space of limited size. The scheme saves on storage requirements and on the amount of computation required to search parameter space for peaks, although it is not able to reduce the total number of votes that have to be accumulated in parameter space. A particular feature of the scheme is that it is essentially two-stage—the second stage being needed (when appropriate) for shape disambiguation and for size analysis. Two-stage procedures are commonly used to make image analysis more efficient, as will be seen in other chapters of this book (see especially Chapter 12 on ellipse detection).

10.4 The Problem of Accurate Center Location

The previous section analyzed the problem of how to cope efficiently with images containing circles of unknown or variable size. This section examines how circle centers may be located with high (preferably subpixel) accuracy. This problem is relevant because the alternative of increased resolution would entail the processing of many more pixels. Hence, attaining a high degree of accuracy with images of low or moderate resolution will provide an advantage.

There are several causes of inaccuracy in locating the centers of circles using the HT: (1) natural edge width may add errors to any estimates of the location of the center (this applies whether or not a thinning algorithm is employed to thin the edge, for such algorithms themselves impose a degree of inaccuracy even when noise is unimportant); (2) image noise may cause the radial position of the edge to become modified; (3) image noise may cause variation in the estimates of local edge orientation; (4) the object itself may be distorted, in which case the most accurate and robust estimate of the circle center is required (i.e., the algorithm should be able to ignore the distortion and take a useful average from the remainder of the boundary); (5) the object may appear distorted because of inadequacies in the method of illumination; and (6) the Sobel or other edge orientation operator will have a significant inherent inaccuracy of its own (typically 1°—see Chapter 5), which then leads to inaccuracy in estimating center location.

Evidently, it is necessary to minimize the effects of all these potential sources of error. In applying the HT, it is usual to remember that all possible center locations that could have given rise to the currently observed edge pixel should be accumulated in parameter space. Thus, we should accumulate in parameter space not a single candidate center point corresponding to the given edge pixel, but a point spread function (PSF) that may be approximated by a Gaussian error function. This will generally have different radial and transverse standard deviations. The radial standard deviation will commonly be 1 pixel—corresponding to the expected width of the edge—whereas the transverse standard deviation will frequently be at least 2 pixels, and for objects of moderate (60 pixels) radius the intrinsic orientation accuracy of the edge detection operator contributes at least 1 pixel to this standard deviation.

It would be convenient if the PSF arising from each edge pixel were isotropic. Then errors could initially be ignored and an isotropic error function could be convolved with the entire contents of parameter space, thus saving computation. However, this is not possible if the PSF is anisotropic, since the information on how to apply the local error function is lost during the process of accumulation.

When the center need only be estimated to within 1 pixel, these considerations hardly matter. However, when locating center coordinates to within 0.1 pixel, each PSF will have to contain 100 points, all of which will have to be accumulated in a parameter space of much increased spatial resolution, or its equivalent,[4] if the required accuracy is to be attained. However, a technique is available that can cut down on the computation, without significant loss of accuracy, as will be seen below.

10.4.1 *Obtaining a Method for Reducing Computational Load*

The HT scheme outlined here for accurately locating circle centers is a completely parallel approach. Every edge pixel is examined independently, and only when this process is completed is a center looked for. Since there is no obvious means of improving the underlying Hough scheme, a sequential approach appears to be required if computation is to be saved. Unfortunately, sequential methods are often less robust than parallel methods. For example, an edge tracking procedure can falter and go wandering off, following shadows and other artifacts around the image. In the current application, if a sequential scheme is to be successful in the sense of being fast, accurate, *and* robust, it must be subtly designed and retain some of the features of the original parallel approach. The situation is now explored in more detail.

First, note that most of the inaccuracy in calculating the position of the center arises from transverse rather than radial errors. Sometimes transverse errors can fragment the peak that appears in the region of the center, thereby necessitating a

---

4 For example, some abstract list structure might be employed that effectively builds up to high resolution only where needed (see also the adaptive HT scheme described in Chapter 11).

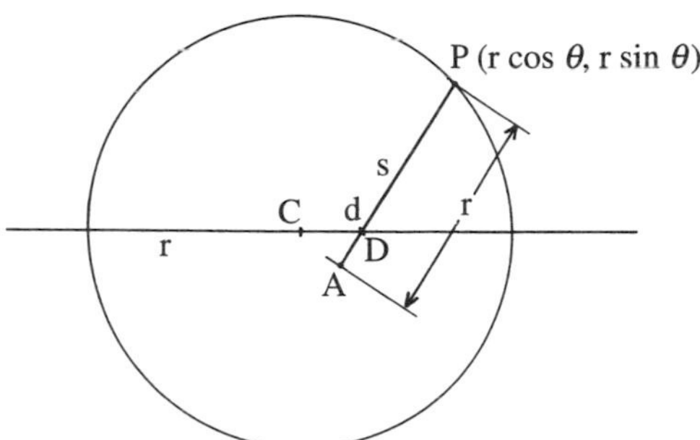

Figure 10.7 Arrangement for obtaining improved center approximation.

considerable amount of computation to locate the real peak accurately. Thus, it is reasonable to concentrate on eliminating transverse errors.

This immediately leads to the following strategy: find a point D in the region of the center and use it to obtain a better approximation A to the center by moving from the current edge pixel P a distance equal to the expected radius r in the direction of D (Fig. 10.7). Then repeat the process one edge pixel at a time until all edge pixels have been taken into account. Although intuition indicates that this strategy will lead to a good final approximation to the center, the situation is quite complex. There are two important questions:

1. Is the process guaranteed to converge?
2. How accurate is the final result?

There is a small range of boundary pixels for which the method described here gives a location A that is *worse* than the initial approximation D (see Fig. 10.8). However, this does not mean that our intuition is totally wrong, and theory confirms the usefulness of the approach when d is small ($d \ll r$). Figure 10.9 shows a set of graphs that map out the loci of points forming second approximations to the center positions for initial approximations at various distances d from the center. In each graph the locus is that obtained on moving around the periphery of the circle. Generally, for small d it is seen that the second approximation is on average a significantly better approximation than the initial approximation. Theory shows that for small d the mean result will be an improvement by a factor close to 1.6 (the limiting value as $d \to 0$ is $\pi/2$ (Davies, 1988e)). In order to prove this result, it is first necessary to show that as d approaches zero, the shape of the locus becomes a circle whose center is halfway between the original approximation D and the center C, and whose radius is $d/2$ (Fig. 10.9).

At first, an improvement factor of only ~ 1.6 seems rather insignificant—until it is remembered that it should be obtainable for *every* circle boundary pixel. Since there are typically 200 to 400 such pixels (e.g., for $r \approx 50$), the overall improvement

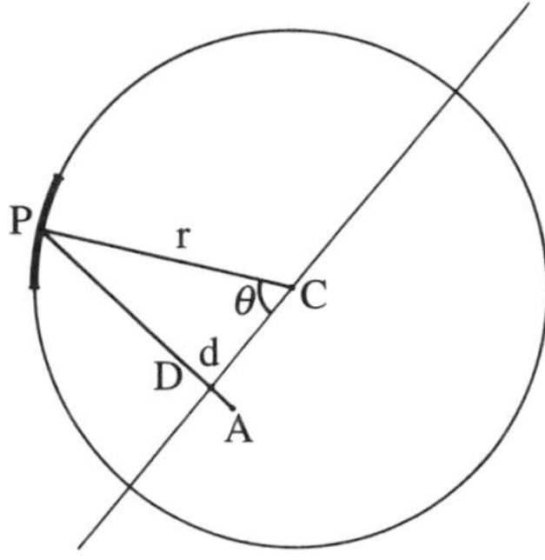

Figure 10.8 Boundary points that lead to a worse approximation.

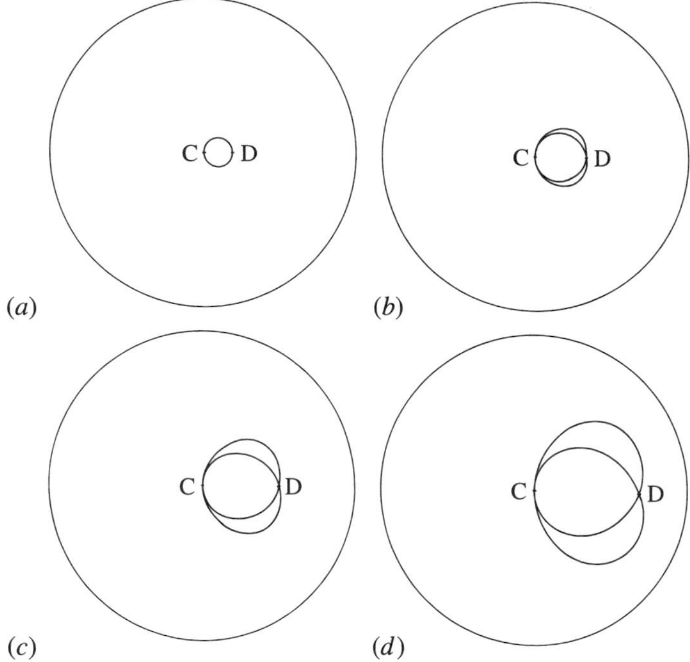

Figure 10.9 Loci obtained for various starting approximations: (*a*) $d=0.17r$; (*b*) $d=0.32r$; (*c*) $d=0.50r$; (*d*) $d=0.68r$.

can in principle be virtually infinite. (Limitations on the actual accuracy are discussed later in this chapter.) However, to maximize the gain in accuracy, the order in which the algorithm processes boundary pixels should be randomized. (If this were not done, two adjacent pixels might often appear consecutively, the second giving negligible additional improvement in center location.)

10.4.2 Improvements on the Basic Scheme

In practice, convergence initially occurs much as theory suggests it does. However, after a time no further improvement occurs, and the system settles down to a situation where it is randomly wandering around the ideal center, sometimes getting better and sometimes getting worse. The reason for this is that the particular approximation obtained at this stage depends more on the noise on individual edge pixels than on the underlying convergence properties discussed above.

To improve the situation further, the noise on the edge pixels must be suppressed by an averaging procedure. Since the positions of the successive approximations produced by this method after a certain stage are essentially noise limited, the best final approximation will result from averaging all the individual approximations by accumulating all the intermediate x and y values and finding the means. Furthermore, just as the best final approximation is obtained by averaging, so the best intermediate approximations are obtained by a similar averaging process—in which the x and y values so far accumulated are divided by the current number. Thus, at this stage of the algorithm it is not necessary to use the previously calculated position at the starting point for the next approximation, but rather some position derived from it. If the kth position that is accumulated is (u_k, v_k), then the kth best estimate is given by:

$$x_k = (1/k) \sum_{i=1}^{k} u_i \tag{10.9}$$

$$x_k = (1/k) \sum_{i=1}^{k} v_i \tag{10.10}$$

Typically, 400 boundary pixels are available, each of which gives an independent measurement of position. On balance we would expect each of the component (x and y) inaccuracies to be averaged and thereby reduced by a factor of $\sqrt{200}$ (Davies, 1987h). Hence, the linear accuracy should improve from around 1 pixel to around 0.1 pixel.

There are three stages of calculation in the algorithm:

1. Find the position of the center to within 2 pixels by the usual Hough technique.
2. Use the procedure as originally outlined until it settles down with an accuracy of 0.5 pixel—that is, until it is limited by radial errors rather than basic convergence problems.
3. Use the averaging version of the procedure (equations (10.9) and (10.10)) to obtain the final result, with an accuracy of 0.1 pixel.

In practice, it is possible to eliminate stage 2, with stage 1 then being used to get within 1 pixel of the center and stage 3 to get down to 0.1 pixel of the center. To retain robustness and accuracy, most of the work is done by parallel processing procedures (stage 1) or procedures that emulate parallel processing (stage 3).

To confirm that stage 3 emulates parallel processing, note that if a set of random points near the center were chosen as initial approximations, then a parallel procedure could be used to accumulate a new set of better approximations to the center, by using radial information from each of the edge pixels. However, the actual stage 3 is better than this as it uses the current best estimate of the position of the center as the starting position for each approximation. This is possible only if the calculation is being done sequentially rather than in parallel. Such a method is an improvement only if it is known that the initial approximation is good enough and if edge pixels that disagree badly with the currently known position of the center can be discarded as being due, for example, to a nearby object. A mechanism has to be provided for this eventuality within the algorithm. A useful strategy is to eliminate from consideration those edge pixels that would on their own point to center locations further than 2 pixels from the approximation indicated by stage 1.

10.4.3 *Discussion*

Based on the forgoing description, we might conclude that instead of using a successive approximation approach we could continue to use the same initial approximation for all boundary pixels and then average the results at the end. This would not be valid, however, since a fixed bias of $0.64d$ in an unknown direction would then result and there would be no way of removing it. Thus, the only real options are a set of random starting approximations or the sequence of approximations outlined above.

One thing that would upset the algorithm would be the presence of gross occlusions in the object, making less than about 30% of it visible, so that the center position might be known accurately in one direction but would be uncertain in a perpendicular direction. Unfortunately, such uncertainties would not be apparent with this method, since the absence of information might well lead to little transverse motion of the apparent center, which would then retain the same incorrect position.

10.4.4 *Practical Details*

The basis of the algorithm for improving on an approximation D (x_i, y_i) to a circle of center C is that of moving a distance r from a general boundary pixel P (x, y) in the direction of D. Thus

$$x_{i+1} = x - (r/s)(x - x_i) \tag{10.11}$$

$$y_{i+1} = y - (r/s)(y - y_i) \tag{10.12}$$

where

$$s = \left[(x - x_i)^2 + (y - y_i)^2\right]^{1/2} \tag{10.13}$$

This process is computationally efficient, since it does not require the evaluation of any trigonometric functions—the most complex operations involved being the calculation of various ratios and the evaluation of a square root.

Generally, it is apparently adequate to randomize boundary pixels (see Section 10.3) by treating them cyclically in the order they appear in a normal raster scan of the image, and then processing every nth one[5] until all have been used (Davies, 1988e). The number n will normally be in the range 50–150 (or for general circles with N boundary pixels, take $n \approx N/4$). One advantage of randomizing boundary pixels is that processing can be curtailed at any stage when sufficient accuracy has been attained, thereby saving computation. This technique is particularly valuable in the case of food products such as biscuits, where accuracy is generally not as much of a problem as speed of processing.

The algorithm is straightforward to set up, parameters being adjusted on a heuristic basis rather than adhering to any specific theory. This does not seem to lead to any real problems or to compromise accuracy. The results of center finding seem to be generally within 0.1 to 0.2 pixel (see below) for accurately made objects where such measurements can meaningfully be made. This rarely applies to food products such as biscuits, which are commonly found to be circular to a tolerance of only 5 to 10%.

When the algorithm was tested using an accurately made dark metal disc, accuracy was found to be in the region of 0.1 pixel (Davies, 1988e). However, some difficulty arose because of uncertainty in the *actual* position of the center. A slight skewness in the original image was found to contribute to such location errors—possibly about 0.05 pixel in this case. The reason for the skewness appeared to lie in a slight nonuniformity in the lighting of the metal disc. It became clear that it will generally be difficult to eliminate such effects and that 0.1-pixel accuracy is close to the limit of measurement with such images. However, other tests using simulated (accurate) circles showed that the method has an intrinsic accuracy of better than 0.1 pixel.

5 The final working algorithm is slightly more complex. If the nth pixel has been used, take the *next* one that has *not* been used. This simple strategy is sufficiently random and also overcomes problems of n sometimes being a factor of N.

Other improvements to the technique involve weighting down the earlier, less accurate approximations in obtaining the average position of the center. Weighting down can be achieved by the heuristic of weighting approximations in proportion to their number in the sequence. This strategy seems capable of finding center locations within 0.05 pixel—at least for cases where radii are around 45 pixels and where some 400 edge pixels are providing information about the position of each center. Although accuracy is naturally data-dependent, the approach fulfills the specification originally set out—of locating the centers of circular objects to 0.1 pixel or better, with minimal computation. As a by-product of the computation, the radius r can be obtained with exceptionally high accuracy—generally within 0.05 pixel (Davies, 1988e).

10.5 Overcoming the Speed Problem

The previous section studied how the accuracy of the HT circle detection scheme could be improved substantially, with modest additional computational cost. This section examines how circle detection may be carried out with significant improvement in speed. There is a definite tradeoff between speed and accuracy, so that greater speed inevitably implies lower accuracy. However, at least the method to be adopted permits graceful degradation to take place.

One way to overcome the speed problem described in Section 10.2 is to sample the image data; another is to attempt to use a more rudimentary type of edge detection operator; and a third is to try to invent a totally new strategy for center location.

Sampling the image data tends to mean looking at, say, every third or fourth line, to obtain a speed increase by a factor of three or so. However, improvements beyond this are unlikely, for we would then be looking at much less of the boundary of the object and the method would tend to lose its robustness. An overall improvement in speed by more than an order of magnitude appears to be unattainable.

Using a more rudimentary edge detection operator would necessarily give much lower accuracy. For example, the Roberts cross operator (Roberts, 1965) could be used, but then the angular accuracy would drop by a factor of five or so (Abdou and Pratt, 1979), and in addition the effects of noise would increase markedly so that the operator would achieve nowhere near its intrinsic accuracy. With the Roberts cross operator, noise in any one of the four pixels in the neighborhood will make estimation of orientation totally erroneous, thereby invalidating the whole strategy of the center finding algorithm.

In this scenario Davies (19871) attempted to find a new strategy that would have some chance of maintaining the accuracy and robustness of the original approach, while improving speed by at least an order of magnitude. In particular, it would be useful to aim for improvement by a factor of three by sampling and by another factor of three by employing a very small neighborhood. By opting for a 2-element neighborhood, the capability of estimating edge orientation is lost. However, this approach still permits an alternative strategy based on bisecting horizontal and vertical chords of a circle. This approach requires two passes over the image to determine the center position fully. A similar amount of computation is needed as for a single application of a 4-element neighborhood edge detector, so it should be possible to gain the desired factor of three in speed relative to a Sobel operator via the reduced neighborhood size. Finally, the overall technique involves much less computation, the divisions, multiplications, and square root calculations being replaced by 2-element averaging operations, thereby giving a further gain in speed. Thus, the gain obtained by using this approach could be appreciably over an order of magnitude.

10.5.1 *More Detailed Estimates of Speed*

This section presents formulas from which it is possible to estimate the gain in speed that would result by applying the above strategy for center location. First, the amount of computation involved in the original Hough-based approach is modeled by:

$$T_0 = N^2 s + S t_0 \tag{10.14}$$

where

T_0 = the total time taken to run the algorithm on an $N \times N$ pixel image
s = the time taken per pixel to compute the Sobel operator and to threshold the intensity gradient g
S = the number of edge pixels that are detected

and

t_0 = the time taken to compute the position of a candidate center point.

Next, the amount of computation in the basic chord bisection approach is modeled as

$$T = 2(N^2 q + Qt) \tag{10.15}$$

where

T = the total time taken to run the algorithm

q = the time taken per pixel to compute a 2-element x or y edge detection operator and to perform thresholding

Q = the number of edge pixels that are detected in one or other scan direction

and

t = the time taken to compute the position of a candidate center point coordinate (either x_c or y_c).

In equation (10.15), the factor of two results because scanning occurs in both horizontal and vertical directions. If the image data are now sampled by examining only a proportion α of all possible horizontal and vertical scan lines, the amount of computation becomes:

$$T = 2\alpha\left(N^2 q + Qt\right) \tag{10.16}$$

The gain in speed from using the chord bisection scheme with sampling is therefore:

$$G = \frac{N^2 s + St_0}{2\alpha(N^2 q + Qt)} \tag{10.17}$$

Typical values of relevant parameters for, say, a biscuit of radius 32 pixels in a 128×128 pixel image are:

$$N^2 = 16{,}384$$

$$S \approx Q \approx 200$$

$$t_0/s \approx 1$$

$$s/q \approx 12/2 = 6$$

$$t_0/t \approx 5$$

$$\alpha \approx 1/3$$

so that

$$\begin{aligned} G &\approx \frac{16{,}384 + 200}{(2/3)(16{,}384/6 + 200/5)} \\ &\approx \frac{16{,}584 \times 1.5}{2731 + 40} \\ &\approx 8.98 \end{aligned}$$

This is a substantial gain, and we now consider more carefully how it could arise. If we assume $N \gg S, Q$, we get:

$$G \approx s/2\alpha q \tag{10.18}$$

This is obviously the product of the sampling factor $1/\alpha$ and the gain from applying an edge detection operator twice in a smaller neighborhood (i.e., applied both horizontally and vertically). In the above example, $1/\alpha = 3$ and $s/2q = 3$, this figure resulting from the ratio of the numbers of pixels involved. This would give an ideal gain of 9, so the actual gain in this case is not much changed by the terms in t_0 and t.

It is important that the strategy also involves reducing the amount of computation in determining the center from the edge points that are found. However, in all cases the overriding factor involved in obtaining an improvement in speed is the sampling factor $1/\alpha$. If this factor could be improved further, then significant additional gains in speed could be obtained—in principle without limit, but in practice the situation is governed by the intrinsic robustness of the algorithm.

10.5.2 *Robustness*

Robustness can be considered relative to two factors: the amount of noise in the image; and the amount of signal distortion that can be tolerated. Fortunately, both the original HT and the chord bisection approach lead to peak finding situations. If there is any distortion of the object shape, then points are thrown into relatively random locations in parameter space and consequently do not have a significant direct impact on the accuracy of peak location. However, they do have an indirect impact in that the signal-to-noise ratio is reduced, so that accuracy is impaired. If a fraction β of the original signal is removed, leaving a fraction $\gamma = 1 - \beta$, due either to such distortions or occlusions or to the deliberate sampling procedures already outlined, then the number of independent measurements of the center location drops to a fraction γ of the optimum. Thus, the accuracy of estimation of the center location drops to a fraction $\sqrt{\gamma}$ of the optimum. Since noise affects the optimum accuracy directly, we have in principle shown the result of both major factors governing robustness.

The important point here is that the effect of sampling is substantially the same as that of signal distortion, so that the more distortion that must be tolerated, the higher the value α has to have. This principle applies both to the original Hough approach and to the chord bisection algorithm. However, the latter does its peak finding in a different way—via 1-D rather than 2-D averaging processes. As a result, it turns out to be somewhat more robust than the standard HT in its capability for accepting a high degree of sampling.

This gain in capability for accepting sampling carries with it a set of related disadvantages: highly textured objects may not easily be located by the method; high noise levels may muddle it via the production of too many false edges; and (since it operates by specific x and y scanning mechanisms) there must be a sufficiently small amount of occlusion and other gross distortion that a significant number of scans (both horizontally and vertically) pass through the object. Ultimately, then, the method will not tolerate more than about one-quarter of the circumference of the object being absent.

10.5.3 *Experimental Results*

Tests (Davies, 1987l) with the image in Fig. 10.10 showed that gains in speed of more than 25 can be obtained, with values of α down to less than 0.1 (i.e., every tenth horizontal and vertical line scanned). To obtain the same accuracy and reliability with smaller objects, a larger value of α was needed. This can be viewed as keeping the number of samples per object roughly constant. The maximum gain in speed was then by a factor of about 10. The results for broken circular products (Figs. 10.11 and 10.12) are self-explanatory; they indicate the limits to which the method can be taken and confirm that it is straightforward to set the algorithm up by eye. An outline of the complete algorithm is given in Fig. 10.14.

Figure 10.13 shows the effect of adjusting the threshold in the 2-element edge detector. The result of setting it too low is seen in Fig. 10.13*a*. Here the surface texture of the object has triggered the edge detector, and the chord midpoints give rise to a series of false estimates of the center coordinates. Figure 10.13*b* shows the result of setting the threshold at too high a level, so that the number of estimates of center coordinates is reduced and sensitivity suffers. The images in Fig. 10.10 were obtained with the threshold adjusted intuitively, but its value is nevertheless close to the optimum. However, a more rigorous approach can be taken by optimizing a suitable criterion function. There are two obvious functions: (1) the number of accurate center predictions n; and (2) the speed–sensitivity product. The speed–sensitivity product can be written in the form $\sqrt{n}/T$, where T is the execution time (the reason for the square root being as in Section 10.4.2). The two methods of optimization make little difference in the example of Fig. 10.13. However, had there

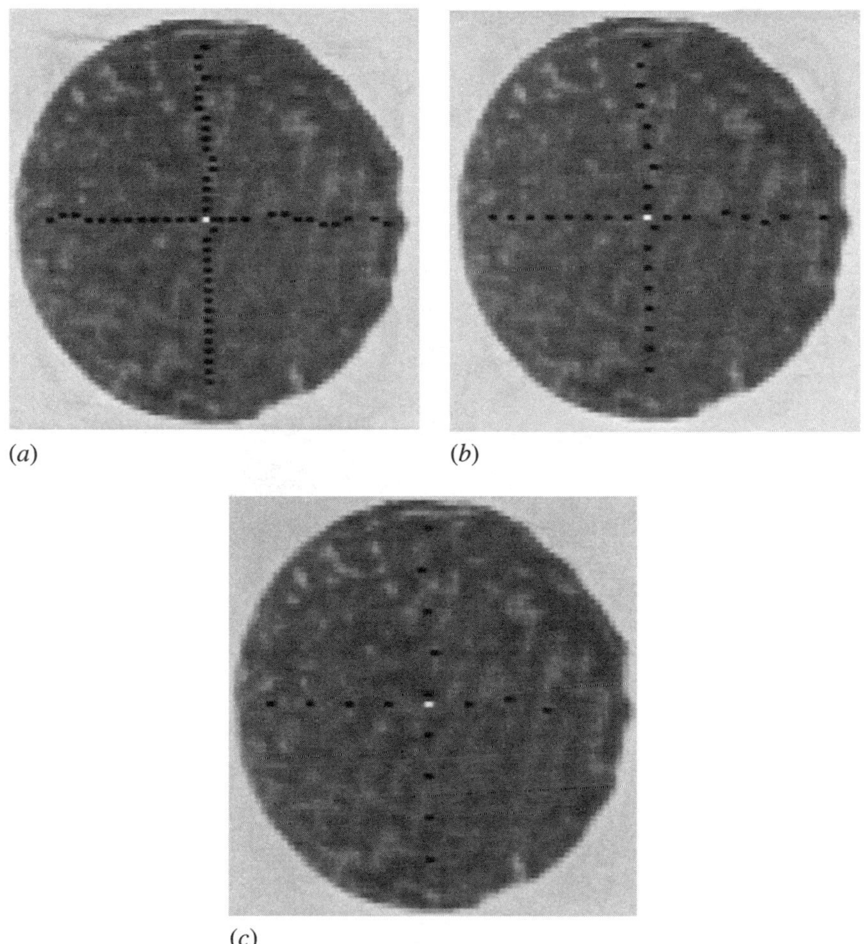

Figure 10.10 Successful object location using the chord bisection algorithm for the same initial image, using successive step sizes of 2, 4, and 8 pixels. The black dots show the positions of the horizontal and vertical chord bisectors, and the white dot shows the position found for the center.

been a strong regular texture on the object, the situation would have been rather different.

10.5.4 *Summary*

The center location procedure described in this chapter is normally over an order of magnitude faster than the standard Hough-based approach. Typically,

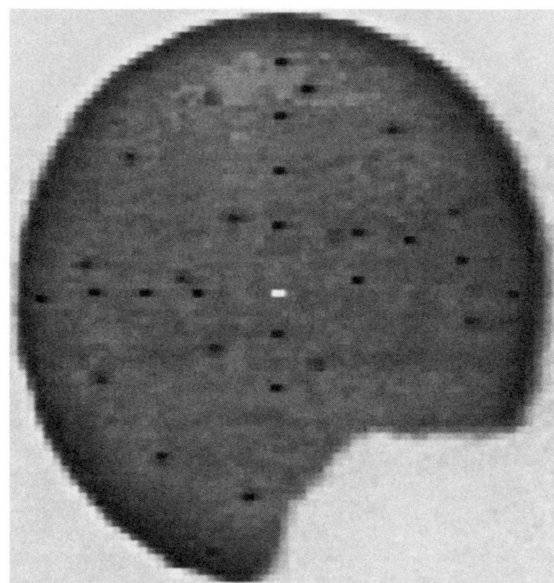

Figure 10.11 Successful location of a broken object using the chord bisection algorithm: only about one-quarter of the ideal boundary is missing.

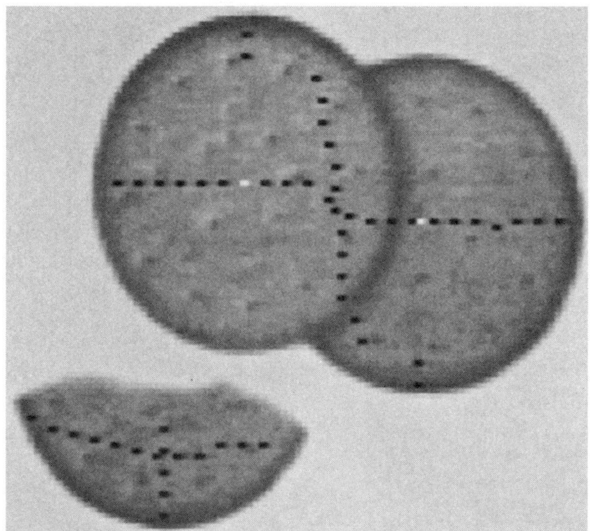

Figure 10.12 A test on overlapping and broken biscuits: the overlapping objects are successfully located, albeit with some difficulty, but there is no chance of finding the center of the broken biscuit since over one-half of the ideal boundary is missing.

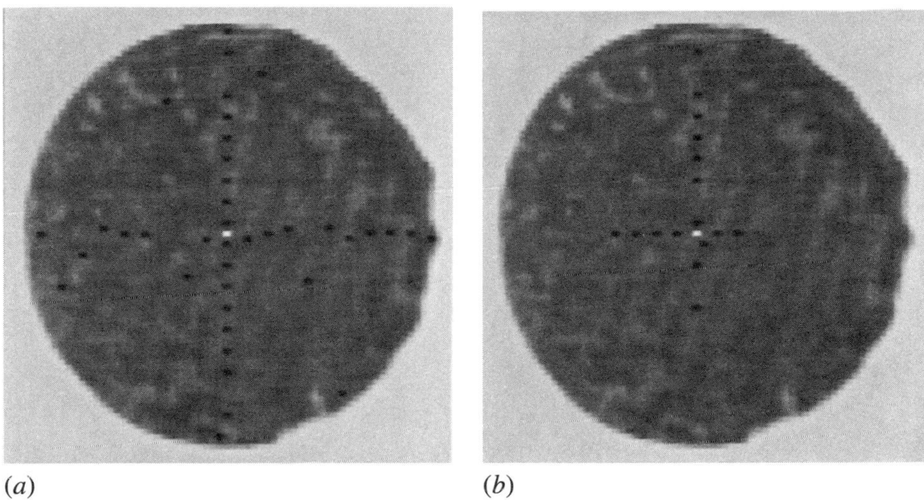

(*a*) (*b*)

Figure 10.13 Effect of misadjustment of the gradient threshold: (*a*) effect of setting the threshold too low, so that surface texture muddles the algorithm; (*b*) loss of sensitivity on setting the threshold too high.

```
y = 0;
do {
    scan horizontal line y looking for start and end of each object;
    calculate midpoints of horizontal object segments;
    accumulate midpoints in 1-D parameter space (x space);
    // note that the same space, x space, is used for all lines y
    y = y + d;
} until y > ymax;

x = 0;
do {
    scan vertical line x looking for start and end of each object;
    calculate midpoints of vertical object segments;
    accumulate midpoints in 1-D parameter space (y space);
    // note that the same space, y space, is used for all lines x
    x = x + d;
} until x > xmax;

find peaks in x space;
find peaks in y space;
test all possible object centers arising from these peaks;
// the last step is necessary only if ∃ > 1 peak in each space
// d is the horizontal and vertical step size (= 1/α)
```

Figure 10.14 Outline of the fast center-finding algorithm.

execution times are reduced by a factor of ~25. In exacting real-time applications, this may be important, for it could mean that the software will operate without special-purpose hardware accelerators. Robustness is good both in noise suppression capability and in the ability to negotiate partial occlusions and breakages of the object being located. The method tolerates at least one-quarter of the circumference of an object being absent for one reason or another. Although this is not as good as for the original HT approach, it is adequate for many real applications. In addition, it is entirely clear on applying the method to real image data whether the real image data would be likely to muddle the algorithm—since a visual indication of performance is available as an integral part of the method.

Robustness is built into the algorithm in such a way as to emulate the standard Hough-based approach. It is possible to interpret the method as being itself a Hough-based approach, since each center coordinate is accumulated in its own 1-D "parameter space." These considerations give further insight into the robustness of the standard Hough technique.

10.6 Concluding Remarks

Although the Hough technique has been found to be effective and highly robust against occlusions, noise, and other artifacts, it incurs considerable storage and computation—especially if it is required to locate circles of unknown radius or if high accuracy is required. Techniques have been described whereby the latter two problems can be tackled much more efficiently. In addition, a method has been described for markedly reducing the computational load involved in circle detection. The general circle location scheme is a type of HT, but although the other two methods are related to the HT, they are distinct methods that draw on the HT for inspiration. As a result, they are rather less robust, although the degree of robustness is quite apparent and there is little risk of applying them when their use would be inappropriate. As remarked earlier, the HT achieves robustness as an integral part of its design. This is also true of the other techniques described, for which considerable care has been taken to include robustness as an intrinsic part of the design rather than as an afterthought.

As in the case of line detection, a trend running through the design of circle detection schemes is the deliberate splitting of algorithms into two or more stages. This consideration is useful for keying into the relevant parts of an image prior to finely discriminating one type of object or feature from another, or prior to measuring dimensions or other characteristics accurately. The concept can be taken further, in that all the algorithms discussed in this chapter have improved their efficiency by searching first for edge features in the image. The concept of two-stage template matching is therefore deep-seated in the methodology of the subject and is developed further in later chapters—notably those on ellipse and corner detection. Although two-stage template matching is a standard means of increasing efficiency (VanderBrug and Rosenfeld, 1977; Davies, 1988i), it is not at all obvious that it is always possible to increase efficiency by this means. It appears to be in the nature of the subject that ingenuity is needed to discover ways of achieving this efficiency.

The Hough transform is straightforwardly applied to circle detection. This chapter has demonstrated that its robustness is impressive in this case, though there are other relevant practical issues such as accuracy, speed, and storage requirements—some of which can be improved by employing parameter spaces of reduced dimension.

10.7 Bibliographical and Historical Notes

The Hough transform was developed in 1962 and first applied to circle detection by Duda and Hart (1972). However, the now standard HT technique, which makes use of edge orientation information to reduce computation, only emerged three years later (Kimme et al., 1975). Davies' work on circle detection for automated inspection required real-time implementation and also high accuracy. This spurred the development of the three main techniques described in Sections 10.3–10.5 of this chapter (Davies, 1987l, 1988b,e). In addition, Davies has considered the effect of noise on edge orientation computations, showing in particular their effect in reducing the accuracy of center location (Davies, 1987k).

Yuen et al. (1989) reviewed various existing methods for circle detection using the HT. In general, their results confirmed the efficiency of the method of Section 10.3 for unknown circle radius, although they found that the two-stage process involved can sometimes lead to slight loss of robustness. It appears that this problem can be reduced in some cases by using a modified version of the algorithm of Gerig and Klein (1986). Note that the Gerig and Klein approach is itself a two-stage procedure: it is discussed in detail in the next chapter. More recently, Pan et al. (1995) have increased the speed of computation of the HT by prior grouping of edge pixels into arcs, for an underground pipe inspection application.

The two-stage template matching technique and related approaches for increasing search efficiency in digital images were known by 1977 (Nagel and Rosenfeld, 1972; Rosenfeld and VanderBrug, 1977; VanderBrug and Rosenfeld, 1977) and have undergone further development since then—especially in relation to particular applications such as those described in this chapter (Davies, 1988i).

Most recently, Atherton and Kerbyson (1999) (see also Kerbyson and Atherton, 1995) showed how to find circles of arbitrary radius in a single parameter space using the novel procedure of coding radius as a phase parameter and then performing accumulation with the resulting phase-coded annular kernel. Using this approach, they attained higher accuracy with noisy images. Goulermas and Liatsis (1998) showed how the Hough transform could be fine-tuned for the detection of fuzzy circular objects such as overlapping bubbles by using genetic algorithms. In effect, genetic algorithms are able to sample the solution space with very high efficiency and hand over cleaner data to the following Hough transform. Toennies et al. (2002) showed how the Hough transform can be used to localize the irises of human eyes for real-time applications, one advantage of the transform being its capability for coping with partial occlusion—a factor that is often quite serious with regard to the iris.

10.8 Problems

1. Prove the result of Section 10.4.1, that as D approaches C and d approaches zero (Fig. 10.9), the shape of the locus becomes a circle on DC as diameter.

2. (a) Describe the use of the Hough transform for circular object detection, assuming that object size is known in advance. Show also how a method for detecting ellipses could be adapted for detecting circles of unknown size.

   (b) A new method is suggested for circle location which involves scanning the image both horizontally and vertically. In each case, the midpoints of chords are determined and their x or y coordinates are accumulated in separate 1-D histograms. Show that these can be regarded as simple types of Hough transform, from which the positions of circles can be deduced. Discuss whether any problems would arise with this approach; consider also whether it would lead to any advantages relative to the standard Hough transform for circle detection.

(c) A further method is suggested for circle location. This again involves scanning the image horizontally, but in this case, for every chord that is found, an estimate is immediately made of the *two* points at which the center could lie, and votes are placed at those locations. Work out the geometry of this method, and determine whether it is faster than the method outlined in (b). Determine whether the method has any disadvantages compared with the method in (b).

The Hough Transform and Its Nature

It has already been seen that the Hough transform can be used to locate straight-line and circle features in digital images. It would be useful to know whether the method can be generalized to cover all shapes and whether it is always as robust as it is for the original two examples. This chapter discusses these questions, showing that the method can be generalized and is broadly able to retain its robustness properties.

Look out for:

- the generalized Hough transform technique.
- its relation to spatial matched filtering.
- how sensitivity is optimized by gradient rather than uniform weighting.
- the effect of applying the generalized Hough transform to line detection.
- how speed has been improved by a variety of methods.
- the value of the Gerig and Klein back-projection technique in cutting down the effects of extraneous clutter.

This chapter not only describes the generalized Hough transform but also generalizes our view of the transform as a generic machine vision technique. The generalized method will be employed in later chapters for the detection of further types of macroscopic image feature.

CHAPTER 11

The Hough Transform and Its Nature

11.1 Introduction

As we have seen, the Hough transform (HT) is of great importance in detecting features such as lines and in circles, and in finding relevant image parameters (cf. the dynamic thresholding problem of Chapter 4). This makes it worthwhile to see the extent to which the method can be generalized so that it can detect arbitrary shapes. The work of Merlin and Farber (1975) and Ballard (1981) was crucial historically and led to the development of the generalized Hough transform (GHT). The GHT is studied in this chapter, first showing how it is implemented and then examining how it is optimized and adapted to particular types of image data. This requires us to go back to first principles, taking spatial matched filtering as a starting point.

Having developed the relevant theory, it is applied to the important case of line detection, showing in particular how detection sensitivity is optimized. Finally, the computational problems of the GHT are examined and fast HT techniques are considered.

11.2 The Generalized Hough Transform

This section shows how the standard Hough technique is generalized so that it can detect arbitrary shapes. In principle, it is trivial to achieve this objective. First, we need to select a localization point L within a template of the idealized shape. Then, we need to arrange it so that, instead of moving from an edge point a *fixed* distance R directly along the local edge normal to arrive at the center, as for circles, we move an appropriate *variable* distance R in a variable direction φ so as

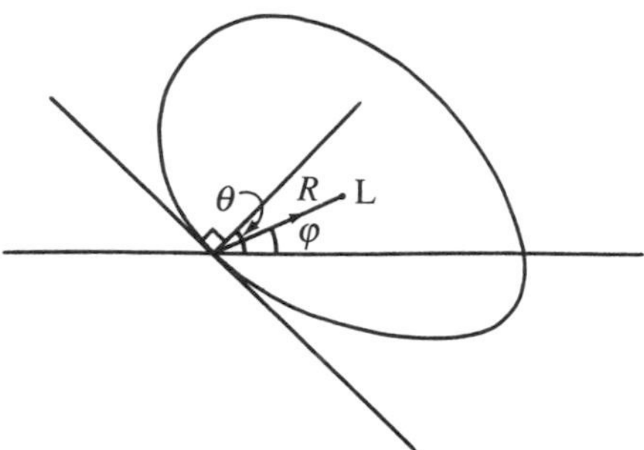

Figure 11.1 Computation of the generalized Hough transform.

to arrive at L. R and φ are now functions of the local edge normal directions θ (Fig. 11.1). Under these circumstances, votes will peak at the preselected object localization point L. The functions $R(\theta)$ and $\varphi(\theta)$ can be stored analytically in the computer algorithm, or for completely arbitrary shapes they may be stored as a lookup table. In either case, the scheme is beautifully simple in principle, but two complications arise in practice. The first develops because some shapes have features such as concavities and holes, so that several values of R and φ are required for certain values of θ (Fig. 11.2). The second arises because we are going from an isotropic shape (a circle) to an anisotropic shape, which may be in a completely arbitrary orientation.

To cope with the first of these complications, the lookup table (usually called the R-table) must contain a list of the positions $\mathbf{r}$, relative to L, of all points on the boundary of the object for each possible value of edge orientation θ (or a similar effect must be achieved analytically). Then, on encountering an edge fragment in the image whose orientation is θ, estimates of the position of L may be obtained by moving a distance (or distances) $\mathbf{R} = -\mathbf{r}$ from the given edge fragment. If the R-table has multivalued entries (i.e., several values of $\mathbf{r}$ for certain values of θ), only

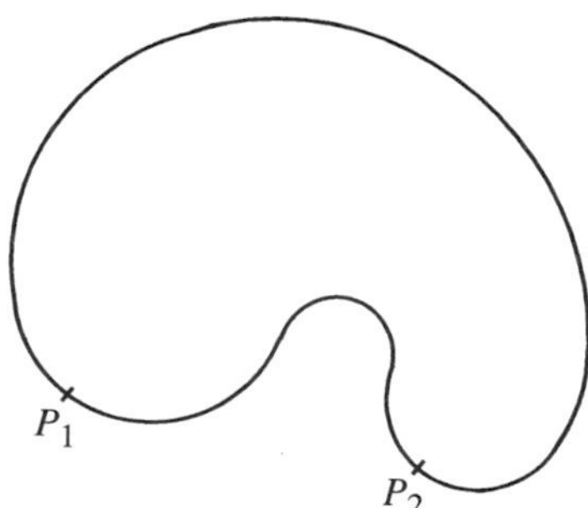

Figure 11.2 A shape exhibiting a concavity: certain values of θ correspond to several points on the boundary and hence require several values of R and φ—as for points P_1 and P_2.

one of these entries (for given θ) can give a correct estimate of the position of L. However, at least the method is guaranteed to give optimum sensitivity, since all relevant edge fragments contribute to the peak at L in parameter space. This property of optimal sensitivity reflects the fact that the GHT is a form of spatial matched filter: this property is analyzed in more detail later in this chapter.

The second complication arises because any shape other than a circle is anisotropic. Since in most applications (including industrial applications such as automated assembly) object orientations are initially unknown, the algorithm has to obtain its own information on object orientation. This means adding an extra dimension in parameter space (Ballard, 1981). Then each edge point contributes a vote in each plane in parameter space at a position given by that expected for an object of given shape and given orientation. Finally, the whole of parameter space is searched for peaks, the highest points indicating both the locations of objects and their orientations. If object size is also a parameter, the problem becomes far worse; this complication is ignored here (although the method of Section 10.3 is clearly relevant).

The changes made in proceeding to the GHT leave it just as robust as the HT circle detector described previously. This gives an incentive to improve the GHT so as to limit the computational problems in practical situations. In particular, the size of the parameter space must be cut down drastically both to save storage and to curtail the associated search task. Considerable ingenuity has been devoted to devising alternative schemes and adaptations to achieve this objective. Important special cases are those of ellipse detection and polygon detection, and in each of these cases definite advances have been made. Ellipse detection is covered in Chapter 12 and polygon detection in Chapter 14. Here we proceed with some more basic studies of the GHT.

11.3 Setting Up the Generalized Hough Transform—Some Relevant Questions

The next few sections explore the theory underpinning the GHT, with the aim of clarifying how to optimize it systematically for specific circumstances. It is relevant to ask what is happening when a GHT is being computed. Although the HT has been shown to be equivalent to template matching (Stockman and Agrawala, 1977) and also to spatial matched filtering (Sklansky, 1978), further clarification is required. In particular, three problems (Davies, 1987c) need to be addressed, as follows:

1. *The parameter space weighting problem.* In introducing the GHT, Ballard mentioned the possibility of weighting points in parameter space according to

the magnitudes of the intensity gradients at the various edge pixels. But when should gradient weighting be used in preference to uniform weighting?

2. *The threshold selection problem.* When using the GHT to locate an object, edge pixels are detected and used to compute candidate positions for the localization point L (see Section 11.2). To achieve this it is necessary to threshold the edge gradient magnitude. How should the threshold be chosen?
3. *The sensitivity problem.* Optimum sensitivity in detecting objects does not automatically provide optimum sensitivity in locating objects, and vice versa. How should the GHT be optimized for these two criteria?

To understand the situation and solve these problems, it is necessary to go back to first principles. A start is made on this in the next section.

11.4 Spatial Matched Filtering in Images

To discuss the questions posed in Section 11.3, it is necessary to analyze the process of spatial matched filtering. In principle, this is the ideal method of detecting objects, since it is well known (Rosie, 1966) that a filter that is matched to a given signal detects it with optimum signal-to-noise ratio under white noise[1] conditions (North, 1943; Turin, 1960). (For a more recent discussion of this topic, see Davies, 1993a.)

Mathematically, using a matched filter is identical to correlation with a signal (or "template") of the same shape as the one to be detected (Rosie, 1966). Here "shape" is a general term meaning the amplitude of the signal as a function of time or spatial location.

When applying correlation in image analysis, changes in background illumination cause large changes in signal from one image to another and from one part of an image to another. The varying background level prevents straightforward peak detection in convolution space. The matched filter optimizes signal-to-noise ratio only in the presence of white noise. Whereas this is likely to be a good approximation in the case of radar signals, this is not generally true in the case of images. For ideal detection, the signal should be passed through a "noise-whitening filter" (Turin, 1960), which in the case of objects in images is usually some form of high-pass filter. This filter must be applied prior to correlation analysis. However, this is likely to be a computationally expensive operation.

1 White noise is noise that has equal power at all frequencies. In image science, white noise is understood to mean equal power at all *spatial* frequencies. The significance of this is that noise at different pixels is completely uncorrelated but is subject to the same gray-scale probability distribution—that is, it has potentially the same range of amplitudes at all pixels.

If we are to make correlation work with near optimal sensitivity but without introducing a lot of computation, other techniques must be employed. In the template matching context, the following possibilities suggest themselves:

1. Adjust templates so that they have a mean value of zero, to suppress the effects of varying levels of illumination in first order.
2. Break up templates into a number of smaller templates, each having a zero mean. Then as the sizes of subtemplates approach zero, the effects of varying levels of illumination will tend to zero in second order.
3. Apply a threshold to the signals arising from each of the subtemplates, so as to suppress those that are less than the expected variation in signal level.

If these possibilities fail, only two further strategies appear to be available:

4. Readjust the lighting system—an important option in industrial inspection applications, although it may provide little improvement when a number of objects can cast shadows or reflect light over each other.
5. Use a more "intelligent" (e.g., context-sensitive) object detection algorithm, although this will almost certainly be computationally intensive.

11.5 From Spatial Matched Filters to Generalized Hough Transforms

To proceed, we note that items (1) to (5) amount to a specification of the GHT. First, breaking up the templates into small subtemplates each having a zero mean and then thresholding them is analogous, and in many cases identical, to a process of edge detection (see, for example, the templates used in the Sobel and similar operators). Next, locating objects by peak detection in parameter space corresponds to the process of reconstructing whole template information from the subtemplate (edge location) data. The important point here is that these ideas reveal how the GHT is related to the spatial matched filter. Basically, *the GHT can be described as a spatial matched filter that has been modified, with the effect of including integral noise whitening, by breaking down the main template into small zero-mean templates and thresholding the individual responses before object detection.*

Small templates do not permit edge orientation to be estimated as accurately as large ones. Although the Sobel edge detector is in principle accurate to about 1° (see Chapter 5), there is a deleterious effect if the edge of the object is fuzzy. In such a case, it is not possible to make the subtemplates very small, and an intermediate

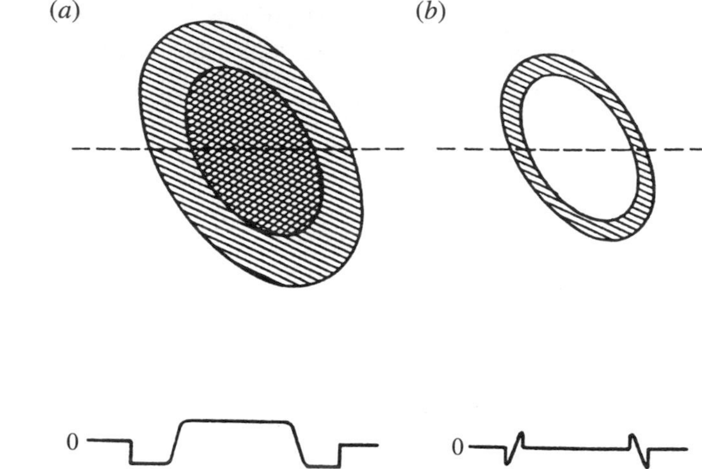

Figure 11.3 The idea of a perimeter template: both the original spatial matched filter template (*a*) and the corresponding "perimeter template" (*b*) have a zero mean (see text). The lower illustrations show the cross sections along the dotted lines.

size should be chosen which gives a suitable compromise between accuracy and sensitivity.

Employing zero-mean templates results in the absolute signal level being reduced to zero and only local relative signal levels being retained. Thus, the GHT is not a true spatial matched filter. In particular, it suppresses the signal from the bulk of the object, retaining only that near its boundary. As a result, the GHT is highly sensitive to object position but is not optimized for object detection.

Thresholding of subtemplate responses has much the same effect as employing zero-mean templates, although it may remove a small proportion of the signal giving positional information. This makes the GHT even less like an ideal spatial matched filter and further reduces the sensitivity of object detection. The thresholding process is particularly important in the present context because it provides a means of saving computational effort *without losing significant positional information*. On its own, this characteristic of the GHT would correspond to a type of perimeter template around the outside of an object (see Fig. 11.3). This must not be taken as excluding all of the interior of the object, since any high-contrast edges within the object will facilitate location.

11.6 Gradient Weighting versus Uniform Weighting

The first problem described in Section 11.1 was that of how best to weight plots in parameter space in relation to the respective edge gradient magnitudes. To find an answer to this problem, it should now only be necessary to go back to the spatial matched filter case to find the ideal solution and then to determine the corresponding solution for the GHT in the light of the discussion in Section 11.4. First, note that the responses to the subtemplates (or to the perimeter template) are proportional to edge gradient magnitude. With a spatial matched filter, signals are detected optimally by templates of the same shape. Each contribution to the spatial matched filter response is then proportional to the local magnitude of the signal and to that of the template. In view of the correspondence between (1) using a spatial matched filter to locate objects by looking for peaks in convolution space and (2) using a GHT to locate objects by looking for peaks in parameter space, we should use weights proportional to the gradients of the edge points *and* proportional to the *a priori* edge gradients.

The choice of weighting is important for two reasons. First, the use of uniform weighting implies that all edge pixels whose gradient magnitudes are above threshold will effectively have them reduced to the threshold value, so that the signal will be curtailed. Accordingly, the signal-to-noise ratio of high-contrast objects could be reduced significantly. Second, the widths of edges of high-contrast objects will be broadened in a crude way by uniform weighting (see Fig. 11.4), but under gradient weighting this broadening will be controlled, giving a roughly Gaussian edge profile. Thus, the peak in parameter space will be narrower and more rounded, and the object reference point L can be located more easily and with

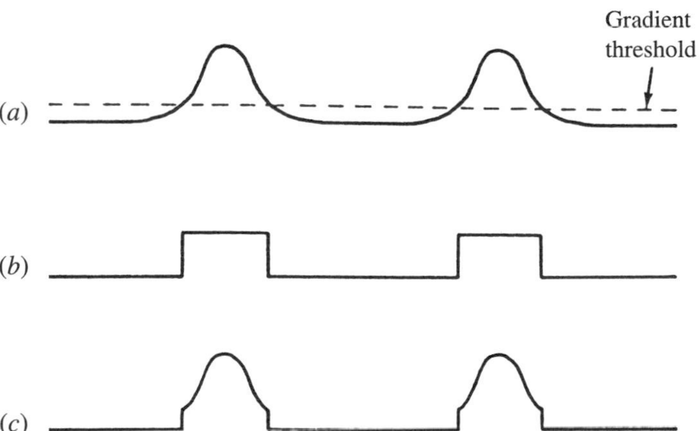

Figure 11.4 Effective gradient magnitude as a function of position within a section across an object of moderate contrast, thresholded at a fairly low level: (*a*) gradient magnitude for original image data and gradient thresholding level; (*b*) uniform weighting: the effective widths of edges are broadened rather crudely, adding significantly to the difficulty of locating the peak in parameter space; (*c*) gradient weighting: the position of the peak in parameter space can be estimated in a manner that is basically limited by the shape of the gradient profile for the original image data.

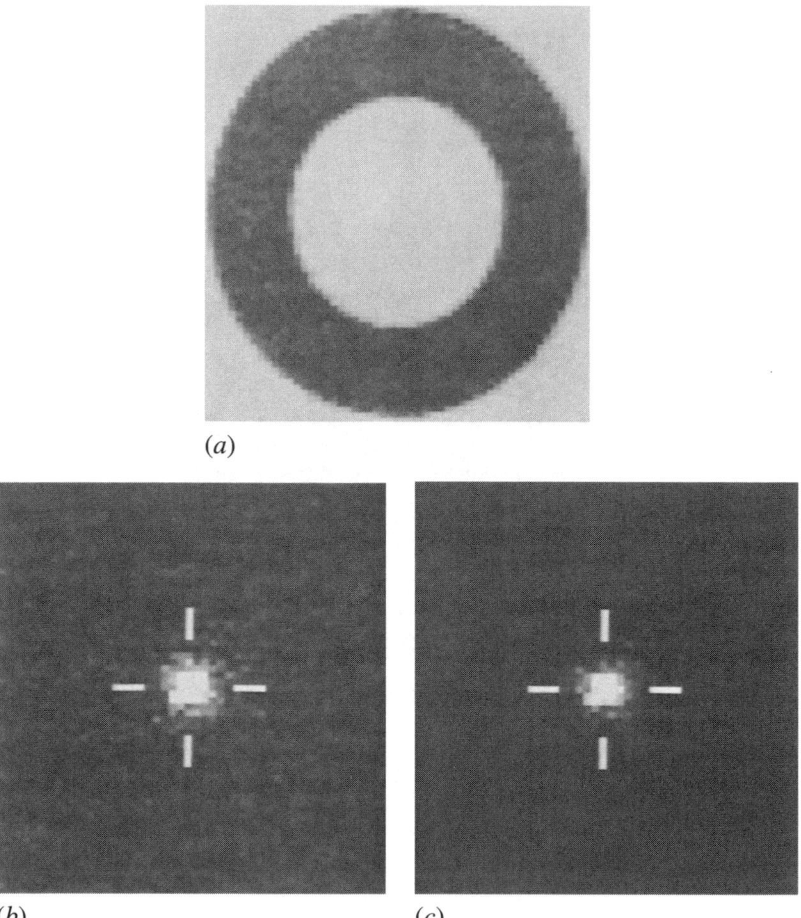

Figure 11.5 Results of applying the two types of weighting to a real image: (*a*) original image; (*b*) results in parameter space for uniform weighting; (*c*) results for gradient weighting. The peaks (which arise from the outer edges of the washer) are normalized to the same levels in the two cases. The increased level of noise in (*b*) is readily apparent. In this example, the gradient threshold is set at a low level (around 10% of the maximum level) so that low-contrast objects can also be detected.

greater accuracy. This effect is visible in Fig. 11.5, which also shows the relatively increased noise level that results from uniform weighting.

Note also that low gradient magnitudes correspond to edges of poorly known location, while high values correspond to sharply defined edges. Thus, the accuracy of the information relevant to object location is proportional to the magnitude of the gradient at each of the edge pixels, and appropriate weighting should therefore be used.

11.6.1 *Calculation of Sensitivity and Computational Load*

In this subsection, we underline the above ideas by working out formulas for sensitivity and computational load. It is assumed that p objects of size around $n \times n$ are being sought in an image of size $N \times N$.

Correlation requires N^2n^2 operations to compute the convolutions for all possible positions of the object in the image. Using the perimeter template, we find that the number of basic operations is reduced to $\sim N^2n$, corresponding to the reduced number of pixels in the template. The GHT requires $\sim N^2$ operations to locate the edge pixels, plus a further pn operation to plot the points in parameter space.

The situation for sensitivity is rather different. With correlation, the results for n^2 pixels are summed, giving a signal proportional to n^2, although the noise (assumed to be independent at every pixel) is proportional to n. This is because of the well-known result that the noise powers of various independent noise components are additive (Rosie, 1966). Overall, this results in the signal-to-noise ratio being proportional to n. The perimeter template possesses only $\sim n$ pixels, and here the overall result is that the signal-to-noise ratio is proportional to $\sqrt{n}$. The situation for the GHT is inherently identical to that for the perimeter template method, as long as plots in parameter space are weighted proportional to edge gradient g multiplied by *a priori* edge gradient G. It is now necessary to compute the constant of proportionality α. Take s as the average signal, equal to the intensity (assumed to be roughly uniform) over the body of the object, and S as the magnitude of a full matched filter template. In the same units, g (and G) is the magnitude of the signal within the perimeter template (assuming the unit of length equals 1 pixel). Then $\alpha = 1/sS$. This means that the perimeter template method and the GHT method lose sensitivity in two ways—first because they look at less of the available signal, and second because they look where the signal is low. For a high value of gradient magnitude, which occurs for a step edge (where most of the variation in intensity takes place within the range of 1 pixel), the values of g and G saturate out, so that they are nearly equal to s and S (see Fig. 11.6). Under these conditions the perimeter template method and the GHT have sensitivities that depend only on the value of n.

Table 11.1 summarizes this situation. The oft-quoted statement that the computational load of the GHT is proportional to the number of perimeter pixels, rather than to the much greater number of pixels within the body of an object, is only an approximation. In addition, this saving is not obtained without cost. In particular, the sensitivity (signal-to-noise ratio) is reduced (at best) as the square root of object area/perimeter. (Note that area and perimeter are measured in the same units, so it is valid to find their ratio.)

Finally, the absolute sensitivity for the GHT varies as gG. As contrast changes so that $g \rightarrow g'$, we see that $gG \rightarrow g'G$. That is, sensitivity changes by a factor of

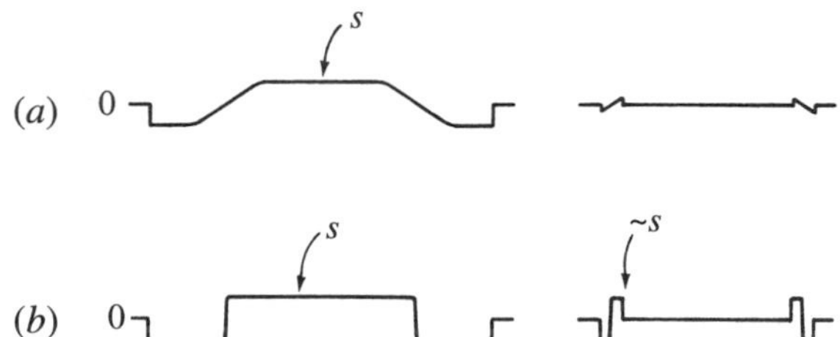

Figure 11.6 Effect of edge gradient on perimeter template signal: (*a*) low edge gradient: signal is proportional to gradient; (*b*) high edge gradient. Signal saturates at value of *s*.

Table 11.1 Formulas for computational load and sensitivity*

	Template Matching	Perimeter Template Matching	Generalized Hough Transform
Number of operations	$O(N^2n^2)$	$O(N^2n)$	$O(N^2) + O(pn)$
Sensitivity	$O(n)$	$O(\sqrt{n}gG/sS)$	$O(\sqrt{n}gG/sS)$
Maximum sensitivity†	$O(n)$	$O(\sqrt{n})$	$O(\sqrt{n})$

*This table gives formulas for computational load and sensitivity when p objects of size $n \times n$ are sought in an image of size $N \times N$. The intensity of the image within the whole object template is taken as s and the value for the ideal template is taken as S: corresponding values for intensity gradient within the perimeter template are g and G.

†Maximum sensitivity refers to the case of a step edge, for which $g \approx s$ and $G \approx S$ (see Fig. 11.6).

g'/g. Hence, theory predicts that sensitivity is proportional to contrast. Although this result might have been anticipated, we now see that it is valid only under conditions of gradient weighting.

11.7 Summary

As described in this chapter, a number of factors are involved in optimizing the GHT, as follows.

1. Each point in parameter space should be weighted in proportion to the intensity gradient at the edge pixel giving rise to it, *and* in proportion to the *a priori* gradient, if sensitivity is to be optimized, particularly for objects of moderate to high contrast.
2. The ultimate reason for using the GHT is to save computation. The main means by which this is achieved is by ignoring pixels having low magnitudes of

intensity gradient. If the threshold of gradient magnitude is set too high, fewer objects are in general detected; if it is set too low, computational savings are diminished. Suitable means are required for setting the threshold, but little reduction in computation is possible if the highest sensitivity in a low-contrast image is to be retained.

3. The GHT is inherently optimized for the location of objects in an image but is not optimized for the detection of objects. It may therefore miss low-contrast objects which are detectable by other methods that take the whole area of an object into account. This consideration is often unimportant in applications where signal-to-noise ratio is less of a problem than finding objects quickly in an uncluttered environment.

Clearly, the GHT is a spatial matched filter only in a particular sense, and as a result it has suboptimal sensitivity. The main advantage of the technique is that it is highly efficient, with overall computational load in principle being proportional to the relatively few pixels on the perimeters of objects rather than to the much greater numbers of pixels within them. In addition, by concentrating on the boundaries of objects, the GHT retains its power to locate objects accurately. It is thus important to distinguish clearly between sensitivity in *detecting* objects and sensitivity in *locating* them.

11.8 Applying the Generalized Hough Transform to Line Detection

This section considers the effect of applying the GHT to line detection because it should be known how sensitivity may be optimized. Straight lines fall into the category of features such as concavities which give multivalued entries in the *R*-table of the GHT—that is, certain values of θ give rise to a whole set of values of $\mathbf{r}$. Here we regard a line as a complete object and seek to detect it by a localization point at its midpoint M. Taking the line as having length S and direction given by unit vector $\mathbf{u}$ along its length, we find then that the transform of a general edge point E at $\mathbf{e}$ on the line is the set of points:[2]

$$\mathrm{P} = \{\mathbf{e} + s\mathbf{u}\} \qquad -S/2 \leq s \leq S/2 \tag{11.1}$$

2 Here it is convenient to employ standard mathematical set notation, the curly brackets being used to represent a set of points.

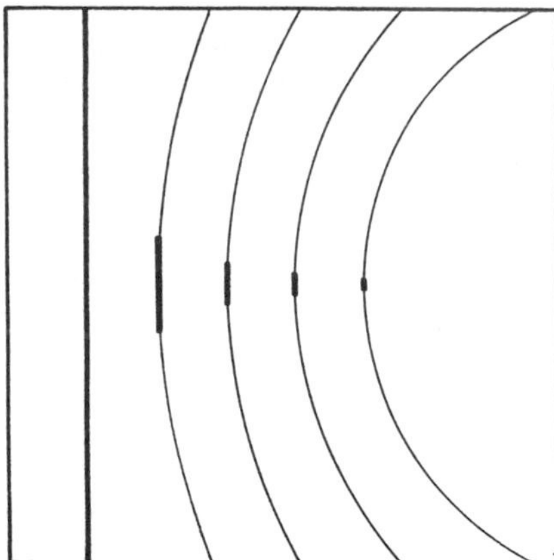

Figure 11.7 Longitudinal accuracy as a function of curvature. Localization points are selected on the lines to which they refer, for clarity of presentation. The sizes of the localization points indicate the relative accuracy in the longitudinal and normal directions.

or, more accurately (since the GHT is constructed so as to accumulate evidence for the existence of objects at particular locations), the transform function $T(\mathbf{x})$ is unity at these points and zero elsewhere:[3]

$$T(\mathbf{x}) = \begin{cases} 1; & \mathbf{x} \in \mathrm{P} \\ 0; & \mathbf{x} \notin \mathrm{P} \end{cases} \tag{11.2}$$

Thus, each edge pixel gives rise to a *line* of votes in parameter space equal in length to the whole length of the line in the original shape. It is profitable to explore briefly why lines should have such an odd PSF, which contrasts strongly with the point PSF for circles. The reason is that a straight line is accurately locatable normal to its length, but inaccurately locatable along its length. (For an infinite line, there is *zero* accuracy of location along the length, and complete precision for location normal to the length.) Hence, lines cannot have a radially symmetrical PSF in a parameter space that is congruent to image space. (This solves a research issue raised by Brown, 1983.)

Circles and lines, however, do not form totally distinct situations but are two extremes of a more general situation. This is indicated in Fig. 11.7, which shows that progressively lowering the curvature for objects within the (limited) field of view leads to progressive lowering of the accuracy with which the object may be located in the direction along the visible part of the perimeter, but to no loss in accuracy in the normal direction.

3 The symbol $\in$ means "is a member of the following set"; the symbol $\notin$ means "is *not* a member of the following set." (In this case, the point **x** is or is not within the given range.)

11.9 The Effects of Occlusions for Objects with Straight Edges

As has been shown, the GHT gives optimum sensitivity of line detection only at the expense of transforming each edge point into a whole line of votes in parameter space. However, there is a resulting advantage, in that the GHT formulation includes a degree of intrinsic longitudinal localization of the line. We now consider this aspect of the situation more carefully.

While studying line transforms obtained using the GHT line parametrization scheme, it is relevant to consider what happens in situations where objects may be partly occluded. It will already be clear that sensitivity for line detection degrades gracefully, as for circle detection. But what happens to the line localization properties? The results are rather surprising and gratifying, as will be seen from Figs. 11.8–11.11 (Davies, 1989d). These figures show that the transforms of partly occluded lines have a distinct peak at the center of the line in all cases where both ends are visible and a flat-topped response otherwise, indicating the exact degree of uncertainty in the position of the line center (see Figs. 11.8 and 11.9). Naturally, this effect is spoiled if the line PSF that is used is of the wrong length, and then an uncertainty in the center position again arises (Fig. 11.10).

Figure 11.11 shows a rectangular component being located accurately despite gross occlusions in each of its sides. The fact that L is independently the highest

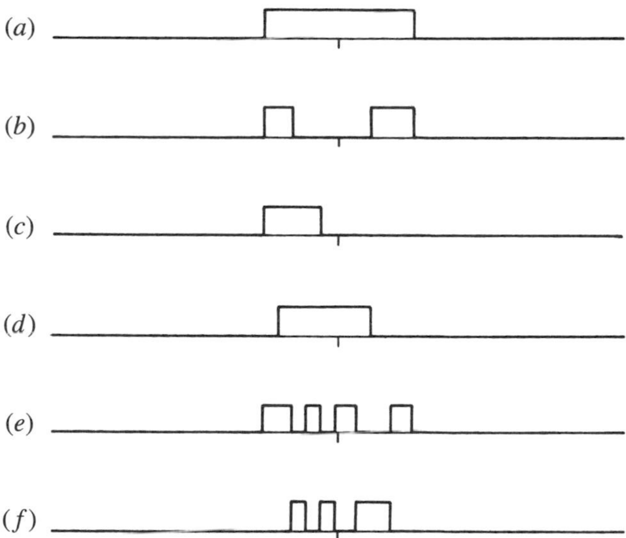

Figure 11.8 Various types of occlusions of a line: (*a*) original line; (*b*)–(*f*) line with various types of occlusion (see text).

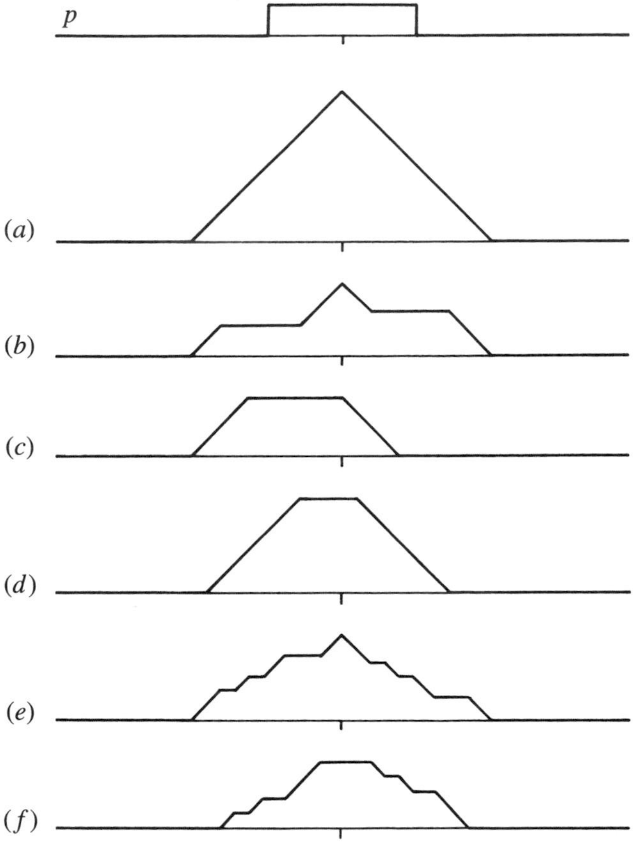

Figure 11.9 Transforms of a straight line and of the respective occluded variants shown in Fig. 11.8 (p, PSF). It is easy to see why response (a) takes the form shown. The result of correlating a signal with its matched filter is the autocorrelation function of the signal. Applying these results to a short straight line (Fig. 11.8a) gives a matched filter with the same shape (p), and the resulting profile in convolution space is the autocorrelation function (a).

point in either branch of the transform at L is ultimately because two ends in each *pair* of parallel sides of the rectangle are visible.

Thus, the GHT parametrization of the line is valuable not only in giving optimum sensitivity in the detection of objects possessing straight edges, but it also gives maximum available localization of these straight edges in the presence of severe occlusions. In addition, if the ends of the lines are visible, then the centers of the lines are locatable precisely—a property that arises automatically *as an integral part of the procedure*, no specific line-end detectors being required. Thus, the GHT with the specified line parametrization scheme can be considered as a

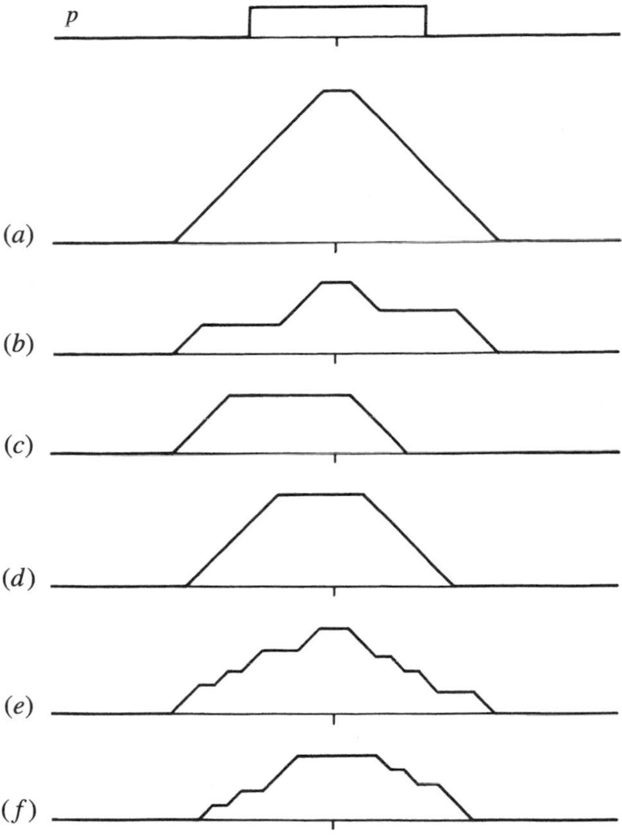

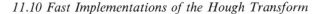

Figure 11.10 Transforms of a straight line and of the respective occluded variants shown in Fig. 11.8, obtained by applying a PSF that is slightly too long. The results should be compared with those in Fig. 11.9 (p, PSF).

process that automatically takes possible occlusions into account in a remarkably "intelligent" way. Table 11.2 summarizes the differences between the normal and the GHT parametrizations of the straight line.

11.10 Fast Implementations of the Hough Transform

As is now abundantly clear, the GHT has a severe weakness in that it demands considerable computation. The problem arises particularly with regard to the

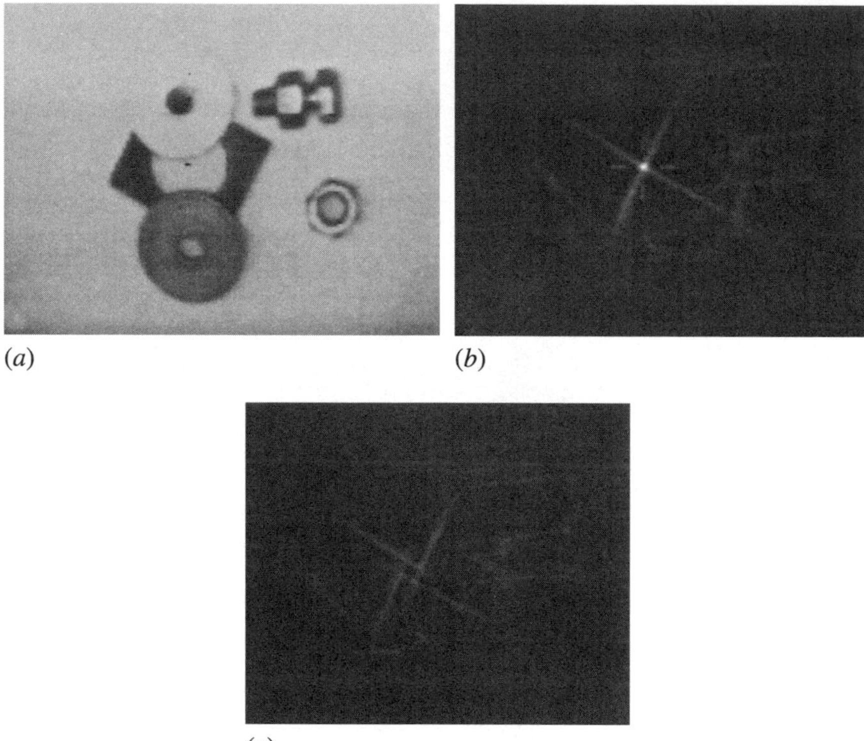

Figure 11.11 Location of a severely occluded rectangular object: (*a*) original image; (*b*) transform assuming that the short sides are oriented between 0 and $\pi/2$; (*c*) transform assuming that the short sides are oriented between $\pi/2$ and π. Note that the rectangular object is precisely located by the peak in (*b*), despite gross occlusions over much of its boundary. The fact that the peak in (*b*) is larger than those in (*c*) indicates the orientation of the rectangular object within $\pi/2$ (Davies, 1989c). (For further details, see Chapter 14.) Notice that occlusions *could* become so great that only one peak was present in either parameter plane *and* the remaining longer side was occluded down to the length of the shorter sides. There would then be insufficient information to indicate which range of orientations of the rectangle was correct, and an inherent ambiguity would remain.

number of planes needed in parameter space to accommodate transforms for different object orientations and sizes. When line detection by the GHT is involved, an additional problem arises in that a whole line of votes is required for each edge pixel. Significant improvements in speed are needed before the GHT can achieve its potential in practical instances of arbitrary shapes. This section considers some important developments in this area.

The source of the computational problems of the HT is the huge size the parameter space can take. Typically, a single plane parameter space has much the same size as an image plane. (This will normally be so in those instances where parameter space is congruent to image space.) But when many planes are required

Table 11.2 Comparison between the two parametrizations of a straight line

Normal parametrization	GHT parametrization
Localized relative to origin of coordinates	Localized relative to the line: independent of any origin
Basic localization point is foot-of-normal from origin	Basic localization point is midpoint of line
Provides no direct information on line's longitudinal position	Provides definite information on line's longitudinal position
Takes no account of length of line: this information not needed	Takes account of length of line: must be given this information, for optimal sensitivity
Not directly compatible with GHT	Directly compatible with GHT
Not guaranteed to give maximum sensitivity	Gives maximum sensitivity even when line is combined with other boundary segments

to cope with various object orientations and sizes, the number of planes is likely to be multiplied by a factor of around 300 for each extra dimension. Hence, the total storage area will then involve some 10,000 million accumulator cells. Reducing the resolution might just make it possible to bring this down to 100 million cells, although accuracy will start to suffer. Further, when the HT is stored in such a large space, the search problem involved in locating significant peaks becomes formidable.

Fortunately, data are not stored at all uniformly in parameter space, which provides a key to solving both the storage and the subsequent search problem. The fact that parameter space is to be searched for the most prominent peaks—in general, the highest and the sharpest ones—means that the process of detection can start during accumulation. Furthermore, initial accumulation can be carried out at relatively low resolution, and then resolution can be increased locally as necessary in order to provide both the required accuracy and the separation of nearby peaks. In this context, it is natural to employ a binary splitting procedure—that is, to repeatedly double the resolution locally in each dimension of parameter space (e.g., the x dimension, the y dimension, the orientation dimension, and the size dimension, where these exist) until resolution is sufficient. A tree structure may conveniently be used to keep track of the situation.

Such a method (the "fast" HT) was developed by Li et al. (1985, 1986). Illingworth and Kittler (1987) did not find this method to be sufficiently flexible in dealing with real data and so produced a revised version (the "adaptive" HT) that permits each dimension in parameter space to change its resolution locally in tune with whatever the data demand, rather than insisting on some previously devised rigid structure. In addition, they employed a 9×9 accumulator array at each resolution rather than the theoretically most efficient 2×2 array, since this was found to permit better judgments to be made on the local data. This approach seemed to work well with fairly clean images, but later, doubts were cast on its effectiveness with complex images (Illingworth and Kittler, 1988). The most serious problem to be overcome here is that, at coarse resolutions, extended patterns of votes from several objects can overlap, giving rise to spurious peaks in parameter space. Since all of these peaks have to be checked at all relevant resolutions, the whole process can consume more computation than it saves. Optimization in multiresolution peak-finding schemes is complex[4] and data-dependent, and so discussion is curtailed here. The reader is referred to the original research papers (Li et al., 1985, 1986; Illingworth and Kittler, 1987) for implementation details.

An alternative scheme is the hierarchical HT (Princen et al., 1989a); so far it has been applied only to line detection. The scheme can most easily be envisaged by considering the foot-of-normal method of line detection described in Section 9.3. Rather small subimages of just 16×16 pixels are taken first of all, and the foot-of-normal positions are determined. Then each foot-of-normal is tagged with its orientation, and an identical HT procedure is instigated to generate the foot-of-normal positions for line segments in 32×32 subimage. This procedure is then repeated as many times as necessary until the whole image is spanned at once. The paper by Princen et al. discusses the basic procedure in detail and also elaborates necessary schemes for systematically grouping separate line segments into full-length lines in a hierarchical context. (In this way the work extends the techniques described in Section 9.4.) Successful operation of the method requires subimages with 50% overlap to be employed at each level. The overall scheme appears to be as accurate and reliable as the basic HT method but does seem to be faster and to require significantly less storage.

11.11 The Approach of Gerig and Klein

The Gerig and Klein approach was first demonstrated in the context of circle detection but was only mentioned in passing in Chapter 10. This is because it is an important *approach* that has much wider application than merely to circle detection. The motivation for it has already been noted in the previous

4 Ultimately, the problem of system optimization for analysis of complex images is a difficult one, since in conditions of low signal-to-noise ratio even the eye may find it difficult to interpret an image and may "lock on" to an interpretation that is incorrect. It is worth noting that in general image interpretation work there are many variables to be optimized—sensitivity, efficiency/speed, storage, accuracy, robustness, and so on—and it is seldom valid to consider any of these individually. Often tradeoffs between just two such variables can be examined and optimized, but in real situations multivariable tradeoffs should be considered. This is a very complex task, and it is one of the purposes of this book to show clearly the serious nature of these types of optimization problems, although at the same time it can only guide the reader through a limited number of basic optimization processes.

section—namely, the problem of extended patterns of votes from several objects giving rise to spurious peaks in parameter space.

Ultimately, the reason for the extended pattern of votes is that each edge point in the original image can give rise to a very large number of votes in parameter space. The tidy case of detection of circles of known radius is somewhat unusual, as will be seen particularly in the chapters on ellipse detection and 3-D image interpretation. Hence, in general *most* of the votes in parameter space are in the end unwanted and serve only to confuse. Ideally, we would like a scheme in which each edge point gives rise only to the single vote corresponding to the localization point of the particular boundary on which it is situated. Although this ideal is not initially realizable, it can be engineered by the "back-projection" technique of Gerig and Klein (1986). Here all peaks and other positions in parameter space to which a given edge point contributes are examined, and a new parameter space is built in which only the vote at the strongest of these peaks is retained. (There is the greatest probability, but no certainty, that it belongs to the *largest* such peak.) This second parameter space thus contains no extraneous clutter, and so weak peaks are found much more easily, giving objects with highly fragmented or occluded boundaries a much greater chance of being detected. Overall, the method avoids many of the problems associated with setting arbitrary thresholds on peak height—in principle no thresholds are required in this approach.

The scheme can be applied to any HT detector that throws multiple votes for each edge point. Thus, it appears to be widely applicable and is capable of improving robustness and reliability at an intrinsic expense of approximately doubling computational effort. (However, set against this is the relative ease with which peaks can be located—a factor that is highly data-dependent.) Note that the method is another example in which a two-stage process is used for effective recognition.

Other interesting features of the Gerig and Klein method must be omitted here for reasons of space, although it should be pointed out that, rather oddly, the published scheme ignores edge orientation information as a means of reducing computation.

11.12 Concluding Remarks

The Hough transform was introduced in Chapter 9 as a line detection scheme and then used in Chapter 10 for detecting circles. In those chapters it appeared as a rather cunning method for aiding object detection. Although it was seen to offer various advantages, particularly in its robustness in the face of noise and occlusion, its rather novel voting scheme appeared to have no real significance. The present chapter has shown that, far from being a trick method, the HT is much more general an approach than originally supposed. Indeed, it embodies the properties of the spatial matched filter and is therefore capable of close-to-optimal sensitivity for object detection. Yet this does not prevent its implementation from entailing considerable computational load, and significant effort and ingenuity have been devoted to overcoming this problem, both in general and in specific cases. The general case is tackled by the schemes discussed in the previous two sections, while two specific cases—those of ellipse and polygon detection—are considered in detail in the following chapters. It is important not to underestimate the value of specific solutions, both because such shapes as lines, circles, ellipses, and polygons cover a large proportion of (or approximations to) manufactured objects, and because methods for coping with specific cases have a habit (as for the original HT) of becoming more general as workers see possibilities for developing the underlying techniques. For further discussion and critique of the whole HT approach, see Chapter 29.

The development of the Hough transform may appear somewhat arbitrary. However, this chapter has shown that it has solid roots in matched filtering. In turn, this implies that votes should be gradient weighted for optimal sensitivity. Back-projection is another means of enhancing the technique to make optimal use of its capability for spatial search.

11.13 Bibliographical and Historical Notes

Although the HT was introduced as early as 1962, a number of developments—including especially those of Merlin and Farber (1975) and Kimme et al. (1975)—were required before the GHT was developed in its current standard form (Ballard, 1981). By that time, the HT was already known to be formally equivalent to template matching (Stockman and Agrawala, 1977) and to spatial matched filtering (Sklansky, 1978). However, the questions posed in Section 11.3 were only answered much later (Davies, 1987c), the necessary analysis being reproduced in Sections 11.4–11.7. Sections 11.8–11.11 reflect Davies' work (1987d,j, 1989d) on line detection by the GHT, which was aimed particularly at optimizing sensitivity of line detection, although deeper issues of tradeoffs between sensitivity, speed, and accuracy are also involved.

By 1985, the computational load of the HT became the critical factor preventing its more general use—particularly as the method could by that time be used for most types of arbitrary shape detection, with well-attested sensitivity and considerable robustness. Preliminary work in this area had been carried out by Brown (1984), with the emphasis on hardware implementations of the HT. Li et al. (1985, 1986) showed the possibility of much faster peak location by using parameter spaces that are not uniformly quantized. This work was developed further by Illingworth and Kittler (1987) and others (see, for example, Illingworth and Kittler, 1988; Princen et al., 1989a,b; Davies, 1992h). The future will undoubtedly bring many further developments. Unfortunately, however, the solutions may be rather complex and difficult (or at least tedious) to program because of the contextual richness of real images. However, an important development has been the randomized Hough transform (RHT), pioneered by Xu and Oja (1993) among others: it involves casting votes until specific peaks in parameter space become evident, thereby saving unnecessary computation.

Accurate peak location remains an important aspect of the HT approach. Properly, this is the domain of robust statistics, which handles the elimination of outliers (of which huge numbers arise from the clutter of background points in digital images—see Appendix A). However, Davies (1992g) has shown a computationally efficient means of accurately locating HT peaks. He has also found why in many cases peaks may appear narrower than *a priori* considerations would indicate (Davies, 1992b). Kiryati and Bruckstein (1991) have tackled the aliasing effects that can arise with the Hough transform and that have the effect of cutting down accuracy.

Over time, the GHT approach has been broadened by geometric hashing, structural indexing, and other approaches (e.g., Lamdan and Wolfson, 1988; Gavrila and Groen, 1992; Califano and Mohan, 1994). At the same time, a probabilistic approach to the subject has been developed (Stephens, 1991) which puts it on a firmer footing (though rigorous implementation would in most cases demand excessive computational load). Finally, Grimson and Huttenlocher (1990) warn against the blithe use of the GHT for complex object recognition tasks because of the false peaks that can appear in such cases, though their paper is probably overpessimistic. For further review of the state of the subject up to 1993, see Leavers (1992, 1993).

In various chapters of Part 2, the statement has been made that the Hough transform[5] carries out a search leading to hypotheses that should be checked before a final decision about the presence of an object can be made. However, Princen et al. (1994) show that the performance of the HT can be improved if it is itself regarded as a hypothesis testing framework. This is in line with the concept that the

5 A similar statement can be made in the case of graph matching methods such as the maximal clique approach to object location (see Chapter 15).

HT is a model-based approach to object location. Other studies have recently been made about the nature of the HT. In particular, Aguado et al. (2000) consider the intimate relationship between the Hough transform and the principle of duality in shape description. The existence of this relationship underlines the importance of the HT and provides a means for a more general definition of it. Kadyrov and Petrou (2001) have developed the trace transform, which can be regarded as a generalized form of the Radon transform, which is itself closely related to the Hough transform.

Other workers have used the HT for affine-invariant search. Montiel et al. (2001) made an improvement to reduce the incidence of wrong evidence in the gathered data, while Kimura and Watanabe (2002) made an extension for 2-D shape detection that is less sensitive to the problems of occlusion and broken boundaries. Kadyrov and Petrou (2002) have adapted the trace transform to cope with affine parameter estimation.

In a generalization of the work of Atherton and Kerbyson (1999), and of Davies (1987c) on gradient weighting (see Section 11.6), Anil Bharath and his colleagues have examined how to optimize the sensitivity of the HT (private communication, 2004). Their method is particularly valuable in solving the problems of early threshold setting that limit many HT techniques. Similar sentiments have emerged in a different way in the work of Kesidis and Papamarkos (2000), which maintains the gray-scale information throughout the transform process, thereby leading to more exact representations of the original images.

Olson (1999) has shown that localization accuracy can be improved efficiently by transferring local error information into the HT and handling it rigorously. An important finding is that the HT can be subdivided into many subproblems without a decrease of performance. This finding is elaborated in a 3-D model-based vision application where it is shown to lead to reduced false positive rates (Olson, 1998). Wu et al. (2002) extend the 3-D possibilities further by using a 3-D HT to find glasses. First, a set of features are located that lie on the same plane, and this is then interpreted as the glasses rim plane. This approach allows the glasses to be separated from the face, and then they can be located in their entirety.

van Dijck and van der Heijden (2003) develop the geometric hashing method of Lamdan and Wolfson (1988) to perform 3-D correspondence matching using full 3-D hashing. This method is found to have advantages in that knowledge of 3-D structure can be used to reduce the number of votes and spurious matches. Tytelaars et al. (2003) describe how invariant-based matching and Hough transforms can be used to identify regular repetitions in planes appearing within visual (3-D) scenes in spite of perspective skew. The overall system has the ability to reason about consistency and is able to cope with periodicities, mirror symmetries, and reflections about a point.

Finally, in an unusual 3-D application, Kim and Han (1998) use the Hough transform to analyze scenes taken from a line-scan camera in linear motion. The images are merged to form a "slit image," which is partitioned into segmented regions, each directly related to a simple surface. In this representation, the correspondence problem (Chapter 16) is reduced to the simpler process of detecting straight lines using the HT.

11.14 Problem

1. (a) Describe the main stages in the application of the Hough transform to locate objects in digital images. What particular advantages does the Hough transform technique offer? Give reasons why these advantages arise.

   (b) It is said that the Hough transform only leads to *hypotheses* about the presence of objects in images and that they should all be checked independently before making a final decision about the contents of any image. Comment on the accuracy of this statement.

Ellipse Detection

Ellipses form a further attractive target for techniques such as the Hough transform, for they appear not only in their own right but also as oblique views of circular objects. This chapter discusses how the Hough transform can be adapted to locate ellipses in three contrasting ways.

Look out for:

- the basic diameter bisection method for ellipse detection using the Hough transform.
- means of testing a shape to confirm that it is an ellipse.
- the chord–tangent method for ellipse detection.
- means of finding all relevant ellipse parameters.
- use of the generalized Hough transform for ellipse detection.
- how speed can be improved by use of a universal lookup table.
- how the computational loads of the various Hough transform techniques can be estimated.

This chapter shows yet again how the Hough transform can detect an important type of macroscopic feature. In addition, it is useful in contrasting different versions of the technique and in providing sound notions about how order calculations of computational loads can be carried out.

CHAPTER 12

Ellipse Detection

12.1 Introduction

As seen in the previous chapter, many manufactured goods are circular or else possess distinctive circular features. This consideration makes it imperative to have an efficient set of algorithms for detecting circular shapes in digital images. However, in many situations such objects are viewed obliquely, and so efficient ellipse detectors are also required. Furthermore, some objects are actually elliptical, which also provides motivation for finding effective ellipse detection algorithms.

This chapter describes some of the algorithms that have been developed for this purpose. We concentrate on parallel algorithms inasmuch as sequential methods have already been covered in Chapter 7. In addition, parallel algorithms such as the Hough transform tend to be more robust. In fact, all the algorithms described here are based on one or other form of the Hough transform. Although Duda and Hart (1972) considered curve detection, the first method that made use of edge orientation information to reduce the amount of computation was that of Tsuji and Matsumoto (1978). Their ingenious method is described in the next section.

12.2 The Diameter Bisection Method

The diameter bisection method of Tsuji and Matsumoto (1978) is very simple in concept. First, a list is compiled of all the edge points in the image. Then, the list is sorted to find those that are antiparallel, so that they can lie at opposite ends of ellipse diameters. Next, the positions of the center points of the connecting lines for all such pairs are taken as voting positions in parameter space (Fig. 12.1). As for circle location, the parameter space that is used for this purpose is

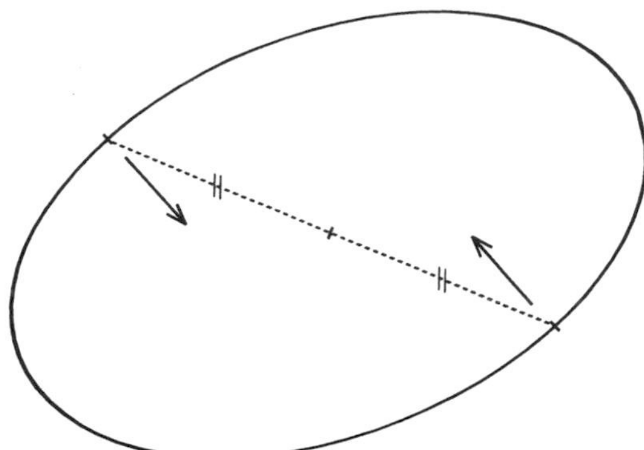

Figure 12.1 Principle of the diameter bisection method. A pair of points is located for which the edge orientations are antiparallel. If such a pair of points lies on an ellipse, the midpoint of the line joining the points will be situated at the center of the ellipse.

congruent to image space. Finally, the positions of significant peaks in parameter space are located to identify possible ellipse centers.

In an image containing many ellipses and other shapes, naturally there will be many pairs of antiparallel edge points, and for most of these the center points of the connecting lines will lead to nonuseful votes in parameter space. Such clutter of course leads to wasted computation. However, it is a principle of the Hough transform that votes must be accumulated in parameter space at all points that *could in principle* lead to correct object center location. It is left to the peak finder to find the voting positions that are most likely to correspond to object centers.

Not only does clutter lead to wasted computation, but the method itself is computationally expensive. This is because it examines all *pairs* of edge points, and there are many more such pairs than there are edge points (m edge points would lead to $\binom{m}{2} \approx m^2/2$ pairs of edge points). Indeed, since there are likely to be at least 1000 edge points in a typical 256×256 image, the computational problems can be formidable. The situation is examined more carefully in Section 12.6.

The basic method is not particularly discriminating about ellipses. It picks out many symmetrical shapes—any indeed that possess 180° rotation symmetry, including rectangles, ellipses, circles, or "ovals" such as ellipses with compressed ends. In addition, the basic scheme sometimes gives rise to a number of false identifications even in an image in which only ellipses are present (Fig. 12.2). However, Tsuji and Matsumoto (1978) also proposed a technique by which true ellipses can be distinguished. The basis of the technique is the property of an ellipse that the lengths of perpendicular semidiameters OP, OQ obey the relation:

$$1/\mathrm{OP}^2 + 1/\mathrm{OQ}^2 = 1/R^2 = \text{constant} \tag{12.1}$$

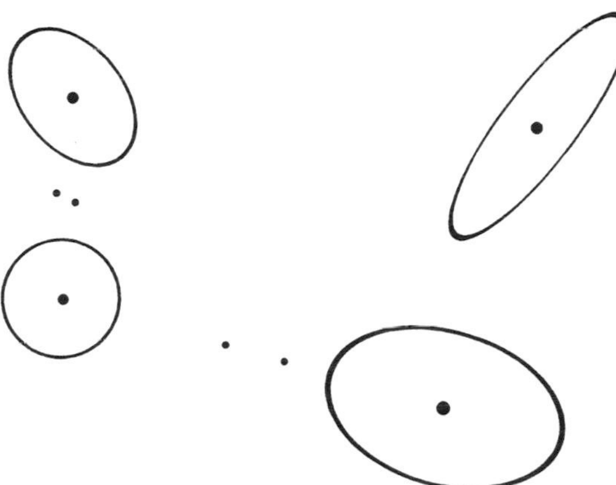

Figure 12.2 Result of using the basic diameter bisection method. The large dots show true ellipse centers found by the method, while the smaller dots show positions at which false alarms commonly occur. Such false alarms are eliminated by applying the test described in the text.

To proceed, the set of edge points that contribute to a given peak in parameter space is used to construct a histogram of R values (the latter being obtained from equation (12.1)). If a significant peak is found in this histogram, then we have clear evidence of an ellipse at the specified location in the image. If two or more such peaks are found, then we have evidence of a corresponding number of concentric ellipses in the image. If, however, no such peaks are found, then a rectangle, oval, or other symmetrical shape may be present, and each of these would need its own identifying test.

Clearly, then, the method relies on the existence of an appreciable number of pairs of edge points on an ellipse lying at opposite ends of diameters. Hence, strict limits are imposed on the amount of the boundary that must be visible (Fig. 12.3). Finally, it should not go unnoticed that the method wastes the signal available from unmatched edge points. These considerations have led to a search for further methods of ellipse detection.

12.3 The Chord–Tangent Method

The chord–tangent method was devised by Yuen et al. (1988) and makes use of another simple geometric property of the ellipse. Again, pairs of edge points are taken in turn, and for each point of the pair, tangents to the ellipse are constructed and found to cross at T; the midpoint of the connecting line is found at M; then the equation of the line TM is found and all points that lie on the portion MK

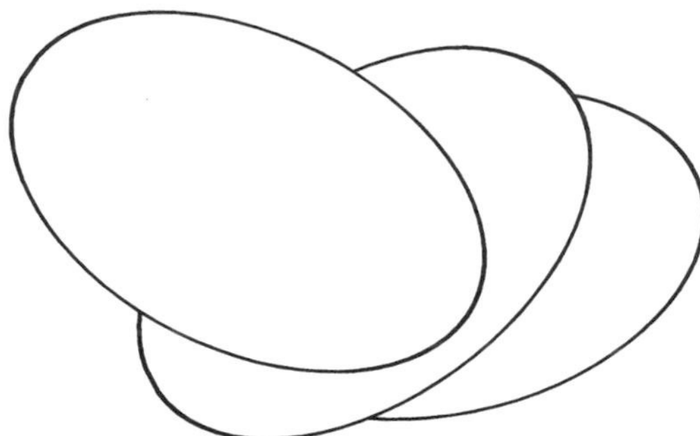

Figure 12.3 Limitations of the diameter bisection method: of the three ellipses shown, only the lowest one cannot be located by the diameter bisection method.

of this line are accumulated in parameter space (Fig. 12.4). (Clearly, T and the center of the ellipse lie on the opposite sides of M.) Finally, peak location proceeds as before.

The proof that this method is correct is trivial. Symmetry ensures that the method works for circles, and projective properties then ensure that it also works for ellipses. Under projection, straight lines project into straight lines, midpoints into midpoints, tangents into tangents, and circles into ellipses; in addition, it is always possible to find a viewpoint such that a circle can be projected into a given ellipse.

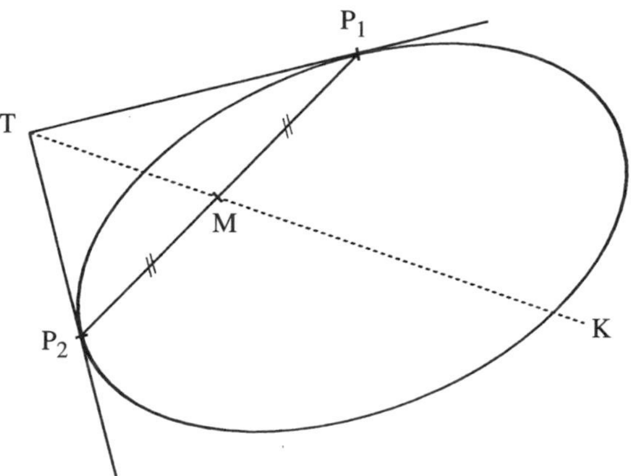

Figure 12.4 Principle of the chord–tangent method. The tangents at P_1 and P_2 meet at T and the midpoint of P_1P_2 is M. The center C of the ellipse lies on the line TM produced. Notice that M lies between C and T. Hence, the transform for points P_1 and P_2 need only include the portion MK of this line.

Unfortunately, this method suffers from significantly increased computation, since so many points have to be accumulated in parameter space. This is obviously the price to be paid for greater applicability. However, computation can be minimized in at least three ways: (1) cutting down the lengths of the lines of votes accumulated in parameter space by taking account of the expected sizes and spacings of ellipses; (2) not pairing edge points initially if they are too close together or too far apart; and (3) eliminating edge points once they have been identified as belonging to a particular ellipse. (The last two techniques are also applicable to the diameter bisection method.)

Section 12.5 explores in more detail whether computation can be saved by reverting to the GHT approach. Meanwhile, it is necessary to consider how ellipse detection can be augmented by suitable procedures for finding ellipse orientation, size, and shape.

12.4 Finding the Remaining Ellipse Parameters

Although it has proved possible to locate ellipses in digital images by simple geometric constructions based on well-known properties of the ellipse, a more formal approach is required to determine other ellipse parameters. Accordingly, we write the equation of an ellipse in the form:

$$Ax^2 + 2Hxy + By^2 + 2Gx + 2Fy + C = 0 \tag{12.2}$$

an ellipse being distinguished from a hyperbola by the additional condition:

$$AB > H^2 \tag{12.3}$$

This condition guarantees that A can never be zero and that the ellipse equation may, without loss of generality, be rewritten with $A = 1$. In any case, only ratios between the parameters have physical significance. This leaves five parameters, which can be related to the position of the ellipse, its orientation, and its size and shape (or eccentricity).

Having located the center of the ellipse, we may select a new origin of coordinates at its center (x_c, y_c). The equation then takes the form:

$$x'^2 + 2Hx'y' + By'^2 + C' = 0 \tag{12.4}$$

where

$$x' = x - x_c; \quad y' = y - y_c \tag{12.5}$$

It now remains to fit to equation (12.4) the edge points that gave evidence for the ellipse center under consideration. The problem will in general be vastly overdetermined. Hence, an obvious approach is the method of least squares. Unfortunately, this technique tends to be very sensitive to outlier points and is therefore liable to be inaccurate. An alternative is to employ some form of Hough transformation. Since Hough parameter spaces require storage that increases exponentially with the number of unknown parameters, it is tempting to try to separate the determination of some of the parameters from that of others. For example, we can follow Tsukune and Goto (1983) by differentiating equation (12.4):

$$x' + By'\mathrm{d}y'/\mathrm{d}x' + H(y' + x'\mathrm{d}y'/\mathrm{d}x') = 0 \tag{12.6}$$

Then $\mathrm{d}y'/\mathrm{d}x'$ can be determined from the local edge orientation at (x', y') and a set of points accumulated in the new (H, B) parameter space. When a peak is eventually located in (H, B) space, the relevant data (a subset of a subset of the original set of edge points) can again be used with equation (12.4) to obtain a histogram of C' values, from which the final parameter for the ellipse can be obtained.

The following formulas must be used to determine the orientation and semiaxes values of an ellipse in terms of H, B, and C':

$$\theta = (1/2)\arctan[2H/(1 - B)] \tag{12.7}$$

$$a^2 = \frac{-2C'}{(B+1) - [(B-1)^2 + 4H^2]^{1/2}} \tag{12.8}$$

$$b^2 = \frac{-2C'}{(B+1) + [(B-1)^2 + 4H^2]^{1/2}} \tag{12.9}$$

Mathematically, θ is the angle of rotation that diagonalizes the second-order terms in equation (12.4). Having performed this diagonalization, we find that the ellipse is essentially in the standard form $\tilde{x}^2/a^2 + \tilde{y}^2/b^2 = 1$, so a and b are determined.

This method finds the five ellipse parameters in *three* stages: first the positional coordinates are obtained, then the orientation, and finally the size and eccentricity.[1] This three-stage calculation involves less computation but compounds any errors—in addition, edge orientation errors, though low, become a limiting factor. For this reason, Yuen et al. (1988) tackled the problem by speeding up the

1 Strictly, the eccentricity is $e = (1 - b^2/a^2)^{1/2}$, but in most cases we are more interested in the ratio of semiminor to semimajor axes, b/a.

HT procedure itself rather than by avoiding a direct assault on equation (12.4). That is, they aimed at a fast implementation of a thoroughgoing second stage, which finds all the parameters of equation (12.4) in one 3-D parameter space. Their fast adaptive HT procedure has already been described in Chapter 11 and is not covered again here.

At this stage it is clear that reasonably optimal means are available for finding the orientation and semiaxes of an ellipse once its position is known. The weak point in the process appears to be that of finding the ellipse initially. Indeed, the two approaches for achieving this that have been described above are particularly computationally intensive, mainly because they examine all pairs of edge points. Hence, we should again consider the GHT approach, which locates objects by taking edge points singly, to see whether any gains can be achieved by this means.

12.5 Reducing Computational Load for the Generalized Hough Transform Method

The complications described in Chapter 11 when the GHT is used to detect anisotropic objects clearly apply to ellipse detection. The various possible orientations of the object lead to the need to employ a large number of planes in parameter space. So far this chapter has avoided considering the GHT for these reasons. However, it will be seen that by accumulating the votes for all possible orientations in a *single* plane in parameter space, significant savings in computation can frequently be made. Basically, the idea is largely to eliminate the enormous storage requirements of the GHT by using only one instead of 360 planes in parameter space, while at the same time reducing by a large factor the computation involved in the final search for peaks. Such a scheme may have concomitant disadvantages such as the production of spurious peaks, and this aspect will have to be examined carefully.

To achieve these aims, it is necessary to analyze the shape of the point spread function (PSF) to be accumulated for each edge pixel, in the case of an ellipse of unknown orientation. We start by taking a general edge fragment at a position defined by ellipse parameter ψ and deducing the bearing of the center of the ellipse relative to the local edge normal (Fig. 12.5). Working first in an ellipse-based axes system, for an ellipse with semimajor and semiminor axes a and b, respectively, it is clear that:

$$x = a \cos \psi \tag{12.10}$$

$$y = b \sin \psi \tag{12.11}$$

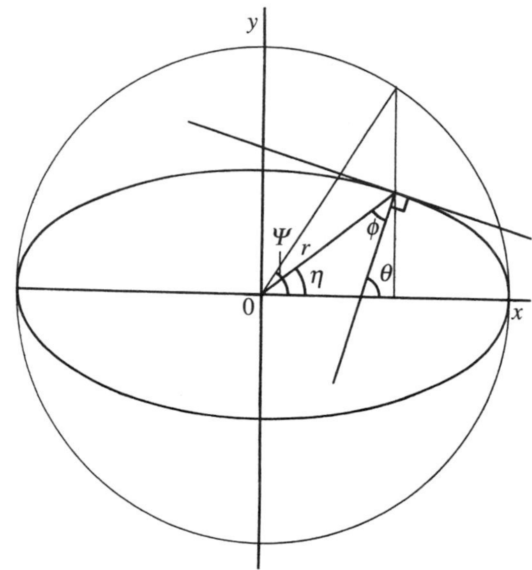

Figure 12.5 Geometry of an ellipse and its edge normal.

Hence,

$$dx/d\psi = -a \sin \psi \tag{12.12}$$

$$dy/d\psi = b \cos \psi \tag{12.13}$$

giving

$$dy/dx = -(b/a) \cot \psi \tag{12.14}$$

Hence, the orientation of the edge normal is given by:

$$\tan \theta = (a/b) \tan \psi \tag{12.15}$$

At this point we wish to deduce the bearing φ of the center of the ellipse relative to the local edge normal. From Fig. 12.5:

$$\varphi = \theta - \eta \tag{12.16}$$

where

$$\tan \eta = y/x = (b/a) \tan \psi \tag{12.17}$$

and

$$\tan\varphi = \tan(\theta - \eta) = \frac{\tan\theta - \tan\eta}{1 + \tan\theta\tan\eta} \tag{12.18}$$

Substituting for tan θ and tan η, and then rearranging, gives:

$$\tan\varphi = \frac{(a^2 - b^2)}{2ab}\sin 2\psi \tag{12.19}$$

In addition:

$$r^2 = a^2\cos^2\psi + b^2\sin^2\psi \tag{12.20}$$

To obtain the PSF for an ellipse of unknown orientation, we now simplify matters by taking the current edge fragment to be at the origin and oriented with its normal along the u-axis (Fig. 12.6). The PSF is then the locus of all possible positions of the center of the ellipse. To find its form, it is necessary merely to eliminate ψ between equations (12.19) and (12.20). This is facilitated by reexpressing r^2 in double angles (the significance of double angles lies in the 180° rotation symmetry of an ellipse):

$$r^2 = \frac{a^2 + b^2}{2} + \frac{a^2 - b^2}{2}\cos 2\psi \tag{12.21}$$

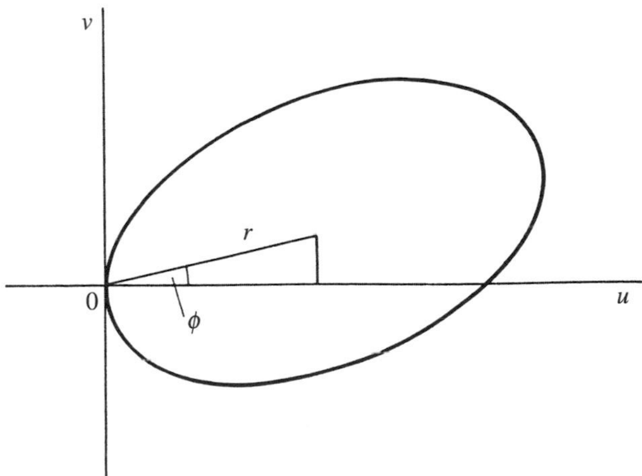

Figure 12.6 Geometry for finding the PSF for ellipse detection by forming the locus of the centers of ellipses touching a given edge fragment.

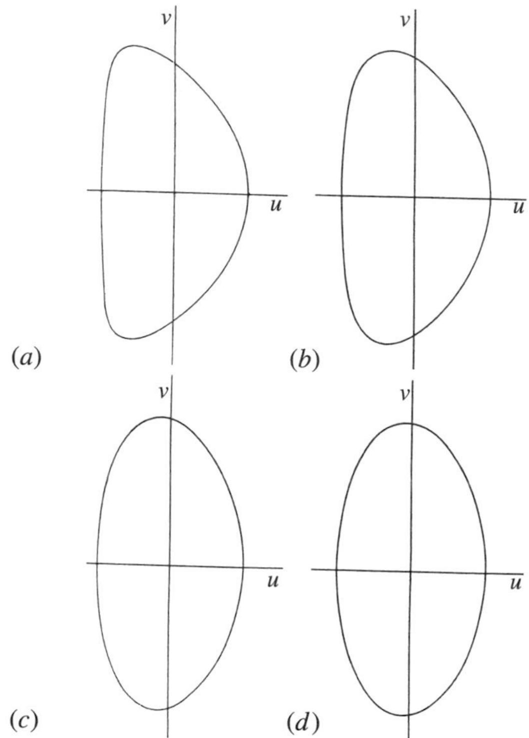

Figure 12.7 Typical PSF shapes for detection of ellipses with various eccentricities: (*a*) ellipse with $a/b = 21.0$, $c/d = 1.1$; (*b*) ellipse with $a/b = 5.0$, $c/d = 1.5$; (*c*) ellipse with $a/b = 2.0$, $c/d = 3.0$; (*d*) ellipse with $a/b = 1.4$, $c/d = 6.0$. Notice how the PSF shape approaches a small ellipse of aspect ratio 2.00 as eccentricity tends to zero. The semimajor and semiminor axes of the PSF are $2d$ and d, respectively.

After some manipulation the locus is obtained as:

$$r^4 - r^2(a^2 + b^2) + a^2b^2 \sec^2\varphi = 0 \tag{12.22}$$

which can, in the edge-based coordinate system, also be expressed in the form:

$$v^2 = (a^2 + b^2) - u^2 - a^2b^2/u^2 \tag{12.23}$$

Unfortunately, this is not a circle and is not even symmetrical about axes through its center. The shape of the PSF for various values of a and b is shown in Fig. 12.7. For *very* highly eccentric ellipses, the PSF is approximated by the arcs of two circles, one of radius a and the other of radius a^2/b. It will be seen that in the other extreme, ellipses of low eccentricity, the PSF is approximated by an ellipse. In general, however, it is easy to see that, since the original ellipse possesses regions along the principal axes where the radii of curvature are approximately constant with values a^2/b and b^2/a, the PSF also possesses regions at either end of one diameter where the radii of curvature are $(a^2/b) - b$ and $a - (b^2/a)$. This makes it clear that the shape is not symmetrical and that it should become approximately symmetrical when $a \approx b$.

Defining two new parameters:

$$c = (a + b)/2 \tag{12.24}$$

$$d = (a - b)/2 \tag{12.25}$$

and then taking an approximation of small d, the locus is obtained in the form:

$$(u + c + d^2/2c)^2/d^2 + v^2/4d^2 = 1 \tag{12.26}$$

Neglecting the term $d^2/2c$ which provides a small correction to u, the PSF is itself approximately an ellipse of semimajor axis $2d$ and semiminor axis d. Making use of this fact simplifies implementation, but it is practicable only for low-eccentricity ellipses where $d \lesssim 0.1c$, so that $a \lesssim 1.2b$. In practice, however, where ellipses are small and the PSF is only a few pixels across, it is a reasonable approximation to insist only that $a < 2b$.

Implementation is simplified since a universal lookup table (ULUT) for ellipse detection can be compiled, which is independent of the size and eccentricity of the ellipse to be detected as long as the eccentricity is not excessive. Since ellipses are common features in industrial and other applications—arising both from elliptical objects and from oblique views of circles—this factor should be an important consideration in many applications. Thus, a single ULUT is compiled and stored, and it then needs only to be scaled and positioned to produce the PSF in a given instance of ellipse detection.

12.5.1 *Practical Details*

Having constructed a ULUT for ellipse detection, we find that the detection algorithm has to scale it, position it and rotate it so that points can be accumulated in parameter space. *A priori*, it would be imagined that a considerable amount of

```
for all pixels in image { // scan over image
    findgx; // apply edge operator
    findgy;
    g = magnitude[gx][gy]; // lookup magnitude
    if (g > gthreshold) // if edge point
        for (i = 1; i <= tablesize; i++) { // run over PSF
            dx = c + d * xtable[i]; // scale and position
            dy = d * ytable[i];
            xc = xedge – (int)((dx * gx – dy * gy)/g);
            yc = yedge – (int)((dx * gy + dy * gx)/g);
            // rotate and position
            vote[xc][yc]++; // accumulate vote
        }
} // end image scan
```

Figure 12.8 Algorithm for ellipse detection. In this section of code, it is arbitrarily (i) that the magnitude of the vector (g_x, g_y) is available from a lookup table, and (ii) that all calculations are carried out in floating-point form until the final "rounding" stage where voting positions are being estimated to the nearest pixel.

trigonometric computation is involved in this process. However, it is possible to avoid calculating angles directly (e.g., using the arctan function) by always working with sines and cosines. This is rendered possible partly because such edge orientation operators as the Sobel give two components (g_x, g_y) for the intensity gradient vector. (This has now been seen to happen in several line and circle detection schemes—see Chapters 9 and 10.) Hence, a considerable amount of computation can be saved, as is demonstrated by the simplicity of the final code for computing a given contribution to the PSF (Fig. 12.8).

Figure 12.9 shows the result of testing the above scheme on an image of some O-rings lying on a slope of arbitrary direction. The O-rings are found accurately and with a fair degree of robustness—that is, despite overlapping and partial occlusion (up to 40% in one case). In several cases, incidental transforms from points on the inner edges of the O-rings overlap other transforms from points on the outer edges, although only the latter are actually employed usefully here for peak finding. Hence, the scheme is able to overcome problems resulting from additional clutter in parameter space.

Figure 12.9 also shows the arrangement of points in parameter space that results from applying the PSF to every edge point on the boundary of an ellipse. The pattern is somewhat clearer in Fig. 12.10. In either case it is seen to contain a high degree of structure. For an ideal transform, there would be no structure apart from the main peak. Looking at the situation in another way, in deducing the locus of points forming the PSF we found all points that could possibly be

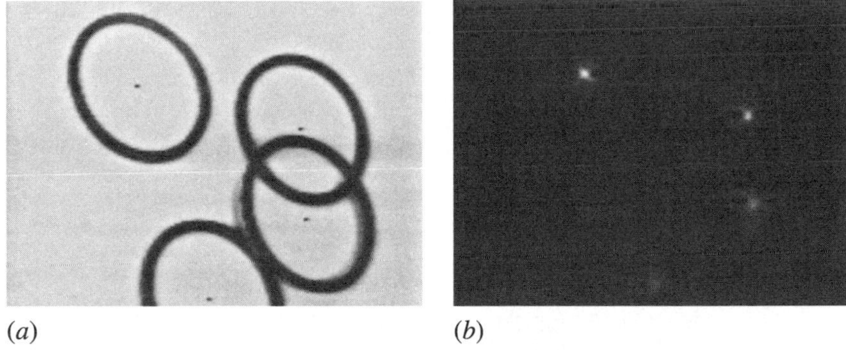

(a) (b)

Figure 12.9 Applying the PSF to detection of tilted circles: (*a*) off-camera 128 × 128 image of a set of circular O-rings on a 45° slope of arbitrary direction; (*b*) transform in parameter space. Notice the peculiar shape of the ellipse transform, which is close to a "four-leaf clover" pattern. (*a*) also indicates the positions of the centers of the O-rings as located from (*b*): accuracy is limited by the presence of noise, shadows, clutter, and available resolution, to an overall standard deviation of about 0.6 pixel.

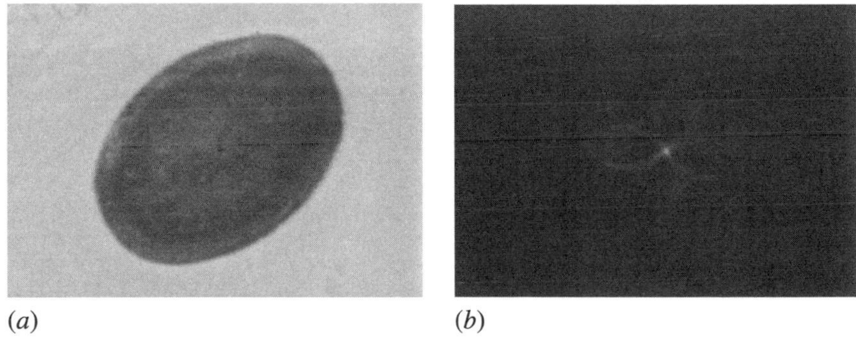

(a) (b)

Figure 12.10 Applying the PSF to detection of elliptical objects: (*a*) off-camera 128 × 128 image of an elliptical bar of soap of arbitrary orientation; (*b*) transform in parameter space. In this case, the cloverleaf pattern is better resolved. Accuracy of location is limited partly by distortions in the shape of the object, but the peak location procedure results in an overall standard deviation of the order of 0.5 pixel.

at the center of an ellipse when we had no knowledge about its orientation. All points on the PSF *not* falling on the peak at the center of the ellipse would hopefully be randomly (and fairly uniformly) distributed nearby. However, for a shape such as an ellipse—which has a particular structure and symmetry of its own—this is not the situation. A full computation of the

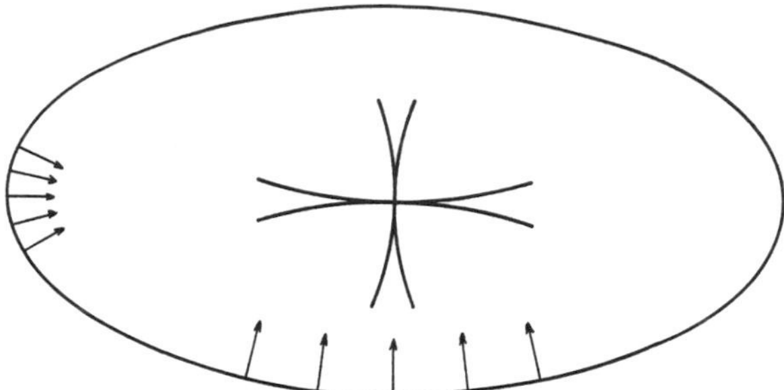

Figure 12.11 Formation of the ellipse transform by moving the regions of the ellipse boundary lying roughly along the principal axes radially inward until they pass through the center of the ellipse. The reason for this peculiar construction lies in the fact that the prominent lines in the transform are the envelopes of all possible PSFs. The necessary coherence is provided by the nearly constant radii of curvature roughly along the directions of the principal axes.

positions of votes in parameter space yields many candidate positions, and to find strong structure in the transform it is necessary to look for *envelopes* to the system of PSFs.

This is a sufficiently complex task to be left to experiment, but one or two observations on the results are useful and relevant. First, where there is a lot of local symmetry in the starting shape (the ellipse)—and specifically where the curvature on the boundary is approximately constant—this leads automatically to strong envelope lines in the transform, as indicated in Fig. 12.11, where the on-axes regions of the original ellipse have essentially been moved radially (and are hence also compressed laterally) until they pass through its center. These lines seem practically to form a "four-leaf clover" pattern, although the loops become more diffuse on moving away from the center, in accordance with the envelope idea expressed above.

The important problem here is whether the structure in the transform is misleading—that is, whether it gives rise to sharply focused peaks at other positions than the main peak. This does not appear to happen, the main and most intense structure being the cross pattern at the center. This pattern is clearly strongest and most focused (both practically and theoretically) near the center, while, in addition, it goes up to double the height quite sharply as the two branches cross at the very center. The overall conclusion is that the additional structure does no particular harm, although it gives rise to interesting complications if two ellipses overlap so that their centers are quite close. Hence, it is more difficult to use the method described here to disentangle two such ellipses than it would be to disentangle two circular objects with a similar degree of overlap. The way around this problem is to check individual high points in the transform by interrogating the original image. See, for example, Bolles and Cain (1982), Davies (1987g), and Chapters 13 and 15.

More help would be obtained in interpreting the transform in such cases if the PSFs were regarded not as mere loci of the centers of ellipses of different orientations touching a given edge fragment, but as loci with proper weights, unit weight corresponding to unit boundary distance on the original ellipse. This is too complex a problem to be pursued here, but it may be noted that curvature information immediately leads to the prediction of larger weights for the part of the PSF corresponding to the two regions of the ellipse along the minor axes.

Finally, note that accuracy will be lost if the PSF is not guaranteed to place a vote at the center of the ellipse for every edge pixel. This means that the PSF should be a connected chain of pixels. Hence, the size of the lookup table (LUT) must vary with the size of the PSF. Or, if a large ULUT is used, a small PSF should be able to skip over a proportion of the entries. In the examples of Figs. 12.9 and 12.10, the PSF contained 50 and 100 votes, respectively.

12.6 Comparing the Various Methods

To compare the computational loads for the methods of ellipse detection discussed above, we concentrate on ellipse location per se and ignore any additional procedures concerned with (1) finding other ellipse parameters, (2) distinguishing ellipses from other shapes, or (3) separating concentric ellipses. We start by examining the GHT method and the diameter bisection method.

First, suppose that an $N \times N$ pixel image contains p ellipses of identical size given by the parameters a, b, c, d as defined in Section 12.3. By ignoring noise and general background clutter, we shall be favoring the diameter bisection method, as will be seen below. Next, the discussion is simplified by supposing that the computational load resides mainly in calculating the positions at which votes should be accumulated in parameter space—the effort involved in locating edge pixels and in locating peaks in parameter space is much smaller.

Under these circumstances, the load for the GHT method may be approximated by the product of the number of edge pixels and the number of points per edge pixel that have to be accumulated in parameter space, the latter being equal to the number of points on the PSF. Hence, the load is proportional to:

$$\begin{aligned} L_A &\approx p \times 2\pi c \times 2\pi(2d + d)/2 = 6\pi^2 pcd \\ &\approx 60pcd \end{aligned} \tag{12.27}$$

where the ellipse has been taken to have relatively low eccentricity so that the PSF itself approximates to an ellipse of semiaxes $2d$ and d.

For the diameter bisection method, the actual voting is a minor part of the algorithm—as indeed it is in the GHT method (see the snippet of code listed in Fig 12.8). In either case, most of the computational load concerns edge orientation calculations or comparisons. Assuming that these calculations and comparisons involve similar inherent effort, we may fairly assess the load for the diameter bisection method as:

$$\begin{aligned} L_B &\approx \binom{p \times 2\pi c}{2} \approx (2\pi pc)^2/2 \\ &\approx 20p^2c^2 \end{aligned} \tag{12.28}$$

Hence,

$$L_B/L_A \approx pc/3d \tag{12.29}$$

When a is close to b, as for a circle, $L_A \rightarrow 0$, and then the diameter bisection method becomes a very poor option. In many cases, however, a is close to $2b$, so that c is close to $3d$. The ratio of the loads then becomes:

$$L_B/L_A \approx p \tag{12.30}$$

It is possible that p will be as low as 1 in some cases. Such cases are likely to be rare, however, and are offset by applications where there is significant background image clutter and noise, or where all p ellipses have other edge detail giving irrelevant signals that can be considered as a type of self-induced clutter (see the O-ring example of Fig. 12.9). The overall image "clutter factor" would normally be expected to be at least two.

It is also possible that some of the pairs of edge points in the diameter bisection method can be excluded before they are considered, for example, by giving every edge point a range of interaction related to the size of the ellipses. This would tend to reduce the computational load by a factor of the order of (but not as small as) p. However, the computational overhead required for this would not be negligible.

Overall, the GHT method will no doubt be significantly faster than the diameter bisection method in most real applications, the diameter bisection method being at a definite disadvantage when image clutter and noise is strong. The only situation where the diameter bisection method might win appears to be when an image contains a minimal number of highly eccentric ellipses. By comparison, the chord–tangent method always requires more computation than the diameter bisection method, since not only does it examine every pair of edge points but also it generates a *line* of votes in parameter space for each pair.

Against these computational limitations must be noted the different characteristics of the methods. First, the diameter bisection method is not particularly discriminating in that it locates many symmetrical shapes, as remarked earlier. The chord–tangent method is selective for ellipses but is not selective about their size or eccentricity. The GHT method is selective about all of these factors. These types of discriminability, or lack of it, can turn out to be advantageous or disadvantageous, depending on the application. Hence, we do no more here than draw attention to these different properties. It is also relevant that the diameter bisection method is rather less robust than the other methods. This is so because if one edge point of an antiparallel pair is not detected, then the other point of the pair cannot contribute to detection of the ellipse—a factor that does not apply for the other two methods since they take all edge information into account.

Ellipse detection is another task that can be tackled using the Hough transform. This chapter has demonstrated three different means for achieving this. Key factors in the choice of method are the degree of robustness, the speed, and the specificity that are required.

12.7 Concluding Remarks

The simplest scheme for ellipse detection in digital images is the diameter bisection method, which, unfortunately, is not very robust because many pairs of diameter ends need to be visible before sufficient signal can be built up. One alternative is the chord–tangent method; this is highly robust, yet at the penalty of considerable additional computation. Another alternative is the GHT, but when applied to anisotropic shapes this method requires that one plane in parameter space be used for each discriminable orientation of the shape. However, if a single plane parameter space is used for this purpose, the GHT approach frequently requires less computational effort than the other two methods. This is because they are based on using pairs of edge points to deduce candidate ellipse centers, whereas the GHT takes edge points one at a time. However, the three methods differ markedly in their shape discriminability characteristics, as should be noted in real applications.

In particular, the diameter bisection method locates ellipses of all eccentricities, sizes, and orientations, and also other symmetrical shapes; the chord–tangent method locates ellipses of all eccentricities, sizes, and orientations but discriminates against other shapes; and the GHT locates ellipses of all orientations but discriminates against ellipses of the wrong eccentricity and size and against other shapes. However, the GHT can be adapted to detect many other shapes, whereas the diameter bisection method is limited and the chord–tangent method is specific to ellipses.

With the GHT, the avoidance of many planes in parameter space is one aspect of a more general problem. For example, we might need to use many planes in parameter space to allow for variation in object size. It is also possible that we might know the size and orientation of an ellipse but not its eccentricity, many planes being needed to cope with this situation. In Chapter 10 it was shown that circles of many sizes can be sought simultaneously in a single plane in parameter space, by a method related to that described here (viz., using a PSF consisting of a radial line in place of a single point). It is clear that this method will carry over to ellipses and other shapes (i.e., replace the $R(\theta)$ in the basic transform in turn by $\alpha R(\theta)$, for all a in a given range, leaving $\varphi(\theta)$ unchanged). Similarly, it is evident from the derivation in Section 12.5 that the GHT approach carries over to other shapes when orientation is unknown.

Finally, a similar extension can be made when eccentricity is unknown but size and orientation are known. Thus, we hypothesize generally that *parameter space can be compressed by one degree of freedom from the ideal* in order to save computation. We also note that *it is not possible to compress parameter space by more than one degree of freedom from the ideal.* If we tried to do this, virtually any shape would be able to trigger the detector and there would be a great many false alarms. As an example, if we try to detect ellipses of unknown orientation and size in a single plane in parameter space, the PSF becomes an extended blob that is unable to discriminate between a great variety of shapes. (In general, the PSF becomes an *area* that is the result of convolving the two basic PSF *curves* in parameter space. Indeed, it is not possible to find two such curves that do not produce an area when convolved.) These observations appear both to extend and ultimately to limit the capabilities of the GHT approach for object location, and they are by no means limited to ellipse detection.

12.8 Bibliographical and Historical Notes

This chapter is based on relatively few papers: first, that of Tsuji and Matsumoto (1978), whose insight led to a totally new ellipse detector; second, that of Tsukune and Goto (1983), which aimed at more accurate determination of ellipse parameters; third, that of Yuen et al. (1988), whose detector made good certain shortcomings of the previous two methods; and fourth, that of Davies (1989a), who developed the GHT idea of Ballard (1981) in order to save computation. The contrasts between these methods are many and intricate, as the chapter has shown. In particular, the idea of saving dimensionality in the implementation of the GHT appears also in a general circle detector (Davies, 1988b). At that point in time, the necessity for a multistage approach to determine ellipse parameters seemed proven, although somewhat surprisingly the optimum number of such stages was just two.

Later algorithms represented moves to greater degrees of robustness with real data by explicit inclusion of errors and error propagation (Ellis et al., 1992). Greater attention was subsequently given to the verification stage of the Hough approach (Ser and Siu, 1995). In addition, work was carried out on the detection of superellipses, which are shapes intermediate in shape between ellipses and rectangles, though the technique used (Rosin and West, 1995) was that of segmentation trees rather than Hough transforms. (Nonspecific detection of superellipses can, of course, be achieved by the diameter bisection method; see Section 12.2.). See also Rosin (2000).

Most recently, ultrafast algorithms were needed for cereal grain inspection—flow rates in excess of 300 grains per second being typical—and the resulting algorithms were limiting cases of chord-based versions of the Hough transform (Davies, 1999b,e). A related approach was adopted by Xie and Ji (2002) for their efficient ellipse detection method. Lei and Wong (1999) employed a method that was based on symmetry and was found to be able to detect parabolas and hyperbolas as well as ellipses.[2] It was also reported as being more stable than other methods since it does not have to calculate tangents or curvatures; Sewisy and Leberl (2001) have also reported this advantage. Wang and Sung

2 Note that while this is advantageous in some applications, the lack of discrimination could prove to be a disadvantage in other applications.

(2001) applied the Hough transform to eye gaze determination by location of irises. They found that the method had to be refined sufficiently to detect the irises as ellipses, not only because of iris orientation but also because the shape of the eye is far from spherical and the horizontal diameter is larger than the vertical diameter.

The fact that even in the 2000s, basic new ellipse detection schemes are being developed says something about the science of image analysis. Even today the toolbox of algorithms is incomplete, and the science of how to choose between items in the toolbox, or how, *systematically*, to develop new items for the toolbox, is immature. Furthermore, although all the parameters for specification of such a toolbox may be known, knowledge about the possible tradeoffs between them is still at an early stage.

12.9 Problems

1. Derive equations (12.7)–(12.9) by the method suggested in the text.

2. Determine which of the methods described in this chapter will detect (a) hyperbolas, (b) curves of the form $Ax^3 + By^3 = 1$, and (c) curves of the form $Ax^4 + Bx + Cy^4 = 1$.

3. Prove equation (12.1) for an ellipse. *Hint*: Write the coordinates of P and Q in suitable parametric forms, and then use the fact that OP $\perp$ OQ to eliminate one of the parameters from the left-hand side of the equation.

4. Describe the *diameter–bisection* and *chord–tangent* methods for the location of ellipses in images, and compare their properties. Justify the use of the chord–tangent method by proving its validity for circle detection and then extending the proof to cover ellipse detection.

5. Round coins of a variety of sizes are to be located, identified, and sorted in a vending machine. Discuss whether the chord–tangent method should be used for this purpose instead of the usual form of Hough transform circle location scheme operating within a 3-D (x, y, r) parameter space.

6. Outline each of the following methods for locating ellipses using the Hough transform: (a) the diameter–bisection method; (b) the chord–tangent method. Explain the principles on which these methods rely. Determine which is the more robust and compare the amounts of computation each requires.

7. For the diameter–bisection method, searching through lists of edge points with the right orientations can take excessive computation. It is suggested that a two-stage

approach might speed up the proccss: (a) load the edge points into a table that may be addressed by orientation; (b) look up the right edge points by feeding appropriate orientations into the table. Estimate how much this would be likely to speed up the diameter–bisection method.

8. It is found that the diameter–bisection method sometimes becomes confused when several ellipses appear in the same image and generate false "centers" that are not situated at the centers of any ellipses. It is also found that certain other shapes are detected by the diameter–bisection method. Ascertain in each case what the method is sensitive to and consider ways in which these problems might be overcome.

Hole Detection

Small holes are important features that can be used very conveniently to locate complex objects. If holes are very small, 3×3 or slightly larger window operators can be used to locate them. If large and circular, the Hough transform may be suitable. If of intermediate size, special fast-acting methods may be useful. This chapter analyzes the situation and compares the available methods and their properties.

Look out for:

- the basic template matching method for hole detection.
- the lateral histogram method for hole detection.
- how ambiguities can be eliminated with the lateral histogram method.
- how speed of processing can be estimated.
- limitations of the lateral histogram method.
- how illumination sets a limit on detection sensitivity and accuracy of location.
- final comparison of available techniques.

This chapter contrasts different methods for detecting an important type of image feature. Speed, sensitivity, and accuracy are all found to be important parameters, and the lighting system is found to be unexpectedly important if accuracy of location is crucial.

CHAPTER 13

Hole Detection

13.1 Introduction

The last few chapters have been particularly concerned with the direct location of objects from their shapes. Where objects have simple shapes this approach is a viable one, but in more complex situations they may have to be located from their features. In the case of larger objects and assemblies, these features may be components of the very shapes discussed so far. However, two important features have not so far been covered: these are small holes and corners. Corners are dealt with in Chapter 14. Here we will consider the location of small holes: not only can these be used to locate objects and to identify them positively, but also they are important in the task of accurately measuring object dimensions. A later chapter considers carefully how the presence of an object may be deduced once its hole features (or a subset of them) have been found.

We start by considering the template matching approach, which seems *a priori* the most obvious means of searching an image for small holes. Later sections examine alternative means of carrying out this task. It should be noted that although the chapter is ostensibly about hole location, it is also about *methods* of searching for such features. In this sense, holes are used as convenient vehicles that allow the relevant principles to be aired.

13.2 The Template Matching Approach

When considering features such as small holes, it is convenient to work with a mental model. We can start by imagining a round, relatively dark region some 1 to 2 pixels in diameter surrounded by a lighter region that is part of some object. The obvious means of detecting such a feature is by template matching. This involves applying a Laplacian type of operator, and in a 3×3 neighborhood this may take the form of one of the following convolution masks:

$$\begin{bmatrix} -1 & -1 & -1 \\ -1 & 8 & -1 \\ -1 & -1 & -1 \end{bmatrix} \qquad \begin{bmatrix} -2 & -3 & -2 \\ -3 & 20 & -3 \\ -2 & -3 & -2 \end{bmatrix}$$

Having applied the convolution, the resulting image is thresholded to indicate regions where holes might be present. Note that if dark holes are regions of negative contrast that are to be located after applying one of the above masks, positions of high *negative* intensity must be sought. It should be noted in passing that the coefficients in each of the above masks sum to zero so that they are insensitive to varying levels of background illumination.

If the holes in question are larger than 1 to 2 pixels in diameter, the above masks will prove inadequate. Indeed, they will indicate the edges of the holes rather than their centers, and they will also indicate the edges of most objects of similar contrast. In any case, the edges are located with low efficiency since the signal originates from only one-half of the mask. Larger masks are required if larger holes are to be detected efficiently at their centers. For very large holes, the masks that are required are so large that the method becomes impracticable. For example, holes 15 pixels in diameter require masks of size around 22×22 pixels. Remembering that these have to be applied at all positions in an image, and assuming that the latter has size 256×256 pixels, over 30 million operations are involved. Hence, a better method must be found for locating larger holes. Similar comments clearly apply if any other type of feature is being sought. Indeed, in many cases the problem is then worse, since most such features are not circular, and templates of many different orientations have to be applied in order to locate them effectively.

This problem has been tackled by many workers. Indeed, several powerful general techniques have been developed for this purpose. They include "ordered search" (Nagel and Rosenfeld, 1972), and "coarse–fine" (Rosenfeld and VanderBrug, 1977) and "two–stage" (VanderBrug and Rosenfeld, 1977) template matching. Considerable interest has also been evinced in the basic (Stockman and Agrawala, 1977) and generalized (Ballard, 1981) HT approaches because of their higher inherent efficiency. The next few sections study the lateral histogram technique, which offers significant advantages in certain applications—particularly for corner and round object detection. A later section appraises the situation, taking account of the odd intensity profiles of some holes, and the fact that large holes may be located by the circle detection procedures discussed in Chapter 10.

13.3 The Lateral Histogram Technique

The lateral histogram technique involves projecting an image on two or more axes by summing pixel intensities (see Fig. 13.1) and using the resulting histograms to identify objects in the image. It has been used previously (Pavlidis, 1968; Ogawa and Taniguchi, 1979) as an aid to pattern recognition in binary images, and it has also been applied to the problem of corner detection in gray-scale images (Wu and Rosenfeld, 1983). Here we consider its application to the detection of small round objects and holes, again in gray-scale images.

The basic technique is inherently attractive in that it requires only about $2N^2$ pixel operations in an $N \times N$ image. However, it turns out that if several similar objects are present in an image, the histograms can be interpreted in a number of ways. Making the method work in practical situations therefore involves finding efficient procedures for resolving ambiguities. It is thus necessary to study the additional computational load these procedures impose. This is done in Sections 13.4–13.5.

13.4 The Removal of Ambiguities in the Lateral Histogram Technique

To resolve the ambiguities that arise with the lateral histogram technique, an obvious strategy is to list all candidate object sites (which arise at the intersections

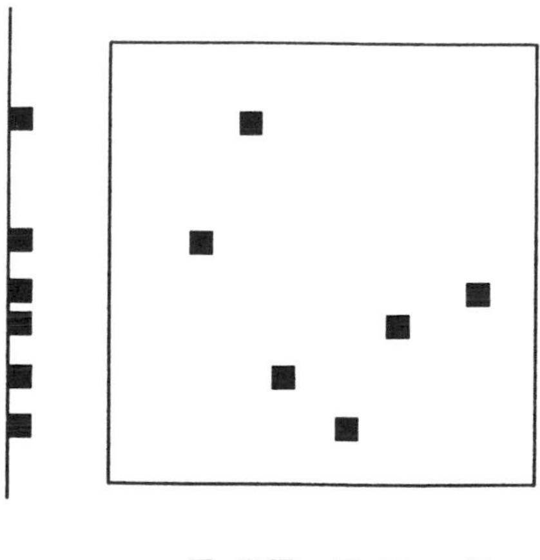

Figure 13.1 The basis of the lateral intensity histogram technique. It is clear that, taken on their own, lateral histograms do not predict object locations uniquely.

of horizontal and vertical lines corresponding to "hits" in the lateral histograms) and to examine them in turn to check rigorously for the presence of an object (Wu and Rosenfeld, 1983). Basically, the lateral histograms are then used to provide rapid screening of the picture data. All relevant locations are checked subsequently (e.g., by normal template matching) to determine their significance. As a result of this procedure, any artifacts due to random combinations of signals are in the end eliminated. The computational implications of this procedure are assessed in this chapter. Meanwhile, it is worth noting that if the initial level of ambiguity in a particular sequence of images turns out to be high, then it is often possible to rectify the situation by computing the lateral histograms of *subimages*.

13.4.1 *Computational Implications of the Need to Check for Ambiguities*

This section shows how many basic computations are required in order to complete the process of object detection and identification. For simplicity, it is assumed that images of size $N \times N$ pixels contain p identical objects of size approximately $n \times n$.

First, the number of operations required to obtain each of the lateral intensity histograms is of the order of N^2; the number of operations required to perform template matching in each of the two histograms is Nn; and the number of operations required to confirm the presence of an object at each possible object site is around n^2. Thus, the total number of operations for the complete algorithm is:

$$R_{\text{lat}} = 2aN^2 + 2bNn + cp^2n^2 \tag{13.1}$$

For convenience we shall take $a \approx b \approx c$, since this should be a reasonable approximation and sufficiently accurate to give a useful indication of the efficiency of the method.

A typical practical case is that of locating holes of diameter 5 pixels in an image of 256×256 pixels. If there are 20 holes and a template of size 7×7 pixels is required to locate a hole of diameter 5 pixels, then the total number of operations is:

$$\begin{aligned} R_{\text{lat}} &= a\left(2 \times 256^2 + 2 \times 256 \times 7 + 20^2 \times 7^2\right) \\ &= 2 \times 256 \times 301.3a = 256^2 \times 2.35a \end{aligned} \tag{13.2}$$

This result should be compared with the result for straight template matching, which is:

$$R_{\text{tem}} = cN^2n^2 \tag{13.3}$$

Since we have already decided to make the approximation $c \approx a$, in the given example it is found that

$$R_{\text{tem}} = 256^2 \times 7^2 a = 256^2 \times 49a \tag{13.4}$$

Here the speedup factor is more than 20, so the saving in computational effort is substantial. The checking and 1-D matching procedures reduce the overall saving achieved by the lateral histogram approach very little. The gain in this instance is effectively $n^2/2$. It becomes even more substantial for larger objects (e.g., around 30 for 20 holes of diameter 7 pixels, 50 for 10 holes of diameter 9 pixels, and 400 for 3 holes of diameter 30 pixels).

It is now useful to examine the relative sizes of the terms in equation (13.1). First, the second term is invariably smaller than the first and can essentially be ignored. However, the third term can become so large that it can dominate over the first. In such a situation, it is even possible that the lateral histogram method would require more computational effort than straightforward template matching. Suppose that the computational load of the lateral histogram technique increases to equal that of pure template matching. Then

$$2N^2 + 2Nn + p^2n^2 = N^2n^2 \tag{13.5}$$

Hence

$$2/n^2 + 2/Nn + p^2/N^2 = 1 \tag{13.6}$$

Since $N \gg 1$ and $n^2 \gg 1$, it follows that $p \approx N$. This shows that the lateral histogram technique does not save computation, and using it is not justifiable, if the number of objects approaches N. The danger sign is perhaps when the third term is just equal to the first. This occurs when

$$p = \sqrt{2}N/n \tag{13.7}$$

This value of p can be interpreted as the point at which object projections start to overlap and interpretation consequently becomes more difficult. It therefore seems that use of the method should be restricted to cases when $p \leq N/n$, so that (1) interpretation is straightforward and (2) a significant saving in computational effort can result. These ideas give a precise indication of when it is profitable to use the subimage method mentioned in Section 13.4. An image should be broken into subimages whenever the image data make $N < np$. This is discussed in more detail below.

Finally, the lateral histogram method incorporates feedback by virtue of its two-stage mode of operation. This permits it to adapt optimally to variations in the image data, so that it is both rapid and accurate in its task of object location. However, image variability needs to be monitored in order to confirm that operation can take place within the limits indicated above.

13.4.2 Further Detail on the Subimage Method

If an image is broken down into subimages, then the number of objects in each subimage should vary approximately as the square of subimage size. Suppose that the size of a subimage is $\tilde{N} \times \tilde{N}$ and that the number of objects it contains is p. Defining

$$\alpha = N/\tilde{N} \tag{13.8}$$

we expect

$$\tilde{p} = p/\alpha^2 \tag{13.9}$$

The equation corresponding to equation (13.1) is then:

$$\tilde{R}_{\text{lat}} = \alpha(2\tilde{N}^2 + 2\tilde{N}n + \tilde{p}^2n^2) \tag{13.10}$$

Multiplying by α^2 to obtain a new equation for the whole image, gives:

$$R'_{\text{lat}} = \alpha(2N^2 + 2\alpha Nn + p^2n^2/\alpha^2) \tag{13.11}$$

Thus, the second term is increased while the third is reduced substantially. Differentiating equation (13.11) with respect to α gives:

$$\frac{\mathrm{d}R'_{\text{lat}}}{\mathrm{d}\alpha} = \alpha(2Nn - 2p^2n^2/\alpha^3) \tag{13.12}$$

This is positive at $\alpha = 1$ for p in the range $p < \sqrt{2}N/n$. Hence, if p is small (typically, $p < 4$), the subimage method incurs additional computation and should not be used. However, the minimum of R_{lat} as α varies is shallow, and little is to be gained from using the method until $p > N/n$, as indicated in Section 13.4.1. Under these circumstances, α should be adjusted so that:

$$\tilde{N} \approx n\tilde{p} \tag{13.13}$$

Then it is found that:

$$\alpha \approx np/N \tag{13.14}$$

and hence:

$$\tilde{N} \approx N^2/np \tag{13.15}$$

Substituting for α in equation (13.11) now gives

$$R'_{\text{lat}} \approx \alpha(2N^2 + 2pn^2 + N^2) \tag{13.16}$$

The significance of this result is that the third term is limited by the subimage method to one-half the first term and is no longer able to exceed it by a large factor. The second term remains relatively unimportant since it varies as p instead of p^2. This means that rarely is there likely to be any penalty from using the subimage method. However, a practical limitation exists, which is now considered.

The practical limitation involves the problem of "edge effects" between subimages. If an object falls on the border between two subimages, then it may escape detection. To overcome this problem, it is necessary to permit subimages to overlap—a factor that will plainly reduce efficiency. The amount by which subimages will have to overlap is equal to the diameter of the objects being detected, since when an object starts to move out of one subimage, it must already be included totally in an adjacent subimage. This means that N must be replaced by $\tilde{N} + n$ in equation (13.10). (N must now be defined by equation (13.8) and is no longer the subimage size.) As a consequence, the additional terms $4N\alpha n + 4\alpha^2 n^2$ appear in equation (13.11). These dominate over the first term if $\alpha \gtrsim N/2n$, i.e. $n \gtrsim \tilde{N}/2$.

Yet there is one way of reducing this problem and getting back substantially to the level indicated by equation (13.16)—namely, by hierarchical addition of rows and columns of pixel intensities while computing the lateral histograms. For example, sub-subimages of some fraction of the size of the subimages may be computed first. These may then be added in appropriate groups. If the size of the subimages is reasonably large ($\tilde{N} + n \gg 1$), the final set of summations consumes relatively little computational effort, and equation (13.16) is a realistic approximation.

This subsection has shown that when $p > N/n$, the amount of computational effort needed to locate objects can be kept within reasonable bounds (of order $3N^2$) by use of the subimage method. These bounds are remarkably close to those (of order $2N^2$) which apply when $p \ll N/n$. Bounds of size N^2 would, of course, be ideal. It is not expected that significantly different results would apply if the coefficients *a*, *b*, *c* were not exactly equal as assumed above.

13.5 Application of the Lateral Histogram Technique for Object Location

We now consider what types of image data can be analyzed conveniently with the aid of lateral histograms. An important feature of the approach is that it is not limited to use in binary images, since all the procedures that are employed (e.g., template matching) are equally applicable to gray-scale images. Another feature of the approach is that objects are identified from their 1-D templates. This means that they may be difficult to locate in the lateral histograms if these are very orientation-dependent. Thus, the method is well adapted to the location of round objects. A great many products, from foodstuffs to machined parts, are almost exactly circular (Davies, 1984c), so the approach is frequently valuable. In addition, many components contain accurately machined circular holes, which can be used to detect the products in which they appear (Chapter 15). Significantly, the method can locate objects whose intensity profiles do not possess well-defined edges (see Fig. 13.2). In such situations the HT method cannot easily be applied because its use requires rapid location and orientation of edges.

Thus, the lateral histogram approach is valuable for the efficient location of (1) small round objects or holes, and (2) objects that are fairly large, round, and somewhat fuzzy or low in contrast. If the number of objects is not excessive, they should be located rapidly and reliably. The author has tested the method on a variety of data from foodstuffs to mechanical parts and has confirmed its value in these applications for images of up to 128×128 pixels. For images of these sizes, the subimage method did not have to be used, but for images of 512×512 or larger, the subimage method will be required.

Figure 13.3 shows the histograms obtained for images of metal bars containing holes (see Fig. 13.4). All the holes were located by the method, despite their high variability and noise content and despite the rather rudimentary templates used to analyze the histograms (see Fig. 13.5). One or two comments should be made about the initial application of lateral histograms to find candidate hole locations (see Fig. 13.4). First, occasional false positives arose from noise or strong corners (Fig. 13.4*a–c*,*f*). Second, on some occasions lateral juxtaposition of holes caused "pulling" and merging of estimated candidate positions (Fig. 13.4*b–e*). Third, in two cases holes were located in only one histogram (Fig. 13.4*e*,*f*). More will be said about this matter in Section 13.5.1. These deficiencies are easily corrected in the final template matching procedure.

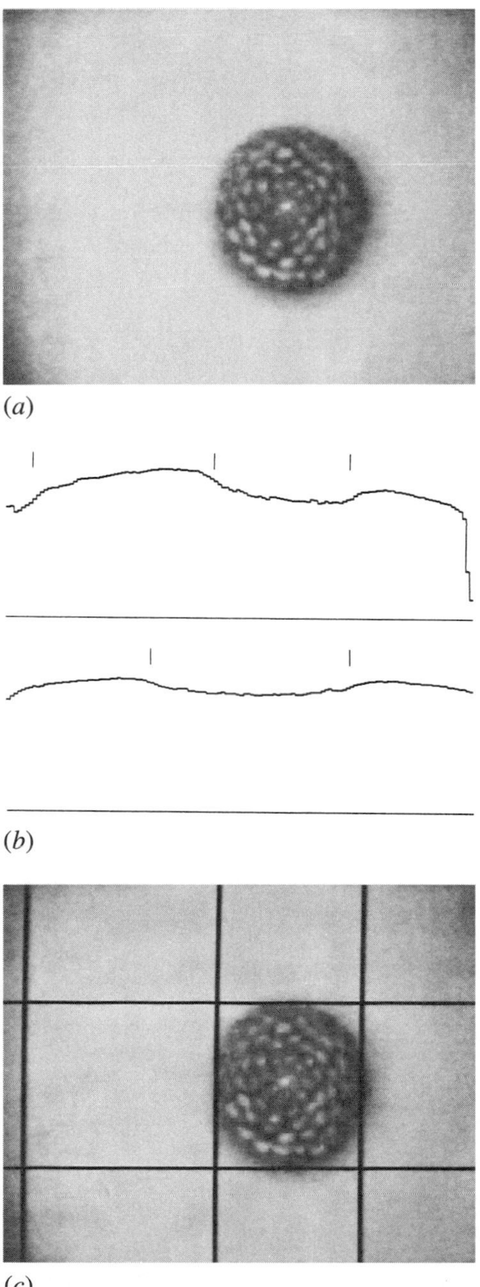

Figure 13.2 (*a*) Image of a decorative candle that appears as a fairly diffuse object with a fuzzy edge; (*b*) lateral histograms; (*c*) edges of the object as detected by differentiating the histograms and locating peaks. The false peak resulting from strongly varying backgrounds is easily eliminated.

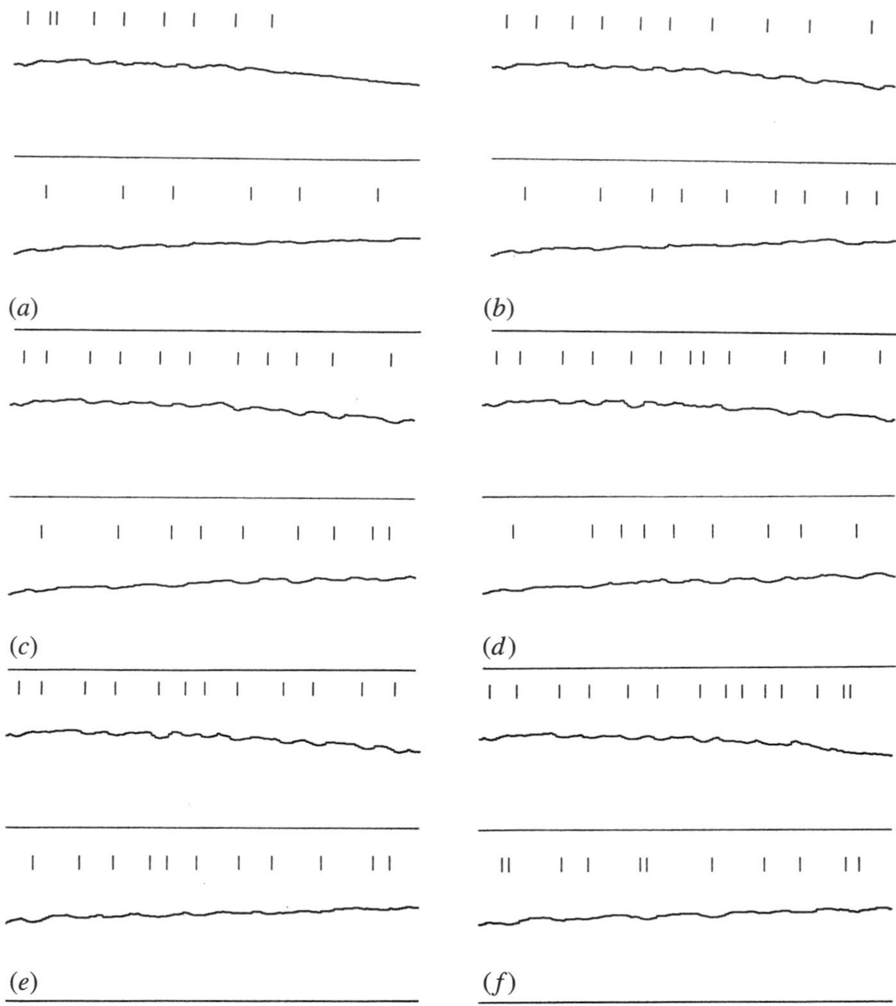

Figure 13.3 Lateral histograms for images of metal bars containing holes for original versions of images in Fig. 13.4.

13.5.1 *Limitations of the Approach*

When applying lateral histograms for object detection in industrial images, the most serious drawback is that some objects may mask others. Ideally, this problem would not arise, because the intensity patterns of all objects are incorporated additively in the histograms, and any suspicion of an object at a particular location would be checked in detail.

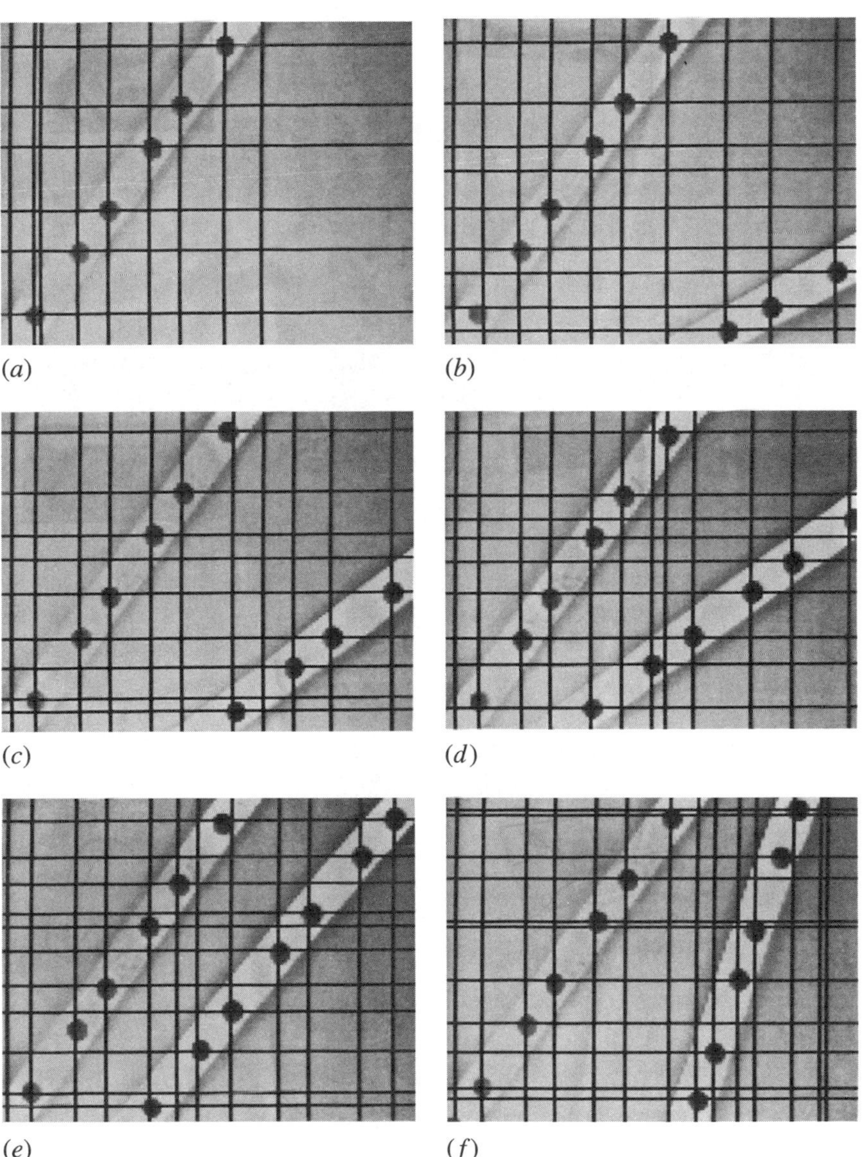

Figure 13.4 Result of applying lateral histograms shown in Fig. 13.3 to find candidate hole positions.

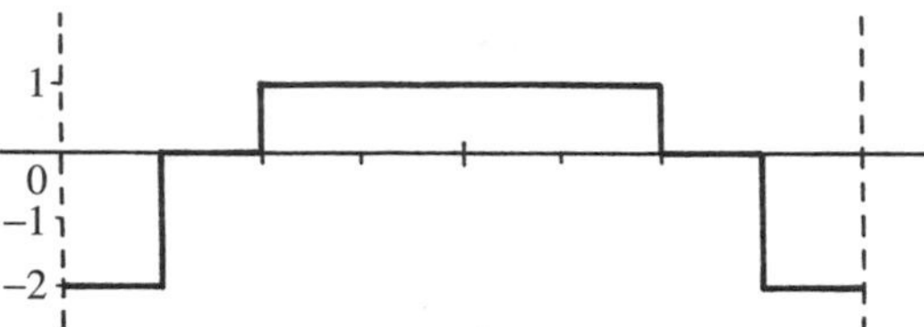

Figure 13.5 Form of the 1-D templates used to locate the hole profiles in the lateral histograms of Fig. 13.3. Greater sophistication was not necessary in this instance.

In practice, the most likely masking error arises when the edges of an object are masked by other objects on both sides. (If only one of an object's edges is so masked, it will almost certainly be detected.) This effect is seen in Fig. 13.4*e* and 13.4*f*, although in Fig. 13.4*f* one of the masking objects is a strong corner in the image. The probability of this occurring is quite low. The probability may be estimated by considering the chance of throwing p balls into N boxes, arranged in line, so that three balls fall into any set of three adjacent boxes. Detailed calculations (Davies, 1987g) show that, if the probability of missing an object must be less than 0.01, the maximum number of objects that can be detected in this way is about 20 in a 128×128 pixel image—though this number will fall off in the presence of noise. These values are generally in accordance with observations for industrial images.

As already seen from Fig. 13.4, false positives can also arise with the lateral histogram technique. However, since the method involves the use of feedback, false positives are rigorously checked in every case, and the final incidence of false predictions can be no worse than for normal template matching. The number of false positives is strongly related to the amount of noise present in the image; further discussion of this is curtailed inasmuch as the effect is so data-dependent.

13.6 Appraisal of the Hole Detection Problem

Much of this chapter has focused on study of the lateral histogram technique because it is a highly efficient means of locating holes, though not its only application. In this section, we return to a more careful appraisal of methods for hole detection.

At this stage, an obvious question is whether the usual methods for circular object detection can be applied to hole detection. In principle, this is undoubtedly practicable; in many cases, the only change arises because holes

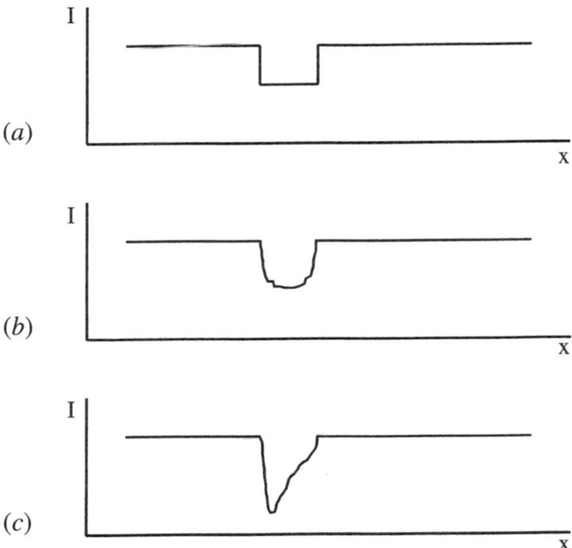

Figure 13.6 (*a*) Cross-sectional intensity pattern of an idealized hole; (*b*) a more commonly found case; (*c*) the effect of the hole being lit obliquely, a significant part of the hole being in shadow.

have negative contrast. Accordingly, if the HT technique is used, votes will have to be accumulated on moving a distance $-R$ along the direction of the local edge normal.

Small holes, however, are rarely found to have an intensity pattern that matches that of the model introduced in Section 13.2. Figure 13.6 shows some of the possibilities. The basic problem is that it is virtually impossible to make lighting totally diffuse. On the contrary, it usually has significant directionality, so there will be a region of shadow on one side of the hole. Assuming that the holes are infinitely deep but narrow, the region of shadow may well stretch halfway across the width of each hole. If the holes are little more than shallow depressions in the surface, there may be no shadows, but luminous intensity will be graded across each hole. Finally, complications arise because holes are often only a few pixels across, so the arrangements of pixel intensities are unlikely to be smoothly varying. In any case, noise can have a significant effect on the appearance of a hole.

In summary, four factors can affect the form of a hole in a digitized image: (1) isotropic variations in intensity over the hole region, (2) anisotropic variations in intensity, (3) digitization effects (both spatial and gray-scale), and (4) noise. Since a perfect working model of a hole is unlikely, even in respect of (1), methods for detecting holes need to include a significant amount of averaging to overcome these effects. For example, when an HT circle detector is applied in such cases, for diameters of 10 pixels or less, edge orientation accuracy and edge position accuracy are frequently poor. In addition, the boundary has very few edge points, so little averaging can occur. Hence, peaks in parameter space are ill-defined and the accuracy of center location is poor. (As a percentage measurement it is very poor, while absolutely it may be little better than ± 1 pixel.) In contrast, the lateral histogram method incorporates significant averaging and usually arrives at reasonable estimates of hole position. The remaining problem is that factor (2) can result in a systematic bias in estimates of hole position, although it is possible that this can be measured and corrected for.

Finally, it should be observed that the HT approach requires use of an edge detector. If a Sobel operator is employed, the total amount of computation involved will be at least $2 \times 8 \times N^2 = 16N^2$, many times that required by the lateral histogram method.

Despite these arguments, a hole must at some point become so large that it is better considered as a circle. This point depends particularly on image resolution, but it also varies with the lighting scheme and with the detailed characteristics (shape, smoothness, etc.) of the hole, as indicated above. Hence, the optimum solution is highly data-dependent. Suffice it to say that if holes are smaller than about 16 pixels in diameter, they are probably best detected by the lateral histogram method; if larger than this, they are probably best treated as normal circular objects and detected by an HT method.

Finally, we consider briefly the algorithm of Kelley and Gouin (1984). This algorithm works by finding the midpoints of horizontal and vertical chords of dark areas in an image and locating holes at the center of the resulting cross. Although the method apparently worked adequately on Kelley and Gouin's data for holes around 12 pixels in diameter, it is liable to fail on less ideal images (e.g., lower contrast or where there are interfering features) since ultimately it relies on the presence of the four edge pixels at the ends of the horizontal and vertical diameters of the holes. Overall, there seems to be no reason to abandon the conclusion that an effective hole-finder needs to incorporate significant averaging, both for general robustness and for accuracy of center location.

13.7 Concluding Remarks

This chapter has aimed to show some of the problems involved in detecting small holes. For all but the smallest holes, template matching is likely to be too computationally intensive. The lateral histogram technique offers a useful solution: it is able to make significant savings in computation since it reduces the dimensionality of the basic search task from two dimensions to one. However, quite large holes are generally best treated as circular objects and detected by an HT technique.

Yet again problems of lighting have arisen, and with them the task of problem specification. Without an accurate problem specification, diagnosis is liable to be inaccurate and algorithm design suboptimal. However, this chapter has been able to show the parameters important in the task of hole detection.

Detection of small holes is a key factor in the location of industrial components. This chapter has shown how this may be achieved with minimal computation using lateral histograms. As with other reduced dimensionality techniques, integrated measures must be employed to avoid ambiguities. In practice, accuracy of hole detection may be limited by lighting problems.

13.8 Bibliographical and Historical Notes

Hole detection is inherently a search problem that would normally involve template matching (correlation). Where template masks are large, some form of speedup is vital and basic methods for achieving this were devised in the 1970s (Nagel and Rosenfeld, 1972; Rosenfeld and VanderBrug, 1977; VanderBrug and Rosenfeld, 1977). At much the same time, it was discovered (Stockman and Agrawala, 1977) that the HT technique (Hough, 1962; Ballard, 1981) is a form of template matching and offers significant savings in computation if edge orientation information is taken into account (Kimme et al., 1975). Lateral intensity histograms were developed for character recognition and later used for corner detection (Pavlidis, 1968; Ogawa and Taniguchi, 1979; Wu and Rosenfeld, 1983). Subsequently, Davies used them for hole detection (1985a, 1987g); Sections 13.3–13.5 embody this work. See also Davies and Barker (1990). Perhaps oddly, the literature gives far greater attention to corner detection than to hole detection (see Chapter 14).

In this chapter holes have been regarded as examples of salient point features that can be used to help locate objects. In principle, they are as suitable as corners for locating industrial components such as brackets and hinges. However, we shall see in Chapter 14 (particularly Section 14.14) that in other applications holes have little part to play—as, for example, when forming correspondences between binocular images of outdoor scenes. In that case "interest" point detectors are needed, and the main characteristics of suitable interest points are that they be distinctive and accurately localizable, not that they approximate to true corners or have the isotropy of holes. For such purposes, the methods of this chapter are not required. This does not represent a failure in any sense—merely the need to find suitable features and detectors to serve the particular purpose to which they will be put (as some would say: "horses for courses").

13.9 Problems

1. Calculate the bias in the estimated position of a hole if it is illuminated from light at an angle 30° to the vertical and has depth three times its diameter.
2. Find a formula giving the accuracy of a Hough-based hole detector as a function of hole diameter. What other information would be needed to choose between this and other hole detection schemes?
3. Estimate how the basic accuracy of a lateral histogram hole detector varies with the size of the image being used. What limits the accuracy of hole detection with the lateral histogram technique? Is there any possibility of improving it?
4. Work out the probability of missing an object if *each* of its edges is masked in the histograms between those of *two* other objects. Hence estimate the maximum number of objects that may be located in an $N \times N$ image if the probability of missing an object has to be less than 0.01. (For further details of the problem, see Section 13.5.1.)
5. (a) Explain why it is more efficient to locate an object by its features, instead of as a whole, using template matching. How does locating the object in this way affect the robustness of the process?

   (b) Show how feature detection masks may be derived for corner detection, and explain why zero-mean masks are normally used for this purpose. How many masks are needed in this case?

   (c) Show that masks such as the following may be used for *line* segment detection in a 3×3 window:

$$\begin{bmatrix} -1 & -1 & 2 \\ -1 & 2 & -1 \\ 2 & -1 & -1 \end{bmatrix}$$

(d) How many masks are needed for this purpose? Why?

(e) Show that line segments may also be detected by small hole detection masks. Explain what is gained by this approach. Show also that holes may be detected by line segment detection masks.

(f) Explain why masks designed for detecting one type of feature may not be quite as effective as tailor-designed masks for detecting other types of feature.

Polygon and Corner Detection

Polygon detection poses further problems, some of which may be solved more conveniently by using corner detection. In any case, corner detection is valuable for locating complex objects and for tracking them in 2-D or 3-D. This chapter discusses these two detection problems and considers whether Hough transforms or other methods are best suited for the task.

Look out for:

- use of the generalized Hough transform for detecting polygons.
- consideration of the minimum numbers of parameter planes needed for the purpose.
- how polygon orientation should be determined.
- why corner features are useful.
- the variety of methods available for corner detection—template matching, the second-order derivative method, the median-based method, the Hough transform, the Plessey detector, and the lateral histogram method.
- how speed of processing can be improved.
- how corner orientation should be determined.

Note again the variety of methods available for performing ostensibly the same tasks. However, different methods have different speeds, accuracies, sensitivities, and degrees of robustness. This chapter seeks to bring out as many of these properties and parameters as possible.

CHAPTER 14

Polygon and Corner Detection

14.1 Introduction

Polygons are ubiquitous among industrial objects and components, and it is of vital importance to have suitable means for detecting them. For example, many manufactured objects are square, rectangular, or hexagonal, and many more have components that possess these shapes. Polygons can be detected by the two-stage process of first finding straight lines and then deducing the presence of the shapes. However, in this chapter we first focus on the *direct* detection of polygonal shapes, for two reasons. First, the algorithms for direct detection are often particularly simple; and second, direct detection is inherently more sensitive (in a signal-to-noise sense), so there is less chance that objects will be missed in a jumble of other shapes. The basic method used is the GHT, and it will be seen how GHT storage requirements and search time can be minimized by exploiting symmetry and other means.

After exploring the possibilities for direct detection of polygons, we move on to more indirect detection, by concentrating on corner features. Although this approach to object detection may be somewhat less sensitive, it is more general, in that it permits objects with curved boundaries to be detected if these objects are interspersed with corners. It also has the advantage of permitting shape distortions to be taken into account if suitable algorithms to infer the presence of objects can be devised (see Chapters 15–17).

Portions of this chapter are reprinted, with permission, from *Proceedings of the 8th International Conference on Pattern Recognition*, Paris (October 27–31, 1986), pp. 495–497 (Davies, 1986d).

14.2 The Generalized Hough Transform

In an earlier chapter, we saw that the GHT is a form of spatial matched filter and is therefore ideal in that it gives optimal detection sensitivity. Unfortunately, when the precise orientation of an object is unknown, votes have to be accumulated in a number of planes in parameter space, each plane corresponding to one possible orientation of the object—a process involving considerable computation. In Chapter 12, it was shown that GHT storage and computational effort can be reduced, in the case of ellipse detection, by employing a single plane in parameter space. This approach appeared to be a useful one which should be worth trying in other cases, including that of polygon detection. To proceed, we first examine straight-line detection by the GHT.

14.2.1 *Straight Edge Detection*

We start by taking a straight edge as an object in its own right and using the GHT to locate it. In this case the PSF of a given edge pixel is simply a line of points of the same orientation and of length equal to that of the (idealized) straight edge (see Chapter 11). The transform is the autocorrelation function of the shape and in this case has a simple triangular profile (Fig. 11.9). The peak of this profile is the position of the localization point (in this case the midpoint M of the line). Thus, the straight edge can be fully localized in two dimensions, which is achieved while retaining the full sensitivity of a spatial matched filter.

We now need to see how best to incorporate this line parametrization scheme into a full object detection algorithm. Basically, the solution is simple—by the addition of a suitable vector to move the localization point from M (the midpoint of the line) to the general point L (Chapter 11).

Thus, the GHT approach is a natural one when searching images for polygonal shapes such as squares, rectangles, hexagons, and so on, since whole objects can be detected directly in one parameter space. However, note that use of the HT (θ, ρ) parametrization would require identification of lines in one parameter space followed by further analysis to identify whole objects. Furthermore, in situations of low signal-to-noise ratio, it is possible that some such objects would not be visible using a feature-based scheme when they *would* be found by a whole object matched filter or its GHT equivalent (Chapter 11).

14.3 Application to Polygon Detection

When applied to an object such as a square whose boundary consists of straight edges, the parametrization scheme outlined in the preceding section takes the following form: construct a transform in parameter space by moving inward from every boundary pixel by a distance equal to half the side S of the square and accumulate a line of votes of length S at each location, parallel to the local edge orientation. Then the pixel having the highest accumulation of votes is the center of the square. This technique was used to produce Fig. 14.1.

The GHT can be used in this way to detect all types of regular polygons, since in all such cases the center of the object can be chosen as the localization point L. Thus, to determine the positions at which lines of votes will be accumulated in parameter space, it is necessary merely to specify a displacement by the same distance along each edge normal. Taking a narrow rectangle as the lowest-order regular polygon (i.e., having two sides) and the circle as the highest-order polygon with an infinite number of short sides, it is apparent from Fig. 14.2 how the transform changes with the number of sides.

In this scheme, notice that votes can be accumulated in parameter space using locally available edge orientation information and that the method is not muddled by changes in object orientation. This is true only for regular polygons, where the sides are *equidistant* from the localization point L. For more general shapes, including the quadrilateral in Fig. 14.3, the standard method of applying the GHT is to have a set of voting planes in parameter space, each corresponding to a possible orientation of the object. To find general shapes and orient them within 1°, 360 parameter planes appear to be required. This would incur a serious storage

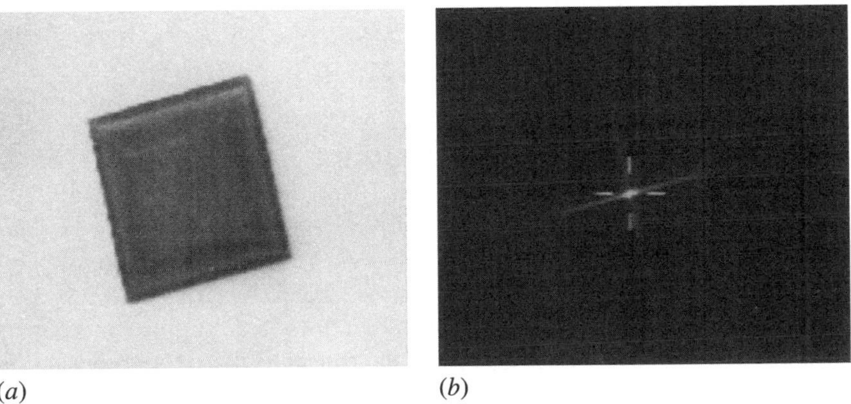

(*a*) (*b*)

Figure 14.1 (*a*) Original off-camera gray-scale image of a square; (*b*) derived transform.

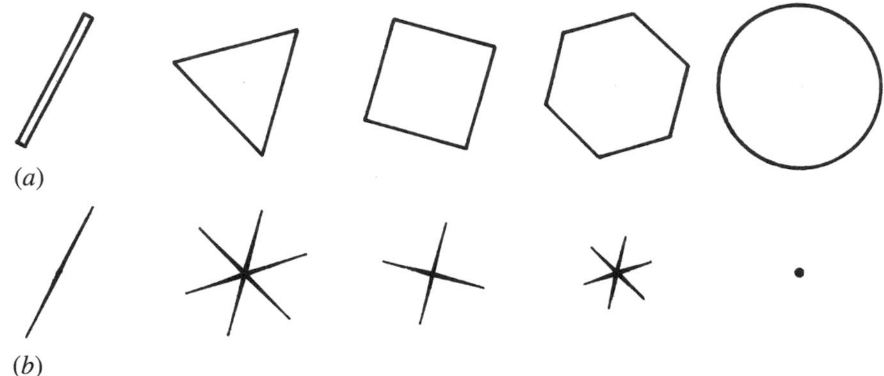

Figure 14.2 (*a*) Set of polygons; (*b*) associated transforms. The widths of the lines in (*b*) represent transform strength: ideally, the transform lines would have zero width.

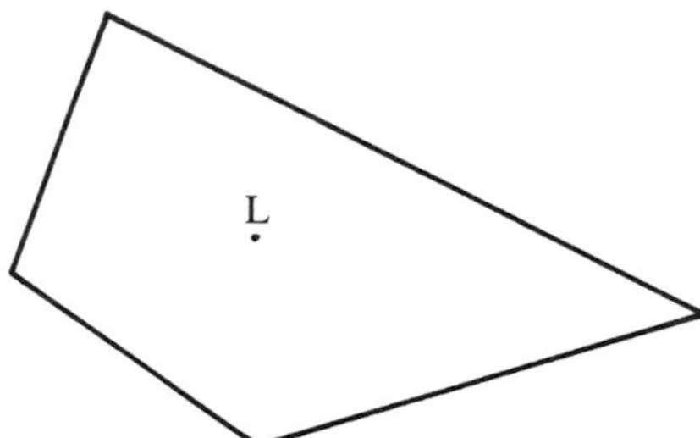

Figure 14.3 Geometry of a typical irregular polygon: the localization point L is at different distances from each of the sides of the polygon.

and search problem. Next we show how this problem may be minimized in the case of irregular polygons.

14.3.1 *The Case of an Arbitrary Triangle*

The case of an arbitrary triangle can be tackled by making use of the well-known property of the triangle that it possesses a unique inscribed circle. Thus, a localization point can be defined which is equidistant from the three sides. This means that the transform of any triangle can be found by moving into the triangle a specified distance from any side and accumulating the results in a single parameter plane (see Fig. 14.4). Such a parametrization can cope with random orientations of a triangle of given size and shape. However, this result is not wholly

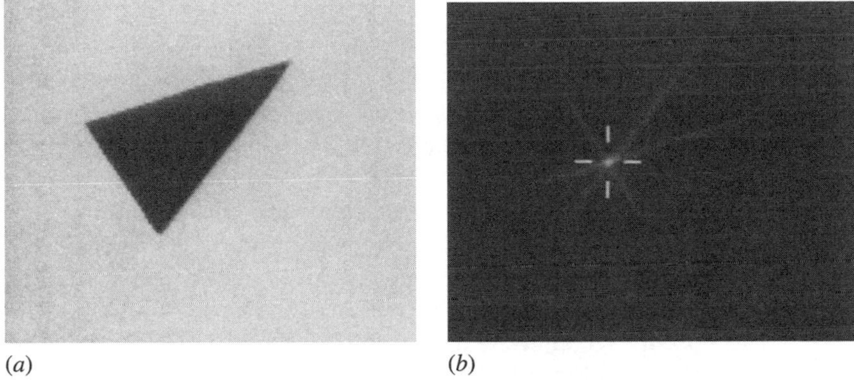

(a) (b)

Figure 14.4 (a) Off-camera image of a triangle; (b) transform in a single parameter plane, making use of a localization point L which is equidistant from the three sides. An optimized transform requires four parameter planes (see Fig. 14.7).

optimal because the separate transforms for each of the three sides do not peak at exactly the same point (except for an equilateral triangle). Neither should the lengths of the edge pixel transforms be the same. Each of these factors results in some loss of sensitivity. Thus, this solution optimizes the situation for storage and search speed but violates the condition for optimal sensitivity. If desired, this tradeoff could be reversed by reverting to the use of a large number of storage planes in parameter space, as described in Section 14.3. A better approach will be described in Section 14.3.3. Finally, note that the method is applicable only to a very restricted class of irregular polygons.

14.3.2 The Case of an Arbitrary Rectangle

The case of a rectangle is characterized by greater symmetry than that for an arbitrary quadrilateral. In this case the number of parameter planes required to cope with variations in orientation may be reduced from a large number (e.g., 360—see Section 14.3) to just two. This is achieved in the following way. Suppose first that the major axis of the rectangle (of sides A, B: $A > B$) is known to be oriented between 0° and 90°. Then we need to accumulate votes by moving a distance $B/2$ from edge pixels oriented between 0° and 90° and accumulating a line of votes of length A in parameter space, and by moving a distance $A/2$ from edge pixels oriented between 90° and 180° and accumulating a line of votes of length B in parameter space. This is very similar to the situation for a square (see Section 14.3). However, suppose that the major axis was instead known to be oriented between 90° and 180°. Then we would need to change the method of transform suitably by interchanging the values of A and B. Since we do not know which supposition is true, we have to try both methods of transform, and for this purpose we need two planes in parameter space. If a strong peak is obtained in the first plane, the first supposition is true and the rectangle major axis is oriented between 0° and 90°—and vice versa if the strong peak is in the second plane in parameter space. In either case, the position of the strong peak indicates the position of the center of the rectangle (see Fig. 14.5).

In summary, the largest peak in either plane indicates (1) which assumption about the orientation of the rectangle was justified and (2) the precise location of the rectangle. Apart from its inherent value, this example is important in giving a clue about how to develop the parametrization as symmetry is gradually reduced.

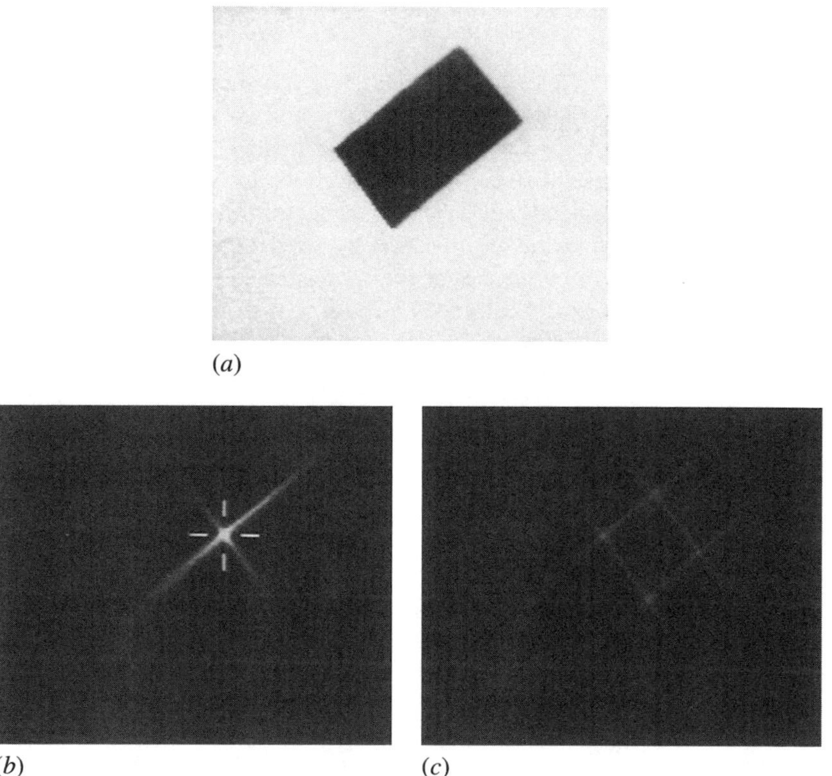

(*a*)

(*b*) (*c*)

Figure 14.5 (*a*) Off-camera image of a rectangle; (*b*), (*c*) optimized transforms in two parameter planes. The first plane (*b*) contains the highest peak, thereby indicating the position of the rectangle and its orientation within 90°.

14.3.3 Lower Bounds on the Numbers of Parameter Planes

This section generalizes the solution of the previous section. First, note that two planes in parameter space are needed for optimal rectangle detection and that each of these planes may be taken to correspond to a range of orientations of the major axis.

For arbitrarily oriented polygons of known size and shape, votes must be accumulated in a number of planes in parameter space. Suppose that the polygon has N sides and that we are using M parameter planes to locate it. Since the complete range of possible orientations of the object is 2π, each of the M planes will have to be taken to represent a range $2\pi/M$ of object orientations. On detecting an edge pixel in the image, its orientation does not make clear which of the N sides of the object it originated from, and it is necessary to insert a line of votes at appropriate positions in N of the parameter planes. When every edge pixel has been processed in this way, the plane with the highest peak again indicates both the location of the object and its approximate orientation. Already it is clear that we must ensure that $M > N$, or else false peaks will start to appear in at least one of the planes, implying some ambiguity in the location of the object. Having obtained some feel for the problem, we now study the situation in more detail in order to determine the precise restriction on the number of planes.

Take $\{\varphi_i: i = 1, \ldots, N\}$ as the set of angles of the edge normals of the polygon in a standard orientation α. Next, suppose an edge of edge normal orientation θ is found in the image. If this is taken to arise from side i of the polygon, we are assuming that the polygon has orientation $\alpha + \theta - \varphi_i$, and we accumulate a line of votes in parameter plane $K_i = (\alpha + \theta - \varphi_i)/(2\pi/M)$. (Here we force an integer result in the range 0 to $M - 1$.) It is necessary to accumulate N lines of votes in the set of N parameter planes $\{K_i: i = 1, \ldots, N\}$, as stated above. However, additional false peaks will start to appear in parameter space if an edge point on any one side of the polygon gives rise to more than one line of votes in the same parameter plane. To ensure that this does not occur it is necessary to ensure that $K_i \neq K_j$ for different i, j. Thus, we must have

$$\left|(\varphi_j - \varphi_i)/(2\pi/M)\right| \geq 1 \tag{14.1}$$

This result has the simple interpretation that the minimum exterior angle of the polygon $\psi_{\min}$ must be greater than $2\pi/M$, the exterior angle being (by definition) the angle through which one side of the polygon has to be rotated before it coincides with an adjacent side (Fig. 14.6). Thus, the minimum exterior angle describes the maximum amount of rotation before an unnecessary ambiguity is generated about the orientation of the object. If no

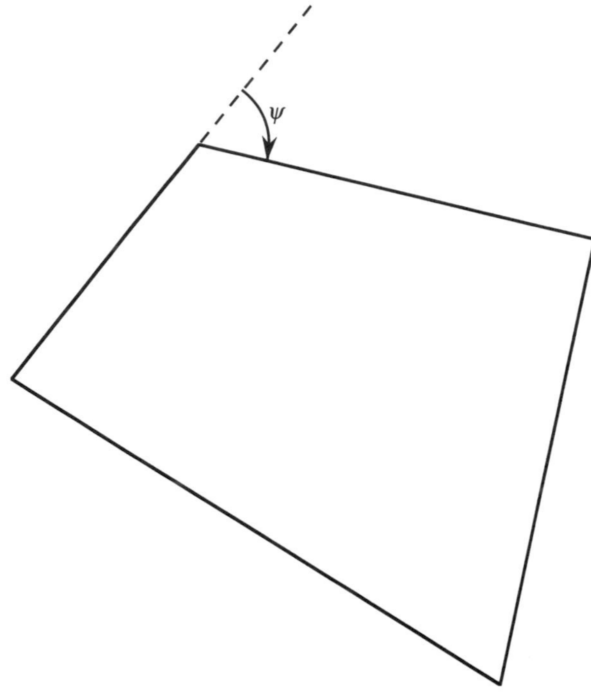

Figure 14.6 The exterior angle of a polygon, ψ angle through which one side has to be rotated before it coincides with an adjacent side.

unnecessary ambiguity is to occur, the number of planes M must obey the condition

$$M \geq \frac{2\pi}{\psi_{\min}} \tag{14.2}$$

We are now in a position to reassess all the results so far obtained with polygon detection schemes. Consider regular polygons first: if there are N sides, then the common exterior angle is $\psi = 2\pi/N$, so the number of planes required is inherently

$$M = 2\pi/\psi = 2\pi/(2\pi/N) = N \tag{14.3}$$

However, symmetry dictates that all N planes contain the same information, so only one plane is required.

Consider next the rectangle case. Here the exterior angle is $\pi/2$, and the inherent number of planes is four. However, symmetry again plays a part, and just two planes are sufficient to contain all the relevant information, as has already been demonstrated.

In the triangle case, for a nonequilateral triangle the minimum exterior angle is typically greater than $\pi/2$ but is never as large as $2\pi/3$. Hence, at least four planes are needed to locate a general triangle in an arbitrary orientation (see Fig. 14.7). Note, however, that for a special choice of localization point fewer planes are required—as shown in Section 14.3.1.

Next, for a polygon that is *almost* regular, the minimum exterior angle is *just* greater than $2\pi/N$, so the number of planes required is $N+1$. In this case, it is at first surprising that as symmetry increases, the requirement jumps from $N+1$ planes, not to N planes but straight to one plane.

Now consider a case where there is significant symmetry but the polygon cannot be considered to be fully regular. Figure 14.8*a* exhibits a semiregular hexagon for which the set of identity rotations forms a group of order 3. Here the minimum exterior angle is the same as for a regular hexagon. Thus, the basic rule predicts that we need a total of six planes in parameter space. However, symmetry shows that only two of these contain different information. Generalizing this result, we find that the number of planes actually required is N/G rather than N, G being the order of the group of identity rotations of the object. To prove this more general result, note that we need to consider only object orientations in the range 0 to $2\pi/G$ instead of 0 to 2π. Hence, condition (14.2) now becomes

$$M \geq (2\pi/G)/\psi_{\min} = (2\pi/G)/(2\pi/N) = N/G \tag{14.4}$$

Suppose instead that the hexagon had threefold rotation symmetry but had two sets of exterior angles with slightly different values (see Fig. 14.8b). Then

$$M \geq (2\pi/G)/\psi_{\min} > N/G \tag{14.5}$$

and at best we will have $M = (N/G)+1$, which is 3 for the hexagon shape of Fig. 14.8*b*.

14.4 Determining Polygon Orientation

Having located a given polygon, it is usually necessary to determine its orientation. One way of proceeding is to find first all the edge points that contribute to a given peak in parameter space and to construct a histogram of their edge orientations. The orientation η of the polygon can be found if the positions of the peaks in this histogram are now determined and matched against the set $\{\varphi_i: i=1, \ldots, N\}$ of the angles of the edge normals for the idealized polygon in standard orientation α (Section 14.3.3).

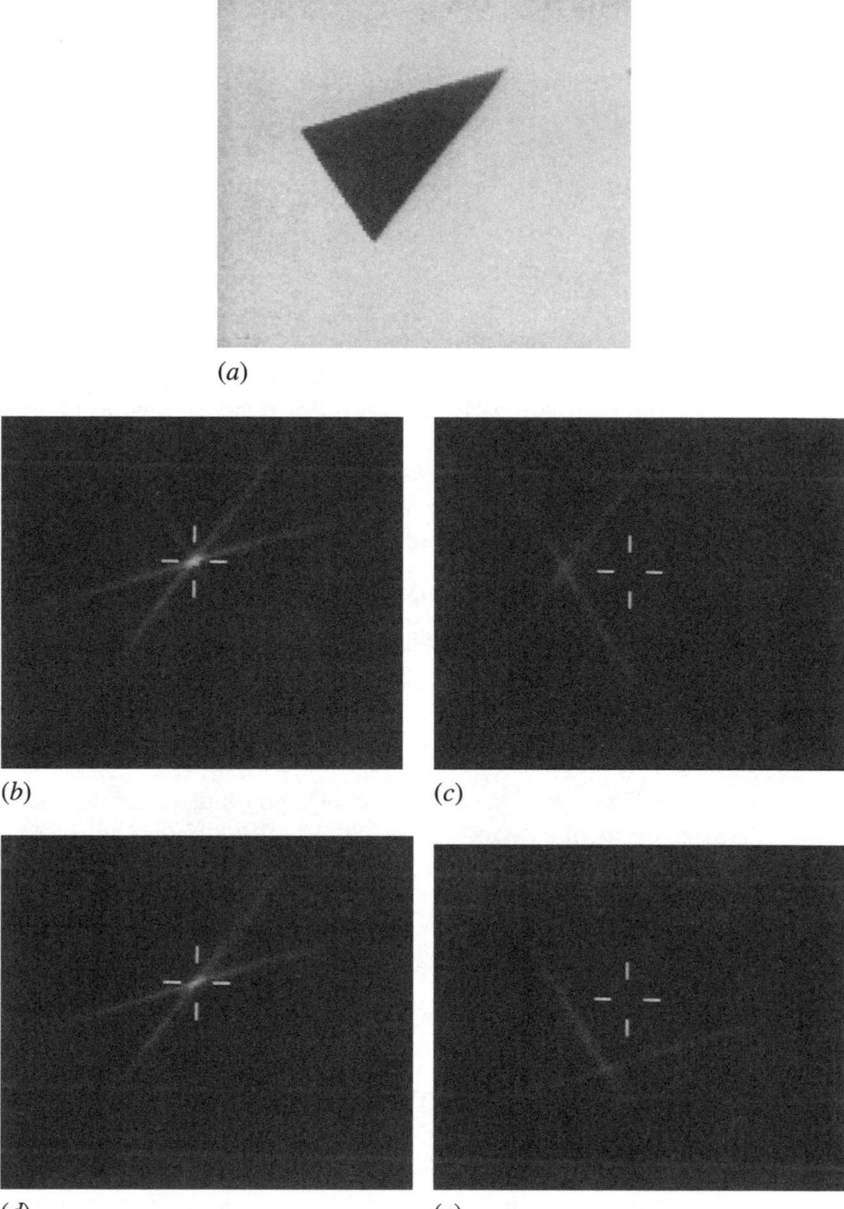

Figure 14.7 An arbitrary triangle and its transform in four planes: (*a*) localization point L selected to optimize sensitivity; (*b*)–(*e*) four planes required in parameter space.

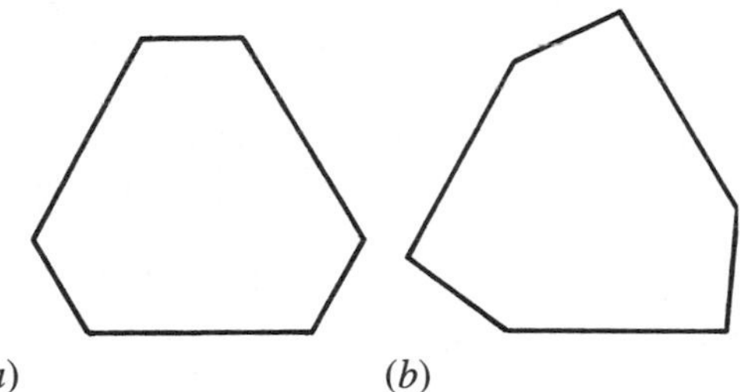

Figure 14.8 (*a*) Semiregular hexagon with threefold rotation symmetry and having all its exterior angles equal: two planes are required in parameter space in this case; (*b*) a similar hexagon with two slightly different exterior angles. Here three planes are needed in parameter space.

This method is wasteful in computation, however, since it does not make use of the fact that the main peak in parameter space has been found to lie in a *particular* plane *K*. Hence, we can take the set of (averaged) polygon side orientations θ from the histogram already constructed, identify which sides had which orientations, deduce from each θ a value for the polygon orientation $\eta = \alpha + \theta - \varphi_i$ (see Section 14.3.3), and enter this value into a histogram of estimated η values. The peak in this histogram will eventually give an accurate value for the polygon orientation η.

Strong similarities, but also significant differences, are evident between the procedure described above and that described in Chapter 13 for the determination of ellipse parameters. In particular, both procedures are two-stage ones, in which it is computationally more efficient to find object location before object orientation.

14.5 Why Corner Detection?

Our analysis has shown the possibilities for direct detection of polygons, but it has also revealed the limitations of the approach: these arise specifically with more complex polygons—such as those having concavities or many sides. In such cases, it may be easier to revert to using line or corner detection schemes coupled with the abstract pattern matching approaches discussed in Chapter 15.

In previous chapters we noted that objects are generally located most efficiently from their features. Prominent features have been seen to include straight lines, arcs, holes, and corners. Corners are particularly important since they may be used to orient objects and to provide measures of their dimensions—a knowledge of orientation being vital on some occasions, for example if a robot is to find the best way of picking up the object, and measurement being necessary in most inspection applications. Thus, accurate, efficient corner detectors are of great relevance in machine vision. In the remainder of the chapter we shall focus on corner detection procedures.

Workers in this field have devoted considerable ingenuity to devising corner detectors, sometimes with surprising results (see, for example, the median-based detector described later in this chapter). We start by considering what is perhaps the most obvious detection scheme—that of template matching.

14.6 Template Matching

Following our experience with template matching methods for edge detection (Chapter 5), it would appear to be straightforward to devise suitable templates for corner detection. These would have the general appearance of corners, and in a 3×3 neighborhood they would take forms such as the following:

$$\begin{bmatrix} -4 & 5 & 5 \\ -4 & 5 & 5 \\ -4 & -4 & -4 \end{bmatrix} \quad \begin{bmatrix} 5 & 5 & 5 \\ -4 & 5 & -4 \\ -4 & -4 & -4 \end{bmatrix}$$

the complete set of eight templates being generated by successive 90° rotations of the first two shown. Bretschi (1981) proposed an alternative set of templates. As for edge detection templates, the mask coefficients are made to sum to zero so that corner detection is insensitive to absolute changes in light intensity. Ideally, this set of templates should be able to locate all corners and to estimate their orientation to within 22.5°.

Unfortunately, corners vary considerably in a number of their characteristics, including in particular their degree of pointedness,[1] internal angle, and the intensity gradient at the boundary. Hence, it is quite difficult to design optimal corner detectors. In addition, generally corners are insufficiently pointed for good results to be obtained with the 3×3 template masks shown above. Another problem is that in larger neighborhoods, not only do the masks become larger but also more of them are needed to obtain optimal corner responses, and it rapidly becomes clear that the template matching approach is likely to involve excessive computation for practical corner detection. The alternative is to approach the problem analytically, somehow deducing the ideal response for a corner at any arbitrary orientation, and thereby bypass the problem of calculating

1 The term *pointedness* is used as the opposite to *bluntness*, the term *sharpness* being reserved for the total angle η through which the boundary turns in the corner region, that is, π minus the internal angle.

many individual responses to find which one gives the maximum signal. The methods described in the remainder of this chapter embody this alternative philosophy.

14.7 Second-order Derivative Schemes

Second-order differential operator approaches have been used widely for corner detection and mimic the first-order operators used for edge detection. Indeed, the relationship lies deeper than this. By definition, corners in gray-scale images occur in regions of rapidly changing intensity levels. By this token, they are detected by the same operators that detect edges in images. However, corner pixels are much rarer[2] than edge pixels. By one definition, they arise where two relatively straight-edged fragments intersect. Thus, it is useful to have operators that detect corners *directly*, that is, *without unnecessarily locating edges*. To achieve this sort of discriminability, it is necessary to consider local variations in image intensity up to at least second order. Hence, the local intensity variation is expanded as follows:

$$I(x, y) = I(0, 0) + I_x x + I_y y + I_{xx}x^2/2 + I_{xy}xy + I_{yy}y^2/2 + \cdots \tag{14.6}$$

where the suffices indicate partial differentiation with respect to x and y and the expansion is performed about the origin X_0 $(0, 0)$. The symmetrical matrix of second derivatives is:

$$\mathcal{I}_{(2)} = \begin{bmatrix} I_{xx} & I_{xy} \\ I_{yx} & I_{yy} \end{bmatrix} \quad \text{where } I_{xy} = I_{yx} \tag{14.7}$$

This gives information on the local curvature at X_0. A suitable rotation of the coordinate system transforms $\mathcal{I}_{(2)}$ into diagonal form:

$$\tilde{\mathcal{I}}_{(2)} = \begin{bmatrix} I_{\tilde{x}\tilde{x}} & 0 \\ 0 & I_{\tilde{y}\tilde{y}} \end{bmatrix} = \begin{bmatrix} \kappa_1 & 0 \\ 0 & \kappa_2 \end{bmatrix} \tag{14.8}$$

where appropriate derivatives have been reinterpreted as principal curvatures at X_0.

---

2 We might imagine a 256×256 image of 64K pixels, of which 1000 ($\sim$2%) lie on edges and a mere 30 ($\sim$0.05%) are situated at corner points.

We are particularly interested in rotationally invariant operators, and it is significant that the trace and determinant of a matrix such as $\mathscr{I}_{(2)}$ are invariant under rotation. Thus we obtain the Beaudet (1978) operators:

$$\text{Laplacian} = I_{xx} + I_{yy} = \kappa_1 + \kappa_2 \tag{14.9}$$

and

$$\text{Hessian} = \det(\mathscr{I}_{(2)}) = I_{xx}I_{yy} - I_{xy}^2 = \kappa_1\kappa_2 \tag{14.10}$$

It is well known that the Laplacian operator gives significant responses along lines and edges and hence is not particularly suitable as a corner detector. On the other hand, Beaudet's "DET" (determinant) operator does not respond to lines and edges but gives significant signals near corners. It should therefore form a useful corner detector. However, DET responds with one sign on one side of a corner and with the opposite sign on the other side of the corner. At the point of real interest—on the corner—it gives a null response. Hence, rather more complicated analysis is required to deduce the presence and exact position of each corner (Dreschler and Nagel, 1981; Nagel, 1983). The problem is clarified by Fig. 14.9. Here the dotted line shows the path of maximum horizontal curvature for various intensity values up the slope. The DET operator gives maximum response at positions P and Q on this line, and the parts of the line between P and Q must be explored to find the "ideal" corner point C where DET is zero.

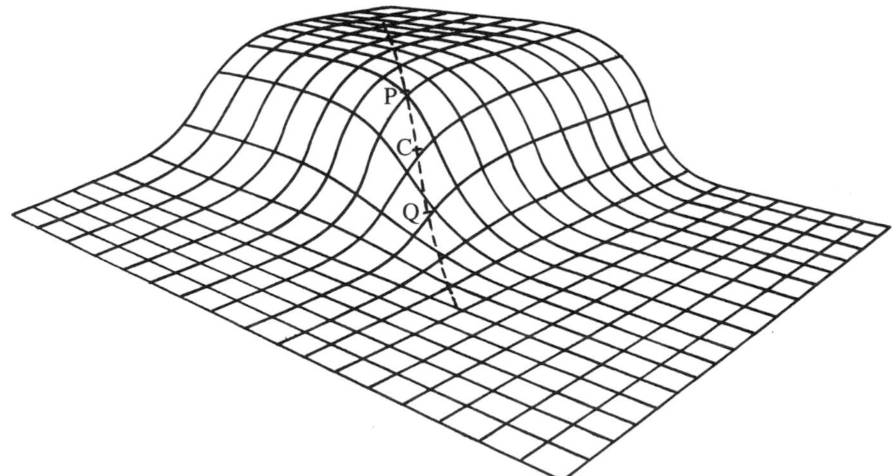

Figure 14.9 Sketch of an idealized corner, taken to give a smoothly varying intensity function. The dotted line shows the path of maximum horizontal curvature for various intensity values up the slope. The DET operator gives maximum responses at P and Q, and it is required to find the ideal corner position C where DET gives a null response.

Perhaps to avoid rather complicated procedures of this sort, Kitchen and Rosenfeld (1982) examined a variety of strategies for locating corners, starting from the consideration of local variation in the directions of edges. They found a highly effective operator that estimates the projection of the local rate of change of gradient direction vector along the horizontal edge tangent direction, and showed that it is mathematically identical to calculating the horizontal curvature κ of the intensity function I. To obtain a realistic indication of the strength of a corner, they multiplied κ by the magnitude of the local intensity gradient g:

$$C = \kappa g = \kappa\left(I_x^2 + I_y^2\right)^{1/2} = \frac{I_{xx}I_y^2 - 2I_{xy}I_xI_y + I_{yy}I_x^2}{I_x^2 + I_y^2} \tag{14.11}$$

Finally, they used the heuristic of nonmaximum suppression along the edge normal direction to localize the corner positions further.

In 1983, Nagel was able to show that mathematically the Kitchen and Rosenfeld (KR) corner detector using nonmaximum suppression is virtually identical to the Dreschler and Nagel (DN) corner detector. A year later, Shah and Jain (1984) studied the Zuniga and Haralick (ZH) corner detector (1983) based on a bicubic polynomial model of the intensity function. They showed that this is essentially equivalent to the KR corner detector. However, the ZH corner detector operates rather differently in that it thresholds the intensity gradient and then works with the subset of edge points in the image, only at that stage applying the curvature function as a corner strength criterion. By making edge detection explicit in the operator, the ZH detector eliminates a number of false corners that would otherwise be induced by noise.

The inherent near-equivalence of these three corner detectors need not be overly surprising, since in the end the different methods would be expected to reflect the same underlying physical phenomena (Davies, 1988d). However, it is gratifying that the ultimate result of these rather mathematical formulations is interpretable by something as easy to visualize as horizontal curvature multiplied by intensity gradient.

14.8 A Median-Filter-Based Corner Detector

Paler et al. (1984) devised an entirely different strategy for detecting corners. This strategy adopts an initially surprising and rather nonmathematical approach based on the properties of the median filter. The technique involves applying a median filter to the input image and then forming another image that is the difference between the input and the filtered images. This difference image contains a set of signals that are interpreted as local measures of corner strength.

Applying such a technique seems risky since its origins suggest that, far from giving a correct indication of corners, it may instead unearth all the noise in the original image and present this as a set of "corner" signals. Fortunately, analysis shows that these worries may not be too serious. First, in the absence of noise, strong signals are not expected in areas of background. Nor are they expected near straight edges, since median filters do not shift or modify such edges significantly (see Chapter 3). However, if a window is moved gradually from a background region until its central pixel is just over a convex object corner, there is no change in the output of the median filter. Hence there is a strong difference signal indicating a corner.

Paler et al. (1984) analyzed the operator in some depth and concluded that the signal strength obtained from it is proportional to (1) the local contrast and (2) the "sharpness" of the corner. The definition of sharpness they used was that of Wang et al. (1983), meaning the angle η through which the boundary turns. Since it is assumed here that the boundary turns through a significant angle (perhaps the whole angle η) within the filter neighborhood, the difference from the second-order intensity variation approach is a major one. Indeed, it is an implicit assumption in the latter approach that first- and second-order coefficients describe the local intensity characteristics with reasonable rigor, the intensity function being inherently continuous and differentiable. Thus, the second-order methods may give unpredictable results with pointed corners where directions change within the range of a few pixels. Although there is some truth in this, it is worth looking at the similarities between the two approaches to corner detection before considering the differences.

14.8.1 *Analyzing the Operation of the Median Detector*

This subsection considers the performance of the median corner detector under conditions where the gray-scale intensity varies by only a small amount within the median filter neighborhood region. This permits the performance of the corner detector to be related to low-order derivatives of the intensity variation, so that comparisons can be made with the second-order corner detectors mentioned earlier.

To proceed, we assume a continuous analog image and a median filter operating in an idealized circular neighborhood. For simplicity, since we are attempting to relate signal strengths and differential coefficients, we ignore noise. Next, recall (Chapter 3) that for an intensity function that increases monotonically with distance in some arbitrary direction $\tilde{x}$ but that does not vary in the perpendicular direction $\tilde{y}$, the median within the circular window is equal to the value at the center of the neighborhood. Thus, the median corner detector gives zero signal if the horizontal curvature is locally zero.

If there is a small horizontal curvature κ, the situation can be modeled by envisaging a set of constant-intensity contours of roughly circular shape and approximately equal curvature within the circular window that will be taken to have radius a (Fig. 14.10). Consider the contour having the median intensity value. The center of this contour does not pass through the center of the window but is displaced to one side along the negative $\tilde{x}$-axis. Furthermore, the

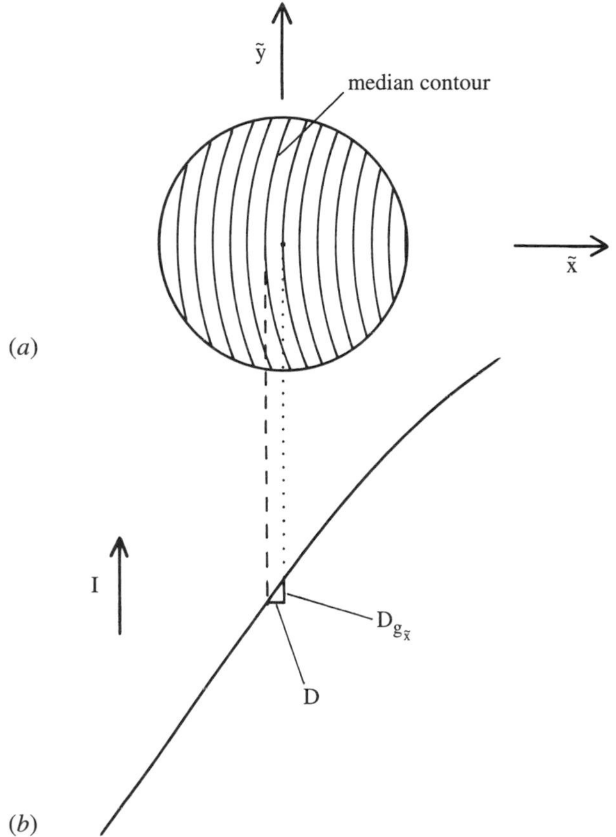

Figure 14.10 (*a*) Contours of constant intensity within a small neighborhood. Ideally, these are parallel, circular, and of approximately equal curvature (the contour of median intensity does not pass through the center of the neighborhood); (*b*) cross section of intensity variation, indicating how the displacement D of the median contour leads to an estimate of corner strength.

signal obtained from the corner detector depends on this displacement. If the displacement is D, it is easy to see that the corner signal is $Dg_{\tilde{x}}$ since $g_{\tilde{x}}$ allows the intensity change over the distance D to be estimated (Fig. 14.10). The remaining problem is to relate D to the horizontal curvature κ. A formula giving this relation has already been obtained in Chapter 3. The required result is:

$$D = \kappa a^2/6 \tag{14.12}$$

so the corner signal is

$$C = Dg_{\tilde{x}} = \kappa g_{\tilde{x}} a^2/6 \tag{14.13}$$

Note that C has the dimensions of intensity (contrast) and that the equation may be reexpressed in the form:

$$C = (g_{\tilde{x}} a) \times (2a\kappa)/12 \tag{14.14}$$

so that, as in the formulation of Paler et al. (1984), corner strength is closely related to corner contrast and corner sharpness.

To summarize, the signal from the median-based corner detector is proportional to horizontal curvature and to intensity gradient. Thus, this corner detector gives an identical response to the three second-order intensity variation detectors discussed in Section 14.7, the closest practically being the KR detector. However, this comparison is valid only when second-order variations in intensity give a complete description of the situation. The situation might be significantly different where corners are so pointed that they turn through a large proportion of their total angle within the median neighborhood. In addition, the effects of noise might be expected to be rather different in the two cases, as the median filter is particularly good at suppressing impulse noise. Meanwhile, for small horizontal curvatures, there ought to be no difference in the positions at which median and second-order derivative methods locate corners, and accuracy of localization should be identical in the two cases.

14.8.2 *Practical Results*

Experimental tests with the median approach to corner detection have shown that it is a highly effective procedure (Paler et al., 1984; Davies, 1988d). Corners are detected reliably, and signal strength is indeed roughly proportional both to local image contrast and to corner sharpness (see Fig. 14.11). Noise is more apparent for 3×3 implementations, which makes it better to use 5×5 or larger

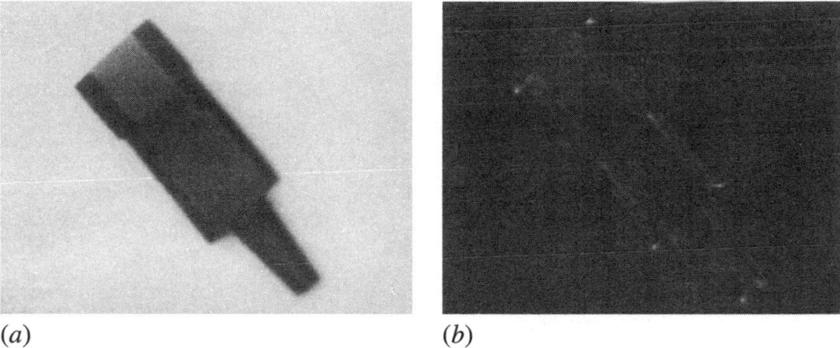

(a) (b)

Figure 14.11 (a) Original off-camera 128×128 6-bit gray-scale image; (b) result of applying the median-based corner detector in a 5×5 neighborhood. Note that corner signal strength is roughly proportional both to corner contrast and to corner sharpness.

neighborhoods to give good corner discrimination. However, because median operations are slow in large neighborhoods and background noise is still evident even in 5×5 neighborhoods, the basic median-based approach gives poor performance by comparison with the second-order methods. However, both of these disadvantages are virtually eliminated by using a "skimming" procedure in which edge points are first located by thresholding the edge gradient, and the edge points are then examined with the median detector to locate the corner points (Davies, 1988d). With this improved method, performance is found to be generally superior to that for, say, the KR method in that corner signals are better localized and accuracy is enhanced. Indeed, the second-order methods appear to give rather fuzzy and blurred signals that contrast with the sharp signals obtained with the improved median approach (Fig. 14.12).

At this stage the reason for the more blurred corner signals obtained using the second-order operators is not clear. Basically, there is no valid rationale for applying second-order operators to pointed corners, since higher derivatives of the intensity function will become important and will at least in principle interfere with their operation. However, it is evident that the second-order methods will probably give strong corner signals when the tip of a pointed corner appears anywhere in their neighborhood, so there is likely to be a minimum blur region of radius a for any corner signal. This appears to explain the observed results adequately. The sharpness of signals obtained by the KR method may be improved by nonmaximum suppression (Kitchen and Rosenfeld, 1982; Nagel, 1983). However, this technique can also be applied to the output of median-based corner detectors. Hence, the fact remains that the median-based method gives inherently better localized signals than the second-order methods (although for low-curvature corners the signals have similar localization).

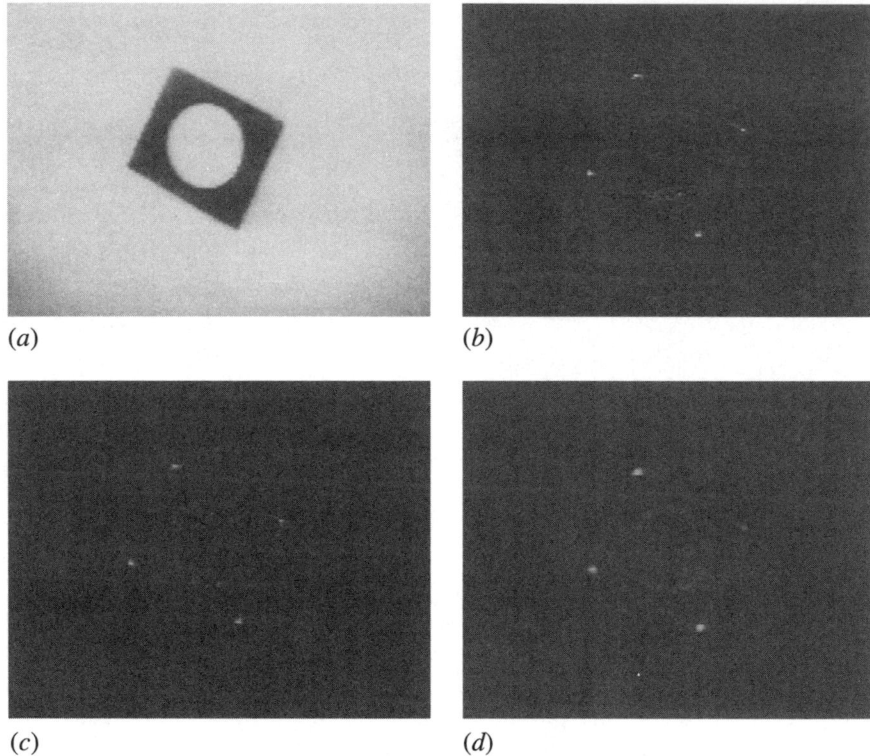

(*a*) (*b*)

(*c*) (*d*)

Figure 14.12 Comparison of the median and KR corner detectors: (*a*) original 128 × 128 gray-scale image; (*b*) result of applying a median detector; (*c*) result of including a suitable gradient threshold; (*d*) result of applying a KR detector. The considerable amount of background noise is saturated out in (*a*) but is evident from (*b*). To give a fair comparison between the median and KR detectors, 5 × 5 neighborhoods are employed in each case, and nonmaximum suppression operations are not applied. The same gradient threshold is used in (*c*) and (*d*).

Overall, the inherent deficiencies of the median-based corner detector can be overcome by incorporating a skimming procedure, and then the method becomes superior to the second-order approaches in giving better localization of corner signals. The underlying reason for the difference in localization properties appears to be that the median-based signal is ultimately sensitive only to the particular few pixels whose intensities fall near the median contour within the window, whereas the second-order operators use typical convolution masks that are in general sensitive to the intensity values of all the pixels within the window. Thus, the KR operator tends to give a strong signal when the tip of a pointed corner is present anywhere in the window.

14.9 The Hough Transform Approach to Corner Detection

In certain applications, corners might rarely appear in the idealized pointed form suggested earlier. Indeed, in food applications corners are commonly chipped, crumbly, or rounded, and are bounded by sides that are by no means straight lines (Fig. 14.13). Even mechanical components may suffer chipping or breakage. Hence, it is useful to have some means of detecting these types of nonideal corners. The generalized HT provides a suitable technique (Davies, 1986b, 1988a).

The method used is a variation on that described in Section 14.3 for locating objects such as squares. Instead of accumulating a line of candidate center points parellel to each side after moving laterally inward a distance equal to half the length of a side, only a small lateral displacement is required when locating corners. The lateral displacement that is chosen must be sufficient to offset any corner bluntness, so that the points that are located by the transform lines are consistently positioned relative to the idealized corner point (Fig. 14.14). In addition, the displacement must be sufficiently large that enough points on the sides of the object contribute to the corner transforms to ensure adequate sensitivity.

There are two relevant parameters for the transform: D, the lateral displacement, and T, the length of the transform line arising from each edge point. These parameters must be related to the following parameters for the object being detected: B, the width of the corner bluntness region (Fig. 14.14), and S, the length of the object side (for convenience the object is assumed to be

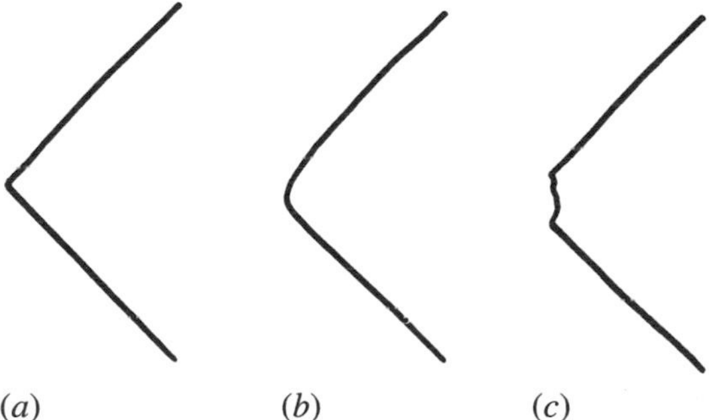

Figure 14.13 Types of corners: (*a*) pointed; (*b*) rounded; (*c*) chipped. Corners of type (*a*) are normal with metal components, those of type (*b*) are usual with biscuits and other food products, whereas those of type (*c*) are common with food products but rarer with metal parts.

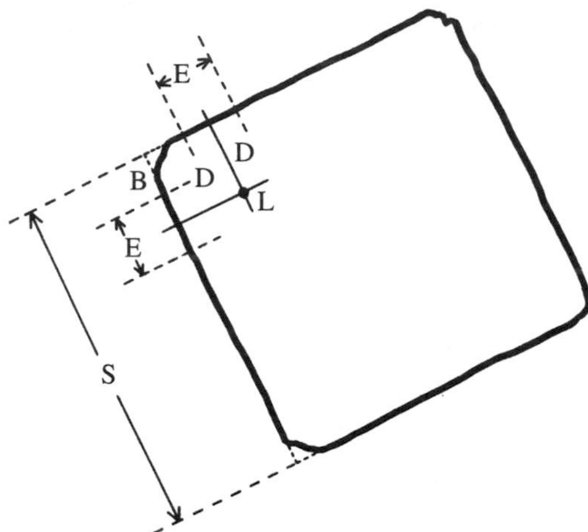

Figure 14.14 Geometry for locating blunt corners: B, region of bluntness; D, lateral displacement employed in edge pixel transform; E, distance over which position of corner is estimated; I, idealized corner point; L, localization point of a typical corner; S, length of a typical side. The degree of corner bluntness shown here is common among food products such as fishcakes.

a square). E, the portion of a side that is to contribute to a corner peak, is simply related to T:

$$T = E \tag{14.15}$$

Since we wish to make corner peaks appear as close to idealized corners as possible, we may also write:

$$D = B + E/2 \tag{14.16}$$

These definitions now allow the sensitivity of corner location to be optimized. Note first that the error in the measurement of each corner position is proportional to $1/\sqrt{E}$, since our measurement of the relevant part of each side is effectively being averaged over E independent signals. Next, note that the purpose of locating corners is ultimately to measure (1) the orientation of the object and (2) its linear dimensions. Hence, the fractional accuracy is proportional to:

$$A = (S - 2D)\sqrt{E} \tag{14.17}$$

Eliminating D now gives:

$$A = (S - 2B - E)\sqrt{E} \tag{14.18}$$

Differentiating, we find:

$$\mathrm{d}A/\mathrm{d}E = (S - 2B - 3E)/2\sqrt{E} \tag{14.19}$$

Accuracy can now be optimized by setting $\mathrm{d}A/\mathrm{d}E = 0$, so that:

$$E = S/3 - 2B/3 \tag{14.20}$$

making

$$D = S/6 + 2B/3 \tag{14.21}$$

Hence, there is a clear optimum in accuracy, and as $B \to 0$ it occurs for $D \approx S/6$, whereas for rather blunt or crumbly types of objects (typified by biscuits) $B \approx S/8$ and then $D \approx S/4$.

If optimal object detection and location had been the main aim, this would have been achieved by setting $E = S - 2B$ and detecting all the corners together at the center of the square. It is also possible that by optimizing object orientation, object detection is carried out with reduced sensitivity, since only a length $2E$ of the perimeter contributes to each signal, instead of a length $4S - 8B$.

The calculations presented here are important for a number of reasons: (1) they clarify the point that tasks such as corner location are carried out not in isolation but as parts of larger tasks that are optimizable; (2) they show that the ultimate purpose of corner detection may be either to locate objects accurately or to orient and measure them accurately; and (3) they indicate that there is an optimum region of interest E for any operator that is to locate corners. This last factor has relevance when selecting the optimum size of neighborhood for the other corner detectors described in this chapter.

Although the calculations above emphasized sensitivity and optimization, the underlying principle of the HT method is to find corners by interpolation rather than by extrapolation, as interpolation is generally a more accurate procedure. The fact that the corners are located away from the idealized corner positions is to some extent a disadvantage (although it is difficult to see how a corner can be located directly if it is missing). However, once the corner peaks have been found, for known objects the idealized corner positions may be deduced (Fig. 14.15). It is as well to note that this step is unlikely to be necessary if the real purpose of corner location is to orient the object and to measure its dimensions.

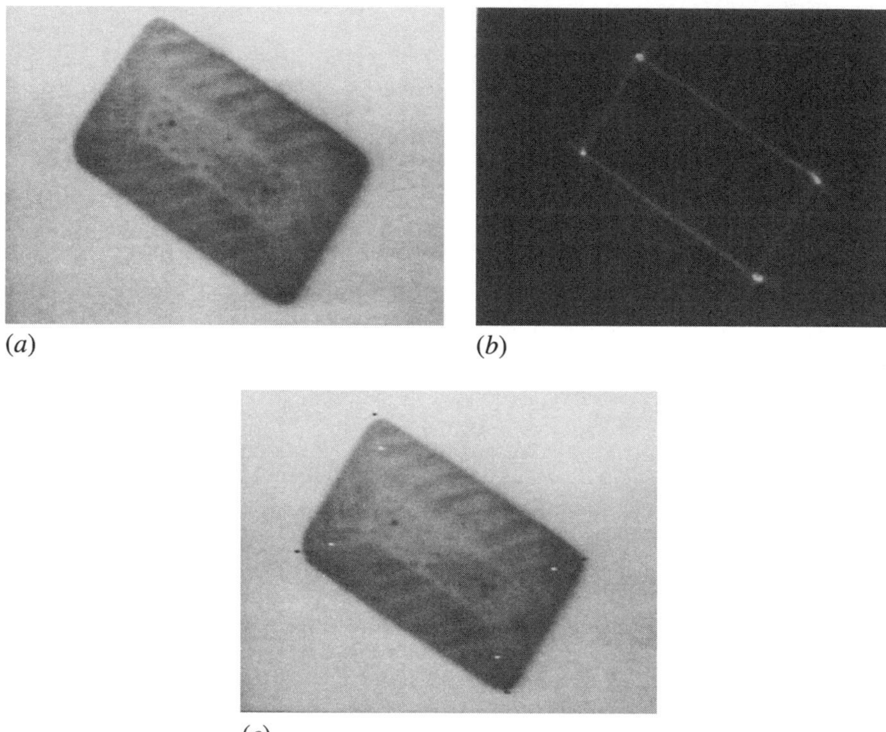

Figure 14.15 Example of the Hough transform approach: (*a*) original image of a biscuit (128×128 pixels, 64 gray levels); (*b*) transform with lateral displacement around 22% of the shorter side; (*c*) image with transform peaks located (white dots) and idealized corner positions deduced (black dots). The lateral displacement employed here is close to the optimum for this type of object.

14.10 The Plessey Corner Detector

At this point we consider whether other strategies for corner detection could be devised. The second-order derivative class of corner detector was devised on the basis that corners are ideal, smoothly varying differentiable intensity profiles, whereas the median-based detector was conceived totally independently on the basis of curved step edges whose profiles are quite likely *not* to be smoothly varying and differentiable. It is profitable to take the latter idea further and to consider corners as the intersection of two step edges.

First, imagine that a small region, which is the support region of such a corner detector, contains both a step edge aligned along the x direction and another aligned along the y direction. To detect these edges, we can use an operator that estimates $<I_x>$ and another that estimates $<I_y>$. Note that a corner detector would have to detect both types of edges simultaneously. To do so, we can use a combined estimator such as $<|I_x|><|I_y|>$ or $<I_x^2><I_y^2>$. Each of these operators leads to corner signals that can be thresholded conveniently using a single positive threshold value.

Next, note that this estimator requires the edge orientations to be known and indeed aligned along the x- and y-axes directions. To obtain more general properties, the operator needs to transform appropriately under rotation of axes. In addition, magnitudes are difficult to handle mathematically, so we start with the squared magnitude estimator and employ a symmetrical matrix of first derivatives in the transformable form:

$$\mathscr{G} = \begin{bmatrix} <I_x^2> & <I_xI_y> \\ <I_xI_y> & <I_y^2> \end{bmatrix} \tag{14.22}$$

where

$$\det(\mathscr{G}) = <I_x^2><I_y^2> - <I_xI_y>^2 \tag{14.23}$$

This expression would clearly be identically zero if evaluated separately at each individual pixel. Nevertheless, our intuition is correct, and the base parameters $<I_x^2>$, $<I_y^2>$, and $<I_xI_y>$ must all be determined as *averages* over all pixels in the detector support region. If this is done, the expression need not average to zero, as the interpixel variations leading to the three base parameter values will introduce deviations. This situation is easily demonstrated, as rotation of axes to permit diagonalization of the matrix yields:

$$\tilde{\mathscr{G}} = \begin{bmatrix} <I_{\tilde{x}}^2> & 0 \\ 0 & <I_{\tilde{y}}^2> \end{bmatrix} = \begin{bmatrix} \gamma_1 & 0 \\ 0 & \gamma_2 \end{bmatrix} \tag{14.24}$$

with eigenvalues $<I_{\tilde{x}}^2> = \gamma_1$, $<I_{\tilde{y}}^2> = \gamma_2$. In this case, the Hessian is $\gamma_1\gamma_2$, and this is a sensible value to take for the corner strength. In particular, it is substantial only if both edges are present within the support region. The Plessey operator (Harris and Stephens, 1988; see also Noble, 1988) normalizes this measure by dividing by $\gamma_1 + \gamma_2$ and forming a square root:[3]

$$\mathscr{C} = [\gamma_1\gamma_2/(\gamma_1 + \gamma_2)]^{1/2} = [< I_{\tilde{x}}^2 >< I_{\tilde{y}}^2 >/(< I_{\tilde{x}}^2 > + < I_{\tilde{y}}^2 >)]^{1/2} \tag{14.25}$$

Interestingly, this does not ruin the transformation properties of the operator because both the determinant and the trace of the relevant symmetrical matrix are invariant under orthogonal transformations.

This operator has some useful properties. In particular, it is better to call it an "interest" operator rather than a corner detector because it also responds to situations where the two constituent edges cross over, so that a double corner results (as happens in many places on a checkerboard). It is really responding to and averaging the effects of all the locations where two edges appear within the operator support region. We can easily show that it has this property, as the fact that the operator embodies $< I_{\tilde{x}}^2 >$ and $< I_{\tilde{y}}^2 >$ means that it is insensitive to the sign of the local edge gradients (Fig. 14.16*a*).

The Plessey operator also has the undesirable property[4] that it can give a bias at T-junctions, to an extent dependent on the relative intensities of the three adjacent regions (Fig. 14.16*b*). This is the effect of averaging the three base parameters over the whole support region. It is difficult to be absolutely sure where the peak response will be as the computation is quite complex. In such cases, symmetry can act as a useful constraining factor—as in the case of a double corner on a checkerboard, where the peak signal necessarily coincides with the double corner location. (This applies to all the edge crossover points in Fig. 14.16*a*.) Clearly, noise will have some effect on the peak locations in off-camera images.

14.11 Corner Orientation

This chapter has so far considered the problem of corner detection as relating merely to corner location. However, corners differ from holes in that they are not isotropic, and hence they are able to provide orientation information. Such information can be used by procedures that collate the information from various features in order to deduce the presence and positions of objects

3 As originally devised, the Plessey operator is inverted relative to this definition, and a square root is not computed. Here we include a square root so that the corner signal is proportional to image contrast, as for the other corner detectors described in this chapter.

4 See, for example, the images in Shen and Wang (2002).

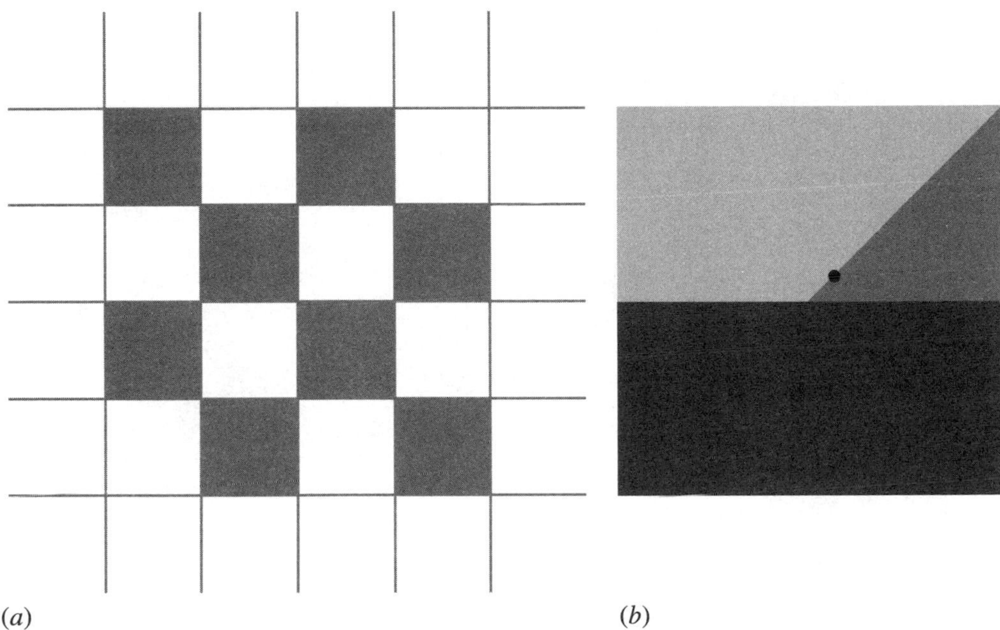

(a) (b)

Figure 14.16 Effect of the Plessey corner detector. (*a*) shows a checkerboard pattern that gives high responses at each of the edge crossover points. Because of symmetry, the location of the peaks is exactly at the crossover locations. (*b*) Example of a T-junction. The black dot shows a typical peak location. In this case there is no symmetry to dictate that the peak must occur exactly at the T-junction.

containing them. In Chapter 15 we will see that orientation information is valuable in immediately eliminating a large number of possible interpretations of an image, and hence of quickly narrowing down the search problem and saving computation.

Clearly, when corners are not particularly pointed, or are detected within rather small neighborhoods, the accuracy of orientation will be somewhat restricted. However, orientation errors will seldom be worse than 90° and will often be less than 20°. Although these accuracies are far worse than those (around 1°) for edge orientation (see Chapter 5), they provide valuable constraints on possible interpretations of an image.

Here we consider only simple means of estimating corner orientation. Basically, once a corner has been located accurately, it is a rather trivial matter to estimate the orientation of the intensity gradient at that location. The main problem that arises is that a slight error in locating the corner gives a bias in estimating its orientation toward that of one or another of the bounding sides—and for a sharp pointed corner, this can mean an error approaching 90°. The simplest solution to this problem is to find the mean orientation over a small region surrounding the estimated corner position. Then, without excessive computation, the orientation accuracy will usually be within the 20° limit suggested above.

Polygonal objects may be recognized by their sides or by their corners. This chapter has studied both approaches, with corner detection probably being more generic—especially if corners are considered as general salient features. Some corner detectors are better classed as interest point detectors, for they may not restrict their attention to true corners.

14.12 Concluding Remarks

When using the GHT for the detection of polygons, it is useful to try to minimize storage and computation. In some cases only one parameter plane is needed to detect a polygonal object: commonly, a number of parameter planes of the order of the number of sides of the polygon is required, but this still represents a significant saving compared with the numbers needed to detect general shapes. It has been demonstrated how optimal savings can be achieved by taking into account the symmetry possessed by many polygon shapes such as squares and rectangles.

The analysis has shown the possibilities for direct detection of polygons, but it has also revealed the limitations of this approach. These limitations arise specifically with more complex polygons—such as those having concavities or many sides. In such cases, it may be easier to revert to using line or corner detection schemes coupled with the abstract pattern matching approaches discussed in Chapter 15. Indeed, the latter approach offers the advantage of greater resistance to shape distortions if suitable algorithms to infer the presence of objects can be found.

Hence, the chapter also appropriately examined corner detection schemes. Apart from the obvious template matching procedure, which is strictly limited in its application, and its derivative, the lateral histogram method (see Problems 14.5 and 14.6), three main approaches have been described. The first was the second-order derivative approach, which includes the KR, DN, and ZH methods—all of which embody the same basic schema. The second was the median-based method, which turned out to be equivalent to the second-order methods—but only in situations where corners have smoothly varying intensity functions; the third was a Hough-based scheme that is particularly useful when corners are liable to be blunt, chipped, or damaged. The analysis of this last scheme revealed interesting general points about the specification and design of corner detection algorithms.

Although the median-based approach has obvious inherent disadvantages—its slowness and susceptibility to noise—these negative aspects can largely be overcome by a skimming (two-stage template matching) procedure. Then the method becomes essentially equivalent to the second-order derivative methods for low-curvature intensity functions, as stated above. However, for pointed corners, higher-order derivatives of the intensity function are bound to be important. As a result the second-order methods are based on an unjustifiable rationale and are found to blur corner signals, hence reducing the accuracy of corner localization. This problem does not arise for the median-based method, which seems to offer rather superior performance in practical situations. Clearly, however, it should not be used when image noise is excessive, as then the skimming procedure will either (with a high threshold) remove a proportion of the corners with the noise, or else (with a low threshold) not provide adequate speedup to counteract the computational load of the median filtering routine.

Overall, the Plessey corner detector described in Section 14.10 and more recently the SUSAN detector (Smith and Brady, 1997) have been among the most widely used of the available corner detectors, though whether this is the result of conclusive tests against carefully defined criteria is somewhat doubtful. For more details and comparisons, see the next section.

14.13 Bibliographical and Historical Notes

The initial sections of this chapter arose from the author's own work which was aimed at finding optimally sensitive means of locating shapes, starting with spatial matched filters. Early work in this area included that of Sklansky (1978), Ballard (1981), Brown (1983), and Davies (1987c), which compared the HT approach with spatial matched filters. These general investigations led to more detailed studies of how straight edges are optimally detected (Davies, 1987d,j), and then to polygon detection (Davies, 1986d, 1989c). Although optimal sensitivity is not the only criterion for effectiveness of object detection, nonetheless it is of great importance, since it governs both effectiveness per se (especially in the presence of "clutter" and occlusion) and the accuracy with which objects may be located.

The emphasis on spatial matched filters seemed to lead inevitably to use of the GHT, with its accompanying computational problems. The present chapter should make it plain that although these problems are formidable, systematic analysis can make a significant impact on them.

Corner detection is a subject that has been developing for well over two decades. Beaudet's (1978) work on rotationally invariant image operators set the scene for the development of parallel corner detection algorithms. This was soon followed by Dreschler and Nagel's (1981) more sophisticated second-order corner detector. The motivation for their research was mapping the motion of cars in traffic scenes, with corners providing the key to unambiguous interpretation of image sequences. One year later, Kitchen and Rosenfeld (1982) had completed their study of corner detectors based mainly on edge orientation and had developed the second-order KR method described in Section 14.7. The years 1983 and 1984 saw the development of the lateral histogram corner detector, the second-order ZH detector, and the median-based detector (Wu and Rosenfeld, 1983; Zuniga and Haralick, 1983; Paler et al., 1984). Subsequently, Davies' work on the detection of blunt corners and on analyzing and improving the median-based detector appeared (1986b, 1988a,d, 1992a). These papers form the basis of Sections 14.8 and 14.9. Meanwhile, other methods had been developed, such as the Plessey algorithm (Harris and Stephens, 1988; see also Noble, 1988), which is covered in Section 14.10. The much later algorithm of Seeger and Seeger (1994) is also of interest because it appears to work well on real (outdoor) scenes in real time without requiring the adjustment of any parameters.

The Smith and Brady (1997) SUSAN algorithm marked a further turning point, needing no assumptions on the corner geometry, for it works by making simple comparisons of local gray levels. This is one of the most often cited of all corner detection algorithms. Rosin (1999) adopted the opposite approach of modeling corners in terms of subtended angle, orientation, contrast, bluntness, and boundary curvature with excellent results. However, parametric approaches such as this tend to be computation intensive. Note also that parametric models sometimes run into problems when the assumptions made in the model do not match the data of real images. It is only much later, when many workers have tested the available operators, that such deficiencies come to light. The late 1990s also saw the arrival of a new boundary-based corner detector with good detection and localization (Tsai et al., 1999). A development of the Plessey detector called a gradient direction detector (Zheng et al., 1999)—which was optimized for good localization—also has reduced complexity.

In the 2000s, further corner detectors have been developed. The Ando (2000) detector, similar in concept to the Tsai et al. (1999) detector, has been found to have good stability, localization, and resolution. Lüdtke et al. (2002) designed a detector based on a mixture model of edge orientation. In addition to being effective in comparison with the Plessey and SUSAN operators, particularly at large opening angles, the method provides accurate angles and strengths for the corners. Olague and Hernández (2002) worked on a unit step-edge function (USEF) concept, which is able to model complex corners well. This innovation resulted in adaptable detectors that are able to detect corners with subpixel accuracy. Shen and Wang (2002) described a Hough-transform-based detector; as this works in a 1-D parameter space, it is fast enough for real-time operation. A useful feature of the Shen and Wang paper is the comparison with, and hence between, the Wang and Brady detector, the Plessey detector, and the SUSAN detector. Several example images show that it is difficult to know exactly what one is looking for in a corner detector (i.e., corner detection is an ill-posed problem), and that even the well-known detectors sometimes inexplicably fail to find corners in some very obvious places. Golightly and Jones (2003) present a practical problem in outdoor country scenery. They discuss not only the incidence of false positives and false negatives but also the probability of correct association in corner matching (e.g., during motion).

Rocket (2003) presents a performance assessment of three corner detection algorithms—the Kitchen–Rosenfeld detector, the median-based detector, and the Plessey detector. The results are complex, with the three detectors all having very different characteristics. Rocket's paper is valuable because it shows how to optimize the three methods (not least showing that a particular parameter k for the Plessey detector should be ~ 0.04) and also because it concentrates on careful research, with no "competition" from detectors invented by its author. Finally, Tissainayagam and Suter (2004) assess the performance of corner detectors, relating this study specifically to point feature (motion) tracking applications. It would appear that whatever the potential value of other reviews, this one has particular validity within its frame of reference. Interestingly, it finds that, in image sequence analysis, the Plessey detector is more robust to noise than the SUSAN detector. A possible explanation is that it "has a built-in smoothing function as part of its formulation."

This review of the literature covers many old and recent developments on corner detection, but it is not the whole story. In many applications, it is not specific corner detectors that are needed but salient point detectors. As indicated in Chapter 13, salient point features can be small holes or corners. However, they may also be characteristic patterns of intensity. Recent literature has referred to such patterns as interest points. Often interest point detectors are essentially corner detectors, but this does not have to be the case. For example, the Plessey detector is sometimes called an interest detector—with good reason, as indicated in Section 14.10. Moravec (1977) was among the first to refer to interest points. He was followed, for example, by Schmid et al. (2000), and later by Sebe and Lew (2003) and Sebe et al. (2003), who revert to calling them salient points but with no apparent change of meaning (except that the points should also be visually interesting). Overall, it is probably safest to use the Haralick and Shapiro (1993) definition: a point being "interesting" if it is both distinctive and invariant—that is, it stands out and is invariant to geometric distortions such as might result from moderate changes in scale or viewpoint. (Note that Haralick and Shapiro also list other desirable properties—stability, uniqueness, and interpretability.) The invariance aspect is taken up by Kenney et al. (2003) who show how to remove from consideration ill-conditioned points in order to make matching more reliable.

The considerable ongoing interest in corner detection and interest point detection underlines the point, made in Section 14.5, that good point feature detectors are vital to efficient image interpretation, as will be seen in more detail in subsequent chapters.

14.14 Problems

1. How would a quadrant of a circle be detected optimally? How many planes would be required in parameter space?
2. Find formulas for the number of planes required in parameter space for optimal detection of (a) a parallelogram and (b) a rhombus.
3. Find how far the signal-to-noise ratio is below the optimum for an isosceles triangle of sides 1, 5, and 5 cm, which is detected in a single-plane parameter space.
4. An N-sided polygon is to be located using the Hough transform. Show that the number of planes required in parameter space can, in general, be reduced from $N+1$ to $N-1$ by a suitable choice of localization point. Show that this rule breaks down for a semicircle, where the circular part of the boundary can be regarded as being composed of an infinite number of short sides. Determine why the rule breaks down and find the minimum number of parameter planes.
5. By examining suitable binary images of corners, show that the median corner detector gives a maximal response within the corner boundary rather than halfway down the edge outside the corner. Show how the situation is modified for gray-scale images. How will this affect the value of the gradient noise-skimming threshold to be used in the improved median detector?
6. Sketch the lateral histograms (defined as in Section 13.3) for an image containing a single convex quadrilateral shape. Show that it should be possible to find the corners of the quadrilateral from the lateral histograms and hence to locate it in the image. How would the situation change for a quadrilateral with a concavity? Why should the lateral histogram method be particularly well suited to the detection of objects with blunt corners?

7. Sketch the response patterns for a [−1 2 −1] type of 1-D Laplacian operator applied to corner detection in lateral histograms for an image containing a triangle, (a) if the triangle is randomly oriented, and (b) if one or more sides are parallel to the x- or y-axis. In case (b), how is the situation resolved? In case (a), how would sensitivity depend on orientation?

8. Prove equation (14.11), starting with the following formula for curvature:

$$\kappa = (\mathrm{d}^2 y/\mathrm{d}x^2)/\left[1 + (\mathrm{d}y/\mathrm{d}x)^2\right]^{3/2}$$

Hint: First express $\mathrm{d}y/\mathrm{d}x$ in terms of the components of intensity gradient, remembering that the intensity gradient vector (I_x, I_y) is oriented along the edge normal; then replace the x, y variation by I_x, I_y variation in the formula for κ.

Abstract Pattern Matching Techniques

Abstract pattern matching involves stepping back from the image itself and working at a higher level, grouping features in an abstract way to infer the presence of objects. Graph matching has long been a conventional method for performing this task, but in the right circumstances a suitable adaptation of the generalized Hough transform can actually outperform it. This chapter discusses inference procedures and considers relational descriptions of scenes and the various types of searches that can be used with image data.

Look out for:

- the match graph approach for identifying objects from their point features.
- how the need for robustness against noise, clutter, and occlusion translates into the requirement for subgraph–subgraph isomorphism.
- the maximal clique paradigm.
- how symmetry can be used to simplify the matching task.
- how the generalized Hough transform can be used for point pattern matching.
- how order calculations can be used to compare the speeds of matching algorithms.
- how relational descriptors may be used for logical analysis of scenes.
- the different types of search algorithms that may be used in scene analysis.

This chapter completes the work of Part 2 by showing how the presence of objects can be inferred from point features as an alternative to edge features. Even with point features it is found that the Hough transform may sometimes be used with advantage. However, all inference techniques need to be analyzed for computational complexity and suitable optimizations made. The need for complexity analysis carries on with even more force in subsequent work—not least the more complex algorithms used for processing 3-D images in Part 3 of this volume.

CHAPTER 15

Abstract Pattern Matching Techniques

15.1 Introduction

We have seen how objects having quite simple shapes may be located in digital images via the Hough transform. For more complex shapes this approach tends to require excessive computation. In general, this problem may be overcome by locating objects from their features. Suitable salient features include small holes, corners, straight, circular, or elliptical segments, and any readily localizable subpatterns. Earlier chapters have shown how such features may be located. However, at some stage it becomes necessary to find methods for collating the information from the various features in order to recognize and locate the objects containing them. This task is studied in the present chapter.

It is perhaps easiest to envisage the feature collation problem when the features themselves are unstructured points carrying no directional information—nor indeed any attributes other than their x, y coordinates in the image. Then the object recognition task is often called *point pattern matching*.[1] The features that are closest to unstructured points are small holes, such as the "docker" holes in many types of biscuits. Corners can also be considered as points if their other attributes—including sharpness, and orientation—are ignored. In the following section we start with point features, and then we see how the attributes of more complex types of features can be included in recognition schemes.

Overall, we should remember that it is highly efficient to use small high-contrast features for object detection, since the computation involved in searching

1 Note that this term is sometimes used not just for object recognition but for initial matching of two stereo views of the same scene.

an image decreases as the template becomes smaller. As will be clear from the preceding chapters, the main disadvantage of such an approach to object detection is the loss in sensitivity (in a signal-to-noise sense) owing to the greatly impoverished information content of the point feature image. However, even the task of identifying objects from a rather small number of point features is far from trivial and frequently involves considerable computation, as we will see in this chapter. We start by studying a graph-theoretic approach to point pattern matching, which involves the "maximal clique" concept.

15.2 A Graph-theoretic Approach to Object Location

This section considers a commonly occurring situation that involves considerable constraints—objects appearing on a horizontal worktable or conveyor at a known distance from the camera. It is also assumed (1) that objects are flat or can appear in only a restricted number of stances in three dimensions, (2) that objects are viewed from directly overhead, and (3) that perspective distortions are small. In such situations, the objects may in principle be identified and located with the aid of very few point features. Because such features are taken to have no structure of their own, it will be impossible to locate an object uniquely from a single feature, although positive identification and location would be possible using two features if these were distinguishable and if their distance apart were known. For truly indistinguishable point features, an ambiguity remains for all objects not possessing 180° rotation symmetry. Hence, at least three point features are in general required to locate and identify objects at a known range. Clearly, noise and other artifacts such as occlusions modify this conclusion. When matching a template of the points in an idealized object with the points present in a real image, we may find the following:

1. A great many feature points may be present because of multiple instances of the chosen type of object in the image.
2. Additional points may be present because of noise or clutter from irrelevant objects and structure in the background.
3. Certain points that should be present are missing because of noise or occlusion, or because of defects in the object being sought.

These problems mean that we should in general be attempting to match a subset of the points in the idealized template to various subsets of the points in the image. If the point sets are considered to constitute *graphs* with the point features as *nodes*, the task devolves into the mathematical problem of subgraph—subgraph isomorphism, that is, finding which subgraphs in the image graph are isomorphic[2] to subgraphs of the idealized template graph. Of course, there may be a large number of matches involving rather few points. These would arise from sets of features that *happen* (see, for example, item (2) above) to lie at valid distances apart in the original image. The most significant matches will involve a fair number of features and will lead to correct object identification and location. A point feature matching scheme will be most successful if it finds the most likely interpretation by searching for solutions with the greatest internal consistency—that is, with the greatest number of point matches per object.

Unfortunately, the scheme of things presented here is still too simplistic in many applications, for it is insufficiently robust against distortions. In particular, optical (e.g., perspective) distortions may arise, or the objects themselves may be distorted, or by resting partly on other objects they may not be quite in the assumed stance. Hence, distances between features may not be exactly as expected. These factors mean that some tolerance has to be accepted in the distances between pairs of features, and it is common to employ a threshold such that interfeature distances have to agree within this tolerance before matches are accepted as potentially valid. Distortions produce more strain on the point matching technique and make it all the more necessary to seek solutions with the greatest possible internal consistency. Thus, as many features as possible should be taken into account in locating and identifying objects. The maximal clique approach is intended to achieve this.

As a start, as many features as possible are identified in the original image, and these are numbered in some convenient order such as the order of appearance in a normal TV raster scan. The numbers then have to be matched against the letters corresponding to the features on the idealized object. A systematic way to do this is to construct a *match graph* (or *association graph*) in which the nodes represent feature assignments and arcs joining nodes represent pairwise compatibilities between assignments. To find the best match, it is necessary to find regions of the match graph where the cross linkages are maximized. To do so, *cliques* are sought within the match graph. A clique is a *complete subgraph*—that is, one for which all pairs of nodes are connected by arcs. However, the previous arguments indicate that if one clique is completely included within another clique, the larger clique likely represents a better match—and indeed *maximal cliques* can be taken as leading to the most reliable matches between the observed image and the object model.

Figure 15.1*a* illustrates the situation for a general triangle: for simplicity, the figure takes the observed image to contain only one triangle and assumes that lengths match exactly and that no occlusions occur. The match graph in this

2 That is, of the same basic shape and structure.

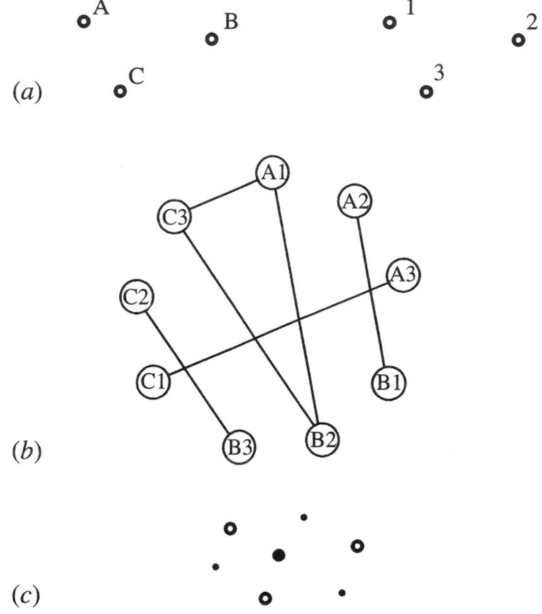

Figure 15.1 A simple matching problem—a general triangle: (*a*) basic labeling of model (*left*) and image (*right*); (*b*) match graph; (*c*) placement of votes in parameter space. In (*b*) the maximal cliques are: (1) A1, B2, C3; (2) A2, B1; (3) B3, C2; and (4) C1, A3. In (*c*) the following notation is used: o, positions of observed features; •, positions of votes, ●, position of main voting peak.

example is shown in Fig. 15.1*b*. There are nine possible feature assignments, six valid compatibilities, and four maximal cliques, only the largest corresponding to an exact match.

Figure 15.2*a* shows the situation for the less trivial case of a quadrilateral, the match graph being shown in Fig. 15.2*b*. In this case, there are 16 possible feature assignments, 12 valid compatibilities, and 7 maximal cliques. If occlusion of a feature occurs, this will (taken on its own) reduce the number of possible feature assignments and also the number of valid compatibilities. In addition, the number of maximal cliques and the size of the largest maximal clique will be reduced. On the other hand, noise or clutter can add erroneous features. If the latter are at arbitrary distances from existing features, then the number of possible feature assignments will be increased, but there will not be any more compatibilities in the match graph. The match graph will therefore have only trivial additional complexity. However, if the extra features appear at *allowed* distances from existing features, this will introduce extra compatibilities into the match graph and make it more tedious to analyze. In the case shown in Fig. 15.3, both types of complications—an occlusion and an additional

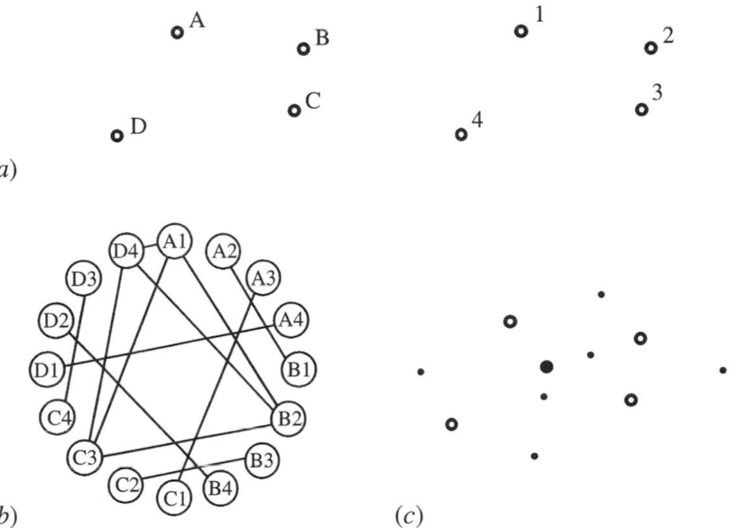

Figure 15.2 Another matching problem—a general quadrilateral: (*a*) basic labeling of model (*left*) and image (*right*); (*b*) match graph; (*c*) placement of votes in parameter space (notation as in Fig. 15.1).

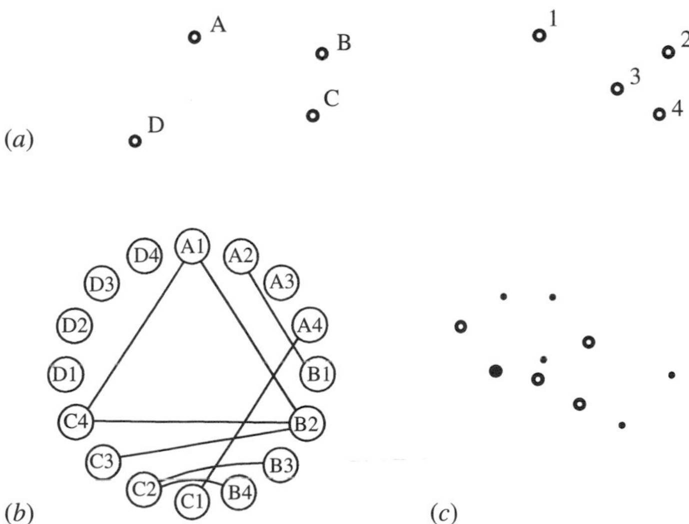

Figure 15.3 Matching when one feature is occluded and another is added: (*a*) basic labeling of model (*left*) and image (*right*); (*b*) match graph; (*c*) placement of votes in parameter space (notation as in Fig. 15.1).

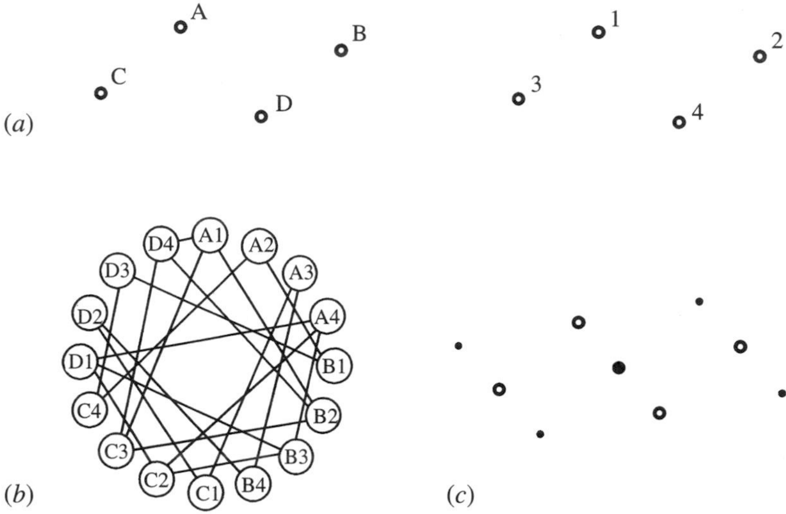

Figure 15.4 Matching a figure possessing some symmetry: (*a*) basic labeling of model (*left*) and image (*right*); (*b*) match graph; (*c*) placement of votes in parameter space (notation as in Fig. 15.1).

feature—arise. There are now eight pairwise assignments and six maximal cliques, rather fewer overall than in the original case of Fig. 15.2. However, the important factors are that the largest maximal clique still indicates the most likely interpretation of the image and that the technique is inherently highly robust.

When using methods such as the maximal clique approach, which involves repetitive operations, it is useful to look for a means of saving computation. When the objects being sought possess some symmetry, economies can be made. Consider the case of a parallelogram (Fig. 15.4). Here the match graph has 20 valid compatibilities, and there are 10 maximal cliques. Of these cliques, the largest two have equal numbers of nodes, and *both* identify the parallelogram within a symmetry operation. This means that the maximal clique approach is doing more computation than absolutely necessary. This can be avoided by producing a new "symmetry-reduced" match graph after relabeling the model template in accordance with the symmetry operations (see Fig. 15.5). This gives a much smaller match graph with half the number of pairwise compatibilities and half the number of maximal cliques. In particular, there is only one nontrivial maximal clique. Note, however, that the application of symmetry does not reduce its size.

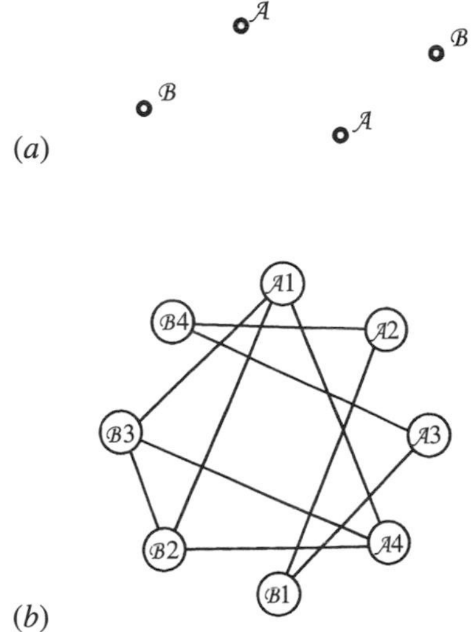

Figure 15.5 Using a symmetry-reduced match graph: (*a*) relabeled model template; (*b*) symmetry-reduced match graph.

15.2.1 *A Practical Example—Locating Cream Biscuits*

Figure 15.6*a* shows one of a pair of cream biscuits that are to be located from their "docker" holes. This strategy is advantageous because it has the potential for highly accurate product location prior to detailed inspection. (In this case, the purpose is to locate the biscuits accurately from the holes, and then to check the alignment of the biscuit wafers and detect any excess cream around the sides of the product.) The holes found by a simple template matching routine are indicated in Fig. 15.6*b*. The template used is rather small, and, as a result, the routine is fairly fast but fails to locate all holes. In addition, it can give false alarms. Hence, an "intelligent" algorithm must be used to analyze the hole location data.

This is a highly symmetrical type of object, and so it should be beneficial to employ the symmetry-reduced match graph described above. To proceed, it is helpful to tabulate the distances between all pairs of holes in the object model (Fig. 15.7*b*). Then this table can be regrouped to take account of symmetry operations (Fig. 15.7*d*). This will help when we come to build the match graph for a particular image. Analysis of the data in the above example shows that there are two nontrivial maximal cliques, each corresponding correctly to one of the two biscuits in the image. Note, however, that the reduced match graph does not give a *complete* interpretation of the image. It locates the two objects, but it does not confirm uniquely which hole is which.

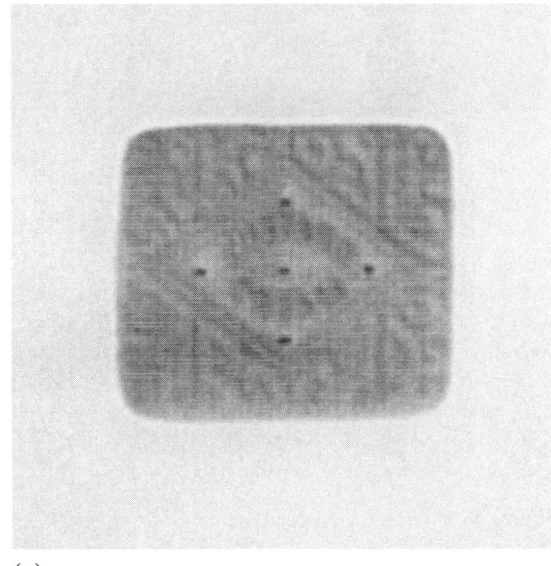

(*a*)

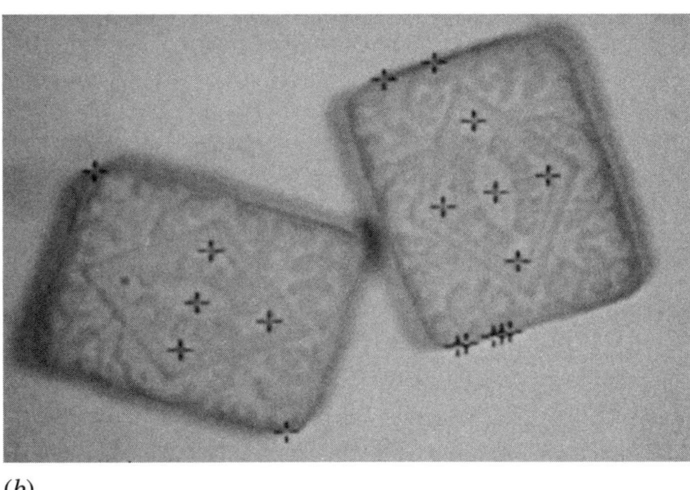

(*b*)

Figure 15.6 (*a*) A typical cream sandwich biscuit; (*b*) a pair of cream sandwich biscuits with crosses indicating the result of applying a simple hole detection routine; (*c*) the two biscuits reliably located by the GHT from the hole data in (*b*). The isolated small crosses indicate the positions of single votes.

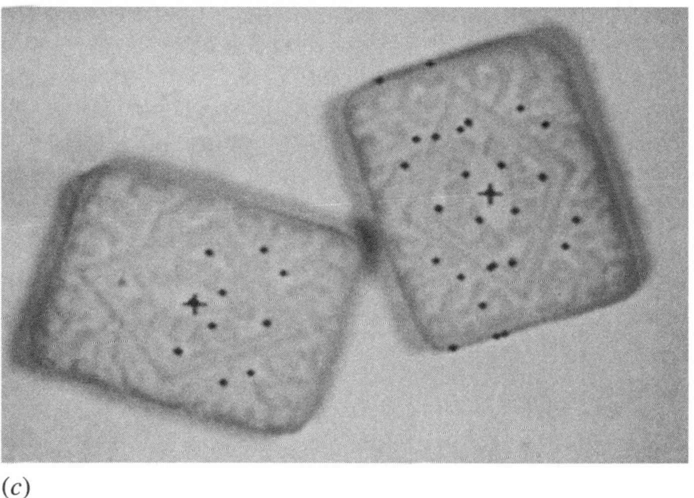

(*c*)

Figure 15.6 Continued.

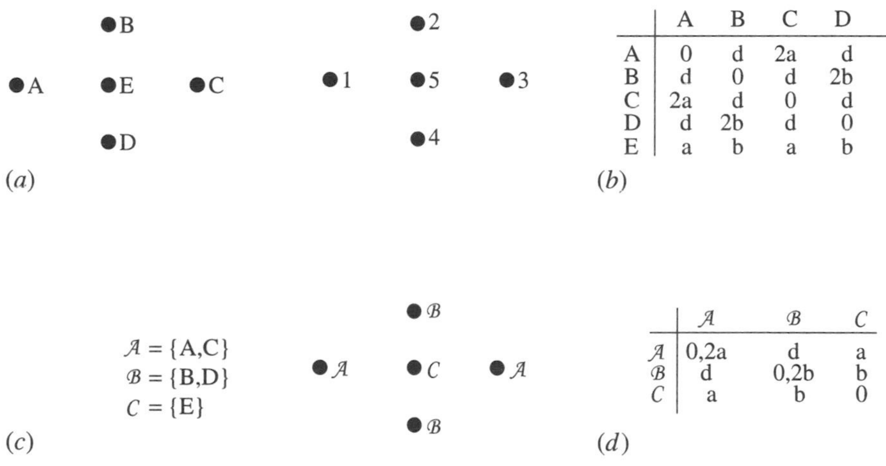

	A	B	C	D	E
A	0	d	2a	d	a
B	d	0	d	2b	b
C	2a	d	0	d	a
D	d	2b	d	0	b
E	a	b	a	b	0

	𝒜	ℬ	𝒞
𝒜	0,2a	d	a
ℬ	d	0,2b	b
𝒞	a	b	0

Figure 15.7 Interfeature distances for holes on cream biscuits: (*a*) basic labeling of model (*left*) and image (*right*); (*b*) allowed distance values; (*c*) revised labeling of model using symmetrical set notation; (*d*) allowed distance values. The cases of zero interfeature distance in the final table can be ignored as they do not lead to useful matches.

In particular, for a given starting hole of type 𝒜, it is not known which is which of the two holes of type ℬ. This can be ascertained by applying simple geometry to the coordinates in order to determine, say, which hole of type ℬ is reached by moving around the center hole E in a clockwise sense.

15.3 Possibilities for Saving Computation

In these examples, it is simple to check which subgraphs are maximal cliques. However, in real matching tasks it can quickly become unmanageable. (The reader is encouraged to draw the match graph for an image containing two objects of seven points!)

Figure 15.8 shows what is perhaps the most obvious type of algorithm for finding maximal cliques. It operates by examining in turn all cliques of a given number of nodes and finding what cliques can be constructed from them by adding additional nodes (bearing in mind that any additional nodes must be compatible with all existing nodes in the clique). This permits all cliques in the match graph to be identified. However, an additional step is needed to eliminate (or relabel) all cliques that are included as subgraphs of a new larger clique before it is known which cliques are maximal.

In view of the evident importance of finding maximal cliques, many algorithms have been devised for the purpose. The best of these is probably now close to the fastest possible speed of operation. Unfortunately, the optimum execution time is known to be bounded not by a polynomial in M (for a match graph containing maximal cliques of up to M nodes) but by a much faster

```
set clique size to 2;
// this is the size already included by the match graph
while (newcliques = true) { // new cliques still being found
        increment clique size;
        set newcliques = false;
        for all cliques of previous size {
            set all cliques of previous size to status maxclique;
            for all possible extra nodes
                  if extra node is joined to all existing nodes in clique {
                     store as a clique of current size;
                     set newcliques = true;
            }
        }
        // the larger cliques have now been found
        for all cliques of current size
            for all cliques of previous size
                 if all nodes of smaller clique are included in current clique
                    set smaller clique to status not maxclique;
        // the subcliques have now been relabeled
}
```

Figure 15.8 A simple maximal clique algorithm.

varying function. Specifically, the task of finding maximal cliques is akin to the well-known traveling salesperson problem and is known to be NP-complete, implying that it runs in exponential time (see Section 15.4.1). Thus, whatever the run-time is for values of M up to about 6, it will typically be 100 times slower for values of M up to about 12 and 100 times slower again for M greater than ~15. In practical situations, this problem can be tackled in several ways:

1. Use the symmetry-reduced match graph wherever possible.
2. Choose the fastest available maximal clique algorithm.
3. Write critical loops of the maximal clique algorithm in machine code.
4. Build special hardware or multiprocessor systems to implement the algorithm.
5. Use the LFF method (see below: this means searching for cliques of small M and then working with an alternative method).
6. Use an alternative sequential strategy, which may, however, not be guaranteed to find all the objects in the image.
7. Use the GHT approach (see Section 15.4).

Of these methods, the first should be used wherever applicable. Methods (2)–(4) amount to improving the implementation and are subject to diminishing returns. It should be noted that the execution time varies so rapidly with M that even the best software implementations are unlikely to give a practical increase in M of more than 2 (i.e., $M \to M + 2$). Similarly, dedicated hardware implementations may only give increases in M of the order of 4 to 6. Method (5) is a "shortcut" approach that proves highly effective in practice. The idea is to search for specific subsets of the features of an object and then to hypothesize that the object exists and go back to the original image to check that it is actually present. Bolles and Cain (1982) devised this method when looking for hinges in quite complex images. In principle, the method has the disadvantage that the particular subset of an object that is chosen as a cue may be missing because of occlusion or some other artifact. Hence, it may be necessary to look for several such cues on each object. This is an example of further deviation from the matched filter paradigm, which reduces detection sensitivity yet again. The method is called the local-feature–focus (LFF) method because objects are sought by cues or local foci.

The maximal clique approach is a type of exhaustive search procedure and is effectively a parallel algorithm. This has the effect of making it highly robust but is also the source of its slow speed. An alternative is to perform some sort of sequential search for objects, stopping when sufficient confidence is attained in the interpretation or partial interpretation of the image. For example, the search process may be terminated when a match has been obtained for a certain minimum number of features on a given number of objects. Such an approach may be useful in some applications and will generally be considerably faster than the full maximal clique procedure when M is greater than about 6. Ullmann (1976) carried out an analysis of several tree-search algorithms for subgraph isomorphism in which he tested algorithms using artificially generated data; it is not clear how they relate to real images. The success or otherwise of all nonexhaustive search algorithms must, however, depend critically on the particular types of image data being analyzed. Hence, it is difficult to give further general guidance on this matter (but see Section 15.7 for additional comments on search procedures).

The final method listed above is based on the GHT. In many ways, this provides an ideal solution to the problem because it presents an exhaustive search technique that is essentially equivalent to the maximal clique approach, while not falling into the NP-complete category. This may seem contradictory, since any approach to a well-defined mathematical problem should be subject to the mathematical constraints known to be involved in its solution. However, although the abstract maximal clique problem is known to be NP-complete, the *subset* of maximal clique problems that arises from 2-D image-based data may well be solved with less computation by other means, and in particular by a 2-D technique. This special circumstance does appear to be valid, but unfortunately it offers no possibility of solving general NP-complete problems by reference to the specific solutions found using the GHT approach! The GHT approach is described in the following section.

15.4 Using the Generalized Hough Transform for Feature Collation

The GHT can be used as an alternative to the maximal clique approach, to collate information from point features in order to find objects. Initially, we consider situations where objects have no symmetries—as for the cases of Figs. 15.1–15.3.

To apply the GHT, we first list all features and then accumulate votes in parameter space at every possible position of a localization point L consistent with each *pair* of features (Fig. 15.9). This strategy is particularly suitable in the present context, for it corresponds to the pairwise assignments used in the maximal clique method. To proceed it is necessary merely to use the interfeature distance as a lookup parameter in the GHT R-table. For indistinguishable point features this means that there must be two entries for the position of L for each value of the interfeature distance. Note that we have assumed that no symmetries exist and that all pairs of features have different interfeature distances. If this were not so, then more than two vectors would have to be stored in the R-table per interfeature distance value.

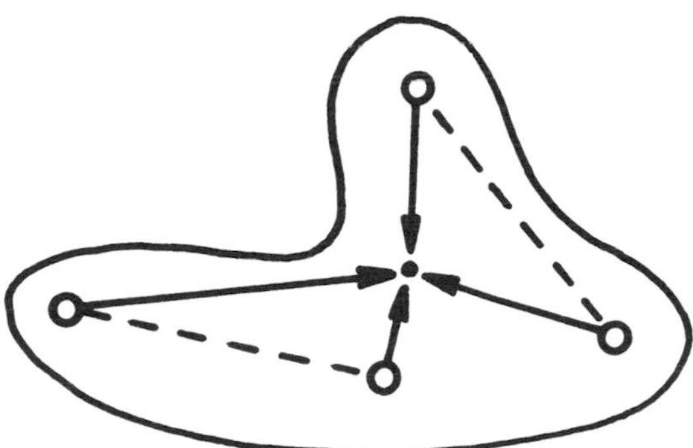

Figure 15.9 Method for locating L from pairs of feature positions. Each pair of feature points gives two possible voting positions in parameter space, when objects have no symmetries. When symmetries are present, certain pairs of features may give rise to up to four voting positions. This is confirmed on careful examination of Fig. 15.6*c*.

To illustrate the procedure, it is applied first to the triangle example of Fig. 15.1. Figure 15.1*c* shows the positions at which votes are accumulated in parameter space. There are four peaks with heights of 3, 1, 1, 1; it is clear that, in the absence of complicating occlusions and defects, the object is locatable at the peak of maximum size. Next, the method is applied to the general quadrilateral example of Fig. 15.2: this leads to seven peaks in parameter space, whose sizes are 6, 1, 1, 1, 1, 1, 1 (Fig. 15.2*c*).

Close examination of Figs. 15.1–15.3 indicates that every peak in parameter space corresponds to a maximal clique in the match graph. Indeed, there is a one-to-one relation between the two. In the uncomplicated situation being examined here, this is bound to be so for any general arrangement of features within an object, since every pairwise compatibility between features corresponds to two potential object locations, one correct and one that can be correct only from the point of view of that pair of features. Hence, the correct locations all add to give a large maximal clique and a large peak in parameter space, whereas the incorrect ones give maximal cliques, each containing two wrong assignments and each corresponding to a false peak of size 1 in parameter space. This situation still applies even when occlusions occur or additional features are present (see Fig. 15.3). The situation is slightly more complicated when symmetries are present, with each of the two methods deviating in a different way. Space does not permit the matter to be explored in depth here, but the solution for the case of Fig. 15.4*a* is presented in Fig. 15.4*c*. Overall, it seems simplest to assume that there is still a one-to-one relationship between the solutions from the two approaches.

Finally, consider again the example of Section 15.2.1 (Fig. 15.6*a*), this time obtaining a solution by the GHT. Figure 15.6*c* shows the positions of candidate object centers as found by the GHT. The small isolated crosses indicate the positions of single votes, and those very close to the two large crosses lead to voting peaks of weights 10 and 6 at these respective positions. Hence, object location is both accurate and robust, as required.

15.4.1 ***Computational Load***

This subsection compares the computational requirements of the maximal clique and GHT approaches to object location. For simplicity, imagine an image that contains just one wholly visible example of the object being sought. Also, suppose that the object possesses n features and that we are trying to recognize it by seeking all possible pairwise compatibilities, whatever their distance apart (as for all examples in Section 15.2).

For an object possessing n features, the match graph contains n^2 nodes (i.e., possible assignments), and there are $\binom{n^2}{2} = n^2(n^2 - 1)/2$ possible pairwise compatibilities to be checked in building the graph. The amount of computation at this stage of the analysis is $O(n^4)$. To this must be added the cost of finding the maximal cliques. Since the problem is NP-complete, the load rises at a rate that is faster than polynomial and probably exponential in n^2 (Gibbons, 1985).

Now consider the cost of getting the GHT to find objects via pairwise compatibilities. As has been seen, the total height of all the peaks in parameter space is in general equal to the number of pairwise compatibilities in the match graph. Thus, the computational load is of the same order, $O(n^4)$. Next comes the problem of locating all the peaks in parameter space. In this case, parameter space is congruent with image space. Hence, for an $N \times N$ image only N^2 points have to be visited in parameter space, and the computational load is $O(N^2)$. Note, however, that an alternative strategy is available in which a running record is kept of the relatively small numbers of voting positions in parameter space. The computational load for this strategy will be $O(n^4)$: although of a higher *order*, this often represents less computation in practice.

The reader may have noticed that the basic GHT scheme as outlined so far is able to locate objects from their features but does not determine their orientations. However, orientations can be computed by running the algorithm a second time and finding all the assignments that contribute to each peak. Alternatively, the second pass can aim to find a different localization point within each object. In either case, the overall task should be completed in little over twice the time, that is, still in $O(n^4 + N^2)$ time.

Although the GHT at first appears to solve the maximal clique problem in polynomial time, what it actually achieves is to solve a real-space template matching problem in polynomial time. It does not solve an *abstract* graph-theoretic problem in polynomial time. The overall lesson is that the graph theory representation is not well matched to real space, not that real space can be used to solve abstract NP-complete problems in polynomial time.

15.5 Generalizing the Maximal Clique and Other Approaches

This section considers how the graph matching concept can be generalized to cover alternative types of features and various attributes of features. We restricted our earlier discussion to point features and in particular to small holes that were supposed to be isotropic. Corners were also taken as point features by ignoring attributes other than position coordinates. Both holes and corners seem to be ideal in that they give maximum localization and hence maximum accuracy for object location. Straight lines and straight edges at first appear to be rather less well suited to the task. However, more careful thought shows that this is not so. One possibility is to use straight lines to deduce the positions of corners that can then be used as point features, although this approach is not as powerful as might be hoped because of the abundance of irrelevant line crossings that are thrown up. (In this context, note that the maximal clique method is inherently capable of sorting the true corners from the false alarms.) A more elegant solution is simply to determine the angles between pairs of lines, referring to a lookup table for each type of object to determine whether each pair of lines should be marked as compatible in the match graph. Once the match graph has been built, an optimal match can be found as before (although ambiguities of scale will arise which will have to be resolved by further processing). These possibilities significantly generalize the ideas of the foregoing sections.

Other types of features generally have more than two specifying parameters, one of which may be contrast and the other size. This applies for most holes and circular objects, although for the smallest (i.e., barely resolvable) holes it is sometimes most practicable to take the central dip in intensity as the measured parameter. For straight lines the relevant size parameter is the length. (We here count the line ends, if visible, as points that will already have been taken into account.) Corners may have a number of attributes, including contrast, color, sharpness, and orientation—none of which is likely to be known with high accuracy. Finally, more complex shapes such as ellipses have orientation, size, and eccentricity, and again contrast or color may be a usable attribute. (Generally, contrast is an unreliable measure because of possible variations in the background.)

So much information is available that we need to consider how best to use it for locating objects. For convenience this is discussed in relation to the maximal clique method. Actually, the answer is very simple. When compatibilities are being considered and the arcs are being drawn in the match graph, *any* available information may be taken into account in deciding whether a pair of features in the image matches a pair of features in the object model. In Section 15.2 the discussion was simplified by taking interfeature distances as the only relevant measurements. However, it is quite acceptable to describe the features in the object model more fully and to insist that they all match within prespecified tolerances. For example, holes and corners may be permitted to lead to a match only if the holes are of the correct size, the corners are of the correct sharpness and orientation, and the distances between these features are also appropriate. All relevant information has to be held in suitable lookup tables. In general, the gains easily outweigh the losses, since a considerable number of potential interpretations will be eliminated—thereby making the match graph significantly simpler and reducing, in many cases by a large factor, the amount of computation that is required to find the maximal cliques. Note, however, that there is a limit to this, since in the absence of occlusions and erroneous tolerances on the additional attributes, the number of nodes in the match graph cannot be less than the square of the number of features in an object, and the number of nodes in a maximal clique will be unchanged.

Thus, extra feature attributes are of great value in reducing computation; they are also useful in making interpretation less ambiguous, a property that is "obvious" but not always realizable. In particular, extra attributes help in this way only if (1) some of the features on an object are missing, through occlusion or for other reasons such as breakage, or (2) the distance tolerances are so large as to make it unclear which features in the image match those in the model.

Suppose next that the distance attributes become very imprecise, either because of shape distortions or because of unforeseen rotations in 3-D. How far can we proceed under these circumstances? In the limit of low-distance accuracy, we may only be able to employ an "adjacent to" descriptor. This parallels the situation in general scene analysis, where use is made of a number of relational descriptors such as "on top of," "to the left of," and so on. Such possibilities are considered in the next section.

15.6 Relational Descriptors

The previous section showed how additional attributes could be incorporated into the maximal clique formalism so that the effects of diminishing accuracy of distance measurement could be accommodated. This section considers what happens when the accuracy of distance measurement drops to zero and we are left just with relational attributes such as "adjacent to," "near," "inside," "underneath," "on top of," and so on. We start by taking *adjacency* as the basic relational attribute. To illustrate the approach, imagine a simple outdoor scene where various rules apply to segmented regions. These rules will be of the type "sky may be adjacent to forest," "forest may be adjacent to field," and so on. Note that "adjacent to" is not transitive; that is, if P is adjacent to Q, and Q is adjacent to R, this does not imply that P is adjacent to R. (In fact, it is quite likely that P will not be adjacent to R, for Q may well separate the two regions completely!)

The rules for a particular type of scene may be summarized as in Table 15.1. Now consider the scene shown in Fig. 15.10. Applying the rules for

Table 15.1 Adjacency table for simple scene

	Sky	Forest	Field
Sky	—	✓	—
Forest	✓	—	✓
Field	—	✓	—

This table is relevant to the scene in Fig. 15.9. Ticks indicate that regions must be adjacent: dashes indicate that regions must not be adjacent (i.e., adjacency is not optional).

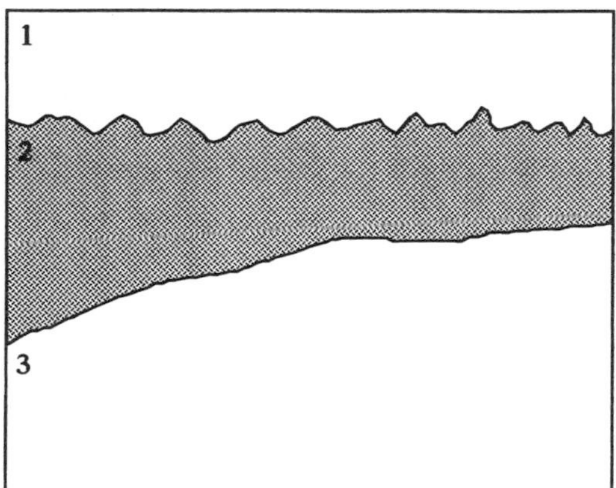

Figure 15.10 Regions in a simple scene: 1, sky; 2, forest; 3, field.

adjacency from Table 15.1 will be seen to permit four different solutions, in which regions 1 to 3 may be interpreted as:

1. Sky, forest, field (the correct interpretation)
2. Sky, forest, sky
3. Field, forest, sky
4. Field, forest, field

Evidently, there are too few constraints. Possible constraints are the following: *sky is above field and forest, sky is blue, field is green*, and so on. Of these, the first is a binary relation like adjacency, whereas the other two are unary constraints. It is easy to see that two such constraints are required to resolve completely the ambiguity in this particular example.

Paradoxically, adding further regions can make the situation inherently less ambiguous. This is because other regions are less likely to be adjacent to all the original regions and in addition may act in such a way as to label them uniquely. This is seen in Fig. 15.11, which is interpreted by reference to Table 15.2, although the keys to unique interpretation are the notions that *any small white objects must be sheep*, and *any small dark objects must be birds*. On analyzing the available data, the whole scene can now be interpreted unambiguously. As expected, the maximal clique technique can be applied successfully, although to achieve this, assignments must be taken to be compatible not only when image regions are adjacent and their interpretations are marked as adjacent, but also when image regions are nonadjacent and their interpretations are marked as nonadjacent. (The reason for this is easily seen by drawing

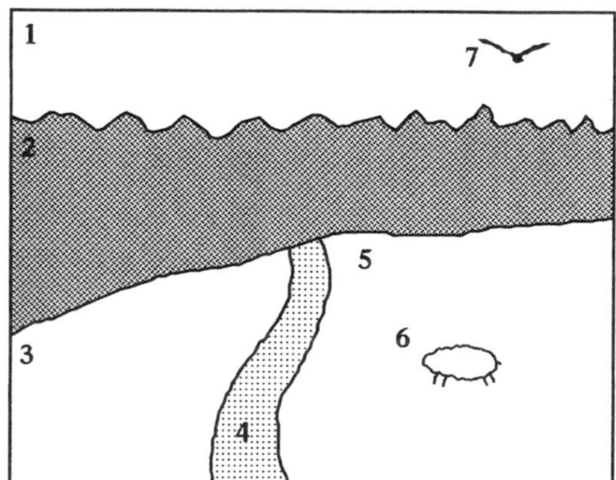

Figure 15.11 Regions in a more complex scene: 1, sky; 2, forest; 3, field; 4, road; 5, field; 6, sheep; 7, bird.

Table 15.2 Adjacency table for more complex scene

	Sky	Forest	Field	Road	Sheep	Bird
Sky	—	✓	—	—	—	✓
Forest	✓	—	✓	✓	—	—
Field	—	✓	—	✓	✓	—
Road	—	✓	✓	—	—	—
Sheep	—	—	✓	—	—	—
Bird	✓	—	—	—	—	—

This table is relevant to the scene in Fig. 15.11. Ticks indicate adjacency and dashes indicate nonadjacency.

an analogy between adjacency and distance, adjacent and nonadjacent corresponding, respectively, to zero and a significant distance apart.) Note that in this scene analysis type of situation, it is *very* tedious to apply the maximal clique method manually because there tend to be a fair number of trivial solutions (i.e., maximal cliques with just a few nodes).

Perhaps oddly, some of the trivial solutions that the maximal clique technique gives rise to are provably *incorrect*. For example, in the above problem, one solution appears to include regions 1 and 2 being forest and sky, respectively, whereas it is clear from the presence of birds that region 1 *cannot* be forest and *must* be sky. Here logic appears to be a stronger arbiter of correctness than an evidence-building scheme such as the maximal clique technique. On the plus side, however, the maximal clique technique will take contradictory evidence and produce the best possible response. For example, if noise appears and introduces an invalid "bird" in the forest, it will simply ignore it, whereas a purely logic-based scheme will be unable to find a satisfactory solution. Clearly, much depends on the accuracy and universality of the assertions on which these schemes are based.

Logical problems such as the ones outlined here are commonly tackled with the aid of declarative languages such as PROLOG, in which the rules are written explicitly in a standard form not dissimilar to IF statements in English. Space does not permit a full discussion of PROLOG here. However, it should be noted that PROLOG is basically designed to obtain a single "logical" solution where one exists. It can also be instructed to search for all available solutions or for any given number of solutions. When instructed to search for all solutions, it will finally arrive at the same main set of solutions as the maximal clique approach (although, as indicated above, noise can make the situation more complicated). Thus, despite their apparent differences, these approaches are quite similar. It is the implementation of the underlying search problem that is different. (PROLOG uses a "depth-first" search strategy—see below.) Next, we move on to another scheme—that of relaxation labeling—which has a quite different formulation.

Relaxation labeling is an iterative technique in which evidence is gradually built up for the proper labeling of the solution space—in this case, of the various regions in the scene. Two main possibilities exist. In the first, called *discrete relaxation*, evidence for pairs of labels is examined and a label discarding rule is instigated that eliminates at each iteration those pairs of labels that are currently inconsistent. This technique is applied until no further change in the labeling occurs. The second possibility is that of *probabilistic relaxation*. Here the labels are set up numerically to correspond to the probabilities of a given interpretation of each region; that is, a table is compiled of regions against possible interpretations, each entry being a number representing the probability of that particular interpretation. After providing a suitable set of starting probabilities (possibly all being weighted equally), these are updated iteratively, and in the ideal case they converge on the values 0 or 1 to give a unique interpretation of the scene. Unfortunately, convergence is by no means guaranteed and can depend on the starting probabilities and the updating rule. A prearranged constraint function defines the underlying process, and this could in principle be either better or worse at matching reality than (for example) the logic programming techniques that are used in PROLOG. Relaxation labeling is a complex optimization process and is not discussed further here. Suffice it to say that the technique often runs into problems of excessive computation. The reader is referred to the seminal paper by Rosenfeld et al. (1976) and other papers mentioned in Section 15.9 for further details.

15.7 **Search**

We have shown how the maximal clique approach may be used to locate objects in an image, or alternatively to label scenes according to predefined rules about what arrangements of regions are expected in scenes. In either case, the basic process being performed is that of a search for solutions that are compatible with the observed data. This search takes place in assignment space—that is, a space in which all combinations of assignments of observed features with possible interpretations exist. The problem is that of finding one or more valid sets of observed assignments.

Generally, the search space is very large, so that an exhaustive search for all solutions would involve enormous computational effort and would take considerable time. Unfortunately, one of the most obvious and appealing methods of obtaining solutions, the maximal clique approach, is NP-complete and can require impracticably large amounts of time to find all the solutions. It is therefore useful to clarify the nature of the maximal clique approach. To achieve this clarification, we first describe the two main categories of search—breadth-first and depth-first search.

Breadth-first search is a form of search that systematically works down a tree of possibilities, never taking shortcuts to nearby solutions. Depth-first search, in contrast, involves taking as direct a path as possible to individual solutions, stopping the process when a solution is found and backtracking up the tree whenever a wrong decision is found to have been made. It is normal to curtail the depth-first search when sufficient solutions have been found, and this means that much of the tree of possibilities will not have been explored. Although breadth-first search can be curtailed similarly when enough solutions have been found, the maximal clique approach as described earlier is in fact a form of breadth-first search that is exhaustive and runs to completion.

In addition to being an exhaustive breadth-first search, the maximal clique approach may be described as being "blind" and "flat"; that is, it involves neither heuristic nor hierarchical means of guiding the search. Faster search methods involve guiding the search in various ways. First, heuristics are used to specify at various stages in which direction to proceed (which node of the tree to expand) or which paths to ignore (which nodes to prune). Second, the search can be made more "hierarchical," so that it searches first for outline features of a solution, returning later (perhaps in several stages) to fill in the details. Details of these techniques are omitted here. However, Rummel and Beutel (1984) used an interesting approach: they searched images for industrial components using features such as corners and holes, alternating at various stages between breadth-first and depth-first search by using a heuristic based on a dynamically adjusted parameter. This parameter was computed on the basis of how far the search was still away from its goal, and the quality of the fit when a "guide factor" based on the number of features required for recognition was adjusted. Rummel and Beutel noted the existence of a tradeoff between speed and accuracy. The problem was that trying to increase speed introduces some risk of not finding the optimum solution.

15.8 Concluding Remarks

In this chapter the maximal clique approach was seen to be capable of finding solutions to the task of recognizing objects from their features and the related task of scene analysis. Although ultimately these are search problems, a much greater range of methods is applicable to each. In particular, blind, flat, exhaustive breadth-first search (i.e., the maximal clique method) involves considerable computation and is often best replaced by guided depth-first search, with suitable heuristics being devised to guide the search. In addition, languages such as PROLOG can implement depth-first search and rather different procedural techniques such as relaxation labeling are available and should be considered (although these are subject to their own complexities and computational problems).

The task of recognizing objects from their features tends to involve considerable computation, and the GHT can provide a satisfactory solution to these problems. This is probably because the graph theory representation is not well matched to the relevant real-space template matching task in the way that the GHT is. Here, recall what was noted in Chapter 11—that the GHT is particularly well suited to object detection in real space as it is one type of spatial matched filter. Indeed, the maximal clique approach can be regarded as a rather inefficient substitute for the GHT form of the spatial matched filter. Furthermore, the LFF method takes a shortcut to save computation, which makes it less like a spatial matched filter, thereby adversely affecting its ability to detect objects among noise and clutter. Note that for scene analysis, where relational descriptors rather than precise dimensional (binary) attributes are involved, the GHT does not provide any very obvious possibilities—probably because we are dealing here with much more abstract data that are linked to real space only at a very high level (but see the work of Kasif et al., 1983).

Finally, note that, on an absolute scale, the graph matching approach takes very little note of detailed image structure, using at most only pairwise feature attributes. This is adequate for 2-D image interpretation but inadequate for situations such as 3-D image analysis where there are more degrees of freedom to contend with (normally three degrees of freedom for position and three for orientation, for each object in the scene). Hence, more specialized and complex approaches need to be taken in such cases; these approaches are examined in the following chapter.

Searching for objects via their features is far more efficient than template matching. This chapter has shown that this raises the need to *infer* the presence of the objects—which can still be computation intensive. Graph matching and generalized Hough transform approaches are robust, though each can lead to ambiguities, so tests of potential solutions need to be made.

15.9 Bibliographical and Historical Notes

Graph matching and clique finding algorithms started to appear in the literature around 1970; for an early solution to the graph isomorphism problem, see Corneil and Gottlieb (1970). The subgraph isomorphism problem was tackled soon after by Barrow et al. (1972; see also Ullmann, 1976). The double subgraph isomorphism (or subgraph–subgraph isomorphism) problem is commonly tackled by seeking maximal cliques in the match graph, and algorithms for achieving this have been described by Bron and Kerbosch (1973), Osteen and Tou (1973), and Ambler et al. (1975). (Note that in 1989 Kehtarnavaz and Mohan reported preferring the algorithm of Osteen and Tou on the grounds of speed.) Improved speed has also been achieved using the minimal match graph concept (Davies, 1991a).

Bolles (1979) applied the maximal clique technique to real-world problems (notably the location of engine covers) and showed how operation could be made more robust by taking additional features into account. By 1982, Bolles and Cain had formulated the local–feature–focus method, which (1) searches for restricted sets of features on an object, (2) takes symmetry into account to save computation, and (3) reconsiders the original image data in order to confirm a valid match. Bolles and Cain give various criteria for ensuring satisfactory solutions with this type of method.

Not satisfied by the speed of operation of maximal clique methods, other workers have tended to use depth-first search techniques. Rummel and Beutel (1984) developed the idea of alternating between depth-first and breadth-first searches as dictated by the data—a powerful approach, although the heuristics they used may well lack generality. Meanwhile, Kasif et al. (1983) showed how a modified GHT (the "relational HT") could be used for graph matching, although their paper gives few practical details. A somewhat different application of the GHT to perform 2-D matching (Davies, 1988g) was described in Section 15.4 and has been extended to optimize accuracy (Davies, 1992d) as well as to improve speed (Davies, 1993b). Geometric hashing has been developed to perform similar tasks on objects with complex polygonal shapes (Tsai, 1996).

Relaxation labeling in scenes dates from the seminal paper by Rosenfeld et al. (1976). For later work on relaxation labeling and its use for matching, see Kitchen and Rosenfeld (1979), Hummel and Zucker (1983), and Henderson (1984); for rule-based methods in image understanding, see, for example, Hwang et al. (1986); and for preparatory discussion and careful contrasting of these approaches, see Ballard and Brown (1982). Discussion of basic artificial intelligence search techniques and rule-based systems may be found in Charniak and McDermott (1985). A useful reference on PROLOG is Clocksin and Mellish (1984).

Over the past three decades, inexact matching algorithms have acquired increasing predominance over exact matching methods because of the ubiquitous presence of noise, distortions, and missing or added feature points, together with inaccuracies and thus mismatches of feature attributes. One class of work on inexact (or "error-tolerant") matching considers how structural representations should be compared (Shapiro and Haralick, 1985). This early work on similarity measures shows how the concept of string edit distance can be applied to graphical structures (Sanfeliu and Fu, 1983). The formal concept of edit distance was later extended by Bunke and Shearer (1998) and Bunke (1999), who considered and rationalized the cost functions for methods such as graph isomorphisms, subgraph isomorphisms, and maximum common subgraph isomorphisms. Choice of cost functions was shown to be of crucial importance to success in each particular data set, though detailed analysis demonstrated important subtleties in the situation (Bunke, 1999). Another class of work on inexact matching involves the development of principled statistical measures of similarity. In this category, the work of Christmas et al. (1995) on probabilistic relaxation and that of Wilson and Hancock (1997) on discrete relaxation, have been much cited.

Yet another class of work is that on optimization. This has included work on simulated annealing (Herault et al., 1990), genetic search (Cross et al., 1997), and neural processing (Pelillo, 1999). The work of Umeyama (1988) develops the least-squares approach using a matrix eigen decomposition method to recover the permutation matrix relating the two graphs being matched. One of the most recent developments has been the use of spectral graph theory[3] to recover the permutation structure. The Umeyama (1988) approach only matches graphs of the same size. Other related methods have emerged (e.g., Horaud and Sossa, 1995), but they have all suffered from an inability to cope with graphs of different sizes. However, Luo and Hancock (2001) have demonstrated how this particular problem can be overcome—by showing how the graph matching task can be posed as maximum likelihood estimation using the EM algorithm formalism. Hence, singular value decomposition is used efficiently to solve correspondence problems. Ultimately, the method is important because it helps to move graph matching away from a discrete process in which a combinatorial search problem exists toward a continuous optimization problem that moves systematically toward the optimum solution. It ought to be added that the method works under considerable levels of structural corruption—such as when 50% of the initial entries in the data-graph adjacency matrix are in error (Luo and Hancock, 2001). In a later development, Robles-Kelly and Hancock (2002) managed to achieve the same end, and to achieve even better performance within the spectral graph formalism itself.

Meanwhile, other developments included a fast, phased approach to inexact graph matching (Hlaoui and Wang, 2002), and an RKHS (reproducing kernel Hilbert space) interpolator-based graph matching algorithm capable of efficiently matching huge graphs of more than 500 vertices (e.g., those extracted from aerial scenes) on a PC (van Wyk et al., 2002). For a more detailed appraisal of inexact matching algorithms, see Lladós et al. (2001), which appears in a Special Section of IEEE Trans. PAMI on *Graph Algorithms and Computer Vision* (Dickinson et al., 2001).

---

3 This subject involves analysis of the structural properties of graphs using the eigenvalues and eigenvectors of the adjacency matrix.

15.10 Problems

1. Find the match graph for a set of features arranged in the form of an isosceles triangle. Find how much simplification occurs by taking account of symmetry and using the symmetry-reduced match graph. Extend your results to the case of a kite (two isosceles triangles arranged symmetrically base to base).

2. Two lino-cutter blades (trapeziums) are to be located from their corners. Consider images in which two corners of one blade are occluded by the other blade. Sketch the possible configurations, counting the number of corners in each case. If corners are treated like point features with no other attributes, show that the match graph will lead to an ambiguous solution. Show further that the ambiguity can in general be eliminated if proper account is taken of corner orientation. Specify how accurately corner orientation would need to be determined for this to be possible.

3. In Problem 15.2, would the situation be any better if the GHT were used?

4. (a) Metal flanges are to be located from their holes using a graph matching (maximal clique) technique. Each bar has four identical holes at distances from the narrow end of the bar of 1, 2, 3, and 5 cm, as shown in Fig. 15.P1. Draw match graphs for the four different cases in which one of the four holes of a given flange is obscured. In each case, determine whether the method is able to locate the metal flange without any error and whether any ambiguity arises.

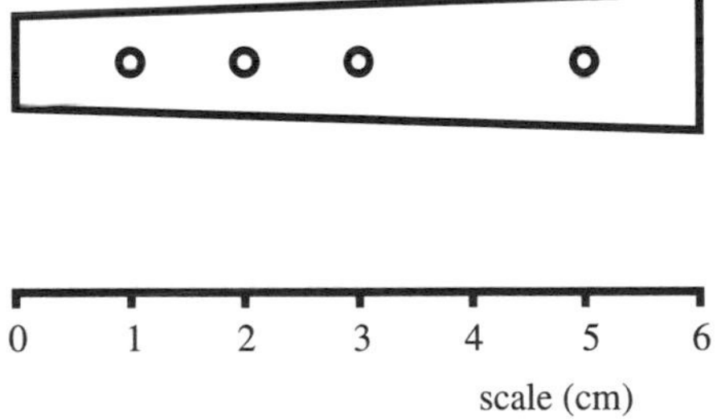

Figure 15.P1

   (b) Do your results tally with the results for human perception? How would any error or ambiguity be resolved in practical situations?

5. (a) Describe the maximal clique approach to object location. Explain why the largest maximal clique will normally represent the most likely solution to any object location task.

   (b) If symmetrical objects with four feature points are to be located, show that suitable labeling of the object template will permit the task to be simplified. Does the type of symmetry matter? What happens in the case of a rectangle?

What happens in the case of a parallelogram? (In the latter case, see points A, B, C, and D in Fig. 15.P2.)

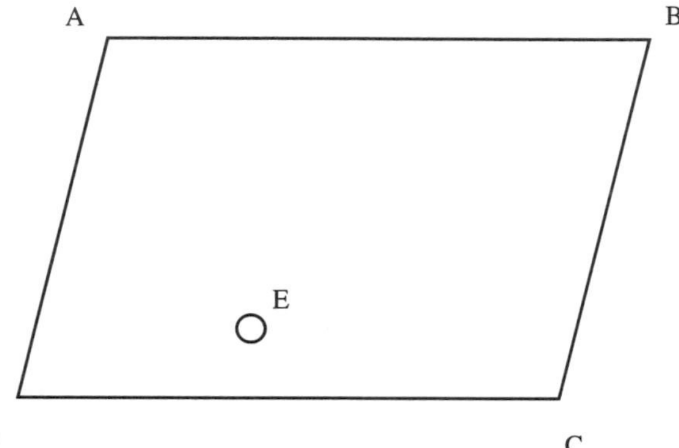

Figure 15.P2

(c) A nearly symmetrical object with *five* feature points (see Fig. 15.P2) is to be located. This is to be achieved by looking initially for the feature points A, B, C, and D and ignoring the fifth point E. Discuss how the fifth point may be brought into play to finally determine the orientation of the object, using the maximal clique approach. What disadvantage might there be in adopting this two-stage approach?

6. (a) What is template matching? Explain why objects are normally located from their features rather than using whole object templates. What features are commonly used for this purpose?

   (b) Describe templates that can be used for corner and hole detection.

   (c) An improved type of lino-cutter blade (Fig. 15.P3) is to be placed into packs of six by a robot. Show how the robot vision system could locate the blades *either* from their corners *or* from their holes by applying the maximal clique method (i.e., show that *both* schemes would work).

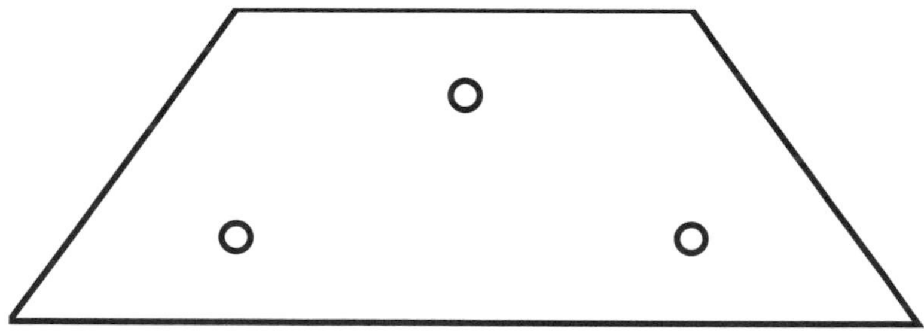

Figure 15.P3

(d) After a time, it appears that the robot is occasionally confused when the blades overlap. It is then decided to locate the blades from their holes *and* their corners. Show why this helps to eliminate any confusion. Show also how finally distinguishing the corners from the holes can help in extreme cases of overlap.

7. (a) A certain type of lino-cutter blade has four corners and two fixing holes (Fig. 15.P4). Blades of this type are to be located using the maximal clique technique. Assume that the objects lie on a worktable and that they are viewed orthogonally at a known distance.

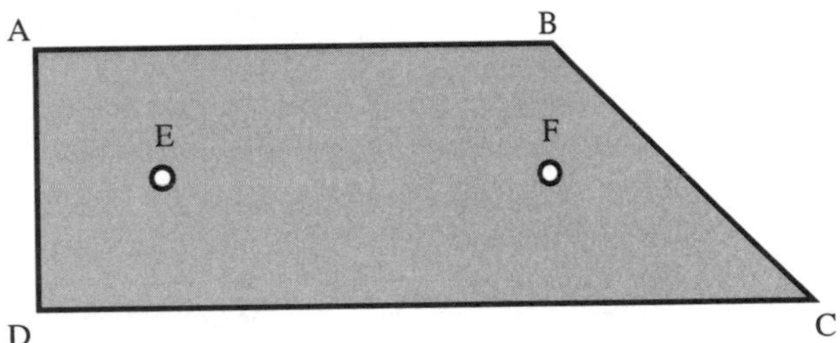

Figure 15.P4

(b) Draw match graphs for the following situations:
   (i) the objects are to be located by their holes and their corners, regarding these as *indistinguishable* point features;
   (ii) the objects are to be located solely by their corners (i.e., matching corners in the image with corners on an idealized object);
   (iii) the objects are to be located solely by their holes;
   (iv) the objects are to be located by their holes *and* their corners, but these are to be regarded as *distinguishable* features.

(c) Discuss your results with particular reference to:
   (i) the robustness that can be achieved;
   (ii) the speed of computation.

(d) In the latter case, distinguish the time taken to build the basic match graph from the time taken to find all the maximal cliques in it. State any assumptions you make about the time taken to find a maximal clique of m nodes in a match graph of n nodes.

8. (a) Decorative biscuits are to be inspected after first locating them from their holes. Show how the maximal clique graph matching technique can be applied to identify and locate the biscuits shown in Fig. 15.P5*a*, which are of the same size and shape.

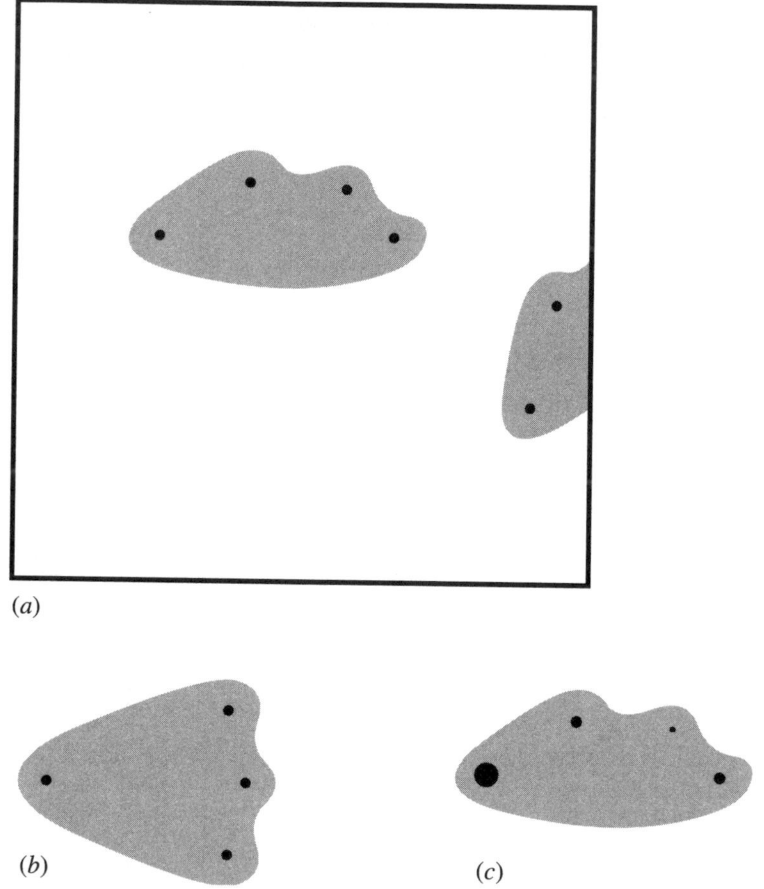

Figure 15.P5

(b) Show how the analysis will be affected for biscuits that have an axis of symmetry, as shown in Fig. 15.P5*b*. Show also how the technique may be modified to simplify the computation for such a case.

(c) A more detailed model of the first type of biscuit shows it has holes of *three* sizes, as shown in Fig. 15.P5*c*. Analyze the situation and show that a much simplified match graph can be produced from the image data, leading to successful object location.

(d) A further matching strategy is devised to make use of the hole size information. Matches are *only* shown in the match graph if they arise between pairs of holes of *different* sizes. Determine how successful this strategy is, and discuss whether it is likely to be generally useful, for example, for objects with increased numbers of features.

(e) Work out an optimal object identification strategy, which will be capable of dealing with cases where holes and/or corners are to be used as point features, where the holes might have different sizes, where the corners might have different angles and orientations, where the object surfaces might have different colors or textures, and where the objects might have larger numbers of features. Make clear what the term *optimal* should be taken to mean in such cases.

9. (a) Figure 15.P6 shows a 2-D view of a widget with four corners. Explain how the maximal clique technique can be used to locate widgets even if they are partly obscured by various types of objects including other widgets.

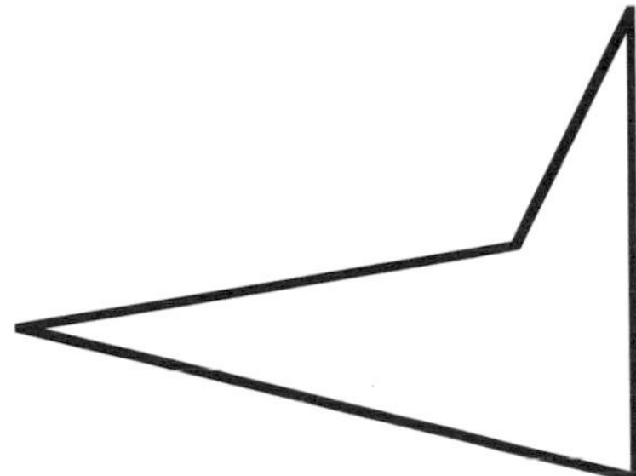

Figure 15.P6 Diagram of a widget.

(b) Explain why the basic algorithm will not distinguish between widgets that are normally presented from those that are upside down. Consider how the basic method could be extended to ensure that a robot only picks up those that are the right way up.

(c) The camera used to view the widgets is accidentally jarred and then reset at a different, unknown height above the worktable. State clearly why the usual maximal clique technique will now be unable to identify the widgets. Discuss how the overall program could be modified to make sense of the data, and make correct interpretations in which all the widgets are identified. Assume *first* that widgets are the only objects appearing in the scene, and *second* that a variety of other objects may appear.

(d) The camera is jarred again, and this time is set at a small, unknown angle to the vertical. To be sure of detecting such situations and of correcting for them, a flat calibration object of known shape is to be stuck on the worktable. Decide on a suitable shape and explain how it should be used to make the necessary corrections.

PART 3

3-D Vision and Motion

Part 3 covers the developments needed for an understanding of real scenes, which necessarily contain 3-D objects—a number of which may be in motion. 3-D vision is considerably more complex than 2-D vision, not least because the number of degrees of freedom of an object will typically have increased from three to six, with an accompanying combinatorial increase in the number of scene configurations to be considered.

This part of the book starts by gently airing the problems (Chapter 16), before concentrating on the complexities, of full perspective projection (Chapter 17). Motion is then considered (Chapter 18), but before this subject can be dealt with in depth (Chapter 20), it is useful to see what shortcuts can be achieved by taking invariants into account (Chapter 19). Finally, Chapter 21 deals with camera calibration but also shows how recent research has attempted to avoid the need for explicit calibration by making careful calculations that interrelate multiple scenes. Here the emphasis is on taking opportunities that permit some of the complexities to be bypassed.

The Three-Dimensional World

Humans are able to employ 3-D vision with consummate ease, and according to conventional wisdom, binocular vision is the key to this success. But the truth is more complex than this, and this chapter demonstrates why.

Look out for:

- what can be achieved using binocular vision.
- how the shading of surfaces can be used in place of binocular vision to achieve similar ends.
- how these basic methods provide dimensional information for 3-D scenes but do not immediately lead to object recognition.
- how the process of 3-D object recognition can be tackled by studies of 3-D geometry.

Note that this is only an introductory chapter on 3-D vision, designed to give the flavor of the subject and to show its origins in human vision. It will be followed by the other five chapters (Chapters 17–21) that comprise Part 3 of this volume.

At a more detailed level, the importance of the epipolar line approach in solving the correspondence problem should be noted. The concept is deservedly taken considerably further in Chapter 21, in conjunction with the required mathematical formulation.

CHAPTER 16

The Three-Dimensional World

16.1 Introduction

In the forgoing chapters, it has been assumed that objects are essentially flat and are viewed from above in such a way that there are only three degrees of freedom—namely, the two associated with position and one concerned with orientation. Although this approach was adequate for carrying out many useful visual tasks, it is totally inadequate for interpreting outdoor or factory scenes, or even for helping with quite simple robot assembly and inspection tasks. Over the past 25 years or so a considerable amount of sophisticated theory has been developed and backed up by experiment, to find how scenes composed of real 3-D objects can be understood in detail.

In general, this means attempting to interpret scenes in which objects may appear in totally arbitrary positions and orientations—corresponding to six degrees of freedom. Interpreting such scenes, and deducing the translation and orientation parameters of arbitrary sets of objects, requires a substantial amount of computation—partly because of the inherent ambiguity in inferring 3-D information from 2-D images.

A variety of approaches are now available for proceeding with 3-D vision. We cannot describe them all in a single chapter. Rather, we will provide an overview, outlining the basic principles and classifying the methods according to generality, applicability, and so on. Although computer vision need not necessarily mimic the capabilities of the human eye–brain system, much research on 3-D vision has been aimed at biological modeling. This type of research shows that the human visual system makes use of a number of different methods simultaneously, taking appropriate cues from the input data and forming hypotheses about the content of a scene, progressively enhancing these hypotheses until a useful working model of what is present is produced. Thus, individual methods are not expected to work in isolation but instead merely need to provide the model generator with whatever data become available. Biological machinery of various types may in principle lie idle for much of the time until triggered by specific input stimuli. Computer vision systems are currently less sophisticated than this and tend to be built on specific processing models, so that they can be applied efficiently to more restricted types of image data. In this chapter, we adopt the pragmatic view that particular methods need to be (or have been) developed for specific types of situations and that they should be used only when appropriate—although we do take some care to elucidate what the appropriate types of applications are.

16.2 Three-dimensional Vision—The Variety of Methods

One of the most obvious characteristics of the human visual system is that it employs two eyes, and it is well known to the layperson that binocular (or "stereo") vision permits depth to be discerned within a scene. However, the loss of vision that results when one eye is shut is relatively insignificant and is by no means a disqualification from driving a car or even flying an airplane. On the contrary, depth can readily be deduced in monocular vision from a plethora of cues buried in an image. Naturally, to achieve depth perception the eye–brain system can call on a huge amount of prestored data about the physical world and about the types of objects in it, be they synthetic or natural entities. For example, the size of any car being viewed is strongly constrained. Similarly, most objects have highly restricted sizes, both absolutely and in their depths relative to their frontal dimensions. Nevertheless, in a single view of a scene, it is normally impossible to deduce absolute sizes—all the objects and their depths can be scaled up or down by arbitrary factors, and this cannot be discerned from a monocular view.

The eye–brain system makes use of a huge database relating to the physical world, but much can be learned with negligible prior knowledge, even from a single monocular view. The main key to this knowledge is the "shape from shading" concept. For 3-D shape to be deducible from shading information (i.e., from the gray-scale intensities in an image), something has to be known about how the scene is lit—the simplest situation being when the scene is illuminated by a single point light source at a known position. Indoors a single overhead tungsten light is still the most usual illuminator, whereas outdoors the sun performs a similar function. In either case, an obvious result is that a single source will illuminate one part of an object and not another—which then remains in shadow—and parts that are oriented in various ways relative to the source and to the observer appear with different brightness values, so that orientation can in principle be deduced. As will be seen later, deduction of orientation and position is not at all trivial and may even be ambiguous. Nevertheless, successful methods have been developed for carrying out this task. One problem that often arises is that the position of the light source is unknown, but this information can generally be extracted (at least by the eye) from the scene being examined. A bootstrapping procedure is therefore able to unlock the image data gradually and proceed to an interpretation.

Although these methods enable the eye to interpret real scenes, it is difficult to say quite to what degree of precision they are carried out. With computer vision, the required precision levels are liable to be higher, although the machine will be aided by knowing the exact source of illumination. However, with computer vision we can go further and arrange artificial lighting schemes that would not appear in nature, so the computer can acquire an advantage over the human visual system. In particular, a set of light sources can be applied in sequence to the scene—an approach known as photometric stereo—which can sometimes help the computer to interpret the scene more rigorously and efficiently. In other cases, structured light may be applied: this means projecting onto the scene a pattern of spots or stripes, or even a grid of lines, and measuring their positions in the resulting image. By this means, depth information can be obtained much as for pairs of stereo images.

Finally, a number of methods have been developed for analyzing images on the basis of readily identifiable sets of features. These methods are the 3-D analogs of the graph matching and the GHT approaches of Chapter 15. However, they are significantly more complex because they generally involve six degrees of freedom in place of the three assumed throughout Chapter 15. Furthermore, such methods make strong assumptions about the particular objects to be located within the scene. In general situations, such assumptions could not likely be made, and so initial analysis of any images must be made on the basis that the entire scene must be mapped out in 3-D, then 3-D models are built up, and finally deductions must be made by noticing the relationship of one part of the scene to another. If a scene is composed from an entirely new set of objects, all that can be done is to *describe* what is present and say perhaps what the set most closely *resembles*. Recognition per se cannot be performed. Scene analysis is—at least from a single monocular image—an inherently ambiguous process. Every scene can have a number of possible interpretations, and there is evidence that the eye looks for the simplest and most probable explanation rather than an absolute interpretation. Indeed, the many illusions to which the eye–brain system is subject underline the notion that decisions must repeatedly be made concerning the most likely interpretation of a scene and that there is some risk that its internal model builder will lock on to an interpretation or part-interpretation that is suboptimal (see the paintings of Escher!).

This section has indicated that methods of 3-D vision can be categorized according to whether they start by mapping out the shapes of objects in 3-D space and then attempt to interpret the resulting shapes, or whether they try to identify objects directly from their features. In either case a knowledge base is ultimately called for. It has also been seen that methods of mapping objects in real space include monocular and binocular methods, although structured lighting can help to offset the deficiencies of employing a single "eye." Laser scanning and ranging techniques must also be included in 3-D mapping methods, although space precludes detailed discussion of these techniques in this book.

16.3 Projection Schemes for Three-dimensional Vision

Engineering drawings generally provide three views of an object to be manufactured—the plan, the side view, and the elevation. Traditionally, these views are simple orthographic (nondistorting) projections of the object. That is, they are made by taking sets of parallel lines from points on the object to the flat plane on which it is being projected.

When objects are viewed by eye or from a camera, however, rays converge to the lens, and so images formed in this way are subject not only to change of scale but also to perspective distortions (Fig. 16.1). This type of projection is called *perspective projection*, although it includes orthographic projection as the special

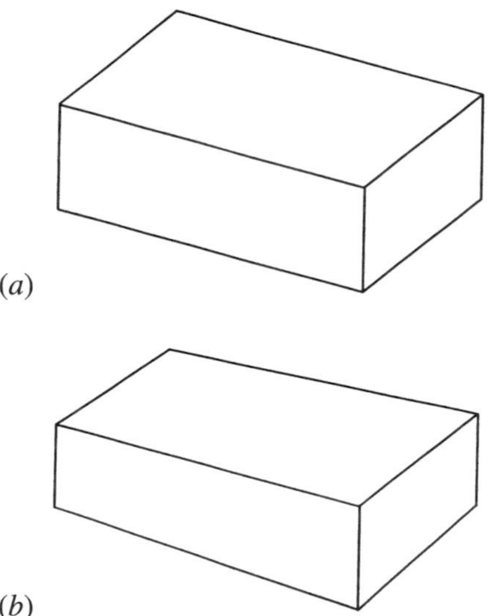

Figure 16.1 (*a*) Image of a rectangular box taken using orthographic projection; (*b*) the same box taken using perspective projection. In (*b*) note that parallel lines no longer appear parallel, although paradoxically the box appears more realistic.

case of viewing from a distant point. Unfortunately, perspective projections have the disadvantage that they tend to make objects appear more complex than they really are by destroying simple relationships between their features. Thus, parallel edges no longer appear parallel, and midpoints no longer appear as such (although many useful geometric properties still hold—for example, a tangent line remains a tangent line, and the order of points on a straight line remains unchanged).

In outdoor scenes, it is very common to see lines that are known to be parallel apparently converging toward a vanishing point on the horizon line (Fig. 16.2). The horizon line is the projection onto the image plane of the line at infinity on the ground plane $\mathscr{G}$. It is the set of all possible vanishing points for parallel lines on $\mathscr{G}$. In general, the vanishing points of a plane $\mathscr{P}$ are defined as the projections onto the image plane corresponding to points at infinity in a given direction on $\mathscr{P}$. Thus, any plane $\mathscr{Q}$ within the field of view may have vanishing

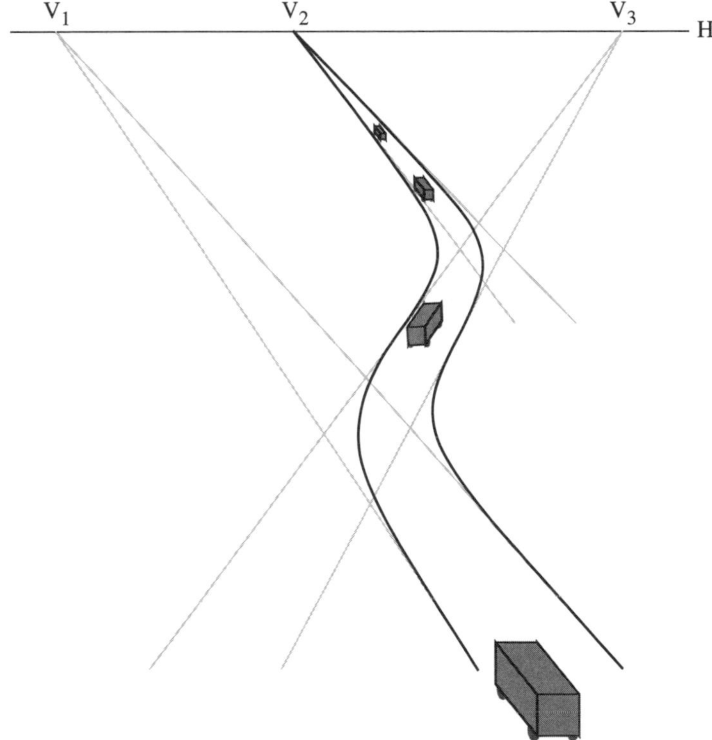

Figure 16.2 Vanishing points and the horizon line. This figure shows how parallel lines on the ground plane appear, under perspective projection, to meet at vanishing points V on the horizon line H. (Note that V and H lie in the *image* plane.) If two parallel lines do not lie on the ground plane, their vanishing point will lie on a different vanishing line. Hence, it should be possible to determine whether any roads are on an incline by computing all the vanishing points for the scene.

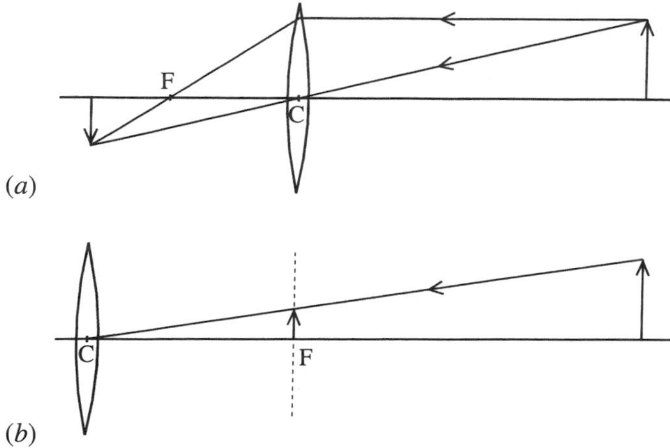

Figure 16.3 (*a*) Projection of an image into the image plane by a convex lens; note that a single image plane only brings objects at a single distance into focus but that for far-off objects the image plane may be taken to be the focal plane, a distance f from the lens; (*b*) a commonly used convention that imagines the projected image to appear noninverted at a focal plane F in front of the lens. The center of the lens is said to be the center of projection for image formation.

points in the image plane, and these will lie on a vanishing line that is the analog of the horizon line for $\mathscr{Q}$.

Figure 16.3*a* shows how an image is projected into the image plane by a convex (eye or camera) lens at the origin. It is inconvenient to have to consider inverted images, and it is a commonly used convention in image analysis to set the center of the lens at the origin (0,0,0) and to imagine the image plane to be the plane $Z = f$, f being the focal length of the lens. With this simplified geometry (Fig. 16.3*b*), images in the image plane appear noninverted. Taking a general point in the scene as (X, Y, Z), which appears in the image as (x_1, y_1), perspective projection now gives:

$$(x_1, y_1) = (fX/Z, fY/Z) \tag{16.1}$$

16.3.1 *Binocular Images*

Figure 16.4 shows the situation when two lenses are used to obtain a stereo pair of images. In general, the two optical systems do not have parallel optical axes but exhibit a "vergence" (which may be variable, as it is for human eyes), so that they intersect at some point within the scene. Then a general point (X, Y, Z) in the scene has two different pairs of coordinates, (x_1, y_1) and (x_2, y_2), in its two

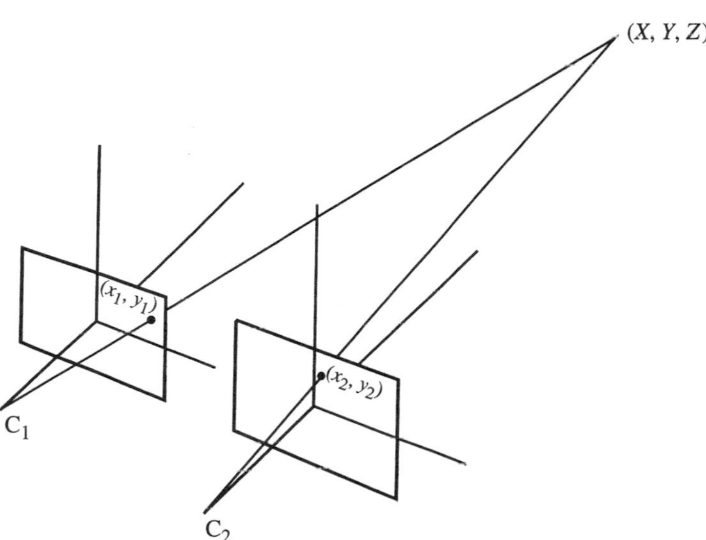

Figure 16.4 Stereo imaging using two lenses. The axes of the optical systems are parallel; that is, there is no "vergence" between the optical axes.

images, which differ both because of the vergence between the optical axes and because the baseline b between the lenses causes relative displacement or "disparity" of the points in the two images.

For simplicity, we now take the vergence to be zero; that is, the optic axes are parallel. Then, with a suitable choice of Z-axis on the perpendicular bisector of the baseline b, we obtain two equations:

$$x_1 = (X + b/2)\,f/Z \tag{16.2}$$

$$x_2 = (X - b/2)f/Z \tag{16.3}$$

so that the disparity is

$$D = x_1 - x_2 = bf/Z \tag{16.4}$$

Rewriting this equation in the form:

$$Z = bf/(x_1 - x_2) \tag{16.5}$$

now permits the depth Z to be calculated. Computation of Z only requires the disparity for a stereo pair of image points to be found and parameters of the optical systems to be known. However, confirming that both points in a stereo pair actually correspond to the same point in the original scene is in general not at all trivial, and much of the computation in stereo vision is devoted to this task. In addition, to obtain good accuracy in the determination of depth, a large baseline b is required. Unfortunately, as b is increased the correspondence between the images decreases, so it becomes more difficult to find matching points.

16.3.2 *The Correspondence Problem*

Two important approaches may be used to find pairs of points that match in the two images of a stereo pair. One approach is that of "light striping" (one form of structured lighting), which encodes the two images so that it is easy to see pairs of corresponding points. If a single vertical stripe is used, for every value of y there is in principle only one light stripe point in each image, and so the matching problem is solved. We return to this problem in a later section.

The second important approach is to employ epipolar lines. To understand this approach, imagine that we have located a distinctive point in the first image and that we are marking all possible points in the object field that could have given rise to it. This will mark out a line of points at various depths in the scene, and, when viewed in the second image plane, a locus of points can be constructed in that plane. This locus is the *epipolar line* corresponding to the original image point in the alternate image (Fig. 16.5). If we now search along the epipolar line for a similarly distinctive point in the second image, the chance of finding the correct match is significantly enhanced. This method has the advantage not only of cutting down the amount of computation required to find corresponding points, but also of significantly reducing the chance of false alarms. Note that the concept of an epipolar line applies to both images—a point in one image gives an epipolar line in the other image. Note also that in the simple geometry of Fig. 16.4, all epipolar lines are parallel to the x-axis, although this is not so in general. (The general situation is that all epipolar lines in one image plane pass through the point that is the image of the projection point of the alternate image plane.)

The correspondence problem is rendered considerably more difficult by the fact that some points in the scene will give rise to points in one image but not in the other. Such points are either occluded in the one image or are so distorted as not to give a recognizable match in the two images (e.g., the different background might mask a corner point in one image while permitting it to stand out in the other). Any attempt to match such points can then only lead to false alarms. Thus, it is necessary to search for consistent sets of solutions in the form of continuous object surfaces in the scene. For this reason iterative "relaxation" schemes are widely used to implement stereo matching.

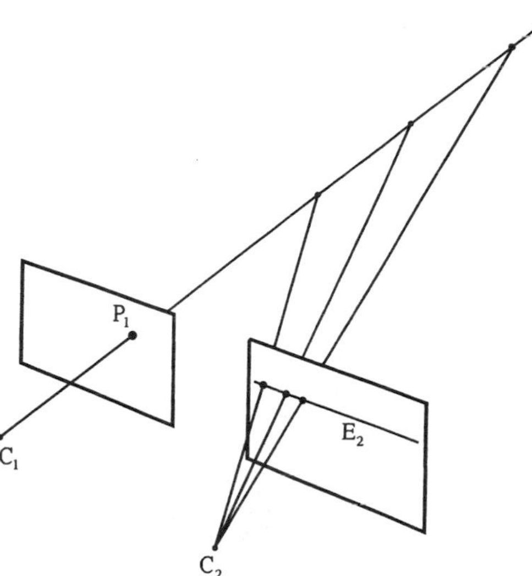

Figure 16.5 Geometry of epipolar lines. A point P_1 in one image plane may have arisen from any one of a line of points in the scene and may appear in the alternate image plane at any point on the so-called epipolar line E_2.

Broadly speaking, correspondences are sought by two methods: one is the matching of near-vertical edge points in the two images (near-horizontal edge points do not give the required precision); the other is the matching of local intensity patterns using correlation techniques. Correlation is an expensive operation, and in this case it is relatively unreliable—principally because intensity patterns frequently appear significantly foreshortened[1] in one or other image and hence are difficult to match reliably. In such cases, the most practical solution is to reduce the baseline. As noted earlier, this has the effect of reducing the accuracy of depth measurement. Further details of these techniques are to be found in Shirai (1987).

Before leaving this topic, it is worth considering in slightly more detail how the problems of visibility mentioned above arise. Figure 16.6 shows a situation in which an object is being observed by two cameras giving stereo images. Much of the object will not be visible in either image because of self-occlusion, whereas some feature points will only be visible in one or the other image. Now consider the order in which the points appear in the two images (Fig. 16.7). The points that are visible appear in the same order as in the scene, and the points that are just going out of sight are those for which the order between the scene and the image is just about to change. Points that provide information about the front surface of the object can thus only bear a simple

1 That is, distorted by the effects of perspective.

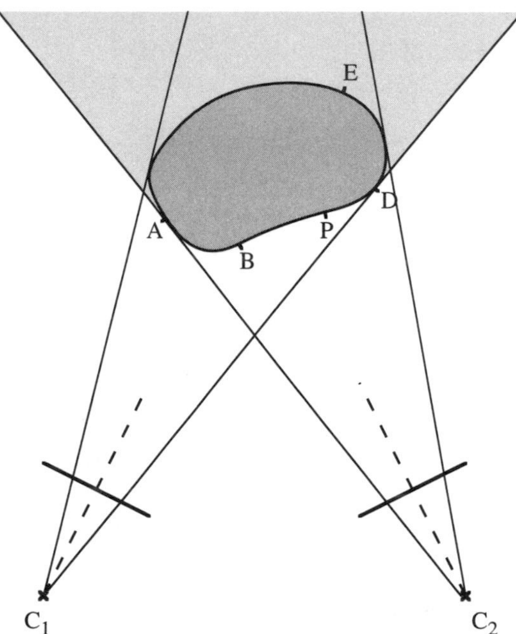

Figure 16.6 Visibility of feature points in two stereo views. Here an object is viewed from two directions. Only feature points that appear in both views are of value for depth estimation. This eliminates all points in the shaded region, such as E, from consideration.

geometrical relation to each other. In particular, for points not to obscure, or be obscured by, a given point P, they must not lie within a double-ended cone region defined by P and the centers of projection C_1, C_2 of the two cameras. This region is shown shaded in Fig. 16.7. A surface passing through P for which full depth information can be retrieved must lie entirely within the nonshaded region. (Of course, a new double-ended cone must be considered for each point on the surface being viewed.) The possibility of objects containing holes, or having transparent sections, must not be forgotten. (Such cases can be detected from differences in the ordering of feature points in the two views—see Fig. 16.7.) Neither must it be ignored that the foregoing figures represent a single horizontal cross section of an object that can have totally different shapes and depths in different cross sections.

16.4 Shape from Shading

As mentioned in Section 16.2, it is possible to analyze the pattern of intensities in a single (monocular) image and to deduce the shapes of objects from the

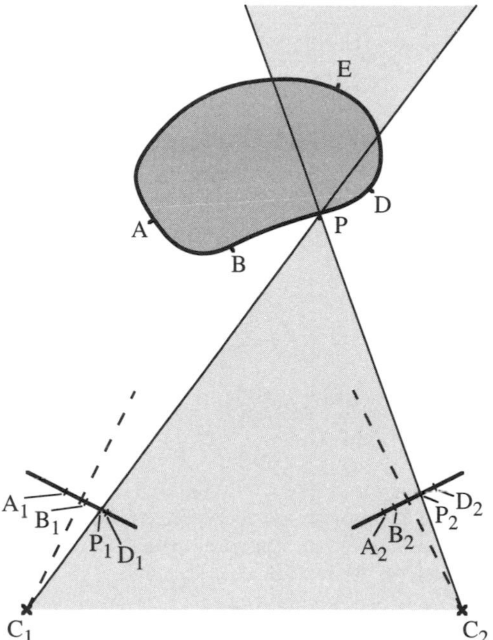

Figure 16.7 Ordering of feature points on an object. In the two views of the object shown here, the feature points all appear in the same order A, B, P, D as on the surface of the object. Points for which this would not be valid, such as E, are behind the object and are obscured from view. Relative to a given visible feature P, there is a double-ended cone (shaded) in which feature points must not appear if they are not to obscure the feature under consideration. An exception to these rules might be if the object had a semitransparent window through which an additional feature T were visible: in that case, interpretation would be facilitated by noting that the orderings of the features seen in the two views were different—for example, A_1, T_1, B_1, P_1, D_1 and A_2, B_2, T_2, P_2, D_2.

shading information. The principle underlying this technique is that of modeling the reflectance of objects in the scene as a function of the angles of incidence i and emergence e of light from their surfaces. A third angle is also involved, and it is called the phase g (Fig. 16.8).

A general model of the situation gives the radiance I (light intensity in the image) in terms of the irradiance E (energy per unit area falling on the surface of the object) and the reflectance R:

$$I(x_1, y_1) = E(x, y, z)R(\mathbf{n}, \mathbf{s}, \mathbf{v}) \tag{16.6}$$

It is well known that a number of matte surfaces approximate reasonably well to an ideal Lambertian surface whose reflectance function depends only on the

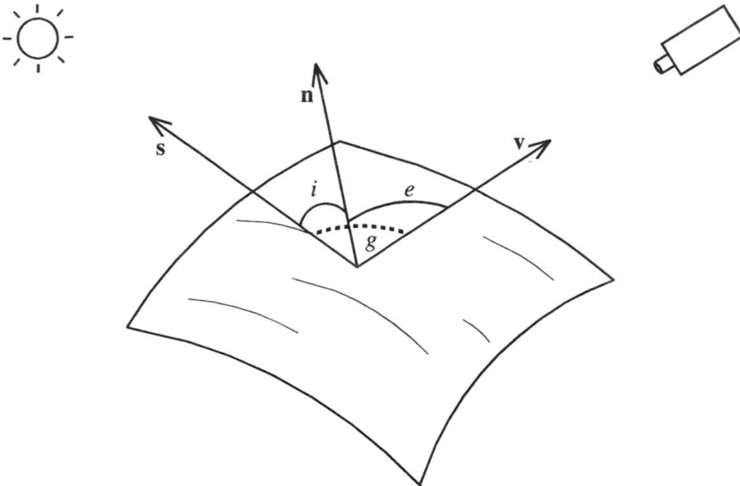

Figure 16.8 Geometry of reflection. An incident ray from source direction **s** is reflected along the viewer direction **v** by an element of the surface whose local normal direction is **n**; *i*, *e*, and *g* are defined, respectively, as the incident, emergent, and phase angles.

angle of incidence *i*—that is, the angles of emergence and phase are immaterial:

$$I = (1/\pi)E\cos i \tag{16.7}$$

For the present purpose E is regarded as a constant and is combined with other constants for the camera and the optical system (including, for example, the *f*-number). In this way a normalized reflectance is obtained, which in this case is:

$$\begin{aligned} R &= R_0 \cos i = R_0\, \mathbf{s}\cdot\mathbf{n} \\ &= \frac{R_0(1 + pp_s + qq_s)}{(1 + p^2 + q^2)^{1/2}\left(1 + p_s^2 + q_s^2\right)^{1/2}} \end{aligned} \tag{16.8}$$

where the commonly used convention is taken of writing orientations in 3-D in terms of p and q values. These are not direction cosines but correspond to the coordinates of the point $(p, q, 1)$ at which a particular direction vector from the origin meets the plane $z = 1$. Hence, they need suitable normalization, as in equation (16.8).

This equation gives a reflectance map in gradient (p, q) space. We now temporarily put the absolute reflectance value R_0 equal to unity. The reflectance map can be drawn as a set of contours of equal brightness, starting with a point having $R = 1$ at $\mathbf{s} = \mathbf{n}$, and going down to zero for $\mathbf{n}$ perpendicular to $\mathbf{s}$. When $\mathbf{s} = \mathbf{v}$, so that the light source is along the viewing direction (here taken to be the direction $p = q = 0$), zero brightness occurs only for infinite distances on the reflectance map ($(p^2+q^2)^{1/2}$ approaching infinity) (Fig. 16.9*a*). In a more general case, when $\mathbf{s} \neq \mathbf{v}$, zero brightness occurs along a straight line in gradient space (Fig. 16.9*b*). To find the exact shapes of the contours we can set R at a constant value a, which results in:

$$a(1+p^2+q^2)^{1/2}(1+p_s^2+q_s^2)^{1/2} = 1+pp_s+qq_s \tag{16.9}$$

Squaring this equation clearly gives a quadratic in p and q, which could be simplified by a suitable change of axes. Thus, the contours must be curves of conic section, namely, circles, ellipses, parabolas, hyperbolas, lines, or points. (The case of a point arises only when $a=1$, when we get $p=p_s$, $q=q_s$; and that of a line only if $a=0$, when we get the equation $1+pp_s+qq_s=0$. Both of these solutions were implied above.)

Unfortunately, object reflectances are not all Lambertian, and an obvious exception is for surfaces that approximate to pure specular reflection. In that case, $e=i$ and $g=i+e$ (**s**, **n**, **v** are coplanar); the only nonzero reflectance position in gradient space is the point representing the bisector of the angle between the source direction $\mathbf{s}(p,q)$ and the viewing direction $\mathbf{v}$ $(0,0)$—that is, $\mathbf{n}$ is along $\mathbf{s}+\mathbf{v}$—and very approximately:

$$p \approx p_s/2 \tag{16.10}$$

$$q \approx q_s/2 \tag{16.11}$$

For less perfect specularity, a peak is obtained around this position. A good approximation to the reflectance of many real surfaces is obtained by modeling them as basically Lambertian but with a strong additional reflectance near the specular reflectance position. Using the Phong (1975) model for the latter component gives:

$$R = R_0 \cos i + R_1 \cos^m \theta \tag{16.12}$$

θ being the angle between the actual emergence direction and the ideal specular reflectance direction.

The resulting contours now have two centers around which to peak: the first is the ideal specular reflection direction ($p \approx p_s/2$, $q \approx q_s/2$), and the second is that of the source direction ($p=p_s$, $q=q_s$). When objects are at all shiny—such as metal, plastic, liquid, or even wood surfaces—the specular peak is quite sharp and rather intense. Casual observation may not even indicate the presence of another peak since Lambertian reflection is so diffuse (Fig. 16.10). In other cases the specular peak can broaden and become more diffuse; hence, it may merge with the Lambertian peak and effectively disappear.

With regards to the Phong model, we wish to point out that it is adapted to different materials by adjusting the values of R_0, R_1, and m. Phong remarks that R_1

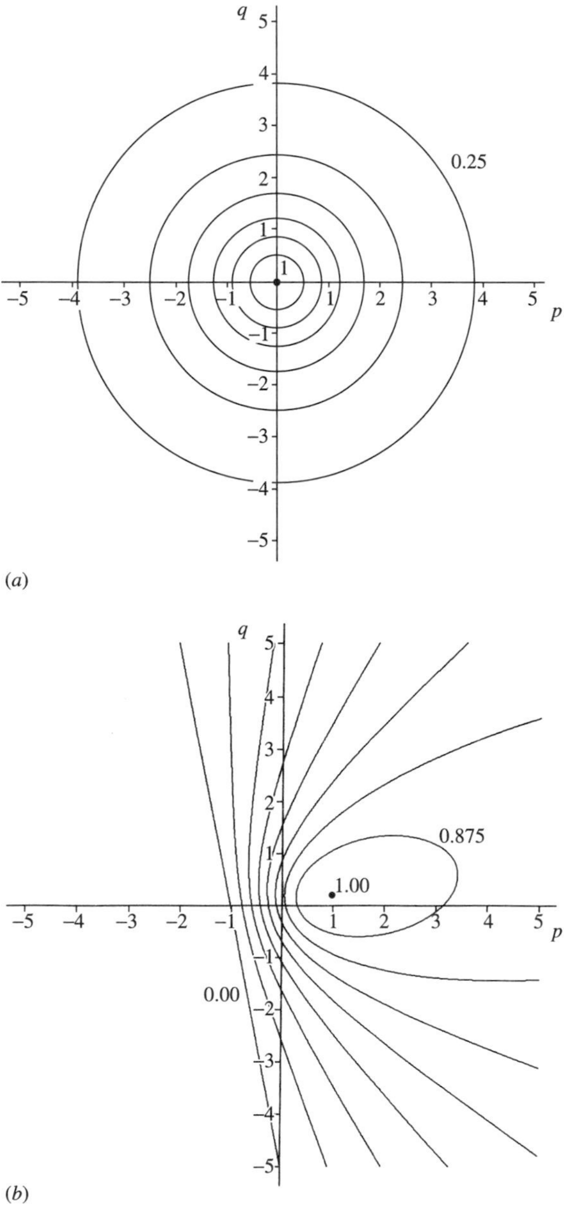

Figure 16.9 Reflectance maps for Lambertian surfaces: (*a*) contours of constant intensity plotted in gradient (p, q) space for the case where the source direction **s** (marked by a black dot) is along the viewing direction **v** $(0, 0)$ (the contours are taken in steps of 0.125 between the values shown); (*b*) the contours that arise where the source direction (p_s, q_s) is at a point (marked by a black dot) in the positive quadrant of (p, q) space. Note that there is a well-defined region, bounded by the straight line $1 + pp_s + qq_s = 0$, for which the intensity is zero. (The contours are again taken in steps of 0.125.)

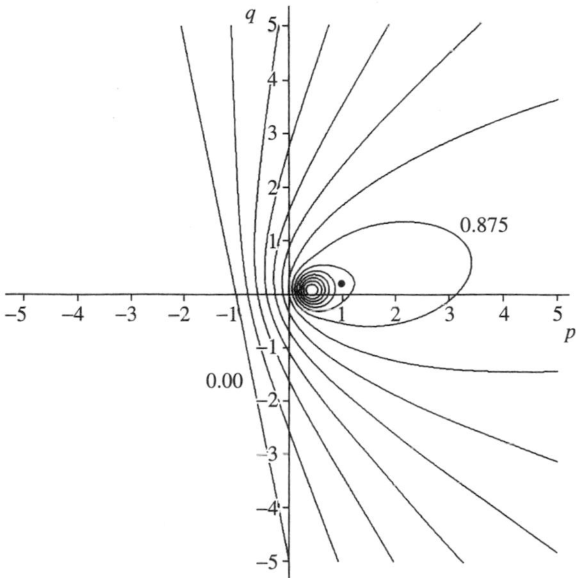

Figure 16.10 Reflectance map for a non-Lambertian surface. A modified form of Fig. 16.9*b* for the case where the surface has a marked specular component ($R_0 = 1.0$, $R_1 = 0.8$). Note that the specular peak can have very high intensity (much greater than the maximum value of unity for the Lambertian component). In this case, the specular component is modeled with a $\cos^8\theta$ variation. (The contours are again taken in steps of 0.125.)

typically lies between 10 and 80%, while m is in the range 1–10. However, Rogers (1985) indicates that m may be as high as 50. There is no physical significance in these numbers—the model is simply a phenomenological one. This being so, care should be taken to prevent the $\cos^m\theta$ term from contributing to reflectance estimates when $|\theta| > 90°$. The Phong model is reasonably accurate but has been improved upon by Cook and Torrance (1982). Finding the right model is important in computer graphics applications, but the improvement is difficult to make use of in computer vision because of lack of data concerning the reflectances of real objects and because of variability in the current state (cleanliness, degree of polish, etc.) of a given surface. However, the method of photometric stereo gives some possibility of overcoming these problems.

16.5 Photometric Stereo

Photometric stereo is a form of structured lighting that increases the information available from surface reflectance variations. Basically, instead of taking a single monocular image of a scene illuminated from a single source, several images are taken, from the same vantage point, with the scene illuminated in turn by separate light sources. Again, these light sources are ideally point sources some distance away in various directions, so that there is in each case a well-defined light source direction from which to measure surface orientation.

The basic idea of photometric stereo is that of cutting down the number of possible positions in gradient space for a given point on the surface of an object. It has already been seen that, for known absolute reflectance R_0, a constant brightness in one image permits the surface orientation to be limited to a curve of conic cross section in gradient space. This would also be true for a second such image, the curve being a new one if the illuminating source is different. In general, two such conic curves meet in two points, so there is now only a single ambiguity in the gradient of the surface at any given point in the image. To resolve this ambiguity, a third source of illumination can be employed (this must not be in the plane containing the first two and the surface point being examined), and the third image gives another curve in gradient space that should pass through the appropriate crossing point of the first two curves (Fig. 16.11). If a third source of illumination cannot be used, it is sometimes possible to arrange that the inclination of each of the sources is so high that $(p^2 + q^2)^{1/2}$ on the surface is always lower than $(p_s^2 + q_s^2)^{1/2}$ for each of the sources,

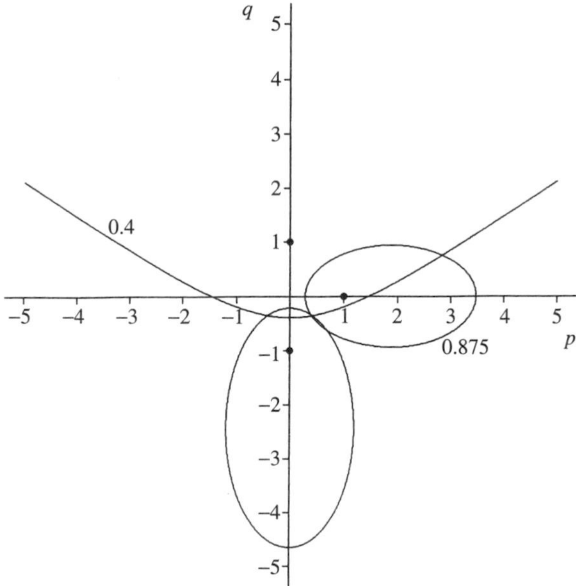

Figure 16.11 Obtaining a unique surface orientation by photometric stereo. Three contours of constant intensity arise for different light sources of equal strength. All three contours pass through a single point in (p,q) space and result in a unique solution for the local gradient.

so that only one interpretation of the data is possible. This method is prone to difficulty, however, since it means that parts of the surface could be in shadow, thereby preventing the gradient for these parts of the surface from being measured. Another possibility is to assume that the surface is reasonably smooth, so that p and q vary continuously over it. This itself ensures that ambiguities are resolved over most of the surface.

Other advantages can be gained from using more than two sources of illumination. One is that information can be gathered on the absolute surface reflectance. Another is that the assumption of a Lambertian surface can be tested. Thus, three sources of illumination ensure that the remaining ambiguity is resolved *and* permits absolute reflectivity to be measured. This is obvious since if the three contours in gradient space do not pass through the same point, then the absolute reflectivity cannot be unity. Thus, corresponding contours should be sought which do pass through the same point. In practice, the calculation is normally carried out by defining a set of nine matrix components of irradiance, s_{ij} being the jth component of light source vector $\mathbf{s}_i$. Then, in matrix notation:

$$\mathbf{E}=R_0\mathcal{S}\mathbf{n} \tag{16.13}$$

where

$$\mathrm{E} = (E_1, E_2, E_3)^\mathsf{T} \tag{16.14}$$

and

$$\mathcal{S} = \begin{bmatrix} s_{11} & s_{12} & s_{13} \\ s_{21} & s_{22} & s_{23} \\ s_{31} & s_{32} & s_{33} \end{bmatrix} \tag{16.15}$$

Provided that the three vectors $\mathbf{s}_1$, $\mathbf{s}_2$, $\mathbf{s}_3$ are not coplanar, so that $\mathcal{S}$ is not a singular matrix, R_0 and $\mathbf{n}$ can now be determined from the formulas:

$$R_0 = \left|\mathcal{S}^{-1}\mathbf{E}\right| \tag{16.16}$$

$$\mathbf{n} = \mathcal{S}^{-1}\mathbf{E}/R_0 \tag{16.17}$$

An interesting special case arises if the three source directions are mutually perpendicular. Taking them to be aligned along the respective major axes' directions, $\mathcal{S}$ is now the unit matrix, so that:

$$R_0 = \left[E_1^2 + E_2^2 + E_3^2\right]^{1/2} \tag{16.18}$$

and

$$\mathbf{n} = (E_1, E_2, E_3)^\mathsf{T}/R_0 \tag{16.19}$$

If four or more images are obtained using further illumination sources, more information can be obtained—for example, the coefficient of specular reflectance, R_1. In practice, this coefficient varies somewhat randomly with the cleanness of the surface, and it may not be relevant to determine it accurately. More probably it will be sufficient to check whether significant specularity is present, so that the corresponding region of the surface can be ignored for absolute reflectance calculations. Nonetheless, finding the specularity peak can itself give important surface orientation information, as will be clear from the previous section. Although the information from several illumination sources should ideally be collated using least squares analysis, this method requires significant computation. Hence, it seems better to use the images resulting from further illumination sources as confirmatory—or, instead, to select the three that exhibit the least evidence of specularity as giving the most reliable information on local surface orientation.

16.6 The Assumption of Surface Smoothness

As suggested earlier, the assumption of a reasonably smooth surface permits ambiguities to be removed in situations where there are two illuminating sources. This method can be used to help analyze the brightness map even for situations where a single source is employed. The fact that the eye can perform this feat of interpretation indicates that it is possible to find computer methods for achieving it. Much research has been carried out on this topic, and a set of methods is available, although the calculations are complex iterative procedures that cannot be carried out in real time on conventional computers. For this reason they are not studied in depth here: the reader is referred to the volume by Horn (1986) for detailed information on this topic. However, one or two remarks are in order.

First, consider the representation to be employed for this type of analysis. It turns out that normal gradient (p, q) space is not very appropriate for the purpose. In particular, it is necessary to average gradient (i.e., the **n**-values) locally within the image. However, (p, q)-space is not "linear" in that a simple average of (p, q) values within a window gives biased results. It turns out that a conformal representation of gradient (i.e., one that preserves small shapes) is closer to the ideal in that the distances between points in such a representation provide better approximations to the relative orientations of surface normals. Averaging in such a representation gives reasonably accurate results. The required representation is obtained by a stereographic projection, which maps the unit (Gaussian) sphere onto a plane $(z = 1)$ through its north pole but this time using as a

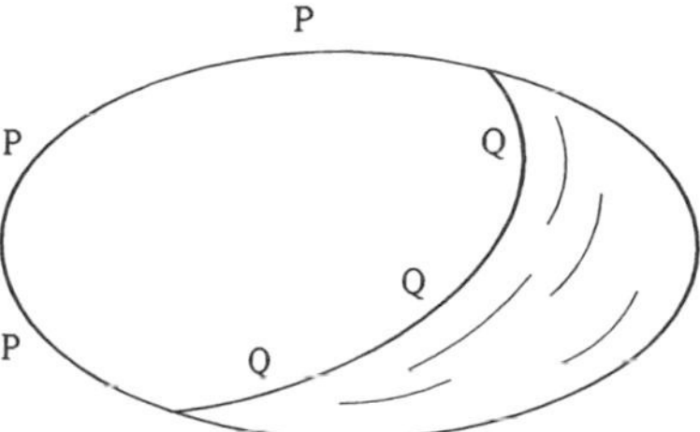

Figure 16.12 The two types of boundary conditions that can be used in shape-from-shading computations of surface orientation: (i) positions P where the surface normal is perpendicular to the viewing direction; (ii) positions Q where the surface normal is perpendicular to the direction of illumination (i.e., shadow boundaries).

projection point not its center but its south pole. This projection has the additional advantage that it projects all possible orientations of a surface onto the plane, not merely those from the northern hemisphere. Hence, backlit objects can be represented conveniently in the same map as used for frontlit objects.

Second, the relaxation methods used to estimate surface orientation have to be provided with accurate boundary conditions. In principle, the more correct the orientations that are initially presented to such procedures, the more quickly and accurately the iterations proceed. Normally, two sets of boundary conditions can be applied in such programs. One is the set of positions in the image where the surface normal is perpendicular to the viewing direction. The other is the set of positions in the image where the surface normal is perpendicular to the direction of illumination. This second set of positions corresponds to the set of shadow edges (Fig. 16.12). Careful analysis of the image must be undertaken to find each set of positions, but once they have been located they provide valuable cues for unlocking the information content of the monocular image and mapping out surfaces in detail.

Finally, it should be remarked that all shape from shading techniques provide information that initially takes the form of surface orientation maps. Dimensions are not obtainable directly, but these can be computed by integration across the image from known starting points. In practice, this tends to mean that absolute dimensions are unknown and that dimensional maps are obtainable only if the size of an object is given or if its depth within the scene is known.

16.7 Shape from Texture

Texture can be very helpful to the human eye in permitting depth to be perceived. Although textured patterns can be very complex, it turns out that even the simplest textural elements can carry depth information. Ohta et al. (1981) showed how circular patches on a flat surface viewed more and more obliquely in the distance become first elliptical and then progressively flatter and flatter. At infinite distance, on the horizon line (here defined as the line at infinity in the given plane), they would clearly become very short line segments. To disentangle such textured images sufficiently to deduce depths within the scene, it is first necessary to find the horizon line reliably. This is achieved by taking all pairs of texture elements and deducing from their areas where the horizon line would have to be. To proceed we make use of the rule:

$$d_1^3/d_2^3 = A_1/A_2 \tag{16.20}$$

which applies since circles at various depths would give a square law, although the progressive eccentricity also reduces the area linearly in proportion to the depth. This information is accumulated in a separate image space, and a line is then fitted to these data. False alarms are eliminated automatically by this Hough-based procedure. At this stage the original data—the ellipse areas—provide direct information on depth, although some averaging is required to obtain accurate results. Although this type of method has been demonstrated in certain instances, it is in practice highly restricted unless considerable amounts of computation are performed. Hence, it is doubtful whether it can be of general practical use in machine vision applications.

16.8 Use of Structured Lighting

Structured lighting has already been considered briefly in Section 16.2 as an alternative to stereo for mapping out depth in scenes. Basically, a pattern of light stripes, or other arrangement of light spots or grids, is projected onto the object field. Then these patterns are enhanced in a (generally) single monocular image and analyzed to extract the depth information. To obtain the maximum information, the light pattern must be close-knit, and the received images must be of very high resolution. When shapes are at all complex, the lines can in places appear so close together that they are unresolvable. It then becomes necessary to separate the elements in the projected pattern, trading resolution and accuracy for reliability of interpretation. Even so, if parts of the objects are along the line of sight, the lines can merge together and even cross back and forth, so unambiguous interpretation is never assured. This is part of a larger problem in that parts of the object will be obscured from the projected pattern by occluding bodies or perhaps by self-occlusion. The method has this feature in common with the shape from shading technique and with stereo vision, which relies on *both* cameras being able to view various parts of the objects simultaneously. Hence, the structured light approach is subject to similar restrictions to those found for other methods of 3-D vision and is not a panacea. Nevertheless, it is a useful technique that is generally simple to set up so as to acquire specific 3-D information that can enable a computer to start the process of cueing into complex images.

Light spots provide perhaps the most obvious form of structured light. However, they are restricted because for each spot an analysis has to be performed to determine which spot is being viewed. Connected lines, in contrast, carry a large amount of coding information with them so that ambiguities are less likely to arise. Grids of lines carry even more coding information but do not necessarily give any more depth information. Indeed, if a pattern of light stripes can be projected (for example) from the left of the camera so that they are parallel to the y-axis in the observed image, then it is clear that there is no point in projecting another set of lines parallel to the x-axis, since these merely replicate information that is already available from the rows of pixels in the image. All the depth information is carried by the vertical lines and their horizontal displacements in the image. This analysis assumes that the camera and projected beams are carefully aligned and that no perspective or other distortions are present. Most practical structured lighting systems in current use appear to employ light stripe patterns rather than spot patterns or full grid patterns.

This section ends with an analysis of the situations that can arise when a single stripe is incident on objects as simple as rectangular blocks. Figure 16.13 shows three types of structure in observed stripes: (1) the effect of a sharp angle being encountered; (2) the effect of "jump edges" at which light stripes jump horizontally and vertically at the same time; and (3) the effect of discontinuous edges at which light stripes jump horizontally but not vertically. The reasons for these circumstances will be obvious from Fig. 16.13. Basically, the problem to be tackled with jump and discontinuous edges is to find whether a given stripe end marks an occluding edge or an occluded edge. The importance of this distinction is that occluding edges mark actual edges of the object being observed, whereas occluded edges may be merely edges of shadow regions and are then not *directly* significant.[2] A simple rule is that, if stripes are projected from the left, the left-hand component of a discontinuous edge will be the occluding edge and the right-hand component will be the occluded edge. Angle edges are located by

2 More precisely, they involve interactions of light with two objects rather than with one and are therefore more complex to interpret.

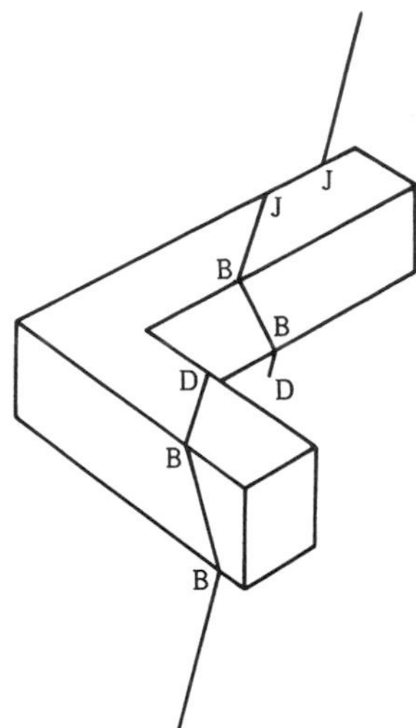

Figure 16.13 Three of the structures that are observed when a light stripe is incident on even quite simple shapes: bends (B), jumps (J), and discontinuities (D).

applying a Laplacian type of operator that detects the change in orientation of the light stripe.

The ideas outlined above correspond to possible 1-D operators that interpret light stripe information to locate nonvertical edges of objects. The method provides no direct information concerning vertical edges. To obtain such information, it is necessary to analyze the information from sets of light stripes. For this purpose 2-D edge operators are required, which collect sufficient data from at least two or three adjacent light stripes. Further details are beyond the scope of this chapter.

Before leaving the topic of light stripe analysis, it should be noted that stripes provide a useful means of recognizing planes forming the faces of polyhedra and other types of manufactured objects. The characteristic sets of parallel lines can be found and demarcated with relative ease, and the fact that the lines usually give rather strong signals means that line tracking techniques can be applied and that algorithms can operate quite rapidly. However, whole-scene interpretation, including inferring the presence and relative positions of different objects, remains a more complex task, as will be seen below.

16.9 Three-dimensional Object Recognition Schemes

The methods described so far in this chapter employ various means for finding depth at all places in a scene and are hence able to map out 3-D surfaces in a fair amount of detail. On the whole, however, they do not give any clue as to what these surfaces represent. In some situations it may be clear that certain planar surfaces are parts of the background, for example, the floor and the walls of a room, but in general individual objects will not be inherently identifiable. Indeed, objects tend to merge with each other and with the background, so specific methods are needed to segment the 3-D space map[3] and finally recognize the objects, giving detailed information on their positions and orientations.

Before proceeding to study this problem, it should be noted that further general processing can be carried out to analyze the 3-D shapes. Agin and Binford (1976) and others have developed techniques for likening 3-D shapes to "generalized cylinders." They are like normal (right circular) cylinders but with additional degrees of freedom so that the axes can bend and the cross sections can vary, both in size and in detailed shape. Thus, even an animal like a sheep can be likened to a distorted cylinder. On the whole, this approach is elegant but may not be well adapted to describe many industrial objects, and it is therefore not pursued further here. A simpler approach may be to model the 3-D surfaces as planar, quadratic, cubic, and quartic surfaces, and then to try to understand these model surfaces in terms of what is known about existing objects. This approach was adopted by Hall et al. (1982) and was found to be viable, at least for certain quite simple objects such as cups. Shirai (1987) has taken the approach even further so that a whole range of objects can be found and identified in quite complex indoor scenes.

At this point let us examine the situation anew to find what it is we are trying to achieve. First, can recognition be carried out *directly* on the mapped out 3-D surfaces, just as it could for the 2-D images of earlier chapters? Second, if we can bypass the 3-D modeling process, and still recognize objects, might it not be possible to save even more computation and omit the stage of mapping out 3-D surfaces, instead identifying 3-D objects directly in 2-D images? It might even be possible to locate 3-D objects from a single 2-D image.

Consider the first of these problems. When we studied 2-D recognition, many instances were found where the HT approach was of great help. It turned out to give trouble in more complex cases, particularly when attempts were made to find objects where there were more than two or at most three degrees of freedom. Here, however, we have situations where objects normally have six degrees of freedom—three degrees of freedom for translation and another three for rotation. This doubling of the number of free parameters on going from 2-D to 3-D makes the situation far worse, since the search space is proportional in size not to the number of degrees of freedom, but to its exponent. For example, if each degree of freedom in translation or rotation can have 256 values, the

3 This may be defined as an imagined 3-D map showing, without interpretation, the surfaces of all objects in the scene and incorporating all the information from depth or range images. Note that it will generally include only the front surfaces of objects seen from the vantage point of the camera.

number of possible locations in parameter space changes from 256^3 in 2-D to 256^6 in 3-D. This has a profound effect on object location schemes and tends to make the HT technique difficult to implement. Ballard and Sabbah (1983) examined the problem for the particular case where space maps are available for 3-D scenes.

16.10 The Method of Ballard and Sabbah

In their paper Ballard and Sabbah (1983) made great play about the order in which various parameters could be determined, concluding that there is a natural precedence—(a) scale, (b) orientation, and (c) translation. Thus, if the scale of an object is unknown, it can in principle be determined without first finding the three orientation parameters, and the latter can in turn be found independently of the three translation parameters. It is easy to see that scale can be found regardless of the other six parameters if a complete space map is available, since the volume of the object[4] can be measured and this immediately determines the scale.[5] Similarly, the orientation and translation parameters can be decoupled merely by ignoring the absolute locations of the features in the space map and hence narrowing down the search problem to that of finding the three object orientation parameters. Once this has been achieved, we can then go back to the original image data and find the three translation parameters, keeping the orientation parameters fixed.

The case Ballard and Sabbah examined was that of a polyhedron with planar faces of known area. To proceed, they first considered the case of a polygon all of whose edges have been located and whose lengths l_i have been measured in a 2-D image (Fig. 16.14).

They then considered the difference between the body-centered frame of axes and the viewer-centered frame. It is assumed that once an edge of length l_{vi} is found in the image, its orientation θ_{bi} relative to the body-centered frame can be obtained from a lookup table $\{(l_{bj}, \theta_{bj})\}$. Then its actual orientation θ_{vi} in the image gives an estimate $\theta_{vi} - \theta_{bi}$ for the orientation of the body-centered frame relative to that of the viewer-centered frame. Thus it is possible to construct a 1-D parameter space for relative orientation $\theta_v - \theta_b$ and to use the individual estimates $\theta_{vi} - \theta_{bi}$ to obtain an accurate measure of the body orientation. (Similar ideas have already been met in Chapters 14 and 15.)

---

4 Or what is temporarily hypothesized to be an object.

5 However, there is little evidence to support the idea that separating scale from orientation is universally possible as Ballard and Sabbah imply. Their paper supports the view that scale and orientation generally have to be determined simultaneously, for example, in the same parameter space.

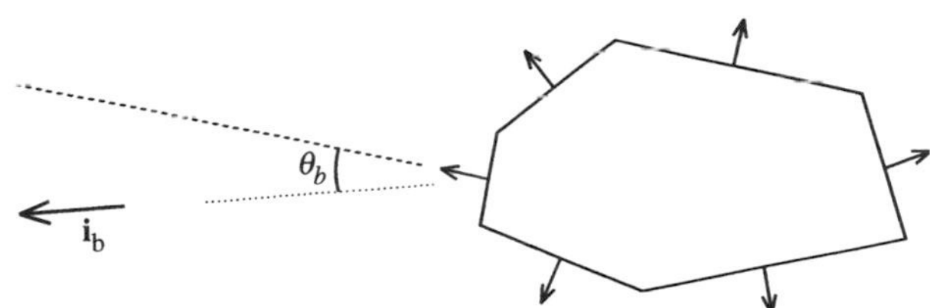

Figure 16.14 Locating the body axis of a polygon. An estimate of one body axis of a polygon may be obtained from the orientation of a single edge.

To generalize the concept to 3-D, the lengths l_{vi} are first generalized to the areas A_{vi} of the planar polygonal faces of a polyhedral shape that is to be located in the 3-D space map (Fig. 16.15). The problem now becomes rather more complicated, since each area A_{vi} (and the direction of its normal $\mathbf{n}_{vi}$) leads to an incomplete estimate of the direction of the body-centered axis $\mathbf{i}_b$. In fact, $\mathbf{i}_b$ is assumed to be inclined at an angle θ_{bi} relative to $\mathbf{n}_{vi}$ Clearly, this defines not a single orientation in space but rather a *cone* of directions whose symmetry axis is $\mathbf{n}_{vi}$ (Fig. 16.15). The parameter space for determining the orientation of body axis $\mathbf{i}_b$ is best pictured on the unit (Gaussian) sphere of direction cosines (l, m, n), where the relation:

$$l^2 + m^2 + n^2 = 1 \tag{16.21}$$

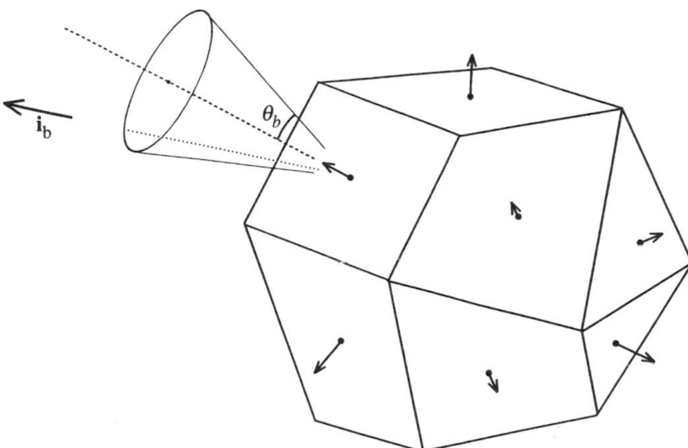

Figure 16.15 Locating the body axis of a polyhedron. The orientation of one body axis of a polyhedron may be narrowed down to a cone of directions, the precise orientation ultimately being determined by the intersection of a number of such cones.

ensures that this apparently 3-D parameter space is actually only 2-D. Details of how to find peaks in this parameter space via 2-D representations are to be found in Ballard and Sabbah (1983).

Next, note another complication—that so far we have only seen how to determine one of the body axes. Determining a second body axis $\mathbf{j}_b$ normal to the first requires a separate lookup table of orientations φ_{bi} to be stored for areas A_{bi}, and a separate parameter space has to be used to accumulate the cones specifying $\mathbf{j}_b$. Once $\mathbf{i}_b$ and $\mathbf{j}_b$ are known, the third axis of a right-angled set is determined uniquely using the relation

$$\mathbf{k}_b = \mathbf{i}_b \times \mathbf{j}_b \tag{16.22}$$

Finally, the method may falsely identify as a polyhedron of the relevant type a set of areas that happen to have the correct orientations but not the correct translations. Hence, it is necessary to proceed by finding the position of the body using another parameter space and to check at that stage that consistent results (sufficiently high peaks) are obtained, indicating a valid body of this type.

The whole procedure involves more computation for 3-D than for 2-D, but it is tractable, supposing that the true areas of the polyhedral faces and the directions of their normals can be measured from the space map. It should be noted that the method has the normal robustness against occlusion and noise that is to be expected with the HT approach, although the maximum peak height for a polyhedron of p faces is p and this reduces to p' when only p' faces are unoccluded. Thus, the method will fail when p' is less than 3. (For two unoccluded faces, there will be two cones specifying a given body axis, and the resulting two circles on the unit sphere intersect in two places, leaving an ambiguity of orientation.)

Many problems of 3-D vision cannot easily be solved by the above approach. Cases where objects have curved faces are intractable by this method, as are situations where simpler bodies such as cubes (for which $p = 3$ but p' may be lower) need to be examined. In addition, depth maps are somewhat more difficult to obtain than intensity images. The method of Silberberg et al. (1984) tackles the problem of locating objects with plane faces in normal 2-D intensity images where depth information is not available. In addition, by working directly from individual surface features, it avoids the stage of segmenting surfaces.

16.11 The Method of Silberberg et al.

This method aims to employ the GHT to locate planar objects in 2-D images by means as close as possible to the normal GHT approach. The problem of too many parameters having to be determined is minimized first by ensuring that the scale of the objects is known in advance, so that six rather than seven parameters have to be determined, and second by arranging that projection is orthographic rather than perspective, so that there is no way in which depth information can affect the image. Hence, only five parameters are active in the image—that is, there are two translational and three rotational parameters.[6]

Of the remaining parameters, two of the rotations are made implicit by assuming a particular viewpoint over the object. Then only its 2-D translation parameters and orientation about the camera axis remain to be found. A normal parameter space is constructed for the latter three parameters. The effect of implicit viewpoint variation is investigated by specifying a fixed large number of viewpoints arranged regularly over the Gaussian sphere, this being achieved by employing a regular icosahedron whose 20 (equilateral) triangular faces are each subdivided into four smaller equilateral triangles. This type of approach has been employed by a number of workers (Koenderink and van Doorn, 1979; Chakravarty and Freeman, 1982).

The overall strategy is to record the height of the main peak in parameter space for each of the 80 viewpoints and then to improve the best set of five viewpoints in several iterations, each time splitting each of the triangles into four smaller equilateral triangles. After two iterations, the best viewpoint is obtained, and the object is located by the method to a suitable degree of accuracy.

The case investigated by Silberberg was that of a cube on whose faces certain letters appear. The faces of a cube are evidently planar, and all relevant features are straight edges. Votes are accumulated in parameter space at every position representing some translation or rotation of a line that leads to a valid position in a model of the object. The criterion for validity also includes matching the lengths of lines in the image and in the projected model, although some fragmentation of lines is taken to be acceptable. Hence lines must, if anything, appear shorter than the lengths deduced from the model. After the initial run, two iterations were found to permit objects to be positively identified, located within 4 pixels and orientated within 5°. Overall, the method is successful for the restricted problem it tackles—objects with planar faces and straight-line features, moderate accuracy requirements, the need to find only five parameters, and the accompanying restriction to orthographic projection (which may often be inconvenient).

With this background of work on orthographic projection, other workers have felt it useful to revert to perspective projection. There are good grounds for doing so since, as stated above, orthographic projection prevents any information on depth within a scene from being deduced using a single monocular view. Unfortunately, there are potential penalties from proceeding with perspective

6 Note that Ballard and Sabbah suggest this idea but then employ space maps that completely specify depth information, thereby bringing the effective parameter count back to six.

projection: (1) the parameter count goes back up from five to six, so a larger parameter space has to be searched, and (2) parallel body lines no longer remain parallel in the image, so a valuable set of cues that can help unlock complex image data is lost. We next show how a powerful strategy developed by Horaud (1987) can be employed, under either projection scheme, to facilitate the recognition of polyhedral objects.

16.12 Horaud's Junction Orientation Technique[7]

Horaud's (1987) technique is special in that it uses as its starting point 2-D images of 3-D scenes and "backprojects" them into the scene, with the aim of making interpretations in 3-D rather than 2-D frames of reference. This has the initial effect of increasing mathematical complexity, although in the end useful and supposedly more accurate results emerge.

Initially, the boundaries of planar surfaces on objects are backprojected. Each boundary line is thus transformed into an "interpretation plane" defined by the center of the camera projection system and the boundary line in the image plane. Clearly, the interpretation plane must contain the line that originally projected into the boundary line in the image. Similarly, angles between boundary lines in the image are backprojected into two interpretation planes, which must contain the original two object lines. Finally, junctions between three boundary lines are backprojected into three interpretation planes that must contain a corner in the space map (Fig. 16.16). The paper focuses on the backprojection of junctions and shows how measurements of the junction angles in the image relate to those of the original corner. It also shows how the space orientation of the corner can be computed. In fact, it is interesting that the orientation of an object in 3-D can in general be deduced from the appearance of just one of its corners in a single image. This is a powerful result and in principle permits objects to be recognized and located from extremely sparse data.

To understand the method, the mathematics first needs to be set up with some care. Assume that lines $\mathbf{L}_1$, $\mathbf{L}_2$, $\mathbf{L}_3$ meet at a junction in an object and appear as lines $\mathbf{l}_1$, $\mathbf{l}_2$, $\mathbf{l}_3$ in the image (Fig. 16.16). Take respective interpretation planes containing the three lines and label them by unit vectors $\mathbf{P}_1$, $\mathbf{P}_2$, $\mathbf{P}_3$ along their normals, so that:

$$\mathbf{P}_1 \cdot \mathbf{L}_1 = 0 \tag{16.23}$$

7 This and related techniques are sometimes referred to as "shape from angle."

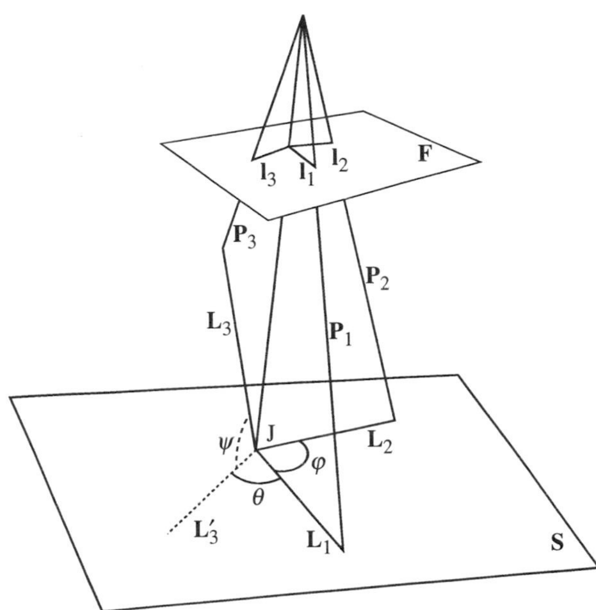

Figure 16.16 Geometry for backprojection from junctions. A junction of three lines in an image may be backprojected into three planes, from which the orientation in space of the original corner J may be deduced.

$$\mathbf{P}_2 \cdot \mathbf{L}_2 = 0 \tag{16.24}$$

$$\mathbf{P}_3 \cdot \mathbf{L}_3 = 0 \tag{16.25}$$

In addition, take the space plane containing $\mathbf{L}_1$ and $\mathbf{L}_2$, and label it by a unit vector $\mathbf{S}$ along its normal, so that:

$$\mathbf{S} \cdot \mathbf{L}_1 = 0 \tag{16.26}$$

$$\mathbf{S} \cdot \mathbf{L}_2 = 0 \tag{16.27}$$

Since $\mathbf{L}_1$ is perpendicular to $\mathbf{S}$ and to $\mathbf{P}_1$, and $\mathbf{L}_2$ is perpendicular to $\mathbf{S}$ and to $\mathbf{P}_2$, it is found that:

$$\mathbf{L}_1 = \mathbf{S} \times \mathbf{P}_1 \tag{16.28}$$

$$\mathbf{L}_2 = \mathbf{S} \times \mathbf{P}_2 \tag{16.29}$$

Note that $\mathbf{S}$ is not in general perpendicular to $\mathbf{P}_1$ and $\mathbf{P}_2$, so $\mathbf{L}_1$ and $\mathbf{L}_2$ are not in general unit vectors. Defining φ as the angle between $\mathbf{L}_1$ and $\mathbf{L}_2$, we now have:

$$\mathbf{L}_1 \cdot \mathbf{L}_2 = L_1 L_2 \cos\varphi \tag{16.30}$$

which can be reexpressed in the form:

$$(\mathbf{S} \times \mathbf{P}_1)\cdot(\mathbf{S} \times \mathbf{P}_2) = |\mathbf{S} \times \mathbf{P}_1||\mathbf{S} \times \mathbf{P}_2|\cos\varphi \tag{16.31}$$

Next we need to consider the junction between $\mathbf{L}_1$, $\mathbf{L}_2$, $\mathbf{L}_3$. To proceed, it is necessary to specify the relative orientations in space of the three lines. θ is the angle between $\mathbf{L}_1$ and the projection $\mathbf{L}_3'$ of $\mathbf{L}_3$ on plane $\mathbf{S}$, while ψ is the angle between $\mathbf{L}_3'$ and $\mathbf{L}_3$ (Fig. 16.16). Thus, the structure of the junction J is described completely by the three angles φ, θ, ψ. $\mathbf{L}_3$ can now be found in terms of other quantities:

$$\mathbf{L}_3 = \mathbf{S}\sin\psi + \mathbf{L}_1\cos\theta\cos\psi + (\mathbf{S} \times \mathbf{L}_1)\sin\theta\cos\psi \tag{16.32}$$

Applying equation (16.25), it may be deduced that:

$$\mathbf{S}\cdot\mathbf{P}_3\sin\psi + \mathbf{L}_1\cdot\mathbf{P}_3\cos\theta\cos\psi + (\mathbf{S} \times \mathbf{L}_1)\cdot\mathbf{P}_3\sin\theta\cos\psi = 0 \tag{16.33}$$

Substituting for $\mathbf{L}_1$ from equation (16.28), and simplifying, we finally obtain:

$$\begin{aligned}(\mathbf{S}\cdot\mathbf{P}_3)|\mathbf{S} \times \mathbf{P}_1|\sin\psi &+ \mathbf{S}\cdot(\mathbf{P}_1 \times \mathbf{P}_3)\cos\theta\cos\psi \\ &+ (\mathbf{S}\cdot\mathbf{P}_1)(\mathbf{S}\cdot\mathbf{P}_3)\sin\theta\cos\psi = (\mathbf{P}_1\cdot\mathbf{P}_3)\sin\theta\cos\psi\end{aligned} \tag{16.34}$$

Equations (16.31) and (16.34) now exclude the unknown vectors $\mathbf{L}_1$, $\mathbf{L}_2$, $\mathbf{L}_3$, but they retain $\mathbf{S}$, $\mathbf{P}_1$, $\mathbf{P}_2$, $\mathbf{P}_3$, and the three angles φ, θ, ψ. $\mathbf{P}_1$, $\mathbf{P}_2$, $\mathbf{P}_3$ are known from the image geometry, and the angles φ, θ, ψ are presumed to be known from the object geometry. In addition, only two components (α, β) of the unit vector $\mathbf{S}$ are independent, so the two equations should be sufficient to determine the orientation of the space plane $\mathbf{S}$. Unfortunately, the two equations are highly nonlinear, and it is necessary to solve them numerically. Horaud (1987) achieved this by reexpressing the formulas in the forms:

$$\cos\varphi = f(\alpha, \beta) \tag{16.35}$$

$$\begin{aligned}\sin\theta\cos\psi = g_1(\alpha, \beta)\sin\psi &+ g_2(\alpha, \beta)\cos\theta\cos\psi \\ &+ g_3(\alpha, \beta)\sin\theta\cos\psi\end{aligned} \tag{16.36}$$

For each image junction, $\mathbf{P}_1$, $\mathbf{P}_2$, $\mathbf{P}_3$ are known, and it is possible to evaluate f, g_1, g_2, g_3. Then, assuming a particular interpretation of the junction, values are assigned to φ, θ, ψ and curves giving the relation between α and β are plotted for each equation. Possible orientations for the space plane $\mathbf{S}$ are then given by positions in (α, β) space where the curves cross. Horaud has shown that, in general, 0, 1, or 2 solutions are possible. The case of no solutions corresponds to trying to make an impossible match between a corner and an image junction when totally the wrong angles φ, θ, ψ are assumed. One solution is the normal situation; and two solutions arise in the interesting special case when orthographic or near-orthographic projection permits perceptual reversals—that is, a convex corner is interpreted as a concave corner or vice versa. Under orthographic projection, the image data from a single corner are insufficient, taken on their own, to give a unique interpretation. In this situation, even the human visual system makes mistakes—as in the case of the well-known Necker cube illusion (see Chapter 17). However, when such cases arise in practical situations, it may be better to take the convex rather than the concave corner interpretation as a working assumption, as it has slightly greater likelihood of being correct.

Horaud has shown that such ambiguities are frequently resolved if the space plane orientation is estimated simultaneously for all the junctions bordering the object face in question, by plotting the α and β values for all such junctions on the same α, β graph. For example, with a cube face on which there are three such junctions, nine curves are coincident at the correct solution, and there are nine points where only two curves cross, indicating false solutions. On the other hand, if the same cube is viewed under conditions approximating very closely to orthographic projection, two solutions with nine coincident curves appear and the situation remains unresolved, as before.

Overall, this technique is important in showing that, although lines and angles individually lead to virtually unlimited numbers of possible interpretations of 3-D scenes, junctions lead individually to at most two solutions and any remaining ambiguity can normally be eliminated if junctions on the same face are considered together. As has been seen, the exception to this rule occurs when projection is accurately orthographic, although this situation can often be avoided in practice.

So far we have considered only how a given hypothesis about the scene may be tested; nothing has been said about how assignments of the angles φ, θ, ψ are made to the observed junctions. Horaud's paper discussed this aspect of the work in some depth. In general, the approach is to use a depth-first search technique in which a match is "grown" from the initial most promising junction assignment. Considerable preprocessing of sample data is carried out to find how to rank image features for their utility during depth-first search interpretation. The idea is to order possible alternatives such as linear or circular arcs, convex or concave junctions, short or long lines, and so on. In this way the tree search becomes more planned and efficient at run-time. Generally, the more frequently occurring types of features should be weighted down in favor of the rarer types of features for greater search efficiency. In addition, it should be remembered that hypothesis generation is relatively expensive in that it demands a stage of backprojection, as described above. Ideally, this stage need be employed only once for each object (in the case that only a single corner is, initially, considered). Subsequent stages of processing then involve hypothesis verification in which other features of the object are predicted and their presence sought in the image. If found they are used to refine the existing match. If the match at any stage becomes worse, then the algorithm backtracks and eliminates one or more features and proceeds with other ones. This process is unavoidable, since more than one image feature may be present near a predicted feature.

One factor that has been found to make the method converge quickly is the use of grouped rather than individual features, since this tends to decrease the combinatorial explosion in the size of the search. In the present context, this means that attempts should be made to match first all junctions or angles bordering a given object face, and further that a face should be selected that has the greatest number of matchable features around it.

In summary, this approach is claimed to be successful since it backprojects from the image and then uses geometrical constraints and heuristic assumptions for matching in 3-D space. It turns out to be suitable for matching objects that possess planar faces and straight-line boundaries, hence giving angle and junction features. However, extending the backprojection technique to situations where object faces are curved and have curved boundaries will almost certainly prove much more difficult.

16.13 An Important Paradigm—Location of Industrial Parts

In this section we consider the location of a common class of industrial part; this constitutes an important example that has to be solved in one way or another. Here we go along with the Bolles and Horaud (1986) approach as it leads to sensible solutions and embodies a number of useful didactic lessons. However, it should be emphasized that many related approaches have been used successfully by other workers over the intervening years. The method starts with a depth map of the scene (obtained in this case using structured lighting).

Figure 16.17 shows in simplified form the type of industrial part being sought in the images. In typical scenes several of these parts may appear jumbled on a worktable, with perhaps three or four being piled on top of each other in some places. In such cases, it is vital that the matching scheme be highly robust if most of the parts are to be found, since even when a part is unoccluded, it appears against a highly cluttered and muddled background. However, the parts themselves have reasonably simple shapes and possess certain salient features. In the particular problem cited, each has a cylindrical base with a concentric cylindrical head, and also a planar shelf attached symmetrically to the base. To locate such objects, it is natural to attempt to search for circular and

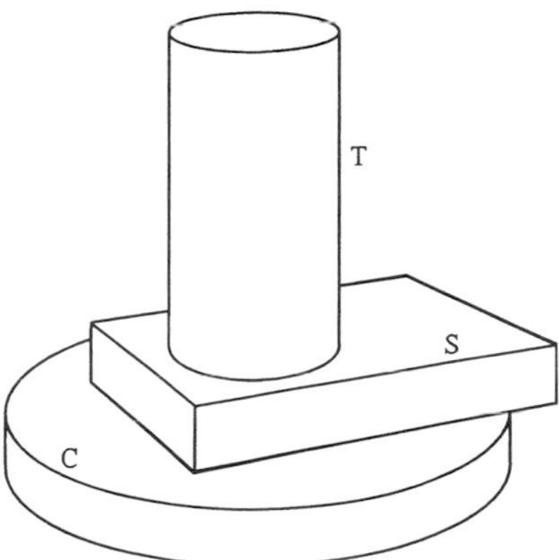

Figure 16.17 The essential features of the industrial components located by the 3DPO system of Bolles and Horaud (1986). S, C, and T indicate, respectively, straight and circular dihedral edges and straight tangential edges, all of which are searched for by the system.

straight dihedral edges. In addition, because of the types of data being used, it is useful to search for straight tangential edges, which appear where the sides of curved cylinders are viewed obliquely.

In general, circular dihedral edges appear elliptical, and parameters for five of the six degrees of freedom of the part can be determined by analyzing these edges. The parameter that cannot be determined in this way corresponds to rotation about the axis of symmetry of the cylinder.

Straight dihedral edges also permit five free parameters to be determined, since location of one plane eliminates three degrees of freedom and location of an adjacent plane eliminates a further two degrees of freedom. The parameter that remains undetermined is that of linear motion along the direction of the edge. However, there is also a further ambiguity in that the part may appear either way around on the dihedral edge.

Straight tangential edges determine only four free parameters, since the part is free to rotate about the axis of the cylinder and can also move along the tangential edge. These edges are the most difficult to locate accurately, since range data are subject to greater levels of noise as surfaces curve away from the sensor.

All three of these types of edges are planar. They also provide useful additional information that can help to identify where they are on a part. For example, both straight and curved dihedral edges provide information on the size of the included angle, and the curved edges in addition give radius values. By now it will be plain that curved dihedral edges provide significantly more parametric information about a part than either of the other two types of edges, and therefore they are of most use to form initial hypotheses about the pose (position and orientation) of a part. Having found such an edge, we need to try out various hypotheses about which edge it is, for example, by searching for other circular dihedral edges at specific relative positions; this is a vital hypothesis verification step. Next, the problem of how to determine the remaining free parameter is solved by searching for the linear straight dihedral edge features from the planar shelf on the part. In principle, this could be done using pure depth-first search, but in this case an exception is made, since all the available line features are very similar and they are best located together by the maximal clique (breadth-first exhaustive search) technique.

At this stage, hypothesis generation is complete, and the part is essentially found, but hypothesis verification is required (1) to confirm that the part is genuine and not an accidental grouping of independent features in the image, (2) to refine the pose estimate, and (3) to determine the configuration of the part, that is, to what extent it is buried under other parts (making it difficult for a robot to pick it up). When the most accurate pose has been obtained, the overall degree of fit can be considered and the hypothesis rejected if some relevant criterion is not met.

In common with other researchers (Faugeras and Hebert, 1983; Grimson and Lozano-Perez, 1984), Bolles and Horaud took a depth-first tree search as the basic matching strategy. Their scheme uses a minimum number of features to key into the data, first generating hypotheses and then taking care to ensure verification. (Note that Bolles and Cain, 1982, had earlier used this technique in a 2-D part location problem.) This contrasts with much work (especially that based on the HT) which makes hypotheses but does not check them. (Forming the initial hypotheses is the difficult and computationally intensive part of the work. Researchers will therefore write about this aspect of their work and perhaps not bother to state the minor amount of computation that went into confirming that objects had indeed been located. Note also that in much 2-D work, images can be significantly simpler and the size of the peak in parameter space can be so large as to make it virtually certain that an object has been located—thus rendering verification unnecessary.)

16.14 Concluding Remarks

To the layperson, 3-D vision is an obvious and automatic result of the fact that the human visual system is binocular and presumes both that binocular vision is the only way to arrive at depth maps and that once they have been obtained the subsequent recognition process is trivial. However, what this chapter has actually demonstrated is that neither of these commonly held views is valid.

First, there are a good many ways of arriving at depth maps, and some of them at least are available using monocular vision. Second, the complexity of the mathematical calculations involved in locating objects and the amount of abstract reasoning involved in obtaining robust solutions—plus the need to ensure that the latter are not ambiguous—are taxing even in simple cases, including those where the objects have well-defined salient features.

Despite the diversity of methods covered in this chapter, certain common themes may be identified: the use of trigger features, the value of combining features into groups that are analyzed together, the need for working hypotheses to be generated at an early stage, the use of depth-first heuristic search (combined where appropriate with more rigorous breadth-first evaluation of the possible interpretations), and the detailed verification of hypotheses. All these can be taken as parts of current methodology. *Details*, however, vary with the dataset. More specifically, if a new type of industrial part is to be considered, some study must be made of its most salient features. Then this causes not only the feature detection scheme to vary but also the heuristics of the search employed—as well as the mathematics of the hypothesis mechanism. The reader is referred to the following chapter for further discussion of object recognition under perspective projection.

Although some of the later sections of this chapter have concentrated on object recognition and have perhaps tended to eschew the value of range measurements and depth maps, it is possible that this might give a misleading impression of the situation. In many situations, recognition is largely irrelevant, but it is mandatory to map out 3-D surfaces in great detail. Turbine blades, automobile body parts, or even food products such as fruit may need to be measured accurately in 3-D. In such cases, it is known in advance what object is in what position, but some inspection or measurement function has to be carried out and a diagnosis made. In such instances, the methods of structured lighting, stereopsis, or photometric stereo come into their own and are highly effective methods. Ultimately, too, one might expect that a robot vision system will have to use all the tricks of the human visual system if it is to be as adaptable and useful when operating in an unconstrained environment rather than at a particular worktable.

This has been a preliminary chapter on 3-D vision, setting the scene for the whole of Part 3. Chapter 17 will be devoted to a careful analysis of the distinction between weak and full perspective projection and how this affects the object recognition process. Chapter 18 will introduce the topic of motion analysis in 3-D scenes. Chapter 19 will aim to show something of the elegance and value of invariants in providing short cuts around some of the complexities of full perspective projection, and Chapter 20 will use this work to take motion analysis one stage further. Chapter 21 will consider camera calibration and will also consider how recent research on interrelating multiple views of a scene has allowed some of the tedium of camera calibration to be bypassed.

Conventional wisdom indicates that binocular vision is the key to understanding the 3-D world. This chapter has shown that the correspondence problem makes the practice of binocular vision tedious, while the solutions it provides are only depth maps and require further intricate analysis before the 3-D world can fully be understood.

16.15 Bibliographical and Historical Notes

As noted earlier in the chapter, the most obvious approach to 3-D perception is to employ a binocular camera system. Burr and Chien (1977) and Arnold (1978) showed how a correspondence could be set up between the two input images by use of edges and edge segments. Forming a correspondence can involve considerable computation: Barnea and Silverman (1972) showed how this problem could be alleviated by passing quickly over unfavorable matches. Similarly, Moravec (1980) devised a coarse-to-fine matching procedure that arrives systematically at an accurate correspondence between images. Marr and Poggio (1979) formulated two constraints—those of uniqueness and continuity—that have to be satisfied in choosing global correspondences: these constraints are important in leading to the simplest available surface interpretation. Ito and Ishii (1986) found that there is something to be gained from three-view stereo in offsetting ambiguity and the effects of occlusions.

The structured lighting approach to 3-D vision was introduced independently by Shirai (1972) and Agin and Binford (1973, 1976) in the form of a single plane of light, while Will and Pennington (1971) developed the grid coding technique. Nitzan et al. (1977) employed an alternative LIDAR (light detecting and ranging) scheme for mapping objects in 3-D. Here short light pulses were timed as they traveled to the object surface and back.

Meanwhile, other workers were attempting monocular approaches to 3-D vision. Some basic ideas underlying shape-from-shading date from as long ago as 1929, with Fesenkov's investigations of the lunar surface. See also van Digellen (1951). However, the first shape-from-shading problem to be solved both theoretically and in an operating algorithm appears to have been that of Rindfleisch (1966), also relating to lunar landscapes. Thereafter Horn systematically tackled the problem both theoretically and with computer investigations, starting with a notable review (1975) and resulting in prominent papers (e.g., Horn, 1977; Ikeuchi and Horn, 1981; Horn and Brooks, 1986), an important book (Horn, 1986), and an edited work (Horn and Brooks, 1989). Interesting papers by other workers in this area include Blake et al. (1985), Bruckstein (1988), and Ferrie and Levine (1989). Woodham (1978, 1980, 1981) must be credited with the photometric stereo idea. Finally, the vital contributions made by workers on computer graphics in this area must not be forgotten—see, for example, Phong (1975) and Cook and Torrance (1982).

The concept of shape-from-texture arose from the work of Gibson (1950) and was developed by Bajcsy and Liebermann (1976), Stevens (1980), and notably by Kender (1981), who carefully explored the underlying theoretical constraints.

The paper by Barrow and Tenenbaum (1981) provides a very readable review of much of this earlier work. It is probably true to say that 1980 marked a turning point, when the emphasis in 3-D vision shifted from mapping out surfaces to interpreting images as sets of 3-D objects. Possibly, this segmentation task could not be tackled easily earlier because basic tools such as the HT were not sufficiently well developed (e.g., the GHT dates from 1981). The work of Koenderink and van Doorn (1979) and Chakravarty and Freeman (1982) was probably also crucial in providing a framework of potential 3-D views of objects enabling interpretation schemes to be developed. The work of Ballard and Sabbah (1983) provided an early breakthrough in segmentation of real objects in 3-D (see Section 16.10), and this was followed by vital further work by Faugeras and Hebert (1983), Silberberg et al. (1984), Bolles and Horaud (1986), Horaud (1987), Pollard et al. (1987), and many others (see foregoing sections).

The review article of Besl and Jain (1984) is particularly useful but is now very dated. Other interesting work includes that of Horaud et al. (1989) on solving the perspective 4-point problem (finding the position and orientation of the camera relative to known points) and complements the work of Section 16.12. For further references on this topic see Section 17.6. Winston (1975) contains several classic papers on 3-D vision—namely, those of Horn, Waltz, and Shirai. Waltz's work on understanding line drawings of polyhedra is interesting in rigorously classifying all types of line junctions and making full use of information available from shadows. However, it assumes an ideal "blocks world," and for this reason its importance is limited.

Though finding vanishing points might be thought a well-worked-through topic, research on it proceeded further in the 1990s (e.g., Lutton et al., 1994; Straforini et al., 1993; Shufelt, 1999). Similarly, stereo correlation matching techniques were still under development, to maintain robustness in real-time applications (Lane et al., 1994).

Trucco and Verri (1998) provide a broad perspective of 3-D vision at a level not dissimilar to that provided in the present volume, while Cipolla and Giblin (2000) develop 3-D vision—together with its relation to motion—adopting a rather more succinct, theoretical stance.

Since 2000, work on stereo vision has continued unabated as a main-line topic (e.g., Lee et al., 2002; Brown et al., 2003), but Horn's approach to shape from shading has been largely superseded. One new technique is the Green's function approach to shape from shading (Torreão, 2001, 2003), while local shape from shading has been used to improve the photometric stereo technique (Sakarya and Erkmen, 2003). Photometric stereo has itself been developed considerably further in a new 4-source technique capable of coping with highlights and shadows (Barsky and Petrou, 2003). Another development is the application of shape from shading to radar data—a translation that required significant new theory (Frankot and Chellappa, 1990; Bors et al., 2003). Finally, a thoroughgoing new approach to the whole study of 3-D vision and its dependence on the light field has been initiated (Baker et al., 2003). This paper starts by comparing what can be learned from (a) stereo vision and (b) a shape from silhouette approach (observing object silhouettes from all directions in the given light field). An important conclusion is that the shapes of Lambertian objects can be uniquely determined with n-camera stereo, without ambiguity, unless there are regions of constant intensity present: indeed, constant intensity is found *always* to lead to ambiguity.[8] This paper is important not only in giving a fresh view of the problems of 3-D vision in general, and shape from shading in particular, but also in demonstrating certain open questions.

16.16 **Problems**

1. Prove that all epipolar lines in one image plane pass through the point that is the image of the projection point of the alternate image plane.

2. What is the physical significance of the straight-line contour in gradient space (see Fig. 16.9*b*)?

3. Find the condition for a contour in gradient space to be an ellipse.

4. Derive a vector formula giving the unit vector **r** of the direction for pure specular reflection in terms of the unit vectors **s** and **n** of the source and surface normal directions (Section 16.4). Hence, show how the total reflectance (equation (16.12)) may be computed as a function of p and q.

---

8 Essentially, this is because there may be a concavity whose light properties outside the concavity hull will be indistinguishable from those of the hull itself (Laurentini, 1994).

5. Sketch a curve of the function $\cos^m\theta$. Estimate what value m would have to have for 90% of the R_1 component to be reflected within 10° of the direction for pure specular reflection.

6. An alien has three eyes. Does this permit it to perceive or estimate depth more accurately than a human? What would be the best placement for a third eye?

7. A cube is viewed in orthographic projection. Show that although the cube is opaque, it is easy to compute the theoretical position of its centroid in the image. Show also that the orientation of the cube can be deduced by considering the apparent areas of its faces. If the contrast between the faces becomes so low that only a hexagonal outline is seen, show that ambiguities will arise in our knowledge of the orientation of the cube. Are ambiguities specific to cubes, or do they arise with other shapes? Why?

8. (a) A feature at (X, Y, Z) appears at locations (x_1, y_1), and (x_2, y_2) in the two images of a binocular imaging system. The image planes of both cameras lie in the same plane; f is the focal length of both camera lenses, and b is the separation of the optical axes of the lenses. Label Fig. 16.P1 appropriately: by considering pairs of similar triangles, show that:

$$\frac{Z}{f} = \frac{(X + b/2)}{x_1} = \frac{(X - b/2)}{x_2}$$

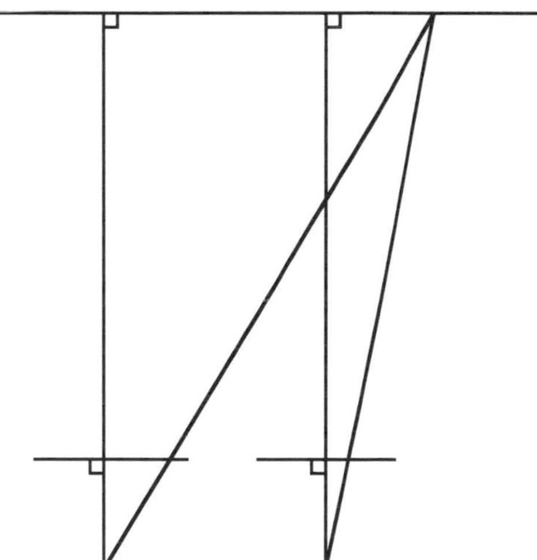

Figure 16.P1

(b) Hence, derive a formula that can be used to determine depth Z from the observed disparity.

9. Give a full proof that the error with which the fractional depth Z in a scene can be computed is: (i) proportional to pixel size, (ii) proportional to Z, and (iii) inversely proportional to the baseline b between the stereo cameras. What other parameter appears in the final formula? Determine under what pair of conditions two very tiny cameras fabricated by nanotechnological methods could still perform viable depth measurement.

10. (a) Draw a diagram that shows that the ordering of visible points is normally the same in both images seen by a binocular vision system.

    (b) An object has a semitransparent front surface through which an interior feature F is just visible. Show that the ordering of the features in the two views of the object may be sufficient to prove that F is inside, or perhaps behind, the object.

11. (a) State the conditions under which matte surfaces may properly be described as "Lambertian." Show that the normal at a point on a Lambertian surface must lie on a cone of directions whose axis points to the point source of illumination. Show that a minimum of three independent light sources will be needed to identify the exact orientation of a matte surface. Why might four light sources help to determine surface orientation for a surface of unknown or nonideal properties?

    (b) Compare the effectiveness of binocular vision and photometric stereo if it is desired to obtain a depth map for each object in a scene. In each case, consider the properties of the object surface and the distance from the observer.

12. (a) Compare the properties of matte surfaces with those that exhibit "normal" specular reflection. Matte surfaces are sometimes described as "Lambertian." Describe how the brightness of the surface varies according to the Lambertian model.

    (b) Show that for a given surface brightness, the orientation of any point on a Lambertian surface must lie on a certain cone of orientations.

    (c) Three images of a surface are obtained on illuminating it in sequence by three independent point light sources. With the aid of a diagram show how this can lead to unambiguous estimates of surface orientation. Would surface orientation of any points on the surface *not* be estimated by this method? Are there any constraints on the allowable positions of the three light sources? Would it help if *four* independent point light sources were used instead of three?

    (d) Discuss whether the surface map that is obtained by shape from shading is identical to that obtained by stereo (binocular) vision. Are the two approaches best applied in the same or different applications? To what extent is the application of structured light able to give better or more accurate information than these basic approaches?

(e) Consider what further processing is required before 3-D objects can be recognized by any of these approaches.

Tackling the Perspective *n*-point Problem

It is possible to recognize 3-D objects from very few point features, even when they are seen in a single view. In fact, the pose of the 3-D object can also be ascertained from a single view. However, ambiguities of interpretation do arise, and this chapter discusses the disambiguation problem.

Look out for:

- the distinction between weak and full perspective projection.
- how the "perspective inversion" type of ambiguity arises under weak perspective projection.
- how more serious ambiguities arise under full perspective projection.
- how full perspective projection has the capability to provide more interpretive information than weak perspective projection.
- how coplanarity can impose quite strong constraints on 3-D data, which can sometimes be helpful and at other times be an impediment.
- how symmetry can help with 3-D image interpretation.

Although this chapter considers only one aspect of 3-D vision, it raises very important issues that are relevant right through the subject of 3-D object recognition.

CHAPTER 17

Tackling the Perspective *n*-point Problem

17.1 Introduction

This chapter tackles a problem of central importance in the analysis of images from 3-D scenes. It has been kept separate and fairly short so as to focus carefully on relevant factors in the analysis. First, we look closely at the phenomenon of perspective inversion, which has already been alluded to several times in Chapter 16. Then we refine our ideas on perspective and proceed to consider the determination of object pose from salient features which are located in the images. It will be of interest to consider how many salient features are required for unambiguous determination of pose.

17.2 The Phenomenon of Perspective Inversion

In this section, we first study the phenomenon of *perspective inversion*. This is actually a rather well-known effect that appears in the following "Necker cube" illusion. Consider a wire cube made from 12 pieces of wire welded together at the corners. Looking at it from approximately the direction of one corner, we have a difficult time telling which way round the cube is, that is, which of the opposite corners of the cube is the nearer (Fig. 17.1). On looking at the cube for a time, one gradually comes to feel one knows which way round it is, but then it suddenly appears to reverse itself. Then that perception remains for some time, until it, too, reverses itself.[1] This illusion reflects the fact that the brain is

1 In psychology, this shifting of attention is known as *perceptual reversal*, which is unfortunately rather similar to the term *perspective inversion*, but is actually a much more general effect that leads to a host of other types of optical illusions. See Gregory (1971) and the many illustrations produced by M.C. Escher.

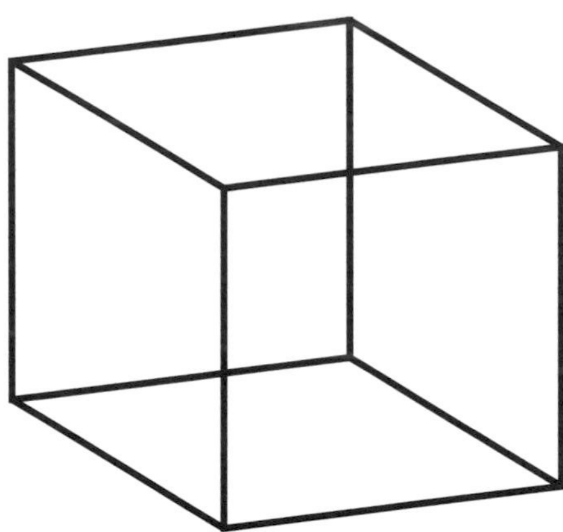

Figure 17.1 The phenomenon of perspective inversion. This figure shows a wire cube viewed approximately from the direction of one corner. The phenomenon of perspective inversion makes it difficult to see which of the opposite corners of the cube is the nearer. There are two stable interpretations of the cube, either of which may be perceived at any moment.

making various hypotheses about the scene, and even making decisions based on incomplete evidence about the situation (Gregory, 1971, 1972).

The wire cube illusion could perhaps be regarded as somewhat artificial. But consider instead an airplane (Fig. 17.2*a*) which is seen in the distance (Fig. 17.2*b*) against a bright sky. The silhouetting of the object means that its surface details are not visible. In that case, interpretation requires that a hypothesis be made about the scene, and it is possible to make the wrong one. Clearly (Fig. 17.2*c*), the airplane could be at an angle α (as for P), though it could equally well be at an angle $-\alpha$ (as for Q). The two hypotheses about the orientation of the object are related by the fact that the one can be obtained from the other by reflection in a plane R normal to the viewing direction D.

Strictly, there is only an ambiguity in this case if the object is viewed under orthographic or scaled orthographic projection.[2] However, in the distance, perspective projection approximates to scaled orthographic projection, and it is often difficult to detect the difference.[3] If the airplane in Fig. 17.2 were quite near, it would be obvious that one part of the silhouette was nearer, as the perspective would distort it in a particular way. In general, perspective

2 *Scaled orthographic projection* is orthographic projection with the final image scaled in size by a constant factor.

3 In this case, the object is said to be viewed under *weak perspective projection*. For weak perspective, the depth ΔZ within the object has to be much less than its depth Z in the scene. On the other hand, the perspective scaling factor can be different for each object and will depend on its depth in the scene. So the perspective can validly be locally weak and globally normal.

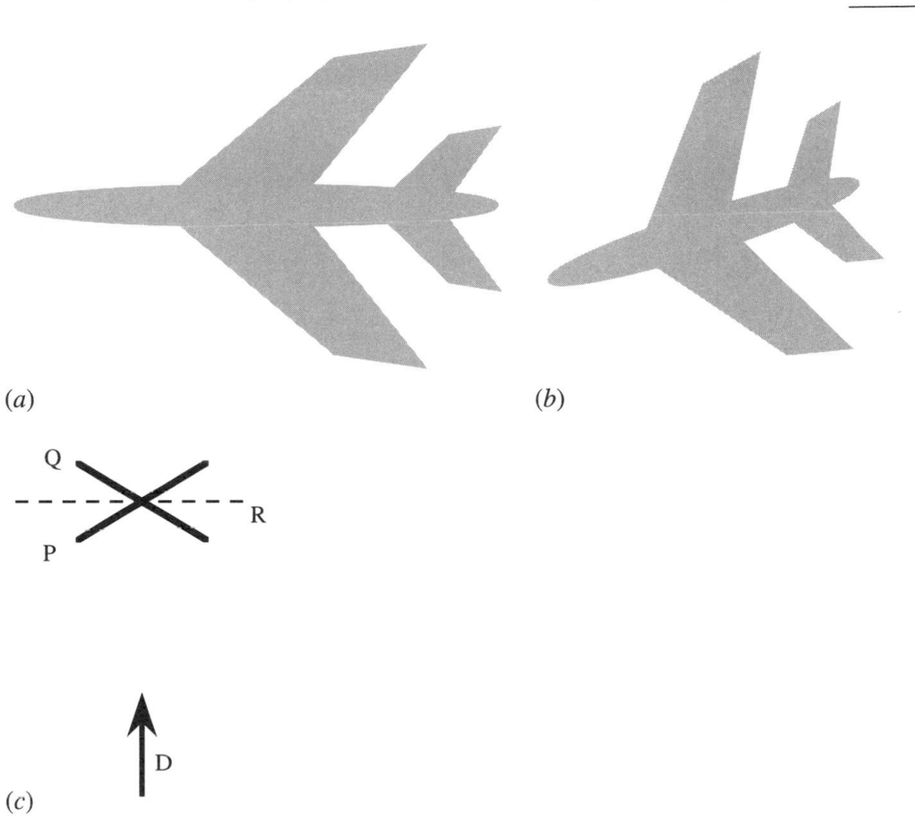

Figure 17.2 Perspective inversion for an airplane. Here an airplane (*a*) is silhouetted against the sky and appears as in (*b*). (*c*) shows the two planes P and Q in which the airplane could lie, relative to the direction D of viewing: R is the reflection plane relating the planes P and Q.

projection will break down symmetries, so searching for symmetries that are known to be present in the object should reveal which way around it is. However, if the object is in the distance, as in Fig. 17.2*b*, it will be virtually impossible to see the breakdown. Unfortunately, short-term study of the motion of the airplane will not help with interpretation in the case shown in Fig. 17.2*b*. Eventually, however, the airplane will appear to become smaller or larger, which will give the additional information needed to resolve the issue.

17.3 Ambiguity of Pose under Weak Perspective Projection

It is instructive to examine to what extent the pose of an object can be deduced under weak perspective projection. We can reduce the above problem to a simple case in which three points have to be located and identified. Any set of three points is coplanar, and the common plane corresponds to that of the silhouette shown in Fig. 17.2*a*. (We assume here that the three points are not collinear, so that they do in fact define a plane.) The problem then is to match the corresponding points on the idealized object (Fig. 17.2*a*) with those on the observed object (Fig. 17.2*b*). It is not yet completely obvious that this is possible or that the solution is unique, even apart from the reflection operation noted earlier. It could be that more than three points will be required—especially if the scale is unknown—or it could be that there are several solutions, even if we ignore the reflection ambiguity. Of particular interest will be the extent to which it is possible to deduce which is which of the three points in the observed image.

To understand the degree of difficulty, let us briefly consider full perspective projection. In this case, any set of three noncollinear points can be mapped into any other three. This means that it may not be possible to deduce much about the original object just from this information. We will certainly not be able to deduce which point maps to which other point. However, we shall see that the situation is rather less ambiguous when we view the object under weak perspective projection.

Perhaps the simplest approach (due to Huang et al. as recently as 1995) is to imagine a circle drawn through the original set of points P_1, P_2, P_3 (Fig. 17.3*a*). We then find the centroid C of the set of points and draw additional lines through the points, all passing through C and meeting the circle in another three points Q_1, Q_2, Q_3 (Fig. 17.3*a*). Now in common with orthographic projection, scaled orthographic projection maintains ratios of distances on the same straight line, and weak perspective projection approximates to this. Thus, the distance ratio $P_iC : CQ_i$ remains unchanged after projection, and when we project the whole figure, as in Fig. 17.3*b*, we find that the circle has become an ellipse, though all lines remain lines and all linear distance ratios remain unchanged. The significance of these transformations and invariances is as follows. When the points P'_1, P'_2, P'_3 are observed in the image, the centroid C' can be computed, as can the positions of Q'_1, Q'_2, Q'_3. Thus, we have six points by which to deduce the position and parameters of the ellipse (in fact, five are sufficient). Once the ellipse is known, the orientation of its major axis gives the axis of rotation of the object; whereas the ratio of the lengths of the minor to major axes immediately gives the value of $\cos\alpha$. (Notice how the ambiguity in the sign of α comes up naturally in this calculation.) Finally, the length of the major axis of the ellipse permits us to deduce the depth of the object in the scene.

We have now shown that observing three projected points permits a unique ellipse to be computed passing through them. When this is back-projected into a circle, the axis of rotation of the object and the angle of rotation can be deduced, but not the sign of the angle of rotation. Two important

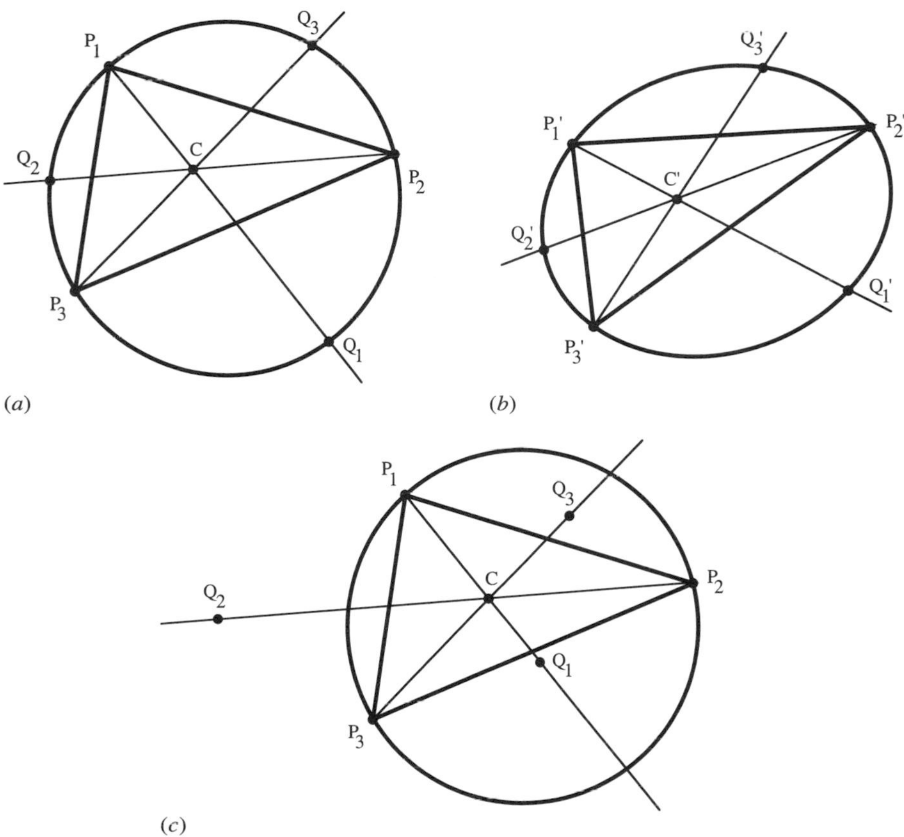

Figure 17.3 Determination of pose for three points viewed under weak perspective projection. (*a*) shows three feature points P_1, P_2, P_3 which lie on a known type of object. The circle passing through P_1, P_2, P_3 is drawn, and lines through the points and their centroid C meet the circle in Q_1, Q_2, Q_3. The ratios $P_iC : CQ_i$ are then deduced. (*b*) shows the three points observed under weak perspective as P'_1, P'_2, P'_3, together with their centroid C′ and the three points Q'_1, Q'_2, Q'_3 located using the original distance ratios. An ellipse drawn through the six points P'_1, P'_2, P'_3, Q'_1, Q'_2, Q'_3 can now be used to determine the orientation of the plane in which P_1, P_2, P_3 must lie, and also (from the major axis of the ellipse) the distance of viewing. (*c*) shows how an erroneous interpretation of the three points does not permit a circle to be drawn passing through P_1, P_2, P_3, Q_1, Q_2, Q_3 and hence no ellipse can be found which passes through the observed and the derived points P'_1, P'_2, P'_3, Q'_1, Q'_2, Q'_3.

comments may be made about the above calculation. The first is that the three distance ratios must be stored in memory, before interpretation of the observed scene can begin. The second is that the order of the three points apparently has to be known before interpretation can be undertaken. Otherwise we will have to perform six computations in which all possible assignments of the distance ratios are tried. Furthermore, it might appear from the earlier introductory remarks that several solutions are possible. Although in some instances feature points might be distinguishable, in many cases they are not (especially in 3-D situations where corner features might vary considerably when viewed from different positions). Thus, the potential ambiguity is important. However, if we can try out each of the six cases, little difficulty will generally arise. For immediately we deduce the positions Q_1', Q_2', Q_3', we will find that it is not possible in general to fit the six resulting points to an ellipse. The reason is easily seen on returning to the original circle. In that case, if the wrong distance ratios are assigned, the Q_i will clearly not lie on the circle, since the only values of the distance ratios for which the Q_i do lie on the circle are the correctly assigned ones (Fig. 17.3*c*). This means that although computation is wasted testing the incorrect assignments, there appears to be no risk of their leading to ambiguous solutions. Nevertheless, there is one contingency under which things could go wrong. Suppose the original set of points P_1, P_2, P_3 forms an almost perfect equilateral triangle. Then the distance ratios will be very similar, and, taking numerical inaccuracies into account, it may not be clear which ellipse provides the best and most likely fit. This mitigates against taking sets of feature points that form approximately isosceles or equilateral triangles. However, in practice more than three coplanar points will generally be used to optimize the fit, making fortuitous solutions rather unlikely.

Overall, it is fortunate that weak perspective projection requires such weak conditions for the identification of unique (to within a reflection) solutions, especially as full perspective projection demands four points before a unique solution can be found (see below). However, under weak perspective projection, additional points lead to greater accuracy but no reduction in the reflection ambiguity. This is because the information content from weak perspective projection is impoverished in the lack of depth cues that could (at least in principle) resolve the ambiguity. To understand this lack of additional information from more than three points under weak perspective projection, note that each additional feature point in the same plane is predetermined once three points have been identified. (Here we are assuming that the model object with the correct distance ratios can be referred to.)

These considerations indicate that we have two potential routes to unique location of objects from limited numbers of feature points. The first is to use noncoplanar points still viewed under weak perspective projection. The second is to use full perspective projection to view coplanar or noncoplanar sets of feature points. We shall see that whichever of these options we take, a unique solution demands that a minimum of four feature points be located on any object.

17.4 Obtaining Unique Solutions to the Pose Problem

The overall situation is summarized in Table 17.1. Looking first at the case of weak perspective projection (WPP), the number of solutions only becomes finite for three or more point features. Once three points have been employed, in the coplanar case there is no further reduction in the number of solutions, since (as noted earlier) the positions of any additional points can be deduced from the existing ones. However, this does not apply when the additional points are noncoplanar since they are able to provide just the right information to eliminate any ambiguity (see Fig. 17.4). (Although this might appear to contradict what was said earlier about perspective inversion, note that we are assuming here that the body is rigid and that all its features are at *known* fixed points on it in three dimensions. Hence, this particular ambiguity no longer applies, except for objects with special symmetries which we shall ignore here—see Fig. 17.4*d*.)

Considering next the case of full perspective projection (FPP), we find that the number of solutions again becomes finite only for three or more point features. The lack of information provided by three point features means that four solutions are in principle possible (see the example in Fig. 17.5

Table 17.1 Ambiguities when estimating pose from point features

Arrangement of the Points	N	WPP	FPP
Coplanar	≤ 2	∞	∞
	3	2	4
	4	2	1
	5	2	1
	≥ 6	2	1
Noncoplanar	≤ 2	∞	∞
	3	2	4
	4	1	2
	5	1	2
	≥ 6	1	1

This table summarizes the number of solutions that will be obtained when estimating the pose of a rigid object from point features located in a single image. It is assumed that N point features are detected and identified correctly and in the correct order. The columns WPP and FPP signify weak perspective projection and full perspective projection, respectively. The upper half of the table applies when all N points are coplanar; the lower half applies when the N points are noncoplanar. Note that when $N \leq 3$, the results strictly apply only in the coplanar case. However, the top two lines in the lower half of the table are retained for easy comparison.

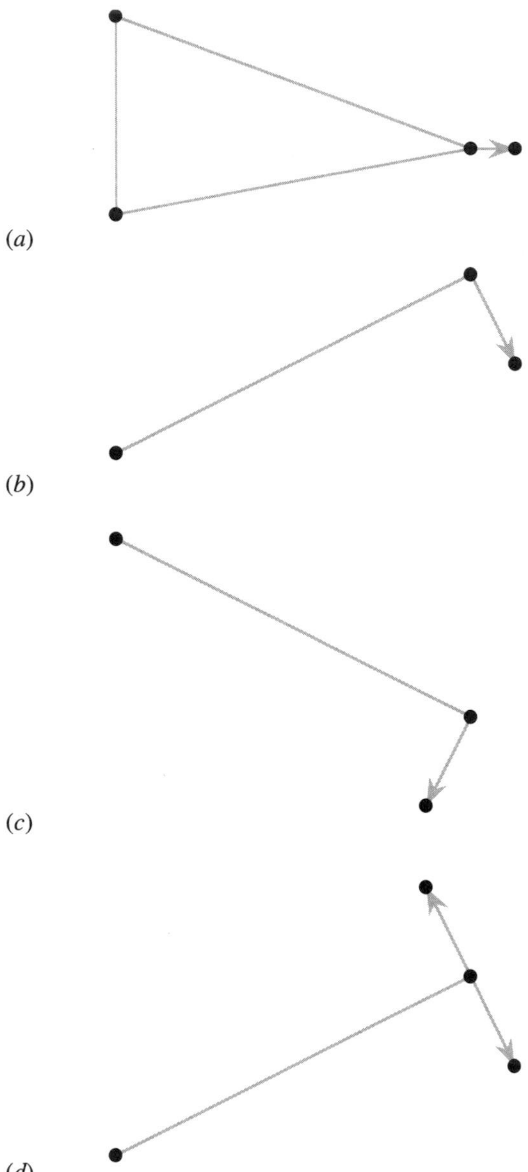

Figure 17.4 Determination of pose for four points viewed under weak perspective projection. (*a*) shows an object containing four noncoplanar points, as seen under weak perspective projection. (*b*) shows a side view of the object. If the first three points (connected by nonarrowed gray lines) were viewed alone, perspective inversion would give rise to a second interpretation (*c*). However, the fourth point gives additional information about the pose, which permits only one overall interpretation. This would not be the case for an object containing an additional symmetry as in (*d*), since its reflection would be identical to the original view (not shown).

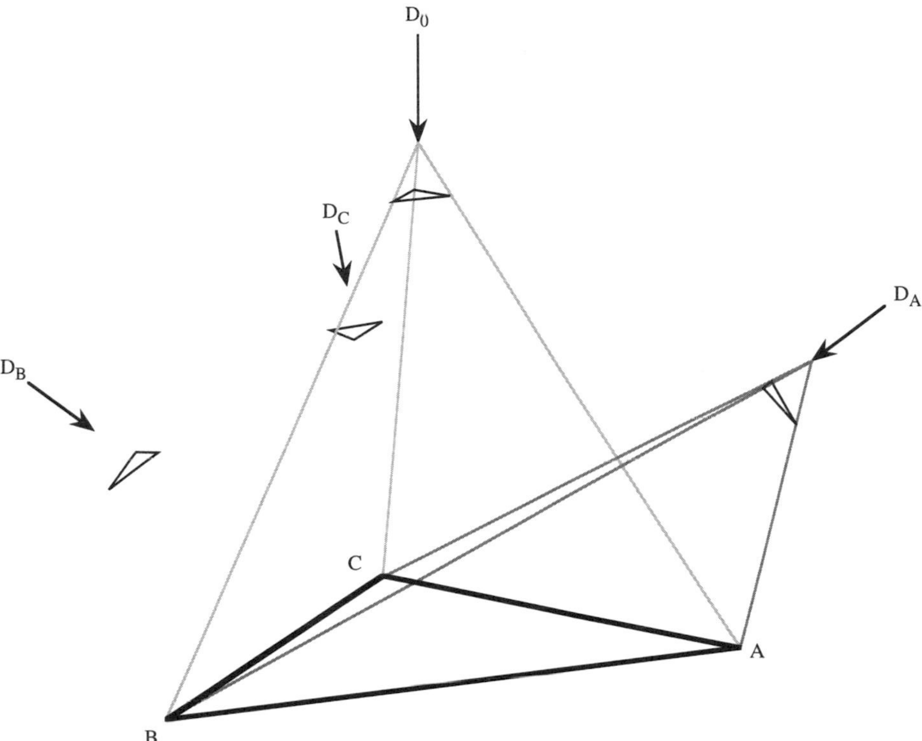

Figure 17.5 Ambiguity for three points viewed under full perspective projection. Under full perspective projection, the camera sees three points A, B, C as three directions in space, and this can lead to a fourfold ambiguity in interpreting a known object. The figure shows the four possible viewing directions and centers of projection of the camera (indicated by the directions and tips of the bold arrows). In each case, the image at each camera is indicated by a small triangle. D_A, D_B, D_C correspond approximately to views from the general directions of A, B, C, respectively.

and the detailed explanation in Section 17.2.1), but the number of solutions drops to one as soon as four coplanar points are employed. (The correct solution can be found by making cross checks between subsets of three points, and eliminating inconsistent solutions.) When the points are noncoplanar, it is only when six or more points are employed that we have sufficient information to unambiguously determine the pose. There is necessarily no ambiguity with 6 or more points, as all 11 camera calibration parameters can be deduced from the 12 linear equations that then arise (see Chapter 21). Correspondingly, it is deduced that 5 noncoplanar points will in general be *insufficient* for all 11 parameters to be deduced, so there will still be some ambiguity in this case.

Next, it should be questioned why the coplanar case is at first ($N = 3$) better[4] under weak perspective projection and then ($N > 3$) better under full perspective projection, although the noncoplanar case is always better, or as good, under weak perspective projection. The reason must be that intrinsically full perspective projection provides more detailed information but is frustrated by lack of data when there are relatively few points. However, the exact stage at which the additional information becomes available is different in the coplanar and noncoplanar cases. In this respect, it is important to note that when coplanar points are being observed under weak perspective projection there is never enough information to eliminate the ambiguity.

Our discussion assumes that the correspondences between object and image features are all known—that is, that N point features are detected and identified correctly and in the correct order. If this is not so, the number of possible solutions could increase substantially, considering the number of possible permutations of quite small numbers of points. This makes it attractive to use the minimum number of features for ascertaining the most probable match (Horaud et al., 1989). Other workers have used heuristics to help reduce the number of possibilities. For example, Tan (1995) used a simple compactness measure (see Section 6.9) to determine which geometric solution is the most likely. Extreme obliqueness is perhaps unlikely, and the most likely solution is taken to be the one with the highest compactness value. This idea follows on from the extremum principle of Brady and Yuille (1984), which states that the most probable solutions are those nearest to extrema of relevant (e.g., rotation) parameters.[5] In this context, it is worth noting that coplanar points viewed under weak or full perspective projection always appear in the same cyclic order. This is not trivial to check given the possible distortions of an object, though if a convex polygon can be drawn through the points, the cyclic order around its boundary will not change on projection.[6] However, for noncoplanar points, the pattern of the perceived points can reorder itself almost randomly. Thus, a considerably greater number of permutations of the points have to be considered for noncoplanar points than for coplanar points.

Finally, to this point we have concentrated on the existence and uniqueness of solutions to the pose problem. The stability of the solutions has not so far been discussed. However, the concept of stability gives a totally different

---

4 In this context, "better" means less ambiguous and leads to fewer solutions.

5 Perhaps the simplest way of understanding this principle is obtained by considering a pendulum, whose extreme positions are also its most probable! However, in this case the extremum occurs when the angle α (see Fig. 16.1) is close to zero.

6 The reason for this is that planar convexity is an invariant of projection.

dimension to the data presented in Table 17.1. In particular, noncoplanar points tend to give more stable solutions to the pose problem. For example, if the plane containing a set of coplanar points is viewed almost head-on ($\alpha \approx 0$), there will be very little information on the exact orientation of the plane because the changes in lateral displacement of the points will vary as $\cos \alpha$ (see Section 17.1) and there will be no linear term in the Taylor expansion of the orientation dependence.

17.4.1 *Solution of the 3-Point Problem*

Figure 17.5 showed how four solutions can arise when three point features are viewed under full perspective projection. Here we briefly explore this situation by considering the relevant equations. Figure 17.5 shows that the camera sees the points as three image points representing three directions in space. This means that we can compute the angles α, β, γ between these three directions. If the distances between the three points A, B, C on the object are the known values D_{AB}, D_{BC}, D_{CA}, we can now apply the cosine rule in an attempt to determine the distances R_A, R_B, R_C of the feature points from the center of projection:

$$D_{BC}^2 = R_B^2 + R_C^2 - 2R_B R_C \cos\alpha \tag{17.1}$$

$$D_{CA}^2 = R_C^2 + R_A^2 - 2R_C R_A \cos\beta \tag{17.2}$$

$$D_{AB}^2 = R_A^2 + R_B^2 - 2R_A R_B \cos\gamma \tag{17.3}$$

Eliminating any two of the variables R_A, R_B, R_C yields an eighth degree equation in the other variable, indicating that eight solutions to the system of equations could be available (Fischler and Bolles, 1981). However, the above cosine rule equations contain only constants and second degree terms. Hence, for every positive solution there is another solution that differs only by a sign change in all the variables. These solutions correspond to inversion through the center of projection and are hence unrealizable. Thus, there are at most four realizable solutions to the system of equations. We can quickly demonstrate that there may sometimes be fewer than four solutions, since in some cases, for one or more of the "flipped" positions shown in Fig. 17.5, one of the features could be on the negative side of the center of projection, and hence would be unrealizable.

Before leaving this topic, we should observe that the homogeneity of equations (17.1)–(17.3) implies that observation of the angles α, β, γ permits the orientation of the object to be estimated independently of any knowledge of its scale. In fact, estimation of scale depends directly on estimation of range, and vice versa. Thus, knowledge of just one range parameter (e.g., R_A) will permit the scale of the object to be deduced. Alternatively, knowledge of its area will permit the remaining parameters to be deduced. This concept provides a slight generalization of the main results of Sections 17.2 and 17.3, which generally start with the assumption that all the dimensions of the object are known.

17.4.2 *Using Symmetrical Trapezia for Estimating Pose*

One more example will be of interest here. That is the case of four points arranged at the corners of a symmetrical trapezium (Tan, 1995). When viewed under weak perspective projection, the midpoints of the parallel sides are easily measured, but under full perspective projection midpoints do not remain midpoints, so the axis of symmetry cannot be obtained in this way. However, producing the skewed sides to meet at S′ and forming the intersection I′ of the diagonals permit the axis of symmetry to be located as the line I′S′ (Fig. 17.6). Thus, we now have not four points but six to describe the perspective view of the trapezium. More important, the axis of symmetry has been located, and this is known to be perpendicular to the parallel sides of the trapezium. This fact is a great help in making the mathematics more tractable and in obtaining solutions quickly so that (for example) object motion can be tracked in real time. Again, this is a case where object orientation can be deduced straightforwardly, even when the situation is one of strong perspective and even when the size of the object is unknown. This result is a generalization from that of Haralick (1989) who noted that a single view of a rectangle of unknown size is sufficient to determine its pose. In either case, the range of the object can be found if its area is known, or its size can be deduced if a single range value can be found from other data (see also Section 17.3.1).

17.5 Concluding Remarks

This chapter has aimed to cover certain aspects of 3-D vision that were not studied in depth in the previous chapter. In particular, the topic of

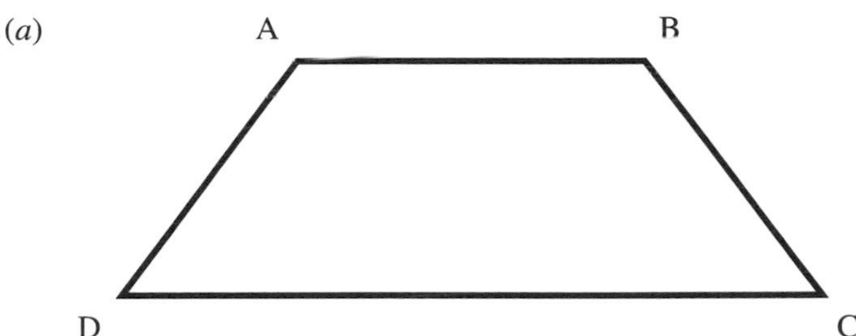

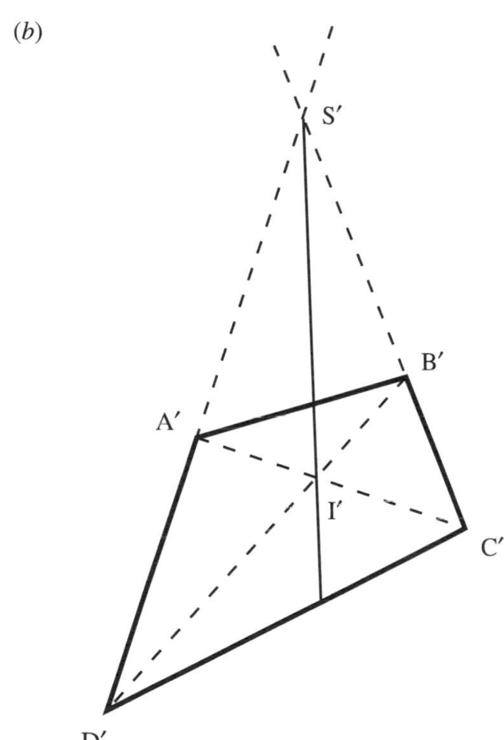

Figure 17.6 Trapezium viewed under full perspective projection. (*a*) shows a symmetrical trapezium, and (*b*) shows how it appears when viewed under full perspective projection. In spite of the fact that midpoints do not project into midpoints under perspective projection, the two points S′ and I′ on the symmetry axis can be located unambiguously as the intersection of two nonparallel sides and two diagonals respectively. This gives six points (from which the two midpoints on the symmetry axis can be deduced if necessary), which is sufficient to compute the pose of the object, albeit with a single ambiguity of interpretation (see text).

perspective inversion was investigated in some detail, and its method of projection was explored. Orthographic projection, scaled orthographic projection, weak perspective projection, and full perspective projection were considered, and the numbers of object points that would lead to correct or ambiguous interpretations were analyzed. It was found that scaled orthographic projection and its approximation, weak perspective projection, led to straightforward interpretation when four or more noncoplanar points were considered, although the perspective inversion ambiguity remained when all the points were coplanar. This latter ambiguity was resolved with four or more points viewed under full perspective projection. However, in the noncoplanar case, some ambiguity remained until six points were being viewed. The key to understanding the situation is the fact that full perspective projection makes the situation more complex, although it also provides more information by which, ultimately, to resolve the ambiguity.

Additional problems were found to arise when the points being viewed are indistinguishable and then a good many solutions may have to be tried before a minimally ambiguous result is obtained. With coplanar points fewer possibilities arise, which leads to less computational complexity. The key to success here is the natural ordering that can arise for points in a plane—as, for example, when they form a convex set that can be ordered uniquely around a bounding polygon. In this context, the role that can be played by the extremum principle of Brady and Yuille (1984) in reducing the number of solutions is significant. (For further insight on the topic, see Horaud and Brady, 1988.)

It is of great relevance to devise methods for rapid interpretation in real-time applications. To do so it is important to work with a minimal set of points and to obtain analytic solutions that move directly to solutions without computationally expensive iterative procedures: for example, Horaud et al. (1989) found an analytic solution for the perspective four-point problem, which works both in the general noncoplanar case and in the planar case. Other low computation methods are still being developed, as with pose determination for symmetrical trapezia (Tan, 1995). Understanding is still advancing, as demonstrated by Huang et al.'s (1995) neat geometrical solution to the pose determination problem for three points viewed under weak perspective projection.

This chapter has covered a specific 3-D recognition problem. Chapter 19 covers another—that of invariants, which provides a speedy and convenient means of bypassing the difficulties associated with full perspective projection. Chapter 21 aims to finalize the study of 3-D vision by showing how camera calibration can be achieved or, to some extent, circumvented.

Perspective makes interpretation of images of 3-D scenes intrinsically difficult. However, this chapter has demonstrated that "weak perspective" views of distant objects are much simplified. As a result, they are commonly located using fewer feature points—though for planar objects an embarrassing pose ambiguity remains, which is not evident under full perspective.

17.6 Bibliographical and Historical Notes

The development of solutions to the so-called *perspective n-point problem* (finding the pose of objects from n features under various forms of perspective) has been proceeding for more than two decades and is by no means complete. Fischler and Bolles summarized the situation as they saw it in 1981, and they described several new algorithms. However, they did not discuss pose determination under weak perspective, and perhaps surprisingly, considering its reduced complexity, pose determination under weak perspective has subsequently been the subject of much research (e.g., Alter, 1994; Huang et al., 1995). Horaud et al. (1989) discussed the problem of finding rapid solutions to the perspective n-point problem by reducing n as far as possible. They also obtained an analytic solution for the case $n = 4$, which should help considerably with real-time implementations. Their solution is related to Horaud's earlier (1987) corner interpretation scheme, as described in Section 16.12, while Haralick et al. (1984) provided useful basic theory for matching wire frame objects.

In a recent paper, Liu and Wong (1999) described an algorithm for pose estimation from four corresponding points under FPP when the points are not coplanar. Strictly, according to Table 17.1, this will lead to an ambiguity. However, Liu and Wong made the point "that the possibility for the occurrence of multiple solutions in the perspective 4-point problem is much smaller than that in the perspective 3-point problem," so that "using a 4-point model is much more reliable than using a 3-point model." They actually only claim "good results." Also, much of the emphasis of the paper is on errors and reliability. Hence, it seems that it is the *scope* for making errors in the sense of misinterpreting the situation that is significantly reduced. Added to this consideration, Liu and Wong's (1999) work involves tracking a known object within a somewhat restricted region of space. This must again cut down the scope for error considerably. Hence, it is not clear that their work violates the relevant (FPP; noncoplanar; $n=4$)[7] entry in Table 17.1, rather than merely making it *unlikely* that a real ambiguity will arise.

Between them, Faugeras (1993), Hartley and Zisserman (2000), Faugeras and Luong (2001), and Forsyth and Ponce (2003) provide good coverage of the whole area of 3-D vision. For an interesting viewpoint on the subject, with particular emphasis on pose refinement, see Sullivan (1992). For further references on specific aspects of 3-D vision, see Sections 16.15, 19.8, and 21.16. (Sections 18.15 and 20.12 give references on motion but also cover aspects of 3-D vision.)

17.7 Problems

1. Draw up a complete table of pose ambiguities that arise for *weak* perspective projection for various numbers of object points identified in the image. Your answer should cover both coplanar points and noncoplanar points, and should make clear in each case how much ambiguity would remain in the limit of an infinite number of object points being seen. Give justification for your results.

2. Distinguish between *full perspective projection* and *weak perspective projection*. Explain how each of these projections presents oblique views of the following real objects: (a) straight lines, (b) several concurrent lines (i.e., lines meeting in a single point), (c) parallel lines, (d) midpoints of lines, (e) tangents to curves, and (f) circles whose centers are marked with a dot. Give justification for your results.

3. Explain each of the following: (a) Why weak perspective projection leads to an ambiguity in viewing an object such as that in Fig. 17.P1*a*. (b) Why the

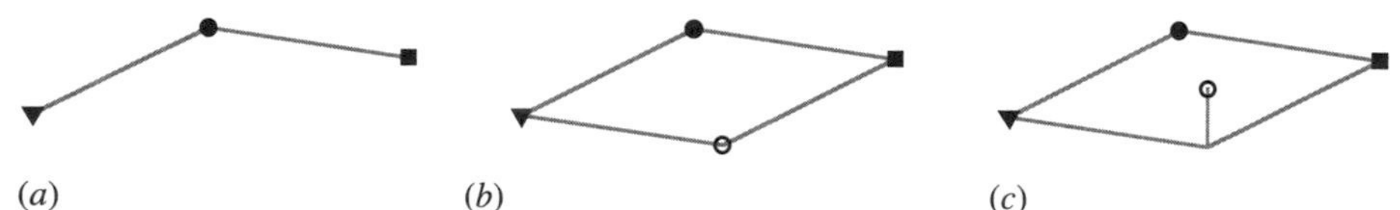

(*a*) (*b*) (*c*)

Figure 17.P1 In this diagram the gray edges are construction lines, *not* parts of the objects. (*a*) and (*b*) are completely planar objects.

7 See Fischler and Bolles (1981) for the original evidence for this particular ambiguity.

ambiguity doesn't disappear for the case of Fig. 17.P1*b*. (c) Why the ambiguity *does* disappear in the case of Fig. 17.P1*c*, if the true nature of the object is known. (d) Why the ambiguity doesn't occur in the case of Fig. 17.P1*b* viewed under full perspective projection. In the last case, illustrate your answer by means of a sketch.

Motion

Motion is another aspect of 3-D vision that humans are able to interpret with ease. This chapter studies the basic theoretical concepts and then tackles real problems where motion is crucial, including the monitoring of traffic flow and the tracking of people.

Look out for:

- the basic concepts of optical flow and its limitations.
- the idea of a focus of expansion.
- identification of the ground plane as an early stage in the analysis of many types of scene incorporating motions.
- the need for "occlusion reasoning" when objects repeatedly pass behind one another and then reemerge.
- how serious studies of the motions of complex objects may have to take into account 3-D articulated models of linked parts.
- the important status of the Kalman filter in motion applications.

Note that although this is only an introductory chapter on 3-D motion, it can demonstrate methods that can be used to perform vital surveillance tasks. Chapter 20 will take this work further—into the area of egomotion—once invariants have provided the means to take useful shortcuts in the analysis.

CHAPTER 18

Motion

18.1 Introduction

In the space available, it will not be possible to cover the whole subject of motion comprehensively; instead, the aim is to provide the flavor of the subject, airing some of the principles that have proved important over the past 20 years or so. Over much of this time, optical flow has been topical, and it is appropriate to study it in fair detail, particularly, as we shall see later from some case studies on the analysis of traffic scenes, because it is being used in actual situations. Later parts of the chapter cover more modern applications of motion anaysis, including people tracking, human gait analysis, and animal tracking.

18.2 Optical Flow

When scenes contain moving objects, analysis is necessarily more complex than for scenes where everything is stationary, since temporal variations in intensity have to be taken into account. However, intuition suggests that it should be possible—even straightforward—to segment moving objects by virtue of their motion. Image differencing over successive pairs of frames should permit motion segmentation to be achieved. More careful consideration shows that things are not quite so simple, as illustrated in Fig. 18.1. The reason is that regions of constant intensity give no sign of motion, whereas edges parallel to the direction of motion also give the appearance of not moving; only edges with a component normal to the direction of motion carry information about the motion. In addition, there is some ambiguity in the direction of the velocity vector. The ambiguity arises partly because too little information is

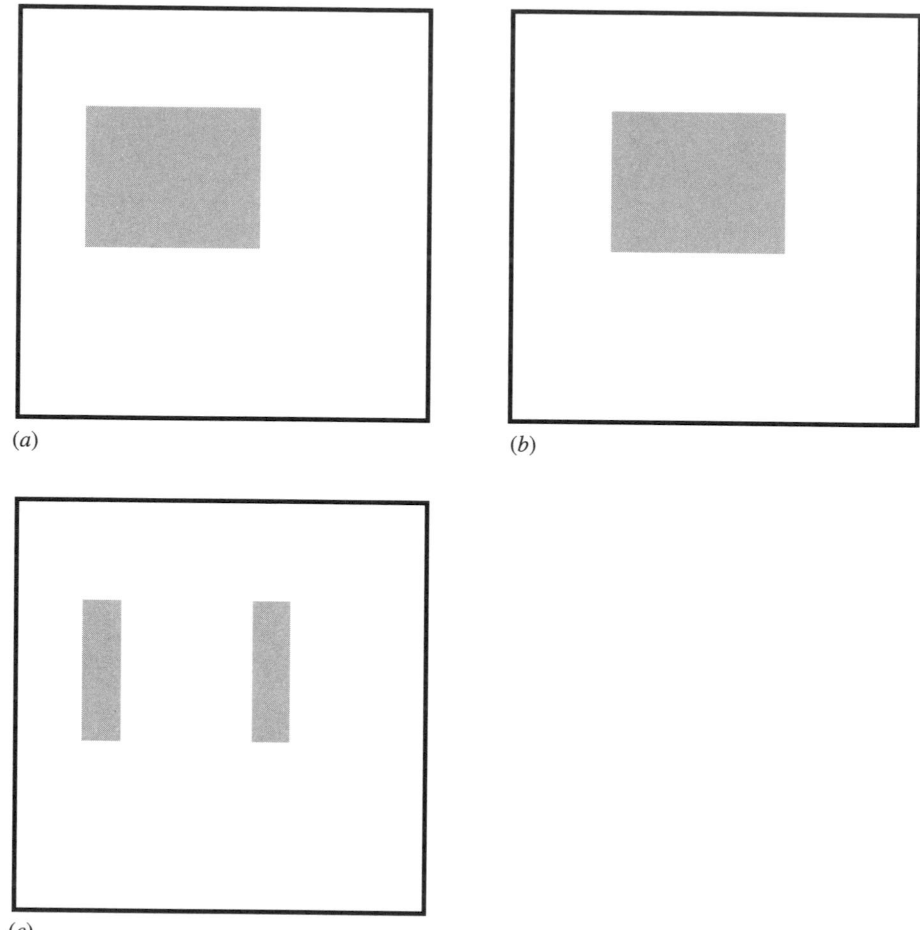

Figure 18.1 Effect of image differencing. This figure shows an object that has moved between frames (*a*) and (*b*). (*c*) shows the result of performing an image differencing operation. Note that the edges parallel to the direction of motion do not show up in the difference image. Also, regions of constant intensity give no sign of motion.

available within a small aperture to permit the full velocity vector to be computed (Fig. 18.2). Hence, this is called the *aperture problem*.

Taken further, these elementary ideas lead to the notion of optical flow, wherein a local operator that is applied at all pixels in the image will lead to a motion vector field that varies smoothly over the whole image. The attraction lies in the use of a local operator, with its limited computational burden. Ideally, it would have an overhead comparable to an edge detector in

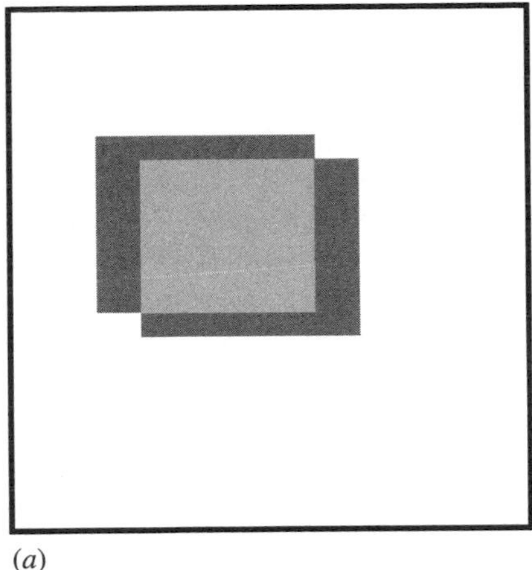

(a)

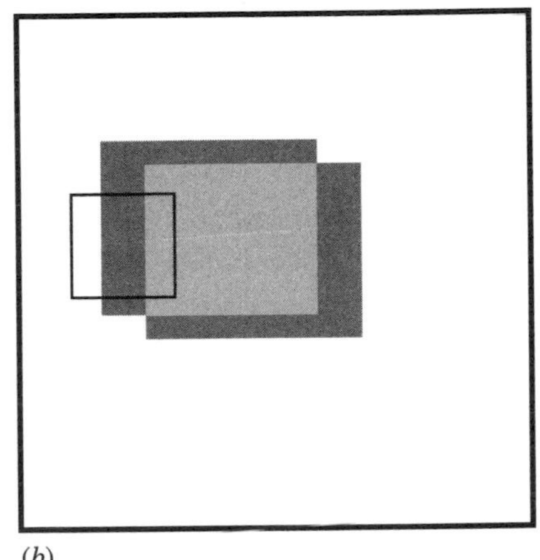

(b)

Figure 18.2 The aperture problem. This figure illustrates the aperture problem. (*a*) shows (dark gray) regions of motion of an object whose central uniform region (light gray) gives no sign of motion. (*b*) shows how little is visible in a small aperture (black border), thereby leading to ambiguity in the deduced direction of motion of the object.

a normal intensity image—though clearly it will have to be applied locally to pairs of images in an image sequence.

We start by considering the intensity function $I(x, y, t)$ and expanding it in a Taylor series:

$$I(x + \mathrm{d}x, y + \mathrm{d}y, t + \mathrm{d}t) = I(x, y, t) + I_x\,\mathrm{d}x + I_y\,\mathrm{d}y + I_t\,\mathrm{d}t + \ldots \tag{18.1}$$

where second and higher order terms have been ignored. In this formula, I_x, I_y, I_t denote respective partial derivatives with respect to x, y, and t.

We next set the local condition that the image has shifted by amount $(\mathrm{d}x, \mathrm{d}y)$ in time $\mathrm{d}t$, so that it is functionally identical[1] at $(x+\mathrm{d}x, y+\mathrm{d}y, t+\mathrm{d}t)$ and (x, y, t):

$$I(x + \mathrm{d}x, y + \mathrm{d}y, t + \mathrm{d}t) = I(x, y, t) \tag{18.2}$$

Hence, we can deduce:

$$I_t = -(I_x\dot{x} + I_y\dot{y}) \tag{18.3}$$

Writing the local velocity $\mathbf{v}$ in the form:

$$\mathbf{v} = (v_x, v_y) = (\dot{x}, \dot{y}) \tag{18.4}$$

1 For further insight into the meaning of "functional identity" in this context, see Section 18.6.

we find:

$$I_t = -(I_x v_x + I_y V_y) = -\nabla I.v \tag{18.5}$$

I_t can be measured by subtracting pairs of images in the input sequence, whereas ∇I can be estimated by Sobel or other gradient operators. Thus, it should be possible to deduce the velocity field $\mathbf{v}(x, y)$ using the above equation. Unfortunately, this equation is a scalar equation and will not suffice for determining the two local components of the velocity field as we require. There is a further problem with this equation—that the velocity value will depend on the values of both I_t and ∇I, and these quantities are only estimated approximately by the respective differencing operators. In both cases significant noise will arise, which will be exacerbated by taking the ratio in order to calculate $\mathbf{v}$.

Let us now return to the problem of computing the full velocity field $\mathbf{v}(x, y)$. All we know about $\mathbf{v}$ is that its components lie on the following line in (v_x, v_y)-space (Fig. 18.3):

$$I_x v_x + I_y v_y + I_t = 0 \tag{18.6}$$

This line is normal to the direction (I_x, I_y), and has a distance from the (velocity) origin which is equal to

$$|\mathbf{v}| = -I_t / \left[I_x^2 + I_y^2\right]^{1/2} \tag{18.7}$$

We need to deduce the component of $\mathbf{v}$ along the line given by equation (18.6). However, there is no purely local means of achieving this with first derivatives of the intensity function. The accepted solution (Horn and Schunck, 1981) is to use relaxation labeling to arrive iteratively at a self-consistent solution that minimizes the global error. In principle, this approach will also minimize the noise problem indicated earlier.

In fact, problems with the method persist. Essentially, these problems arise when there are liable to be vast expanses of the image where the intensity gradient is low. In that case, only very inaccurate information is available about the velocity component parallel to ∇I, and the whole scenario becomes ill-conditioned. On the other hand, in a highly textured image, this situation should not arise, assuming that the texture has a large enough grain size to give good differential signals.

Finally, we return to the idea mentioned at the beginning of this section—that edges parallel to the direction of motion would not give useful motion information. Such edges will have edge normals normal to the direction of motion, so ∇I will be normal to $\mathbf{v}$. Thus, from equation (18.5), I_t will be zero. In addition, regions of constant intensity will have $\nabla I = 0$, so again I_t will be zero. It is interesting and

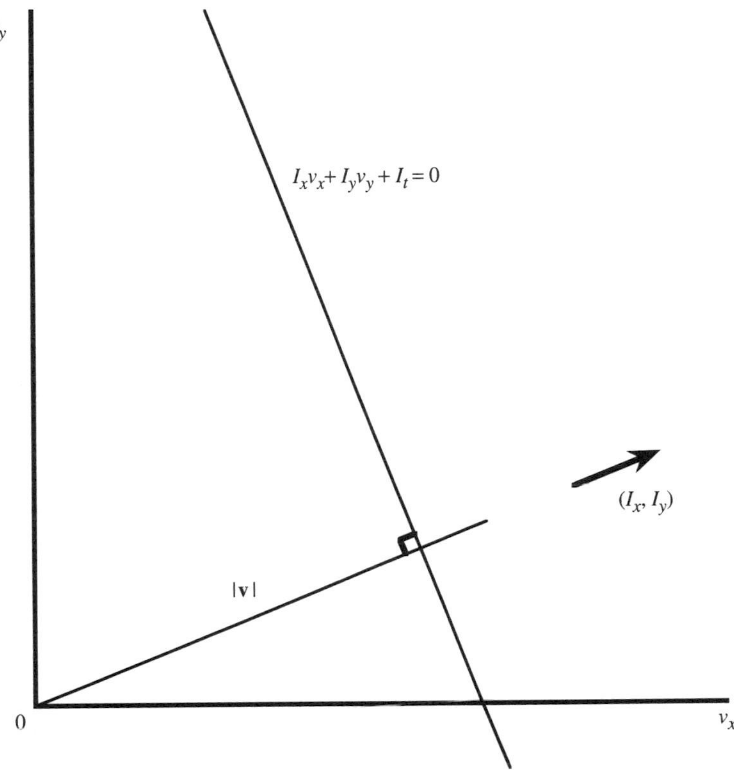

Figure 18.3 Computation of the velocity field. This graph shows the line in velocity space on which the velocity vector $\mathbf{v}$ must lie. The line is normal to the direction (I_x, I_y), and its distance from the origin is known to be $|\mathbf{v}|$ (see text).

highly useful that such a simple equation (18.5) embodies all the cases that were suggested earlier on the basis of intuition.

In the following section, we assume that the optical flow (velocity field) image has been computed satisfactorily, that is, without the disadvantages of inaccuracy or ill-conditioning. It must now be interpreted in terms of moving objects and perhaps a moving camera. We shall ignore motion of the camera by remaining within its frame of reference.

18.3 Interpretation of Optical Flow Fields

We start by considering a case where no motion is visible, and the velocity field image contains only vectors of zero length (Fig. 18.4*a*). Next we take a

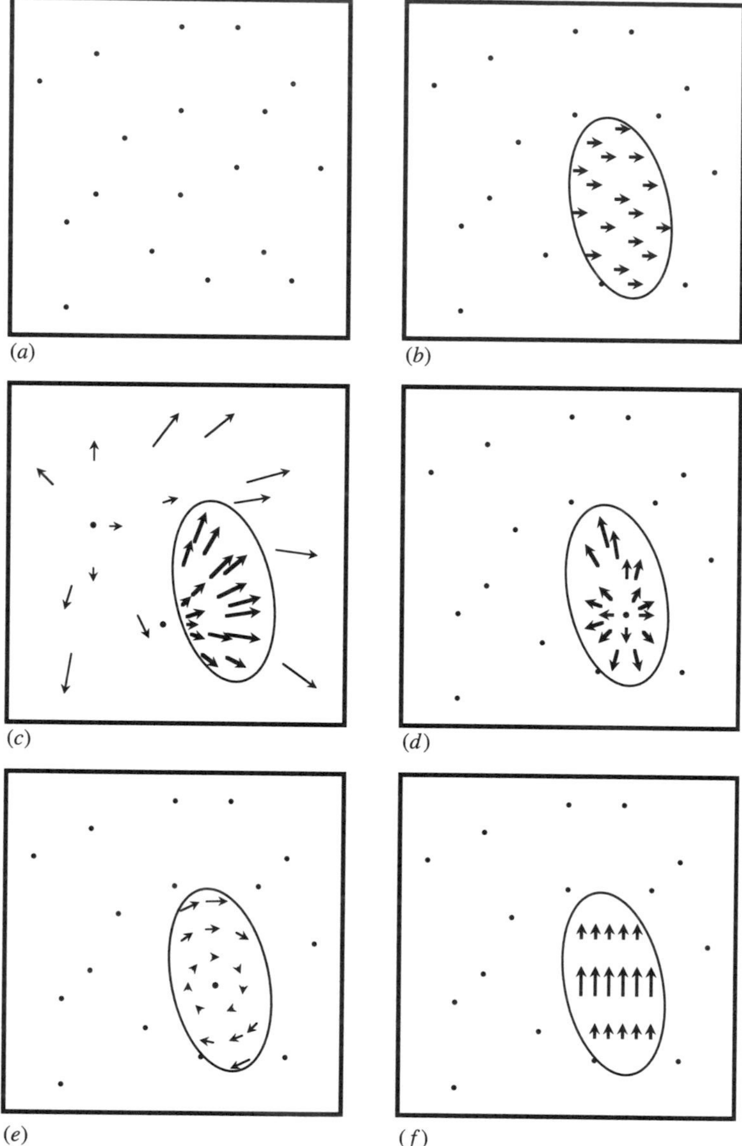

Figure 18.4 Interpretation of velocity flow fields. (*a*) shows a case where the object features all have zero velocity. (*b*) depicts a case where an object is moving to the right. (*c*) shows a case where the camera is moving into the scene, and the stationary object features appear to be diverging from a focus of expansion (FOE), while a single large object is moving past the camera and away from a separate FOE. In (*d*) an object is moving directly toward the camera which is stationary: the object's FOE lies within its outline. In (*e*) an object is rotating about the line of sight to the camera, and in (*f*) the object is rotating about an axis perpendicular to the line of sight. In all cases, the length of the arrow indicates the magnitude of the velocity vector.

case where one object is moving toward the right, with a simple effect on the velocity field image (Fig. 18.4*b*). Next we consider the case where the camera is moving forward. All the stationary objects in the field of view will then appear to be diverging from a point that is called the *focus of expansion* (FOE)—see Fig. 18.4*c*. This image also shows an object that is moving rapidly past the camera and that has its own separate FOE. Figure 18.4*d* shows the case of an object moving directly toward the camera. Its FOE lies within its outline. Similarly, objects that are receding appear to move away from the *focus of contraction*. Next are objects that are stationary but that are rotating about the line of sight. For these the vector field appears as in Fig. 18.4*e*. A final case is also quite simple—that of an object that is stationary but rotating about an axis normal to the line of sight. If the axis is horizontal, then the features on the object will appear to be moving up or down, while paradoxically the object itself remains stationary (Fig. 18.4*f*)—though its outline could oscillate as it rotates.

So far, we have only dealt with cases in which pure translational or pure rotational motion is occurring. If a rotating meteor is rushing past, or a spinning cricket ball is approaching, then both types of motion will occur together. In that case, unraveling the motion will be far more complex. We shall not solve this problem here but will instead refer the reader to more specialized texts (e.g., Maybank, 1992). However, the complexity is due to the way depth (Z) creeps into the calculations. First, note that pure rotational motion with rotation about the line of sight does not depend on Z. All we have to measure is the angular velocity, and this can be done quite simply.

18.4 Using Focus of Expansion to Avoid Collision

We now take a simple case in which a focus of expansion (FOE) is located in an image, and we will show how it is possible to deduce the distance of closest approach of the camera to a fixed object of known coordinates. This type of information is valuable for guiding robot arms or robot vehicles and helping to avoid collisions.

In the notation of Chapter 16, we have the following formulas for the location of an image point (x, y, z) resulting from a world point (X, Y, Z):

$$x = fX/Z \tag{18.8}$$

$$y = fY/Z \tag{18.9}$$

$$z = f \tag{18.10}$$

Assuming that the camera has a motion vector $(-\dot{X}, -\dot{Y}, -\dot{Z}) = (-u, -v, -w)$, fixed world points will have velocity (u, v, w) relative to the camera. Now a point (X_0, Y_0, Z_0) will after a time t appear to move to $(X, Y, Z) = (X_0 + ut, Y_0 + vt, Z_0 + wt)$ with image coordinates:

$$(x, y) = \left(\frac{f(X_0 + ut)}{Z_0 + wt}, \frac{f(X_0 + ut)}{Z_0 + wt} \right) \tag{18.11}$$

and as $t \to \infty$ this approaches the focus of expansion F $(fu/w, fv/w)$. This point is in the image, but the true interpretation is that the actual motion of the center of projection of the imaging system is toward the point:

$$\mathbf{p} = (fu/w, fv/w, f) \tag{18.12}$$

(This is, of course, consistent with the motion vector (u, v, w) assumed initially.) The distance moved during time t can now be modeled as:

$$\mathbf{X}_c = (X_c, Y_c, Z_c) = \alpha t\mathbf{p} = f\alpha t(u/w, v/w, 1) \tag{18.13}$$

where α is a normalization constant. To calculate the distance of closest approach of the camera to the world point $\mathbf{X} = (X, Y, Z)$, we merely specify that the vector $\mathbf{X}_c - \mathbf{X}$ be perpendicular to $\mathbf{p}$ (Fig. 18.5), so that:

$$(\mathbf{X}_c - \mathbf{X}).\mathbf{p} = 0 \tag{18.14}$$

that is,

$$(\alpha t\mathbf{p} - \mathbf{X}).\mathbf{p} = 0 \tag{18.15}$$

$$\therefore\ \alpha t\mathbf{p}.\mathbf{p} = \mathbf{X}.\mathbf{p} \tag{18.16}$$

$$\therefore\ t = (\mathbf{X}.\mathbf{p})/\alpha(\mathbf{p}.\mathbf{p}) \tag{18.17}$$

Substituting in the equation for $\mathbf{X}_c$ now gives:

$$\mathbf{X}_c = \mathbf{p}(\mathbf{X}.\mathbf{p})/(\mathbf{p}.\mathbf{p}) \tag{18.18}$$

Hence, the minimum distance of approach is given by:

$$\begin{aligned} \mathrm{d}_{\min}^2 &= \left[\frac{\mathbf{p}(\mathbf{X}.\mathbf{p})}{(\mathbf{p}.\mathbf{p})} - \mathbf{X} \right]^2 = \frac{(\mathbf{X}.\mathbf{p})^2}{(\mathbf{p}.\mathbf{p})} - \frac{2(\mathbf{X}.\mathbf{p})^2}{(\mathbf{p}.\mathbf{p})} + (\mathbf{X}.\mathbf{X}) \\ &= (\mathbf{X}.\mathbf{X}) - \frac{(\mathbf{X}.\mathbf{p})^2}{(\mathbf{p}.\mathbf{p})} \end{aligned} \tag{18.19}$$

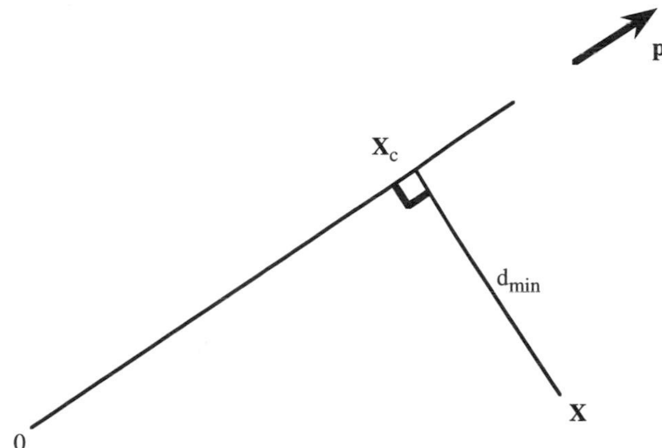

Figure 18.5 Calculation of distance of closest approach. Here the camera is moving from 0 to $\mathbf{X}_c$ in the direction $\mathbf{p}$, not in a direct line to the object at $\mathbf{X}$. d_{min} is the distance of closest approach.

which is naturally zero when $\mathbf{p}$ is aligned along $\mathbf{X}$. Avoidance of collisions requires an estimate of the size of the machine (e.g., robot or vehicle) attached to the camera and the size to be associated with the world point feature $\mathbf{X}$. Finally, it should be noted that while $\mathbf{p}$ is obtained from the image data, $\mathbf{X}$ can only be deduced from the image data if the depth Z can be estimated from other information. This information should be available from time-to-adjacency analysis (see below) if the speed of the camera through space (and specifically w) is known.

18.5 Time-to-Adjacency Analysis

Next we consider the extent to which the depths of objects can be deduced from optical flow. First, note that features on the same object share the same focus of expansion, and this can help us to identify them. But how can we get information on the depths of the various features on the object from optical flow? The basic approach is to start with the coordinates of a general image point (x, y), deduce its flow velocity, and then find an equation linking the flow velocity with the depth Z.

Taking the general image point (x, y) given in equation (18.11), we find:

$$\begin{aligned} \dot{x} &= f[(Z_0 + wt)u - (X_0 + ut)w]/(Z_0 + wt)^2 \\ &= f[(Zu - Xw)]/Z^2 \end{aligned} \tag{18.20}$$

and

$$\dot{y} = f[Zv - Yw]/Z^2 \tag{18.21}$$

Hence:

$$\begin{aligned}\dot{x}/\dot{y} &= [Zu - Xw]/[Zv - Yw] = (u/w - X/Z)/(v/w - Y/Z) \\ &= (x - x_F)/(y - y_F)\end{aligned} \tag{18.22}$$

This result was to be expected, inasmuch as the motion of the image point has to be directly away from the focus of expansion (x_F, y_F). With no loss of generality, we now take a set of axes such that the image point considered is moving along the x-axis. Then we have:

$$\dot{y} = 0 \tag{18.23}$$

$$y_F = y = fY/Z \tag{18.24}$$

Defining the distance from the focus of expansion as Δr (see Fig. 18.6), we find:

$$\Delta r = \Delta x = x - x_F = fX/Z - fu/w = f(Xw - Zu)/Zw \tag{18.25}$$

$$\therefore \Delta r/\dot{r} = \Delta x/\dot{x} = -Z/w \tag{18.26}$$

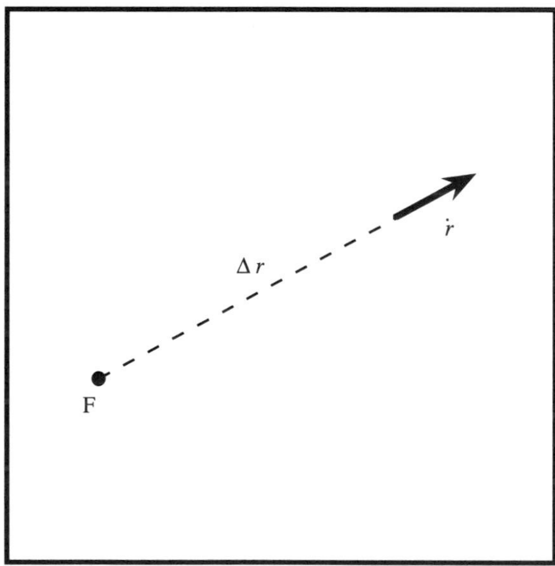

Figure 18.6 Calculation of time to adjacency. Here an object feature is moving directly away from the focus of expansion F with speed $\dot{r}$. At the time of observation, the distance of the feature from F is Δr. These measurements permit the time to adjacency and hence also the relative depth of the feature to be calculated.

Equation (18.26) means that the *time to adjacency*, when the origin of the camera coordinate system will arrive at the object point, is the same (Z/w) when seen in real-world coordinates as when seen in image coordinates $(-\Delta r/\dot{r})$. Hence it is possible to relate the optical flow vectors for object points at different depths in the scene. This is important, for the assumption of identical values of w now allows us to determine the relative depths of object points merely from their apparent motion parameters:

$$\frac{Z_1}{Z_2} = \frac{\Delta r_1/\Delta r_2}{\dot{r}_1/\dot{r}_2} \tag{18.27}$$

This is therefore the first step in the determination of structure from motion. In this context, it should be noted how the implicit assumption that the objects under observation are rigid is included—namely, that all points on the same object are characterized by identical values of w. The assumption of rigidity underlies much of the work on interpreting motion in images.

18.6 Basic Difficulties with the Optical Flow Model

When the optical flow ideas presented above are tried on real images, certain problems arise which are not apparent from the above model. First, not all edge points that should appear in the motion image are actually present. This is due to the contrast between the moving object and the background vanishing locally and limiting visibility. The situation is exactly as for edges that are produced by Sobel or other edge detection operators in nonmoving images. The contrast simply drops to a low value in certain localities, and the edge peters out. This signals that the edge model, and now the velocity flow model, is limited, and such local procedures are ad hoc and too impoverished to permit proper segmentation unaided.

Here we take the view that simple models can be useful, but they become inadequate on certain occasions and so robust methods are required to overcome the problems that then arise. Horn noted some of the problems as early as 1986. First, a smooth sphere may be rotating, but the motion will not show up in an optical flow (difference) image. If we so wish, we can regard this as a simple optical illusion, for the rotation of the sphere may well be invisible to the eye too. Second, a motionless sphere may *appear* to rotate as the light rotates around it. The object is simply subject to the laws of Lambertian optics, and again we may, if we wish, regard this effect as an optical illusion. (The illusion is relative to the baseline provided by the *normally correct* optical flow model.)

We next return to the optical flow model and see where it could be wrong or misleading. The answer is at once apparent; we stated in writing equation (18.2) that we were assuming that the *image* is being shifted. Yet it is not images that shift but the objects imaged within them. Thus, we ought to be considering the images of objects moving against a fixed background (or a variable background if the camera is moving). This will then permit us to see how sections of the motion edge can go from high to low contrast and back again in a rather fickle way, which we must nevertheless allow for in our algorithms. With this in mind, it should be permissible to go on using optical flow and difference imaging, even though these concepts have distinctly limited theoretical validity. (For a more thoroughgoing analysis of the underlying theory, see Faugeras, 1993.)

18.7 Stereo from Motion

An interesting aspect of camera motion is that, over time, the camera sees a succession of images that span a baseline in a similar way to binocular (stereo) images. Thus, it should be possible to obtain depth information by taking two such images and tracking object features between them. The technique is in principle more straightforward than normal stereo imaging in that feature tracking is possible, so the correspondence problem should be nonexistent. However, there is a difficulty in that the object field is viewed from almost the same direction in the succession of images, so that the full benefit of the available baseline is not obtained (Fig. 18.7). We can analyze the effect as follows.

First, in the case of camera motion, the equations for lateral displacement in the image depend not only on X but also on Y, though we can make a simplification in the theory by working with R, the radial distance of an object point from the optical axis of the camera, where:

$$R = [X^2 + Y^2]^{1/2} \tag{18.28}$$

We now obtain the radial distances in the two images as:

$$r_1 = Rf/Z_1 \tag{18.29}$$

$$r_2 = Rf/Z_2 \tag{18.30}$$

so the disparity is:

$$D = r_2 - r_1 = Rf(1/Z_2 - 1/Z_1) \tag{18.31}$$

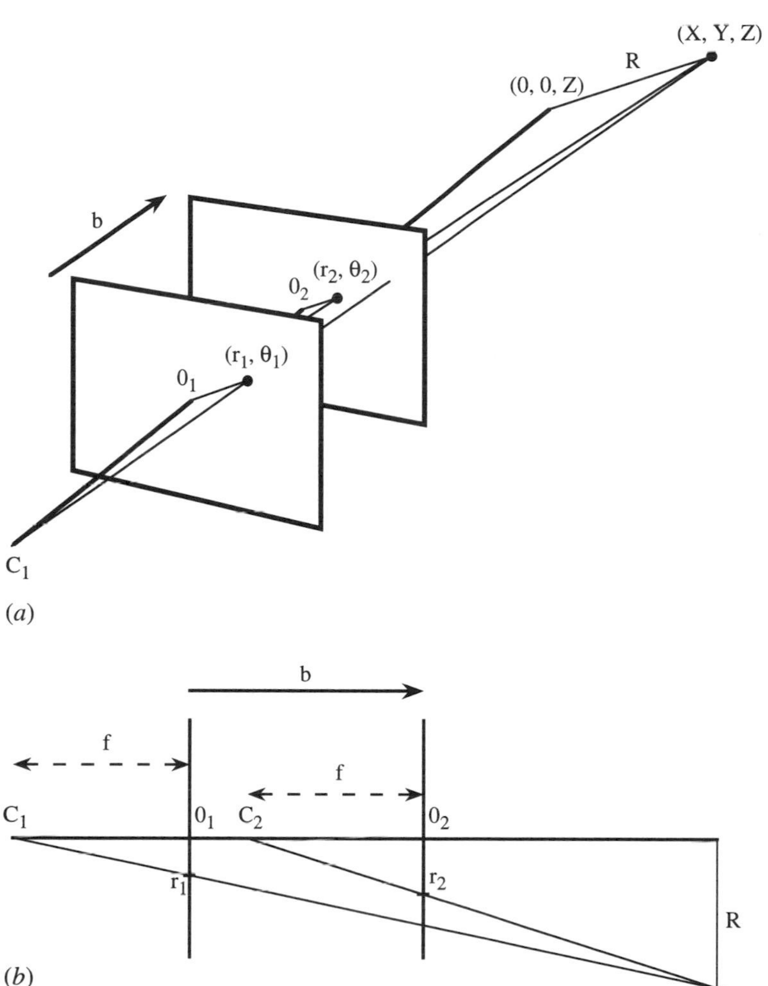

Figure 18.7 Calculation of stereo from camera motion. (*a*) shows how stereo imaging can result from camera motion, the vector **b** representing the baseline. (*b*) shows the simplified planar geometry required to calculate the disparity. It is assumed that the motion is directly along the optical axis of the camera.

Writing the baseline as:

$$b = Z_1 - Z_2 \tag{18.32}$$

and assuming that $b \ll Z_1$, Z_2, and then dropping the suffice, gives:

$$D = Rbf/Z^2 \tag{18.33}$$

Although this would appear to mitigate against finding Z without knowing R, we can overcome this problem by observing that:

$$R/Z = r/f \tag{18.34}$$

where r is approximately the mean value $\frac{1}{2}(r_1 + r_2)$. Substituting for R now gives:

$$D = br/Z \tag{18.35}$$

Hence, we can deduce the depth of the object point as:

$$Z = br/D = br/(r_2 - r_1) \tag{18.36}$$

This equation should be compared with equation (16.5) representing the normal stereo situation. The important point to note is that for motion stereo, the disparity depends on the radial distance r of the image point from the optical axis of the camera, whereas for normal stereo the disparity is independent of r. As a result, motion stereo gives no depth information for points on the optical axis, and the accuracy of depth information depends on the magnitude of r.

18.8 Applications to the Monitoring of Traffic Flow

18.8.1 *The System of Bascle et al.*

Visual analysis of traffic flow provides an important area of 3-D and motion studies. Here we want to see how the principles outlined in the earlier chapters are brought to bear on such practical problems. In one recent study (Bascle et al., 1994) the emphasis was on tracking complex primitives in image sequences, the primitives in question being vehicles. Several aspects of the problem make the analysis easier than the general case; in particular, there is a constraint that the vehicles run on a roadway and that the motion is generally smooth. Nevertheless, the methods that have to be used to make scene interpretation reliable and robust are nontrivial and will be outlined here in some detail.

First, motion-based segmentation is used to initialize the interpretation of the sequence of scenes. The motion image is used to obtain a rough mask of the object, and then the object outline is refined by classical edge detection and linking. B-splines are used to obtain a smoother version of the outline, which is fed to a snake-based tracking algorithm.[2] This algorithm updates the fit of the object outline and proceeds to repeat this for each incoming image.

Snake-based segmentation, however, concentrates on isolation of the object boundary and therefore ignores motion information from the main region of the object. It is therefore more reliable to perform motion-based segmentation of the entire region bounded by the snake and to use this information to refine the description of the motion and to predict the position of the object in the next image. This means that the snake-based tracker is provided with a reliable starting point from which to estimate the position of the object in the next image. The overall process is thus to feed the output of the snake boundary estimator into a motion-based segmenter and position predictor which reinitializes the snake for the next image—so both constituent algorithms perform the operations to which they are best adapted. It is especially relevant that the snake has a good starting approximation in each frame, both to help eliminate ambiguities and to save on computation. The motion-based region segmenter operates principally by analysis of optical flow, though in practice the increments between frames are not especially small. Thus, while true derivatives are not obtained, the result is not as bedeviled by noise as it might otherwise be.

Various refinements have been incorporated into the basic procedure:

- B-splines are used to smooth the outlines obtained from the snake algorithm.
- The motion predictions are carried out using an affine motion model that works on a point-by-point basis. (The affine model is sufficiently accurate for this purpose if perspective is weak so that motion can be approximated locally by a set of linear equations.)
- A multiresolution procedure is invoked to perform a more reliable analysis of the motion parameters.
- Temporal filtering of the motion is performed over several image frames.
- The overall trajectories of the boundary points are smoothed by a Kalman filter.[3]

The affine motion model used in the algorithm involves six parameters:[4]

$$\begin{bmatrix} x(t+1) \\ y(t+1) \end{bmatrix} = \begin{bmatrix} a_{11}(t) & a_{12}(t) \\ a_{21}(t) & a_{22}(t) \end{bmatrix} \begin{bmatrix} x(t) \\ y(t) \end{bmatrix} + \begin{bmatrix} b_1(t) \\ b_2(t) \end{bmatrix} \tag{18.37}$$

2 Snakes are explained in Section 18.12.

3 A basic treatment of Kalman filters is given in Section 18.13.

4 An affine transformation is one that is linear in the coordinates employed. This type of transformation includes the following geometric transformations: translation, rotation, scaling, and skewing (see Chapter 21). An affine motion model is one that takes the motion to lead to coordinate changes describable by affine transformations.

This leads to an affine model of image velocities, also with six parameters:

$$\begin{bmatrix} u(t+1) \\ v(t+1) \end{bmatrix} = \begin{bmatrix} m_{11}(t) & m_{12}(t) \\ m_{21}(t) & m_{22}(t) \end{bmatrix} \begin{bmatrix} u(t) \\ v(t) \end{bmatrix} + \begin{bmatrix} c_1(t) \\ c_2(t) \end{bmatrix} \tag{18.38}$$

Once the motion parameters have been found from the optical flow field, it is straightforward to estimate the following snake position.

An important factor in the application of this type of algorithm is the degree of robustness it permits. In this case, both the snake algorithm and the motion-based region segmentation scheme are claimed to be rather robust to partial occlusions. Although the precise mechanism by which robustness is achieved is not clear, the abundance of available motion information for each object, the insistence on consistent motion, and the recursive application of smoothing procedures, including a Kalman filter, all help to achieve this end. However, no specific nonlinear outlier rejection process is mentioned, which could ultimately cause problems, for example, if two vehicles merged together and became separated later on. Such a situation should perhaps be treated at a higher level in order to achieve a satisfactory interpretation. Then the problem mentioned by the authors of coping with instances of total occlusion would not arise.

Finally, it is of interest that the initial motion segmentation scheme locates the vehicles with their shadows since these are also moving (see Fig. 18.8). Subsequent analysis seems able to eliminate the shadows and to arrive at smooth vehicle boundaries. Shadows can be an encumbrance to practical systems, and systematic rather than ad hoc procedures are ultimately needed to eliminate their effects. Similarly, the algorithm appears able to cope with cluttered backgrounds when tracking faces, though in more complex scenes this might not be possible without specific high-level guidance. In particular, snakes are liable to be confused by background structure in an image, thus, however good the starting approximation, they could diverge from appropriate answers.

18.8.2 *The System of Koller et al.*

Another scheme for automatic traffic scene analysis has been described by Koller et al. (1994). This contrasts with the system described above in placing heavy reliance on high-level scene interpretation through use of belief networks. The basic system incorporates a low-level vision system employing optical flow, intensity gradient, and temporal derivatives. These provide feature extraction and lead to snake approximations to contours. Since convex polygons would be difficult to track from image to image (because the control points would

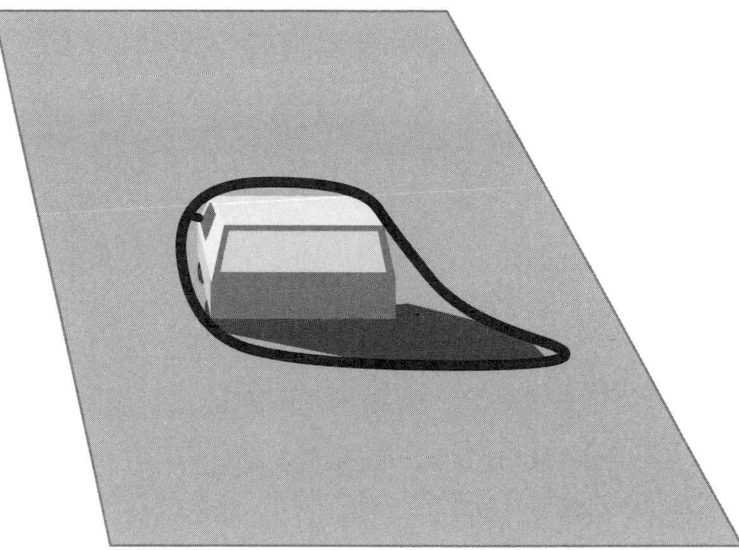

Figure 18.8 Vehicles located with their shadows. In many practical situations, shadows move with the objects that cause them, and simple motion segmentation procedures produce composite objects that include the shadows. Here a snake tracker envelops the car and its shadow.

tend to move randomly), the boundaries are smoothed by closed cubic splines having 12 control points. Tracking can then be achieved using Kalman filters. The motion is again approximated by an affine model, though in this case only three parameters are used, one being a scale parameter and the other two being velocity parameters:

$$\Delta \mathbf{x} = s(\mathbf{x} - \mathbf{x}_{\mathrm{m}}) + \Delta \mathbf{x}_{\mathrm{m}} \tag{18.39}$$

Here the second term gives the basic velocity component of the center of a vehicle region, and the first term gives the relative velocity for other points in the region, s being the change in scale of the vehicle ($s = 0$ if there is no change in scale). The rationale is that vehicles are constrained to move on the roadway, and rotations will be small. In addition, motion with a component toward the camera will result in an increase in size of the object and a corresponding increase in its apparent speed of motion.

Occlusion reasoning is achieved by assuming that the vehicles are moving along the roadway and are proceeding in a definite order, so that later vehicles (when viewed from behind) may partly or wholly obscure earlier ones. This depth ordering defines the order in which vehicles are able to occlude each other, and appears to be the minimum necessary to rigorously overcome problems of occlusion.

As stated earlier, belief networks are employed in this system to distinguish between various possible interpretations of the image sequence. Belief networks are directed acyclic graphs in which the nodes represent random variables and the arcs between them represent causal connections. Each node has an associated list of the conditional probabilities of its various states corresponding to assumed states of its parents (i.e., the previous nodes on the directed network). Thus, observed states for subsets of nodes permit deductions to be made about the probabilities of the states of other nodes. Such networks are used in order to permit rigorous analysis of probabilities of different outcomes when a limited amount of knowledge is available about the system. In like manner, once various outcomes are known with certainty (e.g., a particular vehicle has passed beneath a bridge), parts of the network will become redundant and can be removed. However, before removal their influence must be "rolled up" by updating the probabilities for the remainder of the network. When applied to traffic, the belief network has to be updated in a manner appropriate to the vehicles that are currently being observed. Indeed, each vehicle will have its own belief network, which will contribute a complete description of the entire traffic scene. However, one vehicle will have some influence on other vehicles, and special note will have to be taken of stalled vehicles or those making lane changes. In addition, one vehicle slowing down will have some influence on the decisions made by drivers in following vehicles. All these factors can be encoded into the belief network and can aid in arriving at globally correct interpretations. General road and weather conditions can also be taken into account.

Further sophistication is required to enable the vision part of the system to deal with shadows, brake lights, and other signals, while it remains to be seen whether the system will be able to cope with a wide enough variety of weather conditions. Overall, the system is designed in a very similar manner to that of Bascle et al. (1994), although its use of belief networks places it in a more sophisticated and undoubtedly more robust category.

In a more recent paper (Malik et al., 1995), the same research group has found that a stereo camera scheme can be set up to use geometrical constraints in a way that improves both speed of processing and robustness. Under a number of conditions, interpretation of the disparity map is markedly simplified. The first step is to locate the ground plane (the roadway) by finding road markers such as white lines. Then the disparity map is modified to give the ground plane zero disparity. After zero ground plane disparity has been achieved, vehicles on the roadway are located trivially via their positive disparity values, reflecting their positions above the roadway. The effect is essentially one-dimensional (see Fig. 18.9). It applies only if the following conditions are valid:

1. The camera parameters are identical (e.g., identical focal lengths and image planes, optical axes parallel, etc.), differing only in a lateral baseline **b**.

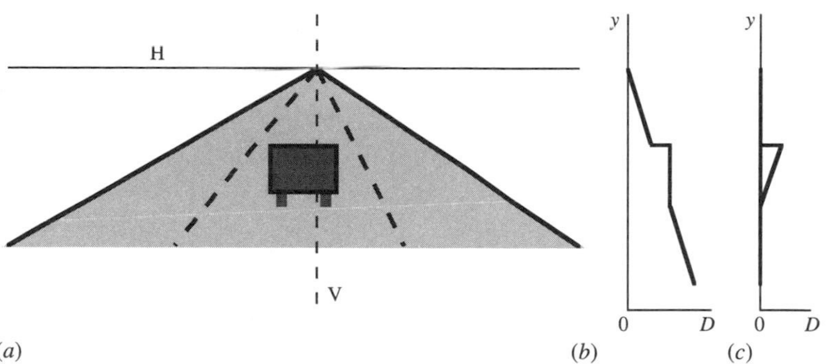

Figure 18.9 Disparity analysis of a road scene. In scene (*a*), the disparity D is greatest near the bottom of the image where depth Z is small, and is reduced to zero at the horizon line H. The vertical line V in the image passes up through the back of a vehicle, which is at constant depth, thereby giving constant disparity in (*b*). (*c*) shows how the disparity graph is modified when the ground plane is reduced to zero disparity by subtracting the Helmholtz shear.

2. The baseline vector **b** has a component only along the image x-axes.
3. The ground plane normal **n** has no component along the image x-axes.

In that case, there is no y-disparity, and the x-disparity on viewing the ground plane reduces to:

$$D_g = (b/h)(y\cos\delta + f\sin\delta) \tag{18.40}$$

where b is the intercamera baseline distance, h is the height of the center of projection of the camera above the ground plane, and δ is the angle of declination of the optical axes below the horizontal. The conditions listed above are reasonable for practical stereo systems, and if they are valid then the disparity will have high values near the bottom of the scene, reducing to $(b/h)f\sin\delta$ for vanishing points on the roadway. The next step in the calculation is to reduce the disparity of the roadway everywhere to zero by subtracting the quantity D_g from the observed disparity D. Thus, we are subtracting what has become known as the Helmholtz shear from the image. (It is a shear because the x-disparity and hence the x-values depend on y.) Here we refrain from giving a full proof of the above formula, and instead we give some insight into the situation by taking the important case $\delta = 0$. In that case we have (cf. equation (16.4)):

$$D_g = bf/Z \tag{18.41}$$

while from Fig. 18.10:

$$y/f = h/Z \tag{18.42}$$

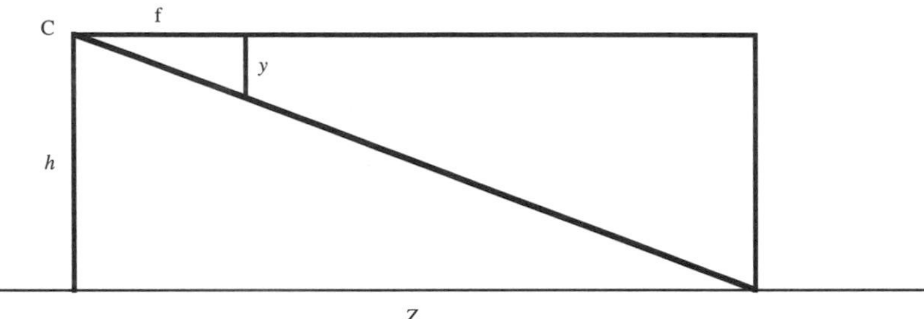

Figure 18.10 Geometry for subtracting the Helmholtz shear. Here the optical center of the camera C is a height h above the ground plane, and the optical axis of the camera is horizontal and parallel to the ground plane. The geometry of similar triangles can then be used to relate declination in the image (measured by y) to depth Z in the scene.

Eliminating Z between these equations leads to:

$$D_g = by/h \tag{18.43}$$

which proves the formula in that case.

Finally, whereas the subtraction of D_g from D is valid below the horizon line (where the ground plane disappears), above that level the definition of D_g is open to question. Equation (18.40) could provide an invalid extrapolation above the horizon line (would zero perhaps be a more appropriate value?). Assuming that equation (18.40) remains valid and that D_g is negative above the horizon line, then we will be subtracting a negative quantity from the observed disparity. Thus, the deduced disparity will increase with height. By way of example, take two telegraph poles equidistant from the camera and placed on either side of the roadway. Their disparity will now increase with height, and as a result they will appear to have a separation that also increases with height. It turns out that this was the form of optical illusion discovered by von Helmholtz (1925), which led him to the hypothesis that a shear (equation (18.40)) is active within the human visual system. In the context of machine vision, it is clearly a construction that has a degree of usefulness, but also carries important warnings about the possibility of making misleading measurements from the resulting sheared images. The situation arises even for the backs of vehicles seen below the horizon line (see Fig. 18.9*c*).

18.9 People Tracking

Visual surveillance is a long-standing area of computer vision, and one of its main early uses was to obtain information on military activities—whether from high-flying aircraft or from satellites. However, with the advent of ever cheaper video cameras, it next became widely used for monitoring road traffic, and most recently it has become ubiquitous for monitoring pedestrians. In fact, its application has actually become much wider than this, the aim being to locate criminals or people acting suspiciously—for example, those wandering around car-parks with the potential purpose of theft. By far the majority of visual surveillance cameras are connected to video recorders and gather miles of videotape, most of which will never be looked at—although following criminal or other activity, hours of videotape may be scanned for relevant events. Further cameras will be attached to closed circuit television monitors where human operators may be able to extract some fraction of the events displayed, though human attentiveness and reliability when overseeing a dozen or so screens will not be high. It would be far better if video cameras could be connected to automatic computer vision monitoring systems, which would at the minimum call human operators' attention to potential hazards or misdemeanors. Even if this were not carried out in real time in specific applications, it would be useful if it could be achieved at high speed with selected videotapes. This could save huge amounts of police time in locating and identifying perpetrators of crime.

Surveillance can cover other useful activities, including riot control, monitoring of crowds at sports events, checking for overcrowding in underground stations, and generally helping in matters of safety as well as crime. To some extent, human privacy must be sacrificed when surveillance is called into play, and a tradeoff between privacy and security then becomes necessary. Suffice it to say that many would consider a small loss of privacy a welcome price to pay to achieve increased security.

Many difficulties must be overcome before the "people-tracking" aspects of surveillance are fully solved. First, in comparison to cars, people are articulated objects that change shape markedly as they move. That their motion is often largely periodic can help visual analysis, though the irregularities in human motion may be considerable, especially if obstacles have to be avoided. Second, human motions are partly self-occluding, one leg regularly disappearing behind another, while arms can similarly disappear from view. Third, people vary in size and apparent shape, having a variety of clothes that can disguise their outlines. Fourth, when pedestrians are observed on a pavement, or on the underground, it is possible to lose track when one person passes behind another, as the two images coalesce before reemerging from the combined object shape.

It could be said that all these problems have been solved, but many of the algorithms that have been applied to these tasks have limited intelligence. Indeed, some employ rather simplified algorithms, as the need to operate continuously in real time frequently overrides the need for absolute accuracy. In any case, given the visual data that the computer actually receives, it is doubtful whether a human operator could always guarantee getting correct answers. For example, on occasions humans turn around in their tracks because they have forgotten something, thereby causing confusion when trying to track every person in a complex scene.

Further complexities can be caused by varying lighting, fixed shadows from buildings, moving shadows from clouds or vehicles, and so on.

18.9.1 *Some Basic Techniques*

The majority of people trackers use a motion detector to start the process of tracking. In principle, a model of the background can be taken with no people present, and then regions of interest (ROIs) can be formed around any substantial regions where movement is detected. However, such an approach will be error-prone, because changes in lighting or new placards, and so on, will appear to be people, and second because after people have criss-crossed the scene many times, error propagation will ensure many erroneous pixel classifications. Instead, it is usual to update the background model either by taking a temporal median of the scene (i.e., updating each pixel independently by using a running temporal median of the image sequence), or by some equivalent procedure that ignores outliers caused by temporary passage of moving objects, including people. Then, image differencing relative to the background model can yield quite reliable ROIs for the various people in the scene. Typically, these ROIs are rectangular in shape and can conveniently be moved with the people around the scene (Fig. 18.11). However, when people's paths cross, complications occur, and the ROIs tend to merge. Such problems have to be overcome by additional region-splitting procedures, often aided by Kalman filters that can help predict the motion over the merger period, thereby efficiently arriving at optimal tracking solutions. Since this approach is somewhat ad hoc, the emphasis has moved to more exact modeling of human shapes, using the ROIs merely as helpful approximations.

The Leeds people tracker (Baumberg and Hogg, 1995) learned a large number of walking pedestrian shapes and represented them as a set of cubic B-spline contours with N control points. These shapes were then analyzed using principal components analysis (Chapter 24) to find the possible "modes of variation" (eigenshapes) for the training data. The principal mode of variation is depicted by the extremes shown in Fig. 18.12. Although to the human eye these shapes do not match the average pedestrian well, they represent a substantial improvement over the basic rectangular ROI concept. In addition, they are well parametrized, in such a way that tracking is facilitated—and of course spatiotemporal shape development can be matched with the shapes that appear within the scene.

The Siebel and Maybank (2002) people tracker incorporates the Leeds people tracker but also includes a head detector. The overall system contains a motion

Figure 18.11 Example of a rectangular region of interest around a pedestrian. When outlined on a picture, the ROI is frequently called a bounding box.

detector (using a temporal median filter), a region tracker, a head detector, and an active shape tracker. Note that the temporal median approach leads to static objects eventually being incorporated into the background—a process that must not be undertaken too quickly, for it could even lead to creation of a negative object when a stationary pedestrian decides to move on. Siebel and Maybank took particular care to avoid this type of problem by keeping alert to objects for some time after they had become static. They also devoted considerable effort to developing a tracking history for the whole system. This permitted the four main modules, including the region tracker and the active shape tracker, to have access to each other in case any individual module lost data or lost track. This allowed the whole system to make up for deficiencies in any of its modules and thus improved overall tracking performance. The overall system employs hypothesis refinement, generating and running several track hypotheses per tracked person. Normally only one track is retained, but if occlusions or dropped frames occur, more are retained on a temporary basis.

The overall system was much more reliable than the active shape tracker at its core, (1) because it did not lose track of a person so easily, (2) because it incorporated more features into the region tracker, and (3) because its history tracking facility meant that it could regain some tracks previously lost.

In summary, as Siebel and Maybank (2002) have observed, there are three main categories of people tracking, of increasing complexity: (1) region- or blob-based tracking, using 2-D segmentation; (2) 2-D appearance modeling of human beings; and (3) full 3-D modeling of human beings. Because the

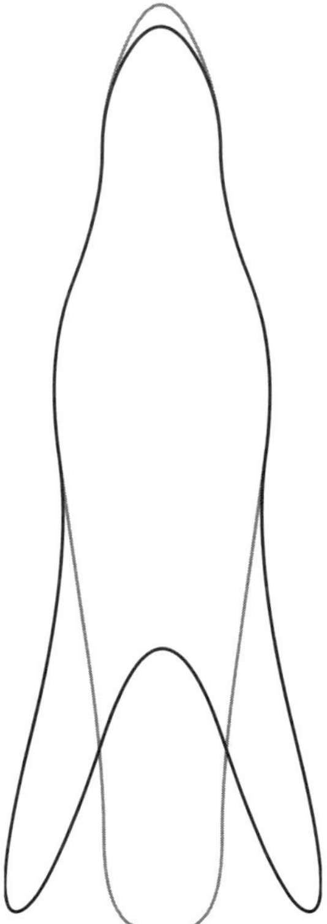

Figure 18.12 Eigenshapes of pedestrians. The two traces show typical extremes of the principal mode of variation of pedestrian shapes, as generated by PCA and characterized by cubic B-splines. Though constrained, this parametrization is actually very useful and far more powerful than the standard rectangular ROI representation (see text).

third category tends to run too slowly for practical cost-effective real-time applications, Siebel and Maybank concentrated on the second approach.

18.9.2 *Within-vehicle Pedestrian Tracking*

Another category of people tracking is that of within-vehicle pedestrian detection. This is a very exacting task, for vehicles travel along busy town roads at up to 30 mph, and it would be impossible for them to spend several minutes building up tracks. Indeed, pedestrians may well suddenly emerge from behind stationary vehicles, leaving the driver of a moving vehicle only a fraction of a second to react and avoid an accident. In these circumstances, it may be necessary to employ more rudimentary detection and tracking schemes than when performing surveillance from a fixed camera in a shopping precinct. The rectangular ROI approach is widely used, but in some more advanced applications (e.g., the ARGO project, Broggi et al., 2000*b*), head detection is also employed. Typically, a pedestrian is approximated by a vertical plane at an appropriate distance, and the ROI for a pedestrian is assumed to have certain size and aspect ratio constraints. Distance can be estimated by the position of the feet on the ground plane, this measure being refined by stereo analysis by reference to a vertical edge map of the scene. Although in general stereo analysis is a relatively slow process, it need not be so if the object in question has already been identified uniquely in both images of the pair—a process that is facilitated if the object is tracked from the moment it first becomes visible. Interestingly, humans often present a strong vertical symmetry, especially when they are seen from in front or behind (as would be the case for pedestrians walking along a pavement), and this can be used to aid image interpretation.

These observations show both that the interpretation of a scene from a vehicle is a highly complex matter and that a great number of cues can help the driver to understand what is going on. Even if only half the pedestrians are walking along the pavement, this would significantly reduce the amount of other information that still has to be analyzed by brute force means. Although we are here looking at the task from the driver's point of view, none of these factors will be any the less important for a future vehicle supervisor robot, whose job it will be to put on the brakes or steer the vehicle at the very last moment, if this will prevent an accident.[5]

Gavrila has published a number of papers on the detection of pedestrians from moving vehicles (e.g., Gavrila, 2000). In particular, he has employed the chamfer system to perform the identification. First, an edge detection algorithm is applied to the image, and then its distance function is obtained using a chamfer transform (which gives a good approximation to a Euclidean distance function). Next a previously derived edge template of a pedestrian is moved over the distance function image. The position giving the least total distance function value corresponds to the best match, as it represents the smallest distance of the (edge) feature points to those in the template. The mean chamfer distance $D(T, I)$ is the most appropriate criterion for a fit:

$$D(T,I) = (1/|T|)\sum_{t\in T} d_I(t) \tag{18.44}$$

5 It will no doubt be some years before a robot will be trusted to drive a car in a built-up area; not least there will be the question of whether the human driver or the vehicle manufacturer is to blame for any accident. Hence last-minute action to prevent an accident is a good compromise.

where $|T|$ is the number of pixels in the template and t is a typical pixel location in the template. Because it involves considerably less computational cost, Gavrila uses the average chamfer distance rather than the Hausdorff distance, even though the latter is more robust, being less affected by missing features in the image data. In addition, use of edge features is claimed to be more effective than general object features, for it gives a much smoother response and allows the search algorithm to lock more readily onto the correct solution. In particular, it allows a degree of dissimilarity between the template and the object of interest in the image (Gavrila, 2000).

To detect pedestrians in images, a wide variety of templates of various shapes is needed. In addition, these templates have to be applied in various sizes, corresponding to a distance parameter. Achieving high speed is still difficult and is attained by (a) using depth-first search over the templates (b) backing this up with a radial-basis function neural classifier for pattern verification, and (c) employing SIMD (MMX) processing.[6] This work (Gavrila, 2000) achieves a "promising" performance, but with upgrading should be of real value for avoiding accidents or at least minimizing their severity.

18.10 Human Gait Analysis

For well over a decade human motion has been studied using conventional cinematography. Often the aim of this work has been to analyze human movements in the context of various sports—in particular tracking the swing of a golf club and thus aiding the player in improving his game. To make the actions clearer, stroboscopic analysis coupled with bright markers judiciously placed on the body have been employed and have resulted in highly effective action displays. In the 1990s, machine vision was applied to the same task. At this point, the studies became much more serious, and there was greater focus on accuracy because of a widening of the area of application not only to other sports but also to medical diagnosis and to animation for modern types of films containing artificial sequences.

Because a high degree of accuracy is needed for many of these purposes—not the least of which is measuring limps or other imperfections of human gait—analysis of the motion of the whole human body in normally lit scenes proved insufficient, and body markers remained important. Typically, two body markers are needed per limb, so that the 3-D orientation of each limb will be deducible.

6 SIMD means "single instruction stream, multiple data stream" (Chapter 28); MMX means "Multimedia Extensions" and is a set of 57 multimedia instructions built into Intel microprocessors and accessed only by special MMX software calls.

Figure 18.13 Stick skeleton model of the lower human body. This model takes the main joints on the skeleton as being universal ball-and-socket joints, which can be approximated by point junctions—albeit with additional constraints on the possible motions (see text). Here a thin line through a joint indicates the single rotational axis of that joint.

Some work has been done to analyze human motions using single cameras, but the majority of the work employs two or more cameras. Multiple cameras are valuable because of the regular occlusion that occurs when one limb passes behind another, or behind the trunk.

To proceed with the analysis, a kinematic model of the human body is required. In general, such models assume that limbs are rigid links between a limited number of ball and socket joints, which can be approximated as point junctions between stick limbs. One such model (Ringer and Lazenby, 2000) employs two rotation parameters for the location where the hips join the backbone and three for the joint where the thigh bone joins the hips, plus one for the knee and another for the ankle. Thus, there are seven joint parameters for in each leg, two of these being common (at the backbone). This leads to a total of 12 parameters covering leg movements (Fig. 18.13). While it is part of the nature of the skeleton that the joints are basically rotational, there is some slack in the system, especially in the shoulders, while the backbone, which is compressed when upright, extends slightly when lying down or when hanging. In addition, the knee has some lateral freedom. A full analysis will be highly complex, though achieving it will be worthwhile for measuring mobility for medical diagnostic purposes. Finally, the whole situation is made more complex by constraints such as the inability of the knee to extend the lower leg too far forward.

Once a kinematic model has been established, tracking can be undertaken. It is relatively straightforward to identify the markers on the body with reasonable accuracy—though quite often one may be missed because of occlusion or because of poor lighting or the effects of noise. The next problem is to distinguish one marker from another and to label them. Considering the huge number of combinations of labels that are possible in the worst case scenario, and the frequency with which occlusions of parts of the farside leg or arm are bound to take place, special association algorithms are required for the purpose. These algorithms include particularly the Kalman filter, which is able to help predict how unseen markers will move until they come back into view. Such methods can be improved by including acceleration parameters as well as position and velocity parameters in the model (Dockstader and Tekalp, 2002). The model of Dockstader and Tekalp (2002) is not merely theoretically deduced: it has to be trained, typically on sequences of 2500 images each separated by 1/30 sec. Even so, the stick model of each human subject has to be initialized manually. A large amount of training is necessary to overcome the slight inaccuracies of measurement and to build up the statistics sufficiently for practical application when testing. Errors are greatest when measuring hand and arm movements because of the frequent occlusions to which they are subject.

Overall, articulated motion analysis involves complex processing and a lot of training data. It is a key area of computer vision, and the subject is evolving rapidly. It has already reached the stage of producing useful output, but accuracy will improve over the next few years. This will set the scene for practical medical monitoring and diagnosis, completely natural animation, detailed help with sports activities at costs that can be afforded by all, not to mention recognition of criminals by their characteristic gaits. Certain requirements—such as multiple cameras—will probably remain, but there will doubtless be a trend to markerless monitoring and minimal training on specific individuals (though there will still need to be bulk training on a sizable number of individuals). This is a fascinating and highly appropriate application of machine vision, and current progress indicates that it will essentially be solved within a decade.[7]

7 As always, it is dangerous to make technological predictions. This particular prediction is based on current progress, the known constraints on the motions of the human skeleton, and the fact that humans have limited limb velocities, accelerations, and reaction times. However, here we discount the interpretation of human facial motion, which involves rather more profound matters.

18.11 Model-based Tracking of Animals—A Case Study

This case study is concerned with animal husbandry—the care of farm animals. Good stockmen notice many aspects of animal behavior, and learn to respond to them. Fighting, bullying, tail biting activity, resting behavior, and posture are useful indicators of states of health, potential lameness, or heat stress, while group behavior may indicate the presence of predators or human intruders. In addition, feeding behavior is all important, as is the incidence of animals giving birth or breaking away from the confinement of the pen. In all these aspects, automatic observation of animals by computer vision systems is potentially useful.

Some animals such as pigs and sheep are lighter than their usual backgrounds of soil and grass, and thus they can in principle be located by thresholding. However, the backgrounds may be cluttered with other objects such as fences, pen walls, drinking troughs, and so on—all of which will complicate interpretation. Whatever the situation on factory conveyors, straightforward thresholding is unlikely to work well in normal outdoor and indoor farm scenes.

McFarlane and Schofield (1995) tackled this problem by image differencing. While differencing between adjacent frames is useful only if specific motion is occurring, far better performance may be achieved if the current frame is differenced against a carefully prepared background image. Employing such an image might seem to be a panacea, but this is not so, as the ambient lighting will vary throughout the day, and care must be taken to ensure that no animals are present when a background image is being taken. McFarlane and Schofield used a background image obtained by finding the median intensity at each pixel for a whole range of images taken over a fair period, in an effort to overcome this problem. Intelligent applications of this procedure were instituted to mask out regions where piglets were known to be resting. In such applications, medians are more effective than mean values, for they robustly eliminate outlier values rather than averaging them in.

The McFarlane and Schofield algorithm modeled piglets as simple ellipses and had fair success with their approach. However, next we examine the more rigorous modeling approach adopted by Marchant and Onyango (1995), and developed further by Onyango and Marchant (1996) and Tillett et al. (1997). These workers aimed to track the movements of pigs within a pen by viewing them from overhead under not too uniform lighting conditions. The main aim of the work at this early stage was tracking the animals, though, as indicated earlier, it was intended to lead on to behavioral analysis in later work. To find the animals, some form of template matching is required. Shape matching is an attractive concept, but with live animals such as pigs, the shapes are highly variable. Specifically, animals that are standing up or walking around will bend from side to side and may also bend their necks sideways or up and down as they feed. It is insufficient to use a small number of template masks to match the shapes, for there is an infinity of shapes related by various values of the shape parameters mentioned. These parameters are in addition to the obvious ones of position, orientation, and size.

Careful trials show that matching with all these parameters is insufficient, for the model is quite likely to be shifted laterally by variations in illumination. If one side of a pig is closer to the source of illumination, it will be brighter, and hence the final template used for matching will also shift in that direction. The resulting fit could be so poor that any reasonable goodness of fit criterion may deny the presence of a pig. Possible variations in lighting therefore have to be taken fully into account in fitting the animal's intensity profile.

A full-blooded approach involves principal components analysis (PCA). The deviation in position and intensity between the training objects and the model at a series of carefully chosen points is fed to a PCA system. The highest energy eigenvalues indicate the main modes of variation to be expected. Then any specific test example is fitted to the model, and amplitudes for each of these modes of variation are extracted, together with an overall parameter representing the goodness of fit. Unfortunately, this rigorous approach is highly computation intensive, because of the large number of free parameters. In addition, the position and intensity parameters are disparate measures that require the use of totally different and somewhat arbitrary scale factors to help the schema work. Some means is therefore required for decoupling the position and intensity information. Decoupling is achieved by performing two independent principal components analyses in sequence—first on the position coordinates and then on the intensity values.

When this procedure was carried out, three significant shape parameters were found, the first being lateral bending of the pig's back, accounting for 78% of the variance from the mean; and the second being nodding of the pig's head, which, because it corresponded to only ~20% of the total variance, was ignored in later analysis. In addition, the gray-level distribution model had three modes of variation, amounting to a total of 77% of the intensity variance. The first two modes corresponded to (1) a general amplitude variation in which the distribution is symmetrical about the backbone, and (2) a more complex variation in which the intensity distribution is laterally shifted relative to the backbone. (This clearly arises largely from lateral illumination of the animal.) See Fig. 18.14.

Although principal components analyses yield the important modes of variation in shape and intensity, in any given case the animal's profile has still to be fitted using the requisite number of parameters—one for shape and two for intensity. The Simplex algorithm (Press et al., 1992) has been found effective for this purpose. The objective function to be minimized to optimize the fit takes account of (1) the average difference in intensity between the rendered (gray-level)

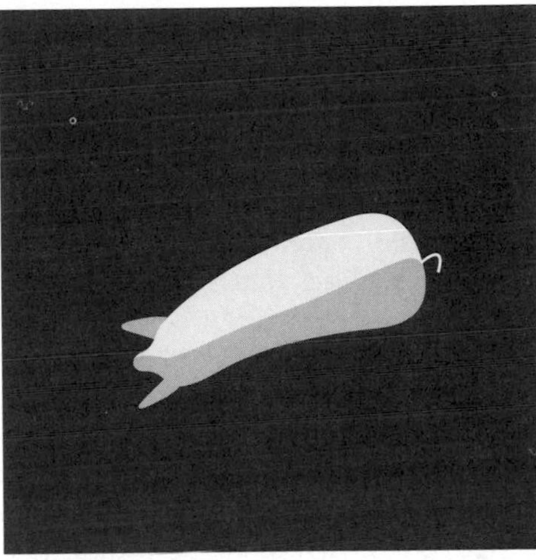

Figure 18.14 Effect of one mode of intensity variation found by PCA. This mode clearly arises from lateral illumination of the pig.

model and the image over the region of the model; and (2) the negative of the local intensity gradient in the image normal to the model boundary averaged along the model boundary. (The local intensity gradient will be a maximum right around the animal if this is correctly outlined by the model.)

One crucial factor has been skirted over in the preceding discussion: that the positioning and alignment of the model to the animal must be highly accurate (Cootes et al., 1992). This applies both for the initial PCA and later when fitting individual animals to the model is in progress. Here we concentrate on the PCA task. It should always be borne in mind when using PCA that the PCA is a method of characterizing deviations. This means that the deviations must already be minimized by referring all variations to the mean of the distribution. Thus, when setting up the data, it is very important to bring all objects to a common position, orientation, and scale before attempting PCA. In the present context, the PCA relates to shape analysis, and it is assumed that prior normalizations of position, orientation, and scale have already been carried out. (Note that in more general cases scaling may be included within PCA if required. However, PCA is a computation-intensive task, and it is best to encumber it as little as possible with unnecessary parameters.)

Overall, the achievements outlined here are notable, particularly in the effective method for decoupling shape and intensity analysis. In addition, the work holds significant promise for application in animal husbandry, demonstrating that animal monitoring and ultimately behavioral analysis should, in the foreseeable future, be attainable with the aid of image processing.

18.12 Snakes

There have been more than a dozen mentions of snakes (also known as deformable contours or active contour models) in this chapter, and it is left to this section to explain what they are and how they are computed.

The basic concept of snakes involves obtaining a complete and accurate outline of an object that may be ill-defined in places, whether through lack of contrast or noise or fuzzy edges. A starting approximation is made either by instituting a large contour that may be shrunk to size, or a small contour that may be expanded suitably, until its shape matches that of the object. In principle, the initial boundary can be rather arbitrary, whether mostly outside or within the object in question. Then its shape is made to evolve subject to an energy minimization process. On the one hand, it is desired to minimize the *external* energy corresponding to imperfections in the degree of fit; on the other hand, it is desired to minimize the *internal* energy, so that the shape of the snake does not become unnecessarily intricate, for example, taking on any of the characteristics of image noise. There are also model constraints that are represented in the formulation as contributions to the external energy. Typical of such constraints is that of preventing the snake from moving into prohibited regions, such as beyond the image boundary, or, for a moving vehicle, off the region of the road.

The snake's internal energy includes elastic energy, which might be needed to extend or compress it, and bending energy. If no bending energy terms were included, sharp corners and spikes in the snake would be free to occur with no restriction. Similarly, if no elastic energy terms were included, the snake would be permitted to grow or shrink without penalty.

The image data are normally taken to interact with the snake via three main types of image features: lines, edges, and terminations (the last-named can be line terminations or corners). Various weights can be given to these features according to the behavior required of the snake. For example, it might be required to hug edges and go around corners, and only to follow lines in the absence of edges. So the line weights would be made much lower than the edge and corner weights.

These considerations lead to the following breakdown of the snake energy:

$$\begin{aligned} E_{\text{snake}} &= E_{\text{internal}} + E_{\text{external}} \\ &= E_{\text{internal}} + E_{\text{image}} + E_{\text{constraints}} \\ &= E_{\text{stretch}} + E_{\text{bend}} + E_{\text{line}} + E_{\text{edge}} + E_{\text{term}} + E_{\text{repel}} \end{aligned} \tag{18.45}$$

The energies are written down in terms of small changes in position $\mathbf{x}(s) = (x(s), y(s))$ of each point on the snake, the parameter s being the arc length distance along the snake boundary. Thus we have:

$$E_{\text{stretch}} = \alpha(s)\left|\mathbf{x}_s(s)\right|^2 \tag{18.46}$$

and

$$E_{\text{bend}} = \beta(s)\left|\mathbf{x}_{ss}(s)\right|^2 \tag{18.47}$$

where the suffices s, ss imply first- and second-order differentiation, respectively. Similarly, E_{edge} is calculated in terms of the intensity gradient magnitude $|\text{grad}I|$, leading to:

$$E_{\text{edge}} = -w_{\text{edge}} \int \left|\text{grad}I\right|^2 ds \tag{18.48}$$

where w_{edge} is the edge-weighting factor.

The overall snake energy is obtained by summing the energies for all positions on the snake; a set of simultaneous differential equations is then set up to minimize the total energy. Space prevents a full discussion of this process here. Suffice it to say that the equations cannot be solved analytically, and recourse has to be made to iterative numerical solution, during which the shape of the snake evolves from some high-energy initialization state to the final low-energy equilibrium state, defining the contour of interest in the image.

In the general case, several possible complications need to be tackled:

1. Several snakes may be required to locate an initially unknown number of relevant image contours.
2. The different snakes will need different initialization conditions.
3. Snakes will occasionally have to split up as they approach contours that turn out to be fragmented.

Procedural problems also exist. The intrinsic snake concept is that of well-behaved differentiability. However, lines, edges, and terminations are usually highly localized, so there is no means by which a snake even a few pixels away could be expected to learn about them and hence to move toward them. In these circumstances the snake would "thrash around" and would fail to systematically zone in on a contour representing a global minimum of energy. To overcome this problem, smoothing of the image is required, so that edges can communicate with the snake some distance away, and the smoothing must gradually be reduced as the snake nears its target position. Ultimately, the problem is that the algorithm has no high-level appreciation of the overall situation, but merely reacts to a conglomerate of local pieces of information in the image. This makes segmentation using snakes somewhat risky despite the intuitive attractiveness of the concept.

In spite of these potential problems, a valuable feature of the snake concept is that, if set up correctly, the snake can be rendered insensitive to minor discontinuities in a boundary. This is important, for this makes it capable of negotiating practical situations such as fuzzy or low-contrast edges, or places where small artifacts get in the way. (This may happen with resistor leads, for example.) This capability is possible because the snake energy is set up *globally*—quite unlike the situation for boundary tracking where error propagation can cause wild deviations from the desired path. The reader is referred to the abundant literature on the subject (e.g., see Section 18.15) not only to clarify the basic theory (Kass and Witkin, 1987; Kass et al., 1988) but also to find how it may be made to work well in real situations.

18.13 The Kalman Filter

When tracking moving objects, it is desirable to be able to predict where they will be in future frames, for this will make maximum use of preexisting information and permit the least amount of search in the subsequent frames. It will also serve to offset the problems of temporary occlusion, such as when one vehicle passes behind another, when one person passes behind another, or even when one limb of a person passes behind another. (There are also many military needs for tracking prediction, and others on the sports field.) The obvious equations to employ for this purpose involve sequentially updating the position and the velocity of points on the object being tracked:

$$x_i = x_{i-1} + v_{i-1} \tag{18.49}$$

$$v_i = x_i - x_{i-1} \tag{18.50}$$

assuming for convenience a unit time interval between each pair of samples.

This approach is too crude to yield the best results. First, it is necessary to make three quantities explicit: (1) the raw measurements (e.g., x), (2) the best estimates of the values of the corresponding variables *before* observation (denoted by $^-$), and (3) the best estimates of these same model parameters *following* observation (denoted by $^+$). In addition, it is necessary to include explicit noise terms, so that rigorous optimization procedures can be derived for making the best estimates.

In the particular case outlined above, the velocity—and possible variations on it, which we shall ignore here for simplicity—constitutes a best estimate model parameter. We include position measurement noise by the parameter

u and velocity (model) estimation noise by the parameter w. The above equations now become:

$$x_i^- = x_{i-1+} + v_{i-1} + u_{i-1} \tag{18.51}$$

$$v_i^- = v_{i-1+} + w_{i-1} \tag{18.52}$$

In the case where the velocity is constant and the noise is Gaussian, we can spot the optimum solutions to this problem:

$$x_i^- = x_{i-1+} \tag{18.53}$$

$$\sigma_i^- = \sigma_{i-1+} \tag{18.54}$$

these being called the *prediction* equations, and

$$x_i^+ = \frac{x_i/\sigma_i^2 + (x_i^-)/(\sigma_i^-)^2}{1/\sigma_i^2 + 1/(\sigma_i^-)^2} \tag{18.55}$$

$$\sigma_i^+ = \left[\frac{1}{1/\sigma_i^2 + 1/(\sigma_i^-)^2}\right]^{1/2} \tag{18.56}$$

these being called the *correction* equations.[8] (In these equations, $\sigma^\pm$ are the standard deviations for the respective model estimates $x^\pm$, and σ is the standard deviation for the raw measurements x.)

These equations show how repeated measurements improve the estimate of the position parameter and the error upon it at each iteration. Notice the particularly important feature—that the noise is being modeled as well as the position itself. This permits all positions earlier than $i-1$ to be forgotten. The fact that there were many such positions, whose values can all be averaged to improve the accuracy of the latest estimate, is of course rolled up into the values of x_i^- and σ_i^-, and eventually into the values for x_i^+ and σ_i^+.

The next problem is how to generalize this result, both to multiple variables and to possibly varying velocity and acceleration. The widely used Kalman filter performs the required function by continuing with a linear approximation and by employing a state vector comprising position, velocity, and acceleration (or other relevant parameters), all in one state vector **s**. This constitutes the dynamic model. The raw measurements **x** have to be considered separately.

In the general case, the state vector is not updated simply by writing:

$$\mathbf{s}_i^- = \mathbf{s}_{i-1+} \tag{18.57}$$

8 The latter are nothing more than the well-known equations for weighted averages (Cowan, 1998).

but requires a fuller exposition because of the interdependence of position, velocity, and acceleration. Hence we have:[9]

$$\mathbf{s}_i^- = K_i \mathbf{s}_{i-1+} \tag{18.58}$$

Similarly, the standard deviations σ_i, $\sigma_i^\pm$ in equations (18.54)–(18.56) (or rather, the corresponding variances) have to be replaced by the covariance matrices Σ_i, $\Sigma_i^\pm$, and the equations become significantly more complicated. We will not go into the calculations fully here because they are nontrivial and would need several pages to iterate. Suffice it to say that the aim is to produce an optimum linear filter by a least-squares calculation (see, for example, Maybeck, 1979).

Overall, the Kalman filter is the optimal estimator for a linear system for which the noise is zero mean, white, and Gaussian, though it will often provide good estimates even if the noise is not Gaussian.

Finally, it will be noticed that the Kalman filter itself works by averaging processes that will give erroneous results if any outliers are present. This will certainly occur in most motion applications. Thus, there is a need to test each prediction to determine whether it is too far away from reality. If this is the case, the object in question will not likely have become partially or fully occluded. A simple option is to assume that the object continues in the same motion (albeit with a larger uncertainty as time goes on) and to wait for it to emerge from behind another object. At the very least, it is prudent to keep a number of such possibilities alive for some time, but the extent of this will naturally vary from situation to situation and from application to application.

18.14 Concluding Remarks

Early in this chapter we described the formation of optical flow fields and showed how a moving object or a moving camera leads to a focus of expansion. In the case of moving objects, the focus of expansion can be used to decide whether a collision will occur. In addition, analysis of the motion taking account of the position of the focus of expansion led to the possibility of determining structure from motion. Specifically, the calculation can be achieved via time-to-adjacency analysis, which yields the relative depth in terms of the motion parameters measurable directly from the image. We then went on to demonstrate some basic difficulties with the optical flow model, which arise since the motion

9 Some authors write K_{i-1} in this equation, but it is only a matter of definition whether the label matches the previous or the new state.

edge can have a wide range of contrast values, making it difficult to measure motion accurately. In practice, larger time intervals may have to be employed to increase the motion signal. Otherwise, feature-based processing related to that of Chapter 15 can be used. Corners are the most widely used feature for this purpose because of their ubiquity and because they are highly localized in 3-D. Space prevents details of this approach from being described here: details may be found in Barnard and Thompson (1980), Scott (1988), Shah and Jain (1984), and Ullman (1979).

Section 18.8 presented a study of two vision systems that have recently been developed with the purpose of monitoring traffic flow. These systems are studied here in order to show the extent to which optical flow ideas can be applied in real situations and to determine what other techniques have to be used to build complete vision systems for this type of application. Interestingly, both systems made use of optical flow, snake approximation to boundary contours, affine models of the displacements and velocity flows, and Kalman filters for motion prediction—confirming the value of all these techniques. On the other hand, the Koller system also employed belief networks for monitoring the whole process and making decisions about occlusions and other relevant factors. In addition, the later version of the Koller system (Malik, 1995) used binocular vision to obtain a disparity map and then applied a Helmholtz shear to reduce the ground plane to zero disparity so that vehicles could be tracked robustly in crowded road scenes. Ground plane location was also a factor in the design of the system.

Later parts of the chapter covered more modern applications of motion analysis, including people tracking, human gait analysis, and animal tracking. These studies reflect the urgency with which machine vision and motion analysis are being pursued for real applications. In particular, studies of human motion reflect medical and film production interests, which can realistically be undertaken on modern high-speed computer platforms.

Further work related to motion will be covered in Chapter 20 after perspective invariants have been dealt with (in Chapter 19), for this will permit progress on certain aspects of the subject to be made more easily.

The obvious way to understand motion is by image differencing and by determination of optical flow. This chapter has shown that the aperture problem is a difficulty that may be avoided by use of corner tracking. Further difficulties are caused by temporary occlusions, thus necessitating techniques such as occlusion reasoning and Kalman filtering.

18.15 Bibliographical and Historical Notes

Optical flow has been investigated by many workers over many years: see, for example, Horn and Schunck (1981) and Heikkonen (1995). A definitive account of the mathematics relating to FOE appeared in 1980 (Longuet-Higgins and Prazdny, 1980). Foci of expansion can be obtained either from the optical flow field or directly (Jain, 1983). The results of Section 18.5 on time-to-adjacency analysis stem originally from the work of Longuet-Higgins and Prazdny (1980), which provides some deep insights into the whole problem of optical flow and the possibilities of using its shear components. Note that numerical solution of the velocity field problem is not trivial; typically, least-squares analysis is required to overcome the effects of measurement inaccuracies and noise and to obtain finally the required position measurements and motion parameters (Maybank, 1986). Overall, resolving ambiguities of interpretation is one of the main problems, and challenges, of image sequence analysis (see Longuet-Higgins (1984) for an interesting analysis of ambiguity in the case of a moving plane).

Unfortunately, the substantial and important literature on motion, image sequence analysis, and optical flow, which impinges heavily on 3-D vision, could not be discussed in detail here for reasons of space. For seminal work on these topics, see, for example, Ullman (1979), Snyder (1981), Huang (1983), Jain (1983), Nagel (1983, 1986), and Hildreth (1984).

Further work on traffic monitoring appears in Fathy and Siyal (1995), and there has been significant effort on automatic visual guidance in convoys (Schneiderman et al., 1995; Stella, et al., 1995). Vehicle guidance and ego-motion control cover much further work (Brady and Wang, 1992; Dickmanns and Mysliwetz, 1992), while several papers show that following vehicles can conveniently be tackled by searching for symmetrical moving objects (e.g., Kuehnle, 1991; Zielke et al., 1993).

Work on the application of snakes to tracking has been carried out by Delagnes et al. (1995); on the use of Kalman filters in tracking by Marslin et al. (1991); on the tracking of plant rows in agricultural vehicles by Marchant and Brivot (1995); and on recognition of vehicles on the ground plane by Tan et al. (1994). For details of belief networks, see Pearl (1988). Note that corner detectors (Chapter 14) have also been widely used for tracking: see Tissainayagam and Suter (2004) for a recent assessment of performance.

Most recently, there has been an explosion of interest in surveillance, particularly in the analysis of human motions (Aggarwal and Cai, 1999; Gavrila, 1999; Collins et al., 2000; Haritaoglu et al., 2000; Siebel and Maybank, 2002; Maybank and Tan, 2004). The latter trend has accelerated the need for studies of articulated motion (Ringer and Lazenby, 2000; Dockstader and Tekalp, 2001), one of the earliest enabling techniques being that of Wolfson (1991).

As a result, a number of workers have been able to characterize or even recognize human gait patterns (Foster et al., 2001; Dockstader and Tekalp, 2002; Vega and Sarkar, 2003). A further purpose for this type of work has been the identification of pedestrians from moving vehicles (Broggi et al., 2000b; Gavrila, 2000). Much of this work has its roots in the early far-sighted paper by Hogg (1983), which was later followed up by crucial work on eigenshape and deformable models (Cootes et al., 1992; Baumberg and Hogg, 1995; Shen and Hogg, 1995). However, it is important not to forget the steady progress on traffic monitoring (Kastrinaki et al., 2003) and related work on vehicle guidance (e.g., Bertozzi and Broggi, 1998)—the latter topic being covered in more depth in Chapter 20.

In spite of the evident successes, only a very limited number of fully automated visual vehicle guidance systems is probably in everyday use. The main problem would appear to be the lack of robustness and reliability required to trust the system in an "all hours–all weathers" situation. Note that there are also legal implications for a system that is to be used for control rather than merely for vehicle-counting functions.

18.16 Problem

1. Explain why, in equation (18.56), the variances are combined in this particular way. (In most applications of statistics, variances are combined by addition.)

Invariants and Their Applications

Invariants are intrinsic to the idea of recognition in both 2-D and 3-D. The basic idea is to identify some parameter or parameters that do not vary between different instances of the same object. Unfortunately, perspective projection makes the issue far harder in the general 3-D case. This chapter explores the problem and demonstrates a number of useful techniques.

Look out for:

- how a ratio of distances between features along the same straight line can act as a convenient invariant under weak perspective projection.
- how a ratio of ratios (or "cross ratio") can act as a convenient invariant under full perspective projection.
- how the cross ratio type of invariant can rather cunningly be generalized to cover many wider possibilities.
- how the cross ratio type of invariant seems largely unable to provide invariance outside any given plane.

Although this chapter considers only one aspect of 3-D vision, it is extremely useful both in helping to cue into complex images (see particularly the egomotion example of Fig. 19.4 and the facial analysis example of Section 24.12) and in taking shortcuts around the tedious analysis of 3-D geometry (see, for example, Sections 20.8 and 20.9).

CHAPTER 19

Invariants and Their Applications

19.1 Introduction

Pattern recognition is a complex task and, as stated in Chapter 1, involves the twin processes of discrimination and generalization. Generalization is in many ways more important than discrimination—especially in the initial stages of recognition—since there is so much redundant information in a typical image. Thus, we need to find ways of helping to eliminate invalid matches. This is where the study of invariants comes into its own.

An invariant is a property of an object or class of objects that does not change with changes of viewpoint or object pose and that can therefore be used to help distinguish it from other objects. The procedure is to search for objects with a specific invariant, so that those that do not possess the invariant can immediately be discarded from consideration. An invariant property can be regarded as a necessary condition for an object to be in the chosen class, though in principle only detailed subsequent analysis will confirm that it is. In addition, if an object is found to possess the correct invariant, it will then be profitable to pursue the analysis further and find its pose, size, or other relevant data. Ideally, an invariant would uniquely identify an object as being of a particular type or class. Thus, an invariant should not merely be a property that leads to further hypotheses being made about the object, but one that fully characterizes it. However, the difference is a subtle one, more a matter of degree and purpose than an absolute criterion. We shall examine the extent of the difference by appealing to a number of specific cases.

Let us first consider an object being viewed from directly overhead at a known distance by a camera whose optical axis is normal to the plane on which the object is lying. We shall assume that the object is flat. Take two point features on the object such as corners or small holes (cf. Chapter 15). If we measure the image distance between these features, then this acts as an invariant, in that:

1. it has a value independent of the translation and orientation parameters of the object.
2. it will be unchanged for different objects of the same type.
3. it will in general be different from the distance parameters of other objects that might be on the object plane.

Thus measurement of distance provides a certain lookup or indexing quality that will ideally identify the object uniquely, though further analysis will be required to fully locate it and ascertain its orientation. Hence, distance has all the requirements of an invariant, though it could also be argued that it is only a feature that helps to classify objects. Here, we are ignoring an important factor—the effect of imprecision in measurement, owing to spatial quantization (or inadequate spatial resolution), noise, lens distortions, and so on. In addition, the effects of partial occlusion or breakage are also being ignored. Most definitely, there is a limit to what can be achieved with a single invariant measure, though in what follows we attempt to reveal what is possible, and we demonstrate the advantages of the way of thinking that involves analysis of invariants.

The above ideas relating to distance as an invariant measure showed it to be useful in suppressing the effects of translations and rotations of objects in 2-D. Hence, it is of little direct value when considering translations and rotations in 3-D. Furthermore, it is not even able to cope with scale variations of objects in 2-D. Moving the camera closer to the object plane and refocusing totally changes the situation: all values of the distance invariant residing in the object indexing table must be changed and the old values ignored. However, a moment's thought will show that this last problem could be overcome. All we need to do is to take *ratios* of distances, which requires a minimum of three point features to be identified in the image and the interfeature distances measured. If we call two of these distances d_1 and d_2, then the ratio d_1/d_2 will act as a scale-independent invariant. That is, we will be able to identify objects using a single indexing operation whatever their 2-D translation, orientation, or apparent size or scale. An alternative to this idea is to measure the angle between pairs of distance vectors, $\cos^{-1}(\mathbf{d}_1.\mathbf{d}_2/|d_1||d_2|)$, which will again be scale invariant.

Of course, this consideration has already been invoked in our earlier work on shape analysis. If objects are subject only to 2-D translations and rotations but not to changes of scale, they can be characterized by their perimeters or areas as well as their normal linear dimensions. Furthermore, parameters such as compactness and aspect ratio, which employ dimensionless ratios of image measurements, were acknowledged in Chapter 6 to overcome the size/scale problem.

Nevertheless, the main motivation for using invariants is to obtain mathematical measures of configurations of object features, which are carefully designed to be independent of the viewpoint or coordinate system used and indeed do not require specific setup or calibration of the image acquisition system. However, it must be emphasized that camera distortions are assumed to be absent or to have been compensated for by suitable postcamera transformations (see Chapter 21).

19.2 Cross Ratios: The "Ratio of Ratios" Concept

It would be most useful if we could extend these ideas to permit indexing for general transformations in 3-D. An obvious question is whether finding ratios of ratios of distances will provide suitable invariants and lead to such a generalization. The answer is that ratios of ratios do provide useful further invariants, though going further than this leads to considerable complication, and there are restrictions on what can be achieved with limited computation. In addition, noise ultimately becomes a limiting factor, since so many parameters become involved in the computation of complex invariants that the method ultimately loses steam. (It becomes just one of many ways of raising hypotheses and therefore it has to compete with other approaches in a manner appropriate to the particular problem application being studied.)

We now consider the ratio of ratios approach. Initially, we only examine a set of four collinear points on an object. Figure 19.1 shows such a set of four points (P_1, P_2, P_3, P_4) and a transformation of them (Q_1, Q_2, Q_3, Q_4) such as that produced by an imaging system with optical center C (c, d). Choice of a suitable pair of oblique axes permits the coordinates of the points in the separate sets to be expressed respectively as:

$$(x_1, 0),\ (x_2, 0),\ (x_3, 0),\ (x_4, 0)$$

$$(0, y_1),\ (0, y_2),\ (0, y_3),\ (0, y_4)$$

Taking points P_i, Q_i, we can write the ratio $CQ_i : PQ_i$ both as $\dfrac{c}{-x_i}$ and as $\dfrac{d - y_i}{y_i}$. Hence:

$$\frac{c}{x_i} + \frac{d}{y_i} = 1 \tag{19.1}$$

which must be valid for all i. Subtraction of the i'th and j'th relations now gives:

$$\frac{c(x_j - x_i)}{x_i x_j} = \frac{-d(y_j - y_i)}{y_i y_j} \tag{19.2}$$

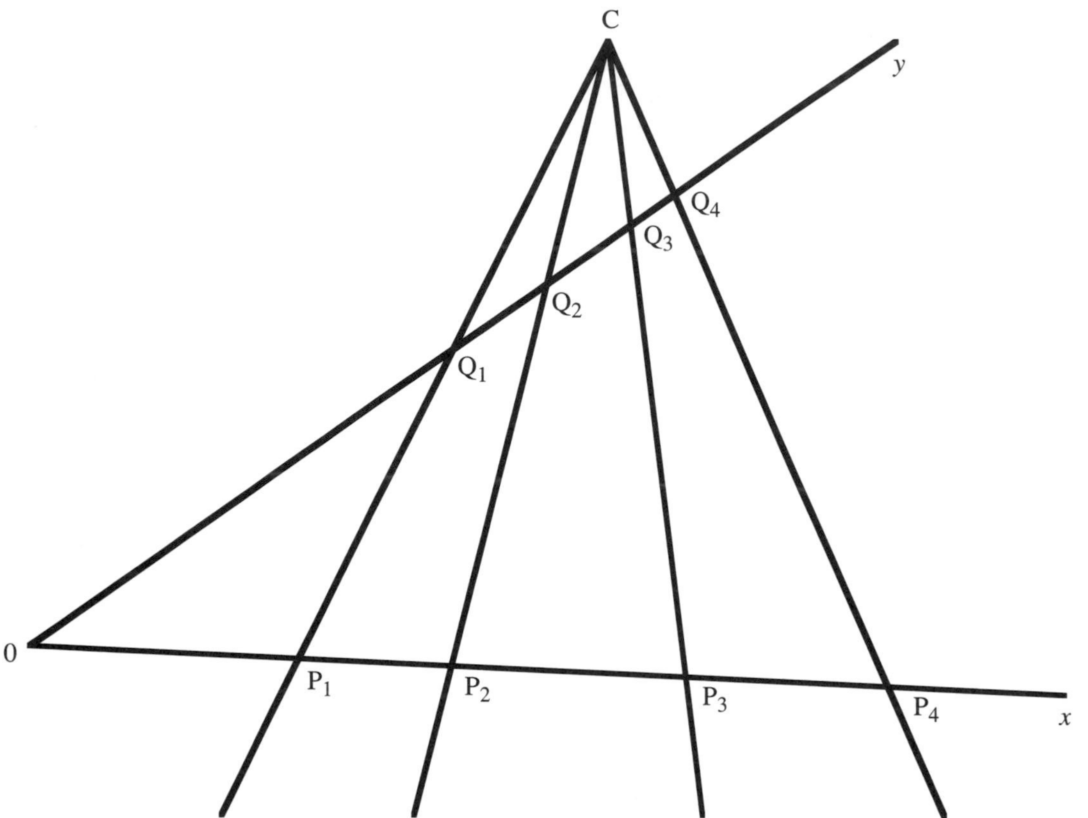

Figure 19.1 Perspective transformation of four collinear points. This figure shows four collinear points (P_1, P_2, P_3, P_4) and a transformation of them (Q_1, Q_2, Q_3, Q_4) similar to that produced by an imaging system with optical center C. Such a transformation is called a *perspective transformation*.

Forming a ratio between two such relations will now eliminate the unknowns c and d. For example, we will have:

$$\frac{x_3(x_2 - x_1)}{x_2(x_3 - x_1)} = \frac{y_3(y_2 - y_1)}{y_2(y_3 - y_1)} \tag{19.3}$$

However, the result still contains factors such as x_3/x_2, which depend on absolute position. Hence, it is necessary to form a suitable ratio of such results, which cancels out the effects of absolute positions:

$$\left(\frac{x_2 - x_4}{x_3 - x_4}\right) \Big/ \left(\frac{x_2 - x_1}{x_3 - x_1}\right) = \left(\frac{y_2 - y_4}{y_3 - y_4}\right) \Big/ \left(\frac{y_2 - y_1}{y_3 - y_1}\right) \tag{19.4}$$

Thus, our original intuition that a ratio of ratios type of invariant might exist which would cancel out the effects of a perspective transformation is correct. In particular, four collinear points viewed from any perspective viewpoint yield the same value of the cross ratio, defined as above. The value of the cross ratio of the four points is written:

$$\mathscr{C}(\mathrm{P}_1, \mathrm{P}_2, \mathrm{P}_3, \mathrm{P}_4) = \frac{(x_3 - x_1)(x_2 - x_4)}{(x_2 - x_1)(x_3 - x_4)} \tag{19.5}$$

For clarity, we shall write this particular cross ratio as κ in what follows. It should be noted that there are $4! = 24$ possible ways in which 4 collinear points can be ordered on a straight line, and hence there could be 24 cross ratios. However, they are not all distinct; in fact, there are only six different values. To verify this, we start by interchanging pairs of points:

$$\mathscr{C}(\mathrm{P}_2, \mathrm{P}_1, \mathrm{P}_3, \mathrm{P}_4) = \frac{(x_3 - x_2)(x_1 - x_4)}{(x_1 - x_2)(x_3 - x_4)} = 1 - \kappa \tag{19.6}$$

$$\mathscr{C}(\mathrm{P}_1, \mathrm{P}_3, \mathrm{P}_2, \mathrm{P}_4) = \frac{(x_2 - x_1)(x_3 - x_4)}{(x_3 - x_1)(x_2 - x_4)} = \frac{1}{\kappa} \tag{19.7}$$

$$\mathscr{C}(\mathrm{P}_1, \mathrm{P}_2, \mathrm{P}_4, \mathrm{P}_3) = \frac{(x_4 - x_1)(x_2 - x_3)}{(x_2 - x_1)(x_4 - x_3)} = 1 - \kappa \tag{19.8}$$

$$\mathscr{C}(\mathrm{P}_4, \mathrm{P}_2, \mathrm{P}_3, \mathrm{P}_1) = \frac{(x_3 - x_4)(x_2 - x_1)}{(x_2 - x_4)(x_3 - x_1)} = \frac{1}{\kappa} \tag{19.9}$$

$$\mathscr{C}(\mathrm{P}_3, \mathrm{P}_2, \mathrm{P}_1, \mathrm{P}_4) = \frac{(x_1 - x_3)(x_2 - x_4)}{(x_2 - x_3)(x_1 - x_4)} = \frac{\kappa}{\kappa - 1} \tag{19.10}$$

$$\mathscr{C}(\mathrm{P}_1, \mathrm{P}_4, \mathrm{P}_3, \mathrm{P}_2) = \frac{(x_3 - x_1)(x_4 - x_2)}{(x_4 - x_1)(x_3 - x_2)} = \frac{\kappa}{\kappa - 1} \tag{19.11}$$

These cases provide the main possibilities, but of course interchanging more points will yield a limited number of further values—in particular:

$$\mathscr{C}(\mathrm{P}_3, \mathrm{P}_1, \mathrm{P}_2, \mathrm{P}_4) = 1 - \mathscr{C}(\mathrm{P}_1, \mathrm{P}_3, \mathrm{P}_2, \mathrm{P}_4) = 1 - \frac{1}{\kappa} = \frac{\kappa - 1}{\kappa} \tag{19.12}$$

$$\mathscr{C}(\mathrm{P}_2, \mathrm{P}_3, \mathrm{P}_1, \mathrm{P}_4) = \frac{1}{(\mathrm{P}_2, \mathrm{P}_1, \mathrm{P}_3, \mathrm{P}_4)} = \frac{1}{1 - \kappa} \tag{19.13}$$

This covers all six cases, and a moment's thought (based on trying further interchanges of points) will show there can be no others. (We can only repeat κ, $1-\kappa$, $\kappa/(\kappa-1)$, and their inverses.) Of particular interest is the fact that numbering the points in reverse (which would correspond to viewing the line from the other side) leaves the cross ratio unchanged. Nevertheless, it is inconvenient that the same invariant has six different manifestations, for this implies that six different index values have to be looked up before the class of an object can be ascertained. On the other hand, if points are labeled in order along the line rather than randomly it should generally be possible to circumvent this situation.

So far we have been able to produce only one projective invariant, and this corresponds to the rather simple case of four collinear points. The usefulness of this measure is augmented considerably when it is noted that four collinear points, taken in conjunction with another point, define a pencil[1] of concurrent coplanar lines passing through the latter point. It will be clear that we can assign a unique cross ratio to this pencil of lines, equal to the cross ratio of the collinear points on any line passing through them. We can clarify the situation by considering the angles between the various lines (Fig. 19.2). Applying the sine rule four times to determine the four distances in the cross ratio $\mathscr{C}(\mathrm{P}_1, \mathrm{P}_2, \mathrm{P}_3, \mathrm{P}_4)$ gives:

$$\frac{x_3 - x_1}{\sin \alpha_{13}} = \frac{\mathrm{OP}_1}{\sin \beta_3} \tag{19.14}$$

$$\frac{x_2 - x_4}{\sin \alpha_{24}} = \frac{\mathrm{OP}_4}{\sin \beta_2} \tag{19.15}$$

$$\frac{x_2 - x_1}{\sin \alpha_{12}} = \frac{\mathrm{OP}_1}{\sin \beta_2} \tag{19.16}$$

$$\frac{x_3 - x_4}{\sin \alpha_{34}} = \frac{\mathrm{OP}_4}{\sin \beta_3} \tag{19.17}$$

Substituting in the cross-ratio formula (equation (19.5)) and canceling the factors OP_1, OP_4, $\sin \beta_2$, and $\sin \beta_3$ now give:

$$\mathscr{C}(\mathrm{P}_1, \mathrm{P}_2, \mathrm{P}_3, \mathrm{P}_4) = \frac{\sin \alpha_{13} \sin \alpha_{24}}{\sin \alpha_{12} \sin \alpha_{34}} \tag{19.18}$$

Thus, the cross ratio depends only on the angles of the pencil of lines. It is interesting that appropriate juxtaposition of the sines of the angles gives

---

1 It is a common nomenclature of projective geometry to call a set of concurrent lines a *pencil* (e.g., Tuckey and Armistead, 1953).

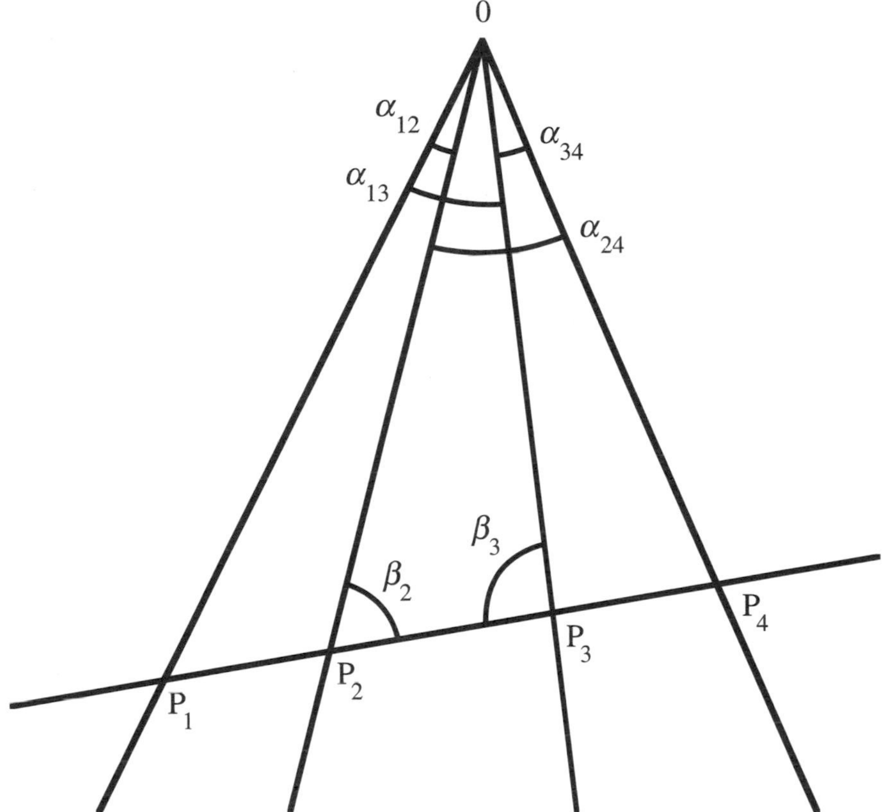

Figure 19.2 Geometry for calculation of the cross ratio of a pencil of lines. The figure shows the geometry required to calculate the cross ratio of a pencil of lines, in terms of the angles between them.

the final formula invariance under perspective projection. Using the angles themselves would not give the desired degree of mathematical invariance. We can immediately see one reason for this: inversion of the direction of any line must leave the situation unchanged, so the formula must be tolerant to adding π to each of the two angles linking the line. This could not be achieved if the angles appeared without suitable trigonometric functions.

We can extend this concept to four concurrent planes, since the concurrent lines can be projected into four concurrent planes once a separate axis for the concurrency has been defined. Because there are infinitely many such axes, there are infinitely many ways in which sets of planes can be chosen. Thus, the original simple result on collinear points can be extended to a much more general case.

We started by trying to generalize the case of four collinear points, but what we achieved was first to find a dual situation in which points become lines also described by a cross ratio and then to find an extension in which planes are described by a cross ratio. We now return to the case of four collinear points, and we will see how we can extend it in other ways.

19.3 Invariants for Noncollinear Points

First, imagine that not all the points are collinear: specifically, let us assume that one point is not in the line of the other three. If this is the case, then there is not enough information to calculate a cross ratio. However, if a further coplanar point is available, we can draw an imaginary line between the noncollinear points to intersect their common line in a unique point, which will then permit a cross ratio to be computed (Fig. 19.3*a*). Nevertheless, this is some way from a general solution to the characterization of a set of noncollinear points. We might inquire how many point features in general position[2] on a plane will be required to calculate an invariant. The answer is five, since the fact that we can form a cross ratio from the angles between four lines immediately means that forming a pencil of four lines from five points defines a cross-ratio invariant (Fig. 19.3*b*).

Although the value of this cross ratio provides a necessary condition for a match between two sets of five general coplanar points, it could be a fortuitous match, as the condition depends only on the relative directions between the various points and the reference point; that is, any of the nonreference points is only defined to the extent that it lies on a given line. Clearly, two cross ratios formed by taking two reference points will define the directions of all the remaining points uniquely (Fig. 19.3*c*).

We can now summarize the general result, which stipulates that for five general coplanar points, no three of which are collinear, two different cross ratios are required to characterize the shape. These cross ratios correspond to taking in turn two separate points and producing pencils of lines passing through them and (in each case) the remaining four points (Fig. 19.3*c*). Although it might appear that at least five cross ratios result from this sort of procedure, there are only two functionally independent cross ratios—essentially because

---

2 Points on a plane are described as being *in general position* if they are chosen at random and are not collinear or in any special pattern such as a regular polygon.

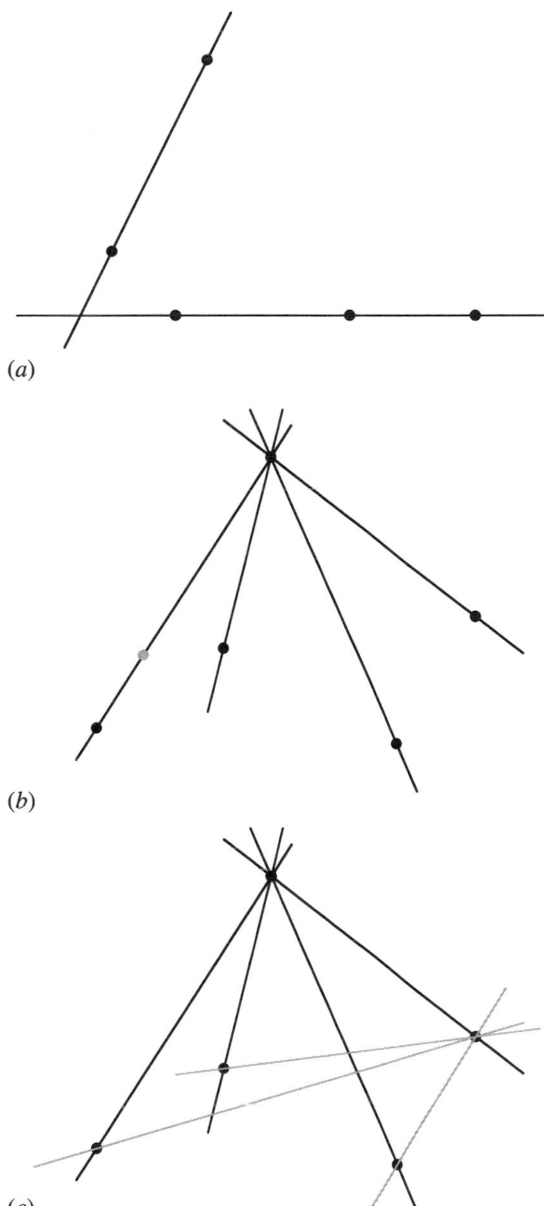

Figure 19.3 Calculation of invariants for a set of noncollinear points. (*a*) shows how the addition of a fifth point to a set of four points, one of which is not collinear with the rest, permits the cross ratio to be calculated. (*b*) shows how the calculation can be extended to any set of noncollinear points; also shown is an additional (gray) point which a single cross ratio fails to distinguish from other points on the same line. (*c*) shows how any failure to identify a point uniquely can be overcome by calculating the cross ratio of a second pencil generated from the five original points.

the position of any point is defined once its direction relative to two other points is known.

Next, we consider the problem of finding the ground plane in practical situations—especially that of egomotion, including vehicle guidance (Fig. 19.4). Here a set of four collinear points can be observed from one frame to the next. If they are on a single plane, then the cross ratio will remain constant, but if one is elevated above the ground plane (as, for example, a bridge or another vehicle), then the cross ratio will vary over time. Taking a larger number of points, it should be possible to deduce by a process of elimination which are on the ground plane and which are not (though the amount of noise and clutter will determine the computational complexity of the task). All this is possible without any calibration of the camera, this being perhaps the main value of concentrating attention on projective invariants. Note that there is a potential problem regarding irrelevant planes, such as the vertical faces of buildings. The cross-ratio test is so resistant to viewpoint and pose that it merely ascertains whether the points being tested are coplanar. Only by using a sufficiently large number of independent sets of points can one plane be discriminated from another. (For simplicity we ignore any subsequent stages of pose analysis that might be carried out.)

19.3.1 *Further Remarks about the 5-Point Configuration*

The above description outlines the principles for solving the 5-point invariance problem but does not show the conditions under which it is guaranteed to operate properly. Actually, these are straightforward to demonstrate. First, the cross ratio can be expressed in terms of the sines of the angles α_{13}, α_{24}, α_{12}, α_{34}. Next, these can be reexpressed in terms of areas of relevant triangles, using equations typified by the following formula to express area:

$$\Delta_{513} = \frac{1}{2} a_{51} a_{53} \sin \alpha_{13} \tag{19.19}$$

Finally, the area can be reexpressed in terms of the point coordinates in the following way:

$$\Delta_{513} = \frac{1}{2} \begin{vmatrix} p_{5x} & p_{1x} & p_{3x} \\ p_{5y} & p_{1y} & p_{3y} \\ p_{5z} & p_{1z} & p_{3z} \end{vmatrix} = \frac{1}{2} | \mathbf{p}_5 \, \mathbf{p}_1 \, \mathbf{p}_3 | \tag{19.20}$$

(*a*)

(*b*)

(*c*)

(*d*)

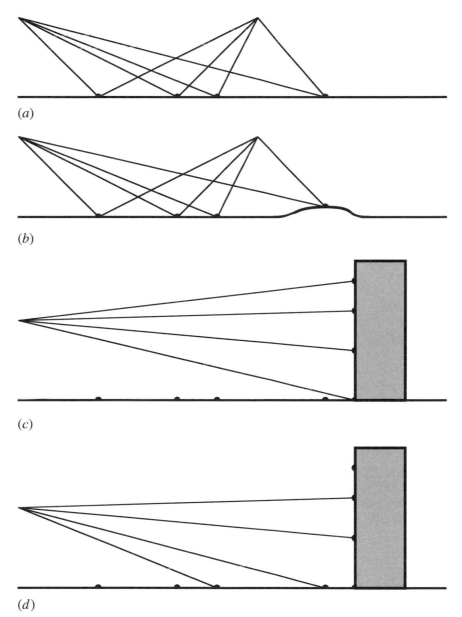

Figure 19.4 Use of cross ratio for egomotion guidance. (*a*) shows how the cross ratio for a set of four collinear points can be tracked to confirm that the points are collinear: this suggests that they lie on the ground plane. (*b*) shows a case where the cross ratio will not be constant. (*c*) shows a case where the cross ratio is constant, though the four points actually lie on a plane which is not the ground plane. (*d*) shows a case where all four points lie on planes, yet the cross ratio will not be constant.

Using this notation, a suitable final pair of cross ratio invariants for the configuration of five points may be written:

$$\mathscr{C}_{\mathrm{a}} = \frac{\Delta_{513}\Delta_{524}}{\Delta_{512}\Delta_{534}} \tag{19.21}$$

$$\mathscr{C}_{\mathrm{b}} = \frac{\Delta_{124}\Delta_{135}}{\Delta_{123}\Delta_{145}} \tag{19.22}$$

Although three more such equations may be written down, these will not be independent of the other two and will not carry any further useful information.

Note that a determinant will go to zero or infinity if the three points it relates to are collinear, corresponding to the situation when the area of the triangle is zero. When this happens any cross ratio containing this determinant will no longer be able to pass on any useful information. On the other hand, there is actually no further information to pass on, for this now constitutes a special case that is describable by a single cross ratio. We have reverted to the situation shown in Fig. 19.3*a*.

Finally, Fig. 19.3 misses one further interesting case: the situation of two points and two lines (Fig. 19.5). Constructing a line joining the two points and producing it until it meets the two lines, we then have four points on a single line. Thus, the configuration is characterized by a single cross ratio. Notice, too, that the two lines can be extended until they join, and further lines can be constructed from the join to meet the two points. This gives a pencil of lines characterized by a single cross ratio (Fig. 19.5*c*); the latter must have the same value as that computed for the four collinear points.

19.4 Invariants for Points on Conics

These discussions promote an understanding of how geometric invariants can be designed to cope with sets of points, lines, and planes in 3-D. Significantly more difficult is the case of curved lines and surfaces, though much headway has been made with regard to the understanding of conics and certain other surfaces (see Mundy and Zisserman, 1992a). It will not be possible to examine all such cases in depth here. However, it will be useful to consider conic sections, particularly ellipses, in more detail.

First, we consider Chasles' (1855) theorem, which dates from the nineteenth century. (The history of projective geometry is quite rich and was initially carried out totally independently of the requirements of machine vision.) Suppose we have four fixed coplanar points F_1, F_2, F_3, F_4 on a conic section

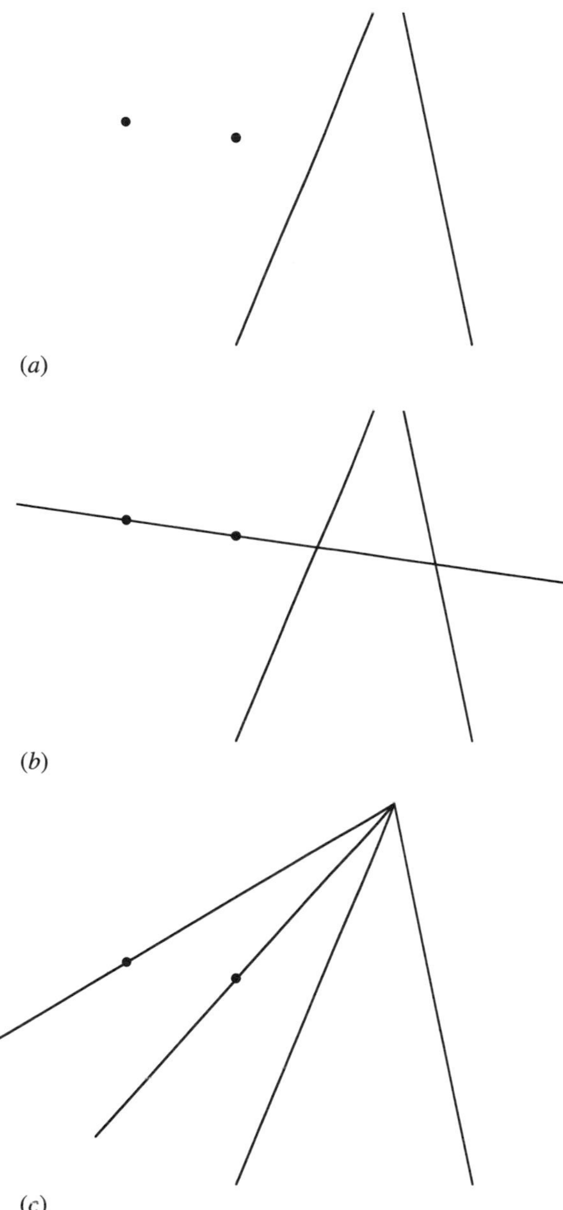

Figure 19.5 Cross ratio for two lines and two points. (*a*) Basic configuration. (*b*) How the line joining the two points introduces four collinear points to which a cross ratio may be applied. (*c*) How joining the two points to the junction of the two lines creates a pencil of four lines to which a cross ratio may be applied.

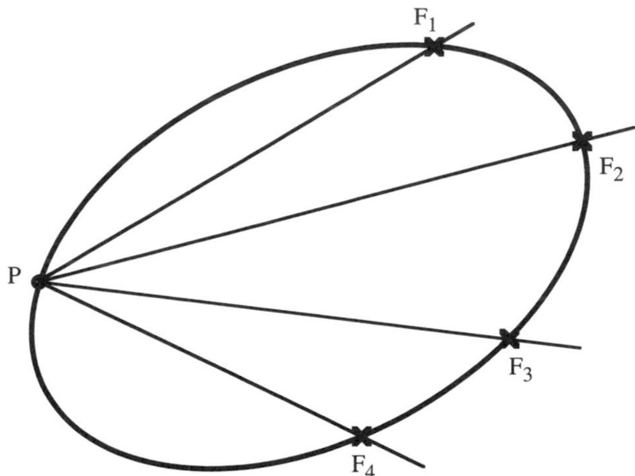

Figure 19.6 Definition of a conic using a cross ratio. Here P is constrained to move so that the cross ratio of the pencil from P to F_1, F_2, F_3, F_4 remains constant. By Chasles' theorem, P traces out a conic curve.

curve and one variable point P in the same plane (Fig. 19.6). Then the four lines joining P to the fixed points form a pencil whose cross ratio will in general vary with the position of P. Chasles' theorem states that if P now moves so as to keep the cross ratio constant, then P will trace out a conic section. This provides a means of checking whether a set of points lies on a planar curve such as an ellipse. Note the close analogy with the problem of ground plane detection already mentioned. Again the amount of computation could become excessive if there were a lot of noise or clutter in the image. When the image contains N boundary features that need to be checked out, the problem complexity is intrinsically $O(N^5)$, since there are $O(N^4)$ ways of selecting the first four points, and for each such selection, $N-4$ points must be examined to determine whether they lie on the same conic. However, choice of suitable heuristics would be expected to limit the computation. Note the problem of ensuring that the first four points are tested in the same order around the ellipse, which is liable to be tedious (1) for point features and (2) for disconnected boundary features.

Although Chasles' theorem gives an excellent opportunity to use invariants to locate conics in images, it is not at all discriminatory. The theorem applies to a general conic, hence, it does not immediately permit circles, ellipses, parabolas, or hyperbolas to be distinguished, a fact that would sometimes be a distinct disadvantage. This is an example of a more general problem in pattern recognition system design—of deciding exactly how and in what sequence one object should be differentiated from another. Space does not permit this point to be considered further here.

Finally, we state without proof that all conic section curves can be transformed under perspective projection to other types of conic section, and thus into ellipses. Subsequently, they can be transformed into circles. Thus, any conic section curve can be transformed projectively into a circle, while the inverse transformation can transform it back again (Mundy and Zisserman, 1992b). This means that simple properties of the circle can frequently be generalized to ellipses or other conic sections. In this context, we should remember that, after perspective projection, lines intersecting curves do so in the same number of points, and thus tangents transform into tangents, chords into chords, three-point contact (in the case of nonconic curves) remains three-point contact, and so on. Returning to Chasles' theorem, we see that a simple proof in the case of circles will automatically generalize to more complex conic section curves.

In response to this assertion, we can derive Chasles' theorem almost trivially for a circle. Appealing to Fig. 19.7, we see that the angles φ_1, φ_2, φ_3 are equal to the respective angles γ_1, γ_2, γ_3 (angles in the same segment of a circle). Thus, the pencils PF_1, PF_2, PF_3, PF_4 and QF_1, QF_2, QF_3, QF_4 have equal angles, their

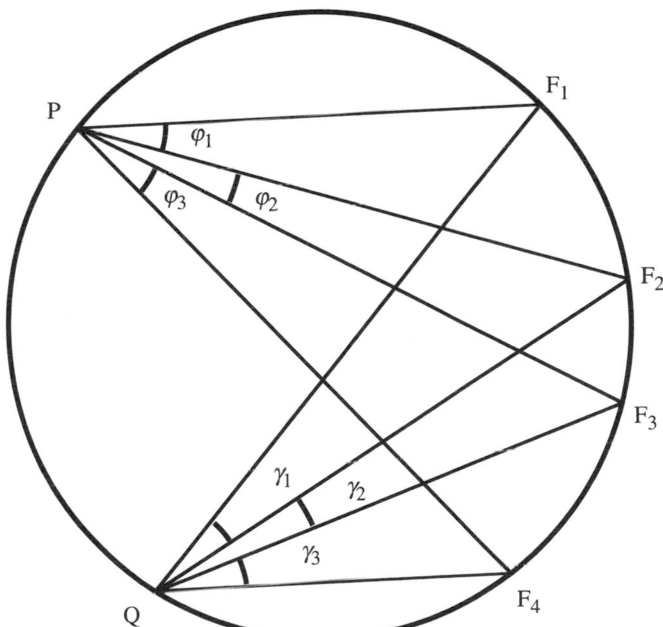

Figure 19.7 Proof of Chasles' theorem. This diagram shows that the four points F_1, F_2, F_3, F_4 subtend the same angles at P as they do at the fixed point Q. Thus, the cross ratio is the same for all points on the circle. This means that Chasles' theorem is valid for a circle.

relative directions being superposable. This means that they will have the same cross ratio, defined by equation (19.18). Hence, the cross ratio of the pencil will remain constant as P traces out the circle. As stated earlier, the property will automatically generalize to any other conic.

19.5 Differential and Semidifferential Invariants

Many attempts have been made to characterize continuous curves by invariants. The obvious way forward is to represent points on a curve in terms of local curve derivatives. If a sufficient number of these can be obtained, invariants can be formed and computed. However, the noise (including digitization noise) that always exists on curves limits the accuracy of higher derivatives, and as a result it is difficult to form useful invariants in this way. In general, the second derivative of the curve function is the highest that can normally be used, and this corresponds to curvature, which is only an invariant for Euclidean transformations (translation and rotation without change of scale).

As a result of this problem, semidifferential invariants are often used instead of differential invariants. They involve considering only a few "distinguished" points on curves and using these to generate invariants. The most common distinguished points to be used in this way are:

1. Points of inflection
2. Sharp corners on curves
3. Cusps on curves (corners where the bounding tangents are coincident)
4. Bitangent points (points of contact of a line that touches the curve twice)
5. Other points whose locations can be derived from existing distinguished points by geometric constructions (see Fig. 19.8).

Tangent points are unlikely to be suitable and so are not included in this list, as a smooth curve will have tangents along its entire length. This is because they are characterized merely by 2-point contact between a limiting chord and the curve. However, a point of inflection represents a 3-point contact, which means that it will be reasonably well localized, and its tangent will have a well-defined direction. On the other hand, bitangent points will be even more accurately represented, as the tangent direction will be accurately defined by two well-separated points on the curve (Fig. 19.8). Nevertheless, bitangent points will still incur some longitudinal error.

Bitangents can be of several sorts. In particular, they can contact the same shape on the same side; they can also cross the body and contact it on

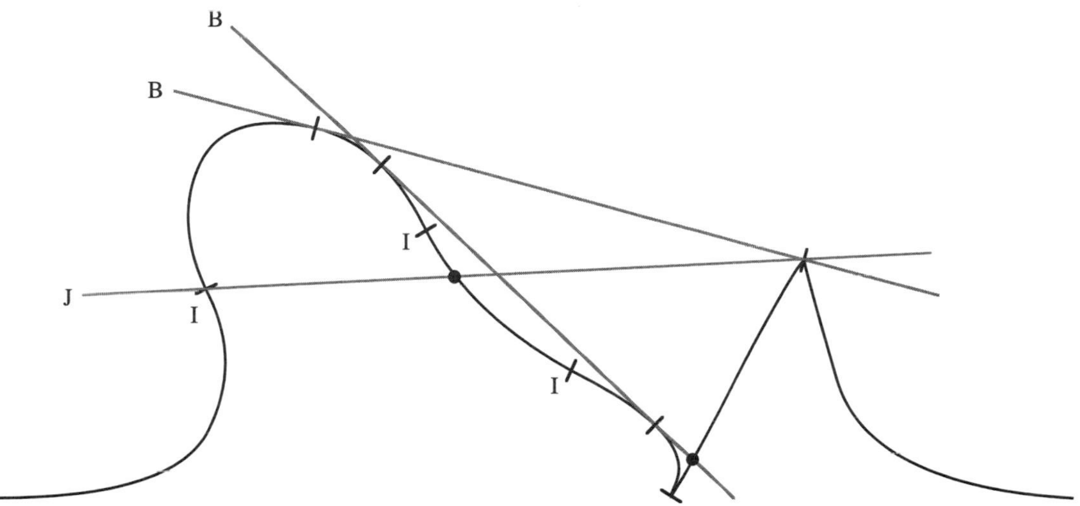

Figure 19.8 Means for finding distinguished points on a curve. The two bitangents contact the curve in a total of four bitangent points. Three points of inflection I provide another three distinguished points. A cusp and a corner provide a further two distinguished points (the latter also being a bitangent point). The line marked J contributes a further distinguished point on the curve, as does one of the bitangents: these are marked as large dots rather than as short lines.

both sides. This latter case is more complex and is therefore sometimes discounted in machine vision applications. Nevertheless, it provides a means of finding further invariant reference points on an object. Note that this happens directly in that the bitangent points are already distinguished points. It also happens indirectly, as the bitangent may cross other reference lines—thereby defining further distinguished points. Figure 19.9 shows several cases of direct and indirect distinguished points, the most accurate of which arise from bitangents, while slightly less accurate ones arise from points of inflection.

Once enough distinguished points and reference lines between them have been found, cross ratio invariants may be obtained (1) from the incidence of distinguished points lying along suitable reference lines and (2) from pencils of lines drawn from distinguished points to line crossings or to other distinguished points.

A remark is needed to confirm that points of inflection can act as suitable distinguished points which are invariant under perspective transformations. Starting from the premise that perspective transformations preserve straight lines and points arising from crossings between curves and lines, we note that a chord that crosses a curve three times will also cross it three times under perspective

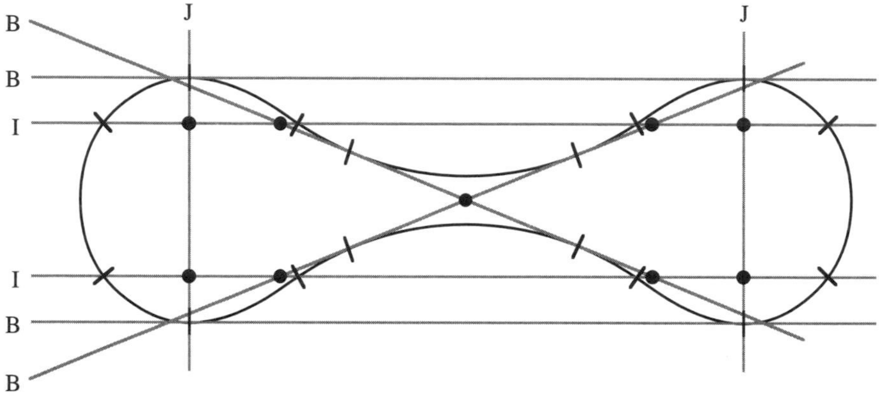

Figure 19.9 Means for finding direct and indirect distinguished points for an object. The four lines marked B are bitangents, which contribute six bitangent points. Two of the bitangents contact the object on opposite sides of its boundary. The two lines marked I arise from points of inflection. The two lines marked J are joins of bitangent points. The nine large dots are indirect distinguished points, which do not lie on the object boundary. Clearly, a good many more indirect distinguished points could be generated, though not all would have accurately defined locations.

projection. This will apply even when the three crossing points merge into 3-point contact.[3] Hence, points of inflection *are* suitable distinguished points and *are* perspective-invariant.

This treatment has only dealt with planar curves and has not covered spatial nonplanar curves. The latter is a significantly more difficult area, as concepts such as bitangents and points of inflection have to be assigned new meaning in this more general domain. It is a subject we cannot broach here.

19.6 Symmetrical Cross Ratio Functions

When applying a cross ratio to a set of points on a line, the order of the points on the line is frequently known. This will almost certainly be the case if feature detection of an image is carried out in a forward raster scan. Hence, the only confusion in the ordering will be the direction in which the line has been traversed. However, the cross ratio is independent of the end from which the line is scanned, since $\mathscr{C}(P_1, P_2, P_3, P_4) = \mathscr{C}(P_4, P_3, P_2, P_1)$. Nevertheless, in some

3 Three-point contact is distinguishable from 2-point contact in that the tangent crosses the curve at the point of contact.

situations the ordering of the cross ratio features will not be known with certainty. This may occur for the situations shown in Figs. 19.3 and 19.5, where the features themselves do not all lie on a single line; or where the features are angles; or where the points lie on a conic whose equation is as yet unknown. In such circumstances, it will be useful to have an invariant that takes in all possible orders of the features.

To derive such an invariant, note first that if there is a confusion in the ordering of the points such that the value could be either κ or $(1-\kappa)$, then we could apply the function $f(\kappa)=\kappa(1-\kappa)$, which has the property $f(\kappa)=f(1-\kappa)$, and this will solve the problem. Alternatively, if confusion exists between κ and $1/\kappa$, then we can apply the function $g(\kappa)=\kappa+1/\kappa$, which has the property $g(\kappa)=g(1/\kappa)$, and again this will solve the problem.

If there is potential confusion between the values κ, $(1-\kappa)$ and $1/\kappa$, however, the situation becomes more complicated. It is difficult to write down any obvious function that satisfies the double condition $h(\kappa)=h(1-\kappa)=h(1/\kappa)$, though we may have soundly based intuition that it will involve symmetrical functions such as $f(\kappa)$ and $g(\kappa)$. The simplest answer seems to be:

$$j(\kappa)=\frac{(1-\kappa+\kappa^2)^3}{\kappa^2(1-\kappa)^2} \tag{19.23}$$

which obeys the symmetry idea as it can be reexpressed in the two forms:

$$j(\kappa)=\frac{[1-\kappa(1-\kappa)]^3}{[\kappa(1-\kappa)]^2}=\frac{(\kappa+1/\kappa-1)^3}{(\kappa+1/\kappa-2)} \tag{19.24}$$

Fortunately, we need go no further in our quest to obey the six conditions required to recognize all six of the cross ratio values κ, $(1-\kappa)$, $1/\kappa$, $1/(1-\kappa)$, $(\kappa-1)/\kappa$, $\kappa/(\kappa-1)$. The reason is that they are all deducible from each other by further applications of the initial negation and inversion rules. (The ultimate reason is that the operations to transform the function from one to another of the six forms form a group of order six, which is generated from the negation and inversion transforms.)

Although this is a powerful result, it does not come without loss. The reason is that a sixfold ambiguity is now inherent in the solution, so that once we have shown that the set of points satisfies the symmetrical cross ratio function, we still have to make tests to determine which of the six possibilities is the correct one. This is reflected by the complexity of the j-function, which contains a sixth degree polynomial and for every value of j there are six possible values of κ (Fig. 19.10).

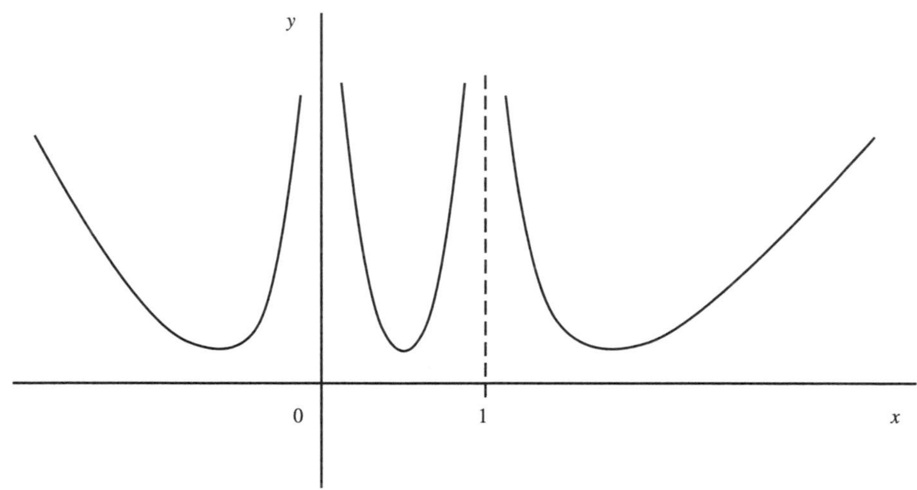

Figure 19.10 Symmetrical cross ratio function. This is the function defined by equation (19.23).

The situation can be described by stating that the function $j(\kappa)$ is not "complete," in the sense that this function alone is insufficient to recognize the set of features unambiguously. To underline this fact, observe that the original cross ratio *is* complete. Once the value of κ is known, we can uniquely determine the position of one of the points from the other three points. This is obvious from the graph of κ as a function of x (where $x = x_{34}$ gives the position of the fourth point[4]), which is a hyperbola:

$$\kappa = \frac{x_{31}x_{24}}{x_{21}x_{34}} = \frac{x_{31}(x_{23} + x)}{x_{21}x} = \frac{x_{31}x_{23}}{x_{21}}\left(\frac{1}{x} + \frac{1}{x_{23}}\right) \tag{19.25}$$

19.7 Concluding Remarks

This chapter seeks to provide some insight into the important subject of invariants and its application in image recognition. The subject begins with

4 In projective geometry it is well known that there are three degrees of freedom on a line: the positions of three points on a line are not predictable from other views of the three points, without further information on the viewpoint.

consideration of ratios of ratios of distances, an idea that leads in a natural way to the cross ratio invariant. While its immediate manifestation lies in its application to recognition of the spacings of points on a line, it generalizes immediately to angular spacings for pencils of lines, as well as angular separations of concurrent planes. A further extension of the idea is the development of invariants that can describe sets of noncollinear points. It turns out that just two cross ratios suffice to characterize a set of four noncollinear points. The cross ratio can also be applied to conics. Indeed, Chasles' theorem describes a conic as the locus of points that maintains a pencil of constant cross ratio with a given set of four points. However, this theorem does not permit one type of conic curve to be distinguished from another.

Many other theorems and types of invariant exist, but space prevents more than a mention of them here. As an extension to the line and conic examples given in this chapter, invariants have been produced which cover a conic and two coplanar nontangent lines, a conic and two coplanar points, and two coplanar conics. Of particular value is the group approach to the design of invariants (Mundy and Zisserman, 1992a). However, certain mathematically viable invariants, such as those that describe local shape parameters on curves, are too unstable for use in their full generality because of image noise. Nevertheless, semidifferential invariants have been shown (Section 19.5) to be capable of fulfilling essentially the same function.

Next, there is the warning of Åström (1995) that perspective transformations can produce such incredible changes in shape that a duck silhouette can be projected arbitrarily closely into something that looks like a rabbit or a circle, hence upsetting invariant-based recognition.[5] Although such reports seem absent from the previous literature, Åström's work indicates that care must be taken to regard recognition via invariants as hypothesis formation, which is capable of leading to false alarms.

Overall, the value of invariants lies in making computationally efficient checks of whether points or other features might form parts of specific objects. In addition, they make these checks without the necessity[6] for camera calibration or knowledge of the camera's viewpoint (though there is an implicit assumption that the camera is Euclidean). Although invariants have been known within the vision community for well over 30 years, only during the last 15 years have they

5 It could, of course, be argued that all recognition methods will be subject to the effects of perspective transformations. However, invariant-based recognition will not flinch from invoking highly extreme transformations that appear to grossly distort the objects in question, whereas more conventional methods are likely to be designed to cope with a reasonable range of expected shape distortions.

6 Here we assume that the aim is location of specific objects in the image. If the objects are then to be located in the world coordinates, camera calibration or some use of reference points will of course be needed. However, there are many applications, such as inspection, surveillance, and identification (e.g., of faces or signatures) where location of objects in the *image* can be entirely adequate.

been systematically developed and applied for machine vision. Such is their power that they will undoubtedly assume a stronger and more central role in the future. Indications of this role can be seen in Chapter 20 where they can help find vanishing points and in Chapter 24 where they are shown to be useful for the analysis of facial features.

Invariants are central to pattern recognition—reflecting within-class constancy vis-à-vis between-class variance. This chapter has shown that in 3-D vision, they are especially valuable in providing a hedge against the problems of perspective projection. Perhaps oddly, the cross ratio invariant always seems to emerge, in one guise or another, in 3-D vision.

19.8 Bibliographical and Historical Notes

The mathematical subject of invariance is very old (cf. the work of Chasles, 1855), but only recently has it been developed systematically for machine vision. Notable in this context is the work of Rothwell, Zisserman, and their co-workers, as reported by Forsyth et al. (1991), Mundy and Zisserman (1992a,b), Rothwell et al. (1992a,b), and Zisserman et al., (1990). In particular, the paper by Forsyth et al. (1991) both shows the range of available invariant techniques and discusses problems of stability that arise in certain cases. The appendix (Mundy and Zisserman, 1992b) on projective geometry for machine vision, which appears in Mundy and Zisserman (1992a), is especially valuable and provides the background needed for understanding the other papers in the volume. This volume has a theoretical flavor that demonstrates what ought to be possible using invariants, though comparisons between invariants and other approaches to recognition are perhaps lacking. Thus, only by examining whether workers choose to use invariants in real applications will the full story emerge. In this respect, the paper by Kamel et al. (1994) on face recognition is of great interest, for it shows how invariants helped to achieve more than had been achieved earlier after many attempts using other approaches—specifically in correcting for perspective distortions during face recognition.

Other more recent work appears in a Special Issue of *Image and Vision Computing* (Mohr and Wu, 1998). In particular, the paper by Van Gool et al.

(1998) shows how shadows can be allowed for in aerial images, and the paper by Boufama et al. (1998) shows how invariants can help with object positioning. Startchik et al. (1998) provides a useful demonstration of the semidifferential invariant methods covered in Section 19.5. Maybank (1996) deals with the problem of accuracy with invariants, making the point that this can be serious even for cross ratios (which contain only four parameters). Another early work, by a totally different set of workers, is Barrett et al. (1991) and contains a number of useful derivations, together with a practical example of aircraft recognition, complete with accuracy assessments.

Rothwell's (1995) book covers the early work on invariance in a thoughtful manner, and the later 3-D books by Hartley and Zisserman (2000) and Faugeras and Luong (2001) integrate the ideas into their structure but are not always easy to understand by students starting out in the subject. Semple and Kneebone (1952) is a standard work on projective geometry, which is still widely used in one or another reprint.

19.9 Problems

1. Find a symmetrical cross ratio function $p(\kappa)$ that has the same value for all cyclic permutations of the four points A, B, C, D on a straight line. Show also that $p(\kappa)$ is incomplete but that it has just two values for each position of point D on the line.

2. Show that the six operations required to transform the cross ratio κ into the six different values for four points on a line form a group G of order six (see Sections 19.2 and 19.6). Show that G is a noncyclic group and has two subgroups of order 2 and 3, respectively. *Hint:* Show that all possible combined operations fall within the same set of six and that this set contains the identity operation and the inverses of all the elements of the set.

3. Show that a conic and two points can be used to define an invariant cross ratio.

4. Show that two conics can be used to define an invariant cross ratio (a) if they intersect in four points, (b) if they intersect in two points, (c) if they do not intersect at all, as long as they have common tangents.[7]

7 The case of nonintersecting conics with no common tangents requires complex algebra: see, for example, Rothwell (1995). The possibility of ambiguity and incompleteness in cases (a)–(c) is also discussed in Rothwell (1995).

5. (a) Perform a geometric calculation based on the sine rule which shows that the angles α, β, γ are related to the distances a, b, c in Fig. 19.P1 by the equation:

$$\frac{a}{\sin\,\alpha} \times \frac{c}{\sin\,\gamma} = \frac{a+b}{\sin\,(\alpha+\beta)} \times \frac{b+c}{\sin\,(\beta+\gamma)}$$

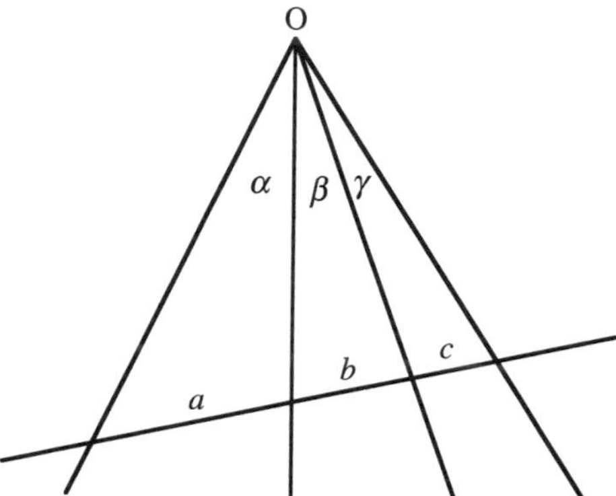

Figure 19.P1 Geometry for cross ratio calculation.

(b) Show that this equation leads to a relation between the cross ratios for various distances on the line and for the sines of various angles. Hence, show that this also leads to the constancy of the cross ratios on any two lines crossing the pencil of four lines passing though O.

6. (a) Explain the value of using *invariants* in relation to pattern recognition systems. Illustrate your answer by considering the value of thinning algorithms in optical character recognition.

(b) The *cross ratio* of four points (P_1, P_2, P_3, P_4) on a line is defined as the ratio:

$$C(P_1, P_2, P_3, P_4) = \frac{(x_3 - x_1)(x_2 - x_4)}{(x_2 - x_1)(x_3 - x_4)}$$

Explain why this is a useful type of invariant for objects viewed under full perspective projection. Show that labeling the points in reverse order will not change the value of the cross ratio.

(c) Give arguments why the cross ratio concept should also be valid for weak perspective projection. Work out a simpler invariant that is valid for straight lines viewed under weak perspective projection.

(d) A flat lino-cutter blade has two parallel sides of different lengths: it is viewed under weak perspective projection. Discuss whether it can be identified from any orientation in 3-D by measuring the lengths of its sides.

7. (a) Flagstones are viewed on a pavement, providing a large number of coplanar feature points. Show that a correspondence can be made between five coplanar feature points in two images—however the camera has been moved between the shots—by checking the values of two cross ratios.

(b) It is required that a panorama of a scene be composed by taking a number of photographs and "stitching" them together after making appropriate image transformations. For this purpose, it is necessary to make correspondences between the images. Show that the two cross ratio types of planar invariants can be used for this purpose, even if the chosen scene features do not lie on a common plane. Determine under what conditions this is possible.

Egomotion and Related Tasks

Humans employ 3-D vision to find their way around in the process called egomotion. The necessary visual processes are carried out with ease and very much in real time. What shortcuts are being taken, and what cues does the brain use to help make the necessary rapid analysis? This chapter considers these issues and finds some interesting methodologies.

Look out for:

- how vanishing points can be found and used to help provide a basic understanding of the scene.
- how a plan view of the ground plane can be obtained and used to help with robot navigation.
- how known measurements of flagstones or other objects can be used to help identify the ground plane and locate the vanishing points.
- how perspective distortions of object center locations may be eliminated.
- how vehicles can be guided using vision to compensate for roll, pitch, and yaw.

This chapter ignores camera calibration and the ultimate need to cope with general 3-D transformations using homogeneous coordinates, both of which are studied in detail in Chapter 21. They have been left until last in order not to demotivate readers who are more orientated toward applications.

CHAPTER 20

Egomotion and Related Tasks

20.1 Introduction

Mechatronics is the study of mechanisms and the electronic systems that are needed to provide them with intelligent control. It has applications in a wide variety of areas, including industry, commerce, transport, space, and even the health service. For instance, robots can be used in principle—and often in practice—for mowing lawns, clearing rubbish, driving vehicles, effecting bomb disposal, carrying out reconnaissance, performing delicate operations, and executing automatic assembly tasks. A related task is that of automated inspection for quality control, though in that case the robots take the form of trapdoors or air jets that divert defective items into reject bins (Chapter 22). There are several categories of mechanisms and robots: mobile robots, which include robot vacuum cleaners and other autonomous vehicles; stationary robots with arms that can spray cars on production lines or assemble television sets; and robots that do not have discernible arms but that insert components into printed circuit boards or inspect food products or brake hubs and reject faulty items.

By definition, all mechatronic systems control the processes they are part of, and to achieve this control they need sensors of various sorts. Sensors can be tactile, ultrasonic, or visual or can involve one of a number of other modalities, some of which provide images that can be handled in the same way as normal visual images. Instances include ultrasonic, infrared, thermographic, X-ray, and ultraviolet images. In this chapter we are concerned primarily with these sorts of visual image, though some mechatronic applications do not involve this general class of input. However, far too many mechatronic applications use visual images for control than we can deal with here. Thus, we shall concentrate on autonomous mobile robots and related egomotion[1] problems.

20.2 Autonomous Mobile Robots

Mobile robots are poised to invade the human living space at an ever increasing rate. The mechanical technology is already in place, and the sensing systems required to guide them have been evolving for many years and are sufficiently mature for serious, though perhaps not ambitious, application. For example, robot lawnmowers, robot vacuum cleaners, robot window cleaners, robot rubbish gatherers, and robot weed-sprayers have been produced, but more ambitious, autonomous vehicles are held back largely because of the risks involved when humans inhabit the same areas—and in case of accidents, litigation could pose serious problems. Indeed, attempting to advance too quickly could set the industry back.

In addition to safety considerations, there are other practical issues such as speed of operation and implementation cost. The problems of applying mobile robots have not all been solved, especially in light of these practical issues. Ignoring the end use of the robot, one of the main things still to be engineered is how the robot is to guide itself safely in territory that will be populated by humans and by other robots. Autonomous guidance involves sensing, scene interpretation, and navigation. On a limited scale, a floor polishing robot could proceed until it bumps into an object and then take evading action. In this case tactile sensing is required. However, ultrasonic devices permit proximity sensing and appreciation that other objects that have to be avoided are nearby. Here the need for planning is beginning to be felt (Kanesalingam et al., 1998). However, serious planning and navigation are probably possible only when vision sensing is used, as vision sensing is remotely instituted and is rich in information content and flow rate.

Although vision sensing provides an extremely rich flow of information that potentially is exceptionally valuable for autonomous control and path planning, the interpretation task is considerable. In addition, it has to be remembered that mobile robots will be moving at walking speed indoors, and at much greater speeds outdoors, so a significant real-time processing problem arises. However, it seems that stereo vision offers limited advantage in that (1) the additional information requires additional processing and (2) it does not offer substantial advantages in 3-D scene interpretation at distances of more than a few meters. On the other hand, stereo can help the vision system to cue into the

1 Although some may consider egomotion to be a synonym for navigation, we interpret the term more widely to include both awareness plus control of motion and navigation per se. Indeed, the need to be sure of remaining upright must be part of the task of egomotion; it hardly falls into the category of navigation.

Figure 20.1 Vanishing point formation. In this case, parallel lines in space appear to converge to a vanishing point to the right of the image. Although the vanishing point V is only marginally outside the image, in many other cases it lies a considerable distance away.

complex information present in the incoming images. Thus, if stereo is not to be used, reliable cues have to be found in monocular images. A widely used means by which such cues can be obtained is by vanishing point (VP) detection. VPs are all too evident to humans in the sort of environment in which they live and work. In particular, factories, office-blocks, hospitals, universities, and outdoor city scenes all tend to contain many structures that are composed not only of straight lines and planes but also of rectangular blocks and towers (Fig. 20.1).

Vanishing points are useful in several ways. They confirm that various lines in the scene arise from parallel lines in the environment; they help to identify where the ground plane is; they permit local scale to be deduced (e.g., objects on the ground plane have width that is referable to, and a known fraction of, the local width of the ground plane); and they permit an estimate to be made of distance along the ground plane, by measuring the distance from the relevant image point to the VP. Thus, they are useful for initiating the process of recognizing and measuring objects, determining their positions and orientations, and helping with the task of navigation. Section 20.4 covers location of VPs, and Section 20.6 examines the process of navigation and path planning in greater detail.

20.3 Active Vision

The term *active vision* has several meanings, and it is important to differentiate carefully between them. First, there is a connotation that comes from radar, active radar being different from passive radar in that the aircraft being sensed deliberately returns a special signal that is able to identify it. This is not a normal connotation in vision. Instead, active vision is taken to mean one of two things: (1) moving the camera around to focus on and attend to a particular item of current interest in the environment; or (2) attending to a particular item or region in the image and interacting dynamically with it until an acceptable interpretation is arrived at. In either case, there is a focus of attention, and the vision system spends its time actively pursuing and analyzing what is present in the scene, rather than blandly interpreting the scene as a whole and attempting to recognize everything. In active vision, it is up to a higher level processor to identify items of interest and to ask, and answer, strategic questions about them.

Although these two modes of active vision may seem very similar, they have evolved to be quite distinct. In the first mode, the activity takes place largely in terms of what the camera may be made to look at. In the second, the activity takes place at the processing level, and active agents such as snakes are used to analyze different parts of the received image. Thus, a snake may initially be looped loosely around the whole region of an object and will then gradually move inward until it has sensed exactly where the object is. This is achieved by interacting intelligently with the object region and interrogating it until a viable solution is reached. In one sense, both schemes are carrying out interrogation, and in that sense active vision is being undertaken in each case. However, each scheme carries a different emphasis. What is important in the case of the moving camera is that recording and memorizing the whole of a scene in all directions around a robot station is necessarily a costly process, and one for which a complete interpretation is unlikely to be fully utilized. It is wasteful to gather information that will not be used later. It is better to concentrate on finding the part of the scene that is currently most useful and to narrow its scope and enhance its quality so that really useful answers can be found (see, for example, Yazdi and King, 1998). This approach mimics the human interrogation of scenes, which involves making "saccades" from one focus of attention to another, thus obtaining a progressive understanding of what is happening in the environment.

The active vision approach is useful for practical control of real robot limbs, but it is even more vital for mobile robots that have to guide themselves around the factory, hospital, office-block, or (in the case of robot vehicles) along the highway. In such cases it is relevant to update only those parts of the scene that give the information needed for the main task—that of safe guidance and navigation.

20.4 Vanishing Point Detection

We next consider how VPs can be detected. Generally, the process is undertaken in two stages: (1) locate all the straight lines in the image; (2) find those lines that pass through common points and deduce which must be VPs. Stage 1 is in principle a straightforward application of the Hough transform, though how straightforward depends on whether some of the lines are texture edges, which are difficult to locate accurately and consistently. Stage 2 is, in effect, another Hough transform in which peaks are located in a parameter space congruent to the image space, in which all points on lines are accumulated. Points on extended lines as well as those on the lines themselves have to be accumulated. Unfortunately, complications arise as VPs may be outside the confines of the initial image space (Fig. 20.1). Extending the image space can help, but there is no guarantee that this will work. In addition, accuracy of location may be poor when the lines leading to a VP converge at a small angle. Finally, the density of votes accumulated in the VP parameter space may (even with the application of suitable point spread functions) be so low that sensitivity of detection will be negligible. Thus, special techniques are necessary to guarantee success.

Perhaps the most important technique for detecting VPs is to use the surface of a Gaussian sphere as the parameter space (Magee and Aggarwal, 1984). Here, a unit sphere S is constructed around the center of projection of the camera, and S is used as a parameter space, instead of an additional image plane. Thus, VPs appear at convenient positions at finite distances, and infinite distances are ruled out (Fig. 20.2). Notice that there is a one-to-one correspondence

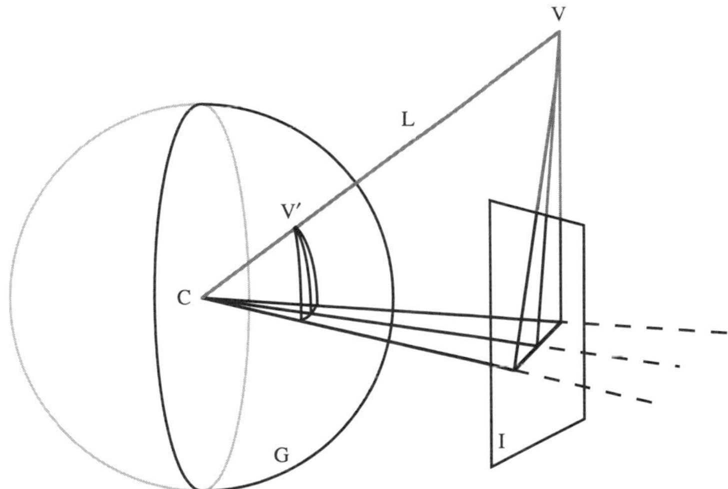

Figure 20.2 Use of Gaussian sphere for vanishing point detection. In this case, the vanishing point V is shown well above the input image I. To be sure of detecting such points, the surface of the Gaussian sphere G is used as the parameter space for accumulating vanishing point votes. Also marked in the figure are the projection line L from the center of projection C to V, and V′, which is the position of the vanishing point on G.

between each point on the extended image plane and each point on the front half of the Gaussian sphere (the back half is not used). Nevertheless, using the Gaussian sphere gives rise to problems when analyzing the vote patterns as these are not deposited uniformly over the sphere. In addition, large numbers of irrelevant votes are cast when lines do not pass through actual VPs as they do not originate from lines that are parallel to other lines in the scene. It would be better if all the votes that were cast at isolated peaks corresponded to likely VPs. To help ensure that only useful votes are cast, *pairs* of lines are taken, and the VPs that would arise if each pair of lines were actually parallel lines in the scene are recorded as votes in the parameter space. If this procedure is implemented for all pairs of lines (or for all pairs of edge segments), all necessary peaks will be recorded in the parameter space, and the amount of unnecessary clutter will be drastically reduced. However, this method for casting votes is not achieved without cost, as the computational load will be proportional to the number of pairs of edge segments, or, alternatively, the number of pairs of lines. If there are N lines, the number of pairs is $\binom{N}{2} = 1/2\,N(N-1)$, which is proportional to N^2 if N is large.

After the Gaussian sphere has been used to locate the peaks corresponding to the VPs, the VPs can, where possible, be recorded in the image space. However, the important step is the initial search to determine which of the initial lines or line segments correspond to the VPs and which correspond to isolated lines or mere clutter. Once the right line assignments have been made, there is considerably greater certainty in the interpretation of the images. By way of confirmation, for a moving robot there should be some correspondence between the VPs seen in successive images.

20.5 Navigation for Autonomous Mobile Robots

When a mobile robot is proceeding down a corridor, the field of view will usually contain at least four lines that converge to a VP situated in the middle of the corridor (Fig. 20.3*a*). In many cases this VP can be used for navigation purposes. For example, the robot can steer so that the VP is in the very center of the field of view, thus giving maximum clearance with the corridor walls. It can also check that other objects do not come between it and the VP. To simplify the latter process, the floor region in front of the robot should be checked for continuity. Such actions can be taken by examining the image space itself. Indeed, both navigation and obstacle detection can be carried out by scrutinizing the image space, without appeal to special navigation models that map out the entire working area.

(*a*)

(*b*)

Figure 20.3 Navigational problems posed by obstacles. (*a*) View of a corridor with four lines converging to a vanishing point, and a narrowed entrance to be negotiated. (*b*) View of a path with various pillars and bollards to be negotiated. (*c*) Plan view of (*b*) showing what can be deduced by examination of the ground plane in (*b*): only walls W that can be seen from the viewpoint (marked △) are drawn; however, the pillars P, bollards B, and litter-bin L (which are easily recognizable) are shown in full.

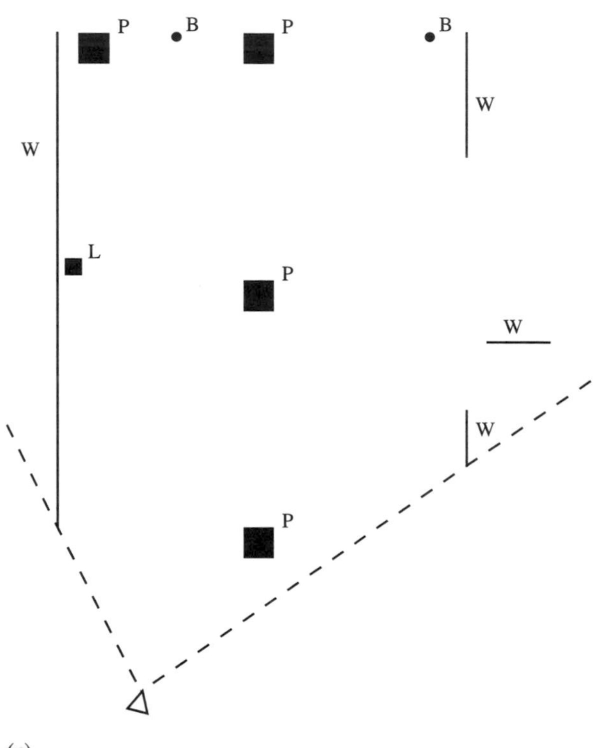

(*c*)

Figure 20.3 Continued.

Although this procedure may be sufficient for guiding a vacuum cleaner, more advanced mobile robots will have to make mapping, path planning, and navigational modeling part of a detailed high-level analysis of the situation (Kortenkamp et al., 1998). This will apply if the corridor has several low obstacles such as chairs, wastepaper bins, and display tables, or if a path has obstacles such as bollards or pillars (Fig. 20.3*b*, *c*); but will *a fortiori* be necessary if the robot is running a maze (as in the "Robot Micromouse" contests). In such cases, the robot will have only limited knowledge and visibility of the working area, and the natural representation for thinking about the situation and path planning is a plan view model of the working area. This plan will gradually be augmented as more and more data come to light. The following section indicates how a (partial) plan view can be calculated for each individual frame obtained by the camera. The overall plan construction algorithm is given in Table 20.1.

Table 20.1 Overall algorithm for constructing plan view of ground plane. This table gives the basic steps for constructing a plan view of the ground plane. For simplicity each input image is treated separately until the last few stages, though more sophisticated implementations would adopt a more holistic strategy.

Steps in Algorithm
• Perform edge detection.
• Locate straight lines using the Hough transform.
• Find VPs using a further Hough transform (see Section 20.4).
• Eliminate all VPs except the primary VP, which lies in the general direction of motion.
• Identify which lines passing through the primary VP lie within the ground plane.[a]
• Mark other lines as irrelevant.
• Find objects that are not related to lines passing through the primary VP.
• Segment these objects and delimit their boundaries: mark them as not being relevant.
• Find shadows and delimit their boundaries: mark them as possibly being on the ground plane.[b]
• Check object and shadow interpretations for consistency between successive images.[c]
• Annotate all feature points in the ground plane with their deduced real-world X and Z coordinates.[d]
• Check coordinates for consistency between successive images.
• Mark significant inconsistencies as due to objects and artifacts such as shadows.
• Mark insignificant inconsistencies as irrelevant noise (to be ignored).[e]
• Produce final map of (X, Z) coordinates.

[a]An initial line of attack is to take the ground plane as having a fair degree of homogeneity, in which case it will have the same general characteristics as the ground immediately in front of the robot. This should lead to identification of a number of ground plane lines. Note also that in a corridor any lines lying above the primary VP cannot be within the ground plane.

[b]Identifying shadows unambiguously is difficult without additional information about the scene illumination. However, dark areas within the ground plane can tentatively be placed in this category. In any case, the purpose of this step is to move toward an understanding of the scene. This means that the later steps in which consistency between successive images is considered are of greater importance.

[c]One example is that of four points lying close together on a flat surface such as the ground plane. These will maintain their spatial relationship in the image during camera motion, the relative positions changing only slowly from frame to frame. However, if one of the points is actually on a separate surface, such as a pillar, it will move in a totally different way, exhibiting a marked stereo motion effect. This inconsistency will allow the disparate nature of the points to be discerned. In this context, it is important to realize that scene interpretation is apparently easy for a human but has to be undertaken by a robot using "hard" and quite extensive deductive processes.

[d]See equations (20.2) and (20.3).

[e]It will often be sufficient to define inconsistent objects as "noise" if they occupy fewer than 3 to 5 pixels. They can often be identified and eliminated by standard median filtering operations.

20.6 Constructing the Plan View of the Ground Plane

This section explains a simple strategy for constructing the plan view of the ground plane, starting with a single view of a scene in which (1) the vanishing point V has been determined, and (2) significant feature points on the boundary of the ground plane have been identified. There is only space here to show how distance can be

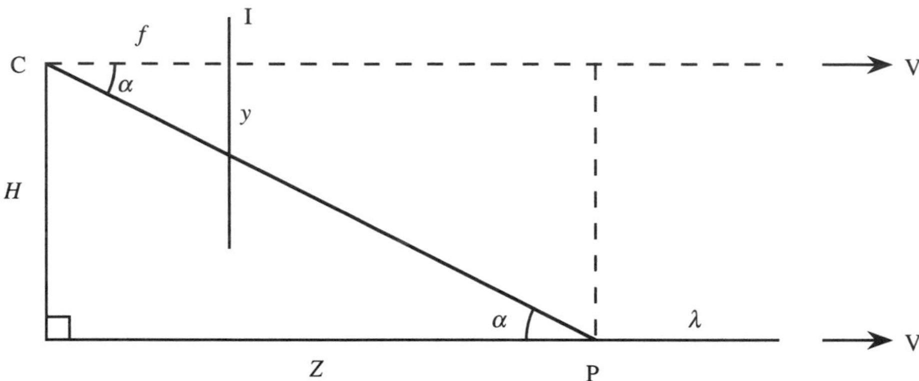

Figure 20.4 Geometry for constructing plan view of ground plane. This diagram is a vertical cross section containing the center of projection C of the camera and the vanishing point V, though vanishing point V can only be indicated by arrows. The cross section also includes a line λ within the ground plane. The projection line from a general point P on λ to the image plane I has a declination angle α, f is the focal length of the camera lens, and H is the height of the optical axis of the camera above the ground plane. Note that the optical axis of the camera is assumed to be horizontal and parallel to λ.

deduced along the ground plane in a line λ defined by the ground plane and a vertical plane containing the center of projection C of the camera and the vanishing point V. In addition, a simplifying assumption is made that the camera has been aligned with its optical axis parallel to the ground plane, and calibrated so that its focal length f and the height H of its axis above the ground plane are both known (see Chapter 21). In that case, Fig. 20.4 shows that the angle of declination α of a general feature point P on λ is given by:

$$\tan\alpha = H/Z = y/f \tag{20.1}$$

where Z is the distance of P along λ (Fig. 20.4), and y is the distance of the projection of P below the vanishing line in the image plane. This leads to the following value for Z:

$$Z = Hf/y \tag{20.2}$$

Similar considerations lead to the lateral distance from λ being given by:

$$X = Hx/y \tag{20.3}$$

where x is the lateral distance in the image plane.

Overall, we now have related image coordinates (x, y) and world (plan view) coordinates (X, Z). Both X and Z vary rapidly with y when y is small because of the distorting (foreshortening) effects of perspective geometry. Thus, as might be expected, X and Z are more difficult to measure accurately in the distance.

More general cases are best dealt with using homogeneous coordinates that lead to linear transformations using 4×4 matrices (Chapter 21).

20.7 Further Factors Involved in Mobile Robot Navigation

We have moved from a simple image view that can successfully indicate a correct way forward to one in which experience of the situation is gradually built up. The first (image view) representation is restricted to providing ad hoc help, whereas the second (plan view) representation is the natural one for storing knowledge and arriving at globally optimal solutions. It is difficult to be sure which representation humans rely on. While it is apparently the first, the second is definitely invoked during conscious, deductive thought processes whenever detailed analysis of what is happening (e.g., in a maze) is undertaken. However, introspection is not a good means of finding out how humans actually carry out the other (subconscious) task. In any case, in a machine vision context, we merely need to see how to set about programming a software system to emulate the human.

It is profitable for us to explore how the navigation task can be undertaken, in the abstract case typified by a maze-running robot. Suppose first that the robot moves over the maze in a search mode, exploring the environment and mapping the allowed and forbidden areas. When a complete map has been built up, some purposive activity can be imposed. How should the robot go from its current position C to a goal position G in a minimum time? We can run the maze abstractly by applying an algorithm to form a distance function (Section 6.7), which starts at G and propagates around the maze, constrained only by the positions of the walls (Fig. 20.5). When the distance function arrives at C, the distance value is noted and the propagation algorithm is terminated. To find the optimum path, it is now only necessary to proceed downhill from C in the direction of the locally greatest gradient, until arriving at G (Kanesalingam et al., 1998). A distance function is used to ensure that the map contains information on which of several alternative routes is the shortest: if it were known that only one route would be possible, simple propagation operations of the type used in connected components analysis would be adequate for locating it.

Thus far we have only considered mobile robots capable of rolling or walking down corridors or paths. However, much of what has been talked about also applies to vehicles driving down city roads because perspective lines are quite

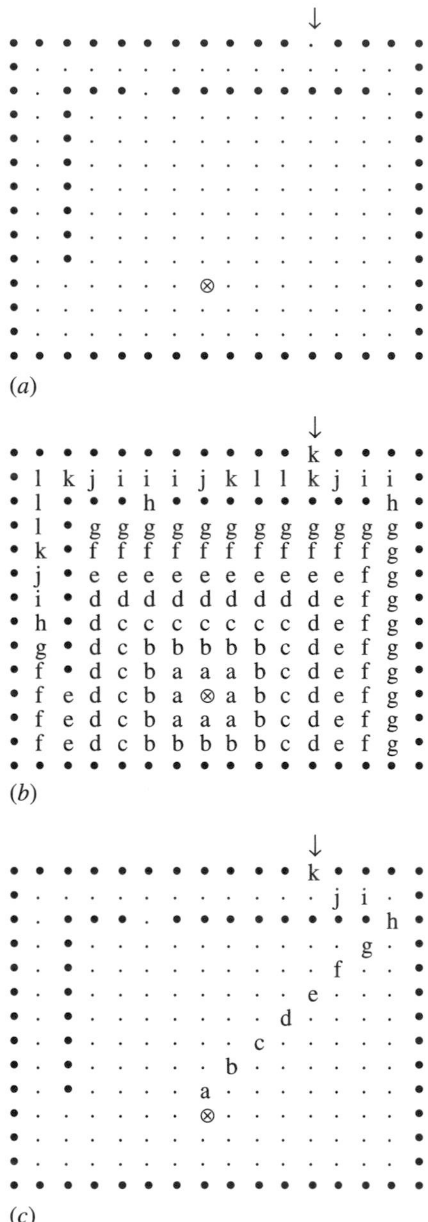

Figure 20.5 Use of distance function for path planning. (*a*) A rudimentary maze. (*b*) Distance function of the maze, with origin at the goal (marked ⊗). (*c*) Optimum path deduced by tracking from the entrance to the maze (marked with an arrow) along directions of maximum gradient. For convenience of display, the distance function is coded in letters, starting with a = 1. Improved forms of distance function would lead to more accurate path optimization.

common in such situations. However, it will be difficult to apply these ideas to country roads which meander along largely unpredictable paths. In such cases, the concept of a VP has to be replaced by a sequence of VPs that move along a vanishing line—typically, the horizon line (Billingsley and Schoenfisch, 1995). Although this complication does not invalidate the general technique, the straight-line detection approach becomes inappropriate, and more suitable methods must be devised. The image-based model can probably still be used for instantaneous steering—especially if a Kalman filter (Section 18.13) is used to maintain continuity. However, practical accident-free country driving will require considerably more realistic scene interpretation than analysis based simply on where the current VP might be. After all, images are rich in cues which the human is able to discern and use—by virtue of the huge database of relevant information stored in the human mind. Robot vision capabilities do not yet match up to this task with the required reliability and speed—though excellent efforts are being made. The next decade will undoubtedly see a revolution in what can be achieved.

20.8 More on Vanishing Points

The cross ratio can turn up in many situations and on each occasion provide yet another neat result. An example is when a road or pavement has flagstones whose boundaries are well demarcated and easily measurable. They can then be used to estimate the position of the vanishing point on the ground plane. Imagine viewing the flagstones obliquely from above, with the camera or the eyes aligned horizontally (Fig. 20.3*b*). Then we have the geometry of Fig. 20.6 where the points O, H_1, H_2 lie on the ground plane, while O, V_1, V_2, V_3 are in the image plane.[2]

If we regard C as a center of projection, the cross ratio formed from the points O, V_1, V_2, V_3 must have the same value as that formed from the points O, H_1, H_2, and infinity in the horizontal direction. Supposing that OH_1 and H_1H_2 have known lengths a and b, equating the cross ratio values gives:

$$\frac{y_1(y_3 - y_2)}{y_2(y_3 - y_1)} = \frac{x_1}{x_2} = \frac{a}{a + b} \tag{20.4}$$

2 Notice that slightly oblique measurement of the flagstone, along a line that is not parallel to the sides of the flagstones, still permits the same cross ratio value to be obtained, as the same angular factor applies to all distances along the line.

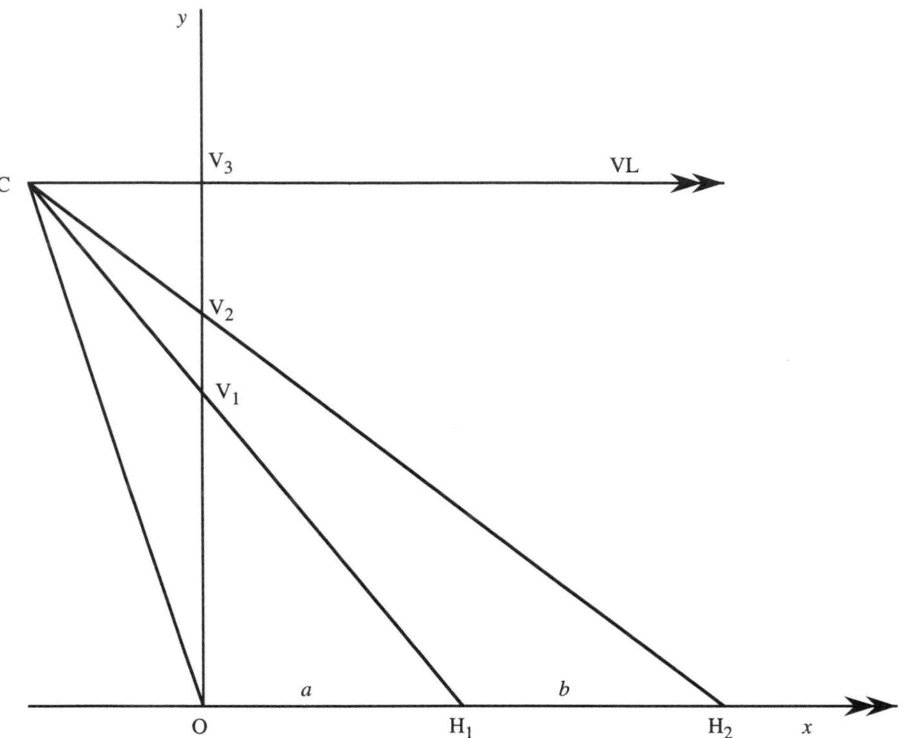

Figure 20.6 Geometry for finding the vanishing line from a known pair of spacings. C is the center of projection. VL is the vanishing line direction, which is parallel to the ground plane OH_1H_2. Although the camera plane $OV_1V_2V_2$ is drawn perpendicular to the ground plane, this is not necessary for successful operation of the algorithm (see text).

This allows us to estimate y_3:

$$(a+b)(y_1y_3 - y_1y_2) = ay_2y_3 - ay_2y_1 \tag{20.5}$$

$$\therefore \quad y_3(ay_1 + by_2 - ay_2) = ay_1y_2 + by_1y_2 - ay_1y_2 \tag{20.6}$$

$$\therefore \quad y_3 = by_1y_2/(ay_1 + by_1 - ay_2) \tag{20.7}$$

If $a = b$ (as is likely to be the case for flagstones):

$$y_3 = \frac{y_1y_2}{2y_1 - y_2} \tag{20.8}$$

Notice that this proof does not actually assume that points V_1, V_2, V_3 are vertically above the origin, or that line OH_1H_2 is horizontal. Rather, these points are seen to lie along two coplanar straight lines, with C in the same plane. Also, note that only the ratio of a to b, not their absolute values, is relevant in this calculation.

Having found y_3, we have calculated the direction of the vanishing point, whether or not the ground plane on which it lies is actually horizontal, and whether or not the camera axis is horizontal.

Finally, notice that equation (20.8) can be rewritten in the simpler form:

$$\frac{1}{y_3} = \frac{2}{y_2} - \frac{1}{y_1} \tag{20.9}$$

The inverse factors give some intuition into the processes involved—not the least of which is considering the inverse relation between distance along the ground plane and image distance from the vanishing line in equation (20.2)—and similarly, the inverse relation between depth and disparity in equation (16.5).

20.9 Centers of Circles and Ellipses

It is well known that circles and ellipses project into ellipses (or occasionally into circles). This statement is widely applicable and is valid for orthographic projection, scaled orthographic projection, weak perspective projection, and full projective projection.

Another factor that is easy to overlook is what happens to the center of the circle or ellipse under these transformations. It turns out that the ellipse (or circle) center does not project into the ellipse (or circle) center under full perspective projection; there will in general be a small offset (Fig. 20.7).[3]

If the position of the vanishing line of the plane can be identified in the image, the calculation of the offset for a circle is quite simple using the theory of Section 20.8, which applies as the center of a circle bisects its diameter (Fig. 20.8). First, let ε be the shift in the center, let d be the distance of the center of the ellipse from the vanishing line, and b be the length of the semiminor axis. Next, identify $b+\varepsilon$ with y_1, $2b$ with y_2, and $b+d$ with y_3. Finally, substitute for y_1, y_2, and y_3 in equation (20.8). We then obtain the result:

$$\varepsilon = b^2/d \tag{20.10}$$

3 The fact that this happens may perhaps suggest that ellipses will be slightly distorted under projection. There is definitely no such distortion, and the source of the shift in the center is merely that full perspective projection does not preserve length ratios—and in particular midpoints do not remain midpoints.

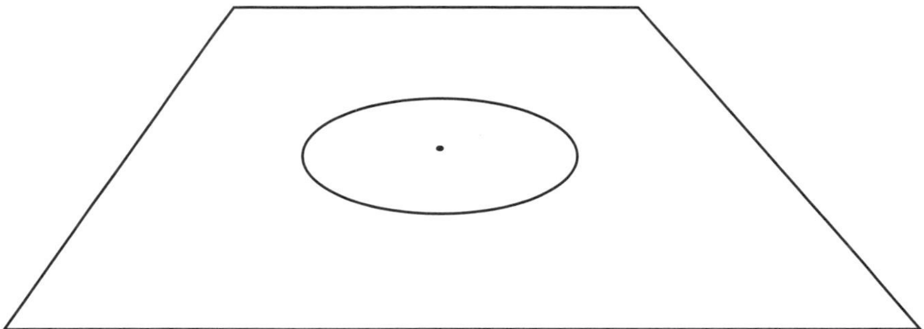

Figure 20.7 Projected position of a circle center under full perspective projection. Note that the projected center is not at the center of the ellipse in the image plane.

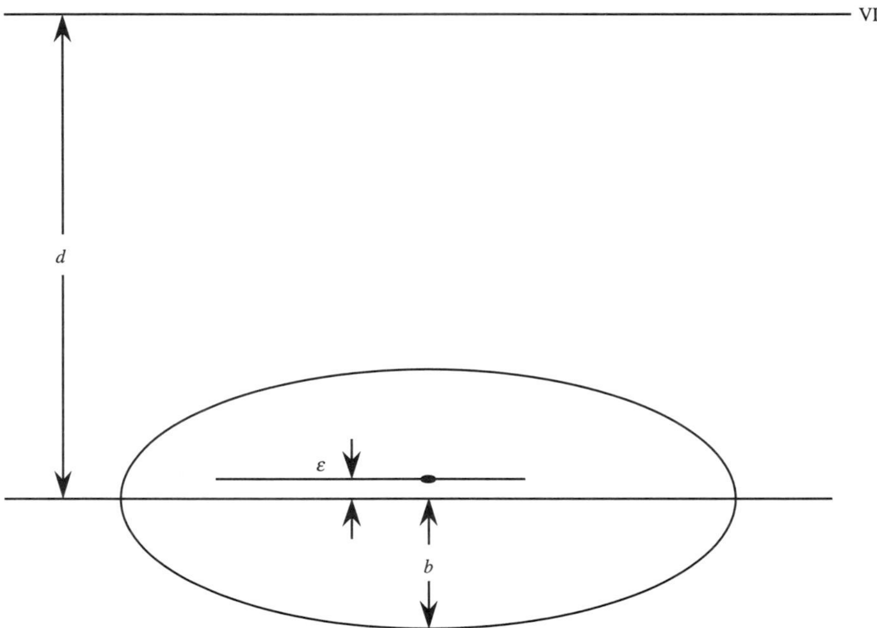

Figure 20.8 Geometry for calculating the offset of the circle center. The projected center of the circle is shown as the elongated dot, and the center of the ellipse in the image plane is a distance d below the vanishing line VL.

Notice that, unlike the situation in Section 20.8, here we are assuming that y_3 is known and we are using its value to calculate y_1 and hence ε.

If the vanishing line is not known, but the orientation of the plane on which the circle lies is known, as is also the orientation of the image plane, then the vanishing

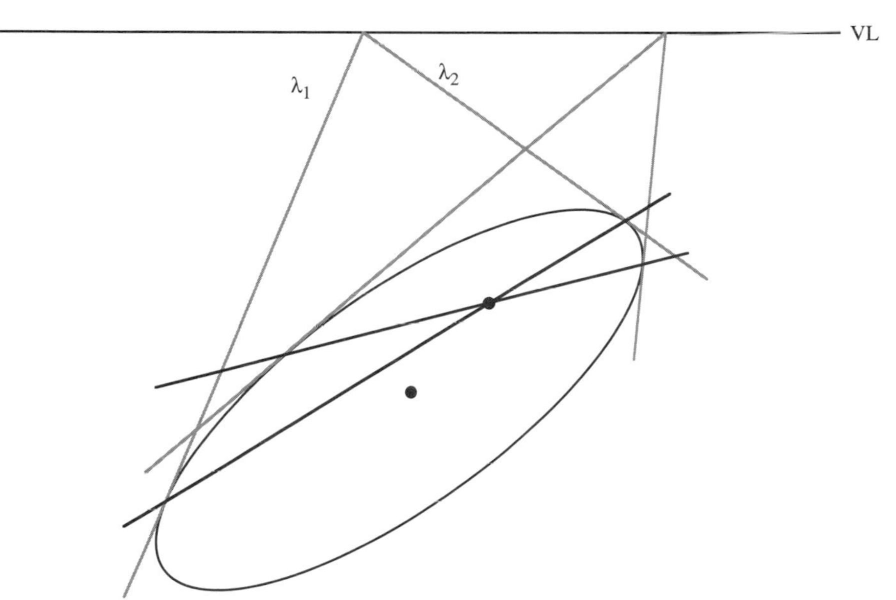

Figure 20.9 Construction for calculating the offsets for a projected ellipse. The two lines λ_1, λ_2 from a point on the vanishing line VL touch the ellipse, and the joins of the points of contact for all such line pairs pass through the projected center. (The figure shows just two chords of contact.)

line can be deduced and the calculation can again proceed as above. However, this approach assumes that the camera has been calibrated (see Chapter 21).

The problem of center determination when an ellipse is projected into an ellipse is a little more difficult to solve: not only is the longitudinal position of the center unknown, but so is the lateral position. Nevertheless, the same basic projective ideas apply. Specifically, let us consider a pair of parallel tangents to the ellipse, which in the image become a pair of lines λ_1, λ_2 meeting on the vanishing line (Fig. 20.9). As the chord joining the contact points of the tangents passes through the center of the original ellipse, and as this property is projectively invariant, so the projected center must lie on the chord joining the contact points of the line-pair λ_1, λ_2. As the same applies for all pairs of parallel tangents to the original ellipse, we can straightforwardly locate the projected center of the ellipse in the image plane.[4] For an alternative, numerical analysis of the situation, see Zhang and Wei (2003).

4 Students familiar with projective geomentry will be able to relate this to the "pole–polar" construction for a conic. In this case, the polar line is the vanishing line, and its pole is the projected center. In general, the pole is not at the center of an ellipse and will not be so unless the polar is at infinite distance. From this point of view, equation (20.9) can be understood in terms of "harmonic ranges," y_2 being the "harmonic mean" of y_1 and y_3.

Both the circle result and the ellipse result are important in cases of inspection of mechanical parts, where accurate results of center positions have to be made regardless of any perspective distortions that may arise. Indeed, circles can also be used for camera calibration purposes, and again high accuracy is required (Heikkilä, 2000).

20.10 Vehicle Guidance in Agriculture—A Case Study

In recent years, farmers have been increasingly pressured to reduce the quantities of chemicals they use for crop protection. This cry has come from both environmentalists and the consumers themselves. The solution to this problem lies in more selective spraying of crops. For example, it would be useful to have a machine that would recognize and spray weeds with herbicides, leaving the vegetable crops themselves unharmed. Alternatively, the individual plants could be sprayed with pesticides. This case study relates to the design of a vehicle that is capable of tracking plant rows and selecting individual plants for spraying (Marchant and Brivot, 1995; Marchant, 1996; Brivot and Marchant, 1996; Sanchiz et al., 1996; Marchant et al., 1998).[5]

The problem would be enormously simplified if plants grew in highly regular placement patterns, so that the machine could tell "at a glance" from their positions whether they were weeds or plants and deal with them accordingly. However, the growth of biological systems is unpredictable and renders such a simplistic approach impracticable. Nevertheless, if plants are grown from seed in a greenhouse and transplanted to the field when they are approaching 100 mm high, they can be placed in straight parallel rows, which will be approximately retained as they grow to full size. There is then the hope (as in the case shown in Fig. 2.2) that the straight rows can be extracted by relatively simple vision algorithms, and the plants themselves located and identified straightforwardly.

At this stage the problems are (1) that the plants will have grown to one side or another and will thus be out of line; (2) that some will have died; (3) that weeds will have appeared near some of the plants; and (4) that some plants will have grown too slowly, and will not be recognized as plants. Thus, a robust algorithm will be required to perform the initial search for the plant rows. The Hough transform approach is well adapted to this type of situation: specifically, it is well suited to looking for line structure in images.

The first step in the process is to locate the plants. This can be achieved with reasonable accuracy by thresholding the input images. (This process is eased if infrared wavelengths are used to enhance contrast.) At this stage, however, the

5 Many of the details of this work are remarkably similar to those for the totally independent project undertaken in Australia by Billingsley and Schoenfisch (1995).

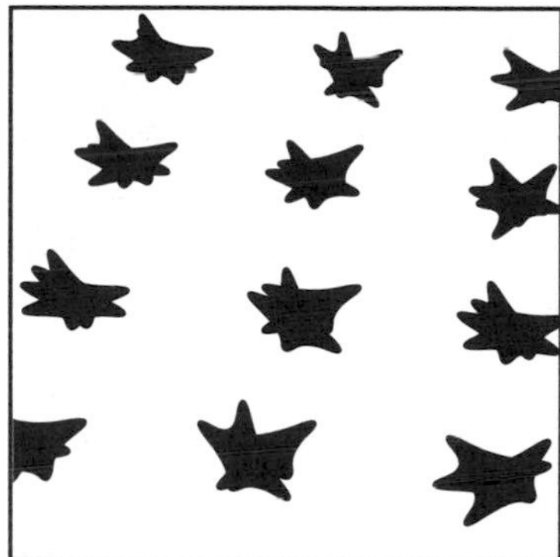

Figure 20.10 Perspective view of plant rows after thresholding. In this idealized sketch, no background clutter is shown.

plant images become shapeless blobs or clumps (Fig. 20.10). These contain holes and lobes (the leaves, in the case of cabbages or cauliflowers), but a certain amount of tidying up can be achieved either by placing a bounding box around the object shape or by performing a dilation of the shape which will regularize it and fill in the major concavities (the real-time solution employed the first of these methods). Then the position of the center of mass of the shape is determined, and it is this that is fed to the Hough transform straight-line (plant row) detector. In common with the usual Hough transform approach, votes are accumulated in parameter space for all possible parameter combinations consistent with the input data. This means taking all possible line gradients and intercepts for lines passing through a given plant center and accumulating them in parameter space. To help get the most meaningful solution, it is useful to accumulate values in proportion to the plant area. In addition, it should be noted that if three rows of plants appear in any image, it will not initially be known which plant is in which row. Therefore, each plant should be allowed to vote for all the row positions. This will naturally be possible only if the inter-row spacing is known and can be assumed in the analysis. However, if this procedure is followed, the method will be far more resistant to missing plants and to weeds that are initially mistakenly assumed to be plants.

The algorithm is improved by preferentially eliminating weeds from the images before applying the Hough transform. Weed elimination is achieved by three techniques: hysteresis thresholding, dilation, and blob size filtering. Dilation refers to the standard shape expansion technique described in Chapter 8 and is used here to fill in the holes in the plant blobs. Filtering by blob area is reasonable since the weeds are seldom as strong as the plants, which were transplanted only when they had become well established.

Hysteresis thresholding is a widely used technique that involves the use of two threshold levels. In this case, if the intensity is greater than the upper level t_u, the object is taken to be plant, if lower than the lower level t_l, it is taken to be weed; if at an intermediate level and next to a region classified as plant, it is taken to be plant. The plant region is allowed to extend sequentially as far as necessary, given only that there is a contiguous region of intensity between t_l and t_u connecting a given point to a true plant ($\geq t_u$) region. This application is unusual in that whole-object segmentation is achieved using hysteresis thresholding. The technique is used more often to help create connected object boundaries (see Section 7.1.1).

Once the Hough transform has been obtained, the parameter space has to be analyzed to find the most significant peak position. Normally, there will be no doubt as to the correct peak—even though the method of accumulation permits plants from adjacent rows to contribute to each peak. With three rows each permitted to contribute to adjacent peaks, the resultant voting patterns in parameter space are: 1,1,1,0,0; 0,1,1,1,0; 0,0,1,1,1—totaling 1,2,3,2,1—thereby making the true center position the most prominent. (Actually, the position is more complicated than this as several plants will be visible in each row, thus augmenting the central position further.) However, the situation could be erroneous if any plants are missing. It is therefore useful to help the Hough transform arrive at the true central position. This can be achieved by applying a Kalman filter (Section 18.13) to keep track of the previous central positions, and anticipate where the next one will be—thereby eliminating false solutions. This concept is taken furthest in the paper by Sanchiz et al. (1996), where all the individual plants are identified on a reliable map of the crop field and allowance is systematically made for errors from any random motions of the vehicle.

20.10.1 *3-D Aspects of the Task*

So far we have assumed that we are looking at simple 2-D images that represent the true 3-D situation in detail. In practice, this is not a valid assumption inasmuch as the rows of plants are being viewed obliquely and therefore appear as straight lines, though with perspective distortions that shift and rotate their positions. The full position can be worked out only if the vehicle motions are kept in mind. Vehicles rolling along the rows of plants exhibit variations in speed and are subject to roll, pitch, and yaw. The first two of these motions correspond, respectively, to rotations about horizontal axes along and perpendicular to the direction of motion. These rotations, being less relevant, are ignored here. The last rotation (yaw) is important as it corresponds to rotation about a vertical axis and affects the immediate direction of motion of the vehicle.

We now have to relate the position (X, Y, Z) of a plant in 3-D with its location (x, y) in an image. We can do so with a general translation:

$$T = (t_x, t_y, t_z)^{\mathrm{T}} \tag{20.11}$$

together with a general rotation:

$$R = \begin{bmatrix} r_1 & r_2 & r_3 \\ r_4 & r_5 & r_6 \\ r_7 & r_8 & r_9 \end{bmatrix} \tag{20.12}$$

giving:

$$\begin{bmatrix} X \\ Y \\ Z \end{bmatrix} = \begin{bmatrix} r_1 & r_2 & r_3 \\ r_4 & r_5 & r_6 \\ r_7 & r_8 & r_9 \end{bmatrix} \begin{bmatrix} x \\ y \\ z \end{bmatrix} + \begin{bmatrix} t_x \\ t_y \\ t_z \end{bmatrix} \tag{20.13}$$

The lens projection formulas are also relevant:

$$x = fX/Z \tag{20.14}$$

$$y = fY/Z \tag{20.15}$$

We shall not give a full analysis here, but assuming that roll and pitch are zero, and that the heading angle (direction of motion relative to the rows of plants) is ψ, and that this is small, we obtain a quadratic equation for ψ in terms of t_x. Thus, two sets of solutions are in general possible. However, it is soon found that only one solution matches the situation, as the wrong solution is not supported by the other feature point positions. This discussion shows the complications introduced by perspective projection—even when highly restrictive assumptions can be made about the geometrical configuration (in particular, ψ being small).

20.10.2 *Real-time Implementation*

Finally, it was found to be possible to implement the vehicle guidance system on a single processor augmented by two special hardware units—a color classifier and a chaincoder. The chaincoder is useful for fast shape analysis following boundary tracking. The overall system was able to process the input images at a rate of 10 Hz, which is sufficient for reliable vehicle guidance. Perhaps more important, the claimed accuracy is in the region of 10 mm and 1° of angle, making the whole guidance system adequate to cope with the particular slightly constrained application considered. A later implementation (Marchant et al., 1998) did a more thorough job of segmenting the individual plants (though still not using the blob size filter), obtaining a final 5 Hz sampling rate—again fast enough for real-time application in the field. All in all, this case study demonstrates the possibility of highly accurate selective spraying of weeds, thereby significantly cutting down on the amount of herbicide needed for crops such as cabbages, cauliflower, and wheat.

20.11 Concluding Remarks

Vanishing point detection is of great importance in helping to obtain a basic understanding of the scene being viewed by the robot camera. The chapter discussed simple means for creating a plan view of the scene, to help with navigation and path planning. It was found that when the ground plane is highly structured (whether with flagstones or other artifacts), cross ratio invariants can help locate vanishing points and provide additional understanding. A case study of vehicle guidance in agriculture was helpful in revealing many more practicalities about motion control.

This chapter has helped focus on mechanistic applications of machine vision and at the same time demonstrated various theoretical aspects of the subject. The circle and ellipse center location ideas of Section 20.9 are a byproduct of the theory and are presented here mainly for convenience and to demonstrate the generality of the concepts.

Egomotion is well mediated by vision, though the problems of 3-D, including perspective, are immense. This chapter has shown how judicious attention to vanishing points can provide a computationally efficient key to solving some of the practical problems. Nevertheless, motion parameters such as roll, pitch, and yaw can make life difficult in a real-time environment.

20.12 Bibliographical and Historical Notes

Mobile robots and the need for path planning are discussed by Kanesalingam et al. (1998) and by Kortenkamp et al. (1998); more recently, DeSouza and Kak (2002) have carried out a survey on vision for mobile robot navigation; see also Davison and Murray (2002). It is interesting how often vanishing point determination has been considered both in relation to egomotion for mobile robots (Lebègue and Aggarwal, 1993; Shuster et al., 1993), and in general, to the vision methodology

(Magee and Aggarwal, 1984; Shufelt, 1999; Almansa et al., 2003), which is prone to suffer from inaccuracy when real off-camera data are used in any context. The seminal paper that gave rise to the crucial Gaussian sphere technique was that by Barnard (1983).

Guidance of outdoor vehicles, particularly on roads, has been discussed by a large number of authors over a period of many years. Most recently, there has been quite an explosive expansion in this area (Bertozzi and Broggi, 1998; Guiducci, 1999; Kang and Jung, 2003; Kastrinaki et al., 2003). De la Escalara et al. (2003) have extended the discussion to recognition of traffic signs. Zhou et al. (2003) have considered the situation for elderly pedestrians—though such work could also be relevant for blind people or wheelchair users. Hofmann et al. (2003) have shown that vision and radar can profitably be used together to combine the excellent spatial resolution of vision with the accurate range resolution of radar. The volume edited by Blake and Yuille (1992) provides a set of key readings on active vision.

Finally, Clark and Mirmehdi (2000, 2002) provide an interesting twist; they show how text in documents can be oriented using vanishing points. This work demonstrates that the techniques described in a volume such as the present one can never be presented in a completely logical order, for the techniques are surprisingly general. Indeed, the very *purpose* of such works is to emphasize what is generic so that future workers can benefit most.

20.13 **Problems**

1. Draw diagrams corresponding to Fig. 20.4 showing the geometry for the lateral distance from the line λ. Prove equation (20.3) and state any assumptions that have to be made in the proof.

2. Redraw Fig. 20.9 using vanishing points aligned along the observed ellipse axes. Show that the problem of finding the transformed center location now reduces to two 1-D cases and that equation (20.10) can be used to obtain the transformed center coordinates.

3. A robot is walking along a path paved with rectangular flagstones. It is able to rotate its camera head so that one set of flagstone lines appears parallel while the other set converges toward a vanishing point. Show that the robot can calculate the position of the vanishing point in two ways: (1) by measuring the varying widths of individual flagstones or (2) by measuring the lengths of adjacent flagstones and proceeding according to equation (20.8). In case (1), obtain a formula that could be used to determine the position of the vanishing point. Which is the more general approach? Which would be applicable if the flagstones appeared in a flower garden in random locations and orientations?

Image Transformations

When setting up a measurement system, it is natural to calibrate it carefully before use. This task has been left to last because (1) it may be more mathematically daunting, (2) there are instances where it can be bypassed, and (3) it is not always possible to perform the calibration entirely in advance, but rather it has to be updated to a sufficient extent as measurements proceed. This chapter outlines some of the problems of calibration and some of the results of recent research that allow the process to be at least partially bypassed.

Look out for:

- the homogeneous coordinates technique for representing general 3-D positions and transformations.
- "extrinsic" (external world) and "intrinsic" (camera) parameters.
- methods of achieving absolute camera calibration.
- the need for correction for camera lens distortions.
- the idea of a generalized epipolar geometry.
- the "essential" and "fundamental" matrix formulations, relating the observed positions of any point in two camera frames of reference.
- the central position of the 8-point algorithm.
- the possibility of image "rectification."
- the possibility of 3-D reconstruction.

This chapter is the last of the six chapters constituting Part 3 of this book. These chapters should be taken together because they involve not merely different topics but also different *aspects* of the subject. In addition, the aim has been to cover them in as gentle an order as possible, considering the mathematical complexities involved in extracting 3-D and motion information from 2-D images.

CHAPTER 21

Image Transformations and Camera Calibration

21.1 Introduction

When images are obtained from 3-D scenes, the exact position and orientation of the camera sensing device are often unknown, and there is a need for it to be related to some global frame of reference. This is especially important if accurate measurements of objects are to be made from their images, for example, in inspection applications. On the other hand, it may sometimes be possible to dispense with such detailed information—as in the case of a security system for detecting intruders, or a system for counting cars on a motorway. There are also more complicated cases, such as those in which cameras can be rotated or moved on a robot arm, or the objects being examined can move freely in space. In such cases, camera calibration becomes a central issue. Before we can consider camera calibration, we need to understand in some detail the transformations that can occur between the original world points and the formation of the final image. We attend to these image transformations in the following section, and we then move on to details of camera parameters and camera calibration in the subsequent two sections. Then, in Section 21.5 we consider how any radial distortions of the image introduced by the camera lens can be corrected.

Section 21.6 signals an apparent break with the previous work and introduces "multiple view" vision. This topic has become important in recent years, for it uses new theory to bypass the need for formal camera calibration, making it possible to update the vision system parameters during actual use. The basis for this work is generalized epipolar geometry. This takes the epipolar line ideas of Section 16.3.2 considerably further. At the core of this new work are the "essential" and "fundamental" matrix formulations, which relate the observed positions of any point in two camera frames of reference.

Short sections on image "rectification" (obtaining a new image as it would be seen from an idealized camera position) and 3-D reconstruction follow.

21.2 Image Transformations

First, we consider the rotations and translations of object points relative to a global frame. After a rotation through an angle θ about the Z-axis (Fig. 21.1), the coordinates of a general point (X, Y) change to:

$$X' = X\cos\theta - Y\sin\theta \tag{21.1}$$

$$Y' = X\sin\theta + Y\cos\theta \tag{21.2}$$

This result is neatly expressed by the matrix equation:

$$\begin{bmatrix} X' \\ Y' \end{bmatrix} = \begin{bmatrix} \cos\theta & -\sin\theta \\ \sin\theta & \cos\theta \end{bmatrix} \begin{bmatrix} X \\ Y \end{bmatrix} \tag{21.3}$$

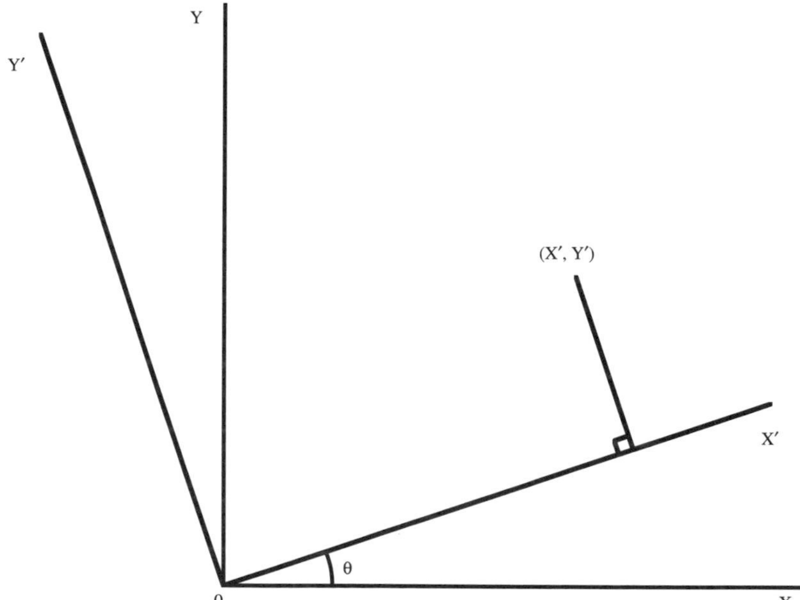

Figure 21.1 Effect of a rotation θ about the origin.

Similar rotations are possible about the X- and Y-axes. To satisfactorily express rotations in 3-D, we require a more general notation using 3×3 matrices, the matrix for a rotation θ about the Z-axis being:

$$\mathbf{Z}(\theta) = \begin{bmatrix} \cos\theta & -\sin\theta & 0 \\ \sin\theta & \cos\theta & 0 \\ 0 & 0 & 1 \end{bmatrix} \tag{21.4}$$

those for rotations ψ about the X-axis and φ about the Y-axis are:

$$\mathbf{X}(\psi) = \begin{bmatrix} 1 & 0 & 0 \\ 0 & \cos\psi & -\sin\psi \\ 0 & \sin\psi & \cos\psi \end{bmatrix} \tag{21.5}$$

$$\mathbf{Y}(\varphi) = \begin{bmatrix} \cos\varphi & 0 & \sin\varphi \\ 0 & 1 & 0 \\ -\sin\varphi & 0 & \cos\varphi \end{bmatrix} \tag{21.6}$$

We can make up arbitrary rotations in 3-D by applying sequences of such rotations. Similarly, we can express arbitrary rotations as sequences of rotations about the coordinate axes. Thus $\mathbf{R} = \mathbf{X}(\psi)\mathbf{Y}(\varphi)\mathbf{Z}(\theta)$ is a composite rotation in which $\mathbf{Z}(\theta)$ is applied first, then $\mathbf{Y}(\varphi)$, and finally $\mathbf{X}(\psi)$. Rather than multiplying out these matrices, we write down here the general result expressing an arbitrary rotation $\mathbf{R}$:

$$\begin{bmatrix} X' \\ Y' \\ Z' \end{bmatrix} = \begin{bmatrix} R_{11} & R_{12} & R_{13} \\ R_{21} & R_{22} & R_{23} \\ R_{31} & R_{32} & R_{33} \end{bmatrix} \begin{bmatrix} X \\ Y \\ Z \end{bmatrix} \tag{21.7}$$

Note that the rotation matrix $\mathbf{R}$ is not completely general. It is orthogonal and thus has the property that $\mathbf{R}^{-1} = \mathbf{R}^{\mathrm{T}}$.

In contrast with rotation, translation through a distance (T_1, T_2, T_3) is given by:

$$X' = X + T_1 \tag{21.8}$$

$$Y' = Y + T_2 \tag{21.9}$$

$$Z' = Z + T_3 \tag{21.10}$$

which is not expressible in terms of a multiplicative 3×3 matrix. However, just as general rotations can be expressed as rotations about various coordinate axes, so general translations and rotations can be expressed as sequences of basic rotations and translations relative to individual coordinate axes. Thus, it would be most useful to have a notation that unified the mathematical treatment so that a generalized displacement could be expressed as a product of matrices. This is indeed possible if so-called *homogeneous coordinates* are used. To achieve this, the matrices must be augmented to 4×4. A general rotation can then be expressed in the form:

$$\begin{bmatrix} X' \\ Y' \\ Z' \\ 1 \end{bmatrix} = \begin{bmatrix} R_{11} & R_{12} & R_{13} & 0 \\ R_{21} & R_{22} & R_{23} & 0 \\ R_{31} & R_{32} & R_{33} & 0 \\ 0 & 0 & 0 & 1 \end{bmatrix} \begin{bmatrix} X \\ Y \\ Z \\ 1 \end{bmatrix} \tag{21.11}$$

while the general translation matrix becomes:

$$\begin{bmatrix} X' \\ Y' \\ Z' \\ 1 \end{bmatrix} = \begin{bmatrix} 1 & 0 & 0 & T_1 \\ 0 & 1 & 0 & T_2 \\ 0 & 0 & 1 & T_3 \\ 0 & 0 & 0 & 1 \end{bmatrix} \begin{bmatrix} X \\ Y \\ Z \\ 1 \end{bmatrix} \tag{21.12}$$

The generalized displacement (i.e., translation plus rotation) transformation clearly takes the form:

$$\begin{bmatrix} X' \\ Y' \\ Z' \\ 1 \end{bmatrix} = \begin{bmatrix} R_{11} & R_{12} & R_{13} & T_1 \\ R_{21} & R_{22} & R_{23} & T_2 \\ R_{31} & R_{32} & R_{33} & T_3 \\ 0 & 0 & 0 & 1 \end{bmatrix} \begin{bmatrix} X \\ Y \\ Z \\ 1 \end{bmatrix} \tag{21.13}$$

We now have a convenient notation for expressing generalized transformations, including operations other than the translations and rotations that account for the normal motions of rigid bodies. First we take a scaling in size of an object, the simplest case being given by the matrix:

$$\begin{bmatrix} S & 0 & 0 & 0 \\ 0 & S & 0 & 0 \\ 0 & 0 & S & 0 \\ 0 & 0 & 0 & 1 \end{bmatrix}$$

The more general case:

$$\begin{bmatrix} S_1 & 0 & 0 & 0 \\ 0 & S_2 & 0 & 0 \\ 0 & 0 & S_3 & 0 \\ 0 & 0 & 0 & 1 \end{bmatrix}$$

introduces a shear in which an object line λ will be transformed into a line that is not in general parallel to λ. Skewing is another interesting transformation, being given by linear translations varying from the simple case:

$$\begin{bmatrix} 1 & B & 0 & 0 \\ 0 & 1 & 0 & 0 \\ 0 & 0 & 1 & 0 \\ 0 & 0 & 0 & 1 \end{bmatrix}$$

to the general case:

$$\begin{bmatrix} 1 & B & C & 0 \\ D & 1 & F & 0 \\ G & H & 1 & 0 \\ 0 & 0 & 0 & 1 \end{bmatrix}$$

Rotations can be regarded as combinations of scaling and skewing and are sometimes implemented as such (Weiman, 1976).

The other simple but interesting case is that of reflection, which is typified by:

$$\begin{bmatrix} 0 & 1 & 0 & 0 \\ 1 & 0 & 0 & 0 \\ 0 & 0 & 1 & 0 \\ 0 & 0 & 0 & 1 \end{bmatrix}$$

This generalizes to other cases of improper rotation where the determinant of the top left 3×3 matrix is -1.

In all the cases discussed above, it will be observed that the bottom row of the generalized displacement matrix is redundant. In fact, we can put this row to good use in certain other types of transformations. Of particular interest in this context is the case of perspective projection. Following Section 15.3, equation (15.1), the equations for projection of object points into image points are:

$$x = fX/Z \tag{21.14}$$

$$y = fY/Z \tag{21.15}$$

$$z = f \tag{21.16}$$

Next we make full use of the bottom row of the transformation matrix by defining the homogeneous coordinates as $(X_h, Y_h, Z_h, h) = (hX, hY, hZ, h)$, where h is a nonzero constant that we can take to be unity. To proceed, we examine the homogeneous transformation:

$$\begin{bmatrix} 1 & 0 & 0 & 0 \\ 0 & 1 & 0 & 0 \\ 0 & 0 & 1 & 0 \\ 0 & 0 & 1/f & 0 \end{bmatrix} \begin{bmatrix} X \\ Y \\ Z \\ 1 \end{bmatrix} = \begin{bmatrix} X \\ Y \\ Z \\ Z/f \end{bmatrix} \tag{21.17}$$

Dividing by the fourth coordinate gives the required values of the transformed Cartesian coordinates $(fX/Z, fY/Z, f)$.

Let us now review this result. First, we have found a 4×4 matrix transformation that operates on 4-D homogeneous coordinates. These do not correspond directly to real coordinates, but real 3-D coordinates can be calculated from them by dividing the first three by the fourth homogeneous coordinate. Thus, there is an arbitrariness in the homogeneous coordinates in that they can all be multiplied by the same constant factor without producing any change in the final interpretation. Similarly, when deriving homogeneous coordinates from real 3-D coordinates, we can employ any convenient constant multiplicative factor h, though we will normally take h to be unity.

The advantage to be gained from use of homogeneous coordinates is the convenience of having a single multiplicative matrix for any transformation, in spite of the fact that perspective transformations are intrinsically nonlinear. Thus, a quite complex nonlinear transformation can be reduced to a more straightforward linear transformation. This eases computer calculation of object coordinate transformations, and other computations such as those for camera calibration (see Section 21.3). We may also note that almost every transformation can be inverted by inverting the corresponding homogeneous transformation matrix. An exception is the perspective transformation, for which the fixed value of z leads merely to Z being unknown, and X, Y only being known relative to the value of Z (hence the need for binocular vision or other means of discerning depth in a scene).

21.3 Camera Calibration

We have shown how homogeneous coordinate systems are used to help provide a convenient linear 4×4 matrix representation for 3-D transformations (including rigid body translations and rotations) and nonrigid operations (including scaling, skewing, and perspective projection). In this last case, it was implicitly assumed that the camera and world coordinate systems are identical, since the image coordinates were expressed in the same frame of reference. However, in general the objects viewed by the camera will have positions that may be known in world coordinates, but that will not *a priori* be known in camera coordinates, since the camera will in general be mounted in a somewhat arbitrary position and will point in a somewhat arbitrary direction. Indeed, it may well be on adjustable gimbals, and may also be motor driven, with no precise calibration system. If the camera is on a robot arm, there are likely to be position sensors that could inform the control system of the camera position and orientation in world coordinates, though the amount of slack may well make the information too imprecise for practical purposes (e.g., to guide the robot toward objects).

These factors mean that the camera system will have to be calibrated very carefully before the images can be used for practical applications such as robot pick-and-place. A useful approach is to assume a general transformation between the world coordinates and the image seen by the camera under perspective projection, and to locate in the image various calibration points that have been placed in known positions in the scene. If enough such points are available, it should be possible to compute the transformation parameters, and then all image points can be interpreted accurately until recalibration becomes necessary.

The general transformation **G** takes the form:

$$\begin{bmatrix} X_H \\ Y_H \\ Z_H \\ H \end{bmatrix} = \begin{bmatrix} G_{11} & G_{12} & G_{13} & G_{14} \\ G_{21} & G_{22} & G_{23} & G_{24} \\ G_{31} & G_{32} & G_{33} & G_{34} \\ G_{41} & G_{42} & G_{43} & G_{44} \end{bmatrix} \begin{bmatrix} X \\ Y \\ Z \\ 1 \end{bmatrix} \tag{21.18}$$

where the final Cartesian coordinates appearing in the image are $(x, y, z) = (x, y, f)$. These are calculated from the first three homogeneous coordinates by dividing by the fourth:

$$x = X_H/H = (G_{11}X + G_{12}Y + G_{13}Z + G_{14})/(G_{41}X + G_{42}Y + G_{43}Z + G_{44}) \quad (21.19)$$

$$y = Y_H/H = (G_{21}X + G_{22}Y + G_{23}Z + G_{24})/(G_{41}X + G_{42}Y + G_{43}Z + G_{44}) \quad (21.20)$$

$$z = Z_H/H = (G_{31}X + G_{32}Y + G_{33}Z + G_{34})/(G_{41}X + G_{42}Y + G_{43}Z + G_{44}) \quad (21.21)$$

However, as we know z, there is no point in determining parameters G_{31}, G_{32}, G_{33}, G_{34}. Accordingly we proceed to develop the means for finding the other parameters. Because only the ratios of the homogeneous coordinates are meaningful, only the ratios of the G_{ij} values need be computed, and it is usual to take G_{44} as unity: this leaves only eleven parameters to be determined. Multiplying out the first two equations and rearranging give:

$$G_{11}X + G_{12}Y + G_{13}Z + G_{14} - x(G_{41}X + G_{42}Y + G_{43}Z) = x \quad (21.22)$$

$$G_{21}X + G_{22}Y + G_{23}Z + G_{24} - y(G_{41}X + G_{42}Y + G_{43}Z) = y \quad (21.23)$$

Noting that a single world point (X, Y, Z) that is known to correspond to image point (x, y) gives us *two* equations of the above form; it requires a minimum of six such points to provide values for all 11 G_{ij} parameters. Figure 21.2 shows a convenient near-minimum case. An important factor is that the world points used for the calculation should lead to independent equations. Thus it is important that they should not be coplanar. More precisely, there must be at least six points, no four of which are coplanar. However, further points are useful in that they lead to overdetermination of the parameters and increase the accuracy with which the latter can be computed. There is no reason why the additional points should not be coplanar with existing points. Indeed, a common arrangement is to set up a cube so that three of its faces are visible, each face having a pattern of squares with 30 to 40 easily discerned corner features (as for a Rubik cube).

Least squares analysis can be used to perform the computation of the 11 parameters (e.g., via the pseudo-inverse method). First, the $2n$ equations have to be expressed in matrix form:

$$\mathbf{A}\mathbf{g} = \boldsymbol{\xi} \quad (21.24)$$

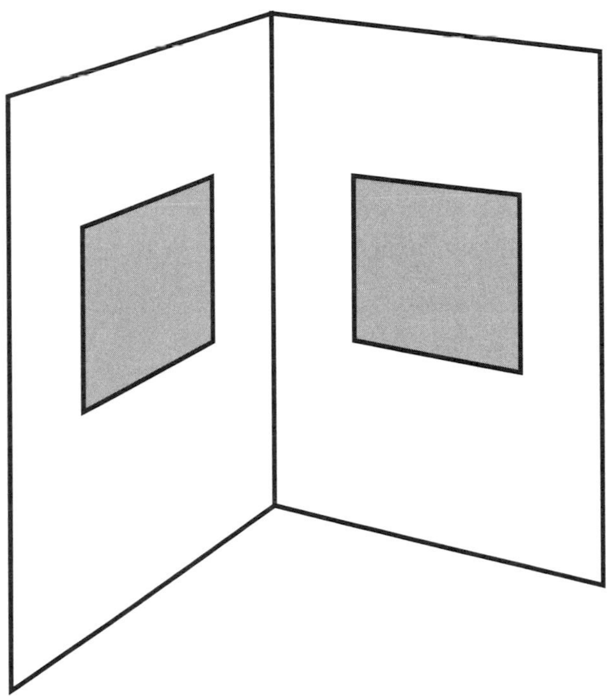

Figure 21.2 A convenient near-minimum case for camera calibration. Here two sets of four coplanar points, each set of four being at the corners of a square, provide more than the absolute minimum number of points required for camera calibration.

where **A** is a $2n \times 11$ matrix of coefficients, which multiplies the G-matrix, now in the form:

$$\mathbf{g} = (G_{11}\, G_{12}\, G_{13}\, G_{14}\, G_{21}\, G_{22}\, G_{23}\, G_{24}\, G_{41}\, G_{42}\, G_{43})^{\mathsf{T}} \tag{21.25}$$

and $\boldsymbol{\xi}$ is a $2n$-element column vector of image coordinates. The pseudo-inverse solution is:

$$\mathbf{g} = \mathbf{A}^{\dagger}\boldsymbol{\xi} \tag{21.26}$$

where:

$$\mathbf{A}^{\dagger} = \left(\mathbf{A}^{\mathsf{T}}\mathbf{A}\right)^{-1}\mathbf{A}^{\mathsf{T}} \tag{21.27}$$

The solution is more complex than might have been expected, since a normal matrix inverse is only defined, and can only be computed, for a square matrix. Note that solutions are only obtainable by this method if the matrix $\mathbf{A}^{\mathrm{T}}\mathbf{A}$ is invertible. For further details of this method, see Golub and van Loan (1983).

21.4 Intrinsic and Extrinsic Parameters

At this point it is useful to look in more detail at the general transformation leading to camera calibration. When we are calibrating the camera, we are actually trying to bring the camera and world coordinate systems into coincidence. The first step is to move the origin of the world coordinates to the origin of the camera coordinate system. The second step is to rotate the world coordinate system until its axes are coincident with those of the camera coordinate system. The third step is to move the image plane laterally until there is complete agreement between the two coordinate systems. (This step is required since it is not known initially which point in the world coordinate system corresponds to the principal point[1] in the image.)

There is an important point to be borne in mind during this process. If the camera coordinates are given by **C**, then the translation **T** required in the first step will be −**C**. Similarly, the rotations that are required will be the inverses of those that correspond to the actual camera orientations. The reason for these reversals is that (for example) rotating an object (here the camera) forward gives the same effect as rotating the axes backwards. Thus, all operations have to be carried out with the reverse arguments to those indicated above in Section 21.1. The complete transformation for camera calibration is hence:

$$\mathbf{G} = \mathbf{PLRT} = \begin{bmatrix} 1 & 0 & 0 & 0 \\ 0 & 1 & 0 & 0 \\ 0 & 0 & 1 & 0 \\ 0 & 0 & 1/f & 0 \end{bmatrix} \begin{bmatrix} 1 & 0 & 0 & t_1 \\ 0 & 1 & 0 & t_2 \\ 0 & 0 & 1 & t_3 \\ 0 & 0 & 0 & 1 \end{bmatrix} \begin{bmatrix} R_{11} & R_{12} & R_{13} & 0 \\ R_{21} & R_{22} & R_{23} & 0 \\ R_{31} & R_{32} & R_{33} & 0 \\ 0 & 0 & 0 & 1 \end{bmatrix} \begin{bmatrix} 1 & 0 & 0 & T_1 \\ 0 & 1 & 0 & T_2 \\ 0 & 0 & 1 & T_3 \\ 0 & 0 & 0 & 1 \end{bmatrix} \tag{21.28}$$

where matrix **P** takes account of the perspective transformation required to form the image. It is usual to group together the transformations **P** and **L** and call

1 The *principal point* is the image point lying on the principal axis of the camera; it is the point in the image that is closest to the center of projection. Correspondingly, the *principal axis* (or *optical axis*) of the camera is the line through the center of projection normal to the image plane.

them internal camera transformations, which include the *intrinsic camera parameters*, while **R** and **T** are taken together as external camera transformations corresponding to *extrinsic camera parameters*:

$$\mathbf{G} = \mathbf{G}_{\text{internal}}\mathbf{G}_{\text{external}} \tag{21.29}$$

where:

$$\mathbf{G}_{\text{internal}} = \mathbf{PL} = \begin{bmatrix} 1 & 0 & 0 & t_1 \\ 0 & 1 & 0 & t_2 \\ 0 & 0 & 1 & t_3 \\ 0 & 0 & 1/f & t_3/f \end{bmatrix} \to \begin{bmatrix} 1 & 0 & t_1 \\ 0 & 1 & t_2 \\ 0 & 0 & 1/f \end{bmatrix} \tag{21.30}$$

$$\mathbf{G}_{\text{external}} = \mathbf{RT} = \begin{bmatrix} \mathbf{R}_1 & \mathbf{R}_1.\boldsymbol{T} \\ \mathbf{R}_2 & \mathbf{R}_2.\boldsymbol{T} \\ \mathbf{R}_3 & \mathbf{R}_3.\boldsymbol{T} \\ \mathbf{0} & 1 \end{bmatrix} \tag{21.31}$$

In the matrix for $\mathbf{G}_{\text{internal}}$ we have assumed that the initial translation matrix **T** moves the camera's center of projection to the correct position, so that the value of t_3 can be made equal to zero: in that case, the effect of **L** will indeed be lateral as indicated above. At that point, we can express the 2-D result in terms of a 3×3 homogeneous coordinate matrix. In the matrix for $\mathbf{G}_{\text{external}}$, we have expressed the result succinctly in terms of the rows $\mathbf{R}_1$, $\mathbf{R}_2$, $\mathbf{R}_3$ of **R**, and have taken dot-products with $\boldsymbol{T}(T_1, T_2, T_3,)$. The 3-D result is a 4×4 homogenous coordinate matrix.

Although the above treatment gives a good indication of the underlying meaning of **G**, it is not general because we have not so far included scaling and skew parameters in the internal matrix. In fact, the generalized form of $\mathbf{G}_{\text{internal}}$ is:

$$\mathbf{G}_{\text{internal}} = \begin{bmatrix} s_1 & b_1 & t_1 \\ b_2 & s_2 & t_2 \\ 0 & 0 & 1/f \end{bmatrix} \tag{21.32}$$

Potentially, $\mathbf{G}_{\text{internal}}$ should include the following:

1. A transform for correcting scaling errors.
2. A transform for correcting translation errors.[2]

2 For this purpose, the origin of the image should be on the principal axis of the camera. Misalignment of the sensor may prevent this point from being at the center of the image.

3. A transform for correcting sensor skewing errors (due to nonorthogonality of the sensor axes).
4. A transform for correcting sensor shearing errors (due to unequal scaling along the sensor axes).
5. A transform for correcting for unknown sensor orientation within the image plane.

Translation errors (item 2) are corrected by adjusting t_1 and t_2. All the other adjustments are concerned with the values of the 2×2 submatrix:

$$\begin{bmatrix} s_1 & b_1 \\ b_2 & s_2 \end{bmatrix}$$

However, it should be noted that application of this matrix performs rotation within the image plane immediately after rotation has been performed in the world coordinates by $\mathbf{G}_{\text{external}}$, and it is virtually impossible to separate the two rotations. This explains why we now have a total of six external and six internal parameters totaling 12 rather than the expected 11 parameters (we return to the factor $1/f$ below). As a result, it is better to exclude item 5 in the above list of internal transforms and to subsume it into the external parameters.[3] Since the rotational component in $\mathbf{G}_{\text{internal}}$ has been excluded, b_1 and b_2 must now be equal, and the internal parameters will be: s_1, s_2, b, t_1, t_2. Note that the factor $1/f$ provides a scaling that cannot be separated from the other scaling factors during camera calibration without specific (i.e., separate) measurement of f. Thus, we have a total of six parameters from $\mathbf{G}_{\text{external}}$ and five parameters from $\mathbf{G}_{\text{internal}}$: this totals 11 and equals the number cited in the previous section.

At this point, we should consider the special case where the sensor is known to be Euclidean to a high degree of accuracy. This will mean that $b = b_1 = b_2 = 0$, and $s_1 = s_2$, bringing the number of internal parameters down to three. In addition, if care has been taken over sensor alignment, and there are no other offsets to be allowed for, it may be known that $t_1 = t_2 = 0$. This will bring the total number of internal parameters down to just one, namely, $s = s_1 = s_2$, or s/f, if we take proper account of the focal length. In this case there will be a total of seven calibration parameters for the whole camera system. This may permit it to be set up unambiguously by viewing a known object having four clearly marked features instead of the six that would normally be required (see Section 21.2).

---

3 While doing so may not be ideal, there is no way of separating the two rotational components by purely optical means. Only measurements on the internal dimensions of the camera system could determine the internal component, but separation is not likely to be a cogent or even meaningful matter. On the other hand, the internal component is likely to be stable, whereas the external component may be prone to variation if the camera is not mounted securely.

21.5 Correcting for Radial Distortions

Photographs generally appear so distortion free that we tend to imagine that camera lenses are virtually perfect. However, it sometimes happens that a photograph will show odd curvatures of straight lines, particularly those appearing around the periphery of the picture. The results commonly take the form of "pincushion" or "barrel" distortion. These terms arise because pincushions have a tendency to be overextended at the corners, whereas barrels usually bulge in the middle. In images of paving stones or brick walls, the amount of distortion is usually not more than a few pixels in a total of the order of 512, that is, typically less than 2%. This explains why, in the absence of particular straight line markers, such distortions can be missed (Fig. 21.3). However, it is important both for recognition and for interimage matching purposes that any distortions be eliminated. Today image interpretation is targeted at, and frequently achieves, subpixel accuracy. In addition, disparities between stereo images are in the first order of small quantities, and single pixel errors would lead to significant errors in depth measurement. Hence, it is more the rule than the exception that 3-D image analysis will need to make corrections for barrel or pincushion distortion.

For reasons of symmetry, the distortions that arise in images tend to involve radial expansions or contractions relative to the optical axis—corresponding, respectively, to pincushion or barrel distortion. As with many types of

Figure 21.3 Photograph of a brick wall showing radial (barrel) distortion.

errors, series solutions can be useful. Thus, it is worthwhile to model the distortions as:

$$\mathbf{r}' = \mathbf{r}f(r) = \mathbf{r}(a_0 + a_2r^2 + a_4r^4 + a_6r^6 + \cdots) \tag{21.33}$$

the odd orders in the brackets canceling to zero, again for reasons of symmetry. It is usual to set a_0 to unity, as this coefficient can be taken up by the scale parameters in the camera calibration matrix.

To fully define the effect, we write the x and y distortions as:

$$x' - x_c = (x - x_c)(1 + a_2r^2 + a_4r^4 + a_6r^6 + \cdots) \tag{21.34}$$

$$y' - y_c = (y - y_c)(1 + a_2r^2 + a_4r^4 + a_6r^6 + \cdots) \tag{21.35}$$

Here x and y are measured relative to the position of the optical axis of the lens (x_c, y_c), so $\mathbf{r} = (x - x_c, y - y_c), \mathbf{r}' = (x' - x_c, y' - y_c)$.

As stated earlier, the errors to be expected are in the 2% or less range. That is, it is normally sufficiently accurate to take just the first correction term in the expansion and disregard the rest. At the very least, this will introduce such a large improvement in the accuracy that it will be difficult to detect any discrepancies, especially if the image dimensions are 512×512 pixels or less.[4] In addition, computation errors in matrix inversion and convergence of 3-D algorithms will add to the digitization errors, tending further to hide higher powers of radial distortion. Thus, in most cases the distortion can be modeled using a single parameter equation:

$$\mathbf{r}' = \mathbf{r}f(r) = \mathbf{r}(1 + a_2r^2) \tag{21.36}$$

Note that the above theory only models the distortion; it has to be corrected by the corresponding inverse transformation.

It is instructive to consider the apparent shape of a straight line that appears, for example, along the top of an image (Fig. 21.3). Take the image dimensions to range over $-x_1 \leq x \leq x_1$, $-y_1 \leq y \leq y_1$, and the optical axis of the camera to be at the center of the image. Then the straight line will have the approximate equation:

$$y' = y_1[1 + a_2(x^2 + y_1^2)] = y_1 + a_2y_1^3 + a_2y_1x^2 \tag{21.37}$$

4 This remark will not apply to many web cameras, which are sold at extremely low prices on the mainly amateur market. While the camera chip and electronics are often a very good value, the accompanying low-cost lens may well require extensive correction to ensure that distortion-free measurements are possible.

Clearly, it is a parabola. The vertical error at the center of the parabola is $a_2 y_1^3$, and the additional vertical error at the ends is $a^2 y_1 x_1^2$. If the image is square ($x_1 = y_1$), these two errors are equal. (The erroneous impression is given by the parabola shape that the error at $x = 0$ is zero.)

Finally, note that digital scanners are very different from single-lens cameras, in that their lenses travel along the object space during acquisition. Thus, longitudinal errors are unlikely to arise to anything like the same extent, though lateral errors could in principle be problematic.

21.6 Multiple-view Vision

In the 1990s, a considerable advance in 3-D vision was made by examining what could be learned from uncalibrated cameras using multiple views. At first glance, considering the efforts made in earlier sections of this chapter to understand exactly how cameras should be calibrated, this may seem nonsensical. Nevertheless, there are considerable potential advantages in examining multiple views—not least, many thousands of videotapes are available from uncalibrated cameras, including those used for surveillance and those produced in the film industry. In such cases, as much must be made of the available material as possible, whether or not any regrets over "what might have been" are entertained. However, the need is deeper than this. In many situations the camera parameters might vary because of thermal variations, or because the zoom or focus setting has been adjusted. It is impracticable to keep recalibrating a camera using accurately made test objects. Finally, if multiple (e.g., stereo) cameras are used, each will have to be calibrated separately, and the results compared to minimize the combined error. It is far better to examine the system as a whole, and to calibrate it on the real scenes that are being viewed.

We have already encountered some aspects of these aspirations, in the form of invariants that are obtained in sequence by a single camera. For example, if a series of four collinear points are viewed and their cross ratio is checked, it will be found to be constant as the camera moves forward, changes orientation, or views the points increasingly obliquely—as long as they all remain within the field of view. For this purpose, all that is required to perform the recognition and maintain awareness of the object (the four points) is an uncalibrated but distortion-free camera. By distortion-free here we mean not the ability to correct perspective distortion—which is, after all, the function of the cross ratio invariant—but the lack of radial distortion, or at least the capability in the following software for eliminating it (see Section 21.5).

To understand how image interpretation can be carried out more generally, using multiple views—whether from the same camera moved to a variety of places, or multiple cameras with overlapping views of the world—we shall need to go back to basics and start afresh with a more general attack on concepts such as binocular vision and epipolar constraints. In particular, two important matrices will be called into play—the "essential" matrix and the "fundamental" matrix. We start with the essential matrix and then generalize the idea to the fundamental matrix. But first we need to look at the geometry of two cameras with general views of the world.

21.7 Generalized Epipolar Geometry

In Section 16.3, we considered the stereo correspondence problem and had already simplified the task by choosing two cameras whose image planes were not only parallel but in the same plane. This made the geometry of depth perception especially simple but suppressed possibilities allowed for in the human visual system (HVS) of having a nonzero vergence angle between the two images. The HVS is special in adjusting vergence so that the current focus of attention in the field of view has almost zero disparity between the two images. It seems likely that the HVS estimates depth not merely by measuring disparity but by measuring the vergence together with remanent small variations in disparity.

Here we generalize the situation to cover the possibility of disparity coupled with substantial vergence. Figure 21.4 shows the revised geometry.

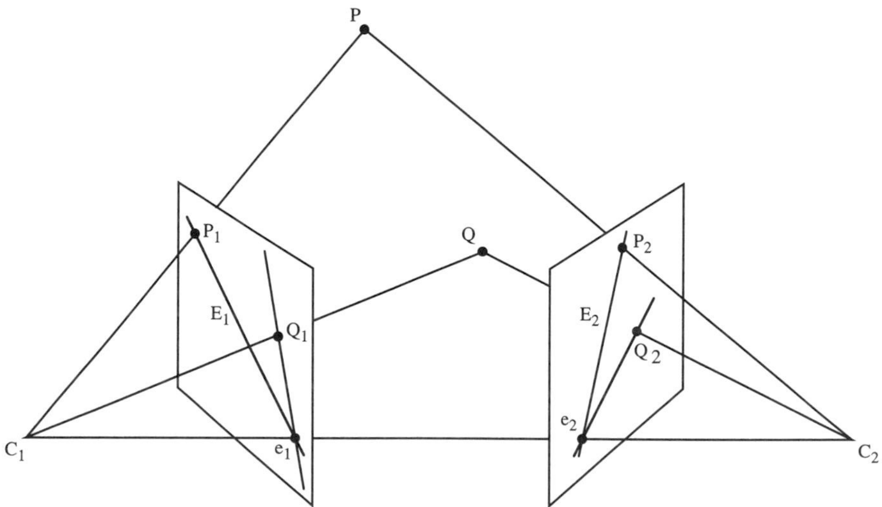

Figure 21.4 Generalized imaging of a scene from two viewpoints. In this case, there is substantial vergence. All epipolar lines in the left image pass through epipole e_1; of these, only E_1 is shown. Similar comments apply for the right image.

Note first that observation of a real point P in the scene leads to points P_1 and P_2 in the two images; that P_1 could correspond to any point on the epipolar line E_2 in image 2; and similarly, that point P_2 could correspond to any point on the epipolar line E_1 in image 1. The so-called epipolar plane of P is the plane containing P and the projection points C_1 and C_2 of the two cameras. The epipolar lines (see Section 16.3) are thus the straight lines in which this plane cuts the two image planes. Furthermore, the line joining C_1 and C_2 cuts the image planes in the so-called epipoles e_1 and e_2. These can be regarded as the images of the alternate camera projection points. Note that all epipolar planes pass through points C_1, C_2 and e_1, e_2. This means that all epipolar lines in the two images pass through the respective epipoles. However, if the vergence angle were zero (as in Fig. 16.5), the epipoles would be at infinity in either direction and all epipolar lines in either image would be parallel, and indeed parallel to the vector **C** from C_1 to C_2.

21.8 The Essential Matrix

In this section we start with the vectors $\mathbf{P}_1$, $\mathbf{P}_2$, from C_1, C_2 to P, and also the vector **C** from C_1 to C_2. Vector subtraction gives:

$$\mathbf{P}_2 = \mathbf{P}_1 - \mathbf{C} \tag{21.38}$$

We also know that $\mathbf{P}_1$, $\mathbf{P}_2$ and **C** are coplanar, the condition of coplanarity being:[5]

$$\mathbf{P}_2 \cdot \mathbf{C} \times \mathbf{P}_1 = 0 \tag{21.39}$$

To progress, we need to relate the vectors $\mathbf{P}_1$ and $\mathbf{P}_2$ when these are expressed relative to their own frames of reference. If we take these vectors as having been defined in the C_1 frame of reference, we now reexpress $\mathbf{P}_2$ in its own (C_2) frame of reference, by applying a translation **C** and a rotation of coordinates expressed as the orthogonal matrix R. This leads to:

$$\mathbf{P}_2' = R\mathbf{P}_2 = R(\mathbf{P}_1 - \mathbf{C}) \tag{21.40}$$

so that:

$$\mathbf{P}_2 = R^{-1}\mathbf{P}_2' = R^{\mathrm{T}}\mathbf{P}_2' \tag{21.41}$$

5 This can be thought of as bringing to zero the volume of the parallelepiped with sides $\mathbf{P}_1$, $\mathbf{P}_2$, and **C**.

Substituting in the coplanarity condition gives:

$$(R^{\mathsf{T}}\mathbf{P}_2') \cdot \mathbf{C} \times \mathbf{P}_1 = 0 \tag{21.42}$$

At this point it is useful to replace the vector product notation by using a skew-symmetrical matrix $C_\times$ to denote $\mathbf{C}\,\times$, where:

$$C_\times = \begin{bmatrix} 0 & -C_z & C_y \\ C_z & 0 & -C_x \\ -C_y & C_x & 0 \end{bmatrix} \tag{21.43}$$

At the same time we observe the correct matrix formulation of all the vectors by transposing appropriately. We now find that:

$$(R^{\mathsf{T}}\mathbf{P}_2')^{\mathsf{T}} C_\times \mathbf{P}_1 = 0 \tag{21.44}$$

$$\therefore \quad \mathbf{P}_2'^{\mathsf{T}} R C_\times \mathbf{P}_1 = 0 \tag{21.45}$$

Finally, we obtain the "essential matrix" formulation:

$$\mathbf{P}_2'^{\mathsf{T}} E \mathbf{P}_1 = 0 \tag{21.46}$$

where the essential matrix has been found to be:

$$E = RC_\times \tag{21.47}$$

Equation (21.46) is actually the desired result: it expresses the relation between the observed positions of the same point in the two camera frames of reference. Furthermore, it immediately leads to formulas for the epipolar lines. To see this, first note that in the C_1 camera frame:

$$\mathbf{p}_1 = (f_1/Z_1)\mathbf{P}_1 \tag{21.48}$$

while in the C_2 camera frame (and expressed in terms of that frame of reference):

$$\mathbf{p}_2' = (f_2/Z_2)\mathbf{P}_2' \tag{21.49}$$

Eliminating $\mathbf{P}_1$ and $\mathbf{P}_2'$, and dropping the prime (as within the respective image planes the numbers 1 and 2 are sufficient to specify the coordinates unambiguously), we find:

$$\mathbf{p}_2^{\mathsf{T}} E \mathbf{p}_1 = 0 \tag{21.50}$$

as Z_1, Z_2 and f_1, f_2 can be cancelled from this matrix equation.

Now note that writing $\mathbf{p}_2^\mathrm{T} E = \mathbf{l}_1^\mathrm{T}$ and $\mathbf{l}_2 = E\,\mathbf{p}_1$ leads to the following relations:

$$\mathbf{p}_1^\mathrm{T}\mathbf{l}_1 = 0 \tag{21.51}$$

$$\mathbf{p}_2^\mathrm{T}\mathbf{l}_2 = 0 \tag{21.52}$$

This means that $\mathbf{l}_2 = E\,\mathbf{p}_1$ and $\mathbf{l}_1 = E^\mathrm{T}\mathbf{p}_2$ are the epipolar lines corresponding to $\mathbf{p}_1$ and $\mathbf{p}_2$, respectively.[6]

Finally, we can find the epipoles from the above formulation. In fact, the epipole lies on every epipolar line within the same image. Thus, $\mathbf{e}_2$ satisfies (can be substituted for $\mathbf{p}_2$ in) equation (21.52), and hence:

$$\mathbf{e}_2^\mathrm{T}\mathbf{l}_2 = 0$$

$$\therefore\ \mathbf{e}_2^\mathrm{T} E\,\mathbf{p}_1 = 0 \text{ for all } \mathbf{p}_1.$$

This means that $\mathbf{e}_2^\mathrm{T} E = 0$, i.e., $E^\mathrm{T}\mathbf{e}_2 = 0$. Similarly, $E\,\mathbf{e}_1 = 0$.

21.9 The Fundamental Matrix

Notice that in the last part of the essential matrix calculation, we implicitly assumed that the cameras are correctly calibrated. Specifically, $\mathbf{p}_1$ and $\mathbf{p}_2$ are corrected (calibrated) image coordinates. However, there is a need to work with uncalibrated images, using the raw pixel measurements[7]—for all the reasons given in Section 21.6. Applying the camera intrinsic matrices G_1, G_2 to the calibrated image coordinates (Section 21.4), we get the raw image coordinates:

$$\mathbf{q}_1 = G_1\mathbf{p}_1 \tag{21.53}$$

$$\mathbf{q}_2 = G_2\mathbf{p}_2 \tag{21.54}$$

6 Consider a line $\mathbf{l}$ and a point $\mathbf{p}$: $\mathbf{p}^\mathrm{T}\mathbf{l} = 0$ means that $\mathbf{p}$ lies on the line $\mathbf{l}$, or dually, $\mathbf{l}$ passes through the point $\mathbf{p}$.

7 However, any radial distortions need to be eliminated, so as to *idealize* the camera, but not to calibrate it in the sense of Sections 21.3 and 21.4.

Here we need to go in the reverse direction, so we use the inverse equations:

$$\mathbf{p}_1 = G_1^{-1}\mathbf{q}_1 \tag{21.55}$$

$$\mathbf{p}_2 = G_2^{-1}\mathbf{q}_2 \tag{21.56}$$

Substituting for $\mathbf{p}_1$ and $\mathbf{p}_2$ in equation (21.50), we find the desired equation linking the raw pixel coordinates:

$$\mathbf{q}_2^{\mathrm{T}}(G_2^{-1})^{\mathrm{T}} E G_1^{-1}\mathbf{q}_1 = 0 \tag{21.57}$$

which can be expressed as:

$$\mathbf{q}_2^{\mathrm{T}} F \mathbf{q}_1 = 0 \tag{21.58}$$

where

$$F = (G_2^{-1})^{\mathrm{T}} E G_1^{-1} \tag{21.59}$$

F is defined as the "fundamental matrix." Because it contains all the information that would be needed to calibrate the cameras, it contains more free parameters than the essential matrix. However, in other respects the two matrices are intended to convey the same basic information, as is confirmed by the resemblance between the two formulations—see equations (21.50) and (21.58).

Finally, just as in the case of the essential matrix, the epipoles are given by $F\mathbf{f}_1 = 0$ and $F^{\mathrm{T}}\mathbf{f}_2 = 0$, though this time in raw image coordinates $\mathbf{f}_1$ and $\mathbf{f}_2$.

21.10 Properties of the Essential and Fundamental Matrices

Next we consider the composition of the essential and fundamental matrices. In particular, note that $C_\times$ is a factor of E and also, indirectly, of F. They are homogeneous in $C_\times$, so the scale of $\mathbf{C}$ will make no difference to the two matrix formulations (equations (21.46) and (21.58)), with only the *direction* of $\mathbf{C}$ being important. The scales of both E and F are immaterial, and as a result only the relative values of their coefficients are of importance. This means that there are at most only eight independent coefficients in E and F. In the case of F there are only *seven*, as $C_\times$ is skew-symmetrical, and this ensures that it has rank 2 rather than rank 3—a property that is passed on to F. The same argument applies for E, but it turns out that the lower complexity of E (by virtue of its not containing the image calibration information) means that it has only *five* free parameters. In the latter case, it is easy to see what they are: they arise from the original three translation (**C**) and three rotation (**R**) parameters, less the one parameter corresponding to scale.

In this context, it should be noted that if **C** arises from a translation of a single camera, the same essential matrix will result whatever the scale of **C**. Only the direction of **C** actually matters, and the same epipolar lines will result from continued motion in the same direction. In this case we can interpret the epipoles as foci of expansion or contraction. This underlines the power of this formulation: specifically, it treats motion and displacement as a single entity.

Finally, we should try to understand why there are seven free parameters in the fundamental matrix. The solution is relatively simple. Each epipole requires two parameters to specify it. In addition, three parameters are needed to map any three epipolar lines from one image to the other. But why do just three epipolar lines have to be mapped? This is because the family of epipolar lines is a pencil whose orientations are related by cross ratios, so once three epipolar lines have been specified the mapping of any other can be deduced. (Knowing the properties of the cross ratio, it is seen that fewer than three epipolar lines would be insufficient and that more than three would yield no additional information.) This fact is sometimes stated in the following form: a homography (a projective transformation) between two 1-D projective spaces has three degrees of freedom.

21.11 Estimating the Fundamental Matrix

In the previous section we showed that the fundamental matrix has seven free parameters. This means that it ought to be possible to estimate it by identifying the same seven features in the two images.[8] Although this is mathematically possible in principle, and a suitable nonlinear algorithm has been devised by Faugeras et al. (1992) to implement it, it has been shown that the computation can be numerically unstable. Essentially, noise acts as an additional variable boosting the effective number of degrees of freedom in the problem to eight. However, a linear algorithm called the *eight-point algorithm* has been devised to overcome the problem. Curiously, Longuet-Higgins (1981) proposed this algorithm many years earlier to estimate the *essential* matrix, but it came into its own when Hartley (1995) showed how to control the errors by first

8 Using the minimum number of points in this way carries the health warning that they must be in general position. Special configurations of points can lead to numerical instabilities in the computations, total failure to converge, or unnecessary ambiguities in the results. In general, coplanar points are to be avoided.

normalizing the values. In addition, by using more than eight points, increased accuracy can be attained, but then a suitable algorithm must be found that can cope with the now overdetermined parameters. Principal component analysis can be used for this purpose, an appropriate procedure being singular value decomposition (SVD).

Apart from noise, gross mismatches in forming trial point correspondences between images can be a source of practical problems. If so, the normal least-squares types of solution can profitably be replaced by the least median of squares robust estimation method (Appendix A).

21.12 Image Rectification

In Section 21.7 we took some pains to generalize the epipolar approach, and we subsequently arrived at general solutions, corresponding to arbitrary overlapping views of scenes. However, there are distinct advantages in special views obtained from cameras with parallel axes—as in the case of Fig. 16.5 where the vergence is zero. Specifically, it is easier to find correspondences between scenes that are closely related in this way. Unfortunately, such well-prepared pairs of images are not in keeping with the aims promoted in Section 21.6, of insisting upon closely aligned and calibrated cameras. The frames taken by a single moving camera certainly do not match well unless the camera's motion is severely constrained by special means. The solution is straightforward: take images with uncalibrated cameras, estimate the fundamental matrix, and then apply suitable linear transformations to compute the images for any desired idealized camera positions. This technique is called image rectification and ensures, for example, that the epipolar lines are all parallel to the baseline **C** between the centers of projection. This then results in correspondences being found by searching along points with the same ordinate in the alternate image. For a point with coordinates (x_1, y_1) in the first image, search for a matching point (x_2, y_1) in the second image.

When rectifying an image, it will in general be rotated in 3-D,[9] and the obvious way of doing so is to transfer each individual pixel to its new location in the rectified image. However, rotations are nonlinear processes and will in some cases have the effect of mapping several pixels into a single pixel. Furthermore, a number of pixels may well not have intensity values assigned to them. Although the first of these problems could be tackled by an intensity averaging process, and

9 Of course, it may also be translated and scaled, in which case the effect described here may be even more significant.

the second problem by applying a median or other type of filter to the transformed image, such techniques are insufficiently thoroughgoing to provide accurate, reliable solutions. The *proper* way of overcoming these intrinsic difficulties is to back-project the pixel locations from the transformed image space to the source image, use interpolation to compute the ideal pixel intensities, and then transfer these intensities to the transformed image space.

Bilinear interpolation is used most often in the transformation process. This works by performing interpolation in the x-direction and then in the y-direction. Thus, if the location to be interpolated is $(x+a, y+b)$ where x and y are integer pixel locations, and $0 \leq a, b \leq 1$, then the interpolated intensities in the x-direction are:

$$I(x+a, y) = (1-a)I(x, y) + aI(x+1, y) \tag{21.60}$$

$$I(x+a, y+1) = (1-a)I(x, y+1) + aI(x+1, y+1) \tag{21.61}$$

and the final result after interpolating in the y-direction is:

$$\begin{aligned} I(x+a, y+b) = (1-a)(1-b)I(x, y) + a(1-b)I(x+1, y) \\ + (1-a)bI(x, y+1) + abI(x+1, y+1) \end{aligned} \tag{21.62}$$

The symmetry of the result shows that it makes no difference which axis is chosen for the first pair of interpolations, and this fact limits the arbitrariness of the method. Note that the method does not assume a locally planar intensity variation in 2-D—as is clear because the value of the $I(x+1, y+1)$ intensity is taken into account as well as the other three intensity values. Nevertheless, bilinear interpolation is not a totally ideal solution, for it takes no account of the sampling theorem. For this reason the bi-cubic interpolation method (which involves more computation) is sometimes used instead. In addition, all such methods introduce slight local blurring of the image as they involve averaging of local intensity values. Overall, transformation processes such as this one are bound to result in slight degradation of the image data.

21.13 3-D Reconstruction

Section 21.10 strongly emphasized that F is determined only up to an unknown scale factor (or equivalently that the actual scales of its coefficients as obtained are arbitrary), thus reflecting the deliberate avoidance of camera calibration in this work. In practice this means that if the results of computations of F are to be related back to the real world, the scaling factor must be reinstated. In principle, scaling can be achieved by viewing a single yardstick. It is unnecessary to view an object such as a Rubik cube, for knowledge of F carries with it a lot of information on relative dimensions in the real world. This factor is important when reconstructing a real scene with a real depth map.

A number of methods are avaliable for image reconstruction, of which perhaps the most obvious is triangulation. This method starts by taking two camera positions containing normalized images, and projecting rays for a given point P back into the real world until they meet. Attempting to do this presents an immediate problem: the inaccuracies in the available parameters, coupled with the pixellation of the images, ensure that in most cases rays will not actually meet, for they are skew lines. The best that can be done with skew lines is to determine the position of closest approach. Once this position has been found, the bisector of the line of closest approach (which is perpendicular to each of the rays) is, in *this* model, the most accurate estimate of the position of P in space.

Unfortunately, this model is not guaranteed to give the most accurate prediction of the position of P. This is because perspective projection is a highly nonlinear process. In particular, slight misjudgment of the orientation of the point from either of the images can cause a substantial depth error, coupled with a significant lateral error. So much is indeed obvious from Fig. 21.5. This being so, we must ask where the error might still be linear, so that, at that position at least, error calculation can be based on Gaussian distributions.[10] The errors can be taken to be approximately Gaussian in the images themselves. Thus, the point in space that has to be chosen as representing the most accurate interpretation of the data is that which results in the minimum error (in a least mean square sense) when reprojected onto the image planes. Typically, the error obtained using this approach is a factor of two smaller than that for the triangulation method described earlier (Hartley and Zisserman, 2000).

Finally, it is useful to mention another type of error that can arise with two cameras. This applies when they both view an object with a smoothly varying boundary. For example, if both cameras are viewing the right-hand edge of a vase of circular cross section, each will see a different point on the boundary and a discrepancy will arise in the estimated boundary position (Fig. 21.6). It is left as an exercise (Section 21.17) to determine the exact magnitude of such errors. The error is proportional both to a, where a is the local radius of curvature of the observed boundary, and to Z^{-2}, where Z is the depth in the scene. This means that the error (and the percentage error) tends to zero at large distances, and that the error falls properly to zero for sharp corners.

---

10 Here we ignore the possibility of gross errors arising from mismatches between images, which is the subject of further discussion in Section 21.11 and elsewhere.

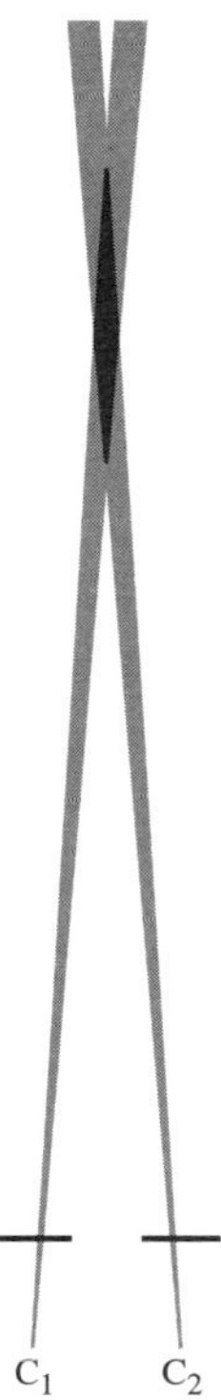

Figure 21.5 Error in locating a feature in space using binocular imaging. The dark shaded regions represent the regions of space that could arise for small errors in the image planes. The crossover region, shaded black, confirms that longitudinal errors will be much larger than lateral errors. A full analysis would involve applying Gaussian or other error functions (see text).

21.14 An Update on the 8-point Algorithm

Section 21.11 outlined the value of the 8-point algorithm for estimating the fundamental matrix. Over a period of about eight years (1995–2003), this became the standard solution to the problem. However, a key contribution by Torr and Fitzgibbon (2003, 2004) has shown that the 8-point algorithm might not after all be the best possible method, since the solutions it obtains depend on the particular coordinate system used for the computation. This is because the normalization generally used, namely, $\sum_i f_i^2 = 1$,[11] is not invariant to shifts in the coordinate

11 Early on, Tsai and Huang (1984) suggested the normalization $f_9 = 1$, but this leads to biased solutions, and, for example, excludes solutions with $f_9 = 0$.

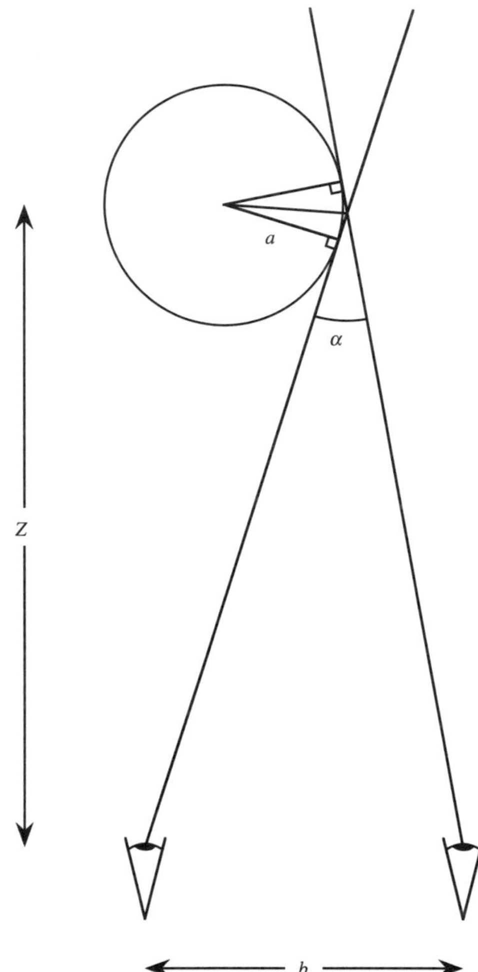

Figure 21.6 Lateral estimation error arising with a smoothly varying boundary. The error arises in estimating the boundary position when information from two views is fused in the standard way. a is the radius of a vase being observed, α is the disparity in direction of its right hand boundary, Z is its depth in the scene, and b is the stereo baseline.

system. It is by no means obvious how to find an invariant normalization. Nevertheless, Torr and Fitzgibbon's logical analysis of the situation, in which they were forced to disregard the affine transform case appropriate for weak perspective, led to the following normalization of F:

$$f_1^2 + f_2^2 + f_4^2 + f_5^2 = K \tag{21.63}$$

where K is a constant and

$$F = \begin{bmatrix} f_1 & f_2 & f_3 \\ f_4 & f_5 & f_6 \\ f_7 & f_8 & f_9 \end{bmatrix} \tag{21.64}$$

Finally, to determine F, equation (21.63) can be applied as a Lagrangian multiplier constraint, which leads to an eigenvector solution for F. Overall, the 8×8 eigenvalue problem solved by the 8-point algorithm is replaced by a 5×5 eigenvalue problem. This approach not only yields the required invariance properties, thus ensuring a more accurate solution, but also gives a much faster computation that loses significantly fewer tracks in image sequence analysis.

21.15 Concluding Remarks

This chapter has discussed the transformations required for camera calibration and has outlined how calibration can be achieved. The camera parameters have been classified as internal and external, thereby simplifying the conceptual problem and throwing light on the origins of errors in the system. It has been shown that a minimum of 6 points are required to perform calibration in the general case where 11 transformation parameters are involved. However, the number of points required might be reduced somewhat in special cases, for example, where the sensor is known to be Euclidean. Nevertheless, it is normally more important to increase the number of points used for calibration than to attempt to reduce it, since substantial gains in accuracy can be obtained via the resulting averaging process.

In an apparent break with the previous work, Section 21.5 introduced multiple-view vision. This important topic was seen to rest on generalized epipolar geometry and led to the essential and fundamental matrix formulations, which relate the observed positions of any point in two camera frames of reference. The importance of the 8-point algorithm for estimating either of these matrices—and particularly the fundamental matrix, which is relevant when the cameras are uncalibrated—was stressed. In addition, the need for accuracy in estimating the fundamental matrix is still a research issue, with at least one new view of how it can be improved having been published as late as 2004 (Torr and Fitzgibbon, 2003, 2004).

The obvious way of tackling vision problems is to set up a camera and calibrate it, and only then to use it in anger. This chapter has shown how, to a large extent, calibration can be avoided or carried out adaptively "on the fly," by performing multiple-view vision and analyzing the various key matrices that arise from the generalized epipolar problems.

21.16 Bibliographical and Historical Notes

One of the first to use the various transformations described in this chapter was Roberts (1965). Important early references for camera calibration are the *Manual of Photogrammetry* (Slama, 1980), Tsai and Huang (1984), and Tsai (1986). Tsai's paper is especially useful in that it provides an extended and highly effective treatment that copes with nonlinear lens distortions. More recent papers on this topic include Haralick (1989), Crowley et al. (1993), Cumani and Guiducci (1995), and Robert (1996): see also Zhang (1995). Note that parametrized plane curves can be used instead of points for the purpose of camera calibration (Haralick and Chu, 1984).

Camera calibration is an old topic that is revisited every time 3-D vision has to be used for measurement, and otherwise when rigorous analysis of 3-D scenes is called for. The calibration scenario started to undergo a metamorphosis in the early 1990s, when it was realized that much could be learned without overt calibration, but rather by *comparing* images taken from moving sequences or from multiple views (Faugeras, 1992; Faugeras et al., 1992; Hartley, 1992; Maybank and Faugeras, 1992). Although it was appreciated that much could be learned without overt calibration, at that stage it was not known how much *might* be learned, and there ensued a rapid sequence of developments as the frontiers were progressively pushed back (e.g., Hartley, 1995, 1997; Luong and Faugeras, 1997). By the late 1990s, the fast evolution phase was over, and definitive, albeit quite complex, texts appeared covering these developments (Hartley and Zisserman, 2000; Faugeras and Luong, 2001; Gruen and Huang, eds., 2001). Nevertheless, many refinements of the standard methods were still emerging (Faugeras et al., 2000; Heikkilä, 2000; Sturm, 2000; Roth and Whitehead, 2002). It is in this light that the innovative insights of Torr and Fitzgibbon (2003, 2004) and Chojnacki et al. (2003) expressing similar but not identical sentiments relating to the 8-point algorithm should be considered.

In retrospect, it is amusing that the early, incisive paper by Longuet-Higgins (1981) presaged many of these developments. Although his 8-point algorithm applied specifically to the essential matrix, it was only much later (Faugeras, 1992; Hartley, 1992) that it was applied to the fundamental matrix, and even later, in a crucial step, that its accuracy was greatly improved by prenormalizing the image data (Hartley, 1997). As we have already noted, the 8-point algorithm continues to be a focus for new research.

21.17 Problems

1. For a two-camera stereo system, obtain a formula for the depth error that arises for a given error in disparity. Hence, show that the percentage error in depth is numerically equal to the percentage error in disparity. What does this result mean in practical terms? How does the pixellation of the image affect the result?

2. A cylindrical vase with a circular cross section of local radius a is viewed by two cameras (Fig. 21.6). Obtain a formula giving the error δ in the estimated position of the boundary of the vase. Simplify the calculation by assuming that the boundary is on the perpendicular bisector of the line joining the centers of projection of the two cameras, and hence find α (Fig. 21.6) in terms of b and Z. Determine δ in terms of α and then substitute for α from the previous formula. Hence, justify the statements made at the end of Section 21.13.

3. Discuss the potential advantages of trinocular vision in the light of the theory presented in Section 21.8. What would be the best placement for a third camera? Where should the third camera *not* be placed? Would any gain be achieved by incorporating even more views of a scene?

PART 4

Toward Real-Time Pattern Recognition Systems

Part 4 aims to cover the family of needs associated with the production of practical visual pattern recognition systems. Not least among these needs are the requirements of error minimization, real-time operation, and system integration. Hence, this part of the book must cover a disparate variety of topics ranging from analysis of 2-D and 3-D scenes, and the methods of statistical pattern recognition, to applications such as automated visual inspection and assembly, and consideration of the lighting and hardware systems required to acquire the images and perform the necessary processing in real time. The chosen topics are so variegated, and interact with each other at so many different levels, that there can be no ideal order of presentation. Nevertheless, the intelligent reader should not have too much difficulty finding relevant information.

Automated Visual Inspection

Humans are good at searching for the unusual and locating faults in manufactured products, but they rapidly tire when large numbers of items have to be scrutinized. Automated visual inspection seeks to achieve 100% untiring inspection and control of quality. This chapter describes the processes and principles needed to achieve this end, the main limitation being the impossibility of covering the full variety of products in a single chapter.

Look out for:

- the variety of products to be inspected.
- the main categories of inspection.
- how deviations relative to a standard template can be measured.
- the methodology for scrutinizing circular objects.
- the problems of inspecting products exhibiting high levels of variability.
- the principles of X-ray inspection.
- the importance of color in inspection.

This chapter seeks to give a broad view of inspection and is counterbalanced by the following chapter which covers a particular application area in more depth. This approach is appropriate because case studies provide one of the best ways to extend the work to the wide variety of products.

CHAPTER 22

Automated Visual Inspection

22.1 Introduction

Over the past two decades, machine vision has consolidated its early promise and has become a vital component in the design of advanced manufacturing systems. On the one hand, it provides a means of maintaining control of quality during manufacture, and on the other, it is able to feed the assembly robot with the right sort of information to construct complex products from sets of basic components. Automated assembly helps to make flexible manufacture a reality, so that costs due to underuse of expensive production lines are virtually eliminated.

These two major applications of vision—automated visual inspection and automated assembly—have many commonalities and can on the whole be performed by similar hardware vision systems employing closely related algorithms. Perhaps the most obvious use of visual inspection is to check products for quality so that defective ones may be rejected. This is easy to visualize in the case of a line making biscuits or washers at rates of the order of one million per day. Another area of application of inspection is to measure a specific parameter for each product and to feed its value back to an earlier stage in the plant in order to "close the loop" of the manufacturing process. A typical application is to adjust the temperature of jam or chocolate in a biscuit factory when the coating is found not to be spreading correctly. A third use of inspection is simply to gather statistics on the efficiency of the manufacturing process, finding, for example, how product diameters vary, in order to provide information that will help management with advance planning.

Another aspect of inspection is what can be learned via other modalities such as X-rays, even if the acquired images often ressemble visible light images. Similarly, it is relevant to ask what additional information color can provide that is useful for inspection. Sections 22.10 and 22.11 aim to give answers to these questions.

In automated assembly, vision can provide feedback to control the robot arm and wrist. For this purpose it needs to provide detailed information on the positions and orientations of objects within the field of view. It also needs to be able to distinguish individual components within the field of view. In addition, a well-designed vision system will be able to check components before assembly, for example, so as to prevent the robot from trying to fit a screw into a nonexistent hole.

Inspection and assembly require virtually identical vision systems, the most notable difference often being that a linescan camera is used for inspecting components on a conveyor, whereas an area (whole picture) camera is required for assembly operations on a worktable. The following discussion centers on inspection, although, because of the similarity of the two types of application, many of the concepts that are developed are also useful for automated assembly tasks.

22.2 The Process of Inspection

Inspection is the process of comparing individual manufactured items against some preestablished standard with a view to maintenance of quality. Before proceeding to study inspection tasks in detail, it is useful to note that the process of inspection commonly takes place in three definable stages:

1. Image acquisition
2. Object location
3. Object scrutiny and measurement

We defer detailed discussion of image acquisition until Chapter 27 and comment here on the relevance of separating the processes of location and scrutiny. This is important because (either on a worktable or on a conveyor) large numbers of pixels usually have to be examined before a particular product is found, whereas once it has been located, its image frequently contains relatively few pixels, and so rather little computational effort need be expended in scrutinizing and measuring it. For example, on a biscuit line, products may be separated by several times the product diameter in two dimensions, so that some 100,000 pixels may need to be examined to locate products occupying say 5000 pixels. Product location is therefore likely to be a much more computationally intensive problem than product scrutiny. Although this is generally true, sampling techniques may permit object location to be performed with much increased efficiency (Chapter 10). Under these circumstances, location may be faster than scrutiny, since the latter process, though straightforward, tends to permit no shortcuts and requires all pixels to be examined.

22.3 Review of the Types of Objects to Be Inspected

Before studying methods of visual inspection (including those required for assembly applications), we should consider the types of objects with which such systems may have to cope. As an example, we can take two rather opposing categories: (1) goods such as food products that are subject to wide variation during manufacture but for which physical appearance is an important factor, and (2) those products such as precision metal parts which are needed in the electronics and automotive industries. The problems specific to each category are discussed first; then size measurement and the problem of 3-D inspection are considered briefly.

22.3.1 *Food Products*

Food products are a particularly wide category, ranging from chocolate cream biscuits to pizzas, and from frozen food packs to complete set meals (as provided by airlines). In the food industry, the trend is toward products of high added value. Logically, such products should be inspected at every stage of manufacture, so that further value is not added to products that are already deficient. However, inspection systems are still quite expensive, and the tendency is to inspect only at the end of the product line. This procedure at least ensures (1) that the final appearance is acceptable and (2) that the size of the product is within the range required by the packaging machine.[1] This strategy is reasonable for many products—as for some types of biscuit where a layer of jam is clearly discernible underneath a layer of chocolate. Pizzas exemplify another category wherein many additives appear on top of the final product, all of which are in principle detectable by a vision system at the end of the product line.

When checking the shapes of chocolate products (chocolates, chocolate bars, chocolate biscuits, etc.), a particular complication that arises is the "footing" around the base of the product. This footing is often quite jagged and makes it difficult to recognize the product or to determine its orientation. However,

1 On many food lines, jamming of packaging machines due to oversize products is one of the major problems.

the eye generally has little difficulty with this task, and hence if a robot is to place a chocolate in its proper place in a box it will have to emulate the eye and employ a full gray-scale image; shortcuts with silhouettes in binary images are unlikely to work well. In this context, it should be observed that chocolate is one of the more expensive ingredients of biscuits and cakes. A frequently recurring inspection problem is to check that chocolate cover is sufficient to please the consumer while low enough to maintain adequate profit margins.

Returning to packaged meals, we find that these present both an inspection and an assembly problem. A robot or other mechanism has to place individual items on the plastic tray, and it is clearly preferable that every item should be checked to ensure, for example, that each salad contains an olive or that each cake has the requisite blob of cream.

22.3.2 *Precision Components*

Many other parts of industry have also progressed to the automatic manufacture and assembly of complex products. It is necessary for items such as washers and O-rings to be tested for size and roundness, for screws to be checked for the presence of a thread, and for mains plugs to be examined for the appropriate pins, fuses, and screws. Engines and brake assemblies also have to be checked for numerous possible faults. Perhaps the worst problems arise when items such as flanges or slots are missing, so that further components cannot be fitted properly. It cannot be emphasized enough that what is missing is at least as important as what is present. Missing holes and threads can effectively prevent proper assembly. It is sometimes stated that checking the pitch of a screw thread is unnecessary—if a thread is present, it is bound to be correct. However, there are many industrial applications where this is not true.

Table 22.1 summarizes some of the common features that need to be checked when dealing with individual precision components. Note that measurement of the extent of any defect, together with knowledge of its inherent seriousness, should permit components to be graded according to quality, thereby saving money for the manufacturer. (Rejecting all defective items is a very crude option.)

22.3.3 *Differing Requirements for Size Measurement*

Size measurement is important both in the food industry and in the automotive and small-parts industry. However, the problems in the two cases are often rather

Table 22.1 Features to be checked on precision components

Dimensions within specified tolerances
Correct positioning, orientation, and alignment
Correct shape, especially roundness, of objects and holes
Whether corners are misshapen, blunted, or chipped
Presence of holes, slots, screws, rivets, and so on
Presence of a thread on screws
Presence of burr and swarf
Pits, scratches, cracks, wear, and other surface marks
Quality of surface finish and texture
Continuity of seams, folds, laps, and other joins

different. For example, the diameter of a biscuit can vary within quite wide limits (~5%) without giving rise to undue problems, but when it gets outside this range there is a serious risk of jamming the packing machine, and the situation must be monitored carefully. In contrast, for mechanical parts, the required precision can vary from 1% for objects such as O-rings to 0.01% for piston heads. This variation makes it difficult to design a truly general-purpose inspection system. However, the manufacturing process often permits little variation in size from one item to the next. Hence, it may be adequate to have a system that is capable of measuring to an accuracy of rather better than 1%, as long as it is capable of checking all the characteristics mentioned in Table 22.1.

When high precision is vital, accuracy of measurement should be proportional to the resolution of the input image. Currently, images of up to 512×512 pixels are common, so accuracy of measurement is basically of the order of 0.2%. Fortunately, gray-scale images provide a means of obtaining significantly greater accuracy than indicated by the above arguments, since the exact transition from dark to light at the boundary of an object can be estimated more closely. In addition, averaging techniques (e.g., along the side of a rectangular block of metal) permit accuracies to be increased even further—by a factor $\sqrt{N}$ if N pixel measurements are made. These factors permit measurements to be made to subpixel resolution, sometimes even down to 0.1 pixel.

22.3.4 *Three-dimensional Objects*

All real objects are 3-D, although the cost of setting up an inspection station frequently demands that they be examined from one viewpoint in a single 2-D image. This requirement is highly restrictive and in many cases overrestrictive. Nevertheless, generally an enormous amount of useful checking and measurement can be done from one such image. The clue that this is possible lies in the prodigious capability of the human eye—for example, to detect at a glance from the play of light on a surface whether or not it is flat. Furthermore, in many cases products are essentially flat, and the information that we are trying to find out about them is simply expressible via their shape or via the presence of some other feature that is detectable in a 2-D image. In cases where 3-D information is required, methods exist for obtaining it from one or more images, for example, via binocular vision or structured lighting, as has already been seen in Chapter 16. More is said about this in the following sections.

22.3.5 *Other Products and Materials for Inspection*

This subsection briefly mentions a few types of products and materials that are not fully covered in the foregoing discussion. First, electronic components are increasingly having to be inspected during manufacture, and, of these, printed circuit boards (PCBs) and integrated circuits are subject to their own special problems, which are currently receiving considerable attention. Second, steel strip and wood inspection are also very important. Third, bottle and glass inspection has its own particular intricacies because of the nature of the material. Glints are a relevant factor—as it also is in the case of inspection of cellophane-covered foodpacks. In this chapter, space permits only a short discussion of some of these topics (see Sections 22.7 and 22.8).

22.4 Summary—The Main Categories of Inspection

The preceding sections have given a general review of the problems of inspection but have not shown how they might be solved. This section takes the analysis a stage further. First, note that the items in Table 22.1 may be classified as *geometrical* (measurement of size and shape—in 2-D or 3-D as necessary), *structural* (whether there are any missing items or foreign objects), and *superficial* (whether the surface has adequate quality). As is evident from Table 22.1, these three categories are not completely distinct, but they are useful for the following discussion.

Start by noting that the methods of object location are also inherently capable of providing geometrical measurements. Distances between relevant edges, holes, and corners can be measured; shapes of boundaries can be checked both absolutely and via their salient features; and alignments can usually be checked quite simply, for example, by finding how closely various straight-line segments fit to the sides of a suitably placed rectangle. In addition, shapes of 3-D surfaces can be mapped out by binocular vision, photometric stereo, structured lighting, or other means (see Chapter 16) and subsequently checked for accuracy.

Structural tests can also be made once objects have been located, assuming that a database of the features they are supposed to possess is available. In the latter case, it is necessary merely to check whether the features are present in predicted positions. As for foreign objects, these can be looked for via unconstrained search as objects in their own right. Alternatively, they may be found as differences between objects and their idealized forms, as predicted from templates or other data in the database. In either case, the problem is very data-dependent, and an optimal solution needs to be found for each situation. For example, scratches may be searched for directly as straight-line segments.

Tests of surface quality are perhaps more complex. Chapter 27 describes methods of lighting that illuminate flat surfaces uniformly, so that variations in brightness may be attributed to surface blemishes. For curved surfaces, it might be hoped that the illumination on the surface would be predictable, and then differences would again indicate surface blemishes. In complex cases, however, probably the only alternative is to resort to the use of switched lights coupled with rigorous photometric stereo techniques (see Chapter 16). Finally, it should be noted that the problem of checking quality of surface finish is akin to that of ensuring an attractive physical appearance, and this judgment can be highly subjective. Inspection algorithms therefore need to be "trained" in some way to make the right judgments. (Note that judgments are decisions or classifications, and so the methods of Chapter 24 are appropriate.)

Accurate object location is a prerequisite to reliable object scrutiny and measurement, for all three main categories of inspection. If a CAD system is available, then providing location information permits an image to be generated which should closely match the observed image. Template matching (or correlation) techniques should in many cases permit the remaining inspection functions to be fulfilled. However, this will not always work without trouble, as when object surfaces have a random or textured component. Preliminary analysis of the texture may therefore have to be carried out before relevant templates can be applied—or at least checks made of the maximum and minimum pixel intensities within the product area. More generally, in order to solve this and other problems, some latitude in the degree of fit should be permitted.

The same general technique—that of template matching—arises in the measurement and scrutiny phase as in the object location phase. However (as remarked earlier), this need not consume as much computational effort as in object location. The underlying reason is that template matching is difficult when there are many degrees of freedom inherent in the situation, since comparisons with an enormous number of templates may be required. However, when the template is in a standard position relative to the product and when it has been oriented correctly, template matching is much more likely to constitute a practical solution to the inspection task, although the problem is very data-dependent.

Despite these considerations, it is necessary to find a computationally efficient means of performing the checks of parts. The first possibility is to use suitable algorithms to model the image intensity and then to employ the model to check relevant surfaces for flaws and blemishes. Another useful approach is to convert 2-D to 1-D intensity profiles. This approach leads to the lateral histogram (Chapter 13) and radial histogram techniques. The histogram technique can conveniently be applied for inspecting the very many objects possessing circular symmetry, as will be seen later. First we consider a simple but useful means of checking shapes.

22.5 Shape Deviations Relative to a Standard Template

For food and certain other products, an important factor in 2-D shape measurement is the deviation relative to a standard template. Maximum deviations are important because of the need to fit the product into a standard pack. Another useful measure is the area of overflow or underfill relative to the template (Fig. 22.1). For simple shapes that are bounded by circular arcs or straight lines (a category that includes many types of biscuits or brackets), it is straightforward to test whether a particular pixel on or near the boundary is inside the template or outside it. For straight-line segments this is achieved in the following way. Taking the pixel as (x_1, y_1) and the line as having equation:

$$lx + my + n = 0 \tag{22.1}$$

the coordinates of the pixel are substituted in the expression:

$$f(x, y) = lx + my + n \tag{22.2}$$

The sign will be positive if the pixel lies on one side of the line, negative on the other, and zero if it is on the idealized boundary. Furthermore, the distance on either side of the line is given by the formula:

$$d = \frac{lx_1 + my_1 + n}{(l^2 + m^2)^{1/2}} \tag{22.3}$$

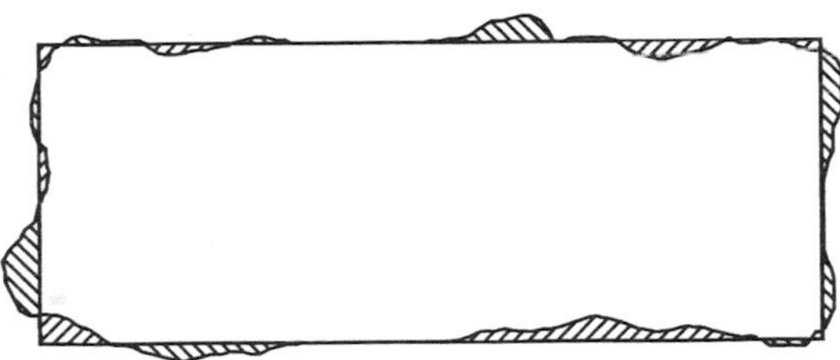

Figure 22.1 Measurement of product area relative to a template. In this example, two measurements are taken, indicating, respectively, the areas of overflow and underfill relative to a prespecified rectangular template.

The same observation about the signs applies to any conic curve if appropriate equations are used. For example, for an ellipse:

$$f(x, y) = \frac{(x - x_c)^2}{a^2} + \frac{(y - y_c)^2}{b^2} - 1 \tag{22.4}$$

where $f(x, y)$ changes sign on the ellipse boundary. For a circle the situation is particularly simple, since the distance to the circle center need only be calculated and compared with the idealized radius value. For more complex shapes, deviations need to be measured using centroidal profiles or the other methods described in Chapter 7. However, the method we have outlined is useful, for it is simple, quick, and reasonably robust, and does not need to employ sequential boundary tracking algorithms; a raster scan over the region of the product is sufficient for the purpose.

22.6 Inspection of Circular Products

Since circular products are so common, it has proved beneficial to develop special techniques for scrutinizing them efficiently. The "radial histogram" technique was developed for this purpose (Davies, 1984c, 1985b). It involves plotting a histogram of intensity as a function of radial distance from the center. Varying numbers of pixels at different radial distances complicate the problem, but smooth histograms result when suitable normalization procedures are applied (Figs. 22.2 and 22.3). These give accurate information on all three major categories of inspection. In particular, they provide values of all relevant radii. For example, both radii of a washer can be estimated accurately using this technique. Surprisingly, accuracy can be as high as 0.3% even when objects have radii as low as 40 pixels (Davies, 1985c), with correspondingly higher accuracies at higher resolutions.

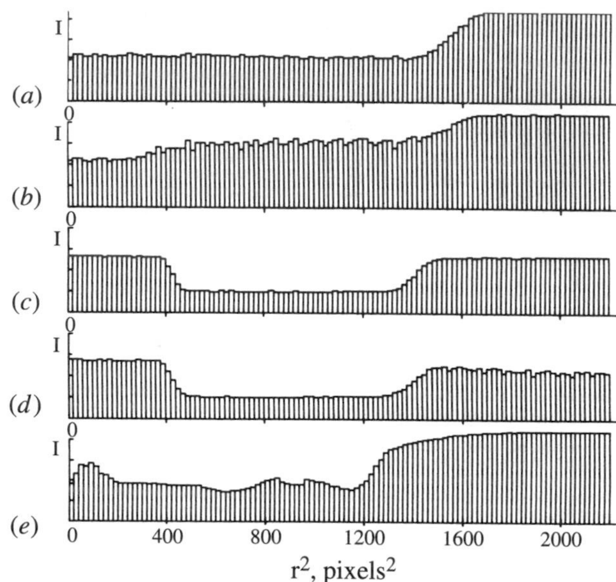

Figure 22.2 Practical applications of the radial histogram approach. In all cases, an r^2 base variable is used, and histogram columns are individually normalized. This corresponds to method (iv) in the text (Section 22.6.1). These histograms were generated from the original images of Fig. 22.3.

Radial histograms are particularly well suited to the scrutiny of symmetrical products that do not exhibit a texture, or for which texture is not prominent and may validly be averaged out. In addition, the radial histogram approach ignores correlations between pixels in the dimension being averaged (i.e., angle). Where such correlations are significant it is not possible to use the approach. An obvious example is in the inspection of components such as buttons, where angular displacement of one of four button holes will not be detectable unless the hole encroaches on the space of a neighboring hole. Similarly, the method does not permit checking the detailed shape of each small hole. The averaging involved in finding the radial histogram mitigates against such detailed inspection, which is then best carried out by separate direct scrutiny of each of the holes.

22.6.1 *Computation of the Radial Histogram: Statistical Problems*

In computing the radial intensity histogram of a region in an image, one of the main variables to be considered is the radial step size s. If s is small, then few pixels contribute to the corresponding column of the histogram, and little averaging of noise and related effects takes place. For textured products,

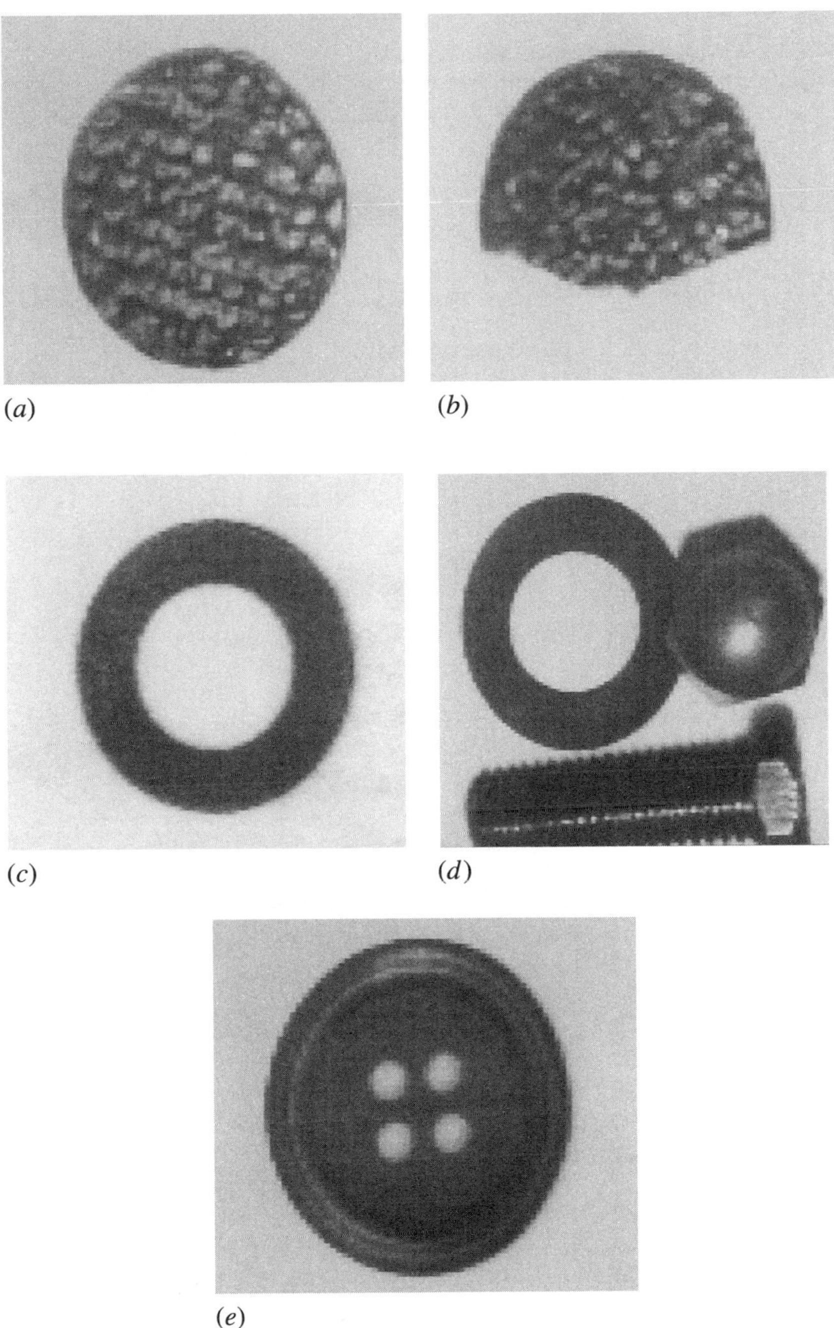

(a) (b) (c) (d) (e)

Figure 22.3 Original images used in generating the radial histograms of Figs. 22.2 and 22.5.

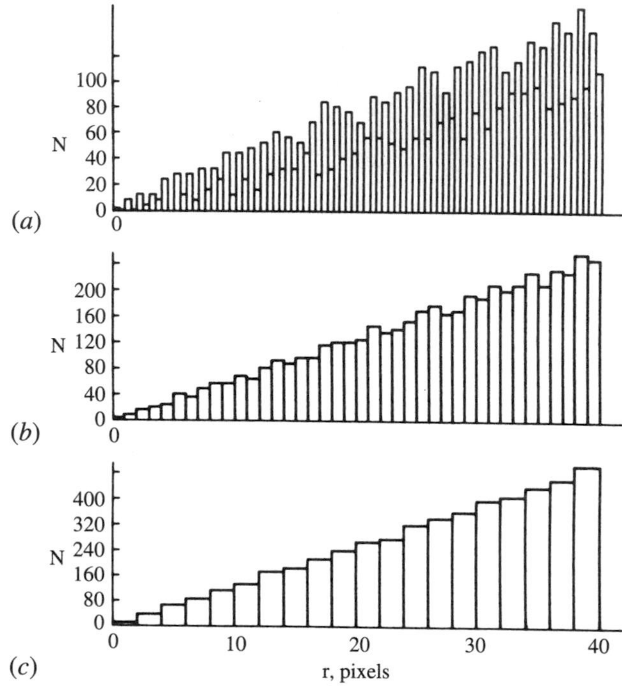

Figure 22.4 Pixel statistics for various radial step sizes; the number of pixels, as a function of radius, that fall into bands of given radial step size. The chosen step sizes are (*a*) 0.50, (*b*) 1.00, and (*c*) 2.00 pixels.

this could be a disadvantage unless it is specifically required to analyze texture by an extension of the method. Ignoring the latter possibility and increasing s in order to obtain a histogram having a statistically significant set of ordinates are bound to result in a loss in radial resolution. To find more about this tradeoff between radial resolution and accuracy, it is necessary to analyze the statistics of interpixel (radial) distances. Assuming a Poisson distribution for the number of pixels falling within a band of a given radius and variable s, we find that histogram column height or "signal" varies, at the given radius, as the step size s, whereas "noise" varies as $\sqrt{s}$ giving an overall signal-to-noise ratio varying (for the given radius) as $\sqrt{s}$. Radial resolution is, of course, inversely proportional to s. For small radii, those of less than about 20 pixels, this simple model is inaccurate, and it is better to follow the actual detailed statistics of a discrete square lattice. Under these circumstances, it seems prudent to examine the pixel statistics plotted in Fig. 22.4 for various values of s and from these to select a value of s which is the best compromise for the application being considered.

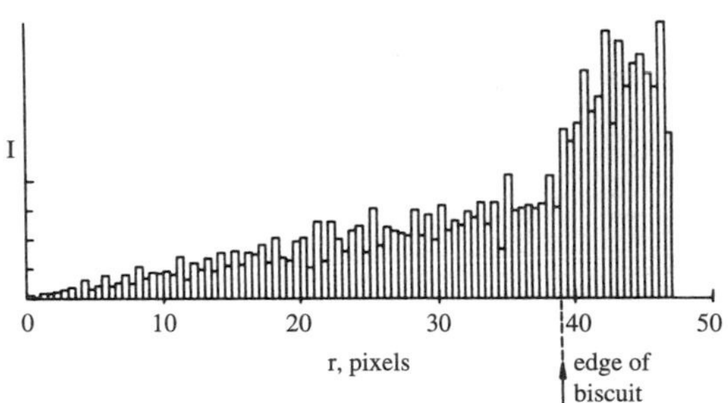

Figure 22.5 Basic form of radial intensity distribution, which varies approximately as the radius. The product being inspected here is the biscuit of Fig. 22.3*a*.

For many applications, it is inconvenient to have radial intensity distributions such as that depicted in Fig. 22.5, in which the column height varies inherently as the radius because the number of pixels in a band of radius r and given s is (approximately) proportional to r. It is more convenient to normalize the distribution so that regions of uniform intensity give rise to a uniform radial intensity distribution. This uniformity may be achieved in two ways. The first is to multiply each column of the radial histogram by a factor that brings it to an inherently uniform value. The other is to plot the histogram as a function of some function $u(r)$ other than r which gives the same effect.

Taking the latter approach first, we find that changing variables from r to $u(r)$ results in

$$\begin{aligned}\int_r^{r+\delta r}\int_0^{2\pi} I(r) r \mathrm{d}r \mathrm{d}\theta &= I(r)\times 2\pi r\delta r \\ &= I(u)\delta u = \text{constant}\end{aligned} \tag{22.5}$$

Hence

$$\delta u = 2\pi r\delta r \tag{22.6}$$

We now require that $I(u) = \text{constant}$ for fixed step size δu. Setting δu equal to a constant leads to

$$\delta r = \gamma/2\pi r \tag{22.7}$$

so that

$$\delta r \propto 1/r \tag{22.8}$$

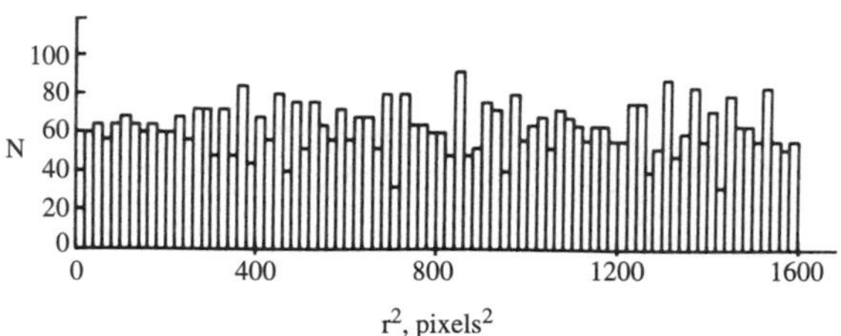

Figure 22.6 Pixel statistics for an r^2 histogram base parameter. The pixel statistics are not exactly uniform even when the radial histogram is plotted with an r^2 base parameter.

as might indeed have been expected. In addition, note that

$$u = \int_0^r 2\pi r \mathrm{d}r = \pi r^2 \tag{22.9}$$

making the histogram base parameter equal to r^2 (within a constant of proportionality).

Unfortunately, this approach is still affected by the pixel statistics problem, as shown in Fig. 22.6. Thus, we are effectively being forced to apply the approach of normalizing the histogram by multiplying by suitable numerical factors, as suggested earlier. There are now four alternatives:

1. Use the basic radial histogram.
2. Use the modified radial histogram, with $u(r) = r^2$ base parameter.
3. Use the basic histogram, with individual column normalization.
4. Use the $u(r) = r^2$ histogram, with individual column normalization.

As a rule, it appears to be necessary to normalize each column of the histogram individually to obtain the most accurate results. For ease of interpretation, it would also be preferable to use the modified base parameter $u(r) = r^2$ rather than r.

Note that the square-law approach is trivial to implement, since r^2 rather than r is the value that is actually available during computation (r has to be derived from r^2 if it is needed). Furthermore, the r^2 base parameter gives a nonlinear stretch to the histogram which matches the increased radial accuracy that is actually available at high values of r. This means that the $u(r) = r^2$ histogram permits greater accuracy in the estimation of radius at large r than it does at small r, thus making $u(r) = r^2$ the natural base variable to employ (Davies, 1985b).

These considerations make method (4) the most generally suitable one, although method (2) is useful when speed of processing is a particular problem. Finally, it is relevant to note that, to a first approximation, all types of radial intensity histograms mentioned earlier can eliminate the effects of a uniform gradient of background illumination over an object.

22.6.2 Application of Radial Histograms

Figure 22.2 shows practical applications of the above theory to various situations, depicted in Fig. 22.3: see also Fig. 22.7, which relates to the biscuits shown in Fig. 10.1. In particular, notice that the radial histogram approach is able to give vital information on various types of defects: the presence or absence of holes in a product such as a washer or a button, whether circular objects are in contact or overlapping, broken objects, "show-through" of the biscuits where there are gaps in a chocolate or other coating, and so on. In addition, it is straightforward to derive dimensional measurements from radial histograms. In particular, radii of discs or washers can be obtained to significantly better than 1 pixel accuracy because of the averaging effect of the histogram approach. However, the method is limited here by the accuracy with which the center of the circular region is first located. This underlines the value of the high-accuracy center finding technique of Chapter 10. To a certain extent, accuracy is limited by the degree of roundness of the product feature being examined. Radius can be measured only to the extent that it is meaningful! In this context, it is worth emphasizing that the combination of techniques described here is not only accurate but also sparing in computation.

It might be imagined that the radial histogram technique is applicable only for symmetrical objects. However, it is also possible to use radial histograms as signatures of intensity patterns in the region of specific salient features. Small

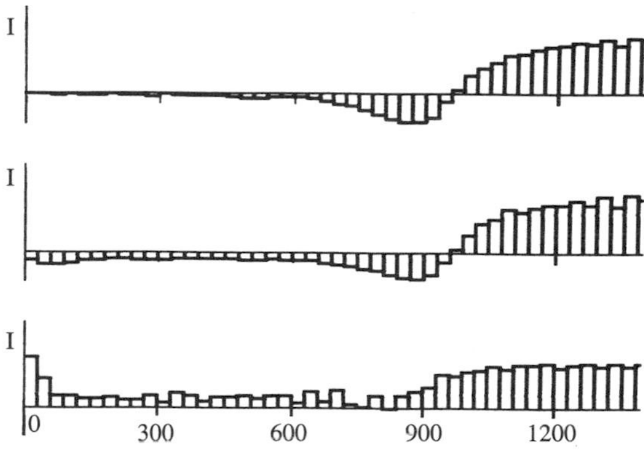

Figure 22.7 Radial intensity histograms for the three biscuits of Fig. 10.1. The order is from top to bottom in both figures. Intensity here is measured relative to that at the center of an ideal product.

holes are suitable salient features, but corners are less suitable unless the background is saturated out at a constant value. Otherwise, too much variation arises from the background, and the technique does not prove viable.

Finally, note that the radial histogram approach has the useful characteristic that it is trainable, since the relevant 1-D templates may be accumulated, averaged over a number of ideal products, and stored in memory, ready for comparisons with less ideal products. This characteristic is valuable, not only because it provides convenience in setting up but also because it permits inspection to be adaptable to cover a range of products.

22.7 Inspection of Printed Circuits

Over the past two or three decades, machine vision has been used increasingly in the electronics industry, notably for inspecting printed circuit boards (PCBs). First, PCBs may be inspected before components are inserted; second, they may be inspected to check that the correct components have been inserted the correct way round; and third, all the soldered joints may be scrutinized. The faults that have to be checked for on the boards include touching tracks, whisker bridges, broken tracks (including hairline cracks), and mismatch between pad positions and holes drilled for component insertion. Controlled illumination is required to eliminate glints from the bare metal and to ensure adequate contrast between the metal and its substrate. With adequate control over the lighting, most of the checks (e.g., apart from reading any print on the substrate) may be carried out on a binarized image, and the problem devolves into that of checking shape. The checking may be tackled by gross template matching—using a logical exclusive-or operation—but this approach requires large data storage and precise registration has to be achieved.

Difficulties with registration errors can largely be avoided if shape analysis is performed by connectedness measurement (using thinning techniques) and morphological processing. For example, if a track disappears or becomes broken after too few erosion operations, then it is too narrow. A similar procedure will check whether tracks are too close together. Similarly, hairline cracks may be detected by dilations followed by tests to check for changes in connectedness.

Alignment of solder pads with component holes is customarily checked by employing a combination of back and front lighting. Powerful back lighting (i.e., from behind the PCB) gives bright spots at the hole positions, while front lighting gives sufficient contrast to show the pad positions. It is then necessary to confirm that each hole is, within a suitable tolerance, at the center of its pad. Counting of the bright spots from the holes, plus suitable measurements around hole positions (e.g., via radial histogram signatures), permits this process to be performed satisfactorily.

The main problem with PCB inspection is the resolution required. Typically, images have to be digitized to at least 4000 × 4000 pixels—as when a 20 × 20 cm board is being checked to an accuracy of 50 μm. In addition, a suitable inspection system normally has to be able to check each board fully in less than one minute. It also has to be trainable, to allow for upgrades in the design of the circuit or improvements in the layout. Finally, it should cost no more than $60,000. However, considerable success has been attained with these aims.

To date, the bulk of the work in PCB inspection has concerned the checking of tracks. Nevertheless, useful work has also been carried out on the checking of soldered joints. Here, each joint has to be modeled in 3-D by structured light or other techniques. In one such case, light stripes were used (Nakagawa, 1982), and in another, surface reflectance was measured with a fixed lighting scheme (Besl et al., 1985). Note that surface brightness says something about the quality of the soldered joint. This type of problem is probably completely solvable (at least up to the subjective level of a human inspector), but detailed scrutiny of each joint at a resolution of say 64 × 64 pixels may well be required to guarantee that the process is successful. This implies an enormous amount of computation to cope with the several hundreds (or in some cases thousands) of joints on most PCBs. Hence, special architectures were needed to handle the information in the time available.

Similar work is under way on the inspection of integrated circuit masks and die bonds, but space does not permit discussion of this rapidly developing area. For a useful review, see Newman and Jain (1995).

22.8 Steel Strip and Wood Inspection

The problem of inspecting steel strip is very exacting for human operators. First, it is virtually impossible for the human eye to focus on surface faults when the strip is moving past the observer at rates in excess of 20 m/s; second, several years of experience are required for this sort of work; and third, the conditions in a steel mill are far from congenial, with considerable heat and noise constantly being present. Hence, much work has been done to automate the inspection process (Browne and Norton-Wayne, 1986). At its simplest, this requires straightforward optics and intensity thresholding, although special laser scanning devices have also been developed to facilitate the process (Barker and Brook, 1978).

The problem of wood inspection is more complex, since this natural material has variable characteristics. For example, the grain varies markedly from sample to sample. As a result of this variation, the task of wood inspection is still in its infancy, and many problems remain. However, the purpose of wood inspection is reasonably clear: first, to look for cracks, knots, holes, bark inclusions, embedded pine needles, miscoloration, and so on; and ultimately to make full use of this natural resource by identifying regions of the wood where strength or appearance is substandard. In addition, the timber may have to be classified as appropriate for different categories of use—furniture, building, outdoor, and so on. Overall, wood inspection is something of an art—that is, it is a highly subjective process—although valiant attempts have long been made to solve the problems (e.g., Sobey, 1989). At least one country (Australia) now has a national standard for inspecting wooden planks.

22.9 Inspection of Products with High Levels of Variability

Thus far we have concentrated on certain aspects of inspection—particularly dimensional checking of components ranging from precision parts to food products, and the checking of complex assemblies to confirm the presence of holes, nuts, springs, and so on. These could be regarded as the geometrical aspects of inspection. For the more imprecisely made products such as foodstuffs and textiles, there are greater difficulties because the template against which any product has to be compared is not fixed. Broadly, there are two ways of tackling this problem: through use of a range of templates, each of which is acceptable to some degree; and through specification of a variety of descriptive parameters. In either case there will be a number of numerical measurements whose values have to be within prescribed tolerances. Overall, it seems inescapable that variable products demand greater amounts of checking and computation, and that inspection is significantly more demanding. Nowhere is this clearer than for food and textiles, for which the relevant parameters are largely textural. However, "fuzzy" inspection situations can also occur for certain products that might initially be considered as precision components: for example, for electric lamps the contour of the element and the solder pads on the base have significant variability. Thus, this whole area of inspection involves checking that a range of parameters do not fall outside certain prespecified limits on some relevant distribution which *may* be reasonably approximated by a Gaussian.

We have seen that the inspection task is made significantly more complicated by natural valid variability in the product, though in the end it seemed best to regard inspection as a process of making measurements that have to be checked statistically. Defects could be detected relative to the templates, either as gross mismatches or as numerical deviations. Missing parts could likewise be detected since they do not appear at the appropriate positions relative to the templates. Foreign objects would also appear to fit into this pattern, being essentially defects under another name. However, this view is rather too simple for several reasons:

1. Foreign objects are frequently unknown in size, shape, material, or nature.
2. Foreign objects may appear in the product in a variety of unpredictable positions and orientations.
3. Foreign objects may have to be detected in a background of texture which is so variable in intensity that they will not stand out.

The unpredictability of foreign objects can make them difficult to see, especially in textured backgrounds (Fig. 22.8). If one knew their nature in advance, then a special detector could perhaps be designed to locate them. But in many practical situations, the only way to detect them is to look for the unusual. In fact, the human eye is well tuned to search for the unusual. On the other hand, few obvious techniques can be used to seek it out automatically in digital images. Simple thresholding would work in a variety of practical cases, especially where plain surfaces have to be inspected for scratches, holes, swarf, or dirt. However, looking for extraneous vegetable matter (such as leaves, twigs, or pods) among a sea of peas on a conveyor may be less easy, as the contrast levels may be similar, and the textural cues may not be able to distinguish the shapes with sufficient accuracy. Of course, in the latter example, it could be imagined that every pea could be identified by its intrinsic circularity. However, the incidence of occlusion, and the very computation-intensive nature of this approach to inspection inhibit such an approach. In any case, a method that would detect pods among peas might not detect round stones or small pieces of wood—especially in a gray-scale image.

Ultimately, the problem is difficult because the paradigm means of designing sensitive detectors—the matched filter—cannot be used, simply because of the high degree of variability in what has to be detected. With so many degrees of freedom—shape dimensions, size, intensity, texture, and so on—foreign objects can be highly difficult to detect successfully in complex images. Naturally, a lot depends on the nature of the substrate, and while a plain background might render the task trivial, a textured product substrate may make it impossible, or at least practically impossible in a real-time factory inspection milieu. In general, the solution devolves into not trying to detect the foreign object directly by means of carefully designed matched filters, but in trying to model the intensity pattern of the substrate with sufficient accuracy that any deviation due to the presence of a foreign object is detected and rendered visible. As hinted, the approach is to search for the unusual. Accordingly, the basic technique is to identify the 3σ or

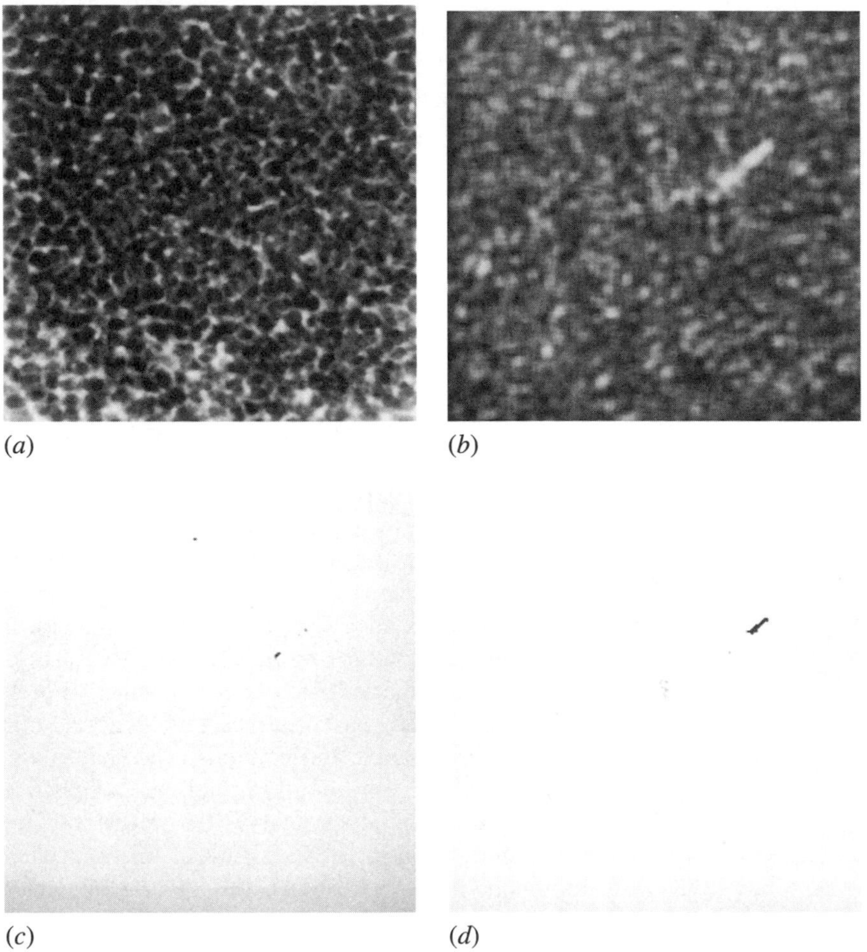

Figure 22.8 Foreign object detection in a packet of frozen vegetables (in this case sweet corn). (*a*) shows the original X-ray image, (*b*) shows an image in which texture analysis procedures have been applied to enhance any foreign objects, and (*c*) and (*d*) show the respective thresholded images. Notice the false alarms that are starting to arise in (*c*), and the increased confidence of detection of the foreign object in (*d*). For further details, see Patel et al. (1995).

other appropriate points on all available measures and to initiate rejection when they are exceeded.

There is a fundamental objection to this procedure. If some limit (e.g., 3σ) is assumed, optimization cannot easily be achieved, since the proper method for achieving this is to find both the distributions—for the background substrate and for the foreign objects—and to obtain a minimum error decision boundary between them. (Ideally, this would be implemented in a multidimensional feature space appropriate to the task.) However, in this case we do not have the distribution corresponding to the foreign objects; we therefore have to fall back on "reasonable" acceptance limits based on the substrate distribution.

It might appear that this argument is flawed in that the proportion of foreign objects coming along the conveyor is well known. While it might occasionally be known, the levels of detectability of the foreign objects in the received images will be unknown and will be less than the actual occurrence levels. Hence, arriving at an optimal decision level will be very difficult.

A far worse problem often exists in practice. The occurrence rates for foreign objects might be almost totally unknown because of (1) their intrinsic variability and (2) their rarity. We might well ask, "How often will an elastic band fall onto a conveyor of peas?", but this is a question that is virtually impossible to answer. Maybe it is possible to answer somewhat more accurately the more general question of how often a foreign object of some sort will fall onto the conveyor, but even then the response may well be that somewhere between 1 in 100,000 and 1 in 10 million of bags of peas contain a foreign object. With low levels of risk, the probabilities are extremely difficult to estimate; indeed, there are few available data or other basis on which to calculate them. Consumer complaints can indicate the possible levels of risk, but they arise as individual items, and in any case many customers will not make any fuss and by no means all instances come to light.

With food products, the penalties for not detecting individual foreign objects are not usually especially great. Glass in baby food may be more apocryphal than real and may be unlikely to cause more than alarm. Similarly, small stones among the vegetables are more of a nuisance than a harm, though cracked teeth could perhaps result in litigation in the $2000 bracket. Far more serious are the problems associated with electric lamps, where a wire emerging from the solder pads is potentially lethal and is a substantive worry for the manufacturer. Litigation for deaths arising from this source could run to millions of dollars (corresponding to the individual's potential lifetime earnings).

This discussion reveals that the cost rate[2] rather than the error rate is the important parameter when there is even a remote risk to life and limb. Indeed, it concerned Rodd and his co-workers so much in relation to the inspection of electric lamps that they decided to develop special techniques for ensuring that their algorithms were tested sufficiently (Thomas and Rodd, 1994; Thomas et al., 1995). Computer graphics techniques were used to produce large numbers of images with automatically generated variations on the basic defects, and it was confirmed that the inspection algorithms would always locate them.

2 The *perceived* cost rate may be even more important, and this can change markedly with reports appearing in the daily press.

22.10 X-ray Inspection

In inspection applications, a tension exists between inspecting products early on, before significant value has been added to a potentially defective product, and at the end of the line, so that the quality of the final product is guaranteed. This consideration applies particularly with food products, where additives such as chocolate can be very expensive and constitute substantial waste if the basic product is broken or misshapen. In addition, inspection at the end of the line is especially valuable as oversized products (which may arise if two normal products become stuck together) can jam pack machines. Ideally, two inspection stations should be placed on the line in appropriate positions, but if only one can be afforded (the usual situation) it will generally have to be placed at the end of the line.

With many products, it is useful to be absolutely sure about the final quality as the customer will receive it. Thus, there is special value in inspecting the packaged products. Since the packets will usually be opaque, it will be necessary to inspect them under X-radiation. This results in significant expense, since the complete system will include not only the the X-ray source and sensing system but also various safety features including heavy shielding. As a result commercial X-ray food inspection systems rarely cost less than \$60,000 and \$200,000 is a more typical figure. Such figures do not take into account maintenance costs. It is also important to note that the X-ray source and the sensors deteriorate with time, so that sensitivity falls, and special calibration procedures have to be invoked.

Fortunately, the X-ray sensors can now take the form of linear photodiode packages constructed using integrated circuit technology.[3] These packages are placed end to end to span the width of a food conveyor which may be 30 to 40 cm wide. They act as a line-scan camera that grabs images as the product moves along the conveyor. The main adjustments to be made in such a system are the voltage across the X-ray tube and the current passing through it. In food inspection applications, the voltage will be in the range of 30 to 100 kV, and the current will be in the range 3 to 10 mA. A *basic* commercial system will include thresholding and pixel counting, permitting the detection of small pieces of metal or other hard contaminants but not soft contaminants. In general, soft contaminants can be detected only if more sophisticated algorithms are used that examine the contrast levels over various regions of the image and arrive at a consensus that foreign objects are present—typically, with the help of texture analysis procedures.

The sensitivity of an X-ray detection system depends on a number of factors. Basically, it is highly dependent on the number of photons arriving at the sensors, and this number is proportional to the electron current passing through the X-ray

3 The X-ray photons are first converted to visible light by passage through a layer of scintillating material.

tube. There are stringent rules on the intensities of X-radiation to which food products may be exposed, but in general these limits are not approached because of good sensitivity at moderate current levels. However, we shall not delve into these matters further here.

Sensitivity also depends critically on the voltages that are applied to X-ray tubes. It is well known that the higher the voltage, the higher the electron energies, the higher the energies of the resulting photons, and the greater the penetrating power of the X-ray beam. However, greater penetrating power is not necessarily an advantage, for the beam will tend to pass through the food without attenuation, and therefore without detecting any foreign objects. Although a poorly setup system may well be able to detect quite small pieces of metal without much trouble, detection of small stones and other hard contaminants will be less easy, and detection of soft contaminants will be virtually impossible. Thus, it is necessary to optimize the contrast in the input images.

Unfortunately, X-ray sources provide a wide range of wavelengths, all of which are scattered or absorbed to varying degrees by the intervening substances. In a thick sample, scattering can cause X-radiation to arrive at the detector after passing through material that is not in a direct line between the X-ray tube and the sensors. This makes a complete analysis of sensitivity rather complicated. In the following discussion we will ignore this effect and assume that the bulk of the radiation reaching the sensors follows the direct path from the X-ray source. We will also assume that the radiation is gradually absorbed by the intervening substances, in proportion to its current strength. Thus, we obtain the standard exponential formula for the decay of radiation through the material which we shall temporarily take to be homogeneous and of thickness z:

$$I = I_0 \exp\left[-\int \mu \, \mathrm{d}z\right] = I_0 \exp\left[-\mu \int \mathrm{d}z\right] = I_0 \mathrm{e}^{-\mu z} \tag{22.10}$$

Here μ depends on the type of material and the penetrating power of the X-radiation. For monochromatic radiation of energy E, we have:

$$\mu = (\rho N/A)\left[k_{\mathrm{P}} Z^a / E^b + k_{\mathrm{C}} Z/E\right] \tag{22.11}$$

where ρ is the density of the material, A is its atomic weight, N is Avogadro's number, a and b are numbers depending on the type of material, and k_{P} and k_{C} are decay constants resulting from photoelectric and Compton scattering, respectively (see e.g., Eisberg, 1961). It will not be appropriate to examine all the implications of this formula. Instead, we proceed with a rather simplified model that nevertheless shows how to optimize sensitivity:

$$\mu = \alpha/E \tag{22.12}$$

Substituting in equation (22.10), we find:

$$I = I_0 \exp(-\alpha z/E) \tag{22.13}$$

If a minute variation in thickness or a small foreign object is to be detected sensitively, we need to consider the change in intensity resulting from a change in z or in αz. (Ultimately, it is the integral of $\mu\,\mathrm{d}z$ that is important—see equation (22.10).) It will be convenient to relabel the latter quantity as a generalized distance X, and the inverse energy factor as f:

$$I = I_0 \exp(-Xf) \tag{22.14}$$

$$\mathrm{d}I/\mathrm{d}X = -I_0 f \exp(-Xf) \tag{22.15}$$

so that:

$$\Delta I = -\Delta X\, I_0 f \exp(-Xf) \tag{22.16}$$

The contrast due to the variation in generalized distance can now be expressed as:

$$\Delta I/I = -\Delta X\, I_0 f \exp(-Xf)/\, I_0 \exp(-Xf) = -\Delta Xf = -\Delta X/E \tag{22.17}$$

This calculation shows that contrast should improve as the energy of the X-ray photons decreases. However, this result appears wrong, as reducing the photon energy will reduce the penetrating power, and in the end no radiation will pass through the sample. First, the sensors will not be sufficiently sensitive to detect the radiation. Specifically, noise (including quantization noise) will become the dominating factor. Second, we have ignored the fact that the X-radiation is not monochromatic. We shall content ourselves here with modeling the situation to take account of the latter factor. Assume that the beam has two energies, one fairly low (as above) and the other rather high and penetrating. This high-energy component will add a substantially constant value to the overall beam intensity and will result in a modified expression for the contrast:

$$\Delta I/I = -\Delta X\, I_0 f \exp(-Xf)/[I_1 + I_0 \exp(-Xf)] = -\Delta X\, I_0 f/[I_1 \exp(Xf) + I_0] \tag{22.18}$$

To optimize sensitivity, we differentiate with respect to f:

$$\mathrm{d}(\Delta I/I)/\mathrm{d}f = -\Delta X\, I_0 \{[I_1 \exp(Xf) + I_0] - XfI_1 \exp(Xf)\}/[I_1 \exp(Xf) + I_0]^2 \tag{22.19}$$

This is zero when:

$$XfI_1 = I_1 + I_0 \exp(-Xf) \tag{22.20}$$

that is, $E = X/[1 + (I_0/I_1)\exp(-X/E)]$ (22.21)

When $I_1 \ll I_0$, we have the previous result, that optimum sensitivity occurs for low E. However, when $I_1 \gg I_0$, we have the result that optimum sensitivity occurs when $E = X$. In general, this formula gives an optimum X-ray that is above zero, in accordance with intuition. In passing, we note that graphical or iterative solutions of equation (22.21) are easily obtained.

Finally, we consider the exponential form of the signal given by equations (22.10), (22.13), and (22.14). These are nonlinear in X (i.e., αz), and meaningful image analysis algorithms would tend to require signals that are linear in the relevant physical quantity, namely, X. Thus, it is appropriate to take the logarithm of the signal from the input sensor, before proceeding with texture analysis or other procedures:

$$I' = \log I + \log[I_0 \exp(-Xf)] = A - Xf \tag{22.22}$$

where

$$A = \log I_0 \tag{22.23}$$

In this way, doubling the width of the sample doubles the change in intensity, and subsequent (e.g., texture analysis) algorithms can once more be designed on an intuitive basis. (In fact, there is a more fundamental reason for performing this transformation—that it performs an element of noise whitening, which should ultimately help to optimize sensitivity.)

22.11 The Importance of Color in Inspection

In many applications of machine vision, it is not necessary to consider color, because almost all that is required can be achieved using gray-scale images. For example, many processes devolve into shape analysis and subsequently into statistical pattern recognition. This situation is exemplified by fingerprint analysis and by handwriting and optical character recognition. However, there is one area where color has a big part to play. This is in the picking, inspection, and sorting of fruit. For example, color is very important in determining apple quality. Not only is it a prime indicator of ripeness, but it also contributes greatly to physical attractiveness and thus encourages purchase and consumption.

Whereas color cameras digitize color into the usual RGB (red, green, blue) channels, humans perceive color differently. As a result, it is better to convert the RGB representation to the HSI (hue, saturation, intensity) domain before assessing the colors of apples and other products.[4] Space prevents a detailed study of the question of color; the reader is referred to more specialized texts for detailed information (e.g., Gonzalez and Woods, 1992; Sangwine and Horne, 1998). However, some brief comments will be useful. Intensity I refers to the total light intensity and is defined by:

$$I = (R + G + B)/3 \tag{22.24}$$

Hue H is a measure of the underlying color, and saturation S is a measure of the degree to which it is *not* diluted by white light (S is zero for white light). S is given by the simple formula:

$$S = 1 - \frac{\min(R, G, B)}{I} = 1 - \frac{3\min(R, G, B)}{R + G + B} \tag{22.25}$$

which makes it unity along the sides of the color triangle and zero for white light ($R = G = B = 1$). Note how the equation for S favors none of the R, G, B components. It does not express color but a measure of the proportion of color and differentiation from white.

Hue is defined as an angle of rotation about the central white point W in the color triangle. It is the angle between the pure red direction (defined by the vector $\mathbf{R} - \mathbf{W}$) and the direction of the color C in question (defined by the vector $\mathbf{C} - \mathbf{W}$). The derivation of a formula for H is fairly complex and will not be attempted here. Suffice it to say that it may be determined by calculating $\cos H$, which depends on the dot product $(\mathbf{C} - \mathbf{W}) \cdot (\mathbf{R} - \mathbf{W})$. The final result is:

$$H = \cos^{-1}\left(\frac{\frac{1}{2}[(R - G) + (R - B)]}{\left[(R - G)^2 + (R - B)(G - B)\right]^{1/2}}\right) \tag{22.26}$$

or 2π minus this value if $B > G$ (Gonzalez and Woods, 1992).

When checking the color of apples, the hue is the important parameter. A rigorous check on the color can be achieved by constructing the hue distribution and comparing it with that for a suitable training set. The most straightforward way to carry out the comparison is to compute the mean and standard deviation of the two distributions to be compared and to perform discriminant analysis assuming Gaussian distribution functions. Standard theory (Section 4.5.3, equations (4.41) to (4.44)) then leads to an optimum hue decision threshold.

4 Usually, a more important reason for use of HSI is to employ the hue parameter, which is independent of the intensity parameter, as the latter is bound to be particularly sensitive to lighting variations.

In the work of Heinemann et al. (1995), discriminant analysis of color based on this approach gave complete agreement between human inspectors and the computer following training on 80 samples and testing on another 66 samples. However, a warning about maintaining the lighting intensity levels identical to those used for training was given. In any such pattern recognition system, it is crucial that the training set be representative in every way of the eventual test set.

Full color discrimination would require an optimal decision surface to be ascertained in the overall 3-D color space. In general, such decision surfaces are hyperellipses and have to be determined using the Mahalanobis distance measure (see, for example, Webb, 2002). However, in the special case of Gaussian distributions with equal covariance matrices, or more simply with equal isotropic variances, the decision surfaces become hyperplanes.

22.12 Bringing Inspection to the Factory

The relationship between the producer of vision systems and the industrial user is more complex than might appear at first sight. The user has a need for an inspection system and states his need in a particular way. However, subsequent tests in the factory may show that the initial statement was inaccurate or imprecise. For example, the line manager's requirements[5] may not exactly match those envisaged by the factory management board. Part of the problem lies in the relative importance given to the three disparate functions of inspection mentioned earlier. Another lies in the change of perspective once it is seen exactly what defects the vision system is able to detect. It may be found immediately that one or more of the major defects that a product is subject to may be eliminated by modifications to the manufacturing process. In that case, the need for vision is greatly reduced, and the very process of trying out a vision system may end in its value being undermined and its not being taken up after a trial period. This does not detract from the inherent capability of vision systems to perform 100% untiring inspection and to help maintain strict control of quality. However, it must not be forgotten that vision systems are not cheap and that they can in some cases be justified only if they replace a number of human operators. Frequently, a payback period of two to three years is specified for installing a vision system.

Textural measurements on products are an attractive proposition for applications in the food and textile industries. Often textural analysis is written

5 In many factories, line managers have the responsibility of maintaining production at a high level on an hour-by-hour basis, while at the same time keeping track of quality. The tension between these two aims, and particularly the underlying economic constraints, means that on occasion quality is bound to suffer.

into the prior justification for, and initial specification of, an inspection system. However, what a vision researcher understands by texture and what a line inspector in either of these industries means by it tend to be different. Indeed, what is required of textural measurements varies markedly with the application. The vision researcher may have in mind higher-order[6] statistical measures of texture, such as would be useful with a rough irregular surface of no definite periodicity[7]—as in the case of sand or pebbles on a beach, or grass or leaves on a bush. However, the textile manufacturer would be very sensitive to the periodicity of his fabric and to the presence of faults or overly large gaps in the weave. Similarly, the food manufacturer might be interested in the number and spatial distribution of pieces of pepper on a pizza, while for fish coatings (e.g., batter or breadcrumbs) uniformity will be important and "texture measurement" may end by being interpreted as determining the number of holes per unit area of the coating. Thus, texture may be characterized not by higher-order statistics but by rather obvious counting or uniformity checks. In such cases, a major problem is likely to be that of reducing the amount of computation to the lowest possible level, so that considerable expanses of fabric or large numbers of products can be checked economically at production rates. In addition, it is frequently important to keep the inspection system flexible by training on samples, so that maximum utility of the production line can be maintained.

With this backcloth to factory requirements, it is vital for the vision researcher to be sensitive to actual rather than idealized needs or the problem as initially specified. There is no substitute for detailed consultation with the line manager and close observation in the factory before setting up a trial system. Then the results from trials need to be considered very carefully to confirm that the system is producing the information that is really required.

22.13 Concluding Remarks

The number of relevant applications of computer vision to industrial tasks, and notably to automated visual inspection, is exceptionally high. For that reason, it

6 The zero-order statistic is the mean intensity level. First-order statistics such as variance and skewness are derived from the histogram of intensities; second- and higher-order statistics take the form of gray-level co-occurrence matrices, showing the number of times particular gray values appear at two or more pixels in various relative positions. For more discussion on textures and texture analysis, see Chapter 26.

7 More rigorously, the fabric is intended to have a long-range periodic order that does not occur with sand or grass. There is a close analogy here with the long- and short-range periodic order for atoms in a crystal and in a liquid, respectively.

has been necessary to concentrate on principles and methods rather than on individual cases. The repeated mention of hardware implementation has been necessary because the economics of real-time implementation is often the factor that ultimately decides whether a given system will be installed on a production line. However, processing speeds are also heavily dependent on the specific algorithms employed, and these in turn depend on the nature of the image acquisition system—including both the lighting and the camera setup. (The decision of whether to inspect products on a moving conveyor or to bring them to a standstill for more careful scrutiny is perhaps the most fundamental one for implementation.) Hence, image acquisition and real-time electronic hardware systems are the main topics of later chapters (Chapters 27 and 28).

More fundamentally, the reader will have noticed that a major purpose of inspection systems is to make instant decisions on the adequacy of products. Related to this purpose are the often fluid criteria for making such decisions and the need to train the system in some way, so that the decisions that are made are at least as good as those that would have arisen with human inspectors. In addition, training schemes are valuable in making inspection systems more general and adaptive, particularly with regard to change of product. Hence, the pattern recognition techniques of Chapter 24 are highly relevant to the process of inspection.

Ideally, the present chapter would have appeared after Chapters 24, 27 and 28. However, it was felt that it was better to include it as early as possible, in order to maintain the reader's motivation; also, Chapters 27 and 28 are perhaps more relevant to the implementation than to the rationale of image analysis.

On a different tack, it is worth noting that automated visual inspection falls under the general heading of computer-aided manufacture (CAM), of which computer-aided design (CAD) is another part. Today many manufactured parts can in principle be designed on a computer, visualized on a computer screen, made by computer-controlled (e.g., milling) machines, and inspected by a computer—all without human intervention in handling the parts themselves. There is much sense in this computer integrated manufacture (CIM) concept, since the original design dataset is stored in the computer, and therefore it might as well be used (1) to aid the image analysis process that is needed for inspection and (2) as the actual template by which to judge the quality of the goods. After all, why key in a separate set of templates to act as inspection criteria when the specification already exists on a computer? However, some augmentation of the original design information to include valid tolerances is necessary before the dataset is sufficient for implementing a complete CIM system. In addition, the purely dimensional input to a numerically controlled milling machine is not generally sufficient—as the frequent references to surface quality in the present chapter indicate.

Automated industrial inspection is a well-worn application area for vision that severely exercises the reliability, robustness, accuracy, and speed of vision software and hardware. This chapter has discussed the practicalities of this topic, showing how color and other modalities such as X-rays impinge on the basic vision techniques.

22.14 Bibliographical and Historical Notes

It is very difficult to provide a bibliography of the enormous number of papers on applications of vision in industry or even in the more restricted area of automated visual inspection. In any case, it can be argued that a book such as the present one ought to concentrate on principles and to a lesser extent on detailed applications and "mere" history. However, the review article by Newman and Jain (1995) gives a useful overview to 1995. For a review of PCB inspection, see Moganti et al. (1996).

The overall history of industrial applications of vision has been one of relatively slow beginnings as the potential for visual control became clear, followed only in recent years by explosive growth as methods and techniques evolved and as cost-effective implementations became possible as a result of cheaper computational equipment. In this respect, the year 1980 marked a turning point, with the instigation of important conferences and symposia, notably, that on "Computer Vision and Sensor-Based Robots" held at General Motors Research Laboratories during 1978 (see Dodd and Rossol, 1979), and the ROVISEC (Robot Vision and Sensory Controls) series of conferences organized annually by IFS (International Fluidic Systems) Conferences Ltd, UK, from 1981. In addition, useful compendia of papers (e.g., Pugh, 1983) and books outlining relevant principles and practical details (e.g., Batchelor et al., 1985) have been published. The reader is also referred to a special issue of *IEEE Transactions on Pattern Analysis and Machine Intelligence* (now dated) which is devoted to industrial machine vision and computer vision technology (Sanz, 1988).

Noble (1995) presents an interesting and highly relevant view of the use of machine vision in manufacturing. Davies (1995) develops the same topic by presenting several case studies together with a discussion of some major problems that remain to be addressed in this area.

The past few years have seen considerable interest in X-ray inspection techniques, particularly in the food industry (Boerner and Strecker, 1988; Wagner,

1988; Chan et al., 1990; Penman et al., 1992; Graves et al., 1994; Noble et al., 1994). In the case of X-ray inspection of food, the interest is almost solely in the detection of foreign objects, which could in some cases be injurious to the consumer. This has been a prime motivation for much work in the author's laboratory (Patel et al., 1994, 1995; Patel and Davies, 1995).

Also of growing interest is the automatic visual control of materials such as lace during manufacture; in particular, high-speed laser scalloping of lace is set to become fully automated (King and Tao, 1995).

A cursory examination of inspection publications over the past few years reveals particular growth in emphasis on surface defect inspection, including color assessment, and X-ray inspection of bulk materials and baggage, (e.g., at airports). Two fairly recent journal special issues (Davies and Ip, eds., 1998; Nesi and Trucco, eds., 1999) cover defect inspection and include a paper on lace inspection (Yazdi, and King, 1998), while Tsai and Huang (2003) and Fish et al. (2003) further emphasize the point regarding surface defects.

Work on color inspection includes both food (Heinemann et al., 1995) and pharmaceutical products (Derganc et al., 2003). Work on X-rays includes the location of foreign bodies in food (Patel et al., 1996; Batchelor et al., 2004), the internal inspection of solder joints (Roh et al., 2003), and the examination of baggage (Wang and Evans, 2003). Finally, a recent volume on inspection of natural products (Graves and Batchelor, 2003) has articles on inspection of ceramics, wood, textiles, food, live fish and pigs, and sheep pelts, and embodies work on color and X-ray modalities. In a sense, such work is neither adventurous nor glamorous. Indeed, it involves significant effort to develop the technology and software sufficiently to make it useful for industry—which means this is an exacting type of task, not tied merely to the production of academic ideas.

With time details often become less relevant, and the fundamental theory of vision limits what is practically possible in real applications. Hence, it does not seem fruitful to dwell further on applications here, beyond noting that the possibilities are continually expanding and should eventually cover most aspects of inspection, assembly automation, and vehicle guidance that are presently handled by humans. (Note, however, that for safety, legal, and social reasons, it may not be acceptable for machines to take control in all situations that they could theoretically handle.) Applications are particularly limited by available techniques for image acquisition and, as suggested above, by possibilities for cost-effective hardware implementation. For this reason, Chapters 27 and 28 should be read in conjunction with the present chapter. It is curious how many books on image analysis almost or even totally ignore one or another of these aspects (especially the former) as if they had no real importance.

Inspection of Cereal Grains

Inspection of cereal grains is an exceptionally mundane and repetitive task, and to some extent contrasts with many of the examples given in the previous chapter in that the emphasis is on detecting contaminants rather than finding manufacturing faults. This chapter presents three case studies that air and solve important theoretical questions.

Look out for:

- the limitations of the immediately obvious technique—global thresholding.
- the value of morphological filtering.
- the unusual need for median filtering as a morphological operation.
- the use of bar (linear feature) detectors for locating insects.
- how the vectorial bar detector operator is optimized.
- how sampling can be used to speed up object location.
- how the outputs of oblique template masks can optimally be combined.

This chapter focuses on the problems of achieving real-time operation and largely solves these problems for less demanding situations (flow rates up to $\sim$300 items per second). For more demanding situations, dedicated hardware accelerators are needed, as discussed in Chapter 28. At a more detailed level, the balance between false positives (false alarms) and false negatives needs to be optimized, as discussed in Section 24.7.

CHAPTER 23

Inspection of Cereal Grains

23.1 Introduction

Cereal grains are among the most important of the foods we grow. A large proportion of cereal grains is milled and marketed as flour, which is then used to produce bread, cakes, biscuits, and many other commodities. Cereal grains can also be processed in a number of other ways (such as crushing), and thus they form the basis of many breakfast cereals. In addition, some cereal grains, or cereal kernels, are eaten with a minimum of processing. Rice is an obvious member of this category—though whole wheat and oat grains are also consumed as decorative additives to "granary" loaves.

Wheat, rice, and other cereal grains are shipped and stored in huge quantities measured in millions of tons, and a very large international trade exists to market these commodities. Transport also has to be arranged between farmers, millers, crushers, and the major bakers and marketers. Typically, transit by road or rail is in relatively small loads of up to twenty tons, and grain depots, warehouses, and ports are not unlikely to receive lorries containing such consignments at intervals ranging from 20 minutes to as little as 3 minutes. As a result of all the necessary transportation and storage, grains are subject to degradation, including damage, molds, sprouting, and insect infestation. In addition, because grains are grown on the land and threshed, potential contamination by rodent droppings, stones, mud, chaff, and foreign grains arises. Finally, the quality of grain from various sources will not be uniformly high, and varietal purity is an important concern.

These factors mean that, ideally, the grain that arrives at any depot should be inspected for many possible causes of degradation. This chapter is concerned with grain inspection. In the space of one chapter we cannot cover all possible methods and modes of inspection, especially since changes in this area are accelerating. Not only are inspection methods evolving rapidly, but so are the standards against which inspection is carried out. To some extent, the improvement of automatic inspection methods and the means of implementing them efficiently in hardware are helping to drive the process onward. We shall explore the situation with the aid of three main case studies, and then we shall look at the overall situation.

The first case study involves the examination of grains to locate rodent droppings and molds such as ergot; the second considers how grains may be scrutinized for insects such as the saw-toothed grain beetle; and the third deals with inspection of the grains themselves. However, this last case study will be more general and will concern itself less with scrutiny of individual grains than with how efficiently they can be located. This is an important factor when truckloads of grains are arriving at depots every few minutes—leading to the need to sample on the order of 300 grains per second. As observed in Chapter 22, object location can involve considerably more computation than object scrutiny and measurement.

23.2 Case Study 1: Location of Dark Contaminants in Cereals

Grain quality needs to be monitored before processing to produce flour, breakfast cereals, and myriad other derived products. So far, automated visual inspection has been applied mainly to the determination of grain quality (Ridgway and Chambers, 1996), with concentration on varietal purity (Zayas and Steele, 1990; Keefe, 1992) and the location of damaged grains (Liao et al., 1994). Relatively little attention has been paid to the detection of contaminants. There is the need for a cheap commercial system that can detect insect infestations and other important contaminants in grain, while not being confused by "permitted admixture" such as chaff and dust. (Up to ~2% is normally permitted.) The inspection work described in this case study (Davies et al., 1998a) pays particular attention to the detection of certain important noninsect contaminants. Relevant contaminants in this category include rodent (especially rat and mouse) droppings, molds such as ergot, and foreign seeds. (Note that ergot is poisonous to humans, so locating any instances of it is of special importance.) In this case study, the substrate grain is wheat, and foreign seeds such as rape would be problematic if present in too great a concentration.

Many of the potential contaminants for wheat grains (and for grains of similar general appearance such as barley or oats) are quite dark in color. This means that thresholding is the most obvious approach for locating them. However, a number of problems arise in that shadows between grains, dark patches on the grains, chaff, and other admixture components, together with rapeseeds, appear as false alarms. Further recognition procedures must therefore be invoked to distinguish between the various possibilities. As a result, the thresholding approach is not as attractive as it might have *a priori* been thought (Fig. 23.1*a*, *b*). This problem is exacerbated by the extreme speed of processing

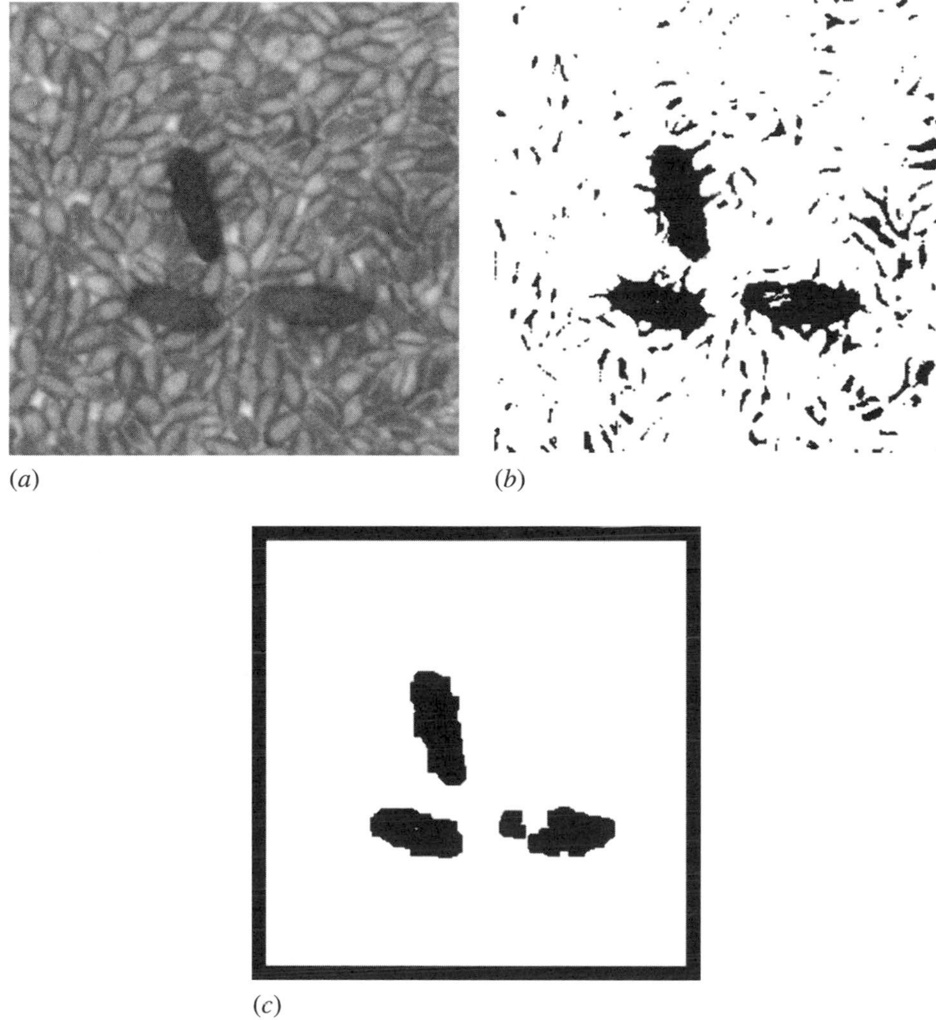

Figure 23.1 Effects of various operations and filters on a grain image. (*a*) Grain image containing several contaminants (rodent droppings). (*b*) Thresholded version of (*a*). (*c*) Result of erosion and dilation on (*b*). (*d*) Result of dilation and erosion on (*b*). (*e*) Result of erosion on (*d*). (*f*) Result of applying 11 × 11 median filter to (*b*). (*g*) Result of erosion on (*f*). In all cases, "erosion" means three applications of the basic 3 × 3 erosion operator, and similarly for "dilation."

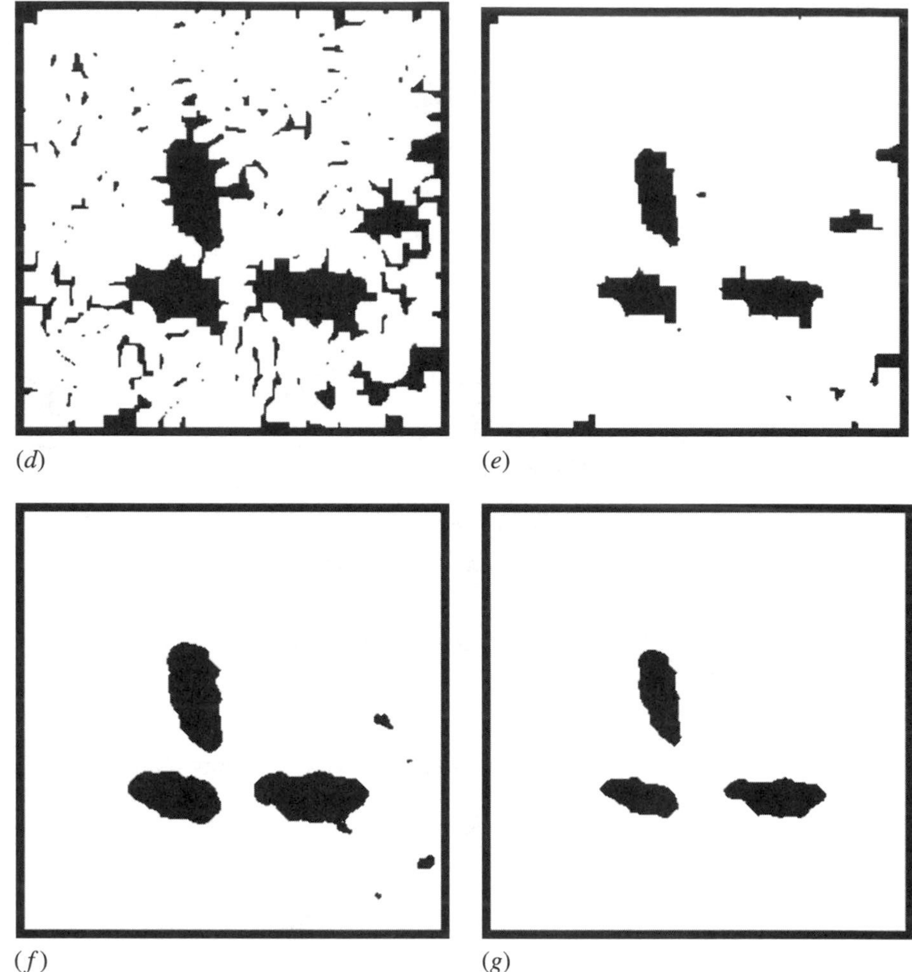

Figure 23.1 Continued.

required in real applications. For example, a successful device for monitoring truckloads of grain might well have to analyze a 3-kg sample of grain in three minutes (the typical time between arrival of lorries at a grain terminal), and this would correspond to some 60,000 grains having to be monitored for contaminants in that time. This places a distinct premium on rapid, accurate image analysis.

This case study concentrates on monitoring grain for rodent droppings. As already indicated, this type of contaminant is generally darker than the grain background, but it cannot simply be detected by thresholding since there are significant shadows between the grains, which themselves often have dark patches. In addition, the contaminants are speckled because of their inhomogeneous constitution and because of lighting and shadow effects. Despite these problems, human operators can identify the contaminants because they are relatively large and their shape follows a distinct pattern (e.g., an aspect ratio of three or four to one). Thus, the combination of size, shape, relative darkness, and speckle characterizes the contaminants and differentiates them from the grain substrate.

Designing efficient, rapidly operating algorithms to identify these contaminants presents a challenge, but an obvious way of meeting it, as we shall see in the next subsection, is via mathematical morphology.

23.2.1 *Application of Morphological and Nonlinear Filters to Locate Rodent Droppings*

The obvious approach to locating rodent droppings is to process thresholded images by erosion and dilation. (Erosion followed by dilation is normally termed opening, but we will not use that term here in order to retain the generality of not insisting on exactly the same number of erosion and dilation operations. In inspection it is the final recognition that must be emphasized rather than some idealized morphological process.) In this way, the erosions would eliminate shadows between grains and discoloration of grains, and the subsequent dilations would restore the shapes and sizes of the contaminants. The effect of this procedure is shown in Fig. 23.1*c*. Note that the method has been successful in eliminating the shadows between the grains but has been decidedly weak in coping with light regions on the contaminants. Remembering that although considerable uniformity might be expected between grains, the same cannot be said about rodent droppings, whose size, shape, and color vary quite markedly. Hence, the erosion–dilation schema has limited value, though it would probably be successful in most instances. Accordingly, other methods of analysis were sought (Davies et al., 1998a).

The second approach to locating rodent droppings is to compensate for the deficiency of the previous approach by ensuring that the contaminants are consolidated even if they are speckled or light in places. Thus, an attempt was made to apply dilation before erosion. The effect of this approach is shown in Fig. 23.1*d*. Notice that the result is to consolidate the shadows between grains even more than the shapes of the contaminants. Even when an additional few erosions are applied (Fig. 23.1*e*) the consolidated shadows do not disappear and are of comparable sizes to the contaminants. Overall, the approach is not viable, for it creates more problems than it solves. (The number of false positives far exceeds the number of false negatives, and similarly for the total areas of false positives and false negatives.) One possibility is to use the results of the earlier erosion–dilation schema as "seeds" to validate a subset of the dilation–erosion schema. However, this would be far more computation intensive, and the results would not be especially impressive (see Fig. 23.1*c*, *e*). Instead, a totally different approach was adopted.

The new approach was to apply a large median filter to the thresholded image, as shown in Fig. 23.1*f*. This gives good segmentation of the contaminants, retaining their intrinsic shape to a reasonable degree, and it suppresses the shadows between grains quite well. In fact, the shadows immediately around the contaminants enhance the sizes of the contaminants in the median filtered image, whereas some shadows further away are consolidated and retained by the median filtering. A reasonable solution is to perform a final erosion operation (Fig. 23.1*g*). This approach eliminates the extraneous shadows and brings the contaminants back to something like their proper size and shape; although the lengths are slightly curtailed, this is not a severe disadvantage. The median filtering–erosion schema easily gave the greatest fidelity to the original contaminants, while being particularly successful at eliminating other artifacts (Davies et al., 1998a).

Finally, although median filtering is intrinsically more computation intensive than erosion and dilation operations, many methods have been devised for accelerating median operations (see Chapter 3). Indeed, the speed aspect was found to be easily soluble within the specification set out above.

23.2.2 *Appraisal of the Various Schemas*

In the previous section, we showed that two basic schemas that aimed in the one case to consolidate the background and then the contaminants and in the other to consolidate the contaminants and then the background failed because each interfered with the alternate region. The median-based schema apparently avoided this problem because it was able to tackle both regions at once without prejudicing either. It is interesting to speculate that in this case the median filter is acting as an analytical device that carefully meditates and obtains the final result in a single rigorous stage, thereby avoiding the error propagation inherent in a two-stage process. Curiously, there seems to be no prior indication of this particular advantage of the median filter in the preexisting and rather wide morphological literature.

23.2.3 Problems with Closing

This case study required the use of a median filter coupled with a final erosion operation to eliminate artifacts in the background. An earlier test similarly involved a closing operation (a dilation followed by an erosion), followed by a final erosion. In other applications, grains or other small particles are often grouped by applying a closing operation to locate the regions where the particles are situated (Fig. 8.6). It is interesting to speculate whether, in the latter type of approach, closing should occasionally be followed by erosion, and also whether the final erosions used in our tests on grain were no more than ad hoc procedures or whether they were vital to achieve the defined goal.

Davies (2000c) analyzed the situation, which is summarized in Sections 8.6.1 and 8.6.2. The analysis starts by considering two regions containing small particles with occurrence densities ρ_1, ρ_2, where $\rho_1 > \rho_2$ (Figs. 8.7 and 8.8). It finds that a final erosion is indeed required to eliminate a shift in the region boundary, the estimated shift δ being[1]

$$\delta = 2ab\rho_2(a + b) \tag{23.1}$$

where a is the radius of the dilation kernel and b is the width of the particles in region 2.

If $b = 0$, no shift will occur, but for particles of measurable size this is not so. If b is comparable to a or if a is much greater than 1, a substantial final erosion may be required. On the other hand, if b is small, the two-dimensional shift may be less than 1 pixel. In that case it will not be correctable by a subsequent erosion, though it should be remembered that a shift has occurred, and any corrections relating to the size of region 1 can be made during subsequent analysis. Although in this work the background artifacts were induced mainly by shadows around and between grains, in other cases impulse noise or light background texture could give similar effects. Care must be taken to select a morphological process that limits any overall shifts in region boundaries. In addition, the whole process can be modeled and the extent of any necessary final erosion can be estimated accurately. More particularly, the final erosion is a necessary measure and is not merely an ad hoc procedure (Davies, 2000c).

23.3 Case Study 2: Location of Insects

As noted in the first case study, a cheap commercial system that can detect a variety of contaminants in grain is needed. The present case study pays

1 A full derivation of this result appears in Section 8.6.1.

particular attention to the need to detect insects. Insects present an especially serious threat because they can multiply alarmingly in a short span of time, so greater certainty of detection is vital in this case. A highly discriminating method was therefore required for locating adult insects (Davies et al., 1998b).

Not surprisingly, thresholding initially seemed to be the approach that offered the greatest promise for locating insects, which appear dark relative to the light brown color of the grain. However, early tests on these lines showed that many false alarms would result from chaff and other permitted admixture, from less serious contaminants such as rapeseeds, and even from shadows between, and discolorations on, the grains themselves (Fig. 23.2). These considerations led to the use of the linear feature detection approach for detecting small adult insects. This proved possible because these insects appear as dark objects with a linear (essentially bar-shaped) structure. Hence, attempts to detect them by applying bar-shaped masks ultimately led to a linear feature detector that had good sensitivity for a reasonable range of insect sizes. Before proceeding, we consider the problems of designing linear feature detector masks.

23.3.1 *The Vectorial Strategy for Linear Feature Detection*

In earlier chapters we found that location of features in 3×3 windows typically required the application of eight template masks in order to cope with arbitrary orientation. In particular, to detect corners, eight masks were required, while only one was needed to locate small holes because of the high degree of symmetry present in the latter case (see Chapters 13 and 14). To detect edges by this means, four masks were required because 180° rotations correspond to a sign change that effectively eliminates the need for half of the masks. However, edge detection is a special case, for edges are vector quantities with magnitude and direction. Hence, they are fully definable using just two component masks. In contrast, a typical line segment detection mask would have the form:

$$\begin{bmatrix} -1 & 2 & -1 \\ -1 & 2 & -1 \\ -1 & 2 & -1 \end{bmatrix}$$

This indicates that four masks are sufficient, the total number being cut from eight to four by the particular rotation symmetry of a line segment. Curiously, it has been shown that although only edges qualify as strict vectors, requiring

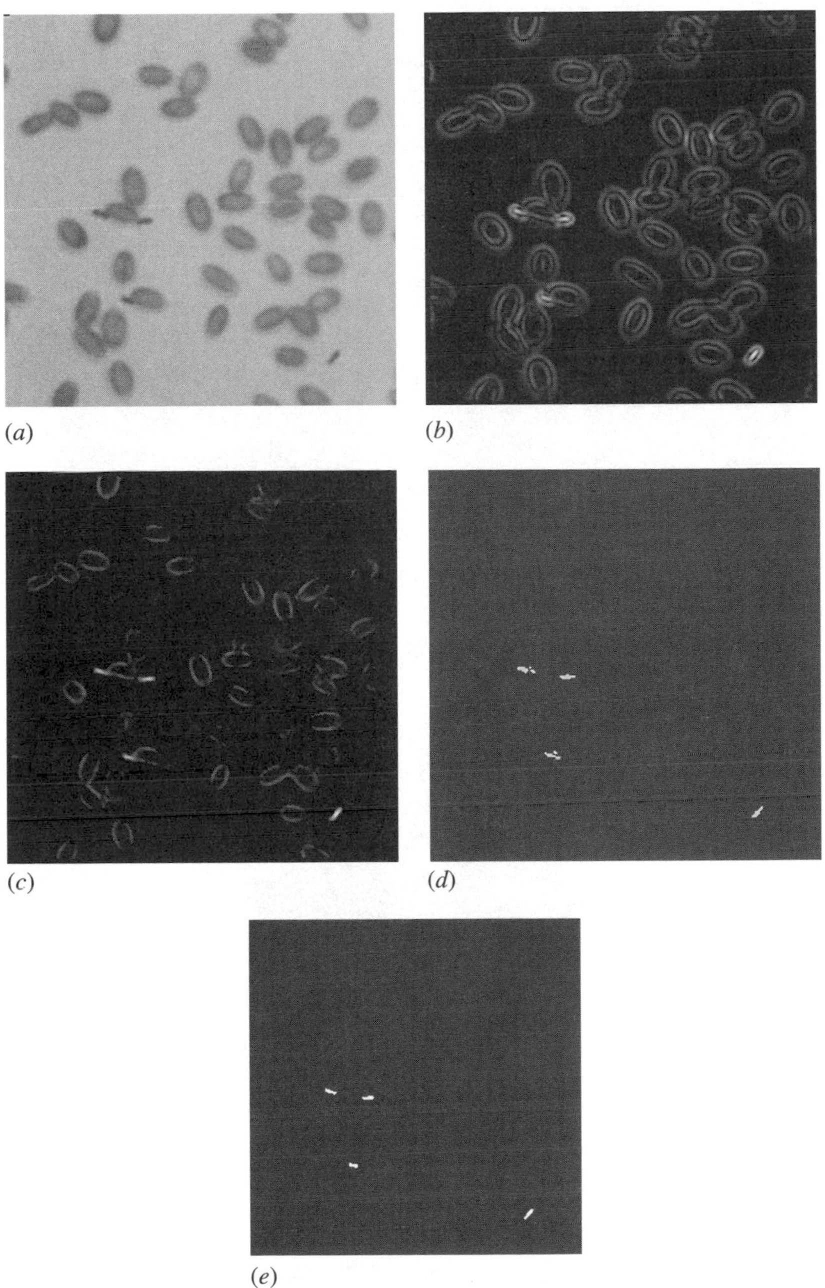

Figure 23.2 Insects located by line segment detector. (*a*) Original image. (*b*) Result of applying vectorial operator. (*c*) Result of masking (*b*) using an intensity threshold at a standard high level. (*d*) Result of thresholding (*c*). (*e*) Result of applying TM operation within (*d*).

just two component masks, the same can be made to apply for line segments. To achieve this, a rather unusual set of masks has to be employed:[2]

$$\mathrm{L}_0 = A\begin{bmatrix} 0 & -1 & 0 \\ 1 & 0 & 1 \\ 0 & -1 & 0 \end{bmatrix} \qquad \mathrm{L}_{45} = B\begin{bmatrix} -1 & 0 & 1 \\ 0 & 0 & 0 \\ 1 & 0 & -1 \end{bmatrix} \tag{23.2}$$

The two masks are given different weights so that some account can be taken of the fact that the nonzero coefficients are, respectively, 1 and $\sqrt{2}$ pixels from the center pixel in the window. These masks give responses g_0, g_{45} leading to:

$$g = (g_0^2 + g_{45}^2)^{1/2} \tag{23.3}$$

$$\theta = \frac{1}{2}\arctan(g_{45}/g_0) \tag{23.4}$$

where the additional factor of one-half in the orientation arises as the two masks correspond to basic orientations of 0° and 45° rather than 0° and 90° (Davies, 1997e). Nevertheless, these masks can cope with the full complement of orientations because line segments have 180° rotational symmetry and the final orientation should only be determined within the range 0° to 180°. In Section 23.5 we comment further on the obvious similarities and differences between equations (23.3), (23.4), and those pertaining to edge detection

Next, we have to select appropriate values of the coefficients A and B. Applying the above masks to a window with the intensity pattern

a	*b*	*c*
d	*e*	*f*
g	*h*	*i*

leads to the following responses at 0° and 45°:

$$g_0 = A(d + f - b - h) \tag{23.5}$$

$$g_{45} = B(c + g - a - i) \tag{23.6}$$

---

2 The proof of the concepts described here involves describing the signal in a circular path around the current position and modeling it as a sinusoidal variation, which is subsequently interrogated by two quadrature sinusoidal basis functions set at 0° and 45° in real space (but by definition at 90° in the quadrature space). The odd orientation mapping between orientation θ in the quadrature space and orientation φ in real space leads to the additional factor of one half in equation (23.3).

Using this model, Davies (1997e) carried out detailed calculations for the case of a thin line of width w passing through the center of the 3×3 window, pixel responses being taken in proportion to the area of the line falling within each pixel. However, for lines of substantial width, theory proved intractable, and simulations were needed to determine the outcome. The remarkable aspect of the results was the high orientation accuracy that occurred for $w = 1.4$, in the case when $B/A = 0.86$, which gave a maximum error of just 0.4°. In the following discussion, we will not be concerned with obtaining a high orientation accuracy, but will be content to use equation (23.3) in order to obtain an accurate estimate of line contrast using just two masks, thereby saving on computation. The reason for this is the need to attain high sensitivity in the detection of image features that correspond to insect contaminants rather than to pass results on to high-level interpretation algorithms, as orientation is not an important output parameter in this application.

23.3.2 Designing Linear Feature Detection Masks for Larger Windows

When larger windows are used, it is possible to design the masks more ideally to match the image features that have to be detected, because of the greater resolution then permitted. However, there are many more degrees of freedom in the design, and there is some uncertainty as to how to proceed. The basic principle (Davies et al., 1998b) is to use masks whose profile matches the intensity profile of a linear feature around a ring of radius R centered on the feature, and at the same time a profile that follows a particular mathematical model—namely, an approximately sinusoidal amplitude variation. For a given linear feature of width w, the sinusoidal model will achieve the best match for a thin ring of radius R_0 for which the two arc lengths within the feature are each one quarter of the circumference $2\pi R_0$ of the ring. Simple geometry (Fig. 23.3) shows that this occurs when:

$$2R_0 \sin(\pi/4) = w \tag{23.7}$$

$$\therefore \quad R_0 = w/\sqrt{2} \tag{23.8}$$

The width ΔR of the ring should in principle be infinitesimal, but in practice, considering noise and other variations in the dataset, ΔR can validly be up to about 40% of R_0. The other relevant factor is the intensity profile of the ring, and how accurately this has to map to the intensity profile of the linear features that are to be located in the image. In many applications, such linear

Figure 23.3 Geometry for application of a thin ring mask. Here two quarters of the ring mask lie within, and two outside, a rectangular bar feature.

features will not have sharp edges but will be slightly fuzzy, and the sides will have significant and varying intensity gradient. Thus, the actual intensity profile is likely to correspond reasonably closely to a true sinusoidal variation. Masks designed on this basis proved close to optimal when experimental tests were made (Davies et al., 1998b). Figure 23.4 shows masks that resulted from this type of design process for one specific value of R (2.5 pixels).

23.3.3 *Application to Cereal Inspection*

The main class of insect targeted in this study (Davies et al., 1998b) was *Oryzaephilus surinamensis* (saw-toothed grain beetle). Insects in this class approximate to rectangular bars of about 10×3 pixels, and masks of size 7×7 (Fig. 23.4) proved to be appropriate for identifying the pixels on the centerlines of these insects. In addition, the insects could appear in any orientation, so potentially quite a large number of template masks might be required: this would have the disadvantage that their sequential implementation would be highly computation intensive and not conducive to real-time operation. Accordingly, it was natural to employ the vectorial approach to template matching, in which two orthogonal masks would be used as outlined above

·	·	−1	−2	−1	·	·		·	·	−1	·	1	·	·
·	·	−1	−2	−1	·	·		·	−2	−2	·	2	2	·
1	1	·	·	·	1	1		−1	−2	·	·	·	2	1
2	2	·	•	·	2	2		·	·	·	•	·	·	·
1	1	·	·	·	1	1		1	2	·	·	·	−2	−1
·	·	−1	−2	−1	·	·		·	2	2	·	−2	−2	·
·	·	−1	−2	−1	·	·		·	·	1	·	−1	·	·

Figure 23.4 Typical (7 × 7) linear feature detection masks.

instead of the much larger number needed when using a more conventional approach.

Preliminary decisions on the presence of the insects were made by thresholding the output of the vectorial operator. However, the vectorial operator responses for insectlike bars have a low-intensity surround, which is joined to the central response at each end. Between the low-intensity surround and the central response, the signal drops close to zero except near the ends (Fig. 23.2*b*). This response pattern is readily understandable because symmetry demands that the signal be zero when the centers of the masks are coincident with the edge of the bar. To avoid problems from the parts of the response pattern outside the object, intensity thresholding (applied separately to the original image) was used to set the output to zero where it indicates no insect could be present (Fig. 23.2*c*). The threshold level was deliberately set high so as to avoid overdependence on this limited procedure.

23.3.4 *Experimental Results*

In laboratory tests, the vectorial operator was applied to 60 grain images containing a total of 150 insects. The output of the basic enhancement operator (described in Section 23.1) was thresholded to perform the actual detection and then passed to an area discrimination procedure. In accordance with our other work on grain images, objects having fewer than 6 pixels were eliminated, as experience had shown that these are most likely to be artifacts due, for example, to dark shadows or discolorations on the grains.

The detection threshold was then adjusted to minimize the total number of errors.[3] This operation resulted in one false positive and three false negatives, so the great majority of the insects were detected. One of the false negatives was due to two insects that were in contact being interpreted as one. The other false

3 In the tests described in this section, all the images were processed with the same version of the algorithm; no thresholds were adjusted to optimize the results for any individual image.

2	2	2	2	2	2	2		2	2	3	2	1	0	0		4	2	2	0	0	-1	-2
0	0	0	0	0	0	0		2	2	0	0	0	-1	-1		2	2	0	0	-1	-2	-1
-1	-1	-1	-1	-1	-1	-1		0	0	0	-2	-2	-2	-1		2	0	0	-1	-3	-1	0
-2	-2	-2	-2	-2	-2	-2		0	-2	-2	-2	-2	-2	0		0	0	-1	-2	-1	0	0
-1	-1	-1	-1	-1	-1	-1		-1	-2	-2	-2	0	0	0		0	-1	-3	-1	0	0	2
0	0	0	0	0	0	0		-1	-1	0	0	0	2	2		-1	-2	-1	0	0	2	2
2	2	2	2	2	2	2		0	0	1	2	3	2	2		-2	-1	0	0	2	2	4

Figure 23.5 7×7 template matching masks. There are eight masks in all, oriented at multiples of 22.5° relative to the x-axis. Only three are shown because the others are straightforwardly generated by 90° rotation and reflection (symmetry) operations.

negatives were due to insects being masked by lying along the edges of grains. The single false positive was due to the dark edge of a grain appearing (to the algorithm) similar to an insect.

Although adjustment of the detection threshold was found to eliminate either the false positive or two of the false negatives (but not the one in which two insects were in contact), the ultimate reason the total number of errors could not be reduced further lay in the limited sensitivity of the vectorial operator masks, which after all contain relatively few coefficients. This was verified by carrying out a test with a set of conventional template matching (TM) masks (Fig. 23.5), which led to a single false negative (due to the two insects being in contact) and no false positives—and a much increased computational load (Davies et al., 2003a). It was accepted that the case of two touching insects would remain a problem that could be eliminated only by introducing a greater level of image understanding—an aspect beyond the scope of this work (though curiously, the problem could easily be eliminated in this particular instance by noting the excessive length and/or area of this one object).

Next, it was desired to increase the interpretation accuracy of the vectorial operator in order to limit computation, and it seemed worth trying to devise a two-stage system, which would combine the speed of the vectorial operator with the accuracy of TM. Accordingly, the vectorial operator was used to create regions of interest for the TM operator. To obtain optimum performance, it was found necessary to decrease the vectorial operator threshold level to reduce the number of false negatives, leaving a slightly increased number of false positives that would subsequently be eliminated by the TM operator. The final result was an error rate exactly matching that of the TM operator, but a considerably reduced overall execution time composed of the adjusted vectorial operator execution time plus the TM execution time for the set of regions of interest. Further speed improvements, and no loss in accuracy, resulted from employing a dilation operation within the vectorial operator to recombine fragmented insects before applying the final TM operator. In this case, the vectorial operator was used without reducing its threshold value. At this stage, the speedup factor on the original TM operator was around seven (Davies et al., 2003a).

Finally, intensity thresholding was used as a preliminary skimming operation on the vectorial operator—where it produced a speedup by a factor of about 20. This brought the overall speedup factor on the original TM algorithm into the range 50–100, being close to 100 where insects are rare. Perhaps more important, the combined algorithm was in line with the target to inspect 3 kg of grain for a variety of contaminants within three minutes.

23.4 Case Study 3: High-speed Grain Location

As already mentioned, object location often requires considerable computation, for it involves unconstrained search over the image data. As a result, the whole process of automated inspection can be slowed down significantly, which can be of crucial importance in the design of real-time inspection systems. If the scrutiny of particular types of objects requires quite simple dimensional measurements to be made, the object location routine can be the bottleneck. This case study is concerned with the high-speed location of objects in 2-D images, a topic on which relatively little systematic work has been carried out—at least on the software side. Many studies have, however, been made on the design of hardware accelerators for image processing. Because hardware accelerators represent the more expensive option, this case study concentrates on software solutions to the problem and then specializes it to the case of cereal grains in images.

23.4.1 *Extending an Earlier Sampling Approach*

In an earlier project, Davies (1987l) addressed the problem of finding the centers of circular objects such as coins and biscuits significantly more rapidly than for a conventional Hough transform, while retaining as far as possible the robustness of that approach. The best solution appeared to be to scan along a limited number of horizontal lines in the image, recording and averaging the x-coordinates of midpoints of chords of any objects, and repeating the process in the vertical direction to complete the process of center location. The method was successful and led to speedup factors as high as 25 in practical situations—in particular for the rapid location of round biscuits (Chapter 10).

In the present project (Davies, 1998b), extreme robustness was not necessary, and it seemed worth finding how much faster the scanning concept

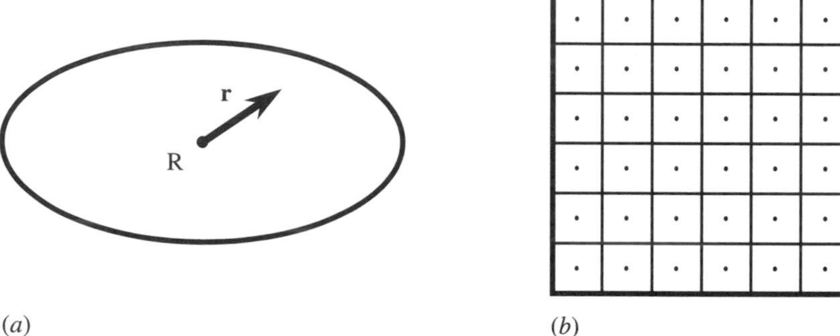

Figure 23.6 Object shape and method of sampling. (*a*) Object shape, showing reference point R and vector $\mathbf{r}$ pointing to a general location $\mathbf{x}_{\mathrm{R}} + \mathbf{r}$. (*b*) Image and sampling points, with associated tiling squares.

could be taken. It was envisaged that significant improvement might be achieved by taking a minimum number of sampling points in the image rather than by scanning along whole lines.

Suppose that we are looking for an object such as that shown in Fig. 23.6*a*, whose shape is defined relative to a reference point R as the set of pixels $A = \{\mathbf{r}_i: i = 1 \text{ to } n\}$, n being the number of pixels within the object. If the position of R is $\mathbf{x}_{\mathrm{R}}$, pixel i will appear at $\mathbf{x}_i = \mathbf{x}_{\mathrm{R}} + \mathbf{r}_i$. Accordingly, when a sampling point $\mathbf{x}_{\mathrm{S}}$ gives a positive indication of an object, the location of its reference point R will be $\mathbf{x}_{\mathrm{R}} = \mathbf{x}_{\mathrm{s}} - \mathbf{r}_i$. Thus, the reference point of the object is known to lie at one of the set of points $U_{\mathrm{R}} = \cup_i(\mathbf{x}_{\mathrm{s}} - \mathbf{r}_i)$, so knowledge of its location is naturally incomplete. The map of possible reference point locations has the same shape as the original object but rotated through 180°—because of the minus sign in front of $\mathbf{r}_i$. Furthermore, the fact that reference point positions are only determined within n pixels means that many sampling points will be needed, the minimum number required to cover the whole image clearly being N/n, if there are N pixels in the image. The optimum speedup factor will therefore be $N/(N/n) = n$, as the number of pixels visited in the image is N/n rather than N (Davies, 1997f).

Unfortunately, it will not be possible to find a set of sampling point locations such that the "tiling" produced by the resulting maps of possible reference point positions covers the whole image without overlap. Thus, there will normally be some overlap (and thus loss of efficiency in locating objects) or some gaps (and thus loss of effectiveness in locating objects). The set of tiling squares shown in Fig. 23.6*b* will only be fully effective if square objects are to be located.

A more serious problem arises because objects may appear in any orientation. This prevents an ideal tiling from being found. It appears that the best that can be

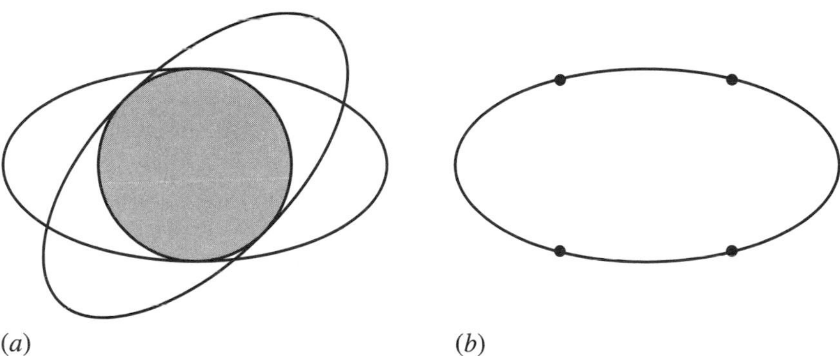

Figure 23.7 Geometry for location of ellipses by sampling. (*a*) Ellipse in two orientations and maximal rotationally invariant subset (shaded). (*b*) Horizontal ellipse and geometry showing size relative to largest permitted spacing of sampling points.

achieved is to search the image for a maximal *rotationally invariant* subset of the shape, which must be a circle, as indicated in Fig. 23.7*a*. Furthermore, as no perfect tiling for circles exists, the tiling that must be chosen is either a set of hexagons or, more practically, a set of squares. This means that the speedup factor for object location will be significantly less than n, though it will still be substantial.

23.4.2 Application to Grain Inspection

A prime application for this technique is that of fast location of grains on a conveyor in order to scrutinize them for damage, varietal purity, sprouting, molds, and the like. Under these circumstances, it is best to examine each grain in isolation; specifically, touching or overlapping grains would be more difficult to cope with. Thus, the grains would need to be spread out with at most 25 grains being visible in any 256×256 image. With so much free image space there would be an intensive search problem, with far more pixels having to be considered than would otherwise be the case. Hence, a very fast object location algorithm would be of special value.

Wheat grains are well approximated by ellipses in which the ratio of semimajor (a) to semiminor (b) axes is almost exactly two. The deviation is normally less than 20%, though there may also be quite large apparent differences between the intensity patterns for different grains. Hence, it was deemed useful to employ this model as an algorithm optimization target. First, the (nonideal) $L \times L$

square tiles would appear to have to fit inside the circular maximal rotationally invariant subset of the ellipse, so that $\sqrt{2}L = 2b$, i.e., $L = \sqrt{2}b$. This value should be compared with the larger value $L = (4/\sqrt{5})b$ that could be used if the grains were constrained to lie parallel to the image x-axis—see Fig. 23.7*b*. (Here we are ignoring the dimensions $2\sqrt{2}b \times \sqrt{2}b$ for optimal *rectangular* sampling tiles.)

Another consequence of the difference in the shape of the objects being detected (here ellipses) and the tile shape (square) is that the objects may be detected at several sample locations, thereby wasting computation (see Section 23.4.1). A further consequence is that we cannot merely count the samples if we wish to count the objects. Instead, we must relate the multiple object counts together and find the centers of the objects. This also applies if the main goal is to locate the objects for inspection and scrutiny. In the present case, the objects are convex, so we only have to look along the line joining any pair of sampling points to determine whether there is a break and thus whether they correspond to more than one object. We shall return later to the problem of systematic location of object centers.

For ellipses, it is relevant to know how many sample points could give positive indications for any one object. Now the maximum distance between one sampling point and another on an ellipse is $2a$, and for the given eccentricity this is equal to $4b$, which in turn is equal to $2\sqrt{2}L$. Thus, an ellipse of this eccentricity could overlap three sample points along the x-axis direction if it were aligned along this direction. Alternatively, it could overlap just two sample points along the 45° direction if it were aligned along this direction, though it could in that case also overlap just one laterally placed sample point. In an intermediate direction (e.g., at an angle of around arctan 0.5 to the image x-axis), the ellipse could overlap up to five points. Similarly, it is easy to see that the minimum number of positive sample points per ellipse is two. The possible arrangements of positive sample points are presented in Fig. 23.8*a*.

Fortunately, this approach to sampling is overrigorous. Specifically, we have insisted upon containing the sampling tile within the ideal (circular) maximal rotationally invariant subset of the shape. However, what is required is that the sampling tile must be of such a size that allowance is made for all possible orientations of the shape. In the present example, the limiting case that must be allowed for occurs when the ellipse is oriented parallel to the x-axis. It must be arranged that it can just pass through four sampling points at the corners of a square, so that on any infinitesimal displacement, at least one sampling point is contained within it. For this to be possible, it can be shown that $L = (4/\sqrt{5})b$, the same situation as already depicted in Fig. 23.7*b*. This leads to the possible arrangements of positive sampling points shown in Fig. 23.8*b*—a distinct reduction in the average number of positive sampling points, which leads to useful savings in computation. (The average number of positive sampling points per ellipse is reduced from ~3 to ~2.)

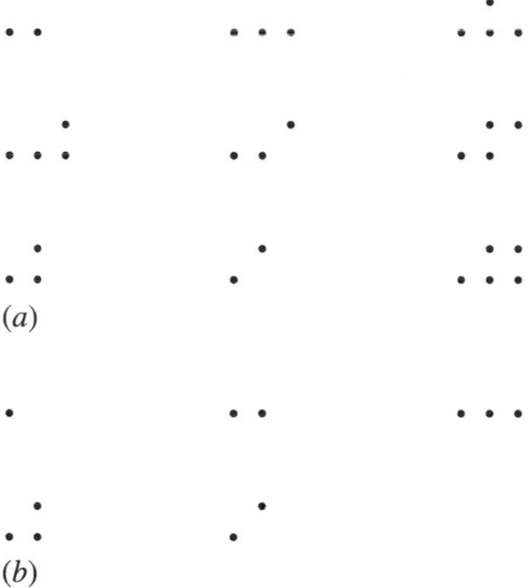

Figure 23.8 Possible arrangements of positive sampling points for ellipse, (*a*) with $L = \surd 2b$, and (*b*) with $L = (4/\surd 5)b$.

Object location normally takes considerable computation because it involves an unconstrained search over the whole image space. In addition, there is normally (as in the ellipse location task) the problem that the orientation is unknown. This contrasts with the other crucial aspect of inspection, that of object scrutiny and measurement, in that relatively few pixels have to be examined in detail, requiring relatively little computation. The sampling approach we have outlined largely eliminates the search aspect of object location, since it quickly eliminates any large tracts of blank background. Nevertheless, the problem of refining the object location phase remains. One approach to this problem is to expand the positive samples into fuller regions of interest and then to perform a restricted search over these regions. For this purpose, we could use the same search tools that we might use over the whole image if sampling were not being performed. However, the preliminary sampling technique is so fast that this approach would not take full advantage of its speed. Instead, we could use the following procedure.

For each positive sample, draw a horizontal chord to the boundary of the object, and find the local boundary tangents. Then use the chord–tangent technique (join of tangent intersection to midpoint of chord: see Chapter 12) to determine one line on which the center of an ellipse must lie. Repeat this procedure for all the positive samples, and obtain all possible lines on which ellipse centers must lie. Finally, deduce the possible ellipse center locations, and check each of them in detail in case some correspond to

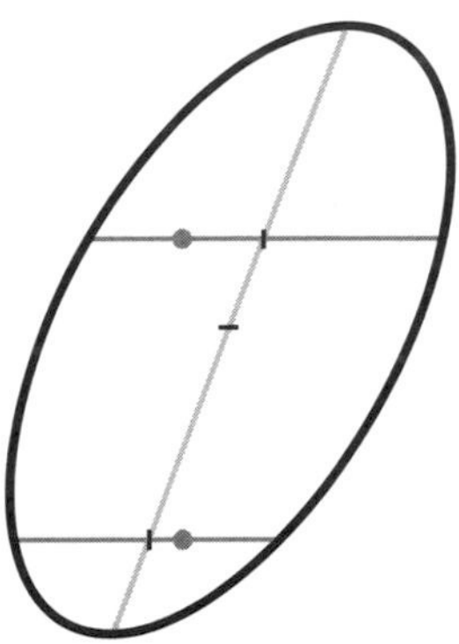

Figure 23.9 Illustration of triple bisection algorithm. The round spots are the sampling points, and the short bars are the midpoints of the three chords, the short horizontal bar being at the center of the ellipse.

false alarms arising from objects that are close together rather than from genuine self-consistent ellipses. Note that when there is a single positive sampling point, another positive sampling point has to be found (say, $L/2$ away from the first).

A significantly faster approach called the *triple bisection* algorithm has recently been developed (Davies, 1998b). Draw horizontal (or vertical) chords through adjacent vertically (or horizontally) separated pairs of positive samples, bisect them, join and extend the bisector lines, and finally find the midpoints of these bisectors (Fig. 23.9). (In cases where there is a single positive sampling point, another positive sampling point has to be found, say, $L/2$ away from the first.) The triple bisection algorithm has the additional advantage of not requiring estimates of tangent directions to be made at the ends of chords, which can prove inaccurate when objects are somewhat fuzzy, as in many grain images. The result of applying this technique to an image containing mostly well-separated grains is shown in Fig. 23.10. This illustrates that the whole procedure for locating grains by modeling them as ellipses and searching for them by sampling and chord bisection approaches is a viable one. In addition, the procedure is very rapid, as the number of pixels that are visited is a small proportion of the total number in each image.

Finally, we show why the triple bisection algorithm is appropriate. First note that it is correct for a circle, for reasons of symmetry. Second, note that in orthographic projection, circles become ellipses, straight lines become

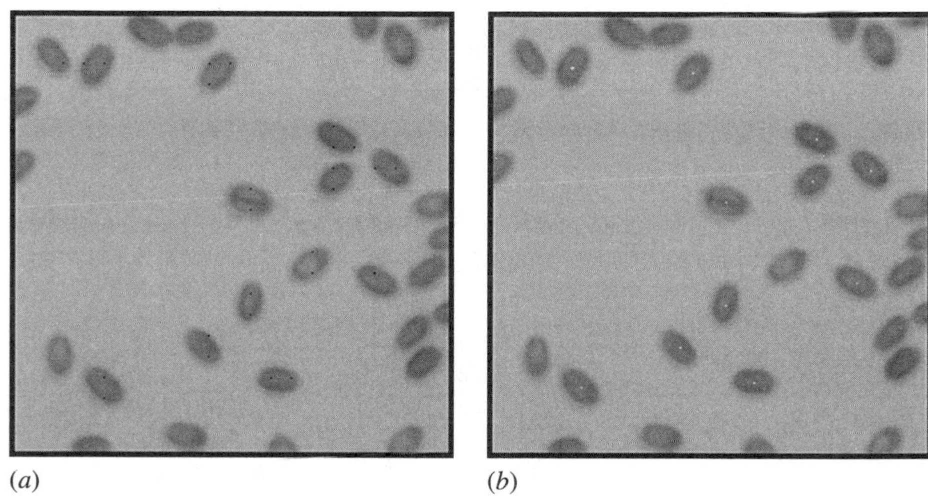

(*a*) (*b*)

Figure 23.10 Image showing grain location using the sampling approach. (*a*) Sampling points. (*b*) Final center locations.

straight lines, parallel lines become parallel lines, chords become chords, and midpoints become midpoints. Hence, choosing the correct orthogonal projection to transform the circle into a correctly oriented ellipse of appropriate eccentricity, we find that the midpoints and center location shown in the diagram of Fig. 23.9 must be validly marked. This proves the algorithm. (For a more rigorous algebraic proof, see Davies, 1999b.)

23.4.3 *Summary*

Motivated by the success of an earlier line-based sampling strategy (Davies, 1987l), Davies has shown that point samples lead to the minimum computational effort when the 180°-rotated object shapes form a perfect tiling of the image space. In practice, imperfect tilings have to be used, but these can be extremely efficient, especially when the image intensity patterns permit thresholding, the images are sparsely populated with objects, and the objects are convex in shape. An important feature of the approach is that detection speed is *improved* for larger objects, though naturally exact location involves some additional effort. In the case of ellipses, this process is considerably aided by the triple bisection algorithm.

The method has been applied successfully to the location of well-separated cereal grains, which can be modeled as ellipses with a 2:1 aspect ratio, ready for scrutiny to assess damage, varietal purity, or other relevant parameters.

23.5 Optimizing the Output for Sets of Directional Template Masks

Several earlier sections of this chapter highlighted the value of low-level operations based on template masks but did not enter into all the details of their design. In particular, they did not breach the problem of how to calculate the signal from a set of n masks of various orientations—a situation exemplified by the case of corner detection, for which eight masks are frequently used in a 3×3 window (see Section 23.2). The standard approach is to take the maximum of the n responses, though clearly this represents a lower bound on the signal magnitude λ and gives only a crude indication of the orientation of the feature. Applying the geometry shown in Fig. 23.11, the true value has components:

$$c_1 = \lambda \cos \alpha \tag{23.9}$$

$$c_2 = \lambda \cos \beta \tag{23.10}$$

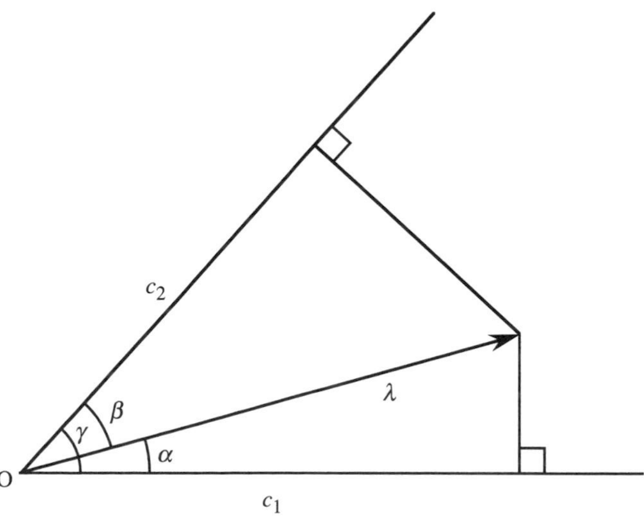

Figure 23.11 Geometry for vector calculation.

Finding the ratio between the components yields the first important result:

$$c_2/c_1 = \frac{\cos\beta}{\cos\alpha} = \cos(\gamma - \alpha)\sec\alpha$$
$$= \cos\gamma + \sin\gamma\,\tan\alpha \tag{23.11}$$

and leads to a formula for tan α:

$$\tan\alpha = (c_2/c_1 - \cos\gamma)/\sin\gamma \tag{23.12}$$

Hence:

$$\alpha = \arctan[(c_2/c_1)\mathrm{cosec}\,\gamma - \cot\gamma] \tag{23.13}$$

Next, we find from equation (23.9):

$$\lambda = c_1 \sec\alpha \tag{23.14}$$

and elimination of α gives:

$$\lambda = c_1\left[1 + \left(\frac{c_2/c_1 - \cos\gamma}{\sin\gamma}\right)^2\right]^{1/2} \tag{23.15}$$

On expansion we finally obtain the required symmetrical formula:

$$\lambda = (c_1^2 + c_2^2 - 2c_1c_2\cos\gamma)^{1/2}\,\mathrm{cosec}\,\gamma \tag{23.16}$$

The closeness to the form of the cosine rule is striking and can be explained by simple geometry (Davies, 2000b).

Interestingly, equations (23.16) and (23.13) are generalizations of the standard vector results for edge detection:

$$\lambda = (c_1^2 + c_2^2)^{1/2} \tag{23.17}$$

$$\alpha = \arctan(c_2/c_1) \tag{23.18}$$

which apply when $\gamma = 90°$. Specifically, the results apply for oblique axes where the relative orientation is γ.

23.5.1 *Application of the Formulas*

Before proceeding further, it will be convenient to define features that have $2\pi/m$ rotation invariance as *m*-vectors. It follows that edges are 1-vectors, and line segments (following the discussion in Section 23.2.1) are 2-vectors. In addition, it would appear that *m*-vectors with $m > 2$ rarely exist in practice.

So far we have established that the formulas given in Section 23.5 apply for 1-vectors such as edges. However, to apply the results to 2-vectors such as line segments, the 1:2 relation between the angles in real and quadrature manifolds has to be acknowledged. Thus, the proper formulas are obtained by replacing α, β, γ, respectively, by 2α, 2β, 2γ in the earlier equations. This leads to the components:

$$c_1 = \lambda \cos 2\alpha \tag{23.19}$$

$$c_2 = \lambda \cos 2\beta \tag{23.20}$$

with $2\gamma = 90°$ replacing $\gamma = 45°$ in the formula giving the final orientation:

$$\begin{aligned} 2\alpha &= \arctan[(\cos 2\beta / \cos 2\alpha)\operatorname{cosec} 90° - \cot 90°] \\ &= \arctan[(\cos 2\beta / \cos 2\alpha)] \end{aligned} \tag{23.21}$$

Hence,

$$\alpha = \frac{1}{2}\arctan(c_2/c_1) \tag{23.22}$$

as found in Section 23.2.1.

Finally, we apply these methods to obtain interpolation formulas for sets of 8 masks for both 1-vectors and 2-vectors. For 1-vectors (such as edges), 8-mask sets have $\gamma = 45°$, and equations (23.16) and (23.13) take the form:

$$\lambda = \surd 2(c_1^2 + c_2^2 - \surd 2 c_1 c_2)^{1/2} \tag{23.23}$$

$$\alpha = \arctan[\surd 2(c_2/c_1) - 1] \tag{23.24}$$

For 2-vectors (such as line segments), 8-mask sets have $\gamma = 22.5°$, so $2\gamma = 45°$, and the relevant equations are:

$$\lambda = \surd 2(c_1^2 + c_2^2 - \surd 2 c_1 c_2)^{1/2} \tag{23.25}$$

$$\alpha = \frac{1}{2}\arctan[\surd 2(c_2/c_1) - 1] \tag{23.26}$$

23.5.2 *Discussion*

This section has examined how the responses to directional template matching masks can be interpolated to give optimum signals. It has arrived at formulas that cover the cases of 1-vector and 2-vector features, typified, respectively, by edge and line segments. Although there are no likely instances of m-vectors with $m > 2$, features can easily be tested for rotation invariance, and it turns out that corners and most other features should be classed as 1-vectors. An exception is a symmetrical S-shape, which should be classed as a 2-vector. Support for classifying corners as 1-vectors arises as they form a subset of the edges in an image. These considerations mean that the solutions arrived at for 1-vectors and 2-vectors will solve all the foreseeable problems of signal interpolation between responses for masks of different orientations. Thus, the main aim of this section is accomplished—to find a useful nonarbitrary solution to determining how to calculate combined responses for n-mask operators.

Oddly, these useful results do not seem to have been reported in the earlier literature of the subject.

23.6 Concluding Remarks

The topic of cereal grain inspection is ostensibly highly specialized. Not only is it just one aspect of automated visual inspection, but also it covers just one sector of food processing. Yet its study has taken these topics to the limits of the capabilities of a number of algorithms, and some very pertinent theory has been exercised and developed. This effort has made the chapter considerably more generic than might have been thought *a priori*. Indeed, this is a major reason for the rather strange inclusion of a section on template matching using sets of directional masks late on in the chapter. It was needed to advance the discussion of the subject, and if included in Part 1 of the book as a further aspect of low-level vision, it might have been regarded as a rather boring, unimportant addition. On the other hand, its presence in the present chapter emphasizes that not all the theory that needs to be known, and might have been expected to be known, is actually available in the literature—and there must surely be many further instances of this waiting to the revealed. In addition, there is a tendency for inspection and other algorithms to be developed ad hoc, whereas there is a need for solid theory to underpin this sort of work and, above all, to ensure that any techniques used are effective and robust. This is a lesson worth including in a volume such as the present one, for which practicalities and theory are intended to go hand in hand.

Food inspection is an important aspect of automated visual inspection. This chapter has shown how bulk cereals may be scrutinized for insects, rodent droppings, and other contaminants. The application is subject to highly challenging speed constraints, and special sampling techniques had to be developed to attain the ultimate speeds of object location.

23.7 Bibliographical and Historical Notes

Food inspection is beset with problems of variability; the expression "Like two peas" gives a totally erroneous indication of the situation. Davies has written much about this over the past many years, the position being summed up in his book and in more recent reviews (Davies, 2001a, 2003d). Graves and Batchelor (2003) have edited a volume that collects much further data on the position.

The work on insect detection described in this chapter concentrates on the linear feature detector approach, which was developed in a series of papers already cited, but founded on theory described in Davies (1997g, 1999f), and Davies et al. (2002, 2003a, 2003b). (For other relevant work on linear feature detection but not motivated by insect detection, see Chauduri et al., 1989; Spann et al., 1989; Koller et al., 1995; and Jang and Hong, 1998.) Zayas and Flinn (1998) offer an alternative strategy based on discriminant analysis, but their work targets the lesser grain borer beetle, which has a different shape, and the data and methods are not really comparable. That different methods are needed to detect different types and shapes of insects is obvious. Indeed, Davies' own work shows that morphological methods are more useful for detecting large insects, as well as rodent droppings and ergot (Davies et al., 2003b; Ridgway et al., 2002). The issue is partly one of sensitivity of detection for the smaller insects versus functionality for the larger ones, coupled with speed of processing—optimization issues typical of those discussed in Chapter 29.

Recent work in this area also covers NIR (near infra-red) detection of insect larvae growing within wheat kernels (Davies et al., 2003c), although the methodology is totally different, being centered on the location of bright (at NIR wavelengths) patches on the surfaces of the grains.

The fast processing issue again arises with regard to the inspection of wheat grains, not least because the 30 tons of wheat that constitute a typical truckload contains some 6000 million grains, and the turnaround time for each lorry can be as little as three minutes. The theory for fast processing by image sampling and the subsequent rapid centering of elliptical shapes is developed in Davies (1997f, 1999b, 1999e, 2001b).

Sensitive feature matching is another aspect of the work described in this chapter. Relevant theory was developed by Davies (2000b), but there are other issues such as the "equal area rule" for designing template masks (Davies, 1999a) and the effect of foreground and background occlusion on feature matching (Davies, 1999c). These theories should all have been developed far earlier in the history of the subject and merely serve to show how little is still known about the basic design rules for image processing and analysis. A summary of much of this work appears in Davies (2000d).

Statistical Pattern Recognition

Humans can achieve pattern recognition "at a glance" with little apparent effort. Much of pattern recognition is structural, being achieved essentially by analyzing shape. In contrast, statistical pattern recognition treats sets of extracted features as abstract entities that can be used to classify objects on a statistical basis, often by mathematical similarity to sets of features for objects with known classes. This chapter explores the subject, presenting relevant theory where appropriate.

Look out for:

- the nearest neighbor algorithm, which is probably the most intuitive statistical pattern recognition technique.
- Bayes' theory, which forms the ideal minimum error classification system.
- the relation linking the nearest neighbor method to Bayes' theory.
- the reason why the optimum number of features will always be finite.
- the ROC (receiver-operator characteristic) curve, which allows an optimum balance between false positives and false negatives to be achieved.
- the distinction between supervised and unsupervised learning.
- the method of principal components analysis and its value.
- some ideas on how face recognition can be initiated.

Statistical pattern recognition is a core methodology in the design of practical vision systems. As such, it has to be used in conjunction with structural pattern recognition methods and many other relevant subjects—as indicated in the heading of Part 4 of this volume. This is the context in which the face recognition ideas of Section 24.12 are included within the present chapter.

CHAPTER 24

Statistical Pattern Recognition

24.1 Introduction

The earlier chapters of this book have concentrated on interpreting images on the basis that when suitable cues have been found, the positions of various objects and their identities will emerge in a natural and straightforward way. When the objects that appear in an image have simple shapes, just one stage of processing may be required—as in the case of circular washers on a conveyor. For more complex objects, location requires at least two stages—as when graph matching methods are used. For situations where the full complexity of three dimensions occurs, more subtle procedures are usually required, as has already been seen in Chapter 16. The very ambiguity involved in interpreting the 2-D images from a set of 3-D objects generally requires cues to be sought and hypotheses to be made before any serious attempt can be made at the task. Thus, cues are vital to keying into the complex data structures of many images. For simpler situations, however, concentration on small features is valuable in permitting image interpretation to be carried out efficiently and rapidly. Neither must it be forgotten that in many applications of machine vision the task is made simpler by the fact that the main interest lies in specific types of objects; for example, the widgets to be inspected on a widget line.

The fact that it is often expedient to start by searching for cues means that the vision system could lock on to erroneous interpretations. For example, Hough-based methods lead to interpretations based on the most prominent peaks in parameter space, and the maximal clique approach supports interpretations based on the largest number of feature matches between parts of an image and the object model. Noise, background clutter, and other artifacts make it possible that an object will be seen (or its presence hypothesized) when none is there, while occlusions, shadows, and so on, may cause an object to be missed altogether.

A rigorous interpretation strategy would ideally attempt to find all possible feature and object matches and should have some means of distinguishing between them. One should not be satisfied with an interpretation until the *whole* of any image has been understood. However, enough has been seen in previous chapters (e.g., Chapter 11) to demonstrate that the computational load of such a strategy generally makes it impracticable, at least for real (e.g., outdoor) scenes. Yet in some more constrained situations the strategy is worth considering and various possible solutions can be evaluated carefully before a final interpretation is reached. These situations arise practically where small relevant parts of images can be segmented and interpreted in isolation. One such case is that of optical character recognition (OCR), which is commonly addressed through statistical pattern recognition (SPR).

The following sections study some of the important principles of SPR. A complete description of all the work that has been carried out in this area would take several volumes to cover. Fortunately, SPR has been researched for well over three decades and has settled down sufficiently so that an overview chapter can serve a useful purpose. The reader is referred to several existing excellent texts for further details (Duda et al., 2001; Webb, 2002). We start by describing the nearest neighbor approach to SPR and then consider Bayes' decision theory, which provides a more general model of the underlying process.

24.2 The Nearest Neighbor Algorithm

The principle of the nearest neighbor (NN) algorithm is that of comparing input image[1] patterns against a number of paradigms and then classifying the input pattern according to the class of the paradigm that gives the closest match (Fig. 24.1). An instructive but rather trivial example is that shown in Fig. 1.1. Here a number of binary patterns are presented to the computer in the training phase of the algorithm. Then the test patterns are presented one at a time and compared bit by bit against each of the training patterns. This approach yields a generally reasonable result, the main problems arising when (1) training patterns of different classes are close together in Hamming distance (i.e., they differ in too few bits to be readily distinguishable), and (2) minor translations, rotations, or noise cause variations that inhibit accurate recognition. More generally, problem (2) means that the training patterns are insufficiently representative of what will appear during the test phase. The latter statement encapsulates an exceptionally important principle and implies that there must be sufficient patterns in the

1 Note that a number of the methods discussed in this chapter are very general and can be applied to recognition in widely different datasets, including, for example, speech and electrocardiograph waveforms.

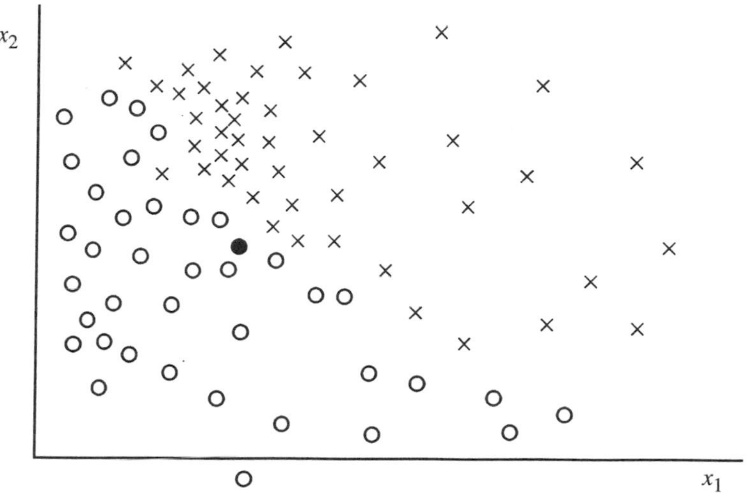

Figure 24.1 Principle of the nearest neighbor algorithm for a two-class problem: ○, class 1 training set patterns; ×, class 2 training set patterns; ●, test pattern.

training set for the algorithm to be able to generalize over all possible patterns of each class. However, problem (1) implies that patterns of two different classes may sometimes be so close as to be indistinguishable by any algorithm, and then it is inevitable that erroneous classifications will be made. As seen later in this chapter, this is because the underlying distributions in feature space overlap.

The example of Fig. 1.1 is rather trivial but nevertheless carries important lessons. General images have many more pixels than those in Fig. 1.1 and also are not merely binary. However, since it is pertinent to simplify the data as far as possible to save computation, it is usual to concentrate on various features of a typical image and classify on the basis of these. One example is provided by a more realistic version of the OCR problem where characters have linear dimensions of at least 32 pixels (although we continue to assume that the characters have been located *reasonably* accurately so that it remains only to classify the subimages containing them). We can start by thinning the characters to their skeletons and making measurements on the skeleton nodes and limbs (see also Chapters 1 and 6). We then obtain the numbers of nodes and limbs of various types, the lengths and relative orientations of limbs, and perhaps information on curvatures of limbs. Thus, we arrive at a set of numerical features that describe the character in the subimage.

The general technique is to plot the characters in the training set in a multidimensional feature space and to tag the plots with the classification index. Then test patterns are placed in turn in the feature space and classified according to the class of the nearest training set pattern. In this way, we generalize the method adopted in Fig. 24.1. Note that in the general case the distance in feature space is no longer Hamming distance but a more general measure such as Mahalanobis distance (Duda et al., 2001). A problem arises since there is no reason why the different dimensions in feature space should contribute equally to distance. Rather, they should each have different weights in order to match the physical problem more closely. The problem of weighting cannot be discussed in detail here, and the reader is referred to other texts such as that by Duda et al. (2001). Suffice it to say that with an appropriate definition of distance, the generalization of the method outlined here is adequate to cope with a variety of problems.

In order to achieve a suitably low error rate, large numbers of training set patterns are normally required, leading to significant storage and computation problems. Means have been found for reducing these problems through several important strategies. Notable among these strategies is that of pruning the training set by eliminating patterns which are not near the boundaries of class regions in feature space, since such patterns do not materially help reduce the misclassification rate.

An alternative strategy for obtaining equivalent performance at lower computational cost is to employ a piecewise linear or other functional classifier instead of the original training set. It will be clear that the NN method itself can be replaced, with no change in performance, by a set of planar decision surfaces that are the perpendicular bisectors (or their analogs in multidimensional space) of the lines joining pairs of training patterns of different classes that are on the boundaries of class regions. If this system of planar surfaces is simplified by any convenient means, then the computational load may be reduced further (Fig. 24.2).

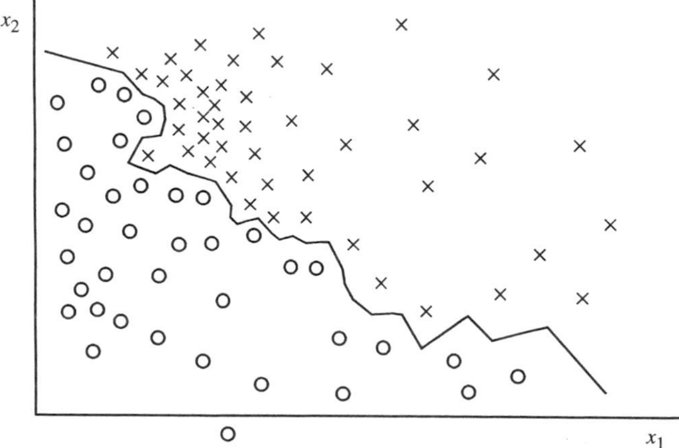

Figure 24.2 Use of planar decision surfaces for pattern classification. In this example the "planar decision surface" reduces to a piecewise linear decision boundary in two dimensions. Once the decision boundary is known, the training set patterns themselves need no longer be stored.

This may be achieved either indirectly by some smoothing process or directly by finding training procedures that act to update the positions of decision surfaces immediately on receipt of each new training set pattern. The second approach is in many ways more attractive than the first since it drastically cuts down storage requirements—although it must be confirmed that a training procedure is selected which converges sufficiently rapidly. Again, discussion of this well-researched topic is left to other texts (Nilsson, 1965; Duda et al., 2001; Webb, 2002).

We now turn to a more generalized approach—Bayes' decision theory which underpins all the possibilities created by the NN method and its derivatives.

24.3 Bayes' Decision Theory

The basis of Bayes' decision theory will now be examined. To get a computer to classify objects, we need to measure some prominent feature of each object such as its length and to use this feature as an aid to classification. Sometimes such a feature may give very little indication of the pattern class—perhaps because of the effects of manufacturing variation. For example, a handwritten character may be so ill formed that its features are of little help in interpreting it. It then becomes much more reliable to make use of the known relative frequencies of letters, or to invoke context: in fact, either of these strategies can give a greatly increased probability of correct interpretation. In other words, when feature measurements are found to be giving an error rate above a certain threshold, it is more reliable to employ the *a priori* probability of a given pattern appearing.

The next step in improving recognition performance is to combine the information from feature measurements and from *a priori* probabilities by applying Bayes' rule. For a single feature x this takes the form:

$$P(C_i|x) = p(x|C_i)P(C_i)/p(x) \tag{24.1}$$

where

$$p(x) = \sum_j p(x|C_j)P(C_j) \tag{24.2}$$

Mathematically, the variables here are (1) the *a priori* probability of class C_i, $P(C_i)$; (2) the probability density for feature x, $p(x)$; (3) the class-conditional probability density for feature x in class C_i, $p(x|C_i)$—that is, the probability that feature x arises for objects known to be in class C_i; and (4) the *a posteriori* probability of class C_i when x is observed, $P(C_i|x)$.

The notation $P(C_i|x)$ is by now a standard one, being defined as the probability that the class is C_i when the feature is known to have the value x. Bayes' rule now says that to find the class of an object we need to know two sets of information about the objects that might be viewed. The first set is the basic probability $P(C_i)$ that a particular class might arise; the second is the distribution of values of the feature x for each class. Fortunately, each of these sets of information can be found straightforwardly by observing a sequence of objects, for example, as they move along a conveyor. Such a sequence of objects is again called the training set.

Many common image analysis techniques give features that may be used to help identify or classify objects. These include the area of an object, its perimeter, the numbers of holes it possesses, and so on. Classification performance may be improved not only by making use of the *a priori* probability but also by employing a number of features simultaneously. Generally, increasing the number of features helps to resolve object classes and reduce classification errors (Fig. 24.3). However, the error rate is rarely reduced to zero merely by adding more and more features, and indeed the situation eventually deteriorates for reasons explained in Section 24.5.

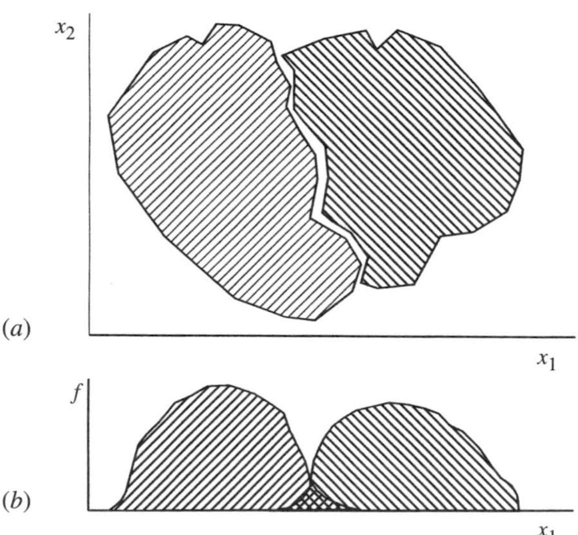

Figure 24.3 Use of several features to reduce classification errors: (*a*) the two regions to be separated in 2-D (x_1, x_2) feature space; (*b*) frequencies of occurrence of the two classes when the pattern vectors are projected onto the x_1-axis. Error rates will be high when either feature is used on its own but will be reduced to a low level when both features are employed together.

Bayes' rule can be generalized to cover the case of a generalized feature **x**, in multidimensional feature space, by using the modified formula:

$$P(C_i|\mathbf{x}) = p(\mathbf{x}|C_i)P(C_i)/p(\mathbf{x}) \tag{24.3}$$

where $P(C_i)$ is the *a priori* probability of class C_i, and $p(\mathbf{x})$ is the overall probability density for feature vector **x**:

$$p(\mathbf{x}) = \sum_j p(\mathbf{x}|C_j)P(C_j) \tag{24.4}$$

The classification procedure is then to compare the values of all the $P(C_j|\mathbf{x})$ and to classify an object as class C_i if:

$$P(C_i|\mathbf{x}) > P(C_j|\mathbf{x}) \qquad \text{for all } j \neq i \tag{24.5}$$

24.4 Relation of the Nearest Neighbor and Bayes' Approaches

When Bayes' theory is applied to simple pattern recognition tasks, it is immediately clear that *a priori* probabilities are important in determining the final classification of any pattern, since these probabilities arise explicitly in the calculation. However, this is not so for the NN type of classifier. The whole idea of the NN classifier appears to be to get away from such considerations, instead classifying patterns on the basis of training set patterns that lie nearby in feature space. However, there must be a definite answer to the question of whether or not *a priori* probabilities are taken into account *implicitly* in the NN formulation, and therefore of whether or not an adjustment needs to be made to the NN classifier to minimize the error rate. Since a categorical statement of the situation is important the next subsection provides such a statement, together with necessary analysis.

24.4.1 *Mathematical Statement of the Problem*

This subsection considers in detail the relation between the NN algorithm and Bayes' theory. For simplicity (and with no ultimate loss of generality), here we take all dimensions in feature space to have equal weight, so that the measure of distance in feature space is not a complicating factor.

For greatest accuracy of classification, many training set patterns will be used and it will be possible to define a density of training set patterns in feature space, $D_i(\mathbf{x})$, for position $\mathbf{x}$ in feature space and class C_i. If $D_k(\mathbf{x})$ is high at position $\mathbf{x}$ in class C_k, then training set patterns lie close together and a test pattern at $\mathbf{x}$ will be likely to fall in class C_k. More particularly, if

$$D_k(\mathbf{x}) = \max_i D_i(\mathbf{x}) \tag{24.6}$$

then our basic statement of the NN rule implies that the class of a test pattern $\mathbf{x}$ will be C_k.

However, according to the outline given above, this analysis is flawed in not showing explicitly how the classification depends on the *a priori* probability of class C_k. To proceed, note that $D_i(\mathbf{x})$ is closely related to the conditional probability density $p(\mathbf{x}|C_i)$ that a training set pattern will appear at position $\mathbf{x}$ in feature space if it is in class C_i. Indeed, the $D_i(\mathbf{x})$ are merely nonnormalized values of the $p(\mathbf{x}|C_i)$:

$$p(\mathbf{x}|C_i) = D_i(\mathbf{x}) / \int D_i(\mathbf{x}) \mathrm{d}\mathbf{x} \tag{24.7}$$

The standard Bayes' formulas (equations (24.3) and (24.4)) can now be used to calculate the *a posteriori* probability of class C_i.

So far it has been seen that the *a priori* probability should be combined with the training set density data before valid classifications can be made using the NN rule. As a result, it seems invalid merely to take the nearest training set pattern in feature space as an indicator of pattern class. However, note that when clusters of training set patterns and the underlying within-class distributions scarcely overlap, there is a rather low probability of error in the overlap region. Moreover, the result of using $p(\mathbf{x}|C_i)$ rather than $P(C_i|\mathbf{x})$ to indicate class often introduces only a very small bias in the decision surface. Hence, although totally invalid *mathematically*, the error introduced need not always be disastrous.

We now consider the situation in more detail, finding how the need to multiply by the *a priori* probability affects the NN approach. Multiplying by the *a priori* probability can be achieved either *directly*, by multiplying the densities of each class by the appropriate $P(C_i)$, or *indirectly*, by providing a suitable amount of additional training for classes with high *a priori* probability. It may now be seen that the amount of additional training required is *precisely* the amount that would be obtained if the training set patterns were allowed to appear with their natural frequencies (see equations below). For example, if objects of different classes are moving along a conveyor, we should not first separate them and then train with equal numbers of patterns from each class.

We should instead allow them to proceed normally and train on them all at their normal frequencies of occurrence in the training stream. If training set patterns do not appear for a time with their proper natural frequencies, this will introduce a bias into the properties of the classifier. Thus, we must make every effort to permit the training set to be representative not only of the *types* of patterns of each class but also of the *frequencies* with which they are presented to the classifier during training.

These ideas for *indirect* inclusion of *a priori* probabilities may be expressed as follows:

$$P(C_i) = \frac{\int D_i(\mathbf{x})\,\mathrm{d}\mathbf{x}}{\sum_j \int D_j(\mathbf{x})\,\mathrm{d}\mathbf{x}} \tag{24.8}$$

Hence

$$P(C_i|\mathbf{x}) = \frac{D_i(\mathbf{x})}{\left(\sum_j \int D_j(\mathbf{x})\,\mathrm{d}\mathbf{x}\right) p(\mathbf{x})} \tag{24.9}$$

where

$$p(\mathbf{x}) = \frac{\sum_k D_k(\mathbf{x})}{\sum_j \int D_j(\mathbf{x})\,\mathrm{d}\mathbf{x}} \tag{24.10}$$

Substituting for $p(\mathbf{x})$ now gives:

$$P(C_i|\mathbf{x}) = D_i(\mathbf{x}) / \sum_k D_k(\mathbf{x}) \tag{24.11}$$

so the decision rule to be applied is to classify an object as class C_i if

$$D_i(\mathbf{x}) > D_j(\mathbf{x}) \text{ for all } j \neq i \tag{24.12}$$

We have now come to the following conclusions:

1. The NN classifier may well not include *a priori* probabilities and hence could give a classification bias.
2. It is in general wrong to train a NN classifier in such a way that an equal number of training set patterns of each class are applied.

3. The correct way to train an NN classifier is to apply training set patterns at the natural rates at which they arise in raw training set data.

The third conclusion is perhaps the most surprising and the most gratifying. Essentially it adds fire to the principle that training set patterns should be representative of the class distributions from which they are taken, although we now see that it should be generalized to the following: *training sets should be fully representative of the populations from which they are drawn*, where "fully representative" includes ensuring that the frequencies of occurrence of the various classes are representative of those in the whole population of patterns. Phrased in this way, the principle becomes a general one that is relevant to many types of trainable classifiers.

24.4.2 *The Importance of the Nearest Neighbor Classifier*

The NN classifier is important in being perhaps the simplest of all classifiers to implement on a computer. In addition, it is guaranteed to give an error rate within a factor of two of the ideal error rate (obtainable with a Bayes' classifier). By modifying the method to base classification of any test pattern on the most commonly occurring class among the k nearest training set patterns (giving the "k-NN" method), the error rate can be reduced further until it is arbitrarily close to that of a Bayes' classifier. (Note that equation (24.11) can be interpreted as covering this case too.) However, both the NN and (*a fortiori*) the k-NN methods have the disadvantage that they often require enormous storage to record enough training set pattern vectors, and correspondingly large amounts of computation to search through them to find an optimal match for each test pattern—hence necessitating the pruning and other methods mentioned earlier for cutting down the load.

24.5 The Optimum Number of Features

Section 24.3 stated that error rates can be reduced by increasing the number of features used by a classifier, but there is a limit to this, after which performance actually deteriorates. In this section we consider why this should happen. Basically, the reason is similar to the situation where many parameters are used to fit a curve to a set of D data points. As the number of parameters P is increased, the fit of the curve becomes better and better, and in general becomes perfect when $P = D$. However, by that stage the significance of the fit is poor, since the parameters are no longer overdetermined and no averaging of their values is taking place. Essentially, all the noise in the raw input data is being transferred to the parameters. The same thing happens with training set patterns in feature space. Eventually, training set patterns are so sparsely packed in feature space that the test patterns have reduced the probability of being nearest to a pattern of the same class, so error rates become very high. This situation can also be regarded as due to a proportion of the features having negligible statistical significances; that is, they add little additional information and serve merely to add uncertainty to the system.

However, an important factor is that the optimum number of features depends on the amount of training a classifier receives. If the number of training set patterns is increased, more evidence is available to support the determination of a greater number of features and hence to provide more accurate classification of test patterns. In the limit of very large numbers of training set patterns, performance continues to increase as the number of features is increased.

This situation was first clarified by Hughes (1968) and verified in the case of n-tuple pattern recognition—a variant of the NN classifier due to Bledsoe and Browning (1959)—by Ullmann (1969). Both workers produced clear curves showing the initial improvement in classifier performance as the number of features increased; this improvement was followed by reduced performance for large numbers of features.

Before leaving this topic, note that the above arguments relate to the number of features that should be used but not to their selection. Some features are more significant than others, the situation being very data-dependent. We will leave it as a topic for experimental tests to determine which subset of features will minimize classification errors (see also Chittineni, 1980).

24.6 Cost Functions and the Error–Reject Tradeoff

The forgoing sections implied that the main criterion for correct classification is that of maximum *a posteriori* probability. Although probability is always a relevant factor, in a practical engineering environment it is often more important to minimize costs. Hence, it is necessary to compare the costs involved in making correct or incorrect decisions. Such considerations can be expressed mathematically by invoking a loss function $L(C_i|C_j)$ that represents the cost involved in making a decision C_i when the true class for feature $\mathbf{x}$ is C_j.

To find a modified decision rule based on minimizing costs, we first define a function known as the conditional risk:

$$R(C_i|\mathbf{x}) = \sum_j L(C_i|C_j)P(C_j|\mathbf{x}) \tag{24.13}$$

This function expresses the expected cost of deciding on class C_i when $\mathbf{x}$ is observed. In order to minimize this function, we now decide on class C_i only if:

$$R(C_i|\mathbf{x}) < R(C_j|\mathbf{x}) \qquad \text{for all } j \neq i \tag{24.14}$$

If we were to choose a particularly simple cost function, of the form:

$$L(C_i|C_j) = \begin{cases} 0 \text{ for } i = j \\ 1 \text{ for } i \neq j \end{cases} \tag{24.15}$$

then the result would turn out to be identical to the previous probability-based decision rule, relation (24.5). Only when certain errors lead to relatively large (or small) costs does it pay to deviate from the normal decision rule. Such cases arise when we are in a hostile environment and must, for example, give precedence to the sound of an enemy tank over that of other vehicles. It is better to be oversensitive and risk a false alarm than to chance not noticing the hostile agent. Similarly, on a production line it may sometimes be better to reject a small number of good products than to risk selling a defective product. Cost functions therefore permit classifications to be biased in favor of a safe decision in a rigorous, predetermined, and controlled manner, and the desired balance of properties obtained from the classifier.

Another way of minimizing costs is to arrange for the classifier to recognize when it is "doubtful" about a particular classification because two or more classes are almost equally likely. Then one solution is to make a safe decision, the decision plane in feature space being biased away from its position for maximum probability classification. An alternative is to reject the pattern, that is, place it into an "unknown" category. In that case, some other means can be employed for making an appropriate classification. Such a classification could be made by going back to the original data and measuring further features, but in many cases it is more appropriate for a human operator to be available to make the final decision. The latter approach is more expensive, and so introducing a "reject" classification can incur a relatively large cost factor. A further problem is that the error rate[2] is reduced only by a fraction of the amount that the rejection rate is increased. In a simple two-class system, the initial decrease in error rate is only one-half the initial increase in reject rate (i.e., a 1% decrease in error rate is obtained only at the expense of a 2% increase in reject rate); the situation gets rapidly worse as progressively lower error rates are attempted (Fig. 24.4). Thus, very careful cost analysis of the error–reject tradeoff curve must be made before an optimal scheme can be developed. Finally, it should be noted that the overall error rate of the

2 All error and reject rates are assumed to be calculated as proportions of the total number of test patterns to be classified.

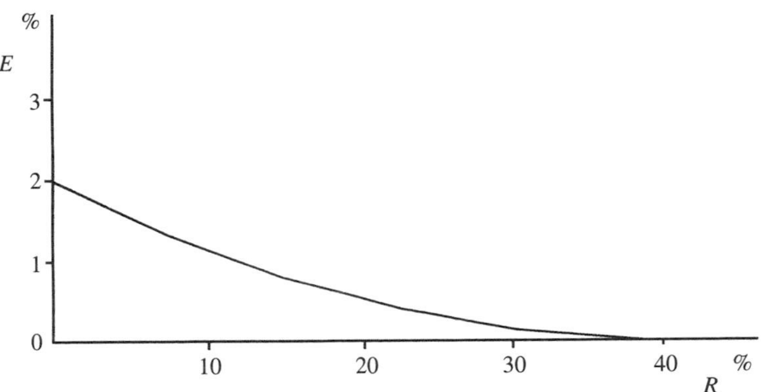

Figure 24.4 An error–reject tradeoff curve (E, error rate; R, reject rate). In this example, the error rate E drops substantially to zero for a reject rate R of 40%. More usually, E cannot be reduced to zero until R is 100%.

classification system depends on the error rate of the classifier that examines the rejects (e.g., the human operator); this consideration needs to be taken into account in determining the exact tradeoff to be used.

24.7 The Receiver–Operator Characteristic

The early sections of this chapter suggest an implicit understanding that classification error rates have to be reduced as far as possible, though in the last section it was acknowledged that it is cost rather than error that is the practically important parameter. It was also found that a tradeoff between error rate and reject rate allows a further refinement to be made to the analysis.

Here we consider another refinement that is required in many practical cases where binary decisions have to be made. Radar provides a good illustration showing that there are two basic types of misclassifications: (1) radar displays may indicate an aircraft or missile when none is present—in which case the error is called a false positive (or in popular parlance, a false alarm); (2) radar may indicate that no aircraft or missile is present when there actually is one—in which case the error is called a false negative. Similarly, in automated industrial inspection, when searching for deficient products, a false positive corresponds to finding one when none is present, whereas a false negative corresponds to missing a deficient product when one is present.

There are four relevant categories: (1) true positives (positives that are correctly classified), (2) true negatives (negatives that are correctly classified), (3) false positives (positives that are incorrectly classified), and (4) false negatives

(negatives that are incorrectly classified). If many experiments are carried out to determine the proportions of these four categories in a given application, we can obtain the four probabilities of occurrence. Using an obvious notation, we can relate these categories by the following formulas:

$$P_{\mathrm{TP}} + P_{\mathrm{FN}} = 1 \tag{24.16}$$

$$P_{\mathrm{TN}} + P_{\mathrm{FP}} = 1 \tag{24.17}$$

(In case the reader finds the combinations of probabilities in these formulas confusing, it is worth noting that an object that is actually a faulty product will either be *correctly* detected as such or else it will be *incorrectly* categorized as acceptable—in which case it is a false negative.)

It will be apparent that the probability of error P_{E} is the sum:

$$P_{\mathrm{E}} = P_{\mathrm{FP}} + P_{\mathrm{FN}} \tag{24.18}$$

In general, false positives and false negatives will have different costs. Thus, the loss function $L(C_1|C_2)$ will not be the same as the loss function $L(C_2|C_1)$. For example, missing an enemy missile or failing to find a glass splinter in baby food may be far more costly than the cost of a few false alarms (which in the case of food inspection merely means the rejection of a few good products). In many applications, there is a prime need to reduce as far as possible the number of false negatives (the number of failures to detect the requisite targets).

But how far should we go to reduce the number of false negatives? This question is an important one that should be answered by systematic analysis rather than by ad hoc procedures. The key to achieving this is to note that the proportions of false positives and false negatives will vary independently with the system setup parameters, though frequently only a single threshold parameter need be considered in any detail. In that case, we can eliminate this parameter and determine how the numbers of false positives and false negatives depend on each other. The result is the receiver–operator characteristic or ROC curve (Fig. 24.5).[3]

The ROC curve will often be approximately symmetrical, and, if expressed in terms of probabilities rather than numbers of items, will pass through the points (1, 0), (0, 1), as shown in Fig. 24.5. It will generally be highly concave, so it will pass well below the line $P_{\mathrm{FP}} + P_{\mathrm{FN}} = 1$, except at its two ends. The point closest to the origin will often be close to the line $P_{\mathrm{FP}} = P_{\mathrm{FN}}$. Thus, if false positives and false negatives are assigned equal costs, the classifier can be optimized simply by minimizing P_{E} with the constraint $P_{\mathrm{FP}} = P_{\mathrm{FN}}$. Note, however, that in general

3 Although this text defines the ROC curve in terms of P_{FP} and P_{FN}, many other texts use alternative definitions, based, for example, on P_{TP} and P_{FP}—in which case the graph will appear inverted.

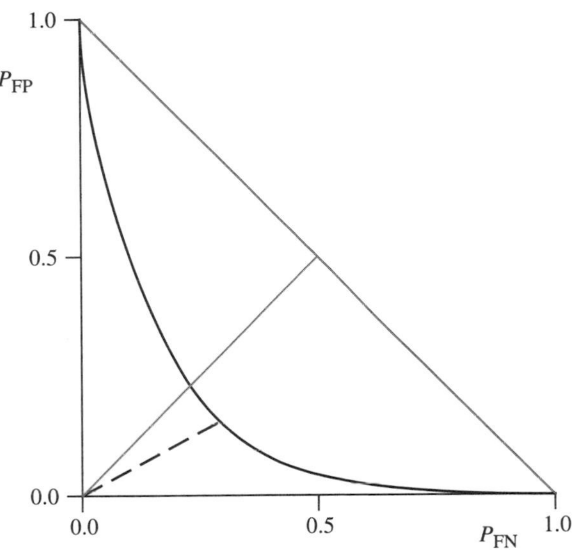

Figure 24.5 Idealized ROC curve. The gray line of gradient +1 indicates the position which *a priori* might be expected to lead to minimum error. The optimum working point is that indicated by the dotted line, where the gradient on the curve is −1. The gray line of gradient −1 indicates the limiting worst case scenario. All practical ROC curves will lie below this line.

the point closest to the origin is *not* the point that minimizes P_E; the point that minimizes total error is actually the point on the ROC curve where the gradient is −1 (Fig. 24.5).

Unfortunately, no general theory can predict the shape of the ROC curve. Furthermore, the number of samples in the training set may be limited (especially in inspection if rare contaminants are being sought), and then it may not prove possible to make an accurate assessment of the shape—especially in the extreme wings of the curve. In some cases, this problem can be handled by modeling, for example, using exponential or other functions—as shown in Fig. 24.6, where the exponential functions lead to reasonably accurate descriptions. However, the underlying shape can hardly be exactly exponential, for this would suggest that the ROC curve tends to zero at infinity rather than at the points (1, 0), (0, 1). Also, there will in principle be a continuity problem at the join of two exponentials. Nevertheless, if the model is reasonably accurate over a good range of thresholds, the relative cost factors for false positives and false negatives can be adjusted appropriately and an ideal working point can be determined systematically. Of course, there may be other considerations: for example, it may not be permissible for the false negative rate to rise above a certain critical level. For examples of the use of the ROC analysis, see Keagy et al. (1995, 1996) and Davies et al. (2003c).

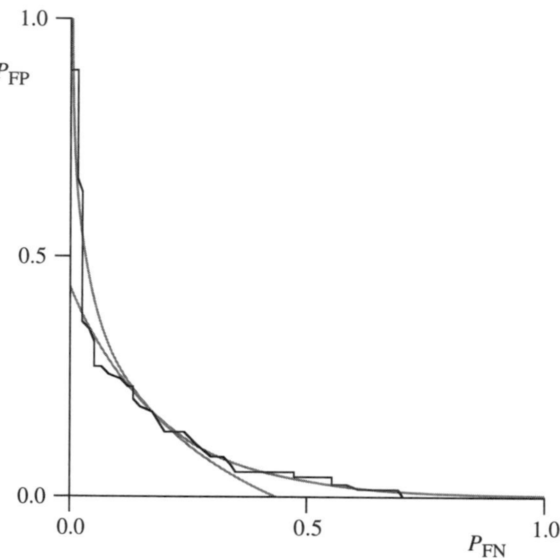

Figure 24.6 Fitting an ROC curve using exponential functions. Here the given ROC curve (see Davies et al., 2003) has distinctive steps resulting from a limited set of data points. A pair of exponential curves fits the ROC curve quite well along the two axes, each having an obvious region that it models best. In this case, the crossover region is reasonably smooth, but there is no real theoretical reason for this. Furthermore, exponential functions will not pass through the limiting points (0, 1) and (1, 0).

24.8 Multiple Classifiers

In recent years, moves have been initiated to make the classification process more reliable by application of multiple classifiers working in cooperation. The basic concept is much like that of three magistrates coming together to make a more reliable judgment than any can make alone. Each is expert in a variety of things, but not in everything, so putting their knowledge together in an appropriate way should permit more reliable judgments to be made. A similar concept applies to expert AI systems. Multiple expert systems should be able to make up for each other's shortcomings. In all these cases, some way should exist to get the most out of the individual classifiers without confusion reigning.

The idea is not just to take all the feature detectors that the classifiers use and to replace their output decision-making devices with a single more complex decision-making unit. Indeed, such a strategy could well run into the problem discussed in Section 24.5—of exceeding the optimum number of features. At best only a minor improvement would result from such a strategy, and at worst the system would be grossly failure-prone. On the contrary, the idea is to take the final classification of a number of complete but totally separate classifiers and to combine their outputs to obtain a substantially improved output. Furthermore, the separate classifiers might use totally different strategies to arrive at their decisions. One may be a nearest neighbor classifier, another may be a Bayes classifier, and yet another may be a neural network classifier (see Chapter 25). Similarly, one might employ structural pattern recognition, another might use statistical pattern recognition, and yet another might use syntactic pattern recognition. Each will be respectable in its own right: each will have strengths, but each will also have weaknesses. Part of the idea is one of convenience: to make use of any soundly based classifier that is available, and to boost its effectiveness by using it in conjunction with other soundly based classifiers.

The next task is to determine how this can be achieved in practice. Perhaps the most obvious way forward is to get the individual classifiers to vote for the class of each input pattern. Although this is a nice idea, it will often fail because the individual classifiers may have more weaknesses than strengths. At the very least, the concept must be made more sophisticated.

Another strategy is again to allow the individual classifiers to vote but this time to make them do so in an exclusive manner, so that as many classes as possible are eliminated for each input pattern. This objective is achievable with a simple intersection rule: a class is accepted as a possibility only if *all* the classifiers indicate that it is a possibility. The strategy is implemented by applying a threshold to each classifier in a special way, which will now be described.

A prerequisite for this strategy to work is that each classifier must give not only a class decision for each input pattern but also the ranks of all possible classes for each pattern. In other words, it must give its first choice of class for any pattern, its second choice for that pattern, and so on. Then the classifier is labeled with the rank it assigned to the true class of that pattern. We apply each classifier to the whole training set and get a table of ranks (Table 24.1). Finally, we find the worst case (largest rank)[4] for each classifier, and we take that as a threshold value that will be used in the final multiple classifier. When using this method for testing input patterns, only those classifiers that are not excluded by the threshold have their outputs intersected to give the final list of classes for the input pattern.

This "intersection strategy" focuses on the worst-case behavior of the individual classifiers, and the result could be that a number of classifiers will hardly reduce the list of possible classes for the input patterns. This tendency can be tackled by an alternative union strategy, which focuses on the specialisms of the individual classifiers. The aim is then to find a classifier that recognizes each particular pattern well. To achieve this we look for the classifier with the

4 In everyday parlance, the worst case corresponds to the lowest rank, which is here the largest numerical rank. Similarly, the highest rank is the smallest numerical rank. It is obviously necessary to be totally unambiguous about this nomenclature.

Table 24.1 Determining a set of classifiers for the intersection strategy

	Classifier Ranks				
	C_1	C_2	C_3	C_4	C_5
D_1	5	3	7	1	8
D_2	4	9	6	4	2
D_3	5	6	7	1	4
D_4	4	7	5	3	5
D_5	3	5	6	5	4
D_6	6	5	4	3	2
D_7	2	6	1	3	8
thr	6	9	7	5	8

In the upper section of this table, the original classifier ranks are shown for each input pattern; in the bottom line of the table, only the worst case rank is retained. When later applying test patterns, this can be used as the threshold (marked "thr") to determine which classifiers should be employed.

Table 24.2 Determining a set of classifiers for the union strategy

	Classifier Ranks					Best Classifiers				
	C_1	C_2	C_3	C_4	C_5	C_1	C_2	C_3	C_4	C_5
D_1	5	3	7	1	8	0	0	0	1	0
D_2	4	9	6	4	2	0	0	0	0	2
D_3	5	6	7	1	4	0	0	0	1	0
D_4	4	7	5	3	5	0	0	0	3	0
D_5	3	5	6	5	4	3	0	0	0	0
D_6	6	5	4	3	2	0	0	0	0	2
D_7	2	6	1	3	8	0	0	1	0	0
min–max threshold						3	0	1	3	2

In the left-hand section of this table, the original classifier ranks are shown for each input pattern; in the right-hand section only one rank is retained, namely, that obtaining the classifier that is best able to recognize that pattern. Note that to facilitate the next piece of analysis—finding the thresholds on the classifier ranks—all remaining places in the table are packed with zeros. A zero final threshold then indicates a classifier that is of no help in analyzing the input data.

smallest rank (classifier rank being defined exactly as already defined above for the intersection strategy) for each individual pattern (Table 24.2). Having found the smallest rank for the individual input patterns, we determine the largest of these ranks that arises for each classifier as we go right through all the input patterns. Applying this value as a threshold now determines whether the output of the classifier should be used to help determine the class of a pattern. Note that the threshold is determined in this way using the training set and is later used to decide which classifiers to apply to individual test patterns. Thus, for any pattern a restricted set of classifiers is identified that can best judge its class.

To clarify the operation of the union strategy, let us examine how well it will work on the training set. It is guaranteed to retain enough classifiers to ensure that the true class of any pattern is not excluded (though naturally this is not guaranteed for any member of the test set). Hence, the aim of employing a classifier that recognizes each particular pattern well is definitely achieved.

Unfortunately, this guarantee is not obtained without cost. Specifically, if a member of the training set is actually an outlier, the guarantee will still apply, and the overall performance may be compromised. This problem can be addressed in many ways, but a simple possibility is to eliminate excessively bad exemplars from the training set. Another way is to abandon the union strategy altogether and go for a more sophisticated voting strategy. Other approaches involve reordering the data to improve the rank of the correct class (Ho et al., 1994).

24.9 Cluster Analysis

24.9.1 *Supervised and Unsupervised Learning*

In the earlier parts of this chapter, we made the implicit assumption that the classes of all the training set patterns are known, and in addition that they should be used in training the classifier. Although this assumption might be regarded as inescapable, classifiers may actually use two approaches to learning—*supervised learning* (in which the classes are known and used in training) and *unsupervised learning* (in which they are either unknown or known and not used in training). Unsupervised learning can frequently be advantageous in practical situations. For example, a human operator is not required to label all the products coming along a conveyor, as the computer can find out for itself both how many classes of product there are and which categories they fall into. In this way, considerable operator effort is eliminated. In addition, it is not unlikely that a number of errors would thereby be circumvented. Unfortunately, unsupervised learning involves a number of difficulties, as will be seen in the following subsections.

Before proceeding, we should state two other reasons why unsupervised learning is useful. First, when the characteristics of objects vary with time—for example, beans changing in size and color as the season develops—it will be necessary to track these characteristics within the classifier. Unsupervised learning

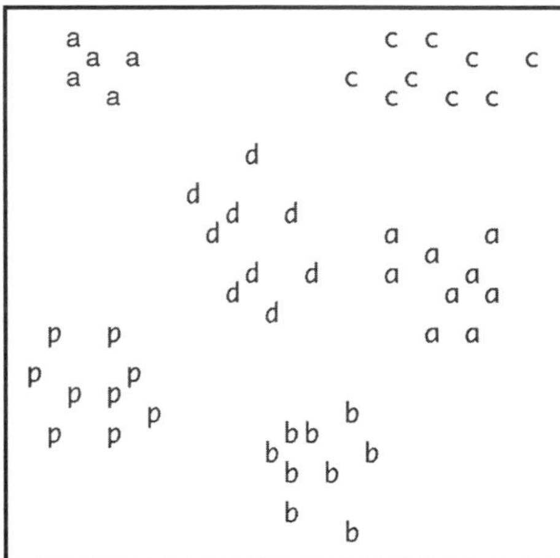

Figure 24.7 Location of clusters in feature space. Here the letters correspond to samples of characters taken from various fonts. The small cluster of a's with strokes bent over the top from right to left appear at a separate location in feature space. This type of deviation should be detectable by cluster analysis.

provides an excellent means of approaching this task. Second, when setting up a recognition system, the characteristics of objects, and in particular their most important parameters (e.g., from the point of view of quality control) may well be unknown, and it will be useful to gain some insight into the nature of the data. Thus, types of faults will need to be logged, and permissible variants on objects will need to be noted. As an example, many OCR fonts (such as Times Roman) have a letter "a" with a stroke bent over the top from right to left, though other fonts (such as Monaco) do not have this feature. An unsupervised classifier will be able to flag this difference by locating a cluster of training set patterns in a totally separate part of feature space (see Fig. 24.7). In general, unsupervised learning is about the location of clusters in feature space.

24.9.2 *Clustering Procedures*

As we have said, an important reason for performing cluster analysis is characterization of the input data. However, the underlying motivation is normally to classify test data patterns reliably. To achieve these aims, it will be necessary both to partition feature space into regions corresponding to significant clusters and to label each region (and cluster) according to the types of data involved. In practice, this can happen in two ways:

1. By performing cluster analysis and then labeling the clusters by specific queries to human operators on the classes of a small number of individual training set patterns.
2. By performing supervised learning on a small number of training set patterns and then performing unsupervised learning to expand the training set to realistic numbers of examples.

In either case, we see that ultimately we do not escape from the need for supervised classification. However, by placing the main emphasis on unsupervised learning, we reduce the tedium and help prevent preconceived ideas on possible classes from affecting the final recognition performance.

Before proceeding further, we should note in some cases we may have absolutely no idea in advance about the number of clusters in feature space. This occurs in classifying the various regions in satellite images. Such cases are in direct contrast with applications such as OCR or recognizing chocolates being placed in a chocolate box.

Unfortunately, cluster analysis involves some very significant problems. Not least is the visualization problem. First, in one, two, or even three dimensions, we can easily visualize and decide on the number and location of any clusters, but this capability is misleading. We cannot extend this capability to feature spaces of many dimensions. Second, computers do not visualize as we do, and special algorithms will have to be provided to enable them to do so. Now computers could be made to emulate our capability in low-dimensional feature spaces, but a combinatorial explosion would occur if we attempted this for high-dimensional spaces. We will therefore have to develop algorithms that operate on *lists* of feature vectors, if we are to produce automatic procedures for cluster location.

Available algorithms for cluster analysis fall into two main groups—*agglomerative* and *divisive*. Agglomerative algorithms start by taking the individual feature points (training set patterns, excluding class) and progressively grouping them together according to some similarity function until a suitable target criterion is reached. Divisive algorithms start by taking the whole set of feature points as a single large cluster and progressively dividing it until some suitable target criterion is reached. Let us assume that there are P feature points. Then, in the worst case, the number of comparisons between pairs of individual feature point positions that will be required to decide whether to combine a pair of clusters in an agglomerative algorithm will be:

$$\binom{P}{2} = \frac{1}{2}P(P-1) \tag{24.19}$$

while the number of iterations required to complete the process will be of order $P-k$. (Here we are assuming that the final number of clusters to be found is k, where $k \leq P$.) On the other hand, for a divisive algorithm, the number of comparisons between pairs of individual feature point positions will be reduced to:

$$\binom{k}{2} = \frac{1}{2}k(k-1) \tag{24.20}$$

while the number of iterations required to complete the process will be of order k.

Although it would appear that divisive algorithms require far less computation than agglomerative algorithms, this is not so. This is because any cluster containing p feature points will have to be examined for a huge number of potential splits into subclusters, the actual number being of the order:

$$\sum\nolimits_{q=1}^{p} \binom{p}{q} = \sum\nolimits_{q=1}^{p} \frac{p!(p-q)!}{q!} \tag{24.21}$$

In general, the agglomerative approach will have to be adopted. In fact, the type of agglomerative approach outlined above is exhaustive and rigorous, and a less exacting, iterative approach can be used. First, a suitable number k of cluster centers are set. (These can be decided from *a priori* considerations or by making arbitrary choices.) Second, each feature vector is assigned to the closest cluster center. Third, the cluster centers are recalculated. This process is repeated if any feature points have moved from one cluster to another during the iteration, though termination can also be instituted if the quality of clustering ceases to improve. The overall algorithm, which was originally due to Forgy (1965), is given in Fig. 24.8.

The effectiveness of this algorithm will be highly data-dependent, in particular with regard to the order in which the data points are presented. In addition, the result could be oscillatory or nonoptimal (in the sense of not arriving at the best solution). This could happen if at any stage a single cluster center arose near the center of a pair of small clusters. In addition, the method gives no indication of the most appropriate number of clusters. Accordingly, a number of variant and alternative algorithms have been devised. One such algorithm is the ISODATA algorithm (Ball and Hall, 1966). This is similar to Forgy's method but is able to merge clusters that are close together and to split elongated clusters.

Another disadvantage of iterative algorithms is that it may not be obvious when to get them to terminate. As a result, they are liable to be too computation intensive. Thus, there has been some support for noniterative algorithms. MacQueen's *k-means* algorithm (MacQueen, 1967) is one of the best known noniterative clustering algorithms. It involves two runs over the data points, one being required to find the cluster centers and the other being required to finally classify the patterns (see Fig. 24.9). Again, the choice of which data points

```
choose target number k of clusters;
set initial cluster centers;
calculate quality of clustering;
do {
    assign each data point to the closest cluster center;
    recalculate cluster centers;
    recalculate quality of clustering;
} until no further change in the clusters or the quality of the clusters;
```

Figure 24.8 Basis of Forgy's algorithm for cluster analysis.

```
choose target number k of clusters;
set the k initial cluster centers at k data points;
for all other data points { // first pass
    assign data point to closest cluster center;
    recalculate relevant cluster center;
}
for all data points // second pass
    re-assign data point to closest cluster center;
```

Figure 24.9 Basis of MacQueen's k-means algorithm.

are to act as the initial cluster centers can be either arbitrary or made on some more informed basis.

Noniterative algorithms are, as indicated earlier, largely dependent on the order of presentation of the data points. With image data this is especially problematic, as the first few data points are quite likely to be similar (e.g., all derived from sky or other background pixels). A useful way of overcoming this problem is to randomize the choice of data points, so that they can arise from anywhere in the image. In general, noniterative clustering algorithms are less effective than iterative algorithms because they are overinfluenced by the order of presentation of the data.

Overall, the main problem with the algorithms described above is the lack of indication they give of the most appropriate value of k. However, if a range of possible values for k is known, all of them can be tried (using one of the above algorithms), and the one giving the best performance in respect to some suitable target criterion can be taken as providing an optimal result. In that case, we will have found the set of clusters which, in some specified sense, gives the best overall description of the data. Alternatively, some method of analyzing the data to determine k can be used before final cluster analysis, for example, using the k-means algorithm; the Zhang and Modestino (1990) approach falls into this category.

24.10 Principal Components Analysis

Closely related to cluster analysis is the concept of data representation. One powerful way of approaching this task is that of principal components analysis. This approach involves finding the mean of a cluster of points in feature space and then finding the principal axes of the cluster in the following way. First an axis is found that passes through the mean position and which gives the maximum variance when the data are projected onto it. Then a second such axis is found which maximizes variance in a direction normal to the first. This process is carried out until a total of N principal axes have been found for an N-dimensional feature space. The process is illustrated in Fig. 24.10. The process is entirely mathematical and need not be undertaken in the strict sequence indicated here. It merely involves finding a set of orthogonal axes that diagonalize the covariance matrix.

The covariance matrix for the input population is defined as:

$$\mathbf{C} = \mathrm{E}\left\{(\mathbf{x}_{(p)} - \mathbf{m})(\mathbf{x}_{(p)} - \mathbf{m})^{\mathrm{T}}\right\} \tag{24.22}$$

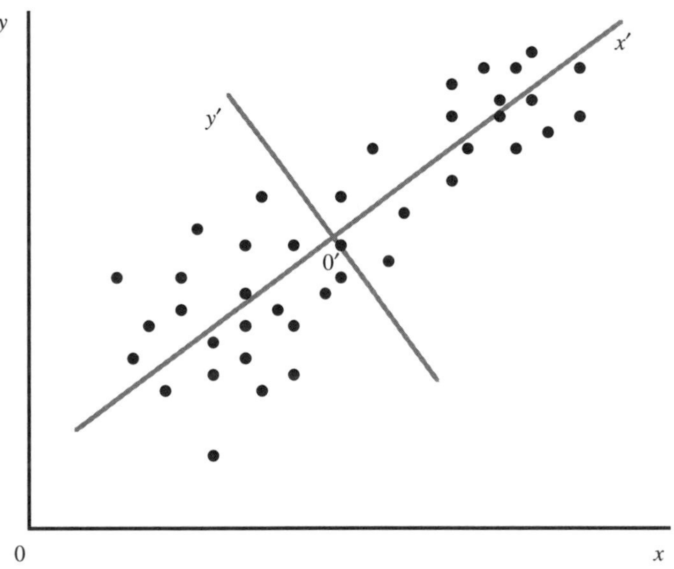

Figure 24.10 Illustration of principal components analysis. Here the dots represent patterns in feature space and are initially measured relative to the x- and y-axes. Then the sample mean is located at $0'$, and the direction $0'x'$ of the first principal component is found as the direction along which the variance is maximized. The direction $0'y'$ of the second principal component is normal to $0'x'$; in a higher dimensional space, it would be found as the direction normal to $0'x'$ along which the variance is maximized.

where $\mathbf{x}_{(p)}$ is the location of the pth data point, $\mathbf{m}$ is the mean of the P data points, and E{...} indicates expectation value for the underlying population. We can estimate $\mathbf{C}$ from the equations:

$$\mathbf{C} = \frac{1}{P}\sum_{p=1}^{P} \mathbf{x}_{(p)}\mathbf{x}_{(p)}^{\mathrm{T}} - \mathbf{m}\mathbf{m}^{\mathrm{T}} \tag{24.23}$$

$$\mathbf{m} = \frac{1}{P}\sum_{p=1}^{P} \mathbf{x}_{(p)} \tag{24.24}$$

Since $\mathbf{C}$ is real and symmetrical, we can diagonalize using a suitable orthogonal transformation matrix $\mathbf{A}$, obtaining a set of N orthonormal eigenvectors $\mathbf{u}_i$ with eigenvalues λ_i given by:

$$\mathbf{C}\mathbf{u}_i = \lambda_i \mathbf{u}_i \qquad (i = 1, 2, \ldots N) \tag{24.25}$$

The vectors $\mathbf{u}_i$ are derived from the original vectors $\mathbf{x}_i$ by:

$$\mathbf{u}_i = \mathbf{A}(\mathbf{x}_i - \mathbf{m}) \tag{24.26}$$

and the inverse transformation needed to recover the original data vectors is:

$$\mathbf{x}_i = \mathbf{m} + \mathbf{A}^{\mathrm{T}}\mathbf{u}_i \tag{24.27}$$

Here we have recalled that, for an orthogonal matrix:

$$\mathbf{A}^{-1} = \mathbf{A}^{\mathrm{T}} \tag{24.28}$$

It may be shown that $\mathbf{A}$ is the matrix whose rows are formed from the eigenvectors of $\mathbf{C}$, and that the diagonalized covariance matrix $\mathbf{C}'$ is given by:

$$\mathbf{C}' = \mathbf{A}\mathbf{C}\mathbf{A}^{\mathrm{T}} \tag{24.29}$$

so that:

$$\mathbf{C}' = \begin{bmatrix} \lambda_1 & & & 0 \\ & \lambda_2 & & \\ & & \cdot & \cdot \\ & & \cdot & \cdot \\ 0 & & \cdot & \lambda_N \end{bmatrix} \tag{24.30}$$

Note that in an orthogonal transformation, the trace of a matrix remains unchanged. Thus, the trace of the input data is given by:

$$\text{trace}\,\mathbf{C} = \text{trace}\,\mathbf{C}' = \sum_{i=1}^{N} \lambda_i = \sum_{i=1}^{N} \sigma_i^2 \tag{24.31}$$

where we have interpreted the λ_i as the variances of the data in the directions of the principal component axes. (Note that for a real symmetrical matrix, the eigenvalues are all real and positive.)

In the following discussion, we shall assume that the eigenvalues have been placed in an ordered sequence, starting with the largest. In that case, λ_1 represents the most significant characteristic of the set of data points, with the later eigenvalues representing successively less significant characteristics. We could even go so far as to say that, in some sense, λ_1 represents the most interesting characteristic of the data, while λ_N would be largely devoid of interest. More practically, if we ignored λ_N, we might not lose much useful information, and indeed the last few eigenvalues would frequently represent characteristics that are not statistically significant and are essentially noise. For these reasons, principal components analysis is commonly used for reduction in the dimensionality of the feature space from N to some lower value N'. In some applications, this would be taken as leading to a useful amount of data compression. In other applications, it would be taken as providing a reduction in the enormous redundancy present in the input data.

We can quantify these results by writing the variance of the data in the reduced dimensionality space as:

$$\text{trace}(\mathbf{C}')_{\text{reduced}} = \sum_{i=1}^{N'} \lambda_i = \sum_{i=1}^{N'} \sigma_i^2 \tag{24.32}$$

Not only is it now clear why this leads to reduced variance in the data, but also we can see that the mean square error obtained by making the inverse transformation (equation (24.27)) will be:

$$\overline{\varepsilon^2} = \sum_{i=1}^{N} \sigma_i^2 - \sum_{i=1}^{N'} \sigma_i^2 = \sum_{i=N'+1}^{N} \sigma_i^2 \tag{24.33}$$

One application in which principal components analysis has become especially important is the analysis of multispectral images, for example, from earth-orbiting satellites. Typically, there will be six separate input channels (e.g., three color and three infrared), each providing an image of the same ground region. If these images are 512×512 pixels in size, there will be about a quarter of a million data points, and these will have to be inserted into a six-dimensional feature space. After finding the mean and covariance matrix for these data points, it is diagonalized and a total of six principal component images can be formed. Commonly, only two or three of these will contain immediately useful information, and the rest can be ignored. (For example, the first three of the six principal component images may well possess 95% of the variance of the input images.) Ideally, the first few principal component images in such a case will highlight such areas as fields, roads, and rivers, and these will be precisely the data required for map-making or other purposes. In general, the vital pattern recognition tasks can be aided *and* considerable savings in storage can be achieved on the incoming image data by attending to just the first few principal components.

Finally, it is as well to note that principal components analysis provides a particular form of data representation. In itself it does not deal with pattern classification, and methods that are required to be useful for the latter type of task must possess useful discrimination. Thus, selection of features simply because they possess the highest variability does not mean that they will necessarily perform well in pattern classifiers. Another important factor relevant to the whole study of data analysis in feature space is the scale of the various features. Often, these will be an extremely variegated set, including length, weight, color, and numbers of holes. Such a set of features will have no special comparability and are unlikely even to be measurable in the same units. Placing them in the same feature space and assuming that the scales on the various axes should have the same weighting factors must therefore be invalid. One way to tackle this problem is to normalize the individual features to some standard scale given by measuring their variances. Such a procedure will totally change the results of principal components calculations and further mitigates against principal components methodology being used thoughtlessly. On the other hand, on some occasions different features can be compatible, and principal components analysis can be performed without such worries. One such situation is where all the features are pixel intensities in the same window. (This case is discussed in Section 26.5.)

24.11 The Relevance of Probability in Image Analysis

Having seen the success of Bayes' theory in pointing to apparently absolute answers in the interpretation of certain types of images, we should now consider complex scenes in terms of the probabilities of various interpretations and the likelihood of a particular interpretation being the correct one. Given a sufficiently large number of such scenes and their interpretations, it seems that it ought to be possible to use them to train a suitable classifier. However, practical interpretation in real time is quite another matter. Next, note that a well-known real-time scene understanding machine, the eye–brain system, does not appear to operate in a manner corresponding to the algorithms we have studied. Instead, it appears to pay attention to various parts of an image in a nonpredetermined sequence of "fixations" of the eye, interrogating various parts of the scene in turn and using the newly acquired information to work out the origin of the next piece of relevant information. It is employing a process of *sequential pattern recognition*, which saves effort overall by progressively building up a store of knowledge about relevant parts of the scene—and at the same time forming and testing hypotheses about its structure.

This process can be modeled as one of modifying and updating the *a priori* probabilities as analysis progresses. This process is inherently powerful, since the eye is thereby not tied to "average" *a priori* probabilities for *all* scenes but is able to use information in a particular scene to improve on the average *a priori* probabilities. Another factor making this a powerful process is that it is not possible to know the *a priori* probabilities for sequences of real, complex scenes at all accurately. Hence the methods of this chapter are unlikely to be applicable in more complex situations. In such cases, the concept of probability is a nice idea but should probably be treated lightly.

The concept of probability is useful, however, when it can validly be applied. At present, we can sum up the possible applications as those where a restricted range of images can arise, such that the consequent image description contains very few bits of information—viz. those forming the various pattern class names. Thus, the image data are particularly impoverished, containing relatively little structure. Structure can here be defined as relationships between relevant parts of an image that have to be represented in the output data. If such a structure is to be present, then it implies recognition of several parts of an image, coupled with recognition of their relation. When only a rudimentary structure is present, the output data will contain only a few times the amount of output data of a set of simple classes, whereas more complicated cases will be subject to a combinatorial explosion in the amount of output data. SPR is a topic characterized by probabilistic interpretation when there is a many-to-few relation between image and interpretation rather than a many-to-many relation. Thus, the input data need to be highly controlled before probabilistic interpretation is possible and SPR can be applied. It is left to other works to consider situations where images are best interpreted as verbal descriptions of their contents and structures. Note, however, that the nature of such descriptions is influenced strongly by the *purpose* for which the information is to be used.

24.12 The Route to Face Recognition

In a book of this type it is important to consider the problem of face recognition, which has long been a target for many practitioners of machine vision. Some would say that by now this is a solved problem, but criminologists repeatedly confirm that there is still some distance to go before reliable identification of people can be attained in practical circumstances. Not least are the problems of hairstyles, hats, glasses, beards, degree of facial stubble, wildly variable facial expressions, and of course variations in lighting and shadow—and this does not even touch on the problem of deliberate disguise. In addition, one must never forget that the face is not flat but part of a solid, albeit malleable, object—the head—which can appear in a variety of orientations and positions in space.

These remarks make it clear that full analysis and solution of the face recognition problem would require a whole book on the topic, and several such books have appeared over time (e.g., Gong et al., 2000). However, the topic of facial recognition is itself dated: not only are workers interested in face recognition, but also there is pressure to measure facial expression, for reasons as diverse as determining whether a person is telling the truth and finding how to mimic real people as accurately as possible in films. (Over the next few years it is possible to anticipate that a good proportion of films will contain no human actors as this has the potential for making them quicker and cheaper to produce.) Medical diagnosis or facial reconstruction can also benefit from facial measurement algorithms, while person verification may well be at least as important as face recognition, as long as it can be done quickly and with minimum error. In this latter respect, recent efforts have moved in the direction of identifying people with great accuracy from their iris patterns (e.g., Daugman, 1993, 2003), and even more accurately from their retinal blood vessels[5] (using the methods of retinal angiography). Although the retinal method would be rather expensive to implement, for example, on all-weather ATR machines, the iris method need not be, and much progress has been made in this direction.

These considerations show that a narrow view of face recognition would be rather inappropriate. For this reason we concentrate here on one or two important aspects. Among these are (1) the task of analyzing the face for key features, which can then at least provide a proper framework for further work on facial recognition, facial expression, facial verification, and so on; and (2) the task of locating that key feature, the iris, which will be used both for detailed verification and as an important starting point for facial analysis.

5 Here some of the main signals are commercial rather than academic, though an important message is that the technical difficulty is only viable where there is a need for the highest security, but in that case "the false acceptance rate for a correctly installed retina scan system falls below 0.0001 percent" (ru.computers.toshiba-europe.com: website accessed May 19, 2004).

We can tackle the iris location and recognition task reasonably straightforwardly. First, if the head has been located with reasonable accuracy, then it can form a region of interest, inside which the iris can be sought. In a front view with the eyes looking ahead, the iris will be seen as a round often high-contrast object, and can in principle be located quite easily with the aid of a Hough transform (Chapter 10; see also Ma et al., 2003). In some cases, this will be less easy because the iris is relatively light and the color may not be distinctive—though a lot will depend on the quality of the illumination. Perhaps more important, in some human subjects, the eyelid and lower contour of the eye may partially overlap the iris (Fig. 24.11*a*), making it more difficult to identify, though a Hough transform is capable of coping with quite a large degree of occlusion. The other factor is that the eyes may not be facing directly ahead. Nevertheless, the iris should appear elliptical and thus should be detectable by another form of the Hough transform (Chapter 12). If this can be done, it should be possible to estimate the direction of gaze with a reasonable degree of accuracy (Gong et al., 2000), thereby taking us further than mere recognition. (The fact that measurement of ellipse eccentricity would lead to an ambiguity in the gaze direction can of course be offset by measuring the position of the ellipse on the eyeball.)

Other facial features, such as the corners of the eye and mouth, can be found by tracking, by snake algorithms, or simply by corner detection. Similarly, the upper and lower contours of the ear and of the nose can be ascertained, thus yielding fixed points that can be used for a multitude of purposes ranging from person identification to recognition of facial expressions. At this point it is useful to reconsider the fact that the face is part of the head and that this is a 3-D object.

24.12.1 *The Face as Part of a 3-D Object*

To start an analysis of the head and face, note that we can define a plane Π containing the outer corners of the eyes and the outer corners of the mouth, if we assume that some odd facial expression has not been adopted. To a very good approximation, it can also be assumed that the inner corners of the eyes will be in the same plane (Fig. 24.11*a*–*c*). The next step is to estimate the position of the vanishing point V for the three horizontal lines λ_1, λ_2, λ_3 joining these three pairs of features (Fig. 24.11*d*). Once this has been done, it is possible to use the relevant cross-ratio invariants to determine the points that in 3-D lie midway between the two features of each pair. This will give the symmetry line λ_s of the face. It will also be possible to determine the horizontal orientation θ of the facial plane Π, that is, the angle through which it has been rotated, about a vertical axis, from a full frontal view. The geometry for these calculations is shown in Fig. 24.11*e*. Finally, it will be possible to convert the interfeature distances along

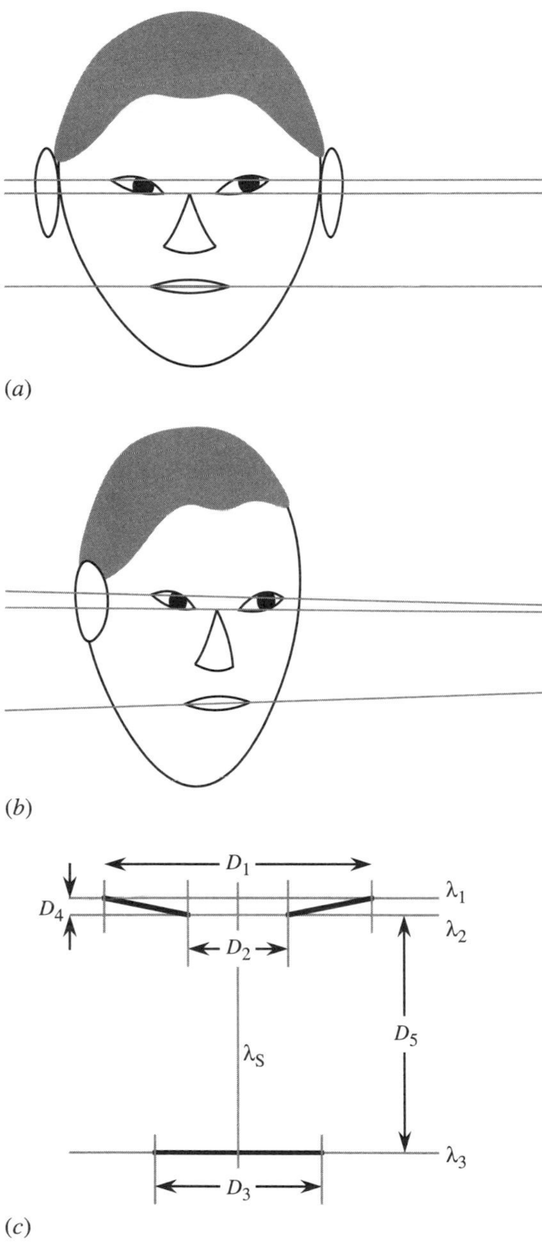

Figure 24.11 3-D analysis of facial parameters. (*a*) Front view of face. (*b*) Oblique view of face, showing perspective lines for corners of the eye and mouth. (*c*) Labeling of eye and mouth features and definition of five interfeature distance parameters. (*d*) Position of vanishing point under an oblique view. (*e*) Positions of one pair of features and their midpoint on the facial plane Π. Note how the midpoint is no longer the midpoint in the image plane I when viewed under perspective projection. Note also how the vanishing point V gives the horizontal orientation of Π.

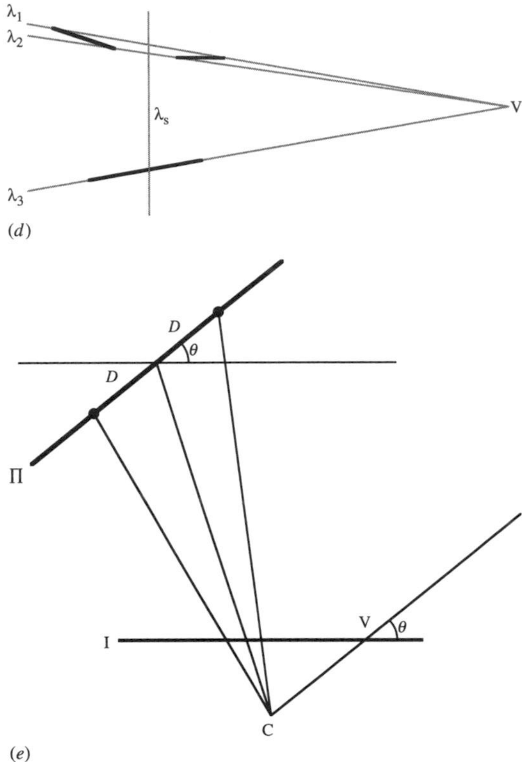

Figure 24.11 Continued.

λ_1, λ_2, λ_3 to the corresponding full frontal values, taking proper account of perspective, but not yet taking account of the vertical orientation φ of Π, which is still unknown.[6]

There is insufficient information to estimate the vertical orientation φ, without making further assumptions; ultimately, this is because the face has no horizontal axis of symmetry. If we can assume that φ is zero (i.e., the head is held neither up nor down, and the camera is on the same level), then we can gain some information on the relative vertical distances of the face, the raw measurements for these being obtained from the intercepts of λ_1, λ_2, λ_3 with the symmetry line λ_s. Alternatively, we can assume average values for the interfeature distances and deduce the vertical orientation of the face. A further alternative is to make other estimates based on the chin, nose, ears, or hairline, but as these are not guaranteed to be in the facial plane Π, the whole assessment of facial pose may not then be accurate and invariant to perspective effects.

6 Note that the theory underlying these procedures is closely related to that of Section 20.8: see also Fig. 20.6.

Overall, we are moving toward measurements either of facial pose or of facial interfeature measurements, with the possibility of obtaining some information on both, even when allowance has to be made for perspective distortions (Kamel et al., 1994, Wang et al., 2003). The analysis will be significantly simpler in the absence of perspective distortions, when the face is viewed from a distance or when a full frontal view is guaranteed. The bulk of the work on facial recognition and pose estimation to date has been done in the context of weak perspective, making the analysis altogether simpler. Even then, the possibility of wide varieties of facial expressions brings in a great deal of complexity. It should also be noticed that the face is not merely a rubber mask (or deformable template), which can be distorted "tidily." The capability for opening and closing the mouth and eyes creates additional nonlinear effects that are not modeled merely by variable stretching of rubber masks.

24.13 Another Look at Statistical Pattern Recognition: The Support Vector Machine

The support vector machine (SVM) is a new paradigm for statistical pattern recognition and emerged during the 1990s as an important contender for practical applications. The basic concept relates to linearly separable feature spaces and is illustrated in Fig. 24.12*a*. The idea is to find the pair of parallel hyperplanes that leads to the maximum separation between two classes of feature so as to provide the greatest protection against errors. In Fig. 24.12*a*, the dashed set of hyperplanes has lower separation and thus represents a less ideal choice, with reduced protection against errors. Each pair of parallel hyperplanes is characterized by specific sets of feature points—the so-called support vectors. In the feature space shown in Fig. 24.12*a*, the planes are fully defined by three support vectors, though clearly this particular value only applies for 2-D feature spaces. In N dimensions the number of support vectors required is $N+1$. This provides an important safeguard against overfitting, since however many data points exist in a feature space, the maximum number of vectors used to describe it is $N+1$.

For comparison, Fig. 24.12*b* shows the situation that would exist if the nearest neighbor method were employed. In this case the protection against errors would be higher, as each position on the separating surface is optimized to the highest local separation distance. However, this increase in accuracy comes at quite a high cost in the much larger number of defining example patterns. As indicated above, much of the gain of the SVM comes from its use of the smallest possible number of defining example patterns (the support vectors). The disadvantage is that the basic method only works when the dataset is linearly separable.

To overcome this problem, it is possible to transform the training and test data to a feature space of higher dimension where the data do become

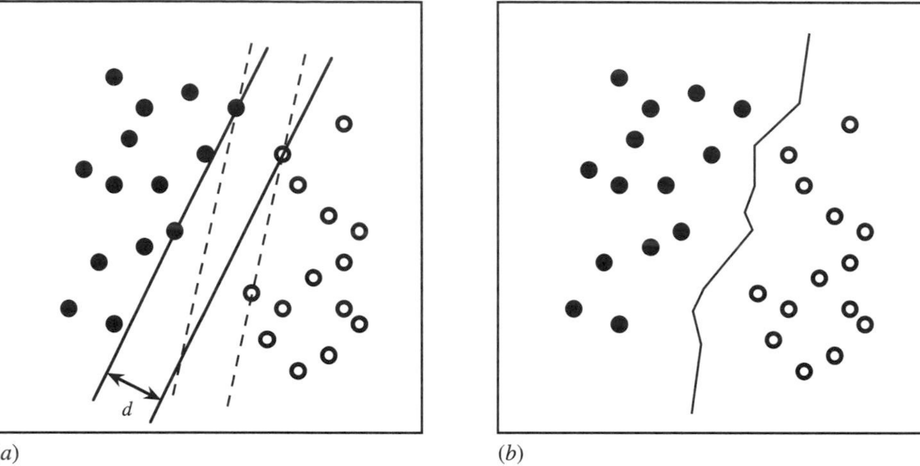

Figure 24.12 Principle of the support vector machine. (*a*) shows two sets of linearly separable feature points: the two parallel hyperplanes have the maximum possible separation d and should be compared with alternatives such as the pair shown dashed. (*b*) shows the optimal piecewise linear solution that would be found by the nearest neighbor method.

linearly separable. However, this approach will tend to reduce or even eliminate the main advantage of the SVM and lead to overfitting of the data plus poor generalizing ability. However, if the transformation that is employed is nonlinear, the final (linearly separable) feature space could have a manageable number of dimensions, and the advantage of the SVM may not be eroded. Nevertheless, there comes a point where the basic restriction of linear separability has to be questioned. At that point, it has been found useful to build "slack" variables s_i into the optimization equations to represent the amount by which the separability constraint can be violated. This is engineered by adding a cost term $C \ \Sigma_i \ s_i$ to the normal error function. C is adjustable and acts as a regularizing parameter, which is optimized by monitoring the performance of the classifier on a range of training data.

For further information on this topic, the reader should consult either the original papers by Vapnik, including Vapnik (1998), the specialized text by Cristianini and Shawe-Taylor (2000), or other texts on statistical pattern recognition, such as Webb (2002).

24.14 Concluding Remarks

The methods of this chapter make it rather surprising that so much of image processing and analysis is possible without any reference to *a priori* probabilities. This situation is likely due to several factors: (1) expediency, and in particular the need for speed of interpretation; (2) the fact that algorithms are designed by humans who have knowledge of the types of input data and thereby incorporate *a priori* probabilities implicitly, for example, via the application of suitable threshold values; and (3) tacit recognition of the situation outlined in the last section, that probabilistic methods have limited validity and applicability. In practice (following on from the ideas of the last section), it is only at the stage of simple image structure and contextual analysis that probabilistic interpretations come into their own. This explains why SPR is not covered in more depth in this book.

Nonetheless, SPR is extremely valuable within its own range of utility. This includes identifying objects on conveyors and making value judgments of their quality, reading labels and codes, verifying signatures, checking fingerprints, and so on. The number of distinct applications of SPR is impressive, and it forms an essential counterpart to the other methods described in this book.

This chapter has concentrated mainly on the supervised learning approach to SPR. However, unsupervised learning is also vitally important, particularly when training on huge numbers of samples (e.g., in a factory environment) is involved. The section on this topic should therefore not be ignored as a minor and insignificant perturbation. Much the same comments apply to the subject of principal components analysis which has had an increasing following in many areas of machine vision—see Section 24.10. Neither should it go unnoticed that these topics link strongly with those of the following chapter on biologically motivated pattern recognition methods, with artificial neural networks playing a very powerful role.

Section 24.12 is entitled "The Route to Face Recognition" to emphasize that facial analysis involves more than mere statistical pattern recognition. Indeed, this topic is something of a maverick, involving structural pattern recognition and the need to cope with the effects of perspective distortions, variations due to facial plasticity, and the enormous number of individual and interracial variations. A number of these variations can be analyzed with the help of PCA (Gong et al., 2000). As facial analysis is such a wide-ranging topic, we are just considering how some of the crucial features could be identified and related to one another. It is left to the now voluminous literature on the subject to take the discussion further.

Vision is largely a recognition process with both structural and statistical aspects. This chapter has reviewed statistical pattern recognition, emphasizing fundamental classification error limits, and has shown the part played by Bayes' theory and the nearest neighbor algorithm, and by ROC and PCA analysis, with face recognition reflecting an ever-changing scenario.

24.15 Bibliographical and Historical Notes

Although the subject of statistical pattern recognition does not tend to be at the center of attention in image analysis work,[7] it provides an important background—especially in the area of automated visual inspection where decisions continually have to be made on the adequacy of products. Most of the relevant work on this topic was already in place by the early 1970s, including the work of Hughes (1968) and Ullmann (1969) relating to the optimum number of features to be used in a classifier. At that stage, a number of important volumes appeared—see, for example, Duda and Hart (1973) and Ullmann (1973) and these were followed a little later by Devijver and Kittler (1982).

The use of SPR for image interpretation dates from the 1950s. For example, in 1959 Bledsoe and Browning developed the n-tuple method of pattern recognition, which turned out (Ullmann, 1973) to be a form of NN classifier. However, it has been useful in leading to a range of simple hardware machines based on RAM (n-tuple) lookups (see for example, Aleksander et al., 1984), thereby demonstrating the importance of marrying algorithms and readily implementable architectures.

Many of the most important developments in this area have probably been those comparing the detailed performance of one classifier with another, particularly with respect to cutting down the amount of storage and computational effort. Papers in these categories include those by Hart (1968) and Devijver and Kittler (1980). Oddly, there appeared to be no overt mention in the literature of how *a priori* probabilities should be used with the NN algorithm until the author's paper on this topic (Davies, 1988f): see Section 24.4.

On the unsupervised approach to SPR, Forgy's (1965) method for clustering data was soon followed by the now famous ISODATA approach of Ball and Hall (1966), and then by MacQueen's (1967) k-means algorithm. Much related work ensued, which has been summarized by Jain and Dubes (1988), a now classic text. However, cluster analysis is an exacting process, and various workers have felt the need to push the subject further forward. For example, Postaire and Touzani (1989) required more accurate cluster boundaries; Jolion and Rosenfeld (1989) wanted better detection of clusters in noise; Chauduri (1994) needed to cope with time-varying data; and Juan and Vidal (1994) required faster k-means clustering. All this work can be described as conventional and did not involve the use of robust statistics per se. However, elimination of outliers is central to the problem of reliable cluster analysis. For a discussion of this aspect of the problem, see Appendix A and the references listed therein.

Although the field of pattern recognition has moved forward substantially since 1990, fortunately there are several quite recent texts that cover the subject relatively painlessly (Theodoridis and Koutroumbas, 1999; Duda et al., 2001; Webb, 2002). The reader can also appeal to the review article by Jain et al. (2000), which outlines new areas that appeared in the previous decade.

7 Note, however, that it is vital to the analysis of multispectral data from satellite imagery: see for example, Landgrebe (1981).

Multiple classifiers is a major new topic and is well reviewed by Duin (2002), a previous review having been carried out by Kittler et al. (1998). Ho et al. (1994) dates from when the topic was rather younger and lists an interesting set of options as seen at that point; some of these are covered in Section 24.8.

Bagging and *boosting* are further variants on the multiple classifier theme. They were developed by Breiman (1996) and Freund and Schapire (1996). Bagging (short for "bootstrap aggregating") means sampling the training set, with replacement, n times, generating b bootstrap sets to train b subclassifiers, and assigning any test pattern to the class most often predicted by the subclassifiers. The method is particularly useful for unstable situations (such as when classification trees are used), but is almost valueless when stable classification algorithms are used (such as the nearest neighbor algorithm). Boosting is useful for aiding the performance of weak classifiers. In contrast with bagging, which is a parallel procedure, boosting is a sequential deterministic procedure. It operates by assigning different weights to different training set patterns according to their intrinsic (estimated) accuracy. For further progress with these techniques, see Rätsch et al. (2002), Fischer and Buhmann (2003), and Lockton and Fitzgibbon (2002). Finally, Beiden et al. (2003) discuss a variety of factors involved in the training and testing of competing classifiers. In addition, much of the discussion relates to multivariate ROC analysis.

Support vector machines (SVMs) also came into prominence during the 1990s and are finding an increasing number of applications. The concept was invented by Vapnik, and the historical perspective is covered in Vapnik (1998). Cristianini and Shawe-Taylor (2000) provide a student-oriented text on the subject.

Progress has also been made in other areas, such as unsupervised learning and cluster analysis (e.g., Mitra et al., 2002; Veenman et al., 2002), but space does not permit further discussion here. Note that, for a more complete view of the modern state of the subject, we must take into account recent work on artificial neural networks (see Chapter 25).

24.16 Problems

1. Show that if the cost function of equation (24.15) is chosen, then the decision rule (24.14) can be expressed in the form of relation (24.5).
2. Show that in a simple two-class system, introducing a reject classification to reduce the number of errors by R in fact requires $2R$ test patterns to be rejected, assuming that R is small. What is likely to happen as R increases?
3. Why is the point on an ROC curve closest to the origin *not* the point that minimizes total error? Prove that the point that minimizes the total error on an ROC curve is actually the point where the gradient is -1 (see Section 24.7).

Biologically Inspired Recognition Schemes

Analogies with the processes being used in the human brain have led to a number of biologically inspired recognition methods—notably genetic algorithms and artificial neural networks. The resulting techniques have proceeded far enough to be widely useful and nowadays must at least be considered when producing new vision algorithms. This chapter outlines the situation regarding these techniques.

Look out for:

- how artificial neural networks can be trained systematically.
- the fact that neural networks are alternative implementations of statistical pattern recognition techniques and are therefore subject to the same underlying rules.
- the problem of overfitting to the training data.
- problems that can arise with inadequate training.
- the existence of alternative learning techniques, such as Hebbian learning.
- the methodology and limitations of genetic algorithms.

The fact that neural networks constitute alternative implementations of statistical pattern recognition techniques is important. They, and genetic algorithms, should be used in appropriate situations with appropriate training. Both have their due place in the panoply of practical recognition systems covered in Part 4 of this volume.

CHAPTER 25

Biologically Inspired Recognition Schemes

25.1 Introduction

The artificial intelligence community has long been intrigued by the capability of the human mind for apparently effortless perception of complex scenes. Of even greater significance is the fact that perception does not occur as a result of overt algorithms, but rather as a result of hard-wired processes, coupled with extensive training of the neural pathways in the infant brain. The hard-wiring pattern is largely the result of evolution, both of the neurons themselves and of their interconnections, though training and use must have some effect on them too. By and large, however, it is probably correct to say that the overall architecture of the brain is the result of evolution, while its specific perceptual and intellectual capabilities are the result of training. It also seems possible that complex thought processes can themselves move toward solutions on an evolutionary basis, on timescales of milliseconds rather than millennia.

These considerations leave us with two interesting possibilities for constructing artificial intelligent systems: the first involves building or simulating networks of neurons and training them to perform appropriate nontrivial tasks; and the second involves designing algorithms that can search for viable solutions to complex problems by "evolutionary" or "genetic" algorithms (GAs). In this chapter we give only an introductory study of GAs, since their capabilities for handling vision problems are at a preliminary stage. On the other hand, we shall study artificial neural networks (ANNs) more carefully, since they are closely linked to quite old ideas on statistical pattern recognition, and, in addition, they are starting to be used quite widely in vision applications.

Neurons operate in ways that are quite different from those of modern electronic circuits. Electronic circuits may be of two main types—analog and digital—though the digital type is predominant in the present generation of computers. Specifically, waveforms at the outputs of logic gates and flip-flops are strongly synchronized with clock pulses, whereas those at the outputs of biological neurons exhibit "firing" patterns that are asynchronous, often being considered as random pulse streams for which only the frequency of firing carries useful information. Thus, the representation of the information is totally different from that of electronic circuits. Some have argued that the representation used by biological neurons is a vital feature that leads to efficient information processing, and have even built stochastic computing units on this basis. We shall not pursue this line here, for there seems to be no reason why the vagaries of biological evolution should have led to the most efficient representation for artificial information processing; that is, the types of neurons that are suitable for biological systems are not necessarily appropriate for machine vision systems. Nevertheless, in designing machine vision systems, it seems reasonable to take *some* hints from biological systems on what methodologies might be useful.

This chapter describes ANNs and gives some examples of their application in machine vision, and also gives a brief outline of work on GAs. In considering ANNs, it should be remembered that there are a number of possible architectures with various characteristics, some being appropriate for supervised learning, and others being better adapted for unsupervised learning. We start by studying the perceptron type of classifier designed by Rosenblatt in the 1960s (1962, 1969).

25.2 Artificial Neural Networks

The concept of an ANN that could be useful for pattern recognition started in the 1950s and continued right through the 1960s. For example, Bledsoe and Browning (1959) developed the "*n*-tuple" type of classifier which involved bitwise recording and lookup of binary feature data, leading to the weightless or logical type of ANN. Although this type of classifier has had some following right through to the present day, it is probably no exaggeration to say that it is Rosenblatt's "perceptron" (1958, 1962) that has had the greatest influence on the subject.

The simple perceptron is a linear classifier that classifies patterns into two classes. It takes a feature vector $\mathbf{x} = (x_1, x_2, \ldots, x_N)$ as its input, and it produces a single scalar output $\sum_{i=1}^{N} w_i x_i$, the classification process being completed by applying a threshold (Heaviside step) function at θ (see Fig. 25.1). The mathematics is simplified by writing $-\theta$ as w_0, and taking it to correspond to an input x_0, which

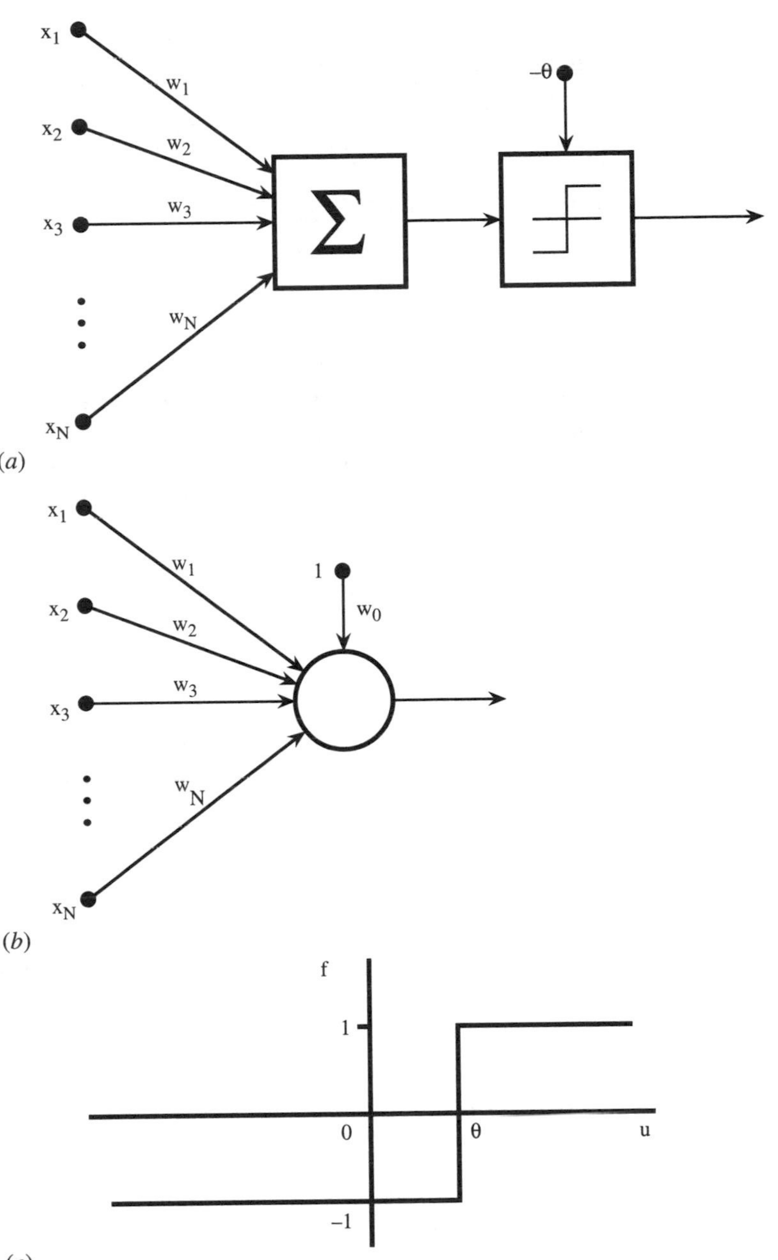

Figure 25.1 Simple perceptron. (*a*) shows the basic form of a simple perceptron. Input feature values are weighted and summed, and the result fed via a threshold unit to the output connection. (*b*) gives a convenient shorthand notation for the perceptron; and (*c*) shows the activation function of the threshold unit.

is maintained at a constant value of unity. The output of the linear part of the classifier is then written in the form:

$$d = \sum_{i=1}^{N} w_i x_i - \theta = \sum_{i=1}^{N} w_i x_i + w_0 = \sum_{i=0}^{N} w_i x_i \tag{25.1}$$

and the final output of the classifier is given by:

$$y = f(d) = f\left(\sum_{i=0}^{N} w_i x_i\right) \tag{25.2}$$

This type of neuron can be trained using a variety of procedures, such as the *fixed increment rule* given in Fig. 25.2. (The original fixed increment rule used a learning rate coefficient η equal to unity.) The basic concept of this algorithm was to try to improve the overall error rate by moving the linear discriminant plane a fixed distance toward a position where no misclassification would occur—but only doing this when a classification error had occurred:

$$w_i(k+1) = w_i(k) \qquad y(k) = \omega(k) \tag{25.3}$$

$$w_i(k+1) = w_i(k) + \eta[\omega(k) - y(k)]x_i(k) \qquad y(k) \neq \omega(k) \tag{25.4}$$

In these equations, the parameter k represents the kth iteration of the classifier, and $\omega(k)$ is the class of the kth training pattern. It is important to know whether this training scheme is effective in practice. It is possible to show that, if the algorithm is modified so that its main loop is applied sufficiently many times, *and* if the feature vectors are linearly separable, then the algorithm will converge on a correct error-free solution.

Unfortunately, most sets of feature vectors are not linearly separable. Thus, it is necessary to find an alternative procedure for adjusting the weights. This is achieved by the Widrow–Hoff delta rule which involves making changes in the weights in proportion to the error $\delta = \omega - d$ made by the classifier. (Note that the error is calculated *before* thresholding to determine the actual class: that is, δ is

```
initialize weights with small random numbers;
select suitable value of learning rate coefficient η in the range 0 to 1;
do {
   for all patterns in the training set {
      obtain feature vector x and class ω;
      compute perceptron output y;
      if (y !=ω) adjust weights according to w_i = w_i + η(ω – y)x_i;
   }
} until no further change;
```

Figure 25.2 Perceptron *fixed increment* algorithm.

calculated using d rather than $f(d)$.) Thus, we obtain the Widrow–Hoff delta rule in the form:

$$w_i(k+1) = w_i(k) + \eta\delta x_i(k) = w_i(k) + \eta[\omega(k) - d(k)]x_i(k) \tag{25.5}$$

There are two important ways in which the Widrow–Hoff rule differs from the fixed increment rule:

1. An adjustment is made to the weights whether or not the classifier makes an actual classification error.
2. The output function d used for training is different from the function $y = f(d)$ used for testing.

These differences underline the revised aim of being able to cope with nonlinearly separable feature data. Figure 25.3 clarifies the situation by appealing to a 2-D case. Figure 25.3(a) shows separable data that are straightforwardly fitted by the fixed increment rule. However, the fixed increment rule is not designed to cope with nonseparable data of the type shown in Fig. 25.3(b) and results in instability during training, as well as inability to arrive at an optimal solution. On the other hand the Widrow–Hoff rule copes satisfactorily with these types of data. An interesting addendum to the case of Fig. 25.3*a* is that although the fixed increment rule apparently reaches an optimal solution, the rule becomes "complacent" once a zero error situation has occurred, whereas an ideal classifier would arrive at a solution that minimizes the probability of error. The Widrow–Hoff rule goes some way to solving this problem.

So far we have considered what can be achieved by a simple perceptron. Although it is only capable of dichotomizing feature data, a suitably trained array of simple perceptrons—the "single-layer perceptron" of Fig. 25.4—should be able to divide feature space into a large number of subregions bounded (in multidimensional space) by hyperplanes. However, in a multiclass application this approach would require a large number of simple perceptrons—up to $\binom{c}{2} = \frac{1}{2}c(c-1)$ for a c-class system. Hence, there is a need to generalize the approach by other means. In particular, multilayer perceptron (MLP) networks (see Fig. 25.5)—which would emulate the neural networks in the brain—seem poised to provide a solution since they should be able to recode the outputs of the first layer of simple perceptrons.

Rosenblatt himself proposed such networks but was unable to propose general means for training them systematically. In 1969, Minsky and Papert published their famous monograph, and in discussing the MLP raised the spectre of "the monster of vacuous generality"; they drew attention to certain problems that apparently would never be solved using MLPs. For example, diameter-limited perceptrons (those that view only small regions of an image within

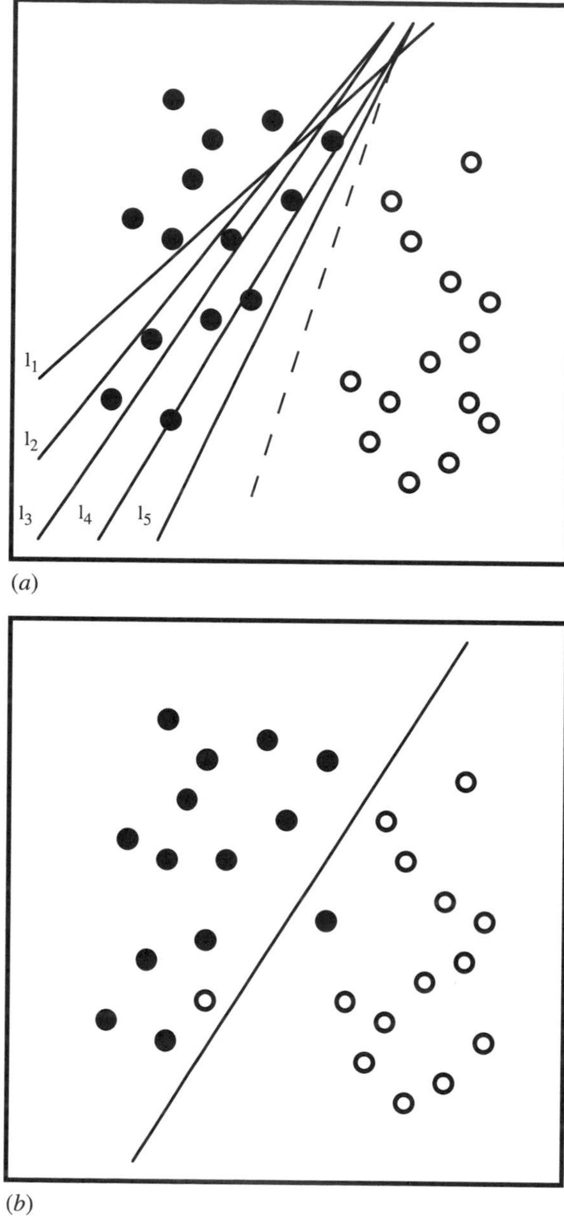

Figure 25.3 Separable and nonseparable data. (*a*) shows two sets of pattern data. Lines l_1–l_5 indicate possible successive positions of a linear decision surface produced by the fixed increment rule. Note that the fixed increment rule is satisfied by the final position l_5. The dotted line shows the final position that would have been produced by the Widrow–Hoff delta rule. (*b*) shows the stable position that would be produced by the Widrow–Hoff rule in the case of nonseparable data. In this case, the fixed increment rule would oscillate over a range of positions during training.

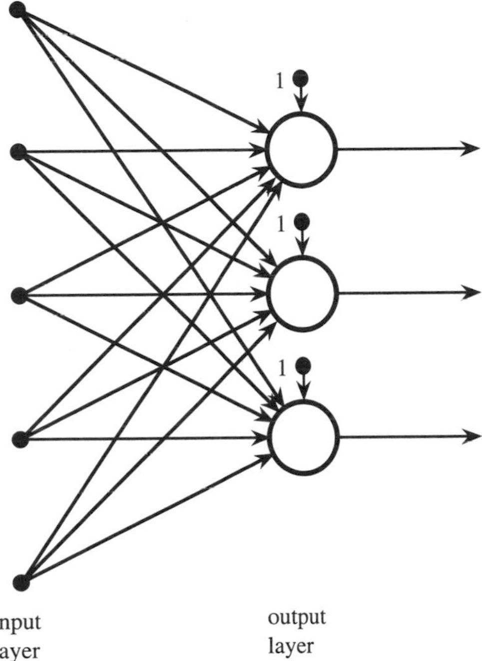

Figure 25.4 Single-layer perceptron. The single-layer perceptron employs a number of simple perceptrons in a single layer. Each output indicates a different class (or region of feature space). In more complex diagrams, the bias units (labeled "1") are generally omitted for clarity.

a restricted diameter) would be unable to measure large-scale connectedness within images. These considerations discouraged effort in this area, and for many years attention was diverted to other areas such as expert systems. It was not until 1986 that Rumelhart et al. were successful in proposing a systematic approach to the training of MLPs. Their solution is known as the backpropagation algorithm.

25.3 The Backpropagation Algorithm

The problem of training an MLP can be simply stated: a general layer of an MLP obtains its feature data from the lower layers and receives its class data from higher layers. Hence, if all the weights in the MLP are potentially changeable, the information reaching a particular layer cannot be relied upon. There is no reason why training a layer in isolation should lead to overall convergence of the MLP toward an ideal classifier (however defined). In addition, it is not evident what the

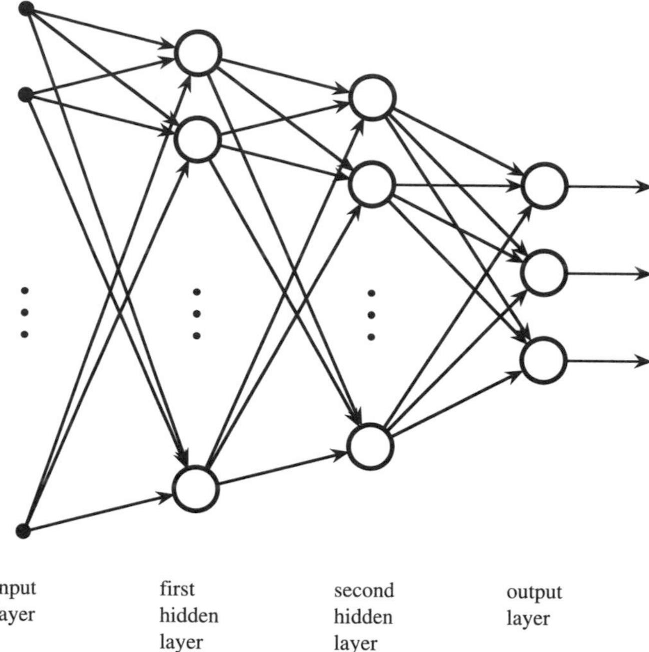

Figure 25.5 Multilayer perceptron. The multilayer perceptron employs several layers of perceptrons. In principle, this topology permits the network to define more complex regions of feature space and thus perform much more precise pattern recognition tasks. Finding systematic means of training the separate layers becomes the vital issue. For clarity, the bias units have been omitted from this and later diagrams.

optimal MLP architecture should be. Although it might be thought that this difficulty is rather minor, in fact this is not so. Indeed, this is but one example of the so-called credit assignment problem.[1]

One of the main difficulties in predicting the properties of MLPs and hence of training them reliably is the fact that neuron outputs swing suddenly from one state to another as their inputs change by infinitesimal amounts. Hence, we might consider removing the thresholding functions from the lower layers of MLP networks to make them easier to train. Unfortunately, this would result in these layers acting together as larger linear classifiers, with far less discriminatory power than the original classifier. (In the limit we would have a set of linear

1 This is not a good first example with which to define the credit assignment problem (in this case it would appear to be more of a deficit assignment problem). The credit assignment problem is the problem of correctly determining the local origins of global properties and making the right assignments of rewards, punishments, corrections, and so on, thereby permitting the whole system to be optimized systematically.

classifiers, each with a single thresholded output connection, so the overall MLP would act as a single-layer perceptron!)

The key to solving these problems was to modify the perceptrons composing the MLP by giving them a less "hard" activation function than the Heaviside function. As we have seen, a linear activation function would be of little use, but one of "sigmoid" shape, such as the tanh(u) function (Fig. 25.6) is effective and

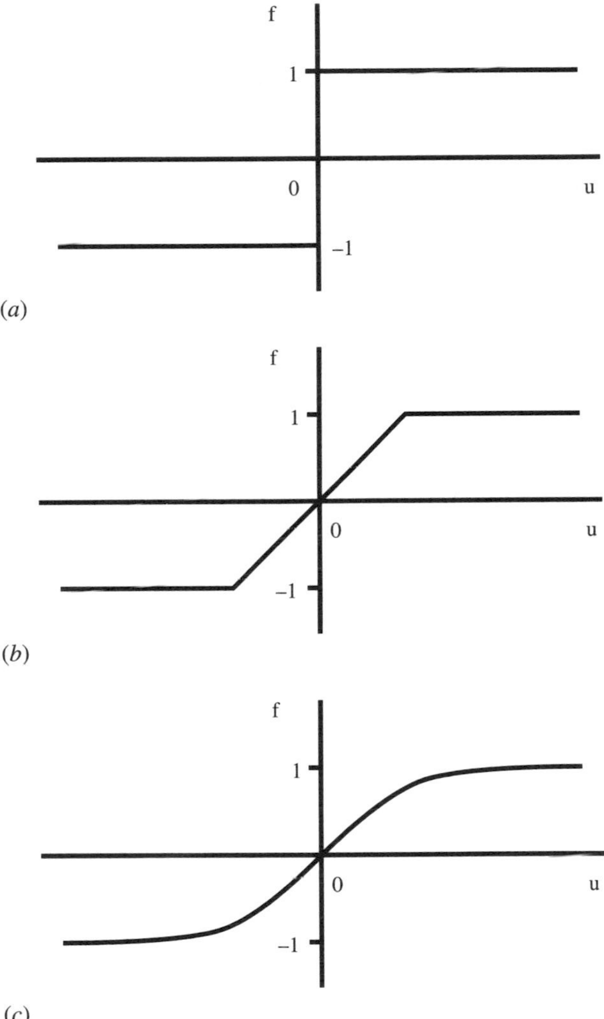

Figure 25.6 Symmetrical activation functions. This figure shows a series of symmetrical activation functions. (*a*) shows the Heaviside activation function used in the simple perceptron. (*b*) shows a linear activation function, which is, however, limited by saturation mechanisms. (*c*) shows a sigmoidal activation function that approximates to the hyperbolic tangent function.

```
initialize weights with small random numbers;
select suitable value of learning rate coefficient η in the range 0 to 1;
do {
   for all patterns in the training set
      for all nodes j in the MLP {
         obtain feature vector x and target output value t;
         compute MLP output y;
         if (node is in output layer)
               δj = yj(1 − yj)(tj − yj);
         else  δj = yj(1 − yj)(Σm δm wjm);
         adjust weights i of node j according to wij = wij + η δj yi;
      }
} until changes are reduced to some predetermined level;
```

Figure 25.7 The backpropagation algorithm.

indeed is almost certainly the most widely used of the available functions.[2] Once these softer activation functions were used, it became possible for each layer of the MLP to "feel" the data more precisely and thus training procedures could be set up on a systematic basis. In particular, the rate of change of the data at each individual neuron could be communicated to other layers, which could then be trained appropriately—though only on an incremental basis. We shall not go through the detailed mathematical procedure, or proof of convergence, beyond stating that it is equivalent to energy minimization and gradient descent on a (generalized) energy surface. Instead, we give an outline of the back-propagation algorithm (see Fig. 25.7). Nevertheless, some notes on the algorithm are in order:

1. The outputs of one node are the inputs of the next, and an arbitrary choice is made to label all variables as output (y) parameters rather than as input (x) variables; all output parameters are in the range 0 to 1.
2. The class parameter ω has been generalized as the target value t of the output variable y.
3. For all except the final outputs, the quantity δ_j has to be calculated using the formula $\delta_j = y_j(1-y_j)(\Sigma_m \delta_m w_{jm})$, the summation having to be taken over all the nodes in the layer *above* node j.

---

2 We do not here make a marked distinction between symmetrical activation functions and alternatives which are related to them by shifts of axes, though the symmetrical formulation seems preferable as it emphasizes bidirectional functionality. In fact, the tanh(u) function, which ranges from –1 to 1, can be expressed in the form:

$$\tanh(u) = (e^u - e^{-u})/(e^u + e^{-u}) = 1 - 2/(1 + e^{2u})$$

and is thereby closely related to the commonly used function $(1 + e^{-v})^{-1}$. It can now be deduced that the latter function is symmetrical, though it ranges from 0 to 1 as v goes from $-\infty$ to ∞.

4. The sequence for computing the node weights involves starting with the output nodes and then proceeding downward one layer at a time.
5. If there are no hidden nodes, the formula reverts to the Widrow–Hoff delta rule, except that the input parameters are now labeled y_i, as indicated above.
6. It is important to initialize the weights with random numbers to minimize the chance of the system becoming stuck in some symmetrical state from which it might be difficult to recover.
7. Choice of value for the learning rate coefficient η will be a balance between achieving a high rate of learning and avoidance of overshoot: normally, a value of around 0.8 is selected.

When there are many hidden nodes, convergence of the weights can be very slow; this is one disadvantage of MLP networks. Many attempts have been made to speed convergence, and a method that is almost universally used is to add a "momentum" term to the weight update formula, it being assumed that weights will change in a similar manner during iteration k to the change during iteration $k-1$:

$$w_{ij}(k+1) = w_{ij}(k) + \eta\delta_j y_i + \alpha[w_{ij}(k) - w_{ij}(k-1)] \tag{25.6}$$

where α is the momentum factor. Primarily, this technique is intended to prevent networks from becoming stuck at local minima of the energy surface.

25.4 MLP Architectures

The preceding sections gave the motivation for designing an MLP and for finding a suitable training procedure, and then outlined a general MLP architecture and the widely used backpropagation training algorithm. However, having a general solution is only one part of the answer. The next question is how best to adapt the general architecture to specific types of problems. We shall not give a full answer to this question here. However, Lippmann attempted to answer this problem in 1987. He showed that a two-layer (single hidden layer) MLP can implement arbitrary convex decision boundaries, and indicated that a three-layer (two-hidden layer) network is required to implement more complex decision boundaries. It was subsequently found that it should never be necessary to exceed two hidden layers, as a three-layer network can tackle quite general situations if sufficient neurons are used (Cybenko, 1988). Subsequently, Cybenko (1989) and Hornik et al. (1989) showed that a two-layer MLP can

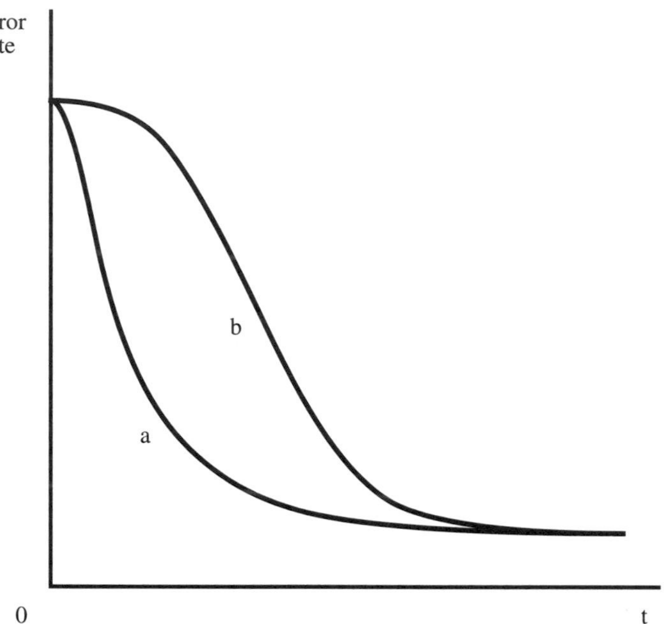

Figure 25.8 Learning curve for the multilayer perceptron. Here (*a*) shows the learning curve for a single-layer perceptron, and (*b*) shows that for a multilayer perceptron. Note that the multilayer perceptron takes considerable time to get going, since initially each layer receives relatively little useful training information from the other layers. Note also that the lower part of the diagram has been idealized to the case of identical asymptotic error rates—though this situation would seldom occur in practice.

approximate any continuous function, though there may sometimes be advantages in using more than two layers.

Although the backpropagation algorithm can train MLPs of any number of layers, in practice, training one layer "through" several others introduces an element of uncertainty that is commonly reflected in increased training times (see Fig. 25.8). Thus, some advantage can be gained from using a minimal number of layers of neurons. In this context, the above findings on the necessary numbers of hidden layers are especially welcome.

25.5 Overfitting to the Training Data

When training MLPs and many other types of ANN there is a problem of overfitting the network to the training data. One basic aim of statistical pattern recognition is for the learning machine to be able to generalize from the particular set of data it is trained on to other types of data it might meet during testing.

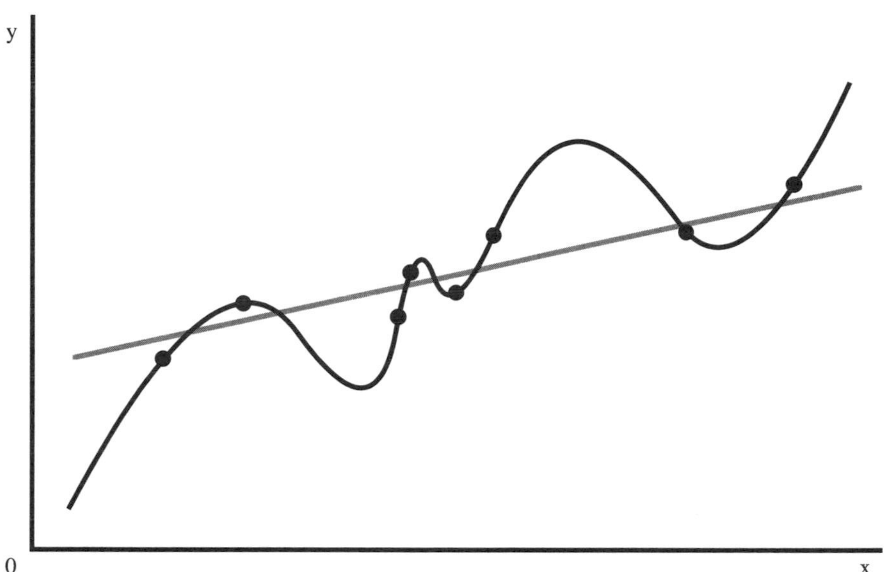

Figure 25.9 Overfitting of data. In this graph, the data points are rather too well fitted by the solid curve, which matches every nuance exactly. Unless there are strong theoretic reasons why the solid curve should be used, the gray line will give a higher confidence level.

In particular, the machine should be able to cope with noise, distortions, and fuzziness in the data, though clearly not to the extent of being able to respond correctly to types of data different from those on which it has been trained. The main points to be made here are (1) that the machine should learn to respond to the underlying population from which the training data have been drawn, and (2) that it must not be so well adapted to the specific training data that it responds less well to other data from the same population. Figure 25.9 shows in a 2-D case both a fairly ideal degree of fit and a situation where every nuance of the set of data has been fitted, thereby achieving a degree of overfit.

Typically, overfitting can arise if the learning machine has more adjustable parameters than are strictly necessary for modeling the training data. With too few parameters, such a situation should not arise. However, if the learning machine has enough parameters to ensure that relevant details of the underlying population are fitted, there may be overmodeling of part of the training set. Thus, the overall recognition performance will deteriorate. Ultimately, the reason for this is that recognition is a delicate balance between capability to discriminate and capability to generalize, and it is most unlikely that any complex learning machine will get the balance right for all the features it has to take account of.

Be this as it may, we need to have some means of preventing overadaptation to the training data. One way of doing so is to curtail the training process before overadaptation can occur.[3] This is not difficult, since we merely need to test the system periodically during training to ensure that the point of overadaptation has not been reached. Figure 25.10 shows what happens when testing is carried out simultaneously on a separate dataset. At first, performance on the test data closely matches that on the training data, being slightly superior for the latter because a small degree of overadaptation is already occurring. But

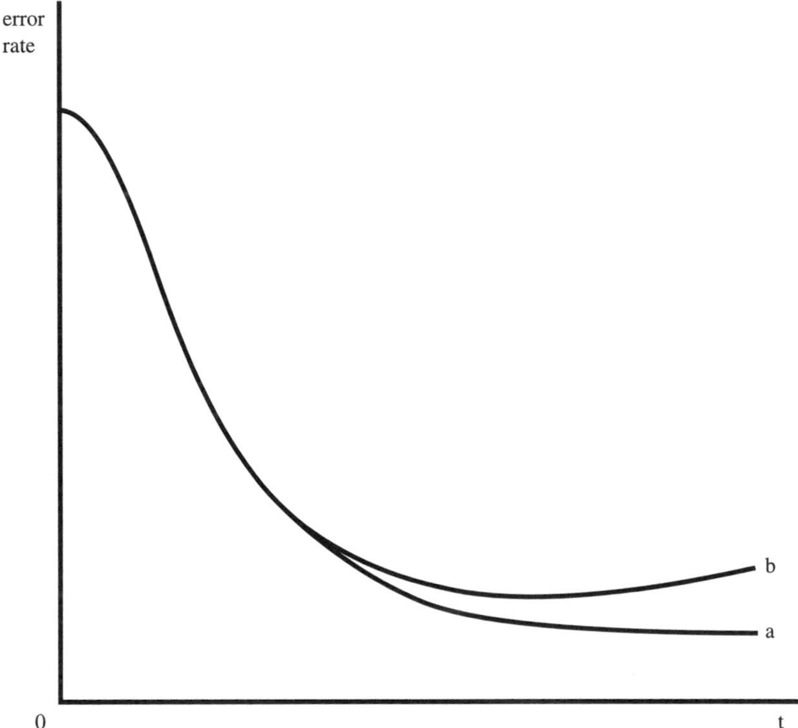

Figure 25.10 Cross-validation tests. This diagram shows the learning curve for a multilayer perceptron (*a*) when tested on the training data and (*b*) when tested on a special validation set. (*a*) tends to go on improving even when overfitting is occurring. However, this situation is detected when (*b*) starts deteriorating. To offset the effects of noise (not shown on curves (*a*) and (*b*)), it is usual to allow 5 to 10% deterioration relative to the minimum in (*b*).

3 It is often stated that this procedure aims to prevent overtraining. However, the term *overtraining* is ambiguous. On the one hand, it can mean recycling through the *same* set of training data until eventually the learning machine is overadapted to it. On the other hand, it can mean using more and more *totally new* data—a procedure that cannot produce overadaptation to the data, and on the contrary is almost certain to improve performance. In view of this ambiguity it seems better not to use the term.

after a time, performance starts deteriorating on the test data while that on the training data appears to go on improving. It is at this point that serious overfitting is occurring, and the training process should be curtailed. The aim, then, is to make the whole training process far more rigorous by splitting the original training set into two parts—the first being retained as a normal training set, and the second being called the *validation set*. Note that the validation set is actually part of the training set in the sense that it is not part of the eventual test set.

The process of checking the degree of training by use of a validation set is called *cross validation* and is vitally important to proper use of an ANN. The training algorithm should include cross validation as a fully integrated part of the whole training schedule. It should not be regarded as an optional extra.

How can overadaptation occur when the training procedure is completely determined by the backpropagation (or other) provably correct algorithms? Overadaptation can occur through several mechanisms. For example, when the training data do not control particular weights closely enough, some can drift to large positive or negative values, while retaining a sufficient degree of cancellation so that no problems appear to arise with the training data. Yet, when test or validation data are employed, the problems become all too clear. The fact that the form of the sigmoid function will permit some nodes to become "saturated out" does not apparently help the situation, for it inactivates parameters and hides certain aspects of the incoming data. Yet it is intrinsic to the MLP architecture and the way it is trained that some nodes are *intended* to be saturated in order to ignore irrelevant features of the training set. The problem is whether inactivation is inadvertent or designed. The answer probably lies in the quality of the training set and how well it covers the available or potential feature space.

Finally, let us suppose that an MLP is being set up, and it is initially unknown how many hidden layers will be required or how many nodes each layer will have to have. It will also be unknown how many training set patterns will be required or how many training iterations will be required—or what values of the momentum or learning parameters will be appropriate. A quite substantial number of tests will be required to decide all the relevant parameters. There is therefore a definite risk that the final system will be overadapted not only to the training set but also to the validation set. In such circumstances, we need a second validation set that can be used after the whole network has been finalized and final training is being undertaken.

25.6 Optimizing the Network Architecture

In the last section we mentioned that the ANN architecture will have to be optimized. In particular, for an MLP the optimum number of hidden layers and numbers of nodes per layer will have to be decided. The optimal degree of connectivity of the network will also have to be determined (it cannot be assumed that full connectivity is necessarily advantageous—it may be better to have some connections that bypass certain hidden layers).

Broadly, there are two approaches to the optimization of network architectures. In the first, the network is built up slowly node by node. In the second, an overly complex network is built, which is gradually pruned until performance is optimized. This latter approach has the disadvantage of requiring excessive computation until the final architecture is reached. It is therefore tempting to consider only the first of the two options—an approach that would also tend to improve generalization. A number of alternative schemes have been described for optimization that improves generalization. There appears to be no universal best method, and research is still proceeding (e.g., Patel and Davies, 1995).

Instead of dwelling on network generation or pruning aspects of architecture optimization, it is helpful to consider an alternative based on use of GAs. In this approach, various trials are made in a space of possible architectures. To apply GAs to this task, two requirements must be fulfilled. Some means of selecting new or modified candidate architectures is needed; and a suitable fitness function for deciding which architecture is optimal must be chosen. The first requirement fits neatly into the GA formalism if each architecture is defined by a binary codeword that can be modified suitably with a meaningful one-to-one correspondence between codeword and architecture. (This is bound to be possible as each actual connection of a potential completely connected network can be determined by a single bit in the codeword.) In principle, the fitness function could be as simple as the mean square error obtained using the validation set. However, some additional weighting factors designed to minimize the number of nodes and the number of interconnections can also be useful. Although this approach has been tested by a number of workers, the computational requirements are rather high. Hence, GAs are unlikely to be useful for optimizing very large networks.

25.7 Hebbian Learning

In Chapter 24, we found how principal components analysis can help with data representation and dimensionality reduction. Principal components analysis is an especially useful procedure, and it is not surprising that a number of attempts have been made to perform it using different types of ANNs. In particular, Oja (1982) was able to develop a method for determining the principal component corresponding to the largest eigenvalue λ_{max} using a single neuron with linear weights.

The basic idea is that of Hebbian learning. To understand this process, we must imagine a large network of biological neurons that are firing according to various input stimuli and producing various responses elsewhere in the network. Hebb (1949) considered how a given neuron could learn from the data it receives.[4] His conclusion was that good pathways should be rewarded so that they become stronger over time; or more precisely, synaptic weights should be strengthened in proportion to the correlation between the firing of pre- and postsynaptic neurons. Here we can visualize the process as "rewarding" the neuron inputs and outputs in proportion to the numbers of input patterns that arrive at them, and modeling this by the equation:[5]

$$\Delta w_i = \eta y x_i \tag{25.7}$$

The problem with this approach is that the weights will grow in an unconstrained manner, and therefore a constraining or normalizing influence is needed. Oja achieved this by adding a weight decay proportional to y^2:

$$\Delta w_i = \eta y x_i - n y^2 w_i = \eta y (x_i - y w_i) \tag{25.8}$$

Linsker (1986, 1988) produced an alternative rule based on clipping at certain maximum and minimum values; and Yuille et al. (1989) used a rule that was designed to normalize weight growth according to the magnitude of the overall weight vector $\mathbf{w}$:

$$\Delta w_i = \eta (y x_i - |\mathbf{w}|^2 w_i) \tag{25.9}$$

We cannot give a complete justification for these rules here: the only really satisfactory justification requires nontrivial mathematical proofs of convergence. However, these ideas led the way to full determination of principal components by ANNs. Both Oja and Sanger developed such methods, which are quite similar and merit close comparison (Oja, 1982; Sanger, 1989).

First, we consider Oja's training rule, which applies to a single-layer feedforward linear network with N input nodes, M processing nodes, and transfer functions:

$$y_i = \sum_{j=1}^{N} w_{ij} x_j \qquad (i = 1 \text{ to } M,\ M < N) \tag{25.10}$$

4 This corresponds to a totally different type of solution to the credit assignment problem than that provided by the backpropagation algorithm.

5 In what follows we suppress the iteration parameter k, and write down merely the increment Δw_i in w_i over a given iteration.

The rule takes the form:

$$\Delta w_{ij} = \eta y_i \left(x_j - \sum_{k=1}^{N} w_{kj} y_k \right) \qquad (i = 1 \text{ to } M,\ j = 1 \text{ to } N) \tag{25.11}$$

This is an extension of the earlier rule for finding the principal component corresponding to $\lambda_{\max}$ (equation (25.5.2)). In particular, it demands that the M nodes, which are all in the same layer, be connected laterally as shown in Fig. 25.11*a*, and it is very plausible that if a single node can determine one principal component, then M modes can find M principal components. However, the symmetry of the situation indicates that it is not clear which mode should correspond to the largest eigenvalue and which to any of the others. The situation is less definite than this, and all we can say is that the output vectors span the space of the largest M eigenvalues. The vectors are mutually orthogonal but are not

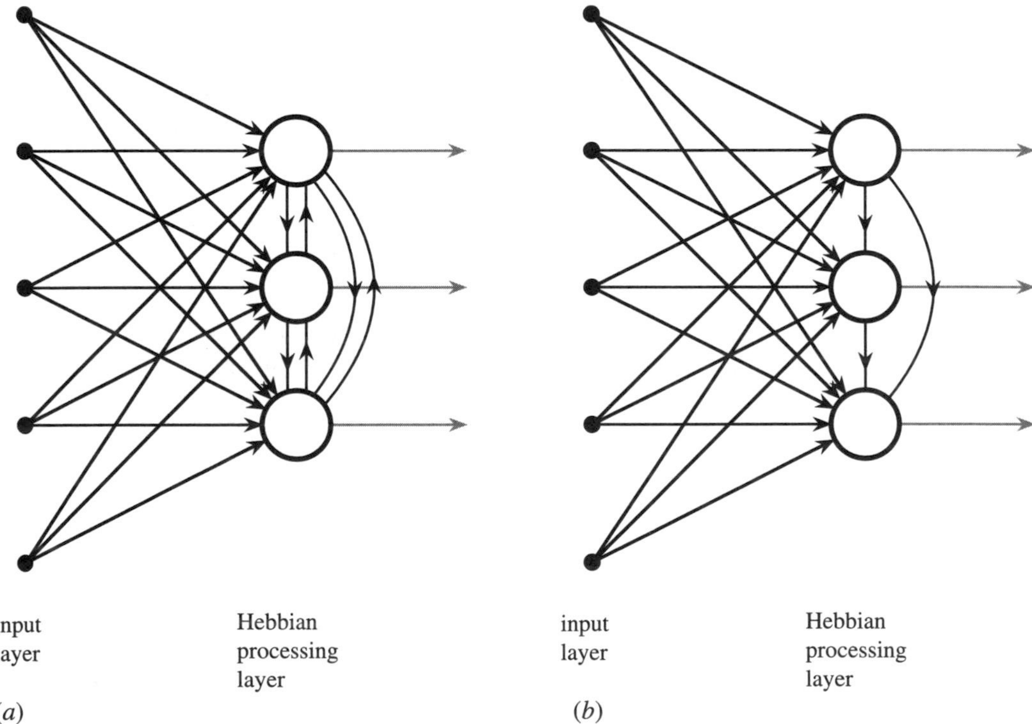

Figure 25.11 Networks employing Hebbian learning. These two networks employ Hebbian learning and evaluate principal components. During training, both networks learn to project the N-dimensional input feature vectors into the space of the M largest principal components. Unlike the Oja network (*a*), the Sanger network (*b*) is able to order the output eigenvalues. For clarity, the output connections are shown in gray.

guaranteed to be oriented along the principal axes directions. It should also be noted that the results obtained by the network will vary significantly with the particular data seen during training and their order of presentation. However, the rule is sufficiently good for certain applications such as data compression, since the principal components corresponding to the smallest $N - M$ eigenvalues are systematically eliminated.

Nevertheless, in many applications it is necessary to force the outputs to correspond to principal components and also to know the order of the eigenvalues. The Sanger network achieves this by making the first node independent of the others; the second dependent only on the first; the third dependent only on the first two; and so on (see Fig. 25.11*b*). Thus, orthogonality of the principal components is guaranteed, whereas independence is maintained at the right level for each node to be sure of producing a specific principal component. It could be argued that the Sanger network is an M-layer network, though this is arguable as the architecture is obtained from the Oja network by *eliminating* some of the connections. The Sanger training rule is obtained from the Oja rule by making the upper limit in the summation i rather than N:

$$\Delta w_{ij} = \eta y_i \left(x_j - \sum_{k=1}^{i} w_{kj} y_k \right) \qquad (i = 1 \text{ to } M,\ j = 1 \text{ to } N) \tag{25.12}$$

At the end of the computation, the outputs of the M processing nodes contain the eigenvalues, while the weights of the M nodes provide the M eigenvectors. An interesting point arises since the successive outputs of the Sanger network have successively lower importance as the eigenvalues (data variances) decrease. Hence, it is valid and sometimes convenient for images or other output data to be computed and presented with progressively fewer bits for the lower eigenvalues. (This applies especially to real-time hardware implementations of the algorithms.)

It might be questioned whether there is any value in techniques such as principal components analysis being implemented using ANNs when a number of perfectly good conventional methods exist for computing eigenvalues and eigenvectors. When the input vectors are rather large, matrix diagonalization involves significant computation, and ANN approaches can arrive at workable approximations very quickly by iterative application of simple linear nodes that are straightforwardly implemented in hardware. Such considerations are especially important in real-time applications, such as those that frequently occur in automated visual inspection or interpretation of satellite images. It is also possible that the approximations made by ANNs will permit them to include other relevant information, hence leading to increased reliability of data classification, for example, by taking into account higher order correlations (Taylor and Coombes, 1993). Such possibilities will have to be confirmed by future research.

We end this section with a simple alternative approach to the determination of principal components. Like the Oja approach, this method projects the input vectors onto a subspace that is spanned by the first M principal components, discarding only minimal information. In this case the network architecture is a two-layer MLP with N inputs, M hidden nodes, and N output nodes, as shown in Fig. 25.12. The special feature is that the network is trained using the backpropagation algorithm to produce the same outputs as the inputs—a scheme commonly known as self-supervised backpropagation. Note that, unusually, the outputs are taken from the hidden layer. Interestingly, it has been found both experimentally and theoretically that nonlinearity of the neuron activation function is of no help in finding principal components (Cottrell et al., 1987; Bourland and Kamp, 1988; Baldi and Hornik, 1989).

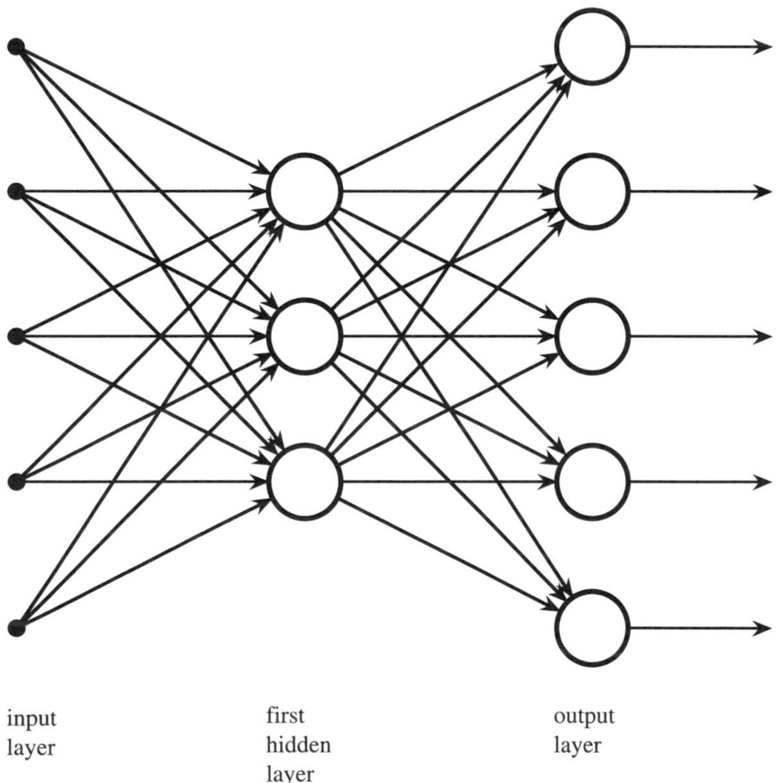

Figure 25.12 Self-supervised backpropagation. This network is a normal two-layer perceptron that is trained by backpropagation so that the output vector tracks the input vector. Once this is achieved, the network can provide information on the M largest principal components. Oddly, this information appears at the outputs of the hidden layer.

25.8 Case Study: Noise Suppression Using ANNs

In what ways can ANNs be useful in applications such as those typical of machine vision? There are three immediate answers to this question. The first is that ANNs are learning machines, and their use should result in greater adaptability than that normally available with conventional algorithms. The second answer is that their concept is that of simple replicable hardware-based computational elements. Hence a parallel machine built using them should in principle be able to operate extremely rapidly, thereby emulating some of the intellectual and perceptual capabilities of the human brain. The third is that their parallel processing capability should lead to greater redundancy and robustness in operation which should again mimic that of the human brain. Although the last two of these possibilities will likely be realized in practice, the position is still not clear, and there is no sign of conventional algorithmic approaches being ousted for these reasons. However, the first point—that of improved adaptability—seems to be at the stage of serious test and practical implementation. It is therefore appropriate to consider it more carefully, in the context of a suitable case study. Here we choose image noise suppression as the vehicle for this study. We use image noise suppression because it is a well researched topic about which a body of relevant theory exists, and any advances made using ANNs should immediately be clear.

As we have seen in Chapter 3, a number of useful conventional approaches are available to image noise suppression. The most widely used, the median filter, is not without its problems, one being that it causes image distortion through the shifting of curved edges. In addition, it has no adjustable parameters other than neighborhood size, and so it cannot be adapted to the different types of noise it might be called upon to eliminate. Indeed, its performance for noise removal is quite limited, as it is largely an ad hoc technique that happens to be best suited to the suppression of impulse noise.

These considerations have prompted a number of researchers to try using ANNs for eliminating image noise (Nightingale and Hutchinson, 1990; Lu and Szeto, 1991; Pham and Bayro-Corrochano, 1992; Yin et al., 1993). Since ANNs are basically recognition tools, they could be used for recognizing noise structures in images, and once a noise structure has been located, it should be possible to replace it by a more appropriate pattern. Greenhill and Davies (1994a,b) have reported a more direct approach in which they trained the ANN how to respond to a variety of input patterns using various example images.

In this work it did not seem appropriate to use an exotic ANN architecture or training method, but rather to test one of the simpler and more widely used forms of ANN. A normal MLP was therefore selected. It was trained using the standard backpropagation algorithm. Preliminary tests were made to find the most appropriate topology for processing the 25 pixel intensities of each

5×5 square neighborhood (the idea being to apply this in turn at all positions in the input images, as with a more conventional filter). The result was a three-layer network with 25 inputs, 25 nodes in the first hidden layer, 5 nodes in the second hidden layer, and one output node, there being 100% connectivity between adjacent layers of the network. All the nodes employed the same sigmoidal transfer function, and a gray-scale output was thus available at the output of the network.

Training was achieved using a normal off-camera image as the low-noise "target" image and a noisy input image obtained by adding artificial noise to the target image. Three experiments were carried out using three different types of noise: (1) impulse noise with intensities 0 and 255; (2) impulse noise with random pixel intensities (taken from a uniform distribution over the range 0 to 255); and (3) Gaussian noise (truncated at 0 and 255). Noise types 1 and 2 were applied to random pixels, whereas noise type 3 was applied to all pixels in the input images. In the experiments, the performances of mean, median, hybrid median (2LH+), and ANN filters were compared, taking particular note of the average absolute deviations in intensity between the output and target images. 3×3 and 5×5 versions of all the filters were used, except that only a 5×5 version of the ANN filter was tried, since it should automatically adapt itself to 3×3 if required by the data seen during training.

As expected, the mean filters produced gross blurring of the images and "blotchy" behavior as a result of partially averaging impulse noise into the surroundings. Although the median filters eliminated noise effectively, they were found to soften these particular images to an unusual degree (Fig. 25.13), making the beans appear rather unnatural and less like beans than those in the original image. Of the conventional filters, the hybrid median filters gave the best overall performance with these images. However, they allowed a small number of "spotty" noise structures to remain, since these essentially emulated the straight line and corner structures that this type of filter is designed to retain intact (see Section 3.8.3). These particular images seemed prone to another artifact: the fairly straight sides of the beans became "bumpy" after processing by the median filters, this being (predictably) less evident on the images processed by the hybrid median filters. Finally, the median filters gave significant filling in of the corners between the beans, the effect being present but barely discernible after processing by the hybrid median filters. The situation is summarized in Table 25.1.

In general, the ANN filters[6] performed better than the conventional filters. First, each was found to perform best with the type of noise it had been trained on and to be more effective than the conventional filters at coping with that particular type of noise. Thus, it exhibited the adaptive capability expected

6 At this stage it is easier to imagine there to be three ANNs; in fact, there was one ANN architecture and three different training processes.

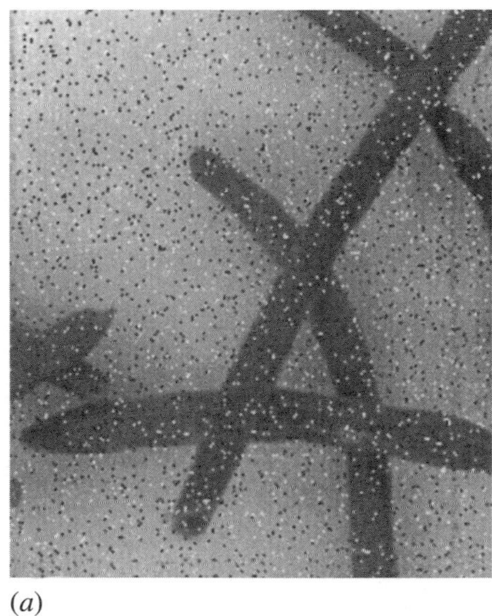

(*a*)

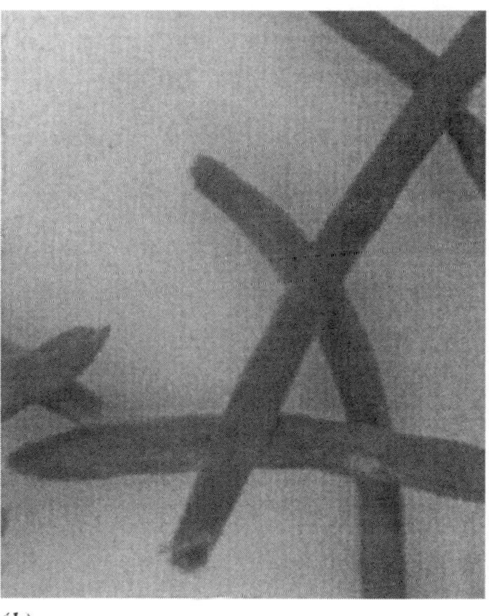

(*b*)

Figure 25.13 Noise suppression by an ANN filter. (*a*) shows an image with added type 1 impulse noise, (*b*) shows the effect of suppressing the noise using an appropriately trained ANN filter, and (*c*) shows the effect of suppressing the noise using a 5×5 median filter. Notice the "softening" effect and the partial filling in of corners in (*c*), neither of which is apparent in (*b*).

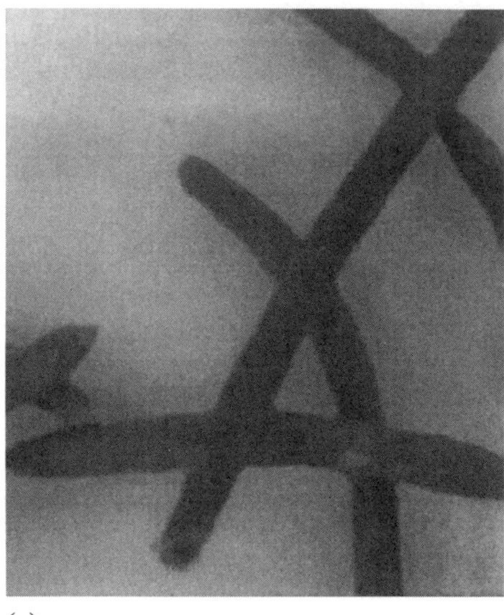

(*c*)

Figure 25.13 Continued.

of it. In addition, the ANN filters produced images without the remanent spottiness, bumpiness, and corner-filling behavior exhibited by the conventional filters.

Nevertheless, the performance of the ANNs was not always exemplary. In one case, an ANN performed less well than a conventional filter when it was tested on an image different from the ones it had been trained on. This must be due to disobeying a basic rule of pattern recognition—namely, that training sets should be fully representative of the classes from which they are drawn (see Chapter 24). However, in this case it could be deemed to have stemmed from the filters being more adapted to the particular type of noise than had been thought possible. Even more surprising was the finding that the training was sensitive not only to the type of noise but also to the types of data in the underlying image. Closer examination of the original images revealed faint vertical camera-induced striations which the ANNs did not remove, but which the median and other filters had eliminated. However, this was because the ANN filters had effectively been "instructed" to keep these striations, since the target images were the original off-camera images, and the input images were produced by adding extra noise to them. Thus, the ANN interpreted the striations as valid image data. A more appropriate training procedure is called for in this problem.

Unfortunately, it is difficult to see how a more appropriate training procedure—or equivalently, a more appropriate set of training images—could be produced. *A priori* it might have been thought that a median filtered image could be used as a target image. If this were done, however, the ANN would, if anything,

Table 25.1 Properties of the various filters investigated

Mean Filter	Median Filter	Hybrid Median Filter	ANN-based Filter
Produces blotches and blur	Does not produce blotches and blur	Does not produce blotches and blur	Does not produce blotches and blur
Does not produce spottiness	Does not produce spottiness	*Produces spottiness*	Does not produce spottiness
Bumps not visible because of blur	*Produces bumps*	Does not produce bumps	Does not produce bumps
Produces corner-filling	*Produces corner-filling*	Produces slight corner-filling	**Does not produce corner-filling**
Cannot adapt to special types of noise	*Cannot adapt to special types of noise*	*Cannot adapt to special types of noise*	**Adapts to the noise seen during training**
Cannot adapt to special types of image data	Cannot adapt to special types of image data	Cannot adapt to special types of image data	**Adapts to the image data seen during training**
Cannot adapt to drifts in the image data	Cannot adapt to drifts in the image data	Cannot adapt to drifts in the image data	Cannot adapt to drifts from the data seen during training
Unlikely to be optimal	Unlikely to be optimal	Unlikely to be optimal	**Potentially optimal**

For clarity of presentation, distinctly good behavior is marked in bold and distinctly poor behavior is marked in italics. In the case of the median and hybrid median filters, these are known to be reasonably well adapted to removal of impulse noise, and in the latter case to coping with straight edges. However, as indicated here, they cannot be adapted to arbitrary types of noise or data.

exhibit worse performance than the median filter. It would hardly be able to *improve* on the median filter and adapt effectively to the particular type of noise it is supposed to remove. Such an approach would therefore negate the main aim of using an ANN for noise suppression.

Overall, the results of these experiments are quite significant. They emphasize that ANN-based filters will be limited by the quality of training. It is also useful to notice that any drift in the characteristics of the incoming image data will tend to cause a mismatch between the ANN filter and its training set, so its performance will be liable to deteriorate if this happens. Although this is now apparent for an ANN-based filter, it must also be true to some extent for other more conventional filters, which are by their nature nonadaptable. For example, as noted earlier, a median filter cannot be adapted conveniently and certainly cannot be matched to its data—except by crudely changing the size of its support region. We must now reflect the question back on conventional algorithms: are they general-purpose, or are they adapted to some specific conditions or data (which are in general not known in detail)? For by and large, conventional algorithms stem from the often fortuitous or even opportunistic inspirations of individuals, and their specifications are decided less in advance than in retrospect. Thus, often it is only later that it is found by chance that they have certain undesirable properties that make them unsuitable for certain applications. Table 25.1 summarizes the properties of the various filters examined in these experiments and attempts to highlight their successes and failures.

25.9 Genetic Algorithms

The introduction to this chapter also included GAs as one of the newer biologically inspired methodologies. GAs are now starting to have significant impact on image recognition, and a growing number of papers reflect their use in this area. We start this section by outlining how GAs operate.

Basically, a GA is a procedure that mimics biological systems in allowing solutions (including complete algorithms) to evolve until they become optimally adapted to their situation. The first requirement is a means of generating random solutions. The second requirement is a means of judging the effectiveness of these solutions. Such a means will generally take the form of a "fitness" function, which should be minimized for optimality, though initially very little information will be available on the optimum value. Next, an initial set of solutions is generated, and the fitness function is evaluated for each of them. The initial solutions are likely to be very poor, but the best of them are selected for "mating" and "procreation." Typically, a pair of solutions is taken, and they are partitioned in some way, normally in a manner that allows corresponding parts to be swapped over (the process is called *crossover*), thereby creating two combination or "infant" solutions. The infants are then subjected to the same procedures—evaluation of their effectiveness using the fitness function, and placement in an order of merit among the set of available solutions.

From this explanation we see that, if both the infant and the original parent solutions are retained, there will be an increasing population of solutions, those at the top of the list being of increasing effectiveness. The storage of solutions will eventually get out of hand, and the worst solutions will have to be killed off (i.e., erased from memory). However, the population of solutions will continue to improve but possibly at a decreasing rate. If evolution stagnates, further random solutions may have to be created, though another widely used mechanism is to *mutate* a few existing solutions. This mimics such occurrences as copying errors or radiation damage to genes in biological systems. However, it corresponds to asexual rather than sexual selection, and may well be less effective at improving the system at the required rate. Normally, crossover is taken to be the basic mechanism for evolution in GAs. Although stagnation remains a possibility, the binary cutting and recombination intrinsic to the basic mechanism are extremely powerful and can radically improve the population over a limited number of generations. Thus, any stagnation is often only temporary.

Overall, the approach can arrive at local minima in a suitably chosen solution space, although it may take considerable computation to improve the situation so that a global minimum is reached. Unfortunately, no universal formula has been derived for guaranteeing finding global minima over a limited number of iterations. Each application is bound to be different, and a good many parameters are to be found and operational decisions are to be made. These include:

1. The number of solutions to be generated initially.
2. The maximum population of solutions to be retained.
3. The number of solutions that are to undergo crossover in each generation.
4. How often mutation should be invoked.
5. Any *a priori* knowledge or rules that can be introduced to guide the system toward a rapid outcome.
6. A termination procedure that can make judgments on (1) the likelihood of further significant improvement and (2) the tradeoff between computation and improvement.

This approach can be made to operate impressively on such tasks as timetable generation. In such cases, the major decisions to be made are (1) the process of deciding a code so that any solution is expressible as a vector in solution space, and (2) a suitable method for assessing the effectiveness of any solution (i.e., the fitness function). Once these two decisions have been made, the situation becomes that of managing the GA, which involves making the operational decisions outlined above.

GAs are well suited to setting up the weights of ANNs. The GAs essentially perform a coarse search of weight space, finding solutions that are relatively near to minima. Then backpropagation or other procedures can be used to refine the weights and produce optimal solutions. There are good reasons for not having the GAs go all the way to these optimal solutions. Basically, the crossover mechanism is rather brute-force, since binary chopping and recombination are radical procedures that will tend to make significant changes. Thus, in the final stages of convergence, if only one bit is crossed over, there is no certainty that this will produce only a minor change in a solution. Indeed, it is the whole value of a GA that such changes can have rather large effects commensurate with searching the whole of solution space efficiently and quickly.

As stated earlier, GAs are starting to have significant impact on image recognition, and more and more papers that reflect their use in this area are appearing (see Section 25.11). Although algorithms unfortunately run slowly and only infrequently lend themselves to real-time implementation, they have the potential for helping with the design and adaptation of algorithms for which long evolution times are less important. They can therefore help to build masks or other processing structures that will subsequently be used for real-time processing, even if they are not themselves capable of real-time processing (Siedlecki and Sklansky, 1989; Harvey and Marshall, 1994; Katz and Thrift, 1994). Adaptation is currently the weak point of image recognition—not least in the area of automated visual inspection, where automatic setup and adaptation procedures are desperately needed to save programming effort and provide the capability for development in new application areas.

25.10 Concluding Remarks

In the recent past there has been much euphoria and glorification of ANNs. Although this was probably inevitable considering the wilderness years before the backpropagation algorithm became widely known, ANNs are starting to achieve a balanced role in vision and other application areas. In this chapter, space has demanded a concentration on three main topics—MLPs trained using backpropagation, Hebbian networks which can emulate principal components analysis, and GAs. MLPs are almost certainly the most widely used of all the biologically inspired approaches to learning, and inclusion here seems amply justified on the basis of wide utility in vision—with the potential for far wider application in the future. Much the same applies to GAs, although they are at a relatively earlier stage of development and application. In addition, GAs tend to involve considerable amounts of computation, and their eventual role in machine vision is not too clear. All the same, it seems most likely (as indicated in Section 25.9) that they will be more useful for setting up vision algorithms than for carrying out the actual processing in real-time systems.

Returning to ANNs that emulate principal components analysis, we find that it is not clear whether they can yet perform a role that would provide a serious threat to the conventional approach. Important here in determining actual use will be the additional adaptivity provided by the ANN solution, and the possibility that higher order correlations (e.g., Taylor and Coombes, 1993) or nonlinear principal component analysis (e.g., Oja and Karhunen, 1993) will lead to far more powerful procedures that *are* able to supersede conventional methodologies.

Overall, ANNs now have greater potential for including automatic setup and adaptivity into vision systems than is possible at present, although the resulting systems then become dependent on the precise training set that is employed. Thus, the function of the system designer changes from code writer to code trainer, raising quite different problems. Although this could be regarded as introducing difficulties, it should be noticed that these difficulties are to a large extent already present. For who can write effective code without taking in (i.e., being trained on) a large quantity of visual and other data, all of which will be subject to diurnal, seasonal, or other changes that are normally totally inexplicit? Similarly, for GAs, who can design a fully effective fitness function appropriate to a given application without taking in similar quantities of data?

Biological inspiration has been a powerful motivator of progress in vision, with artificial neural networks and genetic algorithms being leading contenders. This chapter has dwelled mainly on artificial neural networks, emphasizing that they perform statistical pattern recognition functions and have the same limitations in terms of training, and possibly overfitting any dataset.

25.11 Bibliographical and Historical Notes

ANNs have had a checkered history. After a promising start in the 1950s and 1960s, they fell into disrepute (or at least, disregard) during the 1970s following the pronouncements of Minsky and Papert in 1969. They picked up again in the early 1980s and were subjected to an explosion in interest after the announcement of the backpropagation algorithm by Rumelhart et al. in 1986. It was only in the mid-1990s that they started settling into the role of normal tools for vision and other applications. In addition, it should not be forgotten that the backpropagation algorithm was invented several times (Werbos, 1974; Parker, 1985) before its relevance was finally recognized. In parallel with these MLP developments, Oja (1982) led the field with his Hebbian principal components network.

Useful recent general references on ANNs include the volumes by Hertz et al. (1991), Haykin (1999), and Ripley (1996); and the highly useful review article by Hush and Horne (1993). The reader is also referred to two journal special issues (Backer, 1992; Lowe, 1994) for work specifically related to machine vision. In particular, ANNs have been applied to image segmentation by Toulson and Boyce (1992) and Wang et al. (1992); and to object location by Ghosh and Pal (1992), Spirkovska and Reid (1992), and Vaillant et al. (1994). For work on contextual image labeling, see Mackeown et al. (1994). For references related to visual inspection, see Chapter 22. For work on histogram-based thresholding in which the ANN is trained to emulate the maximal likelihood scheme of Chow and Kaneko (1972), see Greenhill and Davies (1995).

GAs were invented in the early 1970s and have only gained wide popularity in the 1990s. A useful early reference is Holland (1975), and a more recent reference is Michalewicz (1992). A short but highly useful review of the subject appears in Srinivas and Patnaik (1994). These references are general and do not refer to vision applications of GAs. Nevertheless, since the late 1980s, GAs have been applied quite regularly to machine vision. For example, they have been used by Siedlecki and Sklansky (1989) for feature selection, and by Harvey and Marshall (1994) and Katz and Thrift (1994) for filter design; by Lutton and Martinez (1994) and Roth and Levine (1994) for detection of geometric primitives in images; by Pal et al. (1994) for optimal image enhancement; by Ankenbrandt et al. (1990) for scene recognition; and by Hill and Taylor (1992) and Bhattacharya and Roysam (1994) for image interpretation tasks. The IEE Colloquium on *Genetic Algorithms in Image Processing and Vision* (IEE, 1994) presented a number of relevant approaches; see also Gelsema (1994) for a special issue of Pattern Recognition Letters on the subject.

During the early 1990s, papers on artificial neural networks applied to vision were ubiquitous, but by now the situation has settled down. Few see ANNs as being the solution to everything; instead ANNs are taken as alternative means of modeling and estimating, and as such have to compete with conventional, well-documented schemes for performing statistical pattern recognition. Their value lies in their unified approach to feature extraction and selection (even if this necessarily turns into the disadvantage that the statistics are hidden from the user), and their intrinsic capability for finding moderately nonlinear solutions with relative ease. Recent ANN and GA papers include the ANN face detection work of Rowley et al. (1998), among others (Fasel, 2002; Garcia and Delakis, 2002), and the GA object shape matching work of Tsang (2003) and Tsang and Yu (2003). For further information on ANNs and their application to vision, see the books by Bishop (1995), Ripley (1996), and Haykin (1999).

Texture

It is quite easy to understand what a texture is, though somewhat less easy to define it. For many reasons it is useful to be able to classify textures and to distinguish them from one another. It is also useful to be able to determine the boundaries between different textures, as they often signify the boundaries of real objects. This chapter studies the means of achieving these aims.

Look out for:

- basic measures by which textures can be classified—such as regularity, randomness, and directionality.
- problems that arise with "obvious" texture analysis methods, such as autocorrelation.
- the long-standing gray-level co-occurrence matrix method.
- Laws' method and Ade's generalization of it.
- alternative approaches to texture analysis such as fractal-based measures and Markov random field models.
- the fact that textures have to be analyzed statistically because of the random element in their construction.

Texture analysis is a core element in the vision repertoire, just as textures are core components of most images. Thus, this topic had to appear somewhere in Part 4 of the book; its location immediately after the chapters on statistical pattern recognition and neural networks appeared to be the most appropriate.

CHAPTER 26

Texture

26.1 Introduction

In the foregoing chapters many aspects of image analysis and recognition have been studied. At the core of these matters has been the concept of segmentation, which involves the splitting of images into regions that have some degree of uniformity, whether in intensity, color, texture, depth, motion, or other relevant attributes. Care was taken in Chapter 4 to emphasize that such a process will be largely ad hoc, since the boundaries produced will not necessarily correspond to those of real objects. Nevertheless, it is important to make the attempt, either as a preliminary to more accurate or iterative demarcation of objects and their facets, or else as an end in itself—for example, to judge the quality of surfaces.

In this chapter we move on to the study of texture and its measurement. Texture is a difficult property to define. Indeed, in 1979 Haralick reported that no satisfactory definition of it had yet been produced. Perhaps we should not be surprised by this fact, as the concept has rather separate meanings in the contexts of vision, touch, and taste, the particular nuances being understood by different people also being highly individual and subjective. Nevertheless, we require a working definition of texture, and in vision the particular aspect we focus on is the variation in intensity[1] of a particular surface or region of an image. Even with this statement, we are being indecisive about whether we are describing the physical object being observed or the image derived from it. This reflects the fact that it is the roughness of the surface or the structure or composition of the material that gives rise to its visual properties. However, in this chapter we are interested mainly in the interpretation of images, and so we define texture as the characteristic variation in intensity of a region of an image, which should allow us to recognize and describe it and to outline its boundaries (Fig. 26.1).

1 We could at this point generalize the definition to cover variation in color, but this would complicate matters unnecessarily and would not add substantially to coverage of the subject.

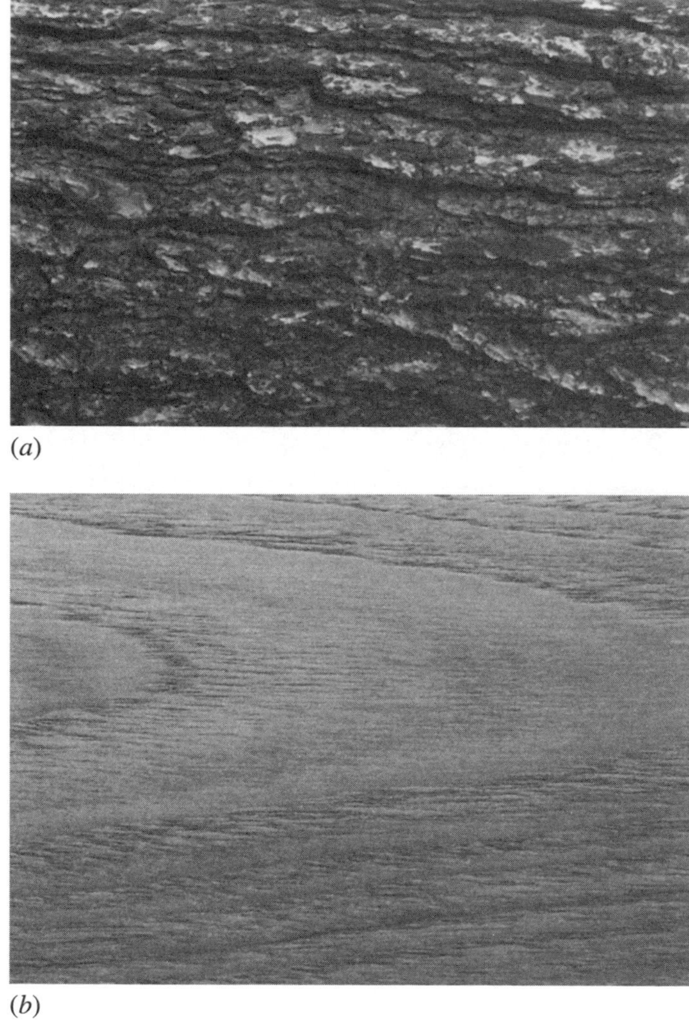

(*a*)

(*b*)

Figure 26.1 A variety of textures. These textures demonstrate the wide variety of familiar textures that are easily recognized from their characteristic intensity patterns.

This definition of texture implies that texture is nonexistent in a surface of uniform intensity, and it does not say anything about how the intensity might be expected to vary or how we might recognize and describe it. Intensity might vary in numerous ways, but if the variation does not have sufficient uniformity, the texture may not be characterized sufficiently closely to permit recognition or segmentation.

We next consider ways in which intensity might vary. It can vary rapidly or slowly, markedly or with low contrast, with a high or low degree of directionality,

(*c*)

(*d*)

Figure 26.1 Continued.

and with greater or lesser degrees of regularity. This last characteristic is often taken as key: either the textural pattern is regular as for a piece of cloth, or it is random as for a sandy beach or a pile of grass cuttings. However, this ignores the fact that a regular textural pattern is often not wholly regular (again, as for a piece of cloth), or not wholly random (as for a mound of potatoes of similar size). Thus, thc degrees of randomness and of regularity will have to be measured and compared when characterizing a texture.

(e)

(f)

Figure 26.1 Continued.

There are more profound things to say about the textures. Often they are derived from tiny objects or components that are themselves similar, but that are placed together in ways ranging from purely random to purely regular—be they bricks in a wall, grains of sand, blades of grass, strands of material, stripes on a shirt, wickerwork on a basket, or a host of other items. In texture analysis it is useful to have a name for the similar textural elements

(g)

(h)

Figure 26.1 Continued.

that are replicated over a region of the image: such textural elements are called *texels*. These considerations lead us to characterize textures in the following ways:

1. The texels will have various sizes and degrees of uniformity.
2. The texels will be oriented in various directions.
3. The texels will be spaced at varying distances in different directions.

4. The contrast will have various magnitudes and variations.
5. Various amounts of background may be visible between texels.
6. The variations composing the texture may each have varying degrees of regularity vis-à-vis randomness.

It is quite clear from this discussion that a texture is a complicated entity to measure. The reason is primarily that many parameters are likely to be required to characterize it. In addition, when so many parameters are involved, it is difficult to disentangle the available data and measure the individual values or decide the ones that are most relevant for recognition. And, of course, the statistical nature of many of the parameters is by no means helpful. However, we have so far only attempted to show how complex the situation can be. In the following paragraphs, we attempt to show that quite simple measures can be used to recognize and segment textures in practical situations.

Before proceeding, it is useful to recall that in the analysis of shape a dichotomy exists between available analysis methods. We could, for example, use a set of measures such as circularity, aspect ratio, and so on which would permit a description of the shape, but which would not allow it to be reconstructed. Or we could use descriptors such as skeletons with distance function values, or moments, which would permit full and accurate reconstruction—though the set of descriptors might have been curtailed so that only limited but predictable accuracy was available. In principle, such a reconstruction criterion should be possible with texture. However, in practice there are two levels of reconstruction. In the first, we could reproduce a pattern that, to human eyes, would be indistinguishable from the off-camera texture until one compared the two on a pixel-by-pixel basis. In the second, we could reproduce a textured pattern exactly. The point is that normally textures are partially statistical in nature, so it will be difficult to obtain a pixel-by-pixel match in intensities. Neither, in general, will it be worth aiming to do so. Thus, texture analysis generally only aims at obtaining accurate statistical descriptions of textures, from which *apparently* identical textures can be reproduced if desired.

At this point it ought to be stated that many workers have contributed to, and used, a wide range of approaches for texture analysis over a period of well over 40 years. The sheer weight of the available material and the statistical nature of it can be daunting for many. It is therefore recommended that those interested in obtaining a quick working view of the subject start by reading Sections 26.2 and 26.4, looking over Sections 26.5 and 26.6, and then proceeding to the end of the chapter. In this way they will bypass much of the literature review material that is part and parcel of a full study of the subject. (Section 26.4 is particularly relevant to practitioners, as it describes the Laws' texture energy approach which is intuitive, straightforward to apply in both software and hardware, and highly effective in many application areas.)

26.2 Some Basic Approaches to Texture Analysis

In Section 26.1 we defined texture as the characteristic variation in intensity of a region of an image that should allow us to recognize and describe it and to outline its boundaries. In view of the likely statistical nature of textures, this prompts us to characterize texture by the variance in intensity values taken over the whole region of the texture.[2] However, such an approach will not give a rich enough description of the texture for most purposes and will certainly not provide any possibility of reconstruction. It will also be especially unsuitable in cases where the texels are well defined, or where there is a high degree of periodicity in the texture. On the other hand, for highly periodic textures such as arise with many textiles, it is natural to consider the use of Fourier analysis. In the early days of image analysis, this approach was tested thoroughly, though the results were not always encouraging.

Bajcsy (1973) used a variety of ring and oriented strip filters in the Fourier domain to isolate texture features—an approach that was found to work successfully on natural textures such as grass, sand, and trees. However, there is a general difficulty in using the Fourier power spectrum in that the information is more scattered than might at first be expected. In addition, strong edges and image boundary effects can prevent accurate texture analysis by this method, though Shaming (1974) and Dyer and Rosenfeld (1976) tackled the relevant image aperture problems. Perhaps more important is the fact that the Fourier approach is a global one that is difficult to apply successfully to an image that is to be segmented by texture analysis (Weszka et al., 1976).

Autocorrelation is another obvious approach to texture analysis, since it should show up both local intensity variations and the repeatability of the texture (see Fig. 26.2). In an early study, Kaizer (1955) examined how many pixels an image has to be shifted before the autocorrelation function drops to $1/e$ of its initial value, and produced a subjective measure of coarseness on this basis. However, Rosenfeld and Troy (1970a,b) later showed that autocorrelation is not a satisfactory measure of coarseness. In addition, autocorrelation is not a very good discriminator of isotropy in natural textures. Hence, workers were quick to take up the co-occurrence matrix approach introduced by Haralick et al. in 1973. This approach not only replaced the use of autocorrelation but during the 1970s became to a large degree the standard approach to texture analysis.

2 We defer for now the problem of finding the region of a texture so that we can compute its characteristics in order to perform a segmentation function. However, some preliminary training of a classifier may be used to overcome this problem for supervised texture segmentation tasks.

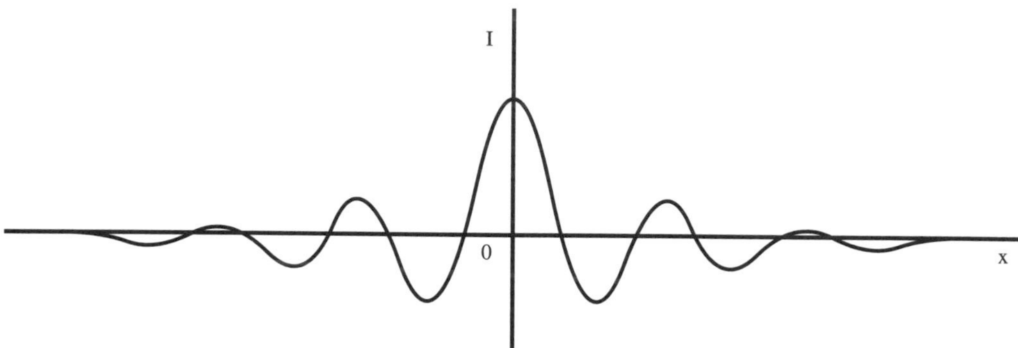

Figure 26.2 Use of autocorrelation function for texture analysis. This diagram shows the possible 1-D profile of the autocorrelation function for a piece of material in which the weave is subject to significant spatial variation. Notice that the periodicity of the autocorrelation function is damped down over quite a short distance.

26.3 Gray-level Co-occurrence Matrices

The gray-level co-occurrence matrix approach[3] is based on studies of the statistics of pixel intensity distributions. As hinted above with regard to the variance in pixel intensity values, single-pixel statistics do not provide rich enough descriptions of textures for practical applications. Thus, it is natural to consider second-order statistics obtained by considering *pairs* of pixels in certain spatial relations to each other. Hence, co-occurrence matrices are used, which express the relative frequencies (or probabilities) $P(i, j|d, \theta)$ with which two pixels having relative polar coordinates (d, θ) appear with intensities i, j. The co-occurrence matrices provide raw numerical data on the texture, though these data must be condensed to relatively few numbers before it can be used to classify the texture. The early paper by Haralick et al. (1973) presented 14 such measures, and these were used successfully for classification of many types of materials (including, for example, wood, corn, grass, and water). However, Conners and Harlow (1980a) found that only five of these measures were normally used, viz. "energy," "entropy," "correlation," "local homogeneity," and "inertia." (Note that these names do not provide much indication of the modes of operation of the respective operators.)

To obtain a more detailed idea of the operation of the technique, consider the co-occurrence matrix shown in Fig. 26.3. This corresponds to a nearly uniform image containing a single region in which the pixel intensities are subject to an approximately Gaussian noise distribution, the attention being on pairs of pixels at a constant vector distance $\mathbf{d} = (d, \theta)$ from each other. Next consider the co-occurrence matrix shown in Fig. 26.4, which corresponds to an almost

3 This is also frequently called the spatial gray-level dependence matrix (SGLDM) approach.

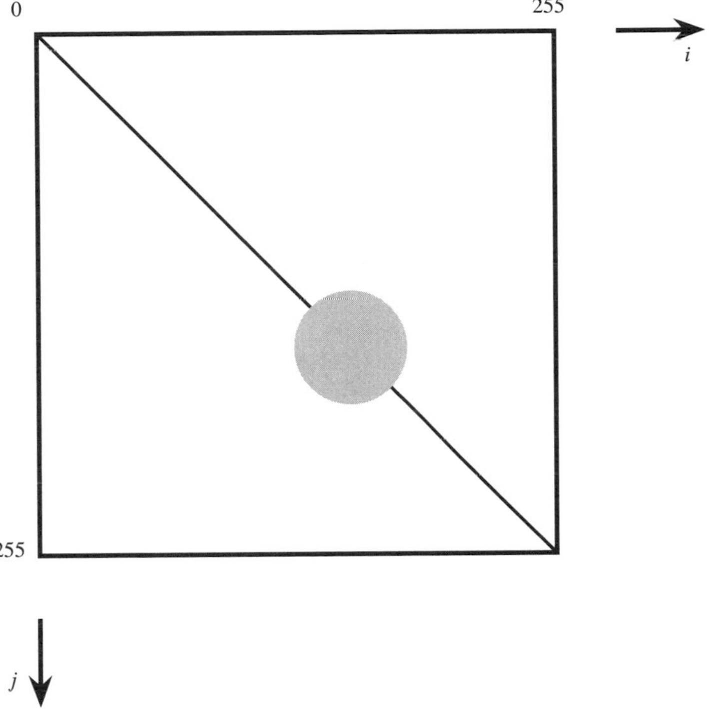

Figure 26.3 Co-occurrence matrix for a nearly uniform gray-scale image with superimposed Gaussian noise. Here the intensity variation is taken to be almost continuous. Normal convention is followed by making the j index increase downwards, as for a table of discrete values (cf. Fig. 26.4).

noiseless image with several nearly uniform image regions. In this case, the two pixels in each pair may correspond either to the same image regions or to different ones, though if d is small they will only correspond to adjacent image regions. Thus, we have a set of N on-diagonal patches in the co-occurrence matrix, but only a limited number L of the possible number M of off-diagonal patches linking them, where $M = \binom{N}{2}$ and $L \leq M$. (Typically, L will be of order N rather than N^2.) With textured images, if the texture is not too strong, it may by modeled as noise, and the $N + L$ patches in the image will be larger but still not overlapping. However, in more complex cases the possibility of segmentation using the co-occurrence matrices will depend on the extent to which $\mathbf{d}$ can be chosen to prevent the patches from overlapping. Since many textures are directional, careful choice of θ will help with this task, though the optimum value of d will depend on several other characteristics of the texture.

As a further illustration, we consider the small image shown in Fig. 26.5*a*. To produce the co-occurrence matrices for a given value of $\mathbf{d}$, we merely

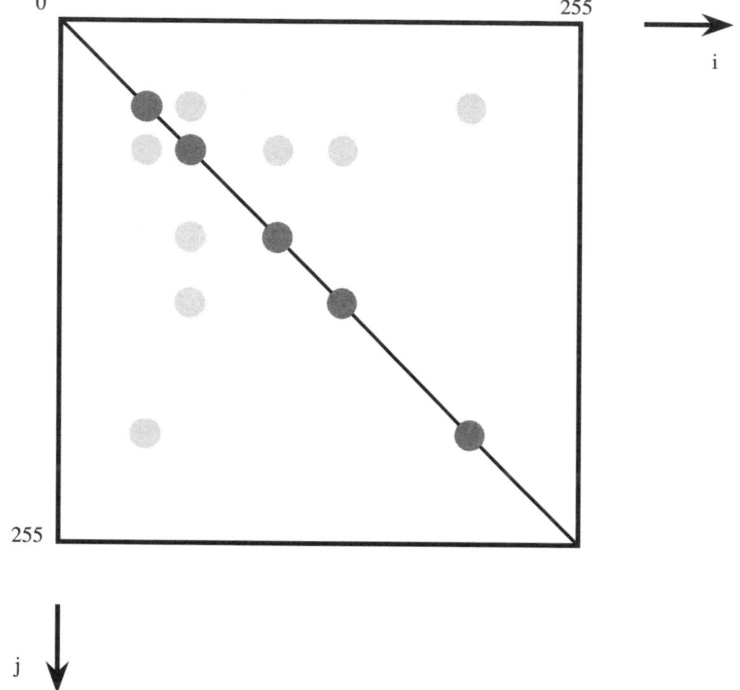

Figure 26.4 Co-occurrence matrix for an image with several distinct regions of nearly constant intensity. Again, the leading diagonal of the diagram is from top left to bottom right (cf. Figs. 26.2 and 26.4).

need to calculate the number of cases for which pixels a distance **d** apart have intensity values i and j. Here, we content ourselves with the two cases $\mathbf{d} = (1, 0)$ and $\mathbf{d} = (1, \pi/2)$. We thus obtain the matrices shown in Fig. 26.5*b* and *c*.

This simple example demonstrates that the amount of data in the matrices is liable to be many times more than in the original image—a situation that is exacerbated in more complex cases by the number of values of d and θ that is required to accurately represent the texture. In addition, the number of gray levels will normally be closer to 256 than to 6, and the amount of matrix data varies as the square of this number. Finally, we should notice that the co-occurrence matrices merely provide a new representation: they do not themselves solve the recognition problem.

These factors mean that the gray-scale has to be compressed into a much smaller set of values, and careful choice of specific sample d, θ values must be made. In most cases, it is not obvious how such a choice should be made, and it is even more difficult to arrange for it to be made automatically. In addition,

0	0	0	1
1	1	1	1
2	2	2	3
3	3	4	5

(*a*)

	0	1	2	3	4	5
0	2	1	0	0	0	0
1	1	3	0	0	0	0
2	0	0	2	1	0	0
3	0	0	1	1	1	0
4	0	0	0	1	0	1
5	0	0	0	0	1	0

(*b*)

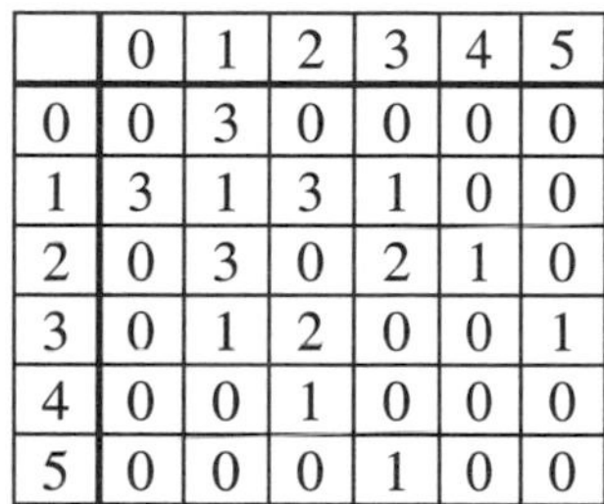

	0	1	2	3	4	5
0	0	3	0	0	0	0
1	3	1	3	1	0	0
2	0	3	0	2	1	0
3	0	1	2	0	0	1
4	0	0	1	0	0	0
5	0	0	0	1	0	0

(*c*)

Figure 26.5 Co-occurrence matrices for a small image. (*a*) shows the original image; (*b*) shows the resulting co-occurrence matrix for $\mathbf{d} = (1, 0)$, and (*c*) shows the matrix for $\mathbf{d} = (1, \pi/2)$. Note that even in this simple case the matrices contain more data than the original image.

various functions of the matrix data must be tested before the texture can be properly characterized and classified.

These problems with the co-occurrence matrix approach have been tackled in many ways: just two are mentioned here. The first is to ignore the distinction between opposite directions in the image, thereby reducing storage by 50%. The second is to work with *differences* between gray levels. This amounts to performing a summation in the co-occurrence matrices along axes parallel to the main diagonal of the matrix. The result is a set of *first-order difference* statistics. Although these modifications have given some additional impetus to the approach, the 1980s saw a highly significant diversification of methods for the analysis of textures. Of these, Laws' approach (1979, 1980a,b) is important in that it has led to other developments that provide a systematic, adaptive means of conducting texture analysis. This approach is covered in the following section.

26.4 Laws' Texture Energy Approach

In 1979 and 1980, Laws presented his novel texture energy approach to texture analysis (1979, 1980a,b). This involved the application of simple filters to digital images. The basic filters he used were common Gaussian, edge detector, and Laplacian-type filters, and were designed to highlight points of high "texture energy" in the image. By identifying these high-energy points, smoothing the various filtered images, and pooling the information from them he was able to characterize textures highly efficiently and in a manner compatible with pipelined hardware implementations. As remarked earlier, Laws' approach has strongly influenced much subsequent work, and it is therefore worth considering it here in some detail.

The Laws' masks are constructed by convolving together just three basic 1×3 masks:

$$\mathrm{L3} = [1\ 2\ 1] \tag{26.1}$$

$$\mathrm{E3} = [-1\ 0\ 1] \tag{26.2}$$

$$\mathrm{S3} = [-1\ 2\ -1] \tag{26.3}$$

The initial letters of these masks indicate *L*ocal averaging, *E*dge detection, and *S*pot detection. These basic masks span the entire 1×3 subspace and form a complete set. Similarly, the 1×5 masks obtained by convolving pairs of these 1×3 masks together form a complete set:[4]

$$\mathrm{L5} = [1\quad 4\quad 6\quad 4\quad 1] \tag{26.4}$$

$$\mathrm{E5} = [-1\quad -2\quad 0\quad 2\quad 1] \tag{26.5}$$

$$\mathrm{S5} = [-1\quad 0\quad 2\quad 0\quad -1] \tag{26.6}$$

$$\mathrm{R5} = [1\quad -4\quad 6\quad -4\quad 1] \tag{26.7}$$

$$\mathrm{W5} = [-1\quad 2\quad 0\quad -2\quad 1] \tag{26.8}$$

4 In principle, nine masks can be formed in this way, but only five of them are distinct.

L3ᵀL3	**L3ᵀE3**	**L3ᵀS3**
1 2 1	–1 0 1	–1 2 –1
2 4 2	–2 0 2	–2 4 –2
1 2 1	–1 0 1	–1 2 –1
E3ᵀL3	**E3ᵀE3**	**E3ᵀS3**
–1 –2 –1	1 0 –1	1 –2 1
0 0 0	0 0 0	0 0 0
1 2 1	–1 0 1	–1 2 –1
S3ᵀL3	**S3ᵀE3**	**S3ᵀS3**
–1 –2 –1	1 0 –1	1 –2 1
2 4 2	–2 0 2	–2 4 –2
–1 –2 –1	1 0 –1	1 –2 1

Figure 26.6 The nine 3×3 Laws masks.

(Here the initial letters are as before, with the addition of *R*ipple detection and *W*ave detection.) We can also use matrix multiplication (see also Section 3.6) to combine the 1×3 and a similar set of 3×1 masks to obtain nine 3×3 masks—for example:

$$\begin{bmatrix} 1 \\ 2 \\ 1 \end{bmatrix} \begin{bmatrix} -1 & 2 & -1 \end{bmatrix} = \begin{bmatrix} -1 & 2 & -1 \\ -2 & 4 & -2 \\ -1 & 2 & -1 \end{bmatrix} \tag{26.9}$$

The resulting set of masks also forms a complete set (Fig. 26.6). Note that two of these masks are identical to the Sobel operator masks. The corresponding 5×5 masks are entirely similar but are not considered in detail here as all relevant principles are illustrated by the 3×3 masks.

All such sets of masks include one whose components do not average to zero. Thus, it is less useful for texture analysis since it will give results dependent more on image intensity than on texture. The remainder are sensitive to edge points, spots, lines, and combinations of these.

Having produced images that indicate local edginess and so on, the next stage is to deduce the local magnitudes of these quantities. These magnitudes are then smoothed over a fair-sized region rather greater than the basic filter mask size (e.g., Laws used a 15×15 smoothing window after applying his 3×3 masks). The effect is to smooth over the gaps between the texture edges and other microfeatures. At this point, the image has been transformed into a vector image, each component of which represents energy of a different type. Although Laws (1980b) used both squared magnitudes and absolute magnitudes

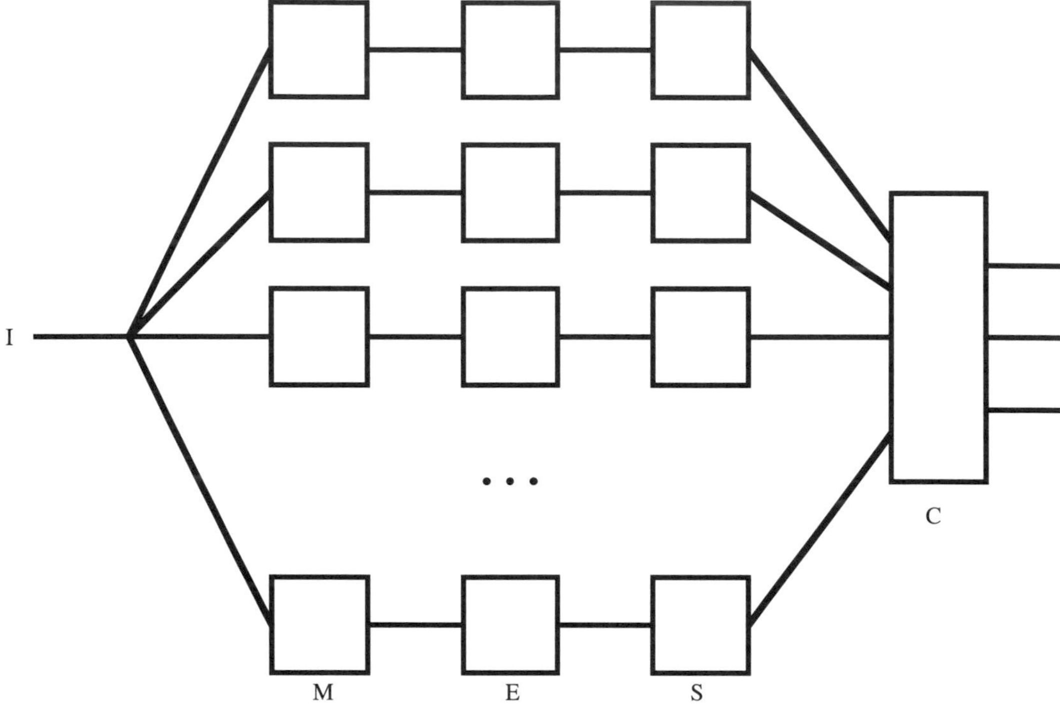

Figure 26.7 Basic form for a Laws texture classifier. Here I is the incoming image, M represents the microfeature calculation, E the energy calculation, S the smoothing, and C the final classification.

to estimate texture energy, the squared magnitude corresponded to true energy and gave a better response, whereas absolute magnitudes require less computation:

$$E(l,m) = \sum_{i=l-p}^{l+p} \sum_{j=m-p}^{m+p} |F(i,j)| \tag{26.10}$$

$F(i, j)$ being the local magnitude of a typical microfeature that is smoothed at a general scan position (l, m) in a $(2p+1) \times (2p+1)$ window.

A further stage is required to combine the various energies in a number of different ways, providing several outputs that can be fed into a classifier to decide upon the particular type of texture at each pixel location (Fig. 26.7). If necessary, principal components analysis is used at this point to help select a suitable set of intermediate outputs.

Laws' method resulted in excellent classification accuracy quoted at (for example) 87% compared with 72% for the co-occurrence matrix method, when applied to a composite texture image of grass, raffia, sand, wool, pigskin, leather, water, and wood (Laws, 1980b). Laws also found that the histogram equalization normally applied to images to eliminate first-order differences in texture field gray-scale distributions produced little improvement in this case.

Pietikäinen et al. (1983) conducted research to determine whether the precise coefficients used in Laws' masks are responsible for the performance of his method. They found that as long as the general forms of the masks were retained, performance did not deteriorate and could in some instances be improved. They were able to confirm that Laws' texture energy measures are more powerful than measures based on pairs of pixels (i.e., co-occurrence matrices), although Unser (1986) later questioned the generality of this result.

26.5 Ade's Eigenfilter Approach

In 1983, Ade investigated the theory underlying Laws' approach and developed a revised rationale in terms of eigenfilters. He took all possible pairs of pixels within a 3×3 window and characterized the image intensity data by a 9×9 covariance matrix. He then determined the eigenvectors required to diagonalize this matrix. These correspond to filter masks similar to Laws' masks (i.e., use of these "eigenfilter" masks produces images that are principal component images for the given texture). Furthermore, each eigenvalue gives that part of the variance of the original image that can be extracted by the corresponding filter. Essentially, the variances give an exhaustive description of a given texture in terms of the texture of the images from which the covariance matrix was originally derived. The filters that give rise to low variances can be taken to be relatively unimportant for texture recognition.

It will be useful to illustrate the technique for a 3×3 window. Here we follow Ade (1983) in numbering the pixels within a 3×3 window in scan order:

1	2	3
4	5	6
7	8	9

This leads to a 9×9 covariance matrix for describing relationships between pixel intensities within a 3×3 window, as stated above. At this point, we recall that we are describing a texture, and assuming that its properties are not synchronous with the pixel tessellation, we would expect various coefficients of the covariance matrix **C** to be equal. For example, C_{24} should equal C_{57}; in addition, C_{57} must equal C_{75}.

Table 26.1 Spatial relationships between pixels in a 3 × 3 window

a	*b*	*c*	*d*	*e*	*f*	*g*	*h*	*i*	*j*	*k*	*l*	*m*
9	6	6	4	4	3	3	1	1	2	2	2	2

This table shows the number of occurrences of the spatial relationships between pixels in a 3 × 3 window. Note that *a* is the diagonal element of the covariance matrix **C**, and that all others appear twice as many times in **C** as indicated in the table.

It is worth pursuing this matter, as a reduced number of parameters will lead to increased accuracy in determining the remaining ones. There are $\binom{9}{2} = 36$ ways of selecting pairs of pixels, but there are only 12 distinct spatial relationships between pixels if we disregard translations of whole pairs—or 13 if we include the null vector in the set (see Table 26.1). Thus, the covariance matrix (see Section 24.10) takes the form:

$$\mathbf{C} = \begin{bmatrix} a & b & f & c & d & k & g & m & h \\ b & a & b & e & c & d & l & g & m \\ f & b & a & j & e & c & i & l & g \\ c & e & j & a & b & f & c & d & k \\ d & c & e & b & a & b & e & c & d \\ k & d & c & f & b & a & j & e & c \\ g & l & i & c & e & j & a & b & f \\ m & g & l & d & c & e & b & a & b \\ h & m & g & k & d & c & f & b & a \end{bmatrix} \tag{26.11}$$

C is symmetrical; the eigenvalues of a real symmetrical covariance matrix are real and positive; and the eigenvectors are mutually orthogonal (see Section 24.10). In addition, the eigenfilters thus produced reflect the proper structure of the texture being studied and are ideally suited to characterizing it. For example, for a texture with a prominent highly directional pattern, there will be one or more high-energy eigenvalues, with eigenfilters having strong directionality in the corresponding direction.

26.6 Appraisal of the Laws and Ade Approaches

At this point, it will be worthwhile to compare the Laws and Ade approaches more carefully. In the Laws approach, standard filters are used, texture energy images are produced, and *then* principal components analysis may be applied to lead to recognition, whereas in the Ade approach, special filters (the eigenfilters) are applied, incorporating the results of principal components analysis, following which texture energy measures are calculated and a suitable number of these are applied for recognition.

The Ade approach is superior to the extent that it permits low-value energy components to be eliminated early on, thereby saving computation. For example, in Ade's application, the first five of the nine components contain 99.1% of the total texture energy, so the remainder can definitely be ignored. In addition, it would appear that another two of the components containing, respectively, 1.9% and 0.7% of the energy could also be ignored, with little loss of recognition accuracy. In some applications, however, textures could vary continually, and it may well *not* be advantageous to fine-tune a method to the particular data pertaining at any one time.[5] In addition, to do so may prevent an implementation from having wide generality or (in the case of hardware implementations) being so cost-effective. There is therefore still a case for employing the simplest possible complete set of masks and using the Laws approach. However, Unser and Ade (1984) have adopted the Ade approach and report encouraging results. More recently, the work of Dewaele et al. (1988) has built solidly on the Ade eigenfilter approach.

In 1986, Unser developed a more general version of the Ade technique, which also covered the methods of Faugeras (1978), Granlund (1980), and Wermser and Liedtke (1982). In this approach not only is performance optimized for texture classification but also it is optimized for discrimination between two textures by simultaneous diagonalization of two covariance matrices. The method has been developed further by Unser and Eden (1989, 1990). This work makes a careful analysis of the use of nonlinear detectors. As a result, two levels of nonlinearity are employed, one immediately after the linear filters and designed (by employing a specific Gaussian texture model) to feed the smoothing stage with genuine variance or other suitable measures, and the other after the spatial smoothing stage to counteract the effect of the earlier filter, and aiming to provide a feature value that is in the same units as the input signal. In practical terms this means having the capability for providing an r.m.s. texture signal from each of the linear filter channels.

Overall, the originally intuitive Laws approach emerged during the 1980s as a serious alternative to the co-occurrence matrix approach. It is as well to note that alternative methods that are potentially superior have also been devised. See, for example, the local rank correlation method of Harwood et al. (1985), and the forced-choice method of Vistnes (1989) for finding edges between different

5 For example, these remarks apply (1) to textiles, for which the degree of stretch will vary continuously during manufacture, (2) to raw food products such as beans, whose sizes will vary with the source of supply, and (3) to processed food products such as cakes, for which the crumbliness will vary with cooking temperature and water content.

textures, which apparently has considerably better accuracy than the Laws approach. Vistnes' (1989) investigation concludes that the Laws approach is limited by (1) the small scale of the masks which can miss larger-scale textural structures, and (2) the fact that the texture energy smoothing operation blurs the texture feature values across the edge. The second finding (or the even worse situation where a third class of texture appears to be located in the region of the border between two textures) has also been noted by Hsiao and Sawchuk (1989, 1990) who applied an improved technique for feature smoothing. They also used probabilistic relaxation for enforcing spatial organization on the resulting data.

26.7 Fractal-based Measures of Texture

Fractals, an important new approach to texture analysis that arose in the 1980s, incorporate the observation due to Mandelbrot (1982) that measurements of the length of a coastline (for example) will vary with the size of the measuring tool used for the purpose, since details smaller than the size of the tool will be missed. If the size of the measuring tool is taken as λ, the measured quantity will be $M = n\lambda^D$, where D is known as the *fractal dimension* and must in general be larger than the immediate geometric dimension if correct measurements are to result. (For a coastline we will thus have $D > 2$.) Thus, when measurements are being made of 2-D textures, it is found that D can take values from 2.0 to at least 2.8 (Pentland, 1984). Interestingly, these values of D have been found to correspond roughly to subjective measures of the roughness of the surface being inspected (Pentland, 1984). The concept of fractal dimension is a mathematical one, which can only apply over a range of scales for physical surfaces. Nevertheless, if the concept is *not* applied over such a range of scales (as is the case for most existing texture analysis methods), then the measures of texture that result cannot be scale invariant. The texture measure will therefore have to be recalibrated for every possible scale at which it is applied. This would be highly inconvenient, so fractal concepts strike at the heart of the texture measurement problem. (Seemingly, ignoring them would only be valid if, for example, textiles or cornfields were being inspected from a fixed height.) At the same time, it is fortunate that fractal dimension appears to represent a quantity as useful as surface roughness. Furthermore, the fact that it reduces the number of textural attributes from the initially large numbers associated with certain other methods of textural measurement (such as co-occurrence matrices) is also something of an advantage.

Since Pentland (1984) put forward these arguments, other workers have had problems with the approach. For example, reducing all textural measurements to the single measure D cannot permit all textures to be distinguished (Keller et al., 1989). Hence, an attempt has been made to define further fractal-based measures. Mandelbrot himself contributed the concept of *lacunarity* and in 1982 provided one definition, whereas Voss (1986) and Keller et al. (1989) provided further definitions. These definitions differ in mathematical detail, but all three appear to measure a sort of texture mark–space ratio that is distinct from average surface roughness. Lacunarity seems to give good separation for most materials having similar fractal dimensions, though Keller et al. proposed further research to combine these measures with other statistical and structural measurements. Ultimately, an important problem of texture analysis will be to determine how many features are required to distinguish different materials and surfaces reliably, while not making the number so large that computation becomes excessive, or (perhaps more important) so that the raw data are unable to support so many features (i.e., so that the features that are chosen are not all statistically significant—see Section 24.5). Meanwhile, it should be noted that the fractal measures currently being used require considerable computation that may not be justified for applications such as automated inspection.

Finally, note that Gårding (1988) found that fractal dimension is not always equivalent to subjective judgments of roughness. In particular, he revealed that a region of Gaussian noise of low amplitude superimposed on a constant gray level will have a fractal dimension that approaches 3.0. This is a rather high value, which is contrary to our judgment of such surfaces as being quite smooth. (An interpretation of this result is that highly noisy textures appear exactly like 3-D landscapes in relief!)

26.8 Shape from Texture

This topic in texture analysis also developed strongly during the 1980s. After early work by Bajcsy and Liebermann (1976) for the case of planar surfaces, Witkin (1981) and Kender (1983) significantly extended this work and at the same time laid the foundations for general development of the whole subject. Many papers have followed (e.g., Aloimonos and Swain, 1985; Aloimonos, 1988; Katatani and Chou, 1989; Blostein and Ahuja, 1989; Stone, 1990); there is no space to cover all this work here. In general, workers have studied how an assumed standard texel shape is distorted and its size changed by 3-D projections; they then relate this to the local orientation of the surface. Since the texel distortion varies as the cosine of the angle between the line of sight and the local normal to the surface plane, essentially similar "reflectance map" analysis is required as in the case of shape-from-shading estimation. An alternative approach adopted by Chang et al. (1987) involves texture discrimination by projective invariants (see Chapter 19). More recently, Singh and Ramakrishna (1990) exploited shadows and integrated the information available from texture and from shadows. The problem is that in many situations insufficient data arise from either source taken on its own. Although the results of Singh and Ramakrishna are only preliminary, they indicate the direction that texture work must develop if it is to be used in practical image understanding systems.

26.9 Markov Random Field Models of Texture

Markov models have long been used for texture synthesis to help with the generation of realistic images. They have also proved increasingly useful for texture analysis. In essence, a Markov model is a one-dimensional construct in which the intensity at any pixel depends only upon the intensity of the previous pixel in a chain and upon a transition probability matrix. For images, this is too weak a characterization, and various more complex constructs have been devised. Interest in such models dates from as early as 1965 (Abend et al., 1965) (see also Woods, 1972), though such work is accorded no great significance in Haralick's extensive review of 1979. However, by 1982 the situation was starting to change (Hansen and Elliott, 1982), and a considerable amount of further work was soon being published (e.g., Cross and Jain, 1983; Geman and Geman, 1984; Derin and Elliott, 1987) as Gibbs distributions came to be used for characterizing Markov random fields. In fact, Ising's much earlier work in statistical mechanics (1925) was the starting point for these developments.

Available space does not permit details of these algorithms to be given here. By 1987 impressive results for texture segmentation of real scenes were being achieved using this approach (Derin and Elliott, 1987; Cohen and Cooper, 1987). In particular, the approach appears to be able to cope with highly irregular boundaries, local boundary estimation errors normally lying between 0 and 3 pixels. Unfortunately, these algorithms depend on iterative processing of the image and so tend to require considerable amounts of computation. Cohen and Cooper (1987) state that little processing is required, though there is an inbuilt assumption that a suitable highly parallel processor will be available. On the other hand, they also showed that a hierarchical segmentation algorithm can achieve considerable computational savings, which are greatest when the texture fields have spatially constant parameters.

Finally, stochastic models are useful not only for modeling texture and segmenting one textured region from another, but also for more general segmentation purposes, including cases where boundary edges are quite noisy (Geman, 1987). The fact that a method such as this effectively takes texture as a mere special case demonstrates the power of the whole approach.

26.10 Structural Approaches to Texture Analysis

It has already been remarked that textures approximate to a basic textural element or primitive that is replicated in a more or less regular manner. Structural approaches to texture analysis aim to discern the textural primitive and to determine the underlying gross structure of the texture. Early work (e.g. Pickett, 1970) suggested the structural approach, though little research on these lines was carried out until the late 1970s. Work of this type has been described by Davis (1979), Conners and Harlow (1980b), Matsuyama et al. (1982), Vilnrotter et al. (1986), and Kim and Park (1990). An unusual and interesting paper by Kass and Witkin (1987) shows how oriented patterns from wood grain, straw, fabric, and fingerprints, as well as spectrograms and seismic patterns, can be analyzed. The method adopted involves building up a flow coordinate system for the image, though the method rests more on edge pattern orientation analysis than on more usual texture analysis procedures. A similar statement may be made about the topologically invariant texture descriptor method of Eichmann and Kasparis (1988), which relies on Hough transforms for finding line structures in highly structured textiles. More recently, pyramidal approaches have been applied to structural texture segmentation (Lam and Ip, 1994).

26.11 Concluding Remarks

In this chapter we have seen the difficulties of analyzing textures. These difficulties arise from the potential of textures, and in many cases the frighteningly real complexities of textures—not least from the fact that their properties are often largely statistical in nature. The erstwhile widely used gray-scale co-occurrence matrix approach has been seen to have distinct computational shortcomings. First, many co-occurrence matrices are in principle required (with different values of d and θ) in order to adequately describe a given texture; second, the co-occurrence matrices can be very large and, paradoxically, may hold more data than the image data they are characterizing—especially if the range of gray-scale values is large. In addition, many sets of co-occurrence matrices may be needed to allow for variation of the texture over the image, and if necessary to initiate segmentation. Hence co-occurrence matrices need to be significantly compressed, though in most cases it is not at all obvious *a priori* how this should be achieved. It is even more difficult to arrange for it to be carried out automatically. This probably explains why attention shifted during the 1980s to other approaches, including particularly Laws' technique and its variations (especially that of Ade). Other natural developments were fractal-based measures and the attention given to Markov approaches. A further important development, which could not be discussed here for space reasons, is the Gabor filter technique (see, for example, Jain and Farrokhnia, 1991).

During the past 15 years or so, considerable attention has also been paid to neural network methodologies, because these are able to extract the rules underlying the construction of textures, apparently without the need for analytic or statistical effort. Although neural networks have been able to achieve certain practical goals in actual applications, there is little firm evidence whether they are actually superior to Markov or other methods because conclusive comparative investigations have still not been performed. However, they do seem to offer solutions requiring minimal computational load.

Textures are recognized and segmented by humans with the same apparent ease as plain objects. This chapter has shown that texture analysis needs to be sensitive to microstructures and then pulled into macrostructures—with PCA being a natural means of finding the optimum structure. The subject has great importance for new applications such as iris recognition.

26.12 Bibliographical and Historical Notes

Early work on texture analysis was carried out by Haralick et al. (1973), and in 1976 Weska and Rosenfeld applied textural analysis to materials inspection. The area was reviewed by Zucker (1976a) and by Haralick (1979), and excellent accounts appear in the books by Ballard and Brown (1982) and Levine (1985).

At the end of the 1970s, the Laws technique (1979, 1980a,b) emerged on the scene (which until then had been dominated by the co-occurrence matrix approach), and led to the principal components approach of Ade (1983), which was further developed by Dewaele et al. (1988), Unser and Eden (1989, 1990), and others. For related work on the optimization of convolution filters, see Benke and Skinner (1987). The direction taken by Laws was particularly valuable as it showed how texture analysis could be implemented straightforwardly and in a manner consistent with real-time applications such as inspection.

The 1980s also saw other new developments, such as the fractal approach led by Pentland (1984), and a great amount of work on Markov random field models of texture. Here the work of Hansen and Elliott (1982) was very formative, though the names Cross, Derin, D. Geman, S. Geman, and Jain come up repeatedly in this context (see Section 26.9). Bajcsy and Liebermann (1976), Witkin (1981), and Kender (1983) pioneered the *shape from texture concept*, which has received much attention ever since. Finally, the last 10 to 15 years have seen much work on the application of neural networks to texture analysis, for example, Greenhill and Davies (1993) and Patel et al. (1994).

A number of reviews and useful comparative studies have been made, including Van Gool et al. (1985), Du Buf et al. (1990), Ohanian and Dubes (1992), and Reed and Du Buf (1993). For further work on texture analysis related to inspection for faults and foreign objects, see Chapter 22.

Recent developments include further work with automated visual inspection in mind (Davies, 2000c; Tsai and Huang, 2003; Ojala et al., 2002; Manthalkar et al., 2003; Pun and Lee, 2003), although several of these papers also cite medical, remote sensing, and other applications, as the authors are aware of the generic nature of their work. Of these papers, the last three are specifically aimed at rotation invariant texture classification, and the last one, Pun and Lee, also aims at scale invariance. In previous years there has not been quite this emphasis on rotation invariance, though it is by no means a new topic. Other work (Clerc and Mallat, 2002) is concerned with recovering shape from texture via a texture gradient equation, while Ma et al. (2003) are particularly concerned with person identification based on iris textures. Mirmehdi and Petrou (2000) describe an in-depth investigation of color texture segmentation. In this context, the importance of "wavelets"[6] as an increasingly used technique of texture analysis with interesting applications (such as human iris recognition) should be noted (e.g., Daugman, 1993, 2003).

Finally, in a particularly exciting advance, Spence et al. (2003) have managed to eliminate texture by using photometric stereo to find the underlying surface shape (or "bump map"), following which they have been able to perform impressive reconstructions, including texture, from a variety of viewpoints. McGunnigle and Chantler (2003) have shown that this kind of technique can also reveal hidden writing on textured surfaces, where only pen pressure marks have been made. Similarly, Pan et al. (2004) have shown how texture can be eliminated from ancient tablets (in particular those made of lead and wood) to reveal clear images of the writing underneath.

6 Wavelets are directional filters reminiscent of the Laws edges, bars, waves, and ripples, but have more rigorously defined shapes and envelopes, and are defined in multiresolution sets (Mallat, 1989).

Image Acquisition

In vision, everything depends on image acquisition, and in image acquisition, everything depends on illumination. Naturally, robust algorithms can be designed to largely overcome any problems of inadequacy on these fronts. On the other hand, care with acquisition often means that simpler, more reliable algorithms can be produced. This chapter considers these important aspects of vision system design.

Look out for:

- lighting effects, reflectance, and the appearance of highlights and shadows.
- the value of soft or diffuse lighting.
- how lighting can systematically be made uniform by use of several point or line sources.
- the types of cameras that are commonly available.
- the sampling theorem and its implications.

The advent of solid state cameras and widely available frame-grabbing devices has made one part of image acquisition completely straightforward. Yet the other aspect—that of providing suitable illumination—is still a rather black art. However, the methods described here at least demonstrate that uniform illumination is subject to design rather than ad hoc experimentation.

This chapter necessarily provides underpinning for all practical vision systems—except perhaps those involving X-rays or other modalities such as ultrasonic imaging. Hence, it was crucial to include it in Part 4 of this volume.

CHAPTER 27

Image Acquisition

27.1 Introduction

When implementing a vision system, nothing is more important than image acquisition. Any deficiencies of the initial images can cause great problems with image analysis and interpretation. An obvious example is that of lack of detail owing to insufficient contrast or poor focusing of the camera. This can have the effect, at best, that the dimensions of objects will not be accurately measurable from the images, and at worst that the objects will not even be recognizable, so the purpose of vision cannot be fulfilled. This chapter examines the problems of image acquisition.

Before proceeding, we should note that vision algorithms are of use in a variety of areas where visual pictures are not directly input. For example, vision techniques (image processing, image analysis, recognition, and so on) can be applied to seismographic maps, to pressure maps (whether these arise from handwriting on pressure pads or from weather data), infrared, ultraviolet, X-ray and radar images, and a variety of other cases. There is no space here to consider methods for acquisition in any of these instances, and so our attention is concentrated on purely optical methods. In addition, space does not permit a detailed study of methods for obtaining range images using laser scanning and ranging techniques, while other methods that are specialized for 3-D work will also have to be bypassed. Instead, we concentrate on (1) lighting systems for obtaining intensity images, (2) technology for receiving and digitizing intensity images, and (3) basic theory such as the Nyquist sampling theorem which underlies this type of work.

First we consider how to set up a basic system that might be suitable for the thresholding and edge detection work of Chapters 2–5.

27.2 Illumination Schemes

The simplest and most obvious arrangement for acquiring images is that shown in Fig. 27.1. A single source provides light over a cluster of objects on a worktable or conveyor, and this scene is viewed by a camera directly overhead. The source is typically a tungsten light that approximates to a point source. Assuming for now that the light and camera are some distance away from the objects, and are in different directions relative to them, it may be noted that:

1. different parts of the objects are lit differently, because of variations in the angle of incidence, and hence have different brightnesses as seen from the camera.
2. the brightness values also vary because of the differing absolute reflectivities[1] of the object surfaces.
3. the brightness values vary with the specularities of the surfaces in places where the incident, emergent, and phase angles are compatible with specular reflection (Chapter 16).

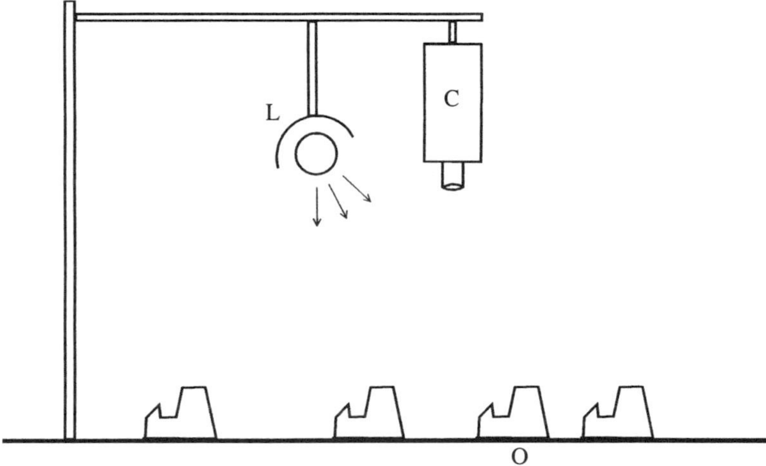

Figure 27.1 Simple arrangement for image acquisition: C, camera; L, light with simple reflector; O, objects on worktable or conveyor.

1 Referring to equation (15.12), R_0 is the absolute surface reflectivity and R_1 is the specularity.

4. parts of the background and of various objects are in shadow, and this again affects the brightness values in different regions of the image.
5. other more complex effects occur because light reflected from some objects will cast light over other objects—factors that can lead to very complicated variations in brightness over the image.

Even in this apparently simple case—one point light source and one camera—the situation can become quite complex. However, (5) is normally a reasonably marginal effect and is ignored in what follows. In addition, effect (3) can often be ignored except in one or two small regions of the image where sharply curved pieces of metal give rise to glints. This still leaves considerable scope for complication due to factors (1), (2), and (4).

Two important reasons may be cited for viewing the surfaces of objects. The first is when we wish to locate objects and their facets, and the second is when we wish to scrutinize the surfaces themselves. In the first instance, it is important to try to highlight the facets by arranging that they are lit differently, so that their edges stand out clearly. In the second instance, it might be preferable to do the opposite—that is, to arrange that the surfaces are lit very similarly, so that any variations in reflectivity caused by defects or blemishes stand out plainly. The existence of effects (1) and (2) implies that it is difficult to achieve both of these effects at the same time: one set of lighting conditions is required for optimum segmentation and location, and another set for optimum surface scrutiny. In most of this book, object location has been regarded as the more difficult task and therefore the one that needs the most attention. Hence, we have imagined that the lighting scheme is set up for this purpose. In principle, a point source of light is well adapted to this situation. However, it is easy to see that if a very diffuse lighting source is employed, then angles of incidence will tend to average out and effect (2) will dominate over (1) so that, *to a first approximation*, the observed brightness values will represent variations in surface reflectance. "Soft" or diffuse lighting also subdues specular reflections (effect (3)), so that for the most part they can be ignored.

Returning to the case of a single point source, recall (effect (4)) that shadows can become important. One special case when this is not so is when the light is projected from exactly the same direction as the camera. We return to this case later in this chapter. Shadows are a persistent cause of complications in image analysis. One problem is that it is not a trivial task to identify them, so they merely contribute to the overall complexity of any image and in particular add to the number of edges that have to be examined in order to find objects. They also make it much more difficult to use simple thresholding. (However, note that shadows can sometimes provide information that is of vital help in interpreting complex 3-D images—see, for example, Section 16.6.)

27.2.1 *Eliminating Shadows*

The above considerations suggest that it would be highly convenient if shadows could be eliminated. A strategy for achieving this elimination is to lower their contrast by using several light sources. Then the region of shadow from one source will be a region of illumination from another, and shadow contrast will be lowered dramatically. Indeed, if there are n lights, many positions of shadow will be illuminated by $n - 1$ lights, and their contrast will be so low that they can be eliminated by straightforward thresholding operations. However, if objects have sharp corners or concavities, there may still be small regions of shadow that are illuminated by only one light or perhaps no light at all. These regions will be immediately around the objects, and if the objects appear dark on a light background, shadows could make the objects appear enlarged, or cause shadow lines immediately around them. For light objects on a dark background this is normally less of a problem.

It seems best to aim for large numbers of lights so as to make the shadows more diffuse and less contrasting, and in the limit it appears that we are heading for the situation of soft lighting discussed earlier. However, this is not quite so. What is often required is a form of diffuse lighting that is still directional—as in the case of a diffuse source of restricted extent directly overhead. This can be provided very conveniently by a continuous ring light around the camera. This technique is found to eliminate shadows highly effectively, while retaining sufficient directionality to permit a good measure of segmentation of object facets to be achieved. That is, it is an excellent compromise, although it is certainly not ideal. For these reasons it is worth describing its effects in some detail. In fact, it is clear that it will lead to good segmentation of facets whose boundaries lie in horizontal planes but to poor segmentation of those whose boundaries lie in vertical planes.

The situation just described is very useful for analyzing the shape profiles of objects with cylindrical symmetry. The case shown in Fig. 27.2 involves a special type of chocolate biscuit with jam underneath the chocolate. If this is illuminated by a continuous ring light fairly high overhead (the proper working position), the region of chocolate above the edge of the jam reflects the light obliquely and appears darker than the remainder of the chocolate. On the contrary, if the ring light is lowered to near the worktable, the region above the edge of the jam appears *brighter* than the rest of the chocolate because it scatters light upward rather than sideways. There is also a particular height at which the ring light can make the jam boundary disappear (Fig. 27.3), this height being dictated by the various angles of incidence and reflection and by the relative direction of the ring light.[2] In comparison, if the lighting were made

2 The latter two situations are described for interest only.

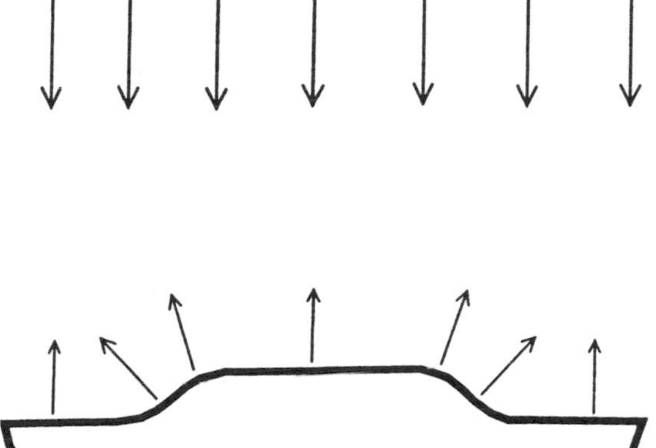

Figure 27.2 Illumination of a chocolate-and-jam biscuit. This figure shows the cross section of a particular type of round chocolate biscuit with jam underneath the chocolate. The arrows show how light arriving from vertically overhead is scattered by the various parts of the biscuit.

completely diffuse, these effects would tend to disappear and the jam boundary would always have very low contrast.

Suitable fluorescent ring lights are readily available and straightforward to use and provide a solution that is more practicable than the alternative means of eliminating shadows mentioned above—that of illuminating objects directly from the camera direction, for example, via a half-silvered mirror.

Earlier, the one case we did not completely solve arose when we were attempting to segment facets whose joining edges were in vertical planes. There appears to be no simple way of achieving a solution to this problem without recourse to switched lights (see Chapter 16). This matter is not discussed further here.

We have now identified various practical forms of lighting that can be used to highlight various object features and that eliminate complications as far as possible. These types of lighting are restricted in what they can achieve (as would clearly be expected from the shape-from-shading ideas of Chapter 16). However, they are exceedingly useful in a variety of applications. A final problem is that two lighting schemes may have to be used in turn, the first for locating objects and the second for inspecting their surfaces. However, this problem can largely be overcome by *not* treating the latter case as a special one requiring its own lighting scheme, but rather noting the direction of lighting and *allowing for* the resulting variation in brightness values by taking account of the known shape of the object. The opposite approach is generally of little use unless other means are used for locating the object. However, the latter situation frequently arises in practice. Imagine that a slab of concrete or a plate of steel is to be inspected for defects. In that case, the position of the object is known, and it is

Figure 27.3 Appearance of the chocolate-and-jam biscuit of Fig. 27.2: (*a*) how the biscuit appears to a camera directly overhead when illuminated as in Fig. 27.2; (*b*) appearance when the lights are lowered to just above table level; (*c*) appearance when the lights are raised to an intermediate level making the presence of the jam scarcely detectable.

clearly best to set up the most uniform lighting arrangement possible, so as to be most sensitive to small variations in brightness at blemishes. This is, then, an important practical problem, to which we now turn.

27.2.2 Principles for Producing Regions of Uniform Illumination

While initially it may seem necessary to illuminate a worktable or conveyor uniformly, a more considered view is that a uniform flat material should *appear* uniform, so that the spatial distribution of the light emanating from its surface is uniform. The relevant quantity to be controlled is therefore the radiance of the surface (light intensity in the image). Following the work of Section 16.4 relating to Lambertian (matte) surfaces, the overall reflectance R of the surface is given by:

$$R = R_0 \mathbf{s}.\mathbf{n} \tag{27.1}$$

where R_0 is the absolute reflectance of the surface and $\mathbf{n}$, $\mathbf{s}$ are, respectively, unit vectors along the local normal to the surface and the direction of the light source.

The assumption of a Lambertian surface can be questioned, since most materials will give a small degree of specular reflection, but in this section we are interested mainly in those nonshiny substances for which equation (27.1) is a good approximation. In any case, special provision normally has to be made for examining surfaces with a significant specular reflectance component. However, notice that the continuous strip lighting systems considered below have the desirable property of largely suppressing any specular components.

Next we recognize that illumination will normally be provided by a set of lights at a certain height h above a worktable or conveyor. We start by taking the case of a single point source at height h. Supposing that this is displaced laterally through a distance a, so that the actual distance from the source to the point of interest on the worktable is d, I will have the general form:

$$I = \frac{c \cos i}{d^2} = \frac{ch}{d^3} \tag{27.2}$$

where c is a constant factor (see Fig. 27.4).

Equation (27.2) represents a distinctly nonuniform region of intensity over the surface. However, this problem may be tackled by providing a suitable distribution of lights. A neat solution is provided by a symmetrical arrangement of

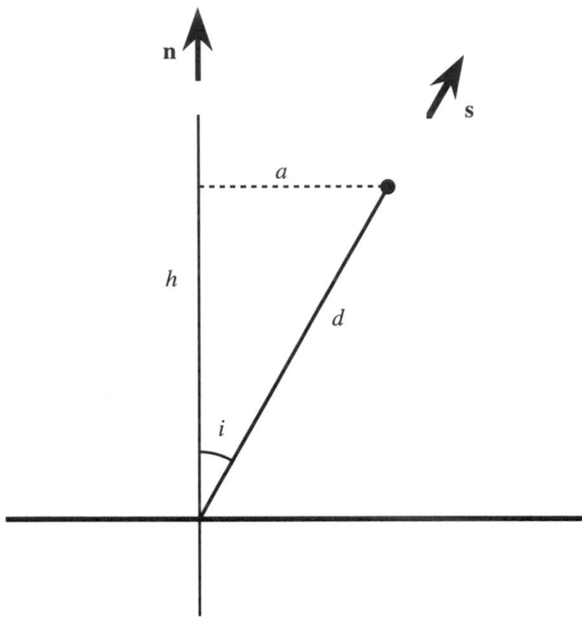

Figure 27.4 Geometry for a single point source illuminating a surface. Here a point light source at a height h above a surface illuminates a general point with angle of incidence i. **n** and **s** are, respectively, unit vectors along the local normal to the surface and the direction of the light source.

two strip lights that will help to make the reflected intensity much more uniform (Fig. 27.5). We illustrate this idea by reference to the well-known arrangement of a pair of Helmholtz coils—widely used for providing a uniform magnetic field, with the separation of the coils made equal to their radius so as to eliminate the second-order variation in field intensity. (Note that in a symmetrical arrangement of coils, all odd orders vanish, *and* a single dimensional parameter can be used to cancel the second order term.)

In a similar way, the separation of the strip lights can be adjusted so that the second-order term vanishes (Fig. 27.5*b*). There is an immediate analogy also with the second-order Butterworth low-pass filter, which gives a maximally flat response, the second-order term in the frequency response curve being made zero, and the lowest order term then being the fourth-order term (Kuo, 1966). The last example demonstrates how the method might be improved further—by aiming for a Chebychev type of response in which there is some ripple in the pass band, yet the *overall* pass-band response is flatter (Kuo, 1966). In a lighting application, we should aim to start with the strip lights not just far enough apart so that the second-order term vanishes, but slightly further apart, so that the intensity is *almost* uniform over a rather larger region (Fig. 27.5*c*).

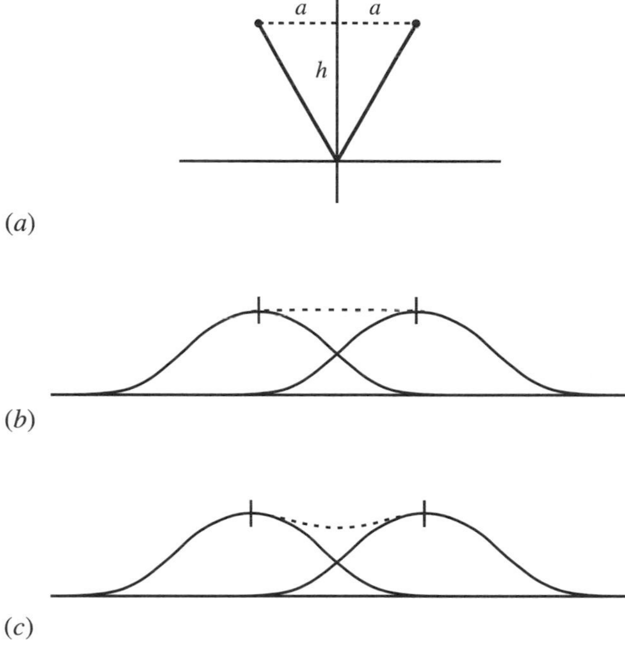

Figure 27.5 Effect of using two strip lights for illuminating a surface. (*a*) shows two strip lights at a height h above a surface, and (*b*) shows the resulting intensity patterns for each of the lights. The dotted line shows the combined intensity pattern. (*c*) shows the corresponding patterns when the separation of the lights is increased slightly.

This demonstrates that in practice the prime aim will be to achieve a given degree of uniformity over the maximum possible size of region.

In principle, it is easy to achieve a given degree of uniformity over a larger region by starting with a given response and increasing the linear dimensions of the *whole* lighting system proportionately. Though valid in principle, this approach will frequently be difficult to apply in practice. For example, it will be limited by convenience and by availability of the strip lights. It must also be noted that as the size of the lighting system increases, so must the power of the lights. Hence in the end we will have only one adjustable geometric parameter by which to optimize the response.

Finally, note that in most practical situations, it will be less useful to have a long narrow working region than one whose aspect ratio is close to unity. We shall consider two such cases—a circular ring light and a square ring light. The first of these is conveniently provided in diameters of up to at least 30 cm by commercially available fluorescent tubes, while the second can readily be constructed— if necessary on a much larger scale—by assembling a set of four linear fluorescent tubes. In this case we take the tubes to be finite in length, and in contact at their ends, the whole being made into an assembly that can be raised or lowered to optimize the system. Thus, these two cases have fixed linear dimensions characterized in each case by the parameter a, and it is h that is adjusted rather than a. To make comparisons easier, we assume in all cases that a is the constant and h is the optimization parameter (Fig. 27.6).

27.2.3 *Case of Two Infinite Parallel Strip Lights*

First we take the case of two infinite parallel strip lights. In this case, the intensity I is given by the sum of the intensities I_1, I_2 for the two tubes:

$$I_1(x) = h \int_{-\infty}^{\infty} [(a-x)^2 + (v-y)^2 + h^2]^{-3/2} dv \tag{27.3}$$

$$I_2(x) = I_1(-x) \tag{27.4}$$

Suitable substitutions permit equation (27.3) to be integrated, and the final result is:

$$I = \frac{2h}{(a-x)^2 + h^2} + \frac{2h}{(a+x)^2 + h^2} \tag{27.5}$$

Differentiating I twice and setting $\mathrm{d}^2I/\mathrm{d}x^2 = 0$ at $x = 0$ eventually (Davies, 1997c) yield the maximally flat condition:

$$h = \sqrt{3}a \tag{27.6}$$

However, as noted earlier, it should be better to aim for minimum *overall* ripple over a region $0 \le x \le x_1$. The situation is shown in Fig. 27.7. We take the ripple ΔI as the difference in height between the maximum intensity I_m and the minimum intensity I_0, at $x = 0$, and on this basis the maximum permissible deviation in x is the value of x where the curve again crosses the minimum value I_0.

A simple calculation shows that the intensity is again equal to I_0 for $x = x_1$, where

$$x_1 = (3a^2 - h^2)^{1/2} \tag{27.7}$$

the graph of h versus x_1 being the circle $x_1^2 + h^2 = 3a^2$ (Fig. 27.8, top curve). It is interesting that the maximally flat condition is a special case of the new one, applying where $x_1 = 0$.

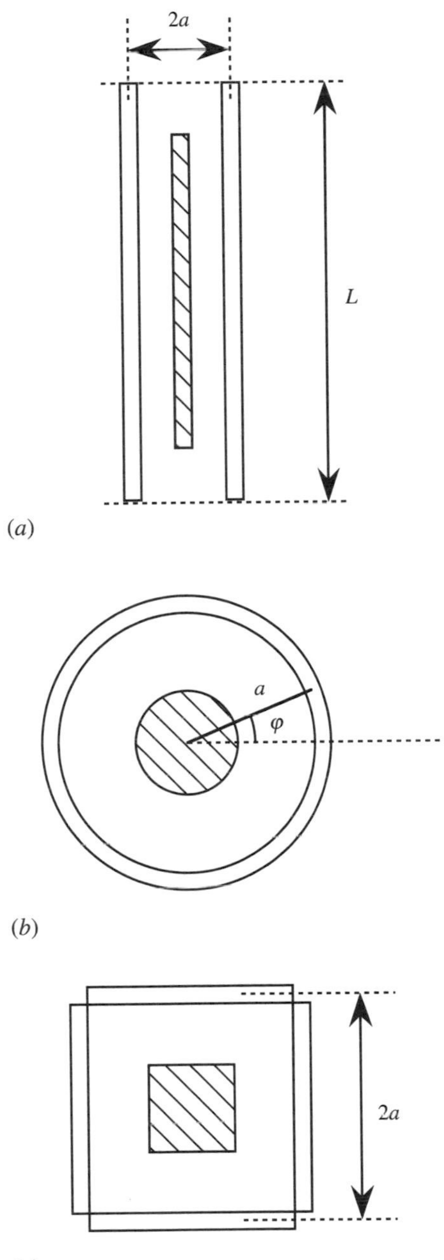

Figure 27.6 Lighting arrangements for obtaining uniform intensity. This diagram shows three arrangements of tubular lights for providing uniform intensity over fairly large regions, shown cross-hatched in each case. (*a*) shows two long parallel strip lights, (*b*) shows a circular ring light, and (*c*) shows four strip lights arranged to form a square "ring." In each case, height *h* above the worktable must also be specified.

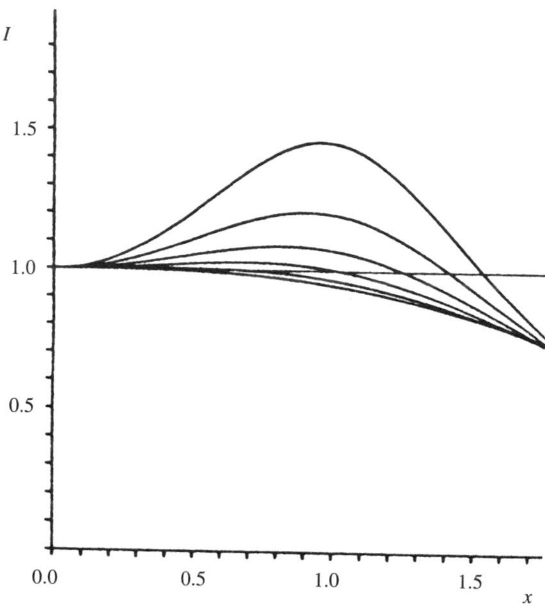

Figure 27.7 Intensity variation for two infinite parallel strip lights. This diagram shows the intensity variation I as a function of the distance x from the center of symmetry for six different values of h. h increases in steps of 0.2 from 0.8 for the top curve to 1.8 for the bottom curve. The value of h corresponding to the maximally flat condition is $h = 1.732$. x and h are expressed in units of a, while I is normalized to give a value of unity at $x = 0$.

Further mathematical analysis of this case is difficult; numerical computation leads to the graphs of Fig. 27.8. The top curve in Fig. 27.8 has already been referred to and shows the optimum height for selected ranges of values of x up to x_1. Taken on its own, this curve would be valueless as the accompanying nonuniformity in intensity would not be known. This information is provided by the left curve in Fig. 27.8. However, for design purposes it is most important first to establish what range of intensities accompanies a given range of values of x, since this information (Fig. 27.8, bottom curve) will permit the necessary compromise between these variables to be made. Having decided on particular values of x_1 and ΔI, the value of the optimization parameter h can then be determined from one of the other two graphs. Both are provided for convenience of reference. (It will be seen that once two of the graphs are provided, the third gives no completely new information.) Maximum acceptable variations in ΔI are assumed to be in the region of 20%, though the plotted variations are taken up to ~50% to give a more complete picture. On the other hand, in most of the practical applications envisaged here, ΔI would be expected not to exceed 2 to 3% if accurate measurements of products are to be made.

The ΔI *v.* x_1 variation varies faster than the fourth power of x_1, there being a very sharp rise in ΔI for higher values of x_1. This means that once ΔI has been

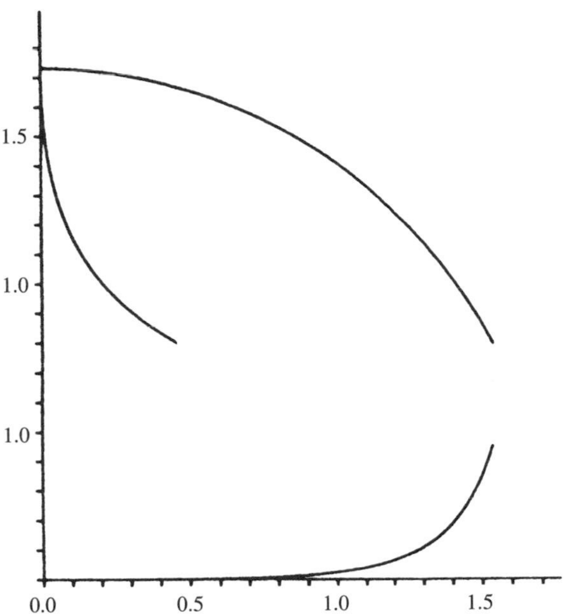

Figure 27.8 Design graphs for two parallel strip lights. Top, h v. x_1. Left, h v. ΔI. Bottom, ΔI v. x_1. The information in these graphs has been extracted from Fig. 27.7. In design work, a suitable compromise working position would be selected on the bottom curve, and then h would be determined from one of the other two curves. In practice, ΔI is the controlling parameter, so the left and bottom curves are the important ones, the top curve containing no completely new information.

specified for the particular application, little is to be gained by trying to squeeze extra functionality through going to higher values of x_1, that is, in practice ΔI is the controlling parameter.

In the case of a circular ring light, the mathematics is more tedious (Davies, 1997c) and it is not profitable to examine it here. The final results are very similar to those for parallel strip lights. They would be used for design in the identical manner to that outlined earlier for the previous case.

In the case of a square ring light, the mathematics is again tedious (Davies, 1997c), but the results follow the same pattern and warrant no special comment.

27.2.4 *Overview of the Uniform Illumination Scenario*

Previous work on optical inspection systems has largely ignored the design of optimal lighting schemes. This section has tackled the problem in a particular case of interest—how to construct an optical system that makes a uniform matte surface appear uniformly bright, so that blemishes and defects can readily be detected with minimal additional computation. Three cases for which calculations have been carried out cover a good proportion of practical lighting schemes, and the design principles described here should be applicable to most other schemes that could be employed.

The results are best presented in the form of graphs. In any one case, the graph shows the tradeoff between variation in intensity and range of position on the working surface, from which a suitable working compromise can be selected. The other two graphs provide data for determining the optimization parameter (the height of the lights above the working surface).

A wide variety of lighting arrangements are compatible with the general principles presented above. Thus, it is not worthwhile to give any detailed dimensional specifications—especially as the lights might be highly directional rather than approximating to point or line sources. However, it is worth underlining that adjustment of just one parameter (the height) permits uniform illumination to be achieved over a reasonable region. Note that (as for a Chebyshev filter) it may be better to arrange a slightly less uniform brightness over a larger region than absolutely uniform brightness over a small region (as is achieved by exact cancellation of the second spatial derivatives of intensity). It is left to empirical tests to finalize the details of the design.

Finally, it should be reiterated that such a lighting scheme is likely to be virtually useless for segmenting object facets from each other—or even for discerning relatively low curvatures on the surface of objects. Its particular value lies in the scrutiny of surfaces via their absolute reflectivities, without the encumbrance of switched lights (see Chapter 16). It should also be emphasized that the aim of the discussion in the past few sections has been to achieve as much as possible with a simple static lighting scheme set up systematically. Naturally, such solutions are compromises, and again are no substitute for the full rigor of switched lighting schemes.

27.2.5 *Use of Line-scan Cameras*

Throughout this discussion it has been assumed implicitly that a conventional "area" camera is employed to view the objects on a worktable. However, when products are being manufactured in a factory they are frequently moved from one stage to another on a conveyor. Stopping the conveyor to acquire an image for inspection would impose unwanted design problems. For this reason use is made of the fact that the speed of the conveyor is reasonably uniform, and an area image is built up by taking successive linear snapshots. This is achieved with a line-scan camera that consists of a row of photocells on a single integrated circuit sensor. The orientation of the line of photocells must, of course, be normal to the direction of motion. More will be said later about the internal design of line-scan and other cameras. However, here we concentrate on the lighting arrangement to be used with such a camera.

When using a line-scan camera it is natural to select a lighting scheme that embodies the same symmetry as the camera. Indeed, the most obvious such scheme is a pair of long fluorescent tubes parallel to the line of the camera (and perpendicular to the motion of the conveyor). We caution against this "obvious" scheme, since a small round object (for example) will not be lit symmetrically. Of course, there are difficulties in considering this problem in that different parts of the object are viewed by the line-scan camera at different moments, but for small objects a linear lighting scheme will not be isotropic. This could lead to small distortions being introduced in measurements of object dimensions. This means that in practice the ring and other symmetrical lighting schemes described above are likely to be more closely optimal even when a line-scan camera is used. For larger objects, much the same situation applies, although the geometry is more complex to work out in detail.

Finally, the comment above that conveyor speeds are "reasonably uniform" should be qualified. The author has come across cases where this is true only as a first approximation. As with many mechanical systems, conveyor motion can be unreliable. For example, it can be jerky, and in extreme cases not even purely longitudinal! Such circumstances frequently arise through a variety of problems that cause slippage relative to the driving rollers—the effects of wear or of an irregular join in the conveyor material, misalignment of the driving rollers, and so on. Furthermore, the motors controlling the rollers may not operate at constant speed, either in the short term (e.g., because of varying load) or in the longer term (e.g., because of varying mains frequency and voltage). While, therefore, it cannot be assumed that a conveyor will operate in an ideal way, careful mechanical design can minimize these problems. However, when high accuracy is required, it will be necessary to monitor the conveyor speed, perhaps by using the optically coded disc devices that are now widely available, and feeding appropriate distance marker pulses to the controlling computer. Even with this method, it will be difficult to match in the longitudinal direction the extremely high accuracy[3] available from the line-scan camera in the lateral direction. However, images of 512×512 pixels that are within 1 pixel accuracy in each direction should normally be available.

3 A number of line-scan cameras are now available with 4096 or greater numbers of photocells in a single linear array. In addition, these arrays are fabricated using very high-precision technology (see Section 27.3), so considerable reliance can be placed on the data they provide.

27.3 Cameras and Digitization

For many years the camera that was normally used for image acquisition was the TV camera with a vidicon or related type of vacuum tube. The scanning arrangements of such cameras became standardized, first to 405 lines, and later to 625 lines (or 525 lines in the United States). In addition, it is usual to interlace the image—that is, to scan odd lines in one frame and even lines in the next frame, then repeat the process, each full scan taking 1/25 second (1/30 second in the United States). There are also standardized means for synchronizing cameras and monitors, using line and frame "sync" pulses. Thus, the vacuum TV camera left a legacy of scanning techniques that are in the process of being eliminated with the advent of digital TV.[4] However, as the result of the legacy is still present, it is worth including a few more details here.

The output of these early cameras is inherently analog, consisting of a continuous voltage variation, although this applies only along the line direction. The scanning action is discrete in that lines are used, making the output of the camera part analog and part digital. Hence, before the image is available as a set of discrete pixels, the analog waveform has to be sampled. Since some of the line scanning time is taken up with frame synchronization pulses, only about 550 lines are available for actual picture content. In addition, the aspect ratio of a standard TV image is 4:3 and it is common to digitize TV pictures as 512×768 or as 512×512 pixels. Note, too, that after the analog waveform has been sampled and pixel intensity values have been established, it is still necessary to digitize the intensity values.

Modern solid-state cameras are much more compact and robust and generate less noise. A very important additional advantage is that they are not susceptible to distortion,[5] because the pixel pattern is fabricated very accurately by the usual integrated circuit photolithography techniques. They have thus replaced vacuum tube cameras in all except special situations.

Most solid-state cameras currently available are of the self-scanned CCD type; therefore, attention is concentrated on these in what follows. In a solid-state CCD camera, the target is a piece of silicon semiconductor that possesses an array of photocells at the pixel positions. Hence, this type of camera digitizes the image from the outset, although in one respect—that signal amplitude represents light intensity—the image is still analog. The analog voltages (or, more accurately, the charges) representing the intensities are systematically passed along analog shift registers to the output of the instrument, where they may be digitized

4 All the vestiges of the old system will not have been swept away until all TV receivers and monitors are digital as well as the cameras themselves.

5 However, this does not prevent distortions from being introduced by other mechanisms—poor optics, poor lighting arrangements, perspective effects, and so on.

to give 6 to 8 bits of gray-scale information (the main limitation here being lack of uniformity among the photosensors rather than noise per se).

The scanning arrangement of a basic CCD camera is not fixed, being fired at any desired rate by externally applied pulses. CCD cameras have been introduced which simulate the old TV cameras, with their line and frame sync pulses. In such cases, the camera initially digitizes the image into pixels, and these are shifted in the standard way to an output connector, being passed en route through a low-pass filter in order to remove the pixellation. When the resulting image is presented on a TV monitor, it appears exactly like a normal TV picture. Some problems are occasionally noticeable, however, when the depixellated image is resampled by a new digitizer for use in a computer frame store. Such problems are due to interference or beating between the two horizontal sampling rates to which the information has been subjected, and take the form of faint vertical lines. Although more effective low-pass filters should permit these problems to be eliminated, the whole problem is being eliminated by memory mapping of CCD cameras to bypass the analog legacy completely.

Another important problem relating to cameras is that of color response. The tube of the vacuum TV camera has a spectral response curve that peaks at much the same position as the spectral pattern of ("daylight") fluorescent lights—which itself matches the response of the human eye (Table 27.1). However, CCD cameras have significantly lower response to the spectral pattern of fluorescent tubes (Table 27.1). In general, this may not matter too greatly, but when objects are moving, the integration time of the camera is limited and sensitivity can suffer. In such cases, the spectral response is an important factor and may dictate against use of fluorescent lights. (This is particularly relevant where CCD line-scan cameras are used with fast-moving conveyors.)

An important factor in the choice of cameras is the delay lag that occurs before a signal disappears and causes problems with moving images. Fortunately, the effect is entirely eliminated with the CCD camera, since the action of reading an image wipes the old image. However, moving images require frequent reading, and this implies loss of integration time and therefore loss of

Table 27.1 Spectral responses

Device	Band (nm)	Peak (nm)
Vidicon	200–800	~550
CCD	400–1000	~800
Fluorescent tube	400–700	~600
Human eye	400–700	~550

In this table, the response of the human eye is included for reference. Note that the CCD response peaks at a much higher wavelength than the vidicon or fluorescent tube, and therefore is often at a disadvantage when used in conjunction with the fluorescent tube.

sensitivity—a factor that normally has to be made up by increasing the power of illuminating sources. Camera "burn-in" is another effect that is absent with CCD cameras but that causes severe problems with certain types of conventional cameras. It is the long-term retention of picture highlights in the light-sensitive material that makes it necessary to protect the camera against bright lights and to take care to make use of lens covers whenever possible. Finally, *blooming* is the continued generation of electron-hole pairs even when the light-sensitive material is locally saturated with carriers, with the result that the charge spreads and causes highlights to envelop adjacent regions of the image. Both CCD and conventional camera tubes are subject to this problem, although it is inherently worse for CCDs. This has led to the production of antiblooming structures in these devices: space precludes detailed discussion of the situation here.

27.3.1 *Digitization*

Digitization is the conversion of the original analog signals into digital form. There are many types of analog-to-digital converters (ADCs), but the ones used for digitizing images have so much data to process—usually in a very short time if real-time analysis is called for—that special types have to be employed. The only one considered here is the "flash" ADC, so called because it digitizes all bits simultaneously, in a flash. It possesses $n-1$ analog comparators to separate n gray levels, followed by a priority encoder to convert the $n-1$ results into normal binary code. Such devices produce a result in a very few nanoseconds and their specifications are generally quoted in megasamples per second (typically in the range 50–200 megasamples/second). For some years these were available only in 6-bit versions (apart from some very expensive parts), but today it is possible to obtain 8-bit versions at almost negligible cost.[6] Such 8-bit devices are probably sufficient for most needs considering that a certain amount of sensor noise, or variability, is usually present below these levels and that it is difficult to engineer lighting to this accuracy.

27.4 The Sampling Theorem

The Nyquist sampling theorem underlies all situations where continuous signals are sampled and is especially important where patterns are to be digitized and analyzed by computers. This makes it highly relevant both with visual patterns and with acoustic waveforms. Hence, it is described briefly in this section.

6 Indeed, as is clear from the advent of cheap web cameras and digital cameras, it is becoming virtually impossible to get noncolor versions of such devices.

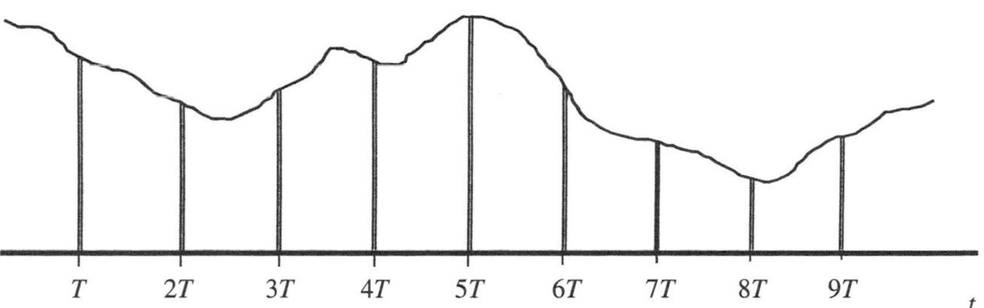

Figure 27.9 The process of sampling a time-varying signal. A continuous time-varying 1-D signal is sampled by narrow sampling pulses at a regular rate $f_r = 1/T$, which must be at least twice the bandwidth of the signal.

Consider the sampling theorem first in respect of a 1-D time-varying waveform. The theorem states that a sequence of samples (Fig. 27.9) of such a waveform contains all the original information and can be used to regenerate the original waveform exactly, but only if (1) the bandwidth W of the original waveform is restricted and (2) the rate of sampling f is at least twice the bandwidth of the original waveform—that is, $f \geq 2W$. Assuming that samples are taken every T seconds, this means that $1/T \geq 2W$.

At first, it may be somewhat surprising that the original waveform can be reconstructed exactly from a set of discrete samples. However, the two conditions for achieving perfect reconstruction are very stringent. What they are demanding in effect is that the signal must not be permitted to change unpredictably (i.e., at too fast a rate) or else accurate interpolation between the samples will not prove possible (the errors that arise from this source are called "aliasing" errors).

Unfortunately, the first condition is virtually unrealizable, since it is nearly impossible to devise a low-pass filter with a perfect cutoff. Recall from Chapter 3 that a low-pass filter with a perfect cutoff will have infinite extent in the time domain, so any attempt at achieving the same effect by time domain operations must be doomed to failure. However, acceptable approximations can be achieved by allowing a "guard-band" between the desired and actual cutoff frequencies. This means that the sampling rate must therefore be higher than the Nyquist rate. (In telecommunications, satisfactory operation can generally be achieved at sampling rates around 20% above the Nyquist rate—see Brown and Glazier, 1974.)

One way to recover the original waveform is to apply a low-pass filter. This approach is intuitively correct because it acts in such a way as to broaden the narrow discrete samples until they coalesce and sum to give a continuous waveform. This method acts in such a way as to eliminate the "repeated" spectra in the transform of the original sampled waveform (Fig. 27.10). This in itself shows why the original waveform has to be narrow-banded before sampling—so that the repeated and basic spectra of the waveform do not cross over each other and become impossible to separate with a low-pass filter. The idea may be taken further

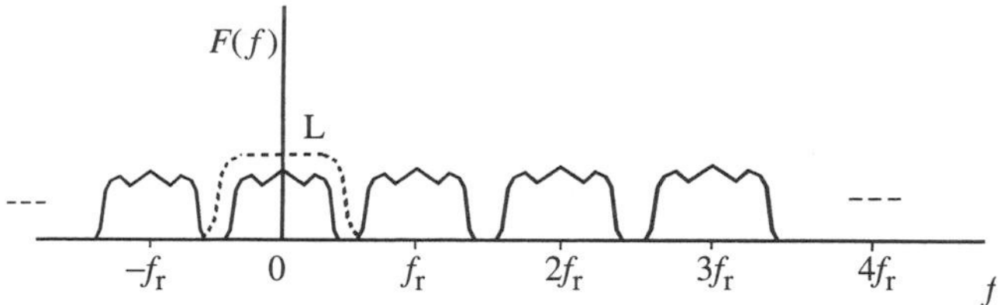

Figure 27.10 Effect of low-pass filtering to eliminate repeated spectra in the frequency domain (f_r, sampling rate; L, low-pass filter characteristic). This diagram shows the repeated spectra of the frequency transform $F(f)$ of the original sampled waveform. It also demonstrates how a low-pass filter can be expected to eliminate the repeated spectra to recover the original waveform.

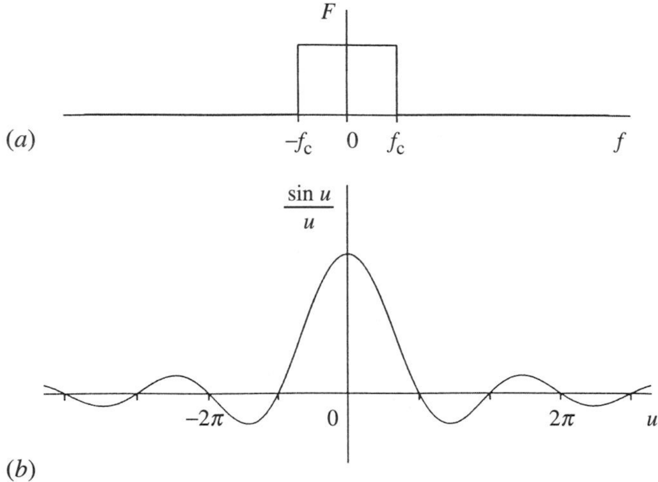

Figure 27.11 The sinc (sin u/u) function shown in (*b*) is the Fourier transform of a square pulse (*a*) corresponding to an ideal low-pass filter. In this case, $u = 2\pi f_c t$, f_c being the cutoff frequency.

because the Fourier transform of a square cutoff filter is the sinc (sin u/u) function (Fig. 27.11). Hence, the original waveform may be recovered by convolving the samples with the sinc function (which in this case means replacing them by sinc functions of corresponding amplitudes). This broadens out the samples as required, until the original waveform is recovered.

So far we have considered the situation only for 1-D time-varying signals. However, recalling that an exact mathematical correspondence exists between time and frequency domain signals on the one hand and spatial and spatial frequency signals on the other, the above ideas may all be applied immediately to each dimension of an image (although the condition for accurate sampling now becomes $1/X \geq 2W_X$, where X is the spatial sampling period and W_X is the spatial bandwidth). Here we accept this correspondence without further discussion and proceed to apply the sampling theorem to image acquisition.

Consider next how the signal from a TV camera may be sampled rigorously according to the sampling theorem. First, it is plain that the analog voltage comprising the time-varying line signals must be narrow-banded, for example, by a conventional electronic low-pass filter. However, how are the images to be narrow-banded in the vertical direction? The same question clearly applies for both directions with a solid-state area camera. Initially, the most obvious solution to this problem is to perform the process optically, perhaps by defocussing the lens. However, the optical transform function for this case is frequently (i.e., for extreme cases of defocusing) very odd, going negative for some spatial frequencies and causing contrast reversals; hence, this solution is far from ideal (Pratt, 2001). Alternatively, we could use a diffraction-limited optical system or perhaps pass the focused beam through some sort of patterned or frosted glass to reduce the spatial bandwidth artificially. None of these techniques will be particularly easy to apply nor will accurate solutions be likely to result. However, this problem is not as serious as might be imagined. If the sensing region of the camera (per pixel) is reasonably large, and close to the size of a pixel, then the averaging inherent in obtaining the pixel intensities will in fact perform the necessary narrow-banding (Fig. 27.12). To analyze the situation in more detail, note that a pixel is essentially square with a sharp cutoff at its borders. Thus, its spatial frequency pattern is a 2-D sinc function, which (taking the central positive peak) approximates to a low-pass spatial frequency filter. This approximation improves somewhat as the border between pixels becomes fuzzier.

The point here is that the worst case from the point of view of the sampling theorem is that of extremely narrow discrete samples, but this worst case is unlikely to occur with most cameras. However, this does not mean that sampling is automatically ideal—and indeed it is not, since the spatial frequency pattern for a sharply defined pixel shape has (in principle) infinite extent in the

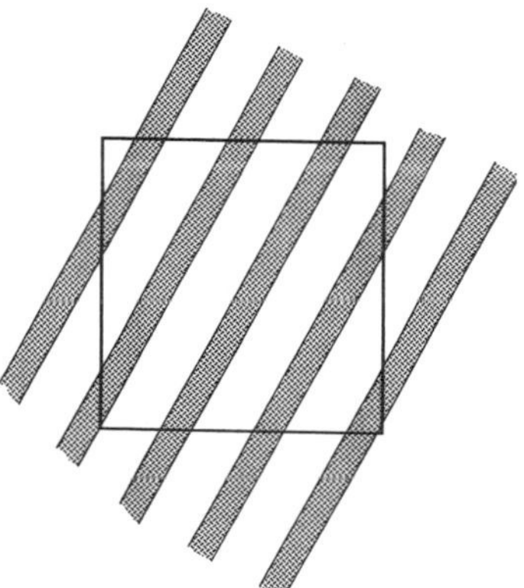

Figure 27.12 Low-pass filtering carried out by averaging over the pixel region. An image with local high-frequency banding is to be averaged over the whole pixel region by the action of the sensing device.

spatial frequency domain. The review by Pratt (2001) clarifies the situation and shows that there is a tradeoff between aliasing and resolution error. Overall, it is underlined here that quality of sampling will be one of the limiting factors if one aims for the greatest precision in image measurement. If the bandwidth of the presampling filter is too low, resolution will be lost; if it is too high, aliasing distortions will creep in; and if its spatial frequency response curve is not suitably smooth, a guard band will have to be included and performance will again suffer.

27.5 Concluding Remarks

This chapter has sought to give some background to the problems of acquiring images, particularly for inspection applications. Methods of illumination were deemed to be worthy of considerable attention since they furnish means by which the practitioner can help to ensure that an inspection system operates successfully—and indeed that its vision algorithms are not unnecessarily complex, thereby necessitating excessive hardware expense for real-time implementation. Means of arranging reasonably uniform illumination and freedom from shadows have been taken to be of significant relevance and are allotted fair attention. (It is of interest that these topics are scarcely mentioned in most books on this subject—a surprising fact that perhaps indicates the importance that most authors ascribe to this vital aspect of the work.) For recent publications on illumination and shadow elimination, see Section 27.6.

By contrast, camera systems and digitization techniques have been taken to be purely technical matters to which little space could be devoted. (To be really useful—and considering that most workers in this area buy commercial cameras and associated frame-grabbing devices ready to plug into a variety of computers—whole chapters would have been required for each of these topics.) Because of its theoretical importance, it appeared to be relevant to give some background to the sampling theorem and its implications, although, considering the applications covered in this book, further space devoted to this topic did not appear to be justified (see Rosie, 1966 and Pratt, 2001 for more details).

Computer vision systems are commonly highly dependent on the quality of the incoming images. This chapter has shown that image acquisition can often be improved, particularly for inspection applications, by arranging regions of nearly uniform illumination, so that shadows and glints are suppressed, and the vision algorithms are much simplified and speeded up.

27.6 Bibliographical and Historical Notes

It is regrettable that very few papers and books give details of lighting schemes that are used for image acquisition, and even fewer give any rationale or background theory for such schemes. Hence, some of the present chapter appears to have broken new ground in this area. However, Batchelor et al. (1985) and Browne and Norton-Wayne (1986) give much useful information on light sources, filters, lenses, light guides, and so on, thereby complementing the work of this chapter. (Indeed, Batchelor et al. (1985) give a wealth of detail on how unusual inspection tasks, such as those involving internal threads, may be carried out.)

Details of various types of scanning systems, camera tubes, and solid-state (e.g., CCD) devices are widely available; see, for example, Biberman and Nudelman (1971), Beynon and Lamb (1980), Batchelor et al. (1985), Browne and Norton-Wayne (1986), and various manufacturers' catalogs. Note that much of the existing CCD imaging device technology dates from the mid-1970s (Barbe, 1975; Weimer, 1975) and is still undergoing development.

Flash (bit-parallel) ADCs are by now used almost universally for digitizing pixel intensities. For a time TRW Inc. (a former USA company, now apparently merged with Northrup Grumman Corp.) was preeminent in this area, and the reader is referred to the catalog of this and other electronics manufacturers for details of these devices.

The sampling theorem is well covered in numerous books on signal processing (see, for example, Rosie, 1966). However, details of how band-limiting should be carried out prior to sampling are not so readily available. Only a brief treatment is given in Section 27.4: for further details, the reader is referred to Pratt (2001) and to references contained therein.

The work of Section 27.2 arose from the author's work on food product inspection, which required carefully controlled lighting to facilitate measurement, improve accuracy, and simplify (and thereby speed up) the inspection algorithms (Davies, 1997d). Similar motivation drove Yoon et al. (2002) to attempt to remove shadows by switching different lights on and off, and then using logic to eliminate the shadows. Under the right conditions, it was only necessary to find the maximum of the individual pixel intensities between the various images. However, in outdoor scenes it is difficult to control the lighting. Instead, various rules must be worked out for minimizing their effect. Prati et al. (2001) have done a comparative evaluation of available methods. Other work on shadow location and elimination has been reported by Rosin and Ellis (1995), Mikic et al. (2000), and Cucchiara et al. (2003). Finally, Koch et al. (2001) have presented results on the use of switched lights to maintain image intensity regardless of changes of ambient illumination and to limit the overall dynamic range of image intensities so that the risk of over- or underexposing the scene is drastically reduced.

Real-Time Hardware and Systems Design Considerations

In general, vision involves huge amounts of computation, as images are two-dimensional and in real-time applications are liable to be delivered at rates of 10 to 20 per second. Although humans can easily cope with these data rates, they are often beyond the processing capabilities of conventional computers. This chapter explores the situation and demonstrates how, using special computational hardware, the processing problems can be addressed and alleviated.

Look out for:

- how parallel processing can radically improve the speeds at which vision algorithms run.
- the concept of an SIMD (single instruction stream, multiple data stream) computer, with one processor per pixel in a 2-D array.
- Flynn's classification of sequential and parallel computers.
- how a vision algorithm may optimally be partitioned between hardware and software.
- modern real-time hardware options.
- the increasingly important status of the FPGA (field-programmable logic array) in real-time hardware design.
- vision systems design considerations and the optimization process.

Much of this volume has been devoted to the systematic design of vision algorithms and of necessity has tended to focus on a large variety of subproblems such as edge detection. However, when embarking upon design of a *complete* vision system, the situation is much less "clean," and indeed is subject to myriad financial and marketing constraints, as well as fiercely nonideal data. In this context, vision system design is as much an art as a science and is more subject to cyclic improvement than in an ideal world—as the last few sections of this chapter have attempted to indicate. Suffice it to say here that the situation must be viewed realistically with an eye to improvement.

CHAPTER 28

Real-Time Hardware and Systems Design Considerations

28.1 Introduction

In Chapter 1 we started by pointing out that of the five senses, vision has the advantage of providing enormous amounts of information at very rapid rates. This ability was observed to be useful to humans and should also be of great value with robots and other machines. However, the input of large quantities of data necessarily implies large amounts of data processing, and this can create problems—especially in real-time applications. Hence, there is no wonder that speed of processing has been alluded to on numerous occasions in earlier chapters of this book.

It is now necessary to examine how serious this situation is and to suggest what can be done to alleviate the problem. Consider a simple situation where it is necessary to examine products of size 64×64 pixels moving at rates of 10 to 20 per second along a conveyor. This amounts to a requirement to process up to 100,000 pixels per second—or typically four times this rate if space between objects is taken into account. The situation can be significantly worse than indicated by these figures. First, even a basic process such as edge detection generally requires a neighborhood of at least 9 pixels to be examined before an output pixel value can be computed. Thus, the number of pixel memory accesses is already 10 times that given by the basic pixel processing rate. Second, functions such as skeletonization or size filtering require a number of basic processes to be applied in turn. For example, eliminating objects up to 20 pixels wide requires 10 erosion operations, while thinning similar objects using simple "north–south–east–west" algorithms requires at least 40 whole-image operations. Third, typical inspection applications require a number of tasks—edge detection, object location, surface scrutiny, and so on—to be carried out. All these factors mean that the overall processing task may involve anything between 1 and 100 million pixels or other memory accesses per second. Finally, this analysis ignores the complex computations required for some types of 3-D modeling, or for certain more abstract processing operations (see Chapters 15 and 16).

These formidable processing requirements imply a need for very carefully thought out algorithm strategies. Accordingly, special hardware will normally be needed[1] for almost all real-time applications. (The exception might be in those tasks where performance rates are governed by the speed of a robot, vehicle, or other slow mechanical device.) Broadly speaking, there are two main strategies for improving processing rates. The first involves employing a single very fast processor that is fabricated using advanced technology, for example, with gallium arsenide semiconductor devices, Josephson junction devices, or perhaps optical processing elements. Such techniques can be expected to yield speed increases by a factor of ten or so, which might be sufficiently rapid for certain applications. However, it is more likely that a second strategy will have to be invoked—that of parallel processing. This strategy involves employing N processors working in parallel, thereby giving the possibility of enhancing speed by a factor N. This second strategy is particularly attractive: to achieve a given processing speed, it should be necessary only to increase the number of processors appropriately—although it has to be accepted that cost will be increased by a factor of around N, as for the speed. It is partly a matter of economics which of the two strategies will be the better choice, but at any time the first strategy will be technology-limited, whereas parallel processing seems more flexible and capable of giving the required speed in any circumstances. Hence, for the most part, parallel processing is considered in what follows.

28.2 Parallel Processing

There are two main approaches to parallel processing. In the first approach, the computational *task* is split into a number of functions, which are then implemented by different processors. In the second, the *data* are split into several parts, and different processors handle the different parts. These two approaches are sometimes called *algorithmic parallelism* and *data parallelism*, respectively. Notice that if the data are split, different parts of the data are likely to be nominally similar, so there is no reason to make the processing elements (PEs) different from each other. However, if the task is split functionally, the functions are liable to be very different, and it is most unlikely that the PEs should be identical.

The inspection example cited in Section 28.1 involved a fixed sequence of processes being applied in turn to the input images. On the whole, this type of

1 Sometime in the future this view will need to be reconsidered; see Section 28.7.

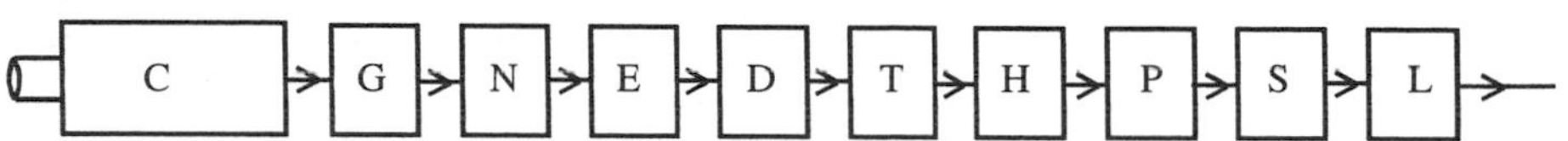

Figure 28.1 Typical pipelined processing system: C, input from camera; G, grab image; N, remove noise; E, enhance edge; D, detect edge; T, thin edge; H, generate Hough transform; P, detect peaks in parameter space; S, scrutinize products; L, log data and identify products to be rejected.

task is well adapted to algorithmic parallelism and indeed to being implemented as a pipelined processing system, each stage in the pipeline performing one task such as edge detection or thinning (Fig. 28.1). (Note that each stage of a thinning task will probably have to be implemented as a single stage in the pipeline.) Such an approach lacks generality, but it is cost-effective in a large number of applications, since it is capable of providing a speedup factor of around two orders of magnitude without undue complexity. Unfortunately, this approach is liable to be inefficient in practice because the speed at which the pipeline operates is dictated by the speed of the slowest device on the pipeline. Faster speeds of the other stages constitute wasted computational capability. Variations in the data passing along the pipeline add to this problem. For example, a wide object would require many passes of a thinning operation, so either thinning would not proceed to completion (and the effect of this would have to be anticipated and allowed for), or else the pipeline would have to be run at a slower rate. Obviously, it is necessary for such a system to be designed in accordance with worst-case rather than "average" conditions—although additional buffering between stages can help to reduce this latter problem.

The design and control of a reliable pipelined processor is not trivial, but, as mentioned above, it gives a generally cost-effective solution in many types of applications. However, both with pipelined processors and with other machines that use algorithmic parallelism significant difficulties are encountered in dividing tasks into functional partitions that match well the PEs on which they are to run. For this and other reasons many attempts have been made at the alternative approach of data parallelism. Image data are, on the whole, reasonably homogeneous, so it is evidently worth searching for solutions incorporating data parallelism. Further consideration then leads to the SIMD type of machine in which each pixel is processed by its own PE. This method is described in the next section.

28.3 SIMD Systems

In the SIMD (single instruction stream, multiple data stream) architecture, a 2-D array of PEs is constructed that maps directly onto the image being processed.

Thus, each PE stores its own pixel value, processes it, and stores the processed pixel value. Furthermore, all PEs run the same program and indeed are subject to the same clock. That is, they execute the same instruction simultaneously (hence the existence of a single instruction stream). An additional feature of SIMD machines that are used for image processing is that each PE is connected to those immediately around it, so that neighborhood operations can conveniently be carried out. The required input data are always available. This means that each PE is typically connected to eight others in a square array. Such machines therefore have the advantage not only of *image parallelism* but also of *neighborhood parallelism*. Data from neighboring pixels are available immediately, and several sequential memory accesses per pixel process are no longer required. (For a useful review of these and other types of parallelism, see Danielsson and Levialdi, 1981.)

The SIMD architecture is extremely attractive in principle because its processing structure seems closely matched to the requirements of many tasks, such as noise removal, edge detection, thinning, size analysis, and so on (although we return to this point below). However, in practice it suffers from a number of disadvantages, some of which are due to the compromises needed to keep costs at reasonable levels. For example, the PEs may not be powerful floating-point processors and may not contain much memory. (This is because available cost is expended on including more PEs rather than making them more powerful.) In addition, the processor array may be too small to handle the whole image at once, and problems of continuity and overlap arise when trying to process subimages separately (Davies et al., 1995). This can also lead to difficulties when global operations (such as finding an accurate convex hull) have to be performed on the whole image. Finally, getting the data in and out of the array can be a relatively slow process.

Although SIMD machines may appear to operate efficiently on image data, this is not always the case in practice, since many processors may be "ticking over" and not doing anything useful. For example, if a thinning algorithm is being implemented, much of the image may be bare of detail for most of the time, since most of the objects will have shrunk to a fraction of their original area. Thus, the PEs are not being kept *usefully* busy. Here the topology of the processing scheme is such that these inactive PEs are unable to get data they can act on, and efficiency drops off markedly. Hence, it is not obvious that an SIMD machine can always carry out the *overall* task any faster than a more modest MIMD machine (see below), or an especially fast but significantly cheaper single processor (SISD) machine.

A more important characteristic is that while the SIMD machine is reasonably well adapted for image processing, it is quite restricted in its capabilities for image analysis. For example, it is virtually impossible to use *efficiently* for implementing Hough transforms, especially when these demand mapping image features into an abstract parameter space. In addition, most serial (SISD) computers are much more efficient at operations such as simple edge tracking, since their single processors are generally much faster than costs will permit for the many processors in an SIMD machine. Overall, these problems should be expected since the SIMD concept is designed for image-to-image transformations via local operators and does not map well to (1) image-to-image transformations that demand *nonlocal* operations, (2) image-to-abstract data transformations (intermediate-level processing), or (3) abstract-to-abstract data (high-level) processing. (Note that some would classify (1) as being a form of intermediate-level processing where *deductions* are made about what is happening in distant parts of an image—that is, higher-level interpretive data are being marked in the transformed image.) This means that unaided SIMD machines are unlikely to be well suited for practical inspection work or for related applications.

Before leaving the topic of SIMD machines, recall that they incorporate two types of parallelism—image parallelism and neighborhood parallelism. Both types contribute to high processing rates. Although it might at first appear that image parallelism contributes mainly through the high processing bandwidth[2] it offers, it also contributes through the high data accessing bandwidth. In contrast, neighborhood parallelism contributes only through the latter mechanism. However, what is important is that this type of parallel machine, in common with any successful parallel machine, incorporates both features. It is of little use to attend to the problem of achieving high processing bandwidth only to run into data bottlenecks through insufficient attention to data structures and data access rates. That is, it is necessary to match the data access and processing bandwidths if full use is to be made of available processor parallelism.

28.4 The Gain in Speed Attainable with *N* Processors

Could the gain in processing rate ever be greater than N, say $2N$ or even N^2? It could in principle be imagined that two robots used to make a bed would operate more efficiently than one, or four more efficiently than two, for a square bed. Similarly, N robots welding N sections of a car body would operate more efficiently than a single one. The same idea should apply to N processors operating in parallel on an N-pixel image. At first sight, it does appear that a gain greater than N could result. However, closer study shows that any task is split between data organization and actual processing. Thus, the maximum gain that could result from the use of N processors is (exactly) N; any other factor is due to the difficulty, either for low or for high N, of getting the right data to the right

2 In this context, it is conventional to use the term *bandwidth* to mean the maximum rate realizable via the stated mechanism.

processor at the right time. Thus, in the case of the bed-making robots, there is an overhead for $N=1$ of having to run around the bed at various stages because the data (the sheets) are not presented correctly. More usually, it is at large N that the data are not available at the right place at the right moment. An immediate practical example of these ideas is that of accessing all eight neighbors in a 3×3 neighborhood where only four are directly connected, and the corner pixels have to be accessed via these four. Then a *threefold* speedup in data access may be obtained by *doubling* the number of local links from four to eight.

Many attempts have been made to model the utilization factor of both SIMD and pipelined machines when operating on branching and other algorithms. Minsky's conjecture (Minsky and Papert, 1971) that the gain in speed from a parallel processor is proportional to $\log_2 N$ rather than N can be justified on this basis and leads to an efficiency $\eta=\log_2 N/N$. Hwang and Briggs (1984) produced a more optimistic estimate of efficiency in parallel systems: $\eta=1/\log_2 N$.

Following Chen (1971), the efficiency of a pipelined processor is usually estimated as $\eta=P/(N+P-1)$, where there are on average P consecutive data points passing through a pipeline of N stages—the reasoning being based on the proportion of stages that are usefully busy at any one time. For imaging applications, such arguments are often somewhat irrelevant since the total delay through the pipelined processor is unimportant compared with the cycle time between successive input or output data values. This is because a machine that does not keep up with the input data stream will be completely useless, whereas one that incorporates a fixed time delay may be acceptable in some cases (such as a conveyor belt inspection problem) though unacceptable in others (such as a missile guidance system).

Broadly speaking, the situation that is being described here involves a speedup factor N coupled with an efficiency η, giving an overall speedup factor of $N'=\eta N$. The loss in efficiency is often due ultimately to frustrated algorithm branching processes but presents itself as underutilization of resources, which cannot be reduced because the incoming data are of variable complexity.

28.5 Flynn's Classification

Early in the development of parallel processing architectures, Flynn (1972) developed a now well-known classification that has already been referred to above: architectures are either SISD, SIMD, or MIMD. Here SI (single instruction stream) means that a single program is employed for all the PEs in the system, whereas MI (multiple instruction stream) means that different programs can be used; SD (single data stream) means that a single stream of data is sent to all the PEs in the system; and MD (multiple data stream) means that the PEs are fed with data independently of each other.

The SISD machine is a single processor and is normally taken to refer to a conventional von Neumann computer. However, the definitions given above imply that SISD falls more naturally under the heading of a Harvard architecture, whose instructions and data are fed to it through separate channels. This gives it a degree of parallelism and makes it generally faster than a von Neumann architecture. (There is almost invariably bit *parallelism* also, the data taking the form of words of data holding several bits of information, and the instructions being able to act on all bits simultaneously. However, this possibility is so universal that it is ignored in what follows.)

The SIMD architecture has already been described with reasonable thoroughness, although it is worth reiterating that the multiple data stream arises in imaging work through the separate pixels being processed by their PEs independently as separate, though similar, data streams. Note that the PEs of SIMD machines invariably embody the Harvard architecture.

The MISD architecture is notably absent from the above classification. It is possible, however, to envisage that pipelined processors fall into this category since a single stream of data is fed through all processors in turn, albeit being modified as it proceeds so the same *data* (as distinct from the same data stream) do not pass through each PE. However, many parties take the MISD category to be null (e.g., Hockney and Jesshope, 1981).

The MIMD category is a very wide one, containing all possible arrangements of separate PEs that get their data and their instructions independently. It even includes the case where none of the PEs is connected together in any way. However, such a wide interpretation does not solve practical problems. We can therefore envisage linking the PEs together by a common memory bus, or every PE being connected to every other one, or linkage by some other means. A common memory bus would tend to cause severe contention problems in a fast-operating parallel system, whereas maintaining separate links between all pairs of processors is at the opposite extreme but would run into a combinatorial explosion as systems become larger. Hence, a variety of other arrangements are used in practice. Crossbar, star, ring, tree, pyramid, and hypercube structures have all been used (Fig. 28.2). In the crossbar arrangement, half of the processors can communicate directly with the other half via N links (and $N^2/4$ switches), although all processors can communicate with each other *indirectly*. In the star there is one central PE, so the maximum communication path has length 2. In the ring all N PEs are placed symmetrically, and the maximum communication path is of length $N-1$. (Note that this figure assumes unidirectional rings, which are easier to implement, and a number of notable examples of this type exist.) In the tree or pyramid, the maximum path length is of order $2(\log_2 N - 1)$, assuming that there are two branches for each node of the tree. Finally, for the hypercube of n dimensions (built in an n-dimensional space with two positions per dimension), the shortest path length between any two PEs is at most n; in this case there are 2^n processors, so the shortest path length is at most $\log_2 N$.

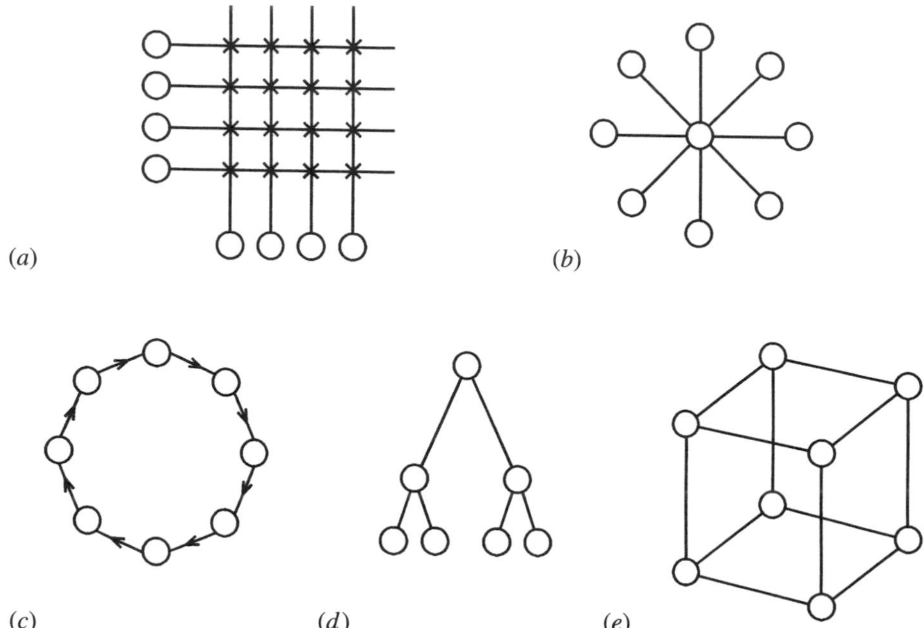

Figure 28.2 Possible arrangements for linking processing elements: (*a*) crossbar; (*b*) star; (*c*) ring; (*d*) tree; (*e*) hypercube. All links are bidirectional except where arrows indicate otherwise.

Overall, the basic principle must be to minimize path length while at the same time reducing the possibility of data bottlenecks. This shows that the star configuration is very limited and explains the considerable attention that has been paid to the hypercube topology. In view of what was said earlier about the importance of matching the data bandwidth to the processing bandwidth, it will be clear that a very careful choice needs to be made concerning a suitable architecture—and there is no lack of possible candidates. (The above set of examples is by no means exhaustive.)

Finally, we should note that factors other than speed enter into the choice of an architecture. For example, many observers have argued that the pyramid type of architecture closely matches the hierarchical data structures most appropriate for scene interpretation.

Much of our discussion has been in the realm of the possible and the ideal (many of the existing systems being expensive experimental ones). It is now necessary to consider more practical issues. How, for example, are we to match the architecture to the data? More particularly, how do we lay down general guidelines for partitioning the tasks and implementing them on practical architectures? In the absence of general guidelines of this type, it is useful to look in detail at a given practical problem to see how to design an optimal hardware system to implement it; this is done in the next section.

28.6 Optimal Implementation of an Image Analysis Algorithm

The particular algorithm considered in this section was examined earlier by Davies and Johnstone (1986, 1989). It involves the inspection of round food products (see also Chapter 22). The purpose of the analysis is to show how to make a systematic selection between available hardware modules (including computers), so that it can be guaranteed that the final hardware configuration is optimal in specific ways and in particular with regard to relevant cost–speed tradeoffs.

In the particular food product application considered, biscuits are moving at rates of up to 20 per second along a conveyor. Since the conveyor is moving continuously, it is natural to use a line-scan camera to obtain images of the products. Before they can be scrutinized for defects and their sizes measured, they have to be located accurately within the images. Since the products are approximately circular, it is straightforward to employ the Hough transform technique for the purpose (see Chapter 10). It is also appropriate to use the radial intensity histogram approach to help with the task of product scrutiny (Chapter 22). In addition, simple thresholding can be used to measure the amount of chocolate cover and certain other product features. The main procedures in the algorithm are summarized in Table 28.1. Note that the Hough transform approach requires the rapid and accurate location of edge pixels, which is achieved using the Sobel operator and thresholding the resulting edge enhanced image. Edge detection is in fact the only 3×3 neighborhood operation in the algorithm and hence it is relatively time-consuming; rather less processing is required by the 1×1 neighborhood operations (Table 28.1). Then come various 1-D processes such as analysis of radial histograms. The fastest operations are those such as logging variables, which are neither 1-D nor 2-D processes.

28.6.1 *Hardware Specification and Design*

On finalization of the algorithm strategy, the overall execution time was found to be about 5 seconds per product (Davies and Johnstone, 1986).[3] With product flow rates on the order of 20 per second, and software optimization subject to severely diminishing returns, a further gain in speed by a factor of around 100 could be obtained only by using special electronic hardware. In this application, a compromise was sought with a single CPU linked to a set of suitable hardware accelerators. In this case, the latter would have had to be designed specially for

3 Although the particular example is dated, the principles involved are still relevant and are worth following here.

Table 28.1 Breakdown of the inspection algorithm

Function	Description	Time (sec)	Cost ($)	c/t ($/ms)
1. Acquire image.	1 × 1	—	1000	—
2. Clear parameter space.	1 × 1	0.017	200	11.8
3. Find edge points.	3 × 3	4.265	3000	0.7
4. Accumulate points in parameter space.	1 × 1	0.086	2000	23.3
5. Find averaged center.	—	0.020	2000	100.0
6. Find area of product.	1 × 1	0.011	100	9.1
7. Find light area (no chocolate cover).	1 × 1	0.019	200	10.5
8. Find dark area (slant on product).	1 × 1	0.021	200	9.5
9. Compute radial intensity histogram.	1 × 1	0.007	400	57.1
10. Compute radial histogram correlation.	1-D	0.013	400	30.8
11. Overheads for functions 6–10.	—	0.415	1200	2.9
12. Calculate product radius.	1-D	0.047	4000	85.1
13. Track parameters and log.	—	0.037	4000	108.1
14. Decide if rejection is warranted.	—	0.002	4000	2000.0
Time for whole algorithm		4.960		

the purpose, and indeed some were produced in this way. However, in the following discussion it is immaterial whether the hardware accelerators are made specially or purchased: the object of the discussion is to present rigorous means for deciding which software functions should be replaced by hardware modules.

As a prerequisite to the selection procedure, Table 28.1 lists execution times and hardware implementation costs of all the algorithm functions. The figures are somewhat notional since it is difficult to divide the algorithm rigorously into completely independent sections. However, they are sufficiently accurate to form the basis for useful decisions on the cost-effectiveness of hardware. Because the aim is to examine the principles underlying cost–speed tradeoffs, the figures presented in Table 28.1 are taken as providing a concrete example of the sort of situation that can arise in practice.

28.6.2 *Basic Ideas on Optimal Hardware Implementation*

The basic strategy that was adopted for deciding on a hardware implementation of the algorithm is as follows:

1. Prioritize the algorithm functions so that it is clear in which order they should be implemented.
2. Find some criterion for deciding when it is *not* worth proceeding to implement further functions in hardware.

From an intuitive point of view, function prioritization seems to be a simple process. Basically, functions should be placed in order of cost-effectiveness; that is, those saving the most execution time per unit cost (when implemented in hardware) should be placed first, and those giving smaller savings should be placed later. Thus, we arrive at the c/t (cost/time) criterion function. Then with limited expenditure we achieve the maximum savings in execution time, that is, the maximum speed of operation.

To decide at what stage it is not worth implementing further functions in hardware is arguably more difficult. Excluding here the practical possibility of strict cost or time limits, the ideal solution results in the optimal balance between total cost and total time. Since these parameters are not expressible in the same units, it is necessary to select a criterion function such as $C \times T$ (cost $\times$ time) which, when minimized, allows a suitable balance to be arrived at automatically.

The procedure outlined here is simple and does not take account of hardware which is common to several modules. This "overhead" hardware must be implemented for the first such module and is then available at zero cost for subsequent modules. In many cases a speed advantage results from the use of overhead hardware. In the example system, it is found that significant economies are possible when implementing functions 6 to 10, since common pixel scanning circuitry may be used. In addition, note that any of these functions that are *not* implemented in hardware engender a time overhead in software. This, and the fact that the time overhead is much greater than the sum of the software times for functions 6 to 10, means that once the initial cost overhead has been paid it is best (at least, in this particular example) to implement all these functions in hardware.

Trying out the design strategy outlined above gives the c/t ratio sequence shown in Table 28.2. A set of overall times and costs resulting from implementing in hardware all functions down to and including the one indicated is now deduced. Examination of the column of $C \times T$ products then shows where the tradeoff between hardware and software is optimized. This occurs here when the first 13 functions are implemented in hardware.

The analysis presented in this section gives only a general indication of the required hardware–software tradeoff. Indeed, minimizing $C \times T$ indicates an overall "bargain package," whereas in practice the system might well have to meet certain cost or speed limits. In this food product application, it was necessary to aim at an overall cost of less than \$15,000. By implementing functions 1, 3, and 6 to 11 in hardware, it was found possible to get to within a factor 3.6 of the optimal $C \times T$ product. Interestingly, by using an upgraded host processor, it proved possible to get within a much smaller factor (1.8) of the optimal tradeoff, with the same number of functions implemented in hardware (Table 28.2). It is a particular advantage of using this criterion function approach that the choice of which processor to use becomes automatic.

Table 28.2 Speed–cost tradeoff figures

Function (see Table 28.1)	c/t (\$/ms)	t (sec)	c (\$)	T (sec)	C (\$)	$C \times T$ (\$–s)	$C' \times T'$ (\$–s)
—		—	6000	4.990	6000	29,940	15,080
3	0.7	4.265	3000	0.725	9000	6530	3190
6–11	5.1	0.486	2500	0.239	11,500	2750	1400
2	11.8	0.017	200	0.222	11,700	2600	1350
4	23.3	0.086	2000	0.136	13,700	1860	1040
12	85.1	0.047	4000	0.089	17,700	1580	1010
5	100.0	0.020	2000	0.069	19,700	1360	860
13	108.1	0.037	4000	0.032	23,700	760	770
14	2000.0	0.002	4000	0.030	27,700	830	860

In this table, the first entry corresponds to the base system cost, including computer, camera, frame store, backplane, and power supply. The other entries are derived from Table 28.1 by ordering the c/t values (see text). The final column shows the figures obtained using an upgraded host processor.

The paper by Davies and Johnstone (1989) goes into some depth concerning the choice of criterion function, showing that more general functions are available and that there is a useful overriding geometrical interpretation—that global concavities are being sought on the chosen curve in $f(C)$, $g(T)$ space. The paper also places the problem of overheads and relevant functional partitions on a more rigorous basis. Finally, a later paper (Davies et al., 1995) emphasizes the value of software solutions, achieved, for example, with the aid of arrays of DSP chips.

28.7 Some Useful Real-time Hardware Options

In the 1980s, real-time inspection systems typically had many circuit boards containing hundreds of dedicated logic chips, though some of the functionality was often implemented in software on the host processor. This period was also a field-day for the Transputer type of microprocessor which had the capability for straightforward coupling of processors to make parallel processing systems. In addition, the "bit-slice" type of microprocessor permitted easy expansion to larger word sizes, though the technology was also capable of use in multipixel parallel processors. Finally, the VLSI type of logic chip was felt by many to present a route forward, particularly for low-level image processing functions.

In spite of all these competing lines, what gradually emerged in the late 1990s as the predominant real-time implementation device was the digital signal processing (DSP) chip. This had evolved earlier in response to the need for fast 1-D signal processors, capable of performing such functions as Fast Fourier Transforms and processing of speech signals. The reason for its success lay in its convenience and programmability (and thus flexibility) and its high speed of operation. By coupling DSP chips together, it was found that 2-D image processing operations could be performed both rapidly and flexibly, thereby to a large extent eliminating the need for dedicated random logic boards, and at the same time ousting Transputers and bit-slices.

Over the same period, however, single-chip Field-Programmable Logic Arrays (FPGAs) were becoming more popular and considerably more powerful, while, in the 2000s, microprocessors are even being embedded on the same chip. At this stage, the concept is altogether more serious, and in the 2000s one has to think very carefully to obtain the best balance between DSP and FPGA chips for implementing practical vision systems.

Another contender in this race is the ordinary PC. While in the 1990s the PC had not normally been regarded as a suitable implementation vehicle for real-time vision, the possibility of implementing some of the slower real-time functions using a PC gradually arose though the relentless progression of Moore's Law. At the same time, other work on special software designs (see Chapter 23) showed how this line of development could be extended to faster running applications. At the time of writing, it can be envisaged that a single unaided PC—with a reliable embedded operating system—will, by the end of the decade, be sufficient to run a large proportion of all machine vision applications. Indeed, this is anticipated by many sectors of industry. Although it is always risky to make predictions, we can at least anticipate that in the next 10 years the whole emphasis of real-time vision will move away from speed being the dominating influence. At that stage effectiveness, accuracy, robustness, and reliability (what one might call "fitness for purpose") will be all that matters. For the first time we will be free to design ideal vision systems. The main word of caution is that this ideal will not necessarily apply for all possible applications—one can imagine exceptions, for example, where ultra–high speed aircraft have to be controlled or where huge image databases have to be searched rapidly.

Overall, we can see a progression amid all the hardware developments outlined here. This is a move from random logic design to use of software-based processing elements, and further, one where the software runs not on special devices with limited capability but on conventional computers for which (1) very long instruction words are not needed, (2) machine code or assembly language programming is not necessary to get the most out of the system (so standard languages can be used), and (3) overt parallel processing (beyond that available in the central processing chip of a standard PC) is no longer crucial.[4] The advantages in terms of flexibility are dramatic compared with the early days of machine vision.

The trends described here are underlined by the publications discussed in Section 28.13.2 and are summarized in Table 28.3.

28.8 Systems Design Considerations

Having focused on the problems of real-time hardware design, we now need to get a clearer idea of the overall systems design process. One of the most important limitations on the rate at which machine vision systems can be produced is the lack of flexibility of existing design strategies. This applies especially for inspection systems. To some extent, this problem stems from lack of understanding of the basic principles of vision upon which inspection systems might optimally be based. In addition, there is the problem of lack of knowledge of what goes into the design process. It is difficult enough designing a complete inspection system, including all the effort that goes into producing a cost-effective real-time hardware implementation, without having to worry at the same time whether the schema used is generic or adaptable to other products. Yet this is a crucial factor that deserves a lot of attention.

A major factor standing in the way of progress in this area is the lack of complete case studies of inspection systems. This problem arises partly from lack of detailed knowledge of existing commercial systems and partly from lack of space in journal or conference papers. In the case of conference papers what suffers is published know-how—particularly on creativity aspects. (Journals see their role as promoting scientific methodology and results rather than subjective design notions.) Nevertheless, subjective notions are part of the whole design process, and if the situation is to develop as it should, the mental background should be aired as part of the overall design procedure. We summarize the situation in the following section.

28.9 Design of Inspection Systems—The Status Quo

As practiced hitherto, the design of an inspection system has a number of stages, much as in the list presented in Table 28.4. Although this list is incomplete (e.g., it includes no mention of lighting systems), it is a useful start, and it does

4 Nevertheless, some functions such as image acquisition and control of mechanical devices will have to be carried out in parallel, to prevent data bottlenecks and other holdups.

Table 28.3 Hardware devices for the implementation of vision algorithms

Device	Function	Summary of properties
PC	Personal computer or more powerful workstation: complete computer with RAM, hard disc, and other peripheral devices. Would need an embedded (restricted) operating system in a real-time application.	• fast • medium cost • extremely flexible • should be envisioned as a software device
MP	Microprocessor: single chip device containing CPU + cache RAM. The core element of a PC.	• fast • low cost • extremely flexible • should be envisioned as a software device
DSP	Digital signal processor: long instruction word MP chip designed specifically for signal processing—high processing speed on a restricted architecture.	• very fast • low cost • highly flexible (some flexibility sacrificed for speed) • should be envisioned as a software device
FPGA	Field programmable gate array: random logic gate array with programmable linkages; may even be dynamically reprogrammable within the application. The latest devices have flip-flops and higher level functions already made up on chip, ready for linking in; some such devices even have one or more MPs on board.	• fast • low to medium cost • extremely flexible • should be envisioned as a hardware device, commonly slaved to a DSP • can be a software device if controlled by on-chip MPs
LUT	Lookup table: RAM or ROM. Useful for fast lookup of crucial functions. Normally slave to a MP or DSP.	• very fast • low cost • extremely flexible, if built using RAM • should be envisioned as a slave software device
ASIC	Application specific integrated circuit: contains devices such as Fourier Transforms, or a variety of specific SP or Vision functions. Normally slave to a MP or DSP.	• very fast • medium cost • inflexible (flexibility sacrificed for speed) • should be envisioned as a slave software device
Vision chip	Vision chips are ASICs that are devised specifically for vision: they may contain several important vision functions, such as edge detectors, thinning algorithms, and connected components analyzers. Normally slave to an MP or DSP.	• very fast • medium cost • inflexible (flexibility sacrificed for speed) • should be envisioned as a slave software device
VLSI	Custom VLSI chip: this commonly has many components from gate level upwards frozen into a fixed circuit with a particular functional application in mind. (Note, however, that the generic name includes MPs, DSPs, though we shall ignore this possibility here.)	• fast • high cost • inflexible • should be envisioned as a hardware device • normally slave to an MP or DSP.

The only high-cost item is the VLSI chip; the cost of producing the masks is only justifiable for high-volume products, such as those used in digital TV. In general, high cost means more than $15,000, medium cost means around $3000, and low cost means less than $150.

If there is a single winner in this table from the point of view of real-time applications, it is the FPGA, supposing only that it contains sufficient on-board raw computing power to make optimum use of its available random logic. Its dynamic reprogrammability is potentially extremely powerful, but it needs to be known how to make best use of it. It has been usual to slave FPGAs to DSPs or MPs,[1] but the picture changes radically for FPGAs containing on-chip MPs.

[1]In fact, the FPGA and the DSP complement each other exceptionally well, and it has been common practice to use them in tandem.

Table 28.4 Stages in the design of a typical inspection system

1. Hearing about the problem
2. Analyzing the situation
3. Looking at the data
4. Testing obvious algorithms
5. Realizing limitations
6. Developing algorithms further
7. Finding things are difficult and to some extent impossible
8. Doing theory to find the source of any limitations
9. Doing further tests
10. Getting an improved approach
11. Reassessing the specification
12. Deciding whether to go ahead
13. Completing a software system
14. Assessing the speed limitations
15. Starting again if necessary
16. Speeding up the software
17. Reaching a reasonable situation
18. Putting through 1000 images
19. Designing a hardware implementation
20. Revamping the software if necessary
21. Putting through another 100,000 images
22. Assessing difficulties regarding rare events
23. Assessing timing problems
24. Validating the final system

reveal something about the creativity aspects—by admitting that reassessments of the efficacy of algorithms may be needed. There are necessarily one or two feedback loops, through which the efficacy can be improved systematically, again and again, until operation is adequate, or indeed until the process is abandoned.[5] The underlying process appears to be:

```
create a basic scheme;
do
   improve current scheme;
   if time up then stop;
   if no further ideas then stop;
until an adequate system is obtained;
```

5 It is in the nature of things that you can't be totally sure whether a venture will be a success without trying it. Furthermore, in the hard world of industrial survival, part of adequacy means producing a working system and part means making a profit out of it. This section must be read in this light.

Table 28.5 Complexities of the design process

- It is not always evident that there is a solution, or at least a cost-effective one.
- Specifications cannot always be made in a nonfuzzy manner.
- There is often no rigorous scientific design procedure to get from specification to solution. (There is certainly no guaranteed way of achieving this optimally.)
- The optimization parameters are not obvious; nor are their relative priorities clear.
- It can be quite difficult to discern whether one solution is better than another.
- Some inspection environments make it difficult to tell whether or not a solution is valid.

The "if no further ideas" clause can be fulfilled if no way is found of making the system fast enough or low enough in cost, or of high enough specification. This would account for most contingencies that could arise.

The problem with the above process is that it represents ad hoc rather than scientific development, and there is no guarantee that the solution that is reached is optimal. Indeed, specification of the problem is not insisted upon (except to the level of adequacy), and if specification and aims are absent, it is impossible to judge whether or not the success that is obtained is optimal. In an engineering environment we ought to insist on problem specification first, solution second. However, things are more complex than this discussion would imply—as will be seen from Table 28.5.

The last case in Table 28.5 illustrates some types of faults[6] that are particularly rare, so that only one may arise in many millions of cases, or, equivalently, every few weeks or months. Thus there is little statistical basis for making judgments of the risk of failure, and there is no proper means of training the system so that it can learn to discriminate these faults. For these reasons, making a rigorous specification and systematically trying to meet it are extremely difficult, though it is still worth trying to do so.

Let us return to the first of the complexities listed in Table 28.5. Although in principle it is difficult to know whether there is a solution to a problem, nevertheless it is frequently possible to examine the computer image data that arise in the application and to see whether the eye can detect the faults or the foreign objects. If it can, this represents a significant step forward, for it means that it should be possible to devise a computer algorithm to do the same thing. What will then be in question will be whether we are creative enough to design such an algorithm and to ensure that it is sufficiently rapid and cost-effective to be useful.

6 An important class of rare faults is that of foreign objects, including the hard and sometimes soft contaminants targeted by X-ray inspection systems.

A key factor at this point is making an appropriate choice of design strategy: with structured types of image data. For example, we can ask the following:

1. Should boundary tracking be employed?
2. Should Hough transform line finding be used?
3. Should corner detection be used?

Which of these alternatives could, *with the particular dataset*, lead to algorithms of appropriate speed and robustness? On the other hand, for data that are fuzzy or where objects are ill-defined, would artificial neural networks be more useful than conventional programming? Overall, the types of data, *and* the types of noise and background clutter that accompany it, are key to the final choice of algorithm.

28.10 System Optimization

To proceed further, we need to examine the optimization parameters that are relevant. There are arguably rather few of these:

1. sensitivity
2. accuracy
3. robustness
4. adaptability
5. reliability
6. speed
7. cost

These one-word parameters are as yet rather imprecise. For example, reliability can mean a multitude of things, including freedom from mechanical failures such as might arise with the camera being shaken loose, or the illumination failing; or where timing problems occur when additional image clutter results in excessive analysis time, so that the computer can no longer keep up with the real-time flow of product. Several of the parameters have been shown to depend upon each other—for example sensitivity and accuracy, cost, and speed: more will be said about this matter in Chapter 29. Thus, we are working in a multidimensional space with various constraining curves and surfaces. In the worst case, all the parameters will be interlinked, corresponding to a single constraining surface, so that adjusting one parameter forces adjustment of at least

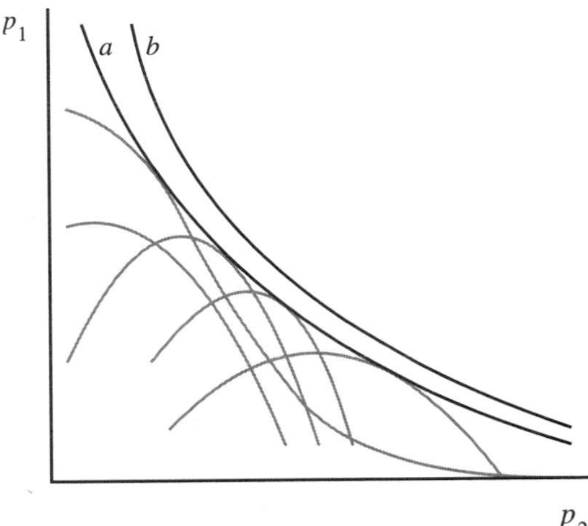

Figure 28.3 Optimization curves for a two-parameter system. The gray curves result from individual algorithms. (*a*) is the envelope of the curves for all known algorithms. (*b*) is the limiting curve for all possible algorithms.

one other. It is also possible that the constraining surface will impose hard limits on the values of some parameters.

Each algorithm will have its own constraining surface, which will in general be separate from that of others. Placing all such surfaces together will create some sort of envelope surface, corresponding to the limits of what is possible with *currently available* algorithms (Fig. 28.3). There will also be an envelope surface, corresponding to what is possible with *all possible* algorithms, that is, including those that have not yet been developed. Thus, there are limits imposed by creativity and ingenuity, and of course it is not known at what stage, if ever, such limits might be overcome.

Returning to the constraining surface, we can see that it provides an element of choice in the situation. Do we prefer a sensitive algorithm or a robust one, a reliable algorithm or a fast one? And so on. However, algorithms are rarely accurately describable as robust or nonrobust, reliable or nonreliable; the multiparameter space concept with its envelope constraining surfaces clearly allows for variations in such parameters. But at the present stage of development of the subject, the problem we have is that these constraining surfaces are, with few exceptions, unknown, and the algorithms that could provide access to their ideal forms are also largely unknown. Even the algorithms that are available are of largely unmapped capabilities. Also, means are not generally known for selecting optimal working points. (An interesting exception is that elucidated by Davies and Johnstone (1989) relating to optimization of cost/speed tradeoffs for image processing hardware.) Thus, there is a long way to go before the algorithm design and selection process can be made fully scientific.

28.11 The Value of Case Studies

In the absence of scientific know-how, we may be best advised to go for a case-law situation, in which we carefully define the regions where given algorithms are thought to be optimal. Thus, one case study will make one claim, and another case study will make another claim. If two algorithms overlap in their regions of applicability, then it may be possible to prove that the first is, in that region, superior to the second, so that the second can be dispensed with. However, methods usually have different regions of applicability so that the one is not a subset of the other, and then both methods have to be retained independently.

It more often happens that the regions of applicability are not so precisely known or specified. At this point, let us consider inherently simple situations where objects are to be located against blank backgrounds. The only interfering factors are noise and other objects of the same type. Even here, we must appreciate that the noise may be partly Gaussian, partly impulse, partly white, partly periodic, it may also vary from one part of the image to another; and it may vary in a complex way with the gray-scale quantization. In addition, the intensity range may be wide or narrow, compressed at one end or the other, subject to exponential (as in X-ray images) or other laws, and subject to shadows or glints as well as the type of object in question. Then the objects themselves may be separate, touching, overlapping, subject to differing variabilities—in size, shape, orientation, distortion, and so on. There may be a great many of these objects or just a few; they may be small, scattered randomly, arranged in piles, appear to form a texture, and may be unsegmentable in the images. Thus, even in this one "simple" example, the potential variability is so great that it may be extremely difficult to write down all the variables and parameters of the input data so that we know exactly when a given algorithm or suite of algorithms will give an optimal solution.

Small wonder, then, that workers often seem to be reinventing the wheel. They may have worked diligently trying to get a previously available algorithm to work with their particular dataset, only to discover some major deficiency. And then, further work will have shown an appropriate route to take, with yet another algorithm which is *a priori* apparently unnecessary. So much for the adaptability that was advocated earlier in this chapter—the matter is far less trivial than might have been imagined.

Exploring the situation further, we can find even low-level algorithms difficult to design and specify rigorously. The noise and intensity range problems referred to earlier still seem not to have been solved satisfactorily. We would be hard pressed to describe these problems quantitatively, and if we cannot do even this, how can we be confident of anything? In particular, if we knew that impulse and Gaussian noise were at specifiably low levels, we could confidently apply thresholding algorithms and then proceed to perform shape analysis by skeletonization techniques. Unfortunately, this degree of confidence may rarely be justifiable. Thus, the odd small hole may appear in a binary shape, thereby significantly altering its skeleton. This will then radically alter the range of solutions that may realistically be adopted for image analysis.[7]

How, then, are we to develop the subject properly and make it more scientific? We must adopt the strategy of gathering as much precise information as possible, mostly with regard to individual cases we will be working on. For each case study we should write out working profiles in as much detail as possible and publish a substantial proportion of the available data. It can't be emphasized enough that a lot of information should be provided defining the type of image data on which any algorithm has worked. By doing all this, we will at least be mapping out the constraint regions for each algorithm in the relevant parameter space in relation to the specific image data. Later, it should be possible to analyze all such mappings carefully. Important lessons will then be learned about when individual algorithms and techniques should properly be used. These ideas make some responses to the important statements in the earlier literature of the subject, which seem, unfortunately, to have yet had relatively little effect on most workers in the field (Jain and Binford, 1991; Haralick, 1992; Pavlidis, 1992). This lack of effect—especially with regard to Haralick's work—is probably due to the workers' awareness of the richness of the data and the algorithmic structures needed to cope rigorously with it, and in the face of such complexity, no one has quite seen how to proceed.

28.12 Concluding Remarks

This chapter has studied the means available for implementing image analysis and inspection algorithms in real time. Fast sequential and various parallel processing architectures have been considered, as well as other approaches

7 Intrinsically, a skeleton gives a good indication of the shape of an object. Furthermore, if one noise point produces an additional loop in a skeleton, this could possibly be coped with by subsequent analysis procedures. However, significant numbers of noise points will add so much additional complexity to a skeleton that a combinatorial explosion of possible solutions will occur, with the result that the skeleton will no longer be a useful representation of the underlying shape of the object.

involving use of DSP and FPGA chips, or even PCs with embedded operating systems. In addition, means of selecting between the various realizable schemes have been studied, these means being based on criterion function optimizations.

The field is characterized on the one hand by elegant parallel architectures that would cost an excessive amount in most inspection applications and on the other by hardware solutions that are optimized to the application and yet are bound to lack generality. It is easy to gain too rosy a view of the elegant parallel architectures that are available. Their power and generality could easily make them an overkill in dedicated applications, while reductions in generality could mean that their PEs spend much of their time waiting for data or performing null operations. Overall, the subject of image analysis is quite variegated and it is difficult to find a set of "typical" algorithms for which an "average" architecture can be designed whose PEs will always be kept usefully busy. Thus, the subject of mapping algorithms to architectures—or, better, of matching algorithms *and* architectures—that is, designing them together as a system—is a truly complex one for which obvious solutions turn out to be less efficient in reality than when initially envisioned. Perhaps the main problem lies in envisaging the nature of image analysis, which in the author's experience turns out to be a far more abstract process than image processing (i.e., image-to-image transformation) ideas would indicate (see especially Chapter 6).

Interestingly, the problems of envisaging the nature of image analysis and of matching hardware to algorithms are gradually being bypassed by the relentless advance in the speed and power of everyday computers. There will be less need for special dedicated hardware for real-time implementation, especially if suitable algorithms such as those outlined in Section 23.4 can be developed. All this represents a very welcome factor for those developing vision systems for industry.

One topic that has been omitted for reasons of space is that of carrying out sufficiently rigorous timing analysis of image analysis and related tasks so that the overall process is *guaranteed* to run in real time. In this context, a particular problem is that unusual image data conditions could arise that might engender additional processing, thereby setting the system behind schedule. This situation could arise because of excessive noise, because more than the usual number of products appear on a conveyor, because the heuristics in a search tree prove inadequate in some situation, or for a variety of other reasons. One way of eliminating this type of problem is to include a watchdog timer, so that the process is curtailed at a particular stage (with the product under inspection being rejected or other suitable action being taken). However, this type of solution is crude, and sophisticated timing analysis methods now exist, based on Quirk analysis. Môtus and Rodd are protagonists of this type of rigorous approach. The reader is referred to Thomas et al. (1995) for further details and to Môtus and Rodd (1994) for an in-depth study.

Finally, the last few sections have highlighted certain weaknesses in the system design process and indicated how the subject can be developed further for the benefit of industry. Perusal of the earlier chapters will quickly show that we can now achieve quite a lot even though our global design base is limited. The key to our ability to solve the problems, and to know that they are solvable, lies in the ability of the human visual system to carry out relevant visual tasks. Note, however, that there is sound reason for replacing the human eye for these visual tasks—so that (for example) 100% tireless inspection can be carried out and so that it can be achieved consistently and reliably to known standards.

The greatest proportion of the real-time vision system design process used to be devoted to the development of dedicated hardware accelerators. This chapter has emphasized that this problem is gradually receding, and that it is already possible to design reconfigurable FPGA systems that in many cases bypass the need for any special parallel processor systems.

28.13 Bibliographical and Historical Notes[8]

28.13.1 *General Background*

The problem of implementing vision in real time presents exceptional difficulties because in many cases megabytes of data have to be handled in times of the order of hundredths of a second, and often repeated scrutiny of the data is necessary. Probably it is no exaggeration to say that hundreds of architectures have been considered for this purpose, all having some degree of parallelism. Of these, many have highly structured forms of parallelism, such as SIMD machines. The idea of an SIMD machine for image parallel operations goes back to Unger (1958) and strongly influenced later work both

8 Because of the rapid ageing process that is particularly evident for computer and hardware development, it has been found necessary to divide this section into two parts "General Background" (including early work) and "Recent Highly Relevant Work."

in the United Kingdom and the United States. For an overview of early SIMD machines, see Fountain (1987).

The SIMD concept was generalized by Flynn (1972) in his well-known classification of computing machines (Section 28.5), and several attempts have been made to devise a more descriptive and useful classification (see Hockney and Jesshope, 1981), although so far none has really caught on (however, see Skillicorn, 1988, for an interesting attempt). Arguably, this is because it is all too easy for structures to become too lacking in generality to be useful for a wide range of algorithmic tasks. There has been a great proliferation of interconnection networks (see Feng, 1981). Reeves (1984) reviewed a number of such schemes that might be useful for image processing. An important problem is that of performing an optimal mapping between algorithms and architectures, but for this to be possible a classification of parallel algorithms is needed to complement that of architectures. Kung (1980) made a start with this task. This work was extended by Cantoni and Levialdi (1983) and Chiang and Fu (1983).

For one period in the 1980s it became an important challenge to design VLSI chips for image processing and analysis. Offen (1985) summarized many of the possibilities, and this work was followed by Fountain's (1987) volume. The problems targeted in that era were defining the most useful image processing functions to encapsulate in VLSI and to understand how best to partition the algorithms, taking account of chip limitations. At a more down-to-earth level, many image processing and analysis systems did not employ sophisticated parallel architectures but "merely" very fast serial processing techniques with hard-wired functions—often capable of processing at video rates (i.e., giving results within one TV frame time).

During the late 1980s, the Transputer provided the system implementer with a powerful processor with good communication facilities (four bidirectional links each with the capability for independent communication at rates of 10 Mbits per second *and* operation independent of the CPU), enabling him or her to build SIMD and MIMD machines on a do-it-yourself basis. The T414 device had a 32-bit processor operating at 10 MIPS with a 5 MHz clock and 2 kB of internal memory, while the T800 device had 4 kB of internal memory and a floating-point processor (Inmos, 1987), yet it cost only around $300. These devices were incorporated in a number of commercially available board-level products plus a few complete rather powerful systems; vision researchers developed parallel algorithms for them, using mainly the Occam parallel processing language (Inmos 1988; Pountain and May, 1987).

Other solutions included the use of bit-slice types of device (Edmonds and Davies, 1991), and DSP chips (Davies et al., 1995), while special-purpose multiprocessor designs were described for implementing multiresolution and other Hough transforms (Atiquzzaman, 1994).

28.13.2 ***Recent Highly Relevant Work***

As indicated in Section 28.7, VLSI solutions to the production of rapid hardware for real-time applications have gradually given way to the much more flexible software solutions permitted by DSP chips and to the highly flexible FPGA type of system that permits random logic to be implemented with relative ease. FPGAs offer the possibility of dynamic reconfigurability, which may ultimately be very useful for space probes and the like, though it may prove to be an unnecessary design burden for most vision applications. Better instead to keep on upgrading a design and extending the number of FPGA chips until the required degree of sophistication and speed has been achieved.[9] In this respect, it is interesting to note that the current trend is toward hybrid CPU/FPGA chips (Andrews et al., 2004), which will have optimal combinations of software and random logic available awaiting software and hardware programming (see also Batlle et al., 2002).

Be this as it may, in the early 2000s considerable attention was still focused on VLSI solutions to vision applications (Tzionas, 2000; Mémin and Risset, 2001; Wiehler et al., 2001; Urriza et al., 2001), and it is clear that there are bound to be high-volume applications (such as digital television) where this will remain the best approach. Similarly, a lot of attention is still focused on SIMD and on linear array processor schemes, as there will always be facilities that offer ultra-high-speed solutions of this type. In any case, small-scale implementations are likely to be cheap and usable for those (mainly low-level vision) applications that match this type of architecture. Examples appear in the papers by Hufnagl and Uhl (2000), Ouerhani and Hügli (2003), and Rabah et al. (2003).

Moving on to FPGA solutions, we find these embodying several types of parallelism and applied to underwater vision applications and robotics (Batlle et al., 2002), subpixel edge detection for inspection (Hussmann and Ho, 2003), and general low-level vision applications, including those implementable as morphological operators (Draper et al., 2003).

DSP solutions, some of which also involve FPGAs, include those by Meribout et al. (2002) and Aziz et al. (2003). It is useful to make comparisons with the Datacube MaxPCI (containing a pipeline of convolvers, histogrammers, and other devices) and other commercially available boards and systems—see Broggi et al. (2000a,b), Yang et al. (2002), and Marino et al. (2001). The Marino et al. paper is exceptional in using a clever architecture incorporating extensive use of lookup tables to perform high-speed matching functions that lead to road following. The matching functions involve location of interest points, followed by

9 However, see Kessal et al. (2003) for a highly interesting investigation of the possibilities—in particular showing that dynamic reconfigurability of a real-time vision system can already be achieved in milliseconds.

high-speed search for matching them between corresponding blocks of adjacent frames. This should be regarded as an application of quite simple low-level operations that *enable* high-level operations to take place quickly. An interesting feature is the use of a residue number system to (effectively) factorize large lookup tables into several much smaller and more manageable lookup tables.

PART 5

Perspectives on Vision

Part 5 aims to draw together and broaden the topics of the previous chapters. In particular, attention is paid to systematizing the subject of machine vision and to summarizing the tradeoffs among the parameters that are relevant in vision algorithms. In addition, the significance and ultimate value of the Hough transform technique is discussed. Appendix A is highly relevant to providing a sound perspective on the subject, for it brings into relief the essential dichotomy between interpretation and accurate measurement (or between robustness "breakdown point" and measurement "efficiency"). It can probably be said without risk of exaggeration that all practical machine vision systems need to incorporate robust measurement in one way or another—whether, on the one hand, making actual measurements on machined parts or, on the other, arriving at robust interpretations of 3-D pose and motion.

CHAPTER 29

Machine Vision: Art or Science?

29.1 Introduction

We have examined how images may be processed to remove noise, how features may be detected, how objects may be located from their features, how to set up lighting schemes, how to design hardware systems for automated visual inspection, and so on. The subject is one that has developed over a period of more than 40 years, and it has clearly come a long way, but it has developed piecemeal rather than systematically. Often, development is motivated by the particular interests of small groups of workers and is relatively ad hoc. Coupled with this is the fact that algorithms, processes, and techniques are all limited by the creativity of the various researchers. The process of design tends to be intuitive rather than systematic, and so again some arbitrariness tends to creep in from time to time. As a result, no means has yet been devised for achieving particular aims, but a number of imperfect methods are available and there is a limited scientific basis for choosing between them.

How then can the subject be placed on a firmer foundation? Time may help, but time can also make things more difficult as more methods and results arise which have to be considered. In any case, there is no shortcut to intellectual analysis of the state of the art. This book has attempted to carry out a degree of analysis at every stage, but in this last chapter it is important to tie it all together, to make some general statements on methodology, and to indicate future directions.

Machine vision is an engineering discipline, and, like all such disciplines, it has to be based on science and an understanding of fundamental processes. However, as an engineering discipline it should involve design based on specifications. Once the specifications for a vision system have been laid down, it can be seen how they match up against the constraints provided by nature and technology. In the following we consider first the parameters of relevance for the specification of vision systems; then we examine constraints and their origins. This leads us to some clues as to how the subject could be developed further.

29.2 Parameters of Importance in Machine Vision

The first requirement of any engineering design is that it should work! This applies as much to vision systems as to other parts of engineering. There is no use in devising edge detectors that do not find edges, corner detectors that do not find corners, thinning algorithms that do not thin, 3-D object detection schemes that do not find objects, and so on. But in what way could such schemes fail? First, ignore the possibility of noise or artifacts preventing algorithms from operating properly. Then what remains is the possibility that at any stage important fundamental factors have not been taken into account.

For example, a boundary tracking algorithm can go wrong because it encounters a part of the boundary that is one pixel wide and crosses over instead of continuing. A thinning algorithm can go wrong because every possible local pattern has not been taken into account in the design, and hence it disconnects a skeleton. A 3-D object detection scheme can go wrong because proper checks have not been made to confirm that a set of observed features is not coplanar. Of course, these types of problems may arise very rarely (i.e., only with highly specific types of input data), which is why the design error is not noticed for a time. Often, mathematics or enumeration of possibilities can help to eliminate such errors, so problems can be removed systematically. However, being absolutely sure no error has been made is difficult—and it must not be forgotten that transcription errors in computer programs can contribute to the problems. These factors mean that algorithms should be put to extensive tests with large datasets in order to ensure that they are correct. In the author's experience, there is no substitute for subjecting algorithms to variegated tests of this type to check out ideas that are "evidently" correct. Although all this is obvious, it is still worth stating, since silly errors continually arise in practice.

At this stage imagine that we have a range of algorithms that all achieve the same results on ideal data and that they really work. The next problem is to compare them critically and, in particular, to find how they react to *real* data and the nasty realities such as noise that accompany it. These nasty realities may be summed up as follows:

1. noise
2. background clutter
3. occlusions

4. object defects and breakages
5. optical and perspective distortions
6. nonuniform lighting and its consequences
7. effects of stray light, shadows, and glints

In general, algorithms need to be sufficiently robust to overcome these problems. However, things are not so simple in practice. For example, HT and many other algorithms are capable of operating properly and detecting objects or features despite considerable degrees of occlusion. But how much occlusion is permissible? Or how much distortion and noise, or how much of any other nasty reality can be tolerated? In each specific case we could state some figures that would cover the possibilities. For example, we may be able to state that a line detection algorithm must be able to tolerate 50% occlusion, and so a particular HT implementation is (or is not) able to achieve this. However, at this stage we end with a lot of numbers that may mean very little on their own; in particular, they seem different and incompatible. This latter problem can largely be eliminated. Each of the defects can be imagined to obliterate a definite proportion of the object. (In the case of impulse noise this is obvious; with Gaussian noise the equivalence is not so clear, but we suppose here that an equivalence can at least in principle be computed.) Hence, we end up by establishing that artifacts in a particular dataset eliminate a certain proportion of the area and perimeter of all objects, or a certain proportion of all small objects.[1] This is a sufficiently clear statement to proceed with the next stage of analysis.

To go further, it is necessary to set up a complete specification for the design of a particular vision algorithm. The specification can be listed as follows (but generality is maintained by not stating any particular algorithmic function):

1. The algorithm must work on ideal data.
2. The algorithm must work on data that is x% corrupted by artifacts.
3. The algorithm must work to p pixels accuracy.
4. The algorithm must operate within s seconds.
5. The algorithm must be trainable.
6. The algorithm must be implemented with failure rate less than 1 per d days.
7. The hardware needed to implement the algorithm must cost less than $\$L$.

No doubt other specifications will arise in practice, but these are ignored here.

1 Certain of the nasty realities (such as optical distortions) tend to act in such a way as to reduce accuracy, but we concentrate here on robustness of object detection.

Before proceeding, note the following points: (a) specification 2 is a statement about the robustness of algorithms, as explained earlier; (b) the accuracy of p pixels in specification 3 may well be fractional; (c) in specification 5, trainability is a characteristic of certain types of algorithms, and in any instance we can only accept the fact that it is or is not achievable—hence, it is not discussed further here; (d) the failure rate referred to in specification 6 will be taken to be a hardware characteristic and is ignored in what follows.

The forgoing set of specifications may at any stage of technological (especially hardware) development be unachievable. This is because they are phrased in a particular way, so they are not compromisable. However, if a given specification is approaching its limit of achievability, a switch to an alternative algorithm might be possible.[2] Alternatively, an internal parameter might be adjusted, which keeps that specification within range while pushing another specification closer to the limits of its range. In general, there will be some hard (nonnegotiable) specifications and others for which a degree of compromise is acceptable. As has been seen in various chapters of the book, this leads to the possibility of tradeoffs—a topic that is reviewed in the next section.

29.3 Tradeoffs

Tradeoffs form one of the most important features of algorithms, because they permit a degree of flexibility subject only to what is possible in the nature of things. Ideally, the tradeoffs that are enunciated by theory provide absolute statements about what is possible, so that if an algorithm approaches these limits it is then probably as "good" as it can possibly be.

Next, there is the problem about where on a tradeoff curve an algorithm should be made to operate. The type of situation was examined carefully in Chapter 28 in a particular context—that of cost–speed tradeoffs of inspection hardware. Generally, the tradeoff curve (or surface) is bounded by hard limits. However, once it has been established that the optimum working point is somewhere within these limits, in a continuum, then it is appropriate to select a criterion function whereby an optimum can be located uniquely. Details will vary from case to case, but the crucial point is that an optimum must exist on a tradeoff curve, and that it can be found systematically once the curve is known. All this implies that the science of the situation has been studied sufficiently so that relevant tradeoffs have been determined.

2 But note that several, or all, relevant algorithms may be subject to almost identical limitations because of underlying technological or natural constraints.

29.3.1 *Some Important Tradeoffs*

Earlier chapters of this book have revealed some quite important tradeoffs that are more than just arbitrary relations between relevant parameters. Here, a few examples will have to suffice by way of summary.

First, in Chapter 5 the DG edge operators were found to have only one underlying design parameter—that of operator radius r. Ignoring here the important matter of the effect of a discrete lattice in giving preferred values of r, it was found that:

1. signal-to-noise ratio varies linearly with r, because of underlying signal and noise averaging effects.
2. resolution varies inversely with r, since relevant linear features in the image are averaged over the active area of the neighborhood: the scale at which edge positions are measured is given by the resolution.
3. the accuracy with which edge position (at the current scale) may be measured depends on the square root of the number of pixels in the neighborhood, and hence varies as r.
4. computational load, and associated hardware cost, is proportional to the number of pixels in the neighborhood, and hence vary as r^2.

Thus, operator radius carries with it four properties that are intimately related—signal-to-noise ratio, resolution (or scale), accuracy, and hardware/computational cost.

Another important problem was that of fast location of circle centers (Chapter 10): in this case, robustness was seen to be measurable as the amount of noise or signal distortion that can be tolerated. For HT-based schemes, noise, occlusions, distortions, and the like all reduce the peak height in parameter space, thereby reducing the signal-to-noise ratio and impairing accuracy. Furthermore, if a fraction β of the original signal is removed, leaving a fraction $\gamma = 1 - \beta$, either by such distortions or occlusions or else by deliberate sampling procedures, then the number of independent measurements of the center location drops to a fraction γ of the optimum. This means that the accuracy of estimation of the center location drops to a fraction around $\sqrt{\gamma}$ of the optimum. (Note, however, that Gaussian noise can act in another way to reduce accuracy—by broadening the peak instead of merely reducing its height—but this effect is ignored here.)

What is important is that the effect of sampling is substantially the same as that of signal distortion, so that the more distortion that must be tolerated, the higher α, the fraction of the total signal sampled, has to be. This means that as the level of distortion increases, the capability for withstanding sampling decreases, and therefore the gains in speed achievable from sampling are reduced—that is, for fixed signal-to-noise ratio and accuracy, a definite robustness–speed tradeoff exists. Alternatively, the situation can be viewed as a three-way relation between accuracy, robustness, and speed of processing. This provides an interesting insight into how the edge operator tradeoff considered earlier might be generalized.

To underline the value of studying such tradeoffs, note that any given algorithm will have a particular set of adjustable parameters which are found to control—and hence lead to tradeoffs between—the important quantities such as speed of processing, signal-to-noise ratio, and attainable accuracy already mentioned. Ultimately, such *practically* realizable tradeoffs (i.e., arising from the given algorithm) should be considered against those that may be deduced on purely theoretical grounds. Such considerations would then indicate whether a better algorithm might exist than the one currently being examined.

29.3.2 *Tradeoffs for Two-stage Template Matching*

Two-stage template matching has been mentioned a number of times in this book as a means whereby the normally slow and computationally intensive process of template matching may be speeded up. In general, it involves looking for easily distinguishable subfeatures, so that locating the features that are ultimately being sought involves only the minor problem of eliminating false alarms. This strategy is useful because the first stage eliminates the bulk of the raw image data, so that only a relatively trivial testing process remains. This latter process can then be made as rigorous as necessary. In contrast, the first "skimming" stage can be relatively crude (see, e.g., the lateral histogram approach of Chapter 13), the main criterion being that it must not eliminate any of the desired features. False positives are permitted but not false negatives. Needless to say, the efficiency of the overall two-stage process is limited by the number of false alarms thrown up by the first stage.

Suppose that the first stage is subject to a threshold h_1 and the second stage to a threshold h_2. If h_1 is set very low, then the process reverts to the normal template matching situation, since the first stage does not eliminate any part of the image. Setting $h_1 = 0$ initially is useful so that h_2 may be adjusted to its normal working value. Then h_1 can be increased to improve efficiency (reduce overall computation); a natural limit arises when false negatives start to occur—that is, some of the desired features are not being located. Further increases in h_1 now have the effect of cutting down available signal, although speed continues to increase. This gives a tradeoff between signal-to-noise ratio, and hence accuracy of location and speed.

In a particular application in which objects were being located by the HT, the number of edge points located were reduced as h_1 increased, so accuracy of object location was reduced (Davies, 1988i). A criterion function approach was then used to determine an optimum working condition. A suitable criterion function turned out to be $C = T/A$, where T is the total execution time and A the achievable accuracy. Although this approach gave a useful optimum, the optimum can be improved further if a mix of normal two-stage template matching and random sampling is used. This turns the problem into a 2-D optimization problem with adjustable parameters h_1 and u (the random sampling coefficient, equal to $1/\alpha$). However, in reality these types of problems are even more complex than indicated so far. In general, this is a 3-D optimization problem, the relevant parameters being h_1, h_2 and u, although in fact a good approximation to the global optimum may be obtained by the procedure of adjusting h_2 first, and then optimizing h_1 and u together—or even of adjusting h_2 first, then h_1 and then u (Davies, 1988i). Further details are beyond the scope of the present discussion.

29.4 Future Directions

It has been indicated once or twice that the constraints and tradeoffs limiting algorithms are sometimes not accidental but rather the result of underlying technological or natural constraints. If so, it is important to determine this in as many cases as possible; otherwise, workers may spend much time on algorithm development only to find their efforts repeatedly being thwarted. Usually, this is more easily said than done, but it underlines the necessity for scientific analysis of fundamentals. At this point we should consider technological constraints in order to see what possibilities exist for advancing machine vision dramatically in the future.

The well-known law due to Moore (Noyce, 1977) relating to computer hardware states that the number of components that can be incorporated into a single integrated circuit increases by a factor of about two per year. Certainly, this was so for the 20 years following 1959, although the rate subsequently decreased somewhat (not enough, however, to prevent the growth from remaining approximately exponential). It is not the purpose of this chapter to speculate on the accuracy of Moore's law. However, it is useful to suppose that computer memory and power will grow by a factor approaching two per year in the foreseeable future. Similarly, computer speeds may also grow at roughly this rate in the foreseeable future. When then of vision?

Unfortunately, many vision processes such as search are inherently NP-complete and hence demand computation that grows exponentially with some internal parameter such as the number of nodes in a match graph. This means that the advance of technology is able to give only a roughly linear improvement in this internal parameter (e.g., something like one extra node in a match graph every two years). It is therefore not solving the major search and other problems but only easing them.

NP-completeness apart, Chapter 28 was able to give an optimistic view that the relentless advance of computer power described by Moore's law is now leading to an era when conventional PCs will be able to cope with a fair proportion of vision tasks. When combined with specially designed algorithms (see Section 23.4), it should prove possible to implement many of the simpler tasks in this way, leading to a much less strenuous life for the vision systems designer.

29.5 Hardware, Algorithms, and Processes

The previous section raised the hope that improvements in hardware systems will provide the key to the development of impressive vision capabilities. However, it seems likely that breakthroughs in vision algorithms will also be required before this can come about. My belief is that until robots can play with objects and materials in the way that tiny children do they will not be able to build up sufficient information and the necessary databases for handling the complexities of real vision. The real world is too complex for all the rules to be written down overtly; these rules have to be internalized by training each brain individually. In some ways this approach is better, since it is more flexible and adaptable and at the same time more likely to be able to correct for the errors that would arise in direct transference of huge databases or programs. Nor should it be forgotten that it is the underlying processes of vision and intelligence that are important. Hardware merely provides a means of implementation. If an idea is devised for a hardware solution to a visual problem, it reflects an underlying algorithmic process that either is or is not effective. Once it is known to be effective, then the hardware implementation can be analyzed to confirm its utility. Of course, we must not segregate algorithms too much from hardware design. In the end, it is necessary to optimize the whole system, which means considering both together. However, the underlying processes should be considered first, before a hardware solution is frozen in. Hardware should not be the driving force since there is a danger that some type of hardware implementation (especially one that is temporarily new and promising) will take over and make workers blind to underlying processes. Hardware should not be the tail that wags the vision dog.

29.6 A Retrospective View

This book has progressed steadily from low-level ideas through intermediate-level methods to high-level processing, covering 3-D image analysis, the necessary technology, and so on—admittedly with its own type of detailed examples and emphasis. Many ideas have been covered, and many strategies described. But where have we got to, and to what extent have we solved the problems of vision referred to in Chapter 1?

Among the worst of all the problems of vision is that of minimizing the amount of processing required to achieve particular image recognition and measurement tasks. Not only do images contain huge amounts of data, but often they need to be interpreted in frighteningly small amounts of time, and the underlying search and other tasks tend to be subject to combinatorial explosions. Yet, in retrospect, we seem to have remarkably few *general* tools for coping with these problems. Indeed, the truly general tools available[3] appear to be the:

1. reducing high-dimensional problems to lower-dimensional problems that can be solved in turn.
2. the Hough transform and other indexing techniques.
3. location of features that are in some sense sparse, and that can hence help to reduce redundancy quickly (obvious examples of such features are edges and corners).
4. two-stage and multistage template matching.
5. random sampling.

These are said to be general tools because they appear in one guise or another in a number of situations, with totally different data. However, it is pertinent to ask to what extent these are genuine tools rather than almost accidental means (or tricks) by which computation may be reduced. Further analysis yields interesting answers to this question, as we will now see.

First, consider the Hough transform, which takes a variety of forms—the normal parametrization of a line in an abstract parameter space, the GHT, which is parametrized in a space congruent to image space, the adaptive thresholding transform (Chapter 4), which is parametrized in an abstract 2-D parameter space, and so on. What is common about these forms is *the choice of a representation in which the data peak naturally at various points*, so that analysis can proceed with improved efficiency. The relation with item (3) above now

3 We here consider only intermediate-level processing, ignoring, for example, efficient AI tree-search methods relevant for purely abstract high-level processing.

becomes clear, making it less likely that either of these procedures is purely accidental in nature.

Next, item (1) appears in many guises—see, for example, the lateral histogram approach (Chapter 13) and the approaches used to locate ellipses (Chapter 12). Thus, item (1) has much in common with item (4). Note also that item (5) can be considered a special case of item (4). (Random sampling is a form of two-stage template matching with a "null" first stage, capable of eliminating large numbers of input patterns with particularly high efficiency; see Davies, 1988i.) Finally, note that the example of so-called two-stage template matching covered in Section 29.3.1 was actually part of a larger problem that was really multistage. The edge detector was two-stage, but this was incorporated in an HT, which was itself two-stage, making the whole problem at least four-stage. It can now be seen that items (1)–(5) are all forms of multistage matching (or sequential pattern recognition), which are potentially more powerful and efficient than a single-stage approach. Similar conclusions are arrived at in Appendix A, which deals with robust statistics and their application to machine vision.

Hence, we are coming to a view that there is just one general tool for increasing efficiency. However, in practical terms this may not itself be too useful a conclusion, since the subject of image analysis is also concerned with the ways in which this underlying idea may actually be realized—how are complex tasks to be broken down into the most appropriate multistage processes, and how then is the most suitable representation found for sparse feature location? Probably the five-point list given above throws the right sort of light onto this particular problem—although in the long run it will hardly turn out to be the whole truth!

29.7 **Just a Glimpse of Vision?**

This book has solved some of the problems it set itself—starting with low-level processing, concentrating on strategies, limitations, and optimizations of intermediate-level processing, going some way with higher-level tasks, and attempting to create an awareness of the underlying processes of vision. Yet this program has taken a lot of space and has thereby had to omit many exciting developments. Notable omissions are advanced work on artificial intelligence and knowledge-based systems and their application in vision; details of much important work on motion and image sequence analysis; practical details of the hardware architectures and techniques; detailed theoretical studies of neural computation, which offer many possibilities for marrying algorithms and architectures; and the huge variety of modern applications in areas such as crime detection and prevention, face recognition and biometrics, and so on. While separate volumes are available on most of these topics, until now a gap appeared to be present, which this book has aimed to fill.

In any field, it is tempting to say that all the main problems have been solved and that only details remain. Indeed, this chapter may inadvertently have implied that only the tradeoffs between optimization parameters need to be determined and the best answers read off. However, as indicated in Chapter 28, there are deeper problems still to be solved—such as finding out how to make valid specifications for image data that are naturally so variegated and varied that setting a detailed specification appears just now to be virtually impossible.

29.8 Bibliographical and Historical Notes

Much of this chapter has summarized the work of earlier chapters and attempted to give it some perspective so the reader is referred back to appropriate sections for further detail. However, two-stage template matching has been highlighted in the current chapter. The earliest work on this topic was carried out by Rosenfeld and VanderBrug (1977) and VanderBrug and Rosenfeld (1977). Later work included that on lateral histograms for corner detection by Wu and Rosenfeld (1983), while the ideas of Section 29.3.2 were developed by Davies (1988i). Two-stage template matching harks back to the spatial matched filtering concept discussed in Chapter 11 and elsewhere. Ultimately, this concept is limited by the variability of the objects to be detected. However, it has been shown that some account can be taken of this problem, for example, in the design of filter masks (see Davies, 1992e).

With regard to the topics that could not be covered in this book (see the previous section), the reader should refer to the bibliographies at the ends of the individual chapters—though this leaves notable omissions on highly relevant AI techniques. Here the reader is referred to Russell and Norvig (2003).

Robust Statistics

At an early stage, science students learn that averaging is an effective way of eliminating noise and improving accuracy. However, Chapter 3 demonstrated unequivocally that median filtering of images is far better than mean filtering, both in retaining the form of the underlying signal and in suppressing impulse noise. Robust statistics is the subject of systematically eliminating outliers from visual or other data. This appendix seeks to give useful insights into this important subject.

Look out for:

- the concepts *breakdown point* and *relative efficiency*.
- M-, R-, and L-estimators.
- the idea of an influence function.
- the least median of squares (LMedS) approach.
- the RANSAC approach.
- the ways these methods can be applied in machine vision.

Although robust statistics is a relatively young discipline, dating largely from the 1980s, it has acquired a considerable following in machine vision and is crucial, for example, in the development of robust 3-D vision algorithms. A basic problem to be tackled is the impossibility of knowing how much of the input data is in the form of outliers.

Appendix A

Robust Statistics

A.1 Introduction

Frequently in this volume we have found that noise can interfere with image signals and result in inaccurate measurements—for example, of object shapes, sizes, and positions. Perhaps more important, however, is the fact that signals other than the particular one being focused upon can lead to gross shape distortions and can thus prevent an object from being recognized or even being discerned at all. In many cases, this will render some obvious interpretation algorithm useless, though algorithms with intrinsic "intelligence" may be able to save the day. For this reason the Hough transform has achieved some prominence. Indeed, this approach to image interpretation has frequently been described as robust, though no rigorous definition of robustness has been attempted so far in this volume. This appendix aims to throw further light on the problem.

Research into robustness did not originate in machine vision but evolved as the specialist area of statistics now known as *robust statistics*. Perhaps the paradigm problem in this area is that of fitting a straight line to a set of points. In the physics laboratory, least squares analysis is commonly used to tackle the task. Figure A.1*a* shows a straightforward situation where all the data points can be fitted with a reasonably uniform degree of exactness, in the sense that the residual errors[1] approximate to the expected Gaussian distribution. Figure A.1*b* shows a less straightforward case, where a particular data point seems not to fall within a Gaussian distribution. Intuition tells us that this particular point represents data that have become corrupted in some way, for example, by misreading an instrument or through a transcription error. Although the wings of a Gaussian distribution stretch out to infinity, the probability that a point will be more than five standard deviations from the center of the distribution is very small,

1 The residual errors or *residuals* are the deviations between the observed values and the theoretical predictions of the current model or current iteration of that model.

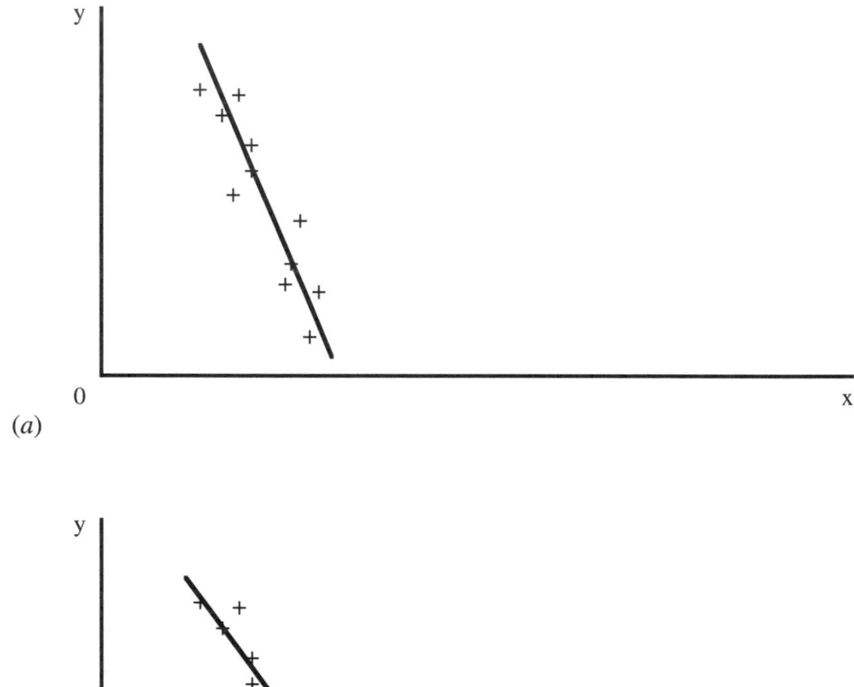

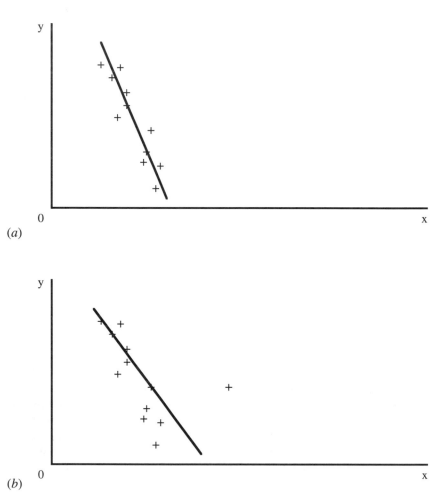

Figure A.1 Fitting of data points to straight lines. (*a*) shows a straightforward situation where all the data points can be fitted with reasonable precision; (*b*) shows a less straightforward case, where a particular data point seems not to fall within a Gaussian distribution; (*c*) shows a situation where the correct solution has been ignored by the numerical analysis procedure; (*d*) shows a situation where there are many rogue points, and it is not clear which points lie on the straight line and which do not. In such cases it may not be known whether there are several, or any, lines to be fitted.

and indeed, $\pm 3\sigma$ limits are commonly taken as demarcating practical limits of correctness. It is taken as reasonable to disregard data points lying outside this range.

Unfortunately, the situation can be much worse than this simple example suggests. For suppose there is a rogue data point that is a very long way off. In least-squares analysis it will have such a large leverage that the correct solution

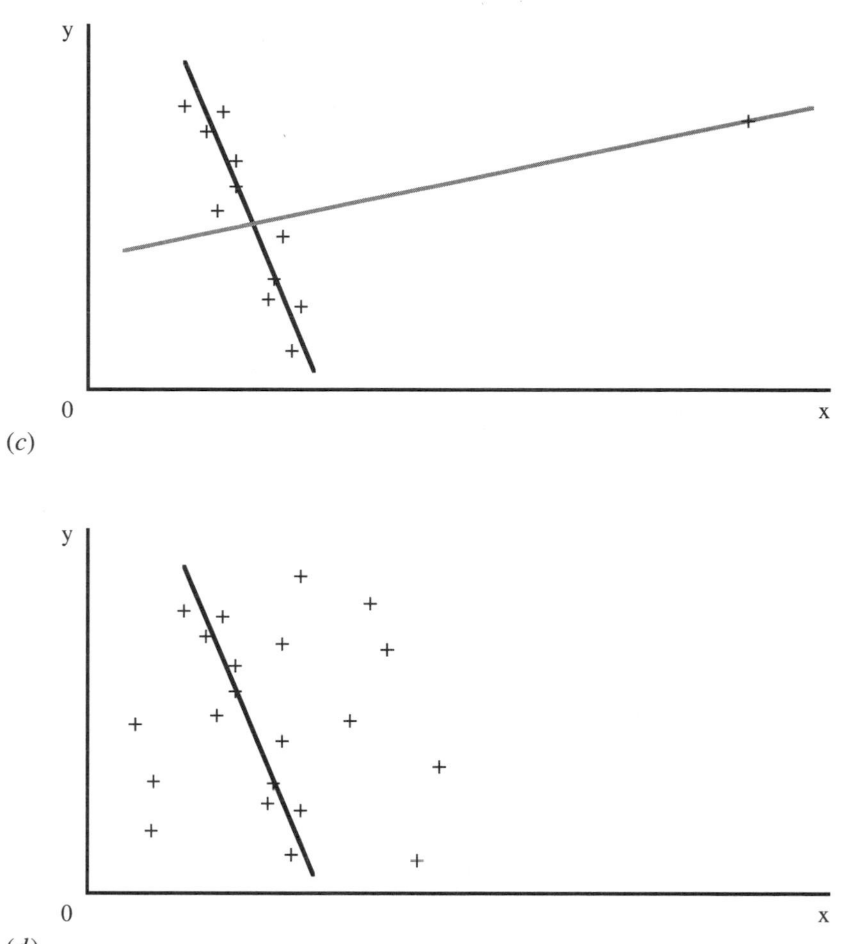

Figure A.1 Continued.

may not be found. And if the correct solution is not found, there will be no basis for excluding the rogue data point. This situation is illustrated in Fig. A.1*c*, where the obviously correct solution has been ignored by the numerical analysis procedure.

A worse case of line fitting occurs when there are many rogue points, and it is not clear which points lie on the straight line and which do not (Fig. A.1*d*). It may not be known whether there are several lines to be fitted, or whether there are *any* lines to be fitted. Whereas this circumstance would appear not to occur while data points are being plotted in physics experiments, it can arise when high-energy particles are being tracked. It also occurs frequently in images of indoor and outdoor scenes where a myriad of straight lines of various lengths can appear in a great many orientations and positions. Thus, it is a real problem for which answers are required. An attempt at a full statement for this type of problem might be: devise a means for finding all the straight lines—of whatever length—in a generalized[2] image, so as to obtain the best overall fit to the dataset. Unfortunately, there are likely to be many solutions to any line-fitting task, particularly if the data points are not especially accurate. (If they are highly accurate, then the number of solutions will be small, and it should be easy to decide intuitively or automatically what the best solution is.) A rigorous answer to the question of which solution provides the best fit requires the definition of a criterion function that in some way takes account of the number of lines and the *a priori* length distribution. We shall not pursue this line of attack here, for the purpose of this appendix is to give a basic account of the subject of robust statistics, not one that is tied to a particular task. Hence, we shall return to the simpler case where there is only one line present in the generalized image, and there are a substantial number of rogue data points or outliers present.

A.2 Preliminary Definitions and Analysis

In the previous section we saw that robustness is an important factor in deciding on a scheme for fitting experimental data to numerical models. It is clearly important to have an exact measure of robustness, and the concept of a *breakdown point* long ago emerged as such a measure. The breakdown point ε of a regression scheme is defined as the smallest proportion of outlier contamination that may force the value of the estimate to exceed an arbitrary range. As we have seen, even a single outlier in a set of plots can cause least squares regression to give completely erroneous results. However, a much simpler example is to hand, namely, a 1-D distribution for which the mean is computed. Here again, a single outlier can cause the mean to exceed any stated bound. This means that the breakdown point for the mean must be zero. On the other hand, the median of a distribution is well known to be highly robust to outliers and remains unchanged if nearly half the data is corrupted. Specifically, for a set of n data points, the median will remain unchanged if the lowest $\lfloor n/2 \rfloor$ points[3] are moved to arbitrary lower values, or the highest $\lfloor n/2 \rfloor$ points are moved to arbitrary higher values, but in either case the median value will be changed to an arbitrary value if $\lfloor n/2 \rfloor + 1$ points are so moved. By definition (see above), this means that the breakdown point of the median is $(\lfloor n/2 \rfloor + 1)/n$. This value should be compared

2 That is, an image that might correspond to off-camera images, or to situations such as data points being plotted on a graph.

3 The function $\lfloor \cdot \rfloor$ denotes the "floor" (rounding down) operation and indicates the largest integer less than or equal to the enclosed value. In the present case we have $\lfloor n/2 \rfloor \leq n/2 < \lfloor n/2 \rfloor + 1$.

Table A.1 Breakdown points for means and medians

n	mean	median
1	1	1
3	1/3	2/3
5	1/5	3/5
11	1/11	6/11
∞	0	0.5

This table shows how the respective breakdown points for the mean and median approach 0 and 0.5 as n tends to infinity, in the case of 1-D data.

with the value $1/n$ for the mean. In the case of the median, the breakdown point approaches 0.5 as n tends to infinity (see Table A.1). Thus, the median attains the apparently maximum achievable breakdown point of 0.5 and is therefore optimal—at least in the 1-D case described in this paragraph.

The breakdown point is not the only relevant parameter for characterizing regression schemes. For example, the *relative efficiency* is also important and is defined as the ratio between the lowest achievable variance and the actual variance achieved by the regression method. It turns out that the relative efficiency depends on the particular noise distribution to which the data are subject. It can be shown that the mean is optimal for elimination of Gaussian noise, having a relative efficiency of unity, while the median has a relative efficiency of $2/\pi = 0.637$. However, when dealing with impulse noise, the median has a higher relative efficiency than the mean, the exact values depending on the nature of the noise. This point will be discussed in more detail below.

Time complexity is a further parameter that is needed for characterizing regression methods. We shall not pursue this aspect further here, beyond making the observation that the time complexity of the mean is $O(n)$, while that for the median varies with the method of computation (e.g., $O(n)$ for the histogram approach of Section 3.3 and $O(n^2)$ when using a bubble sort). In any case, the absolute time for computing a median normally far exceeds that for the mean.

Of the parameters referred to here, the breakdown point has been at the forefront of workers' minds when devising new regression schemes. Although it might appear that the median already provides an optimal approach for robust regression, its breakdown value of 0.5 only applies to 1-D data. It is therefore worth considering what breakdown point could be achieved for tasks such as line fitting, bearing in mind the poor performance of least squares regression. Let us take the method of Theil (1950) in which the slope of each pair of a set of n data points is computed, and the median of the resulting set of $\binom{n}{2} = \frac{1}{2}n(n-1)$ values is taken as the final slope. The intercept can be determined more simply because the problem has at that stage been reduced to one dimension. As the median is used in this procedure, at least half the slopes have to be correct in order to obtain a correct estimate of the actual slope. If we assume that the proportion of outliers in the data is η, the proportion of inliers[4] will be $1-\eta$, and the proportion of correct slopes will be $(1-\eta)^2$, and this has to be at least 0.5. This means that η has to lie in the range:

$$\eta \leq 1 - 1/\sqrt{2} = 1 - 0.707 = 0.293 \tag{A.1}$$

Thus, the breakdown point for this approach to linear regression is less than 0.3. In a 3-D data-space where a best-fit plane has to be found, the best breakdown point will be even smaller, with a value $1-2^{-1/3} \approx 0.2$. The general formula for p dimensions is:

$$\eta_p \leq 1 - 2^{-1/p} \tag{A.2}$$

There is a need for more robust regression schemes that become more urgent for larger values of p.

The development of robust multidimensional regression schemes took place relatively recently, in the 1970s. The basic estimators developed at that time, and classified by Huber in 1981, were the M-, R-, and L-estimators. The M-estimator is by far the most widely used and appears in a variety of forms that encompass median and mean estimators and least squares regression. We shall study this type of estimator in more detail below. The L-estimators employ linear combinations of order statistics, and include the alpha-trimmed mean, with the median and mean as special cases. However, it will be easier to consider the median and the mean under the heading of M-estimators, and in what follows we concentrate on this approach.

A.3 The M-estimator (Influence Function) Approach

M-estimators operate by minimizing the sum of a suitable function ρ of the residuals r_i. Normally, ρ is taken to be a positive definite function, and for least squares (L_2) regression it is the square of the residuals:

$$\rho(r_i) = r_i^2 \tag{A.3}$$

4 Inliers are normal valid data points; the dataset is to be regarded as composed of inliers and outliers.

In general, it is necessary to perform the M-estimation minimization operation iteratively until a stable solution is obtained. (At each iteration the new set of offsets has to be added to the previous set of parameter values.)

To improve upon the poor robustness of L_2 regression, reflected by its zero breakdown point, an improved function ρ must be obtained which is well adapted to the particular noise[5] and outlier content of the data. To understand this process, it is easiest to analyze the situation for 1-D datasets and to consider the influence of each data point. We represent the influence of a data point by an influence function $\psi(r_i)$, where

$$\psi(r_i) = \frac{\mathrm{d}\rho(r_i)}{\mathrm{d}r_i} \tag{A.4}$$

Notice that minimizing $\sum_{i=0}^{n} \rho(r_i)$ is equivalent to reducing $\sum_{i=0}^{n} \psi(r_i)$ to zero, and in the case of L_2 regression:

$$\psi(r_i) = 2r_i \tag{A.5}$$

In one dimension this equation has a simple interpretation—moving the origin of coordinates to a position where $\sum_{i=0}^{n} r_i = 0$, that is, to the position of the mean. Now that we have shown the equivalence of L_2 regression to simple averaging, the source of the lack of robustness becomes all too clear—however far away from the mean a data point is, it still retains a weight proportional to its residual value r_i. Accordingly, a wide range of possible alternative influence functions has been devised to limit the problem by cutting down the weights of distant points that are potential outliers.

An obvious approach is to limit the influence of a distant point to some maximum value. Another is to eliminate its influence altogether once its residual error exceeds a certain limiting value (Fig. A.2). We could achieve this by a variety of schemes, either cutting off the influence suddenly at this limiting distance (as in the case of the $\pm 3\sigma$ points), or letting it approach zero according to a linear profile, or opting for a more mathematically ideal functional form with a smoother profile. There are other considerations, such as the amount of computation involved in dealing with large numbers of data points taken over a fair number of iterations. Thus, it is not surprising that a variety of piecewise linear profiles approximating to

5 At this point, a certain ambiguity creeps into the discussion. *Noise* tends to originate from electronic processes in the image source and typically leads to a Gaussian distribution in the pixel intensity values. By the time positions of objects are being measured, it is, strictly speaking, errors rather than noise that are being considered, and the error distribution is not necessarily identical to the noise distribution that gave rise to it. However, in the later sections of this appendix we usually refer to noise and noise distributions. The term *noise* will be taken to refer either to the original noise source or to the derived errors, as appropriate to the discussion.

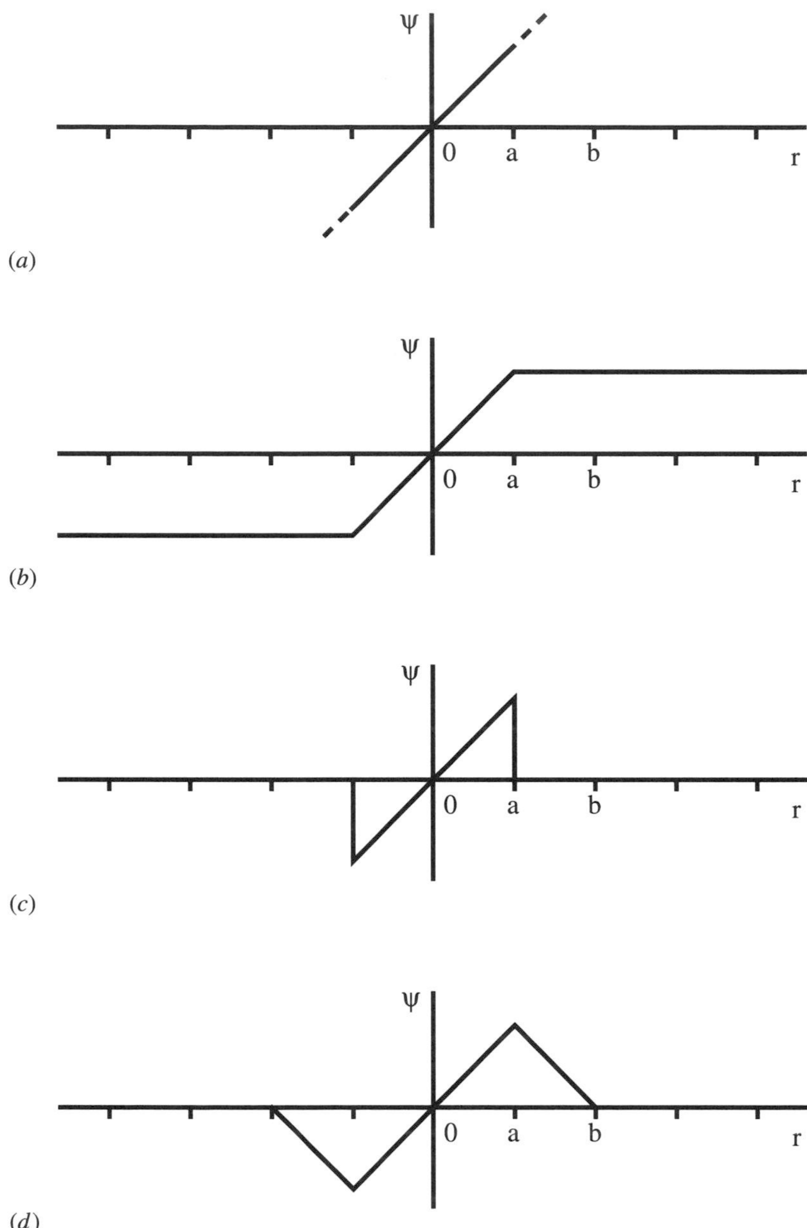

Figure A.2 Influence functions that limit the effects of outliers. (*a*) shows the case where no limit is placed on the influence of distant points; (*b*) shows how the influence is limited to some maximum value; (*c*) shows how the influence is eliminated altogether once the residual exceeds a certain maximum value; (*d*) shows a piecewise-linear profile that gives a less abrupt variation; (*e*) shows a mathematically more well-behaved influence function; (*f*) shows another possible piecewise-linear case, and (*g*) shows a Hampel three-part re-descending M-estimator that approximates the mathematically ideal case (*e*) with reasonable accuracy; (*h*) shows the situation for a median estimator.

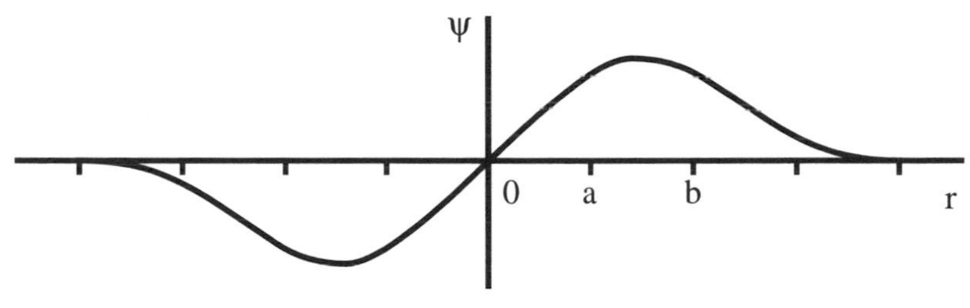

(e)

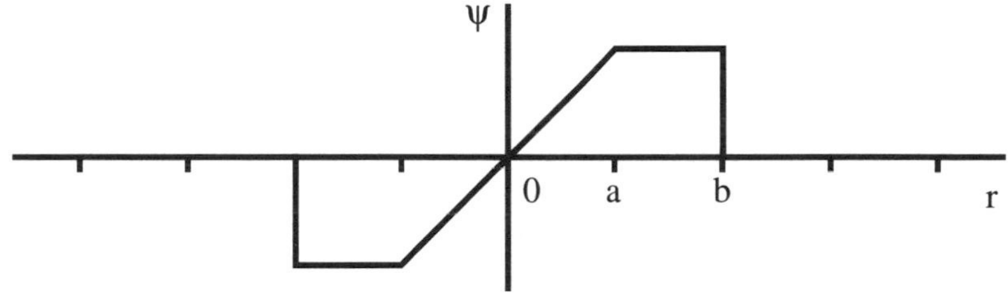

(f)

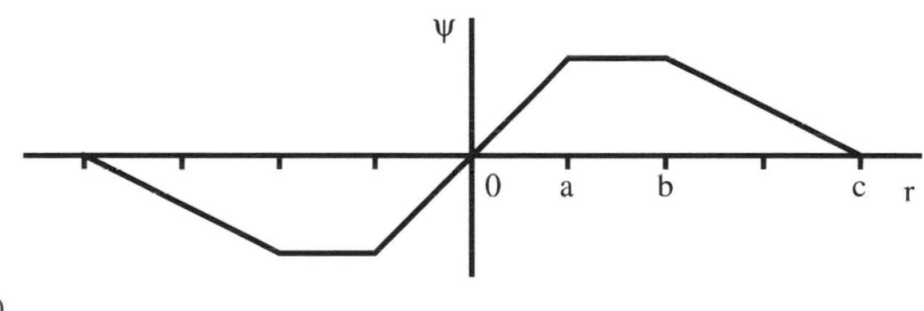

(g)

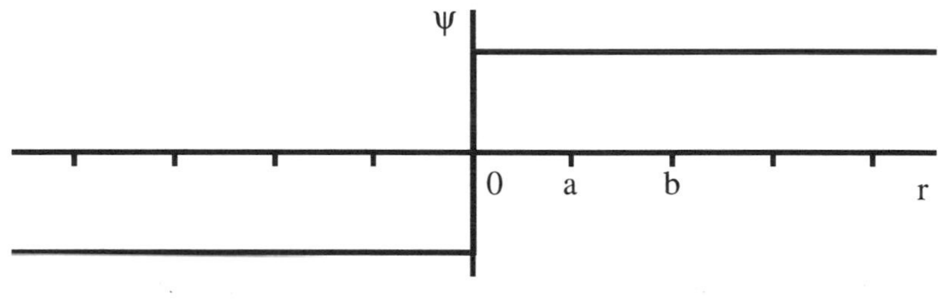

(h)

Figure A.2 Continued.

the smoother ideal profiles have been devised. In general, however, influence functions are linear near the origin, zero at large distances from the origin, and possess a region over which they give significant weight to the data points (Fig. A.2).

Prominent among these possibilities are the Hampel 3-part re-descending M-estimator, whose influence function is composed simply of convenient linear components, and the Tukey biweight estimator (Beaton and Tukey, 1974), which takes a form similar to that shown in Fig. A.2*e*:

$$\begin{aligned} \psi(r_i) &= r_i(\gamma^2 - r_i^2)^2 \quad |r_i| \leq \gamma \\ &= 0 \qquad\qquad\qquad |r_i| > \gamma \end{aligned} \tag{A.6}$$

It was observed earlier that the median operation is a special case of the M-estimator. Here all data points on one side of the origin have a unit positive weight, and all data points on the other side of the origin have unit negative weight:

$$\psi(r_i) = \text{sign}(r_i) \tag{A.7}$$

Thus if more data points are on one side than the other, the solution will be pulled in that direction, iteration proceeding until the median is at the origin.

Although the median has exceptionally useful outlier suppression characteristics, it actually gives outliers significant weight. In fact, the median ignores how far away an outlier is, but it still counts up how many outliers there are on either side of the current origin. As a result, the median is liable to produce a biased estimate. This is good reason for considering other types of influence function for analyzing data. Finally, we remark that the median influence function leads to the value of ρ for L_1 regression:

$$\rho(r_i) = |r_i| \tag{A.8}$$

When selecting an influence function, it is important not only that the function be appropriate but also that its scale must match that of the data. If the width of the influence function is too great, too few outliers will be rejected. If the width is too small, the estimator may be surrounded by a rather homogeneous sea of data points with no guarantee that it will do more than find a locally optimal fit to the data. These factors mean that preliminary measurements must be made to determine the optimal form of the influence function for any application.

At this point it seems clear that there ought to be a more scientific approach than would permit the influence function to be calculated from the noise characteristics. This is indeed so, and if the expected noise distribution is given by $f(r_i)$, the optimal form of the influence function (Huber, 1964) is:

$$\psi(r_i) = -f'(r_i)/f(r_i) = -\sum \frac{\mathrm{d}}{\mathrm{d}r_i} \ln[f(r_i)] \tag{A.9}$$

The existence of the logarithmic form of this solution is interesting and useful, for it simplifies exponential-based noise distributions such as the Gaussian and double exponential functions. For the Gaussian function, $\exp(-r_i^2/2\sigma^2)$, we find:

$$\psi(r_i) = r_i/\sigma^2 \tag{A.10}$$

and for the double exponential function, $\exp(-|r_i|/s)$:

$$\psi(r_i) = \mathrm{sign}(r_i)/s \tag{A.11}$$

Since the constant multipliers may be ignored, we conclude that the mean and median are optimal estimators for signals in Gaussian and double exponential noise, respectively.

Gaussian noise may be expected to arise in many situations (most particularly because of the effects of the central limit theorem), demonstrating the intrinsic value of employing the mean or L_2 regression. On the other hand, the double exponential distribution has no obvious justification in practical situations. However, it represents situations where the wings of the noise distribution stretch out rather widely, and it is good to see under what conditions the widely used median would in fact be optimal. Nevertheless our purpose in wanting an explicit mathematical form for the influence function was to optimize the detection of signals in arbitrary noise conditions and specifically those where outliers may be present.

Let us suppose that the noise is basically Gaussian but that outliers may also be present and that these would be drawn approximately from a uniform distribution. There might, for example, be a uniform (but low-level) distribution of outlier values over a limited range. An overall distribution of this type is shown in Fig. A.3. Near $r_i = 0$, the uniform distribution of outliers will have relatively little effect, and $\psi(r_i)$ will approximate to r_i. For large $|r_i|$, the value of f' will be due mainly to the Gaussian noise contribution, whereas the value of f will arise mainly from the uniform distribution f_u, and the result will be:

$$\psi(r_i) \approx \frac{r_i}{\sigma^2 f_u} \exp(-r_i^2/2\sigma^2) \tag{A.12}$$

a function that peaks at an intermediate value of r_i. This essentially proves that the form shown in Fig. A.2*e* is reasonable. However, there is a severe problem in that outliers are by definition unusual and rare, so it is almost impossible in

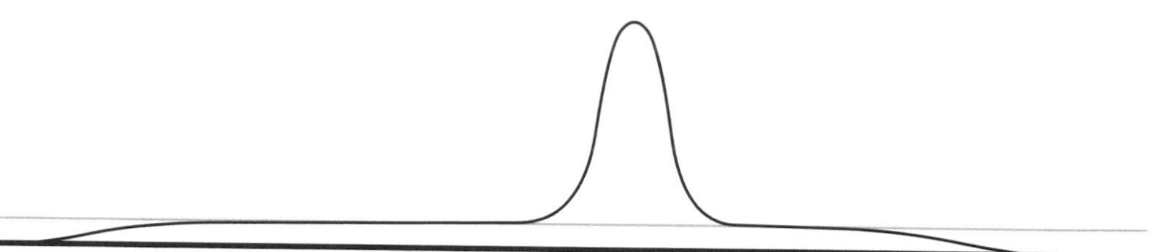

Figure A.3 Distribution resulting from Gaussian noise and outliers. Here the usual Gaussian noise contribution is augmented by a distribution of outliers that is nearly uniform over a limited range.

most cases to be able to produce on optimum form of $\psi(r_i)$ as suggested earlier. Unfortunately, the situation is even worse than this discussion might indicate. Redescending M-estimators are even more limited in that they are sensitive to local densities of data points and are therefore prone to finding false solutions—unique solutions are *not* guaranteed. Non-redescending M-estimators are guaranteed to arrive at unique solutions, though the accuracy of solutions depends on the accuracy of the preliminary scale estimate. In addition, the quality of the initial approximation tends to be of very great importance for M-estimators, particularly for re-descending M-estimators.

Finally, we should point out that the above analysis has concentrated on optimization of accuracy and is ultimately based on maximum likelihood strategies (Huber, 1964). It is really concerned with maximizing relative efficiency on the assumption that the underlying distribution is known. Robustness measured according to the breakdown point criterion is not optimized, and this factor will be of vital importance in any situation where the outliers form part of a totally unexpected distribution, or do not form part of a predictable distribution.[6] Methods must be engineered that are intrinsically highly robust according to the breakdown point criterion. This is what motivated the development of the least median of squares approach to regression during the 1980s.

A.4 The Least Median of Squares Approach to Regression

We have seen that a variety of estimators exist that can be used to suppress noise from numerical data and to optimize the robustness and accuracy of the final result. The M-estimator (or influence function) approach is widely used and is successful in eliminating the main problems associated with the use of least squares

6 It is perhaps a philosophical question whether an outlier distribution does not exist, cannot exist, or cannot be determined by any known experimental means, for example, because of rarity.

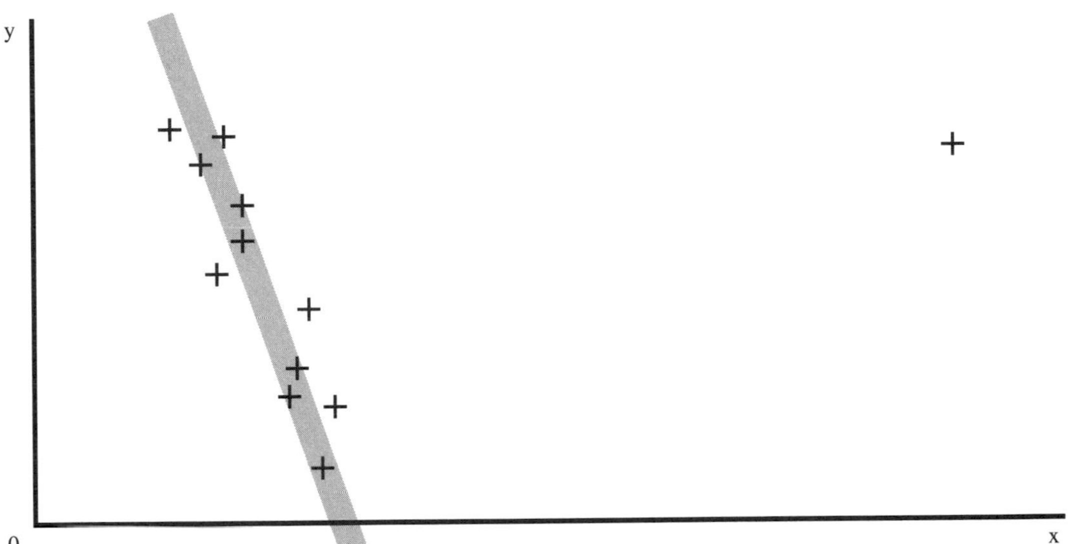

Figure A.4 Application of the least median of squares technique. Here the narrowest parallel-sided strip is found, which includes half the population of the distribution, in an attempt to determine the best-fit line. Notice the effortless superiority in performance when compared with the situation in Fig. A.1*c*.

regression (including, in 1-D, use of the mean). However, it does not in general achieve the ideal breakdown value of 0.5, and it requires careful setting up to give optimal matching to the scale of the variation in the data. Accordingly, much attention has been devoted to a newer approach—least median of squares regression.

The aim of least median of squares (LMedS) regression is to capitalize on the known robustness of the median in a totally different way—by replacing the mean of the least (mean) squares averaging technique by the far more robust median. The effect of this is to ignore errors from the distant parts of the distribution and also from the central parts where the peak is often noisy and ill defined, and to focus on the parts about halfway up and on either side of the distribution. Minimization then balances the contributions from the two sides of the distribution, thereby sensitively estimating the mode position, though clearly this is achieved rather indirectly. Perhaps the simplest view of the technique is that it determines the location of the narrowest width region, which includes half the population of the distribution. In a 2-D straight-line location application, this interpretation amounts to locating the narrowest parallel-sided strip, which includes half the population of the distribution (Fig. A.4). In principle, in such cases the method operates just as effectively if the distribution is sparsely populated—as happens where the best-fit straight line for a set of experimental plots has to be determined.

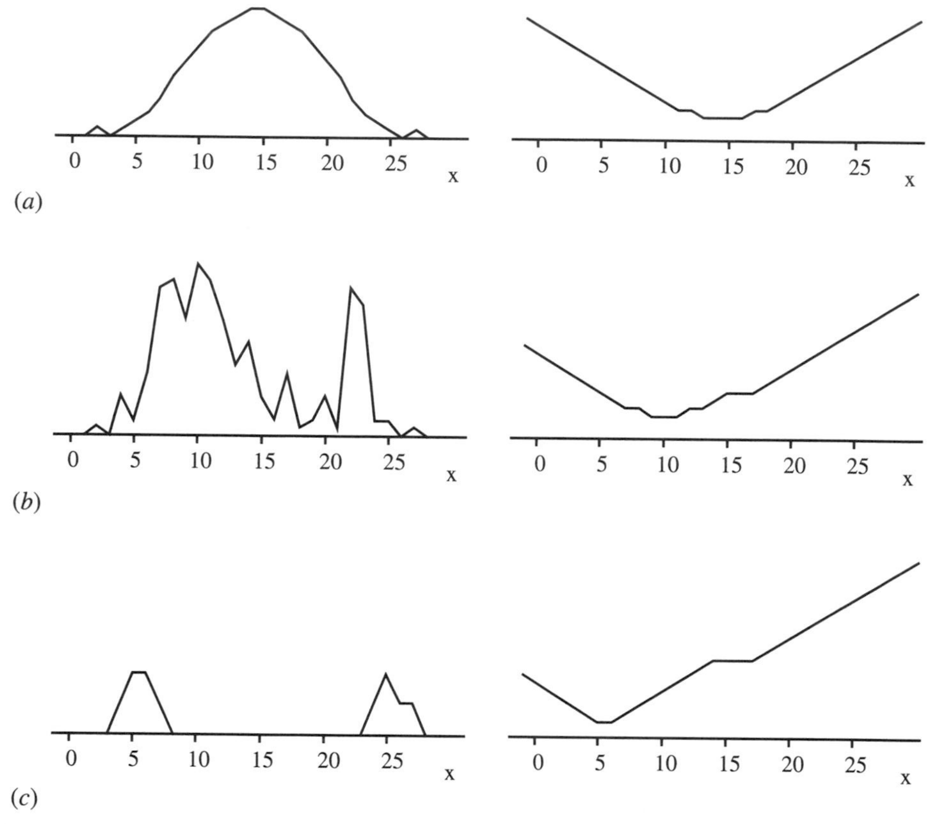

Figure A.5 Minimizing $\mathscr{M}$ for various distributions. This figure shows (left) the original distributions and (right) the resulting response functions $\mathscr{M}$ in the following cases: (*a*) an approximately Gaussian distribution, (*b*) an "untidy" distribution, (*c*) a distribution with two peaks.

The LMedS technique involves minimizing the median of the squares of the residuals r_j for all possible positions in the distribution which are potentially mode positions; that is, it is the position x_i which minimizes $M = \text{med}_j(r_j^2)$. Although it might be thought that minimizing M is equivalent to minimizing $\mathscr{M} = \text{med}_j(|r_j|)$, this is not so if there are two adjacent central positions giving equal responses (as in Fig. A.5 *a–c*). However, the form of M guarantees that a position midway between these two will give an appropriate minimum. For clarity we shall temporarily ignore this technicality and concentrate on $\mathscr{M}$: the reason for doing so is to take advantage of piecewise-linear responses that considerably simplify theoretical analysis.

Figure A.5*a* shows the response $\mathscr{M}$ when the original distribution is approximately Gaussian. There is a clear minimum of $\mathscr{M}$ at the mode position, and the method works perfectly. Figure A.5*b* shows a case where there is a very untidy distribution, and there is a minimum of $\mathscr{M}$ at an appropriate position. Figure A.5*c* shows a more extreme situation in which there are two peaks, and again the response $\mathscr{M}$ is appropriate, except that it is now clear that the technique can only focus on one peak at a time. Nevertheless, it gets an appropriate and robust answer for the case on which it is focusing. If the two peaks are identical, the method will still work but will clearly not give a unique solution.

Although the LMedS approach to regression appeared only two decades ago (Rousseeuw, 1984), it has acquired considerable support, since it has the maximum possible breakdown point of 0.5. In particular, the LMedS approach has been used for pattern recognition and image analysis applications (see for example Kim et al., 1989). In these areas, the method is useful for (1) location of straight lines in digital images, (2) location of Hough transform peaks in parameter space, and (3) location of clusters of points in feature space.

Unfortunately, the LMedS approach is liable to give a biased estimate of the modes if two distributions overlap, and in any case focuses on the main mode of a multimodal distribution. Thus, the LMedS technique has to be applied several times, alternating with necessary truncation processes, to find all the cluster centers, while weighted least squares fitting is required to optimize accuracy. The result is a procedure of some complexity and considerable computational load. Indeed, the load is in general so large that it is normally approximated by taking subsets of the data points, though this aspect cannot be examined in detail here (see, for example, Kim et al., 1989). Once this has been carried out, the method can give quite impressive results.

Ultimately, the value of the LMedS approach lies in its increased breakdown point in situations of multidimensional data. If we have n data points in p dimensions, the LMedS breakdown point is:

$$\varepsilon_{\mathrm{LMedS}} = (\lfloor n/2 \rfloor - p + 2)/n \tag{A.13}$$

which tends to 0.5 as n approaches infinity (Rousseeuw, 1984). This value must be compared with a maximum of

$$\varepsilon = 1/(p+1) \tag{A.14}$$

for standard methods of robust regression such as the M-, R- and L-estimators discussed earlier (Kim et al., 1989). (Equation (A.2) represents the suboptimal solution achieved by the Theil approach to line estimation.) Thus, in these latter cases, 0.33 is the best breakdown point that can be achieved for $p = 2$, while the LMedS approach offers 0.5. However, the relative efficiency of LMedS is relatively low (ultimately because it is a median-based estimator). As already stated, this means that it has to be used with the weighted least squares technique.

We should also point out that the LMedS technique is intrinsically 1-D, so it has to be used in a "projection pursuit" manner (Huber, 1985), concentrating on one dimension at a time. For implementation details the reader is referred to the literature (see Section A.7).

A.5 Overview of the Robustness Problem

For greatest success in solving the robustness and accuracy problems—represented, respectively, by the breakdown point and relative efficiency criteria—it has been found in the foregoing sections that the LMedS technique should be used for finding signals (whether peaks, clusters, lines, or hyperplanes, etc.), and weighted least squares regression should be used for refining accuracy, the whole process being iterated until satisfactory results are achieved. This is a complex and computation-intensive process but reflects an overall strategy that has been outlined several times in earlier chapters—namely, search for an approximate solution, and then refinement to optimize location accuracy. The major question to be considered at this stage is: what is the best method for performing an efficient and effective initial search? There is a further question which is of special relevance: is there any means of achieving a breakdown point of greater than 0.5?

The extent to which the Hough transform tackles and solves these problems is of some interest. First, it is a highly effective search procedure, though in some contexts its computational efficiency has been called into question. (However, in the present context, it must be remembered that the LMedS technique is especially computation intensive.) Second, it seems able to yield breakdown points far higher than 0.5 and even approaching unity. Consider a parameter space where there are many peaks and also a considerable number of randomly placed votes. Then any individual peak includes perhaps only a small fraction of the votes, and the peak location proceeds without difficulty in spite of the presence of 90 to 99% contamination by outliers (the latter arising from noise and clutter). Thus, the strategy of searching for peaks appears to offer significant success at avoiding outliers. Yet this does not mean that the LMedS technique is valueless, since subsequent application of LMedS is essentially able to *verify* the identification of a peak, to locate it more accurately via its greater relative efficiency, and thus to feed reliable information to a subsequent least squares regression stage. Overall, we can see that a staged progression is taking place from a high breakdown point, low relative efficiency procedure, to a procedure of intermediate breakdown point and moderate relative efficiency, and finally to a procedure of low breakdown point and high relative efficiency.

Table A.2 Breakdown points and efficiency values for peak finding

	HT	LMedS	LS	Overall
ε	0.98	0.50	0.2	0.98
η	0.2	0.4	0.95	0.95

This table gives possible breakdown points ε and relative efficiency values η for peak finding. A Hough transform is used to perform an initial search for peaks; then the LMedS technique is employed for validating the peaks and eliminating outliers; finally, least squares regression is used to optimize location accuracy. The result is far higher overall effectiveness than that obtainable by any of the techniques applied alone. However, computational load is not taken into account and is likely to be a major consideration.

We summarize the progression by giving possible figures for the relevant quantities in Table A.2.

A.6 The RANSAC Approach

Over the years, RANSAC has become one of the most widely used outlier rejection and data fitting tools: it has achieved particular value in 3-D vision. RANSAC is an acronym for RANdom SAmple Consensus and involves repeatedly trying to obtain a consensus (set of inliers) from the data until the degree of fit exceeds a given criterion.

To understand the process, let us first return to the LMedS approach, which is useful both in providing a graphic presentation of what it achieves and in requiring no parameters to be set in order to make it work. This last-named feature is in many ways its undoing because if the proportion of outliers in the data exceeds 50%, the resulting fit is liable to be heavily biased. A simple modification of the method is to require a smaller number of inliers—indeed, whatever proportion would be expected in the incoming data. Thus, we may go for 20% inliers, 80% outliers if this seems appropriate. This naturally leads to problems, for ideally we will have to estimate the proportion of inliers in advance, or as part of the fitting process, and then apply the resulting value as part of the technique.

Once the "cleanness" of the LMedS method is lost, a variety of alternative solutions become possible. The RANSAC method involves not taking the proportion of inliers as set and finding how the residual distance (e.g., from a best fit straight line) varies, but rather specifying a threshold residual distance

t and finding how the proportion of inliers varies. Here, the word "inlier" is not a good term to use, for it implies that we already know that these data are acceptable points. Instead, they should be called *consensus points*—at least until the end of the process. In summary, we set a threshold residual distance t and ask how much consensus this gives. Note that in principle at least, t has to be iterated as part of the whole process of finding the best fit. However, it is possible to work on the basis that the experimental uncertainty is known in advance, and if, for example, t is made equal to three standard deviations, this should not lead to too much error in the final fit obtained.

Another aspect of RANSAC is the random extraction of n data points to specify each initial potential fit, following which the hypothesized solution is tested to find how much consensus there is. Then out of k trials, the best solution is the one with the greatest consensus, and at this final stage we can interpret the consensus as the set of inliers.

Finally, the number of data points n needed to specify a potential fit is made equal to the number of degrees of freedom of the data—two for a straight line in a plane, three for a circle, four for a sphere, and so on (Fig. A.6). All that remains to be specified is the number of iterations k of sets of n data points in order to reach the final best fit solution. One way of estimating k is to calculate the risk that all the

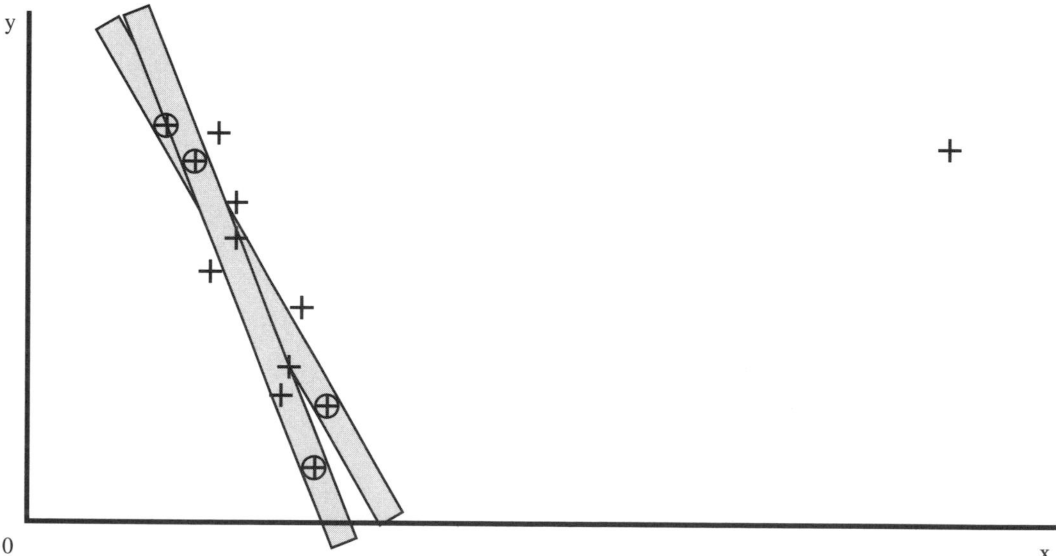

Figure A.6 The RANSAC technique. Here the + signs indicate data points to be fitted, and two instances of pairs of data points (indicated by ⊕ signs) leading to hypothesized lines are also shown. Each hypothesized line has a region of influence of tolerance $\pm t$ within which the support of maximal numbers of data points is sought. The line with the most support indicates the best fit (though weighted least squares analysis may subsequently be applied to improve it further).

k sets of n data points chosen will contain only outliers, so that no good data will be examined. Clearly, k must be sufficiently large to reduce the risk of this eventuality to a low enough level. Formulas to estimate k on this basis appear in several sources, for example, Hartley and Zisserman (2000).

As happens with many other outlier identification processes, improved fits can be obtained by a final stage in which normal or weighted least squares analysis is applied to the remaining (inlier) data.

A.7 Concluding Remarks

This appendix has aimed to place the discussion of robustness on a sounder basis than might have been thought possible in the earlier chapters of the book, where a more intuitive approach was presented. It has been necessary to delve quite deeply into the maturing and highly mathematical subject of robust statistics, and there are certain important lessons to be learned. In particular, three relevant parameters have been found to form the basis for study in this area. The first is the breakdown point of an estimator, which shows the estimator's resistance to outliers and provides the core meaning of robustness. The second is the relative efficiency of an estimator, which provides a measure of how efficiently it will use the inlier data at its disposal to arrive at accurate estimates. The third is the time complexity of the estimator when it is implemented as a computer algorithm. Although this last parameter is a vital consideration in practical situations, available space has not permitted it to be covered in any depth here. It is clear, however, that the most robust techniques (especially LMedS) tend to be highly computation intensive. It is also found that there is a definite tradeoff between the other two parameters—techniques that have high breakdown points have low relative efficiencies and vice versa.[7] These factors make it reasonable, and desirable, to use several techniques in sequence, or iteratively in cycle, in order to obtain the best overall performance. Thus, LMedS is frequently used in conjunction with least squares regression (see for example Kim et al., 1989).

The basis of robust statistics is that of statistical analysis of the available data; the tendency is therefore to presume that outliers are rare events due typically to erroneous readings or transcriptions. Yet in vision the most difficult problems tend to arise from the clutter of irrelevant objects in the background, and only a tiny fraction of the incoming data may constitute the relevant inlier portion. This makes the problem of robustness all the more serious, and in principle *could* mean

7 The reason for this may be summarized as the aim of achieving high robustness requiring considerable potentially outlier data to be discarded, even when this could be accurate data that would contribute to the overall accuracy of the estimate.

that until a whole image has been interpreted satisfactorily no single object can finally be identified and its position and orientation measured accurately. Although this view may be pessimistic, it does indicate that the conventional approach to robust statistics may have significant limitations in its application to machine vision.

Robust statistics are at the core of any practical vision system. This appendix has aimed to cover the intricacies of the subject in an accessible way, dealing with important concepts such as "breakdown point" and measurement "efficiency." What is really in question is *how* robust statistics will be incorporated into any practical vision system, not *whether* they need to be.

A.8 Bibliographical and Historical Notes

This appendix has given a basic introduction to the rapidly maturing subject of robust statistics, which has made a substantial impact on machine vision over the past 15 to 20 years. The most popular and successful approach to robust statistics must still be seen as the M-estimator (influence function) approach, though in high dimensional spaces its robustness is called to question, and it is here that the newer LMedS approach has gathered a firm following. Quite recently, the value of using a sequence of estimators that can optimize the overall breakdown point and relative efficiency has been pointed out (Kim et al., 1989). In particular, the right combination of Hough transform (or other relevant technique), LMedS, and weighted least squares regression would seem especially powerful.

Robust statistics has been applied in a number of areas of machine vision, including robust window operators (Besl et al., 1989), pose estimation (Haralick and Joo, 1988), motion studies (Bober and Kittler, 1993), camera location and calibration (Kumar and Hanson, 1989), and surface defect inspection (Koivo and Kim, 1989), to name but a few.

The original papers by Huber (1964) and Rousseeuw (1984) are still worth reading, and the books by Huber (1981), Hampel et al. (1986), and Rousseeuw and Leroy (1987) are valuable references, containing much insight and readable material. On the application of the LMedS technique, and for a more recent review of robust regression in machine vision, see Meer et al. (1990, 1991).

The RANSAC technique (Fischler and Bolles, 1981) was introduced before LMedS and presaged its possibilities: RANSAC was thus historically of great importance. The work of Siegel (1982) was also important historically in providing the background from which LMedS could take off, while the work of Steele and Steiger (1986) showed how LMedS might be implemented with attainable levels of computation.

Although much of the work on robust statistics dates from the 1980s, one has only to look at the recent book by Hartley and Zisserman (2000) to see how deeply embedded it is in the current methodology and thinking on machine vision. A recent example of its application to 3-D correspondence matching is provided by Hasler et al. (2003): Hasler et al. consider exactly where the outlier data originate and model the whole process. Unexpected motion, occlusion of points in some views, and viewing of convex boundaries from different positions all lead to mismatches and outliers. They arrive at a new way of calculating outliers in image pairs, which helps to put the whole subject area on a more secure footing. Meanwhile, Torr and Davidson (2003) have produced an improved version of RANSAC, which they have called IMPSAC (an acronym for IMPortance SAmpling Consensus). It works in a hierarchical manner and is initialized at the coarsest level by RANSAC, but then it goes on to sample at a finer level to refine relevant *a posteriori* estimates. Although IMPSAC has been applied to 3-D matching tasks, it embodies statistical techniques that can be applied to a wide variety of statistical problems to eliminate outlier corrupted data.

A.9 Problem

1. (a) What is meant by the *breakdown point* of a data analysis method? Show how it is related to the concept of robustness. Consider also how accuracy of measurement is affected by the proportion of data points that are fully utilized by the data analysis method. Discuss the situation in relation to (i) the mean, (ii) the median, and (iii) the result of applying a Hampel three-part re-descending M-estimator.

   (b) A method for locating straight lines in digital images involves taking every pair of edge points and finding where a line through both points of a pair intercepts the x- and y-axes. Then medians for all such intercepts are found, and the positions of any straight lines are deduced. Show that the effect of taking pairs is to reduce the breakdown point from 50% to around 30%, and give an exact answer for the breakdown point. (*Hint:* Start by assuming that the fraction of outliers in the original set of edge points is ε and work out the probability of half the intercept values being correct.)

List of Acronyms and Abbreviations

ACM	Association for Computing Machinery (USA)
AI	artificial intelligence
ANN	artificial neural network
ASCII	American Standard Code for Information Interchange
ASIC	application specific integrated circuit
BMVA	British Machine Vision Association
CAD	computed-aided design
CAM	computer-aided manufacture
CCD	charge-coupled device
CCTV	closed-circuit television
CIM	computer integrated manufacture
CLIP	cellular logic image processor
CPU	central processor unit
DET	Beaudet's (1978) determinant operator
DG	differential gradient
DN	Dreschler and Nagel (1981) corner detector
DSP	digital signal processor
ECL	emitter-coupled logic
FFT	fast Fourier transform
FOE	focus of expansion
FPGA	field programmable gate array
GHT	generalized Hough transform
HT	Hough transform
IDD	integrated directional derivative
IEE	Institution of Electrical Engineers (UK)
IEEE	Institute of Electrical and Electronics Engineers (USA)
k-NN	*k*-nearest neighbor
KR	Kitchen and Rosenfeld (1982) corner detector
LFF	local-feature-focus method (Bolles and Cain, 1982)
LMedS	least median of squares
LS	least squares
LSB	least significant bit

LUT	lookup table
MIMD	multiple instruction stream, multiple data stream
MIPS	millions of instructions per second
MISD	multiple instruction stream, single data stream
MP	microprocessor
MSB	most significant bit
NN	nearest neighbor
OCR	optical character recognition
PC	personal computer
PCB	printed circuit board
PE	processing element
PR	pattern recognition
PSF	point spread function
RAM	random access memory
RANSAC	random sample consensus
RMS	root mean square
ROC	receiver–operator characteristic
ROI	region of interest
SIAM	Society of Industrial and Applicative Mathematics
SIMD	single instruction stream, multiple data stream
SISD	single instruction stream, single data stream
SPIE	Society of Photo-optical Instrumentation Engineers
SPR	statistical pattern recognition
SVM	support vector machine
TM	template matching
TTL	transistor–transistor logic
TV	television
ULUT	universal lookup table
VLSI	very large scale integration
ZH	Zuniga and Haralick (1983) corner detector
1-D	one dimension/one-dimensional
2-D	two dimensions/two-dimensional
3-D	three dimensions/three-dimensional

References

Abdou, I.E. and Pratt, W.K. (1979). Quantitative design and evaluation of enhancement/thresholding edge detectors. *Proc. IEEE* **67**, 753–763.

Abend, K., Harley, T. and Kanal, L.N. (1965). Classification of binary random patterns. *IEEE Trans. Inf. Theory* **11**, 538–544.

Abutaleb, A.S. (1989). Automatic thresholding of gray-level pictures using two-dimensional entropy. *Comput. Vision Graph. Image Process.* **47**, 22–32.

Ade, F. (1983). Characterization of texture by "eigenfilters." *Signal Process.* **5**, no. 5, 451–457.

Aggarwal, J.K. and Cai, Q. (1999). Human motion analysis: a review. *Computer Vision Image Understanding* **73**, no. 3, 428–440.

Agin, G.J. and Binford, T.O. (1973). Computer description of curved objects. *Proc. 3rd Int. Joint Conf. on Artif. Intell., Stanford, California*, pp. 629–640.

Agin, G.J. and Binford, T.O. (1976). Computer description of curved objects. *IEEE Trans. Comput.* **25**, 439–449.

Aguado, A.S., Montiel, M.E. and Nixon, M.S. (2000). On the intimate relationship between the principle of duality and the Hough transform. *Proc. Roy. Soc.* **456**, no. 1995, 503–526

Akey, M.L. and Mitchell, O.R. (1984). Detection and sub-pixel location of objects in digitized aerial imagery. *Proc. 7th Int. Conf. on Pattern Recogn., Montreal* (July 30–August 2), pp. 411–414.

Aleksander, I., Thomas, W.V. and Bowden, P.A. (1984). WISARD: A radical step forward in image recognition. *Sensor Rev.* **4**, 120–124.

Ali, S.M. and Burge, R.E. (1988). A new algorithm for extracting the interior of bounded regions based on chain coding. *Comput. Vision Graph. Image Process.* **43**, 256–264.

Almansa, A., Desolneux, A. and Vamech, S. (2003). Vanishing point detection without any a priori information. *IEEE Trans. Pattern Anal. Mach. Intell.* **25**, no. 4, 502–507.

Aloimonos, J. (1988). Shape from texture. *Biol. Cybern.* **58**, 345–360.

Aloimonos, J. and Swain, M.J. (1985). Shape from texture. *Proc. 9th Int. Joint Conf. on Artif. Intell.* **2**, 926–931.

Alter, T.D. (1994). 3-D pose from 3 points using weak-perspective. *IEEE Trans. Pattern Anal. Mach. Intell.* **16**, no. 8, 802–808.

Ambler, A.P., Barrow, H.G., Brown, C.M., Burstall, R.M. and Popplestone, R.J. (1975). A versatile system for computer-controlled assembly. *Artif. Intell.* **6**, 129–156.

Amit, Y. (2002). *2D Object Detection and Recognition: Models, Algorithms and Networks.* MIT Press, Cambridge, MA.

Ando, S. (2000). Image field categorization and edge/corner detection from gradient covariance. *IEEE Trans. Pattern Anal. Mach. Intell.* **22**, no. 2, 179–190.

Andrews, D., Niehaus, D. and Ashenden, P. (2004). Programming models for hybrid CPU/FPGA chips. *IEEE Computer* **37**, no. 1, 118–120.

Ankenbrandt, C.A., Buckles, B.P. and Petry, F.E. (1990). Scene recognition using genetic algorithms with semantic nets. *Pattern Recogn. Lett.* **11**, no. 4, 285–293.

Arcelli, C. and di Baja, G.S. (1985). A width-independent fast-thinning algorithm. *IEEE Trans. Pattern Anal. Mach. Intell.* **7**, 463–474.

Arcelli, C. and Ramella, G. (1995). Finding grey-skeletons by iterated pixel removal. *Image Vision Comput.* **13**, no. 3, 159–167

Arcelli, C., Cordella, L.P. and Levialdi, S. (1975). Parallel thinning of binary pictures. *Electronics Lett.* **11**, 148–149.

Arcelli, C., Cordella, L.P. and Levialdi, S. (1981). From local maxima to connected skeletons. *IEEE Trans. Pattern Anal. Mach. Intell.* **3**, 134–143.

Arnold, R.D. (1978). Local context in matching edges for stereo vision. *Proc. Image Understanding Workshop, Cambridge, Massachusetts*, pp. 65–72.

Åström, K. (1995). Fundamental limitations on projective invariants of planar curves. *IEEE Trans. Pattern Anal. Mach. Intell.* **17**, no. 1 , 77–81.

Ataman, E., Aatre, V.K. and Wong, K.M. (1980). A fast method for real time median filtering. *IEEE Trans. Acoust. Speech Signal Process.* **28**, 415–420.

Atherton, T.J. and Kerbyson, D.J. (1999). Size invariant circle detection. *Image Vision Comput.* **17**, no. 11, 795–803.

Atiquzzaman, M. (1994). Pipelined implementation of the multiresolution Hough transform in a pyramid multiprocessor. *Pattern Recogn. Lett.* **15**, no. 9, 841–851.

Atiquzzaman, M. and Akhtar, M.W. (1994). Complete line segment description using the Hough transform. *Image Vision Comput.* **12**, no. 5, 267–273.

Aziz, M., Boussakta, S. and McLernon, D.C. (2003). High performance 2D parallel block-filtering system for real-time imaging applications using the Sharc ADSP21060. *Real-Time Imaging* **9**, no. 2, 151–161.

Babaud, J., Witkin, A.P., Baudin, M. and Duda, R.O. (1986). Uniqueness of the Gaussian kernel for scale-space filtering. *IEEE Trans. Pattern Anal. Mach. Intell.* **8**, 26–33.

Backer, E. (ed.) (1992). Special Issue on Artificial Neural Networks. *Pattern Recogn. Lett.* **13**, no. 5.

Bajcsy, R. (1973). Computer identification of visual surface. *Comput. Graph. Image Process.* **2**, 118–130.

Bajcsy, R. and Liebermann, L. (1976). Texture gradient as a depth cue. *Comput. Graph. Image Process.* **5**, no. 1, 52–67.

Baker, S., Sim, T. and Kanade, T. (2003). When is the shape of a scene unique given its light-field: a fundamental theorem of 3D vision? *IEEE Trans. Pattern Anal. Mach. Intell.* **25**, no. 1, 100–109.

Baldi, P. and Hornik, K. (1989). Neural networks and principal component analysis: learning from examples without local minima. *Neural Networks* **2**, 53–58.

Ball, G.H. and Hall, D.J. (1966). ISODATA, an iterative method of multivariate data analysis and pattern classification. *IEEE Int. Communications Conf., Philadelphia, Digest of Techn. Papers II*, pp. 116–117.

Ballard, D.H. (1981). Generalizing the Hough transform to detect arbitrary shapes. *Pattern Recogn.* **13**, 111–122.

Ballard, D.H. and Brown, C.M. (1982). *Computer Vision*. Prentice-Hall, Englewood Cliffs, NJ.

Ballard, D.H. and Sabbah, D. (1983). Viewer independent shape recognition. *IEEE Trans. Pattern Anal. Mach. Intell.* **5**, 653–660.

Bangham, J.A. and Marshall, S. (1998). Image and signal processing with mathematical morphology. *IEE Electronics and Commun. Eng. Journal* **10**, no. 3, 117–128.

Barbe, D.F. (1975). Imaging devices using the charge-coupled principle. *Proc. IEEE* **63**, 38–66.

Barker, A.J. and Brook, R.A. (1978). A design study of an automatic system for on-line detection and classification of surface defects on cold-rolled steel strip. *Optica Acta* **25**, 1187–1196.

Barnard, S. (1983). Interpreting perspective images. *Artif. Intell.* **21**, 435–462.

Barnard, S.T. and Thompson, W.B. (1980). Disparity analysis of images. *IEEE Trans. Pattern Anal. Mach. Intell.* **2**, no. 4, 333–340.

Barnea, D.I. and Silverman, H.F. (1972). A class of algorithms for fast digital image registration. *IEEE Trans. Comput.* **21**. 179–186.

Barrett, E.B., Payton, P.M., Haag, N.N. and Brill, M.H. (1991). General methods for determining projective invariants in imagery. *Computer Vision Graph. Image Process.* **53**, no. 1, 46–65.

Barrow, H.G. and Popplestone, R.J. (1971). Relational descriptions in picture processing. In Meltzer, B. and Michie, D. (eds.), *Machine Intelligence* **6**. Edinburgh University Press, Edinburgh. pp. 377–396.

Barrow, H.G. and Tenenbaum, J.M. (1981). Computational vision. *Proc. IEEE* **69**, 572–595.

Barrow, H.G., Ambler, A.P. and Burstall, R.M. (1972). Some techniques for recognising structures in pictures. In Watanabe, S. (ed.), *Frontiers of Pattern Recognition*. Academic Press, New York, pp. 1–29.

Barsky, S. and Petrou, M. (2003). The 4-source photometric stereo technique for three-dimensional surfaces in the presence of highlights and shadows. *IEEE Trans. Pattern Anal. Mach. Intell.* **25**, no. 10, 1239–1252.

Bartz, M.R. (1968). The IBM 1975 optical page reader. *IBM J. Res. Dev.* **12**, 354–363.

Bascle, B., Bouthemy, P., Deriche, R. and Meyer, F. (1994). Tracking complex primitives in an image sequence. *Proc. 12th Int. Conf. on Pattern Recogn., Jerusalem, Israel (October 9–13)*, Vol. A, pp. 426–431.

Batchelor, B.G. (1979). Using concavity trees for shape description. *Comput. Digital Techniques* **2**, 157–165.

Batchelor, B.G. (1991). *Intelligent Image Processing in Prolog*. Springer-Verlag, London.

Batchelor, B.G. (2003). Like two peas in a pod. Chapter 1 in Graves and Batchelor (2003), ibid., 1–33.

Batchelor, B.G., Davies, E.R. and Graves, M. (2004). Using X-rays to detect foreign bodies in food materials and packs. In Edwards, M. (ed.), *Detecting Foreign Bodies in Food*. Cambridge, UK, Woodhead Publishing Ltd.

Batchelor, B.G., Hill, D.A. and Hodgson, D.C. (1985). *Automated Visual Inspection*. IFS (Publications) Ltd., Bedford, UK/North-Holland, Amsterdam.

Batlle, J., Marti, J., Ridao, P. and Amat, J. (2002). A new FPGA/DSP-based parallel architecture for real-time image processing. *Real-Time Imaging* **8**, no. 5, 345–356.

Baumberg, A. and Hogg, D. (1995). An adaptive eigenshape model. *Proc. British Machine Vision Assoc. Conf.* (*September*), pp. 87–96.

Beaton, A.E. and Tukey, J.W. (1974). The fitting of power series, meaning polynomials, illustrated on band-spectroscopic data. *Technometrics* **16**, no. 2, 147–185.

Beaudet, P.R. (1978). Rotationally invariant image operators. *Proc. 4th Int. Conf. on Pattern Recogn., Kyoto*, pp. 579–583.

Beckers, A.L.D. and Smeulders, A.W.M. (1989). A comment on "A note on 'Distance transformations in digital images'" *Comput. Vision Graph. Image Process.* **47**, 89–91.

Beiden, S.V., Maloof, M.A. and Wagner, R.F. (2003). A general model for finite-sample effects in training and testing of competing classifiers. *IEEE Trans. Pattern Anal. Mach. Intell.* **25**, no. 12, 1561–1569.

Benke, K.K. and Skinner, D.R. (1987). Segmentation of visually similar textures by convolution filtering. *Australian Comput. J.* **19**, no. 3, 134–139.

Bergholm, F. (1986). Edge focusing. *Proc. 8th Int. Conf. on Pattern Recogn., Paris* (*October 27–31*), pp. 597–600.

Berman, S., Parikh, P. and Lee, C.S.G. (1985). Computer recognition of two overlapping parts using a single camera. *IEEE Comput.* (March), 70–80.

Bertozzi, M. and Broggi, A. (1998). GOLD: A parallel real-time stereo vision system for generic obstacle and lane detection. *IEEE Trans. Image Process.* **7**, no. 1, 62–81.

Besl, P.J. and Jain, R.C. (1984). Three-dimensional object recognition. *Comput. Surveys* **17**, 75–145.

Besl, P.J., Birch, J.B. and Watson, L.T. (1989). Robust window operators. *Machine Vision and Applications* **2**, 179–191.

Besl, P.J., Delp, E.J. and Jain, R. (1985). Automatic visual solder joint inspection. *IEEE J. Robot. Automation* **1**, 42–56.

Beun, M. (1973). A flexible method for automatic reading of handwritten numerals. *Philips Techn. Rev.* **33**, 89–101; 130–137.

Beynon, J.D.E. and Lamb, D.R. (eds.) (1980). *Charge-coupled Devices and Their Applications*. McGraw-Hill, London.

Bhanu, B. and Faugeras, O.D. (1982). Segmentation of images having unimodal distributions. *IEEE Trans. Pattern Anal. Mach. Intell.* **4**, 408–419.

Bhattacharya, A.K. and Roysam, B. (1994). Joint solution of low, intermediate and high-level vision tasks by evolutionary optimization: application to computer vision at low SNR. *IEEE Trans. Neural Networks* **5**, no. 1, 83–95.

Biberman, L.M. and Nudelman, S. (eds.) (1971). *Photoelectronic Imaging Devices. Vol. 2. Devices and Their Evaluation.* Plenum Press, New York.

Billingsley, J. (ed.) (1985). *Robots and Automated Manufacture*. IEE Control Engineering Series 28. Peter Peregrinus Ltd., London.

Billingsley, J., and Schoenfisch, M. (1995). Vision-guidance of agricultural vehicles. Autonomous Robots **2**, no. 1, 65–76.

Bishop, C. (1995). *Neural Networks for Pattern Recognition*. Oxford, Oxford University Press.

Blake, A., and Yuille, A. (eds.) (1992). *Active Vision*. MIT Press, Cambridge, MA; London, England.

Blake, A., Zisserman, A. and Knowles, G. (1985). Surface descriptions from stereo and shading. *Image Vision Comput.* **3**, 183–191.

Bledsoe, W.W. and Browning, I. (1959). Pattern recognition and reading by machine. *Proc. Eastern Joint Comput. Conf.*, pp. 225–232.

Blostein, D. and Ahuja, N. (1989). Shape from texture: integrating texture-element extraction and surface estimation. *IEEE Trans. Pattern Anal. Mach. Intell.* **11**, no. 12, 1233–1251.

Blum, H. (1967). A transformation for extracting new descriptors of shape. In Wathen-Dunn, W. (ed.), *Models for the Perception of Speech and Visual Form.* MIT Press, Cambridge, MA, pp. 362–380.

Blum, H., and Nagel, R.N. (1978). Shape description using weighted symmetric axis features. *Pattern Recogn.* **10**, 167–180.

Bober, M. and Kittler, J. (1993). Estimation of complex multimodal motion: an approach based on robust statistics and Hough transform. *Proc. 4th British Machine Vision Assoc. Conf., Univ. of Surrey (September 21–23)*, Vol. 1, pp. 239–248.

Boerner, H. and Strecker, H. (1988). Automated X-ray inspection of aluminium castings. *IEEE Trans. Pattern Anal. Mach. Intell.* **10**, no. 1, 79–91.

Bolles, R.C. (1979). Robust feature matching via maximal cliques. *SPIE, 182. Proc. Technical Symposium on Imaging Applications for Automated Industrial Inspection and Assembly, Washington D.C. (April)*, pp. 140–149.

Bolles, R.C. and Cain, R.A. (1982). Recognizing and locating partially visible objects: the local-feature-focus method. *Int. J. Robot. Res.* **1**, no. 3, 57–82.

Bolles, R.C. and Horaud, R. (1986). 3DPO: a three-dimensional part orientation system. *Int. J. Robot. Res.* **5**, no. 3, 3–26.

Bors, A.G., Hancock, E.R. and Wilson, R.C. (2003). Terrain analysis using Radar shape-from-shading. *IEEE Trans. Pattern Anal. Mach. Intell.*, **25**, no. 8, 974–992.

Boufama, B., Mohr, R. and Morin, L. (1998). Using geometric properties for automatic object positioning. In *Special Issue on Geometric Modelling and Invariants for Computer Vision, Image Vision Comput.* **16**, no. 1, 27–33.

Bourland, H. and Kamp, Y. (1988). Auto-association by multilayer perceptrons and singular value decomposition. *Biol. Cybern.* **59**, 291–294.

Bovik, A.C., Huang, T.S. and Munson, D.C. (1983). A generalization of median filtering using linear combinations of order statistics. *IEEE Trans. Acoustics, Speech Signal Process.* **31**, no. 6, 1342–1349.

Bovik, A.C., Huang, T.S. and Munson, D.C. (1987). The effect of median filtering on edge estimation and detection. *IEEE Trans. Pattern Anal. Mach. Intell.* **9**, 181–194.

Brady, J.M. and Wang, H. (1992). Vision for mobile robots. *Phil. Trans. R. Soc. (London)*, **B337**, 341–350.

Brady, J.M. and Yuille, A. (1984). An extremum principle for shape from contour. *IEEE Trans. Pattern Anal. Mach. Intell.* **6**, 288–301.

Brady, M. (1982). Computational approaches to image understanding. *Comput. Surveys* **14**, 3–71.

Breiman, L. (1996). Bagging predictors. *Machine Learning* **24**, no. 2, 123–140.

Bretschi, J. (1981). *Automated Inspection Systems for Industry.* IFS Publications Ltd., Bedford, UK.

Brink, A.D. (1992). Thresholding of digital images using two-dimensional entropies. *Pattern Recogn.* **25**, 803–808.

Brivot, R. and Marchant, J.A. (1996). Segmentation of plants and weeds for a precision crop protection robot using infrared images. *IEE Proc. Vision Image Signal Process.* **143**, no. 2, 118–124.

Brodatz, P. (1966). *Textures: A Photographic Album for Artists and Designers.* New York, Dover Publications.

Broggi, A., Bertozzi, M. and Fascioli, A. (2000a). Architectural issues on vision-based automatic vehicle guidance: the experience of the ARGO project. *Real-Time Imaging* **6**, no. 4, 313–324.

Broggi, A., Bertozzi, M., Fascioli, A. and Sechi, M. (2000b). Shape-based pedestrian detection. *Proc. IEEE Intell. Vehicles Symp., Dearborn (MI), USA (October 3–5)*, pp. 215–220.

Bron, C. and Kerbosch, J. (1973). Algorithm 457: finding all cliques in an undirected graph [H]. *Comm. ACM* **16**, 575–577.

Brooks, M.J. (1976). Locating intensity changes in digitised visual scenes. *Computer Science Memo-15* (from M.Sc. thesis), University of Essex.

Brooks, M.J. (1978). Rationalising edge detectors. *Comput. Graph. Image Process.* **8**, 277–285.

Brown, C.M. (1983). Inherent bias and noise in the Hough transform. *IEEE Trans. Pattern Anal. Mach. Intell.* **5**, 493–505.

Brown, C.M. (1984). Peak-finding with limited hierarchical memory. *Proc. 7th Int. Conf. on Pattern Recogn., Montreal (July 30–August 2)*, pp. 246–249.

Brown, J. and Glazier, E.V.D. (1974). *Telecommunications.* Chapman and Hall, London (2nd edition).

Brown, M.Z., Burschka, D. and Hager, G.D. (2003). Advances in computational stereo. *IEEE Trans. Pattern Anal. Mach. Intell.* **25**, no. 8, 993–1008.

Browne, A. and Norton-Wayne, L. (1986). *Vision and Information Processing for Automation.* Plenum Press, New York.

Bruckstein, A.M. (1988). On shape from shading. *Comput. Vision Graph. Image Process.* **44**, 139–154.

Bryant, D.J. and Bouldin, D.W. (1979). Evaluation of edge operators using relative and absolute grading. *Proc. IEEE Comput. Soc. Conf. on Pattern Recogn. and Image Process. Chicago (May 31–June 2)*, pp. 138–145.

Bunke, H. (1999). Error correcting graph matching: on the influence of the underlying cost function. *IEEE Trans. Pattern Anal. Mach. Intell.* **21**, no. 9, 917–922.

Bunke, H. and Shearer, K. (1998). A graph distance metric based on the maximal common subgraph. *Pattern Recogn. Lett.* **19**, 255–259.

Burr, D.J. and Chien, R.T. (1977). A system for stereo computer vision with geometric models. *Proc. 5th Int. Joint Conf. on Artif. Intell., Boston*, p. 583.

Califano, A. and Mohan, R. (1994). Multidimensional indexing for recognizing visual shapes. *IEEE Trans. Pattern Anal. Mach. Intell.* **16**, no. 4, 373–392.

Canny, J. (1986). A computational approach to edge detection. *IEEE Trans. Pattern Anal. Mach. Intell.* **8**, 679–698

Cantoni, V. and Levialdi, S. (1983). Matching the task to an image processing architecture. *Comput. Vision Graph. Image Process.* **22**, 301–309.

Chakravarty, I. and Freeman, H. (1982). Characteristic views as a basis for three-dimensional object recognition. *Proc. Soc. Photo-opt. Instrum. Eng. Conf. Robot Vision* **336**, 37–45.

Chakravarty, V.S. and Kompella, B. (2003). The shape of handwritten characters. *Pattern Recogn. Lett.* **24**, no. 12, 1901–1913.

Chan, J.P., Batchelor, B.G., Harris, I.P. and Perry, S.J. (1990). Intelligent visual inspection of food products. *Proc. SPIE Conf. on Machine Vision Systems in Industry*, **1386**, 171–179.

Chang, S., Davis, L.S., Dunn, S.M., Eklundh, J.-O. and Rosenfeld, A. (1987). Texture discrimination by projective invariants. *Pattern Recogn. Lett.* **5**, no. 5, 337–342.

Charles, D. and Davies, E.R. (2003a). Properties of the mode filter when applied to colour images. *Proc. IEE Int. Conf. on Visual Information Engineering*, VIE 2003, Surrey, July 7–9, *IEE Conference Publication* 495, pp. 101–104.

Charles, D. and Davies, E.R. (2003b). Distance-weighted median filters and their application to colour images. *Proc. IEE Int. Conf. on Visual Information Engineering*, VIE 2003, Surrey, July 7–9, *IEE Conference Publication* 495, pp. 117–120.

Charles, D. and Davies, E.R. (2004). Mode filters and their effectiveness for processing colour images. *Imaging Science Journal* **52**, no. 1, 3–25.

Charniak, E. and McDermott, D. (1985). *Introduction to Artificial Intelligence*. Addison-Wesley, Reading, MA.

Chasles, M. (1855). Question no. 296. *Nouv. Ann. Math.* **14**, p. 50.

Chauduri, B.B. (1994). Dynamic clustering for time incremental data. *Pattern Recogn. Lett.* **15**, no. 1, 27–34.

Chauduri, S., Chatterjee, S., Katz, N., Nelson, M. and Goldbaum, M. (1989). Detection of blood vessels in retinal images using two-dimensional matched filters. *IEEE Trans. Medical Imaging* **8**, no. 3, 263–269.

Chen, T.C. (1971). Parallelism, pipelining and computer efficiency. *Comput. Design* (January), 69–74.

Cheng, S.-C. (2003). Content-based image retrieval using moment-preserving edge detection. *Image Vision Comput.* **21**, no. 9, 809–826.

Chiang, Y.P. and Fu, K.-S. (1983). Matching parallel algorithm and architecture. In *IEEE Proceedings of the International Conference on Parallel Processing*. Columbus, Ohio, US, Computer Society Press, pp. 374–380.

Chittineni, C.B. (1980). Efficient feature subset selection with probabilistic distance criteria. *Inf. Sci.* **22**, 19–35.

Chojnacki, W., Brooks, M.J., van den Hengel, A. and Gawley, D. (2003). Revisiting Hartley's normalized eight-point algorithm. *IEEE Trans. Pattern Anal. Mach. Intell.* **25**, no. 9, 1172–1177.

Chow, C.K. and Kaneko, T. (1972). Automatic boundary detection of the left ventricle from cineangiograms. *Comput. Biomed. Res.* **5**, 388–410.

Choy, S.S.O., Choy, C.S.-T. and Siu, W.-C. (1995). New single-pass algorithm for parallel thinning. *Comput. Vision Image Understanding* **62**, no. 1, 69–77.

Christmas, W.J., Kittler, J. and Petrou, M. (1995). Structural matching in computer vision using probabilistic relaxation. *IEEE Trans. Pattern Anal. Mach. Intell.* **17**, no. 8, 749–764.

Cipolla, R. and Giblin, P. (2000). *Visual Motion of Curves and Surfaces*. Cambridge University Press, Cambridge, UK.

Clark, P. and Mirmehdi, M. (2000). Location and recovery of text on oriented surfaces. *Proc. SPIE Conf. on Document Recognition and Retrieval VII* (*January*) **3967**, pp. 267–277.

Clark, P. and Mirmehdi, M. (2002). On the recovery of oriented documents from single images. *Proc. Advanced Concepts for Intelligent Vision Systems* (*ACIVS*), *Ghent, Belgium* (*September 9–11*), 190–197.

Clerc, M. and Mallat, S. (2002). The texture gradient equation for recovering shape from texture. *IEEE Trans. Pattern Anal. Mach. Intell.* **24**, no. 4, 536–549.

Clocksin, W.F. and Mellish, C.S. (1984). *Programming in Prolog*. Springer-Verlag, New York.

Coeurjolly, D. and Klette, R. (2004). A comparative evaluation of length estimators of digital curves. *IEEE Trans. Pattern Anal. Mach. Intell.* **26**, no. 2, 252–258.

Cohen, F.S. and Cooper, D.B. (1987). Simple parallel hierarchical and relaxation algorithms for segmenting noncausal Markovian random fields. *IEEE Trans. Pattern Anal. Mach. Intell.* **9**, no. 2, 195–219.

Coleman, G.B. and Andrews, H.C. (1979). Image segmentation by clustering. *Proc. IEEE* **67**, 773–785.

Collins, R.T., Lipton, A.J. and Kanade, T. (eds.) (2000). Special Section on Video Surveillance. *IEEE Trans. Pattern Anal. Mach. Intell.* **22**, no. 8.

Conners, R.W. and Harlow, C.A. (1980a). A theoretical comparison of texture algorithms. *IEEE Trans. Pattern Anal. Mach. Intell.* **2**, no. 3, 204–222.

Conners, R.W. and Harlow, C.A. (1980b). Toward a structural textural analyzer based on statistical methods. *Comput. Graph. Image Process.* **12**, 224–256.

Cook, R.L. and Torrance, K.E. (1982). A reflectance model for computer graphics. *ACM Trans. Graphics* **1**, 7–24.

Cootes, T.F., Taylor, C.J., Cooper, D.H. and Graham, J. (1992). Training models of shape from sets of examples. *Proc. 3rd British Machine Vision Assoc. Conf., Leeds,* (*September 22–24*) pp. 9–18.

Corneil, D.G. and Gottlieb, C.C. (1970). An efficient algorithm for graph isomorphism. *J. ACM* **17**, 51–64.

Costa, L. da F. and Cesar, R.M. (2000). *Shape Analysis and Classification: Theory and Practice*. CRC Press, Boca Raton, FL.

Cottrell, G.W., Munro, P. and Zipser, D. (1987). Learning internal representations from grey-scale images: an example of extensional programming. *Proc. 9th Annual Conf. of the Cognitive Science Soc.*, Seattle (Erlbaum, Hillsdale), pp. 462–473.

Cowan, G. (1998). *Statistical Data Analysis*. Oxford University Press, Oxford.

Crimmins, T.R. and Brown, W.R. (1985). Image algebra and automatic shape recognition. *IEEE Trans. Aerospace and Electronic Systems* **21**, 60–69.

Cristianini, N. and Shawe-Taylor, J. (2000). *An Introduction to Support Vector Machines*. Cambridge University Press, Cambridge, UK.

Cross, A.D.J., Wilson, R.C. and Hancock, E.R. (1997). Inexact graph matching with genetic search. *Pattern Recogn.* **30**, no. 6, 953–970.

Cross, G.R. and Jain, A.K. (1983). Markov random field texture models. *IEEE Trans. Pattern Anal. Mach. Intell.* **5**, no. 1, 25–39.

Crowley, J.L., Bobet, P. and Schmid, C. (1993). Auto-calibration by direct observation of objects. *Image Vision Comput.* **11**, no. 2, 67–81.

Cucchiara, R., Grana, C., Piccardi, M. and Prati, A. (2003). Detecting moving objects, ghosts, and shadows in video streams. *IEEE Trans. Pattern Anal. Mach. Intell.* **25**, no. 10, 1337–1342.

Cumani, A. and Guiducci, A. (1995). Geometric camera calibration: the virtual camera approach. *Machine Vision and Applications* **8**, no. 6, 375–384.

Cybenko, G. (1988). Continuous valued neural networks with two hidden layers are sufficient. *Techn. Report*, Dept. of Comput. Sci., Tufts University, Medford, MA.

Cybenko, G. (1989). Approximation by superpositions of a sigmoidal function. *Mathematics of Control, Signals and Systems* **2**, no. 4, 303–314.

da Gama Leitão, H.C. and Stolfi, J. (2002). A multiscale method for the reassembly of two-dimensional fragmented objects. *IEEE Trans. Pattern Anal. Mach. Intell.* **24**, no. 9, 1239–1251.

Danielsson, P.-E. (1981). Getting the median faster. *Comput. Graph. Image Process.* **17**, 71–78.

Danielsson, P.-E. and Levialdi, S. (1981). Computer architectures for pictorial information systems. *IEEE Comput.* (*November*) **14**, 53–67.

Daugman, J.G. (1993). High confidence visual recognition of persons by a test of statistical independence. *IEEE Trans. Pattern Anal. Mach. Intell.* **15**, 1148–1161.

Daugman, J.G. (2003). Demodulation by complex-valued wavelets for stochastic pattern recognition. *Int. J. of Wavelets, Multiresolution and Information Processing* **1**, no. 1, 1–17.

Davies, E.R. (1982). Image processing. In Sumner, F.H. (ed.), *State of the Art Report: Supercomputer Systems Technology*. Pergamon Infotech, Maidenhead, pp. 223–244.

Davies, E.R. (1983). Image processing—its milieu, its nature and constraints on the design of special architectures for its implementation. In Duff, M.J.B. (ed.), *Computing Structures for Image Processing*. Academic Press, London, pp. 57–76.

Davies, E.R. (1984a). The median filter: an appraisal and a new truncated version, *Proc. 7th Int. Conf. on Pattern Recogn., Montreal* (*July 30–August 2*), pp. 590–592.

Davies, E.R. (1984b). Circularity—a new principle underlying the design of accurate edge orientation operators. *Image Vision Comput.* **2**, 134–142.

Davies, E.R. (1984c). Design of cost-effective systems for the inspection of certain food products during manufacture. In Pugh, A. (ed.), *Proceedings of the 4th International Conference on Robot Vision and Sensory Controls, London* (*October 9–11*), IFS (Publications) Ltd., Bedford and North-Holland, Amsterdam, pp. 437–446.

Davies, E.R. (1984d). A glance at image analysis—how the robot sees. *Chartered Mech. Eng.* (*December*), 32–35.

Davies, E.R. (1985a). A comparison of methods for the rapid location of products and their features and defects. In McKeown, P.A. (ed.), *Proceedings of the 7th International Conference on Automated Inspection and Product Control, Birmingham, AL* (*March 26–28*), pp. 111–120.

Davies, E.R. (1985b). Radial histograms as an aid in the inspection of circular objects. *IEE Proc. D* **132** (*4, Special Issue on Robotics*), 158–163.

Davies, E.R. (1985c). Precise measurement of radial dimensions in automatic visual inspection and quality control—a new approach. In Billingsley, J. (ed.), *Robots and Automated Manufacture*. IEE Control Engineering Series 28. Peter Peregrinus Ltd., London, pp. 157–171.

Davies, E.R. (1986a). Constraints on the design of template masks for edge detection. *Pattern Recogn. Lett.* **4**, 111–120.

Davies, E.R. (1986b). Corner detection using the generalised Hough transform. *Proceedings of the 2nd International Conference on Image Processing and Its Applications (June 24–26), IEE Conf. Publ.* **265**, 175–179.

Davies, E.R. (1986c). Image space transforms for detecting straight edges in industrial images. *Pattern Recogn. Lett.* **4**, 185–192.

Davies, E.R. (1986d). Reduced parameter spaces for polygon detection using the generalised Hough transform. *Proc. 8th Int. Conf. on Pattern Recogn., Paris (October 27–31)*, pp. 495–497.

Davies, E.R. (1987a). Methods for the rapid inspection of food products and small parts. In McGeough, J.A. (ed.), *Proceedings of the 2nd International Conference on Computer-Aided Production Engineering, Edinburgh (April 13–15)*, pp. 105–110.

Davies, E.R. (1987b). Visual inspection, automatic (robotics). In Meyers, R.A. (ed.), *Encyclopedia of Physical Science and Technology*, Vol. 14. Academic Press, San Diego, pp. 360–377.

Davies, E.R. (1987c). A new framework for analysing the properties of the generalised Hough transform. *Pattern Recogn. Lett.* **6**, 1–7.

Davies, E.R. (1987d). A new parametrisation of the straight line and its application for the optimal detection of objects with straight edges. *Pattern Recogn. Lett.* **6**, 9–14.

Davies, E.R. (1987e). Design of optimal Gaussian operators in small neighbourhoods. *Image Vision Comput.* **5**, 199–205.

Davies, E.R. (1987f). Design of robust algorithms for automated visual inspection. *Proc. R. Swedish Acad. Eng. Sci. Symp. on Machine Vision–the Eyes of Automation, Stockholm (August 27), IVA Rapp.* **336**, 55–81.

Davies, E.R. (1987g). Lateral histograms for efficient object location: speed versus ambiguity. *Pattern Recogn. Lett.* **6**, 189–198.

Davies, E.R. (1987h). The performance of the generalised Hough transform: concavities, ambiguities and positional accuracy. *Proc. 3rd Alvey Vision Conf., Cambridge (September 15–17)*, pp. 327–333.

Davies, E.R. (1987i). Industrial vision systems: segmentation of gray-scale images. In *Encyclopedia of Systems and Control*, Vol. 4. Pergamon, Oxford, pp. 2479–2484.

Davies, E.R. (1987j). Improved localisation in a generalised Hough scheme for the detection of straight edges. *Image Vision Comput.* **5**, 279–286.

Davies, E.R. (1987k). The effect of noise on edge orientation computations. *Pattern Recogn. Lett.* **6**, 315–322.

Davies, E.R. (1987l). A high speed algorithm for circular object location. *Pattern Recogn. Lett.* **6**, 323–333.

Davies, E.R. (1988a). Application of the generalised Hough transform to corner detection. *IEE Proc. E* **135**, 49–54.

Davies, E.R. (1988b). A modified Hough scheme for general circle location. *Pattern Recogn. Lett.* **7**, 37–43.

Davies, E.R. (1988c). On the noise suppression and image enhancement characteristics of the median, truncated median and mode filters. *Pattern Recogn. Lett.* **7**, 87–97.

Davies, E.R. (1988d). Median-based methods of corner detection. In Kittler, J. (ed.), *Proceedings of the 4th BPRA International Conference on Pattern Recognition, Cambridge (March 28–30)*. Lecture Notes in Computer Science, Vol. 301. Springer-Verlag, Heidelberg, pp. 360–369.

Davies, E.R. (1988e). A hybrid sequential-parallel approach to accurate circle centre location. *Pattern Recogn. Lett.* **7**, 279–290.

Davies, E.R. (1988f). Training sets and *a priori* probabilities with the nearest neighbour method of pattern recognition. *Pattern Recogn. Lett.* **8**, 11–13.

Davies, E.R. (1988g). An alternative to graph matching for locating objects from their salient features. *Proc. 4th Alvey Vision Conf., Manchester* (*August 31–September 2*), pp. 281–286.

Davies, E.R. (1988h). Efficient image analysis techniques for automated visual inspection. *Proc. Unicom Semin. on Comput. Vision and Image Process., London* (*November 29–December 1*), pp. 1–19.

Davies, E.R. (1988i). Tradeoffs between speed and accuracy in two-stage template matching. *Signal Process.* **15**, 351–363.

Davies, E.R. (1989a). Finding ellipses using the generalised Hough transform. *Pattern Recogn. Lett.* **9**, 87–96.

Davies, E.R. (1989b). Edge location shifts produced by median filters: theoretical bounds and experimental results. *Signal Process.* **16**, 83–96.

Davies, E.R. (1989c). Minimising the search space for polygon detection using the generalised Hough transform. *Pattern Recogn. Lett.* **9**, 181–192.

Davies, E.R. (1989d). Occlusion analysis for object detection using the generalised Hough transform. *Signal Process.* **16**, 267–277.

Davies, E.R. (1990). *Machine Vision: Theory, Algorithms, Practicalities.* Academic Press, London.

Davies, E.R. (1991a). The minimal match graph and its use to speed identification of maximal cliques. *Signal Process.* **22**, no. 3, 329–343.

Davies, E.R. (1991b). Median and mean filters produce similar shifts on curved boundaries. *Electronics Lett.* **27**, no. 10, 826–828.

Davies, E.R. (1991c). Insight into operation of Kulpa boundary distance measure. *Electronics Lett.* **27**, no. 13, 1178–1180.

Davies, E.R. (1992a). Simple fast median filtering algorithm, with application to corner detection. *Electronics Lett.* **28**, no. 2, 199–201.

Davies, E.R. (1992b). Modelling peak shapes obtained by Hough transform. *IEE Proc. E* **139**, no. 1, 9–12.

Davies, E.R. (1992c). A skimming technique for fast accurate edge detection. *Signal Process.* **26**, no. 1, 1–16.

Davies, E.R. (1992d). Locating objects from their point features using an optimised Hough-like accumulation technique. *Pattern Recogn. Lett.* **13**, no. 2, 113–121.

Davies, E.R. (1992e). Procedure for generating template masks for detecting variable signals. *Image Vision Comput.* **10**, no. 4, 241–249.

Davies, E.R. (1992f). Accurate filter for removing impulse noise from one- or two-dimensional signals. *IEE Proc. E* **139**, no. 2, 111–116.

Davies, E.R. (1992g). Simple two-stage method for the accurate location of Hough transform peaks. *IEE Proc. E* **139**, no. 3, 242–248.

Davies, E.R. (1992h). A framework for designing optimal Hough transform implementations. *Proc. 11th IAPR Int. Conf. on Pattern Recogn., The Hague* (*August 30–September 3*), Vol. III, pp. 509–512.

Davies, E.R. (1993a). *Electronics, Noise and Signal Recovery.* Academic Press, London.

Davies, E.R. (1993b). Computationally efficient Hough transform for 2-D object location. *Proc. 4th British Machine Vision Assoc. Conf., Univ. of Surrey* (*September 21–23*), Vol. 1, pp. 259–268.

Davies, E.R. (1995). Machine vision in manufacturing—what are the real problems? *Proc. 2nd Int. Conf.* on *Mechatronics and Machine Vision in Practice, Hong Kong* (*September 12–14*), pp. 15–24.

Davies, E.R. (1997a). *Machine Vision: Theory, Algorithms, Practicalities*. Academic Press, London (2nd edition).

Davies, E.R. (1997b). Algorithms for inspection: constraints, tradeoffs and the design process. *IEE Digest no. 1997/041, Colloquium on Industrial Inspection, IEE* (*February 10*), pp. 6/1–5.

Davies, E.R. (1997c). Shifts produced by mode filters on curved intensity contours. *Electronics Lett.* **33**, no. 5, 381–382.

Davies, E.R. (1997d). Principles and design graphs for obtaining uniform illumination in automated visual inspection. *Proc. 6th IEE Int. Conf. on Image Processing and its Applications, Dublin* (*July 14–17*), *IEE Conf. Publication* no. 443, pp. 161–165.

Davies, E.R. (1997e). Designing efficient line segment detectors with high orientation accuracy. *Proc. 6th IEE Int. Conf. on Image Processing and Its Applications, Dublin* (*July 14–17*), *IEE Conf. Publication* no. 443, pp. 636–640.

Davies, E.R. (1997f). Lower bound on the processing required to locate objects in digital images. *Electronics Lett.* **33**, no. 21, 1773–1774.

Davies, E.R. (1997g). Vectorial strategy for designing line segment detectors with high orientation accuracy. *Electronics Lett.* **33**, no. 21, 1775–1777.

Davies, E.R. (1998a). Image distortions produced by mean, median and mode filters. *IEE Digest no. 1998/284, Colloquium on Non-Linear Signal and Image Processing, IEE* (*May 22*), pp. 6/1–5.

Davies, E.R. (1998b). Rapid location of convex objects in digital images. *Proc. European Signal Processing Conf. (EUSIPCO'98), Rhodes, Greece* (*September 8–11*), pp. 589–592.

Davies, E.R. (1998c). From continuum model to a detailed discrete theory of median shifts. *Proc. European Signal Processing Conf. (EUSIPCO'98), Rhodes, Greece* (*September 8–11*), pp. 805–808.

Davies, E.R. (1999a). Designing optimal image feature detection masks: equal area rule. *Electronics Lett.* **35**, no. 6, 463–465.

Davies, E.R. (1999b). Chord bisection strategy for fast ellipse location. *Electronics Lett.* **35**, no. 9, 703–705.

Davies, E.R. (1999c). Effect of foreground and background occlusion on feature matching for target location. *Electronics Lett.* **35**, no. 11, 887–889.

Davies, E.R. (1999d). High precision discrete model of median shifts. *Proc. 7th IEE Int. Conf. on Image Processing and Its Applications, Manchester* (*July 13–15*), *IEE Conf. Publication* no. 465, pp. 197–201.

Davies, E.R. (1999e). Algorithms for ultra-fast location of ellipses in digital images. *Proc. 7th IEE Int. Conf. on Image Processing and Its Applications, Manchester* (*July 13–15*), *IEE Conf. Publication* no. 465, pp. 542–546.

Davies, E.R. (1999f). Isotropic masks make efficient linear feature detectors. *Electronics Lett.* **35**, no. 17, 1450–1451.

Davies, E.R. (1999g). Image distortions produced by mean, median and mode filters. *IEE Proc. Vision Image Signal Processing* **146**, no. 5, 279–285.

Davies, E.R. (2000a). *Image Processing for the Food Industry*. World Scientific, Singapore.

Davies, E.R. (2000b). Obtaining optimum signal from set of directional template masks. *Electronics Lett.* **36**, no. 15, 1271–1272.

Davies, E.R. (2000c). Resolution of problem with use of closing for texture segmentation. *Electronics Lett.* **36**, no. 20, 1694–1696.

Davies, E.R. (2000d). Low-level vision requirements. *Electron. Commun. Eng. J.* **12**, no. 5, 197–210.

Davies, E.R. (2000e). Accuracy of multichannel median filter. *Electronics Lett.* **36**, no. 25, 2068–2069.

Davies, E.R. (2000f). A generalized model of the geometric distortions produced by rank-order filters. *Imaging Science Journal* **48**, no. 3, 121–130.

Davies, E.R. (2001a). Some problems in food and cereals inspection and methods for their solution. *Proc. Int. Conf. on Quality Control by Artificial Vision – 2001, Le Creusot, France* (*May 21–23*), pp. 35–46.

Davies, E.R. (2001b). A sampling approach to ultra-fast object location. *Real-Time Imaging*, **7**, no. 4, 339–355.

Davies, E.R. (2003a). Fundamentals of machine vision and their importance for real mechatronic applications. Chapter 7 in Cho, H. (ed.), *Opto-Mechatronic Systems Handbook*, CRC Press, Boca Raton, FL, 7.1–7.34.

Davies, E.R. (2003b). Truncating the Hough transform parameter space can be beneficial. *Pattern Recogn. Lett.* **24**, nos. 1–3, 129–135.

Davies, E.R. (2003c). Formulation of an accurate discrete theory of median shifts. *Signal Process.* **83**, 531–544.

Davies, E.R. (2003d). Design of real-time algorithms for food and cereals inspection. *Imaging Science* **51**, no. 2, 63–78.

Davies, E.R. (2003e). An analysis of the geometric distortions produced by median and related image processing filters. *Advances in Imaging and Electron Physics* **126**, 93–193.

Davies, E.R. and Barker, S.P. (1990). An analysis of hole detection schemes. *Proc. British Machine Vision Assoc. Conf., Oxford* (*September 24–27*), pp. 285–290.

Davies, E.R. and Celano, D. (1993a). Orientation accuracy of edge detection operators acting on binary and saturated grey-scale images. *Electronics Lett.* **29**, no. 7, 603–604.

Davies, E.R. and Celano, D. (1993b). Analysis of skeleton junctions in 3×3 windows. *Electronics Lett.* **29**, no. 16, 1440–1441.

Davies, E.R. and Johnstone, A.I.C. (1986). Engineering trade-offs in the design of a real-time system for the visual inspection of small products. In *Proceedings of the 4th Conference on UK Research in Advanced Manufacture* (*December 10–11*). IMechE Conference Publications, London, pp. 15–22.

Davies, E.R. and Johnstone, A.I.C. (1989). Methodology for optimising cost/speed tradeoffs in real-time inspection hardware. *IEE Proc. E* **136**, 62–69.

Davies, E.R. and Plummer, A.P.N. (1980). A new method for the compression of binary picture data. *Proc. 5th Int. Conf. on Pattern Recogn., Miami Beach, Florida* (*IEEE Comput. Soc.*), pp. 1150–1152.

Davies, E.R. and Plummer, A.P.N. (1981). Thinning algorithms: a critique and a new methodology. *Pattern Recogn.* **14**, 53–63.

Davies, E.R., and Ip, H.H.S. (eds.) (1998). *Special Issue on Real-Time Visual Monitoring and Inspection. Real-Time Imaging* **4**, no. 5.

Davies, E.R., Patel, D. and Johnstone, A.I.C. (1995). Crucial issues in the design of a real-time contaminant detection system for food products. *Real-Time Imaging*, **1**, no. 6, 397–407.

Davies, E.R., Bateman, M., Chambers, J. and Ridgway, C. (1998a). Hybrid nonlinear filters for locating speckled contaminants in grain. *IEE Digest no. 1998/284, Colloquium on Non-Linear Signal and Image Processing, IEE (May 22)*, pp. 12/1–5.

Davies, E.R., Mason, D.R., Bateman, M., Chambers, J. and Ridgway, C. (1998b). Linear feature detectors and their application to cereal inspection. *Proc. European Signal Processing Conf. (EUSIPCO'98), Rhodes, Greece (September 8–11)*, pp. 2561–2564.

Davies, E.R., Chambers, J. and Ridgway, C. (2002). Combination linear feature detector for effective location of insects in grain images. *Measurement Sci. Technology*, **13**, no. 12, 2053–2061.

Davies, E.R., Bateman, M., Mason, D.R., Chambers, J. and Ridgway, C. (2003a). Design of efficient line segment detectors for cereal grain inspection. *Pattern Recogn. Lett.* **24**, nos. 1–3, 421–436.

Davies, E.R., Chambers, J. and Ridgway, C. (2003b). Design of a real-time grain inspection system with high sensitivity for insect detection. *Proc. Int. Conf. on Mechatronics (ICOM 2003), Loughborough, (June 18–20)*, Parkin, R.M. Al-Habaibeh, A. and Jackson, M.R. (eds.), Professional Engineering Publishing, London and Bury St. Edmunds, UK, pp. 377–382.

Davies, E.R., Ridgway, C. and Chambers, J. (2003c). NIR detection of grain weevils inside wheat kernels. *Proc. IEE Int. Conf. on Visual Information Engineering, VIE 2003, Surrey, (July 7–9), IEE Conference Publication* 495, pp. 173–176.

Davis, L.S. (1979). Computing the spatial structure of cellular textures. *Comput. Graph. Image Process.* **11**, no. 2, 111–122.

Davison, A.J. and Murray, D.W. (2002). Simultaneous localization and map-building using active vision. *IEEE Trans. Pattern Anal. Mach. Intell.* **24**, no. 7, 865–880.

de la Escalara, A., Armingol, J.M[a] and Mata, M. (2003). Traffic sign recognition and analysis for intelligent vehicles. *Image Vision Comput.* **21**, no. 3, 247–258.

Deans, S.R. (1981). Hough transform from the Radon transform. *IEEE Trans. Pattern Anal. Mach. Intell.* **3**, 185–188.

Delagnes, P., Benois, J. and Barba, D. (1995). Active contours approach to object tracking in image sequences with complex background. *Pattern Recogn. Lett.* **16**, no. 2, 171–178.

Deravi, F. and Pal, S.K. (1983). Grey level thresholding using second-order statistics. *Pattern Recogn. Lett.* **1**, 417–422.

Derganc, J., Likar, B., Bernard, R., Tomaževič, D. and Pernuš, F. (2003). Real-time automated visual inspection of color tablets in pharmaceutical blisters. *Real-Time Imaging* **9**, no. 2, 113–124.

Derin, H. and Elliott, H. (1987). Modelling and segmentation of noisy and textured images using Gibbs random fields. *IEEE Trans. Pattern Anal. Mach. Intell.* **9**, no. 1, 39–55.

DeSouza, G.N. and Kak, A.C. (2002). Vision for mobile robot navigation: a survey. *IEEE Trans. Pattern Anal. Mach. Intell.* **24**, no. 2, 237–267.

Devijver, P.A. and Kittler, J. (1980). On the edited nearest neighbour rule. *Proc. 5th Int. Conf. on Pattern Recogn., Miami Beach, Florida (IEEE Comput. Soc.)*, pp. 72–80.

Devijver, P.A. and Kittler, J. (1982). *Pattern Recognition: A Statistical Approach*. Prentice-Hall, Englewood Cliffs, NJ.

Dewaele, P., Van Gool, L., Wambacq, P. and Oosterlinck, A. (1988). Texture inspection with self-adaptive convolution filters. *Proc. 9th Int. Conf. on Pattern Recogn.*, pp. 56–60.

Dhome, M., Rives, G. and Richetin, M. (1983). Sequential piecewise linear segmentation of binary contours. *Pattern Recogn. Lett.* **2**, 101–107.

Dickinson, S., Pelillo, M. and Zabih, R. (eds.) (2001). Special Section on *Graph Algorithms and Computer Vision. IEEE Trans. Pattern Anal. Mach. Intell.* **23**, no. 10, 1049–1151.

Dickmanns, E.D. and Mysliwetz, B.D. (1992). Recursive 3-D road and relative ego-state recognition. *IEEE Trans. Pattern Anal. Mach. Intell.* **14**, no. 2, 199–213.

Dockstader, S.L. and Tekalp, A.M. (2001). On the tracking of articulated and occluded video object motion. *Real-Time Imaging* **7**, no. 5, 415–432.

Dockstader, S.L. and Tekalp, A.M. (2002). A kinematic model for human motion and gait analysis. *Proc. Workshop on Statistical Methods in Video Processing (ECCV), Copenhagen, Denmark (June 1–2)*, pp. 49–54

Dodd, G.G. and Rossol, L. (eds.) (1979). *Computer Vision and Sensor-Based Robots*. Plenum Press, New York.

Dorst, L. and Smeulders, A.W.M. (1987). Length estimators for digitized contours. *Comput. Vision Graph. Image Process.* **40**, 311–333.

Dougherty, E.R. (1992). *An Introduction to Morphological Image Processing*. SPIE Press, Bellingham, MA.

Dougherty, E.R. and Giardina, C.R. (1988). Morphology on umbra matrices. *Int. J. Pattern Recogn. and Artif. Intell.* **2**, 367–385.

Dougherty, E.R. and Sinha, D. (1995a). Computational gray-scale mathematical morphology on lattices (a comparator-based image algebra). Part I: Architecture. *Real-Time Imaging* **1**, no. 1, 69–85.

Dougherty, E.R. and Sinha, D. (1995b). Computational gray-scale mathematical morphology on lattices (a comparator-based image algebra). Part II: Image operators. *Real-Time Imaging* **1**, no. 4, 283–295.

Doyle, W. (1962). Operations useful for similarity-invariant pattern recognition. *J. ACM* **9**, 259–267.

Draper, B.A., Beveridge, J.R., Böhm, A.P.W., Ross, C. and Chawathe, M. (2003). Accelerated image processing on FPGAs. *IEEE Trans. Image Processing* **12**, no. 12, 1543–1551.

Dreschler, L. and Nagel, H.-H. (1981). Volumetric model and 3D-trajectory of a moving car derived from monocular TV-frame sequences of a street scene. *Proc. Int. Joint Conf. on Artif. Intell.*, pp. 692–697.

Du Buf, J.M.H., Kardan, M. and Spann, M. (1990). Texture feature performance for image segmentation. *Pattern Recogn.* **23**, 291–309.

Duda, R.O. and Hart, P.E. (1972). Use of the Hough transformation to detect lines and curves in pictures. *Comm. ACM* **15**, 11–15.

Duda, R.O. and Hart, P.E. (1973). *Pattern Classification and Scene Analysis*. Wiley, New York.

Duda, R.O., Hart, P.E. and Stork, D.G. (2001). *Pattern Classification*. Wiley, New York.

Dudani, S.A. and Luk, A.L. (1978). Locating straight-line edge segments on outdoor scenes. *Pattern Recogn.* **10**, 145–157.

Dudani, S.A., Breeding, K.J. and McGhee, R.B. (1977). Aircraft identification by moment invariants. *IEEE Trans. Comput.* **26**, 39–46.

Duin, R.P.W. (2002). The combining classifier: to train or not to train? *Proc. 16th Int. Conf. on Pattern Recogn., Québec, Canada (August 11–15)*, Vol. II, pp. 765–770.

Duin, R.P.W., Haringa, H. and Zeelen, R. (1986). Fast percentile filtering. *Pattern Recogn. Lett.* **4**, 269–272.

Dyer, C.R. and Rosenfeld, A. (1976). Fourier texture features: suppression of aperture effects. *IEEE Trans. Systems Man Cybern.* **6**, no. 10, 703–705.

Edmonds, J.M. and Davies, E.R. (1991). High-speed processor for realtime visual inspection. *Microprocessors and Microsystems*, **15**, no. 1, 11–19.

Eichmann, G. and Kasparis, T. (1988). Topologically invariant texture descriptors. *Comput. Vision Graph. Image Process.* **41**, 267–281.

Eisberg, R.M. (1961). *Fundamentals of Modern Physics*. Wiley, New York.

Ellis, T.J., Abbood, A. and Brillault, B. (1992). Ellipse detection and matching with uncertainty. *Image Vision Comput.* **10**, no. 5, 271–276.

Fasel, B. (2002). Robust face analysis using convolutional neural networks. *Proc. 16th Int. Conf. on Pattern Recogn., Québec, Canada (August 11–15)*, Vol. II, pp. 40–43.

Fathy, M. and Siyal, M.Y. (1995). Real-time image processing approach to measure traffic queue parameters. *IEE Proc. Vision Image Signal Process.* **142**, no. 5, 297–303.

Faugeras, O.D. (1978). Texture analysis and classification using a human visual model. *Proc. 4th Int. Joint Conf. on Pattern Recogn., Kyoto, (November 7–10)*, pp. 549–552.

Faugeras, O. (1992). What can be seen in three dimensions with an uncalibrated stereo rig? Proc. 2nd European Conf. on Computer Vision: Sandini, G. (ed.), *Lecture Notes in Computer Science* **588**, Springer-Verlag, Berlin, Heidelberg, pp. 563–578.

Faugeras, O. (1993). *Three-Dimensional Computer Vision—a Geometric Viewpoint*. MIT Press, Cambridge, MA.

Faugeras, O.D. and Hebert, M. (1983). A 3-D recognition and positioning algorithm using geometrical matching between primitive surfaces. *Proc. 8th Int. Joint Conf. on Artif. Intell.*, pp. 996–1002.

Faugeras, O. and Luong, Q.-T. (2001). *The Geometry of Multiple Images*. MIT press, Cambridge, MA.

Faugeras, O., Luong, Q.-T. and Maybank, S.J. (1992). Camera self-calibration: theory and experiments. Proc. 2nd European Conf. on Computer Vision: Sandini, G. (ed.), *Lecture Notes in Computer Science* **588**, Springer-Verlag, Berlin, Heidelberg, pp. 321–334.

Faugeras, O., Quan, L. and Sturm, P. (2000). Self-calibration of a 1D projective camera and its application to the self-calibration of a 2D projective camera. *IEEE Trans. Pattern Anal. Mach. Intell.* **22**, no. 10, 1179–1185.

Feng, T.-Y. (1981). A survey of interconnection networks. *IEEE Comput.* **14**, 12–17.

Ferrie, F.P. and Levine, M.D. (1989). Where and why local shading analysis works. *IEEE Trans. Pattern Anal. Mach. Intell.* **11**, 198–206.

Fesenkov, V.P. (1929). Photometric investigations of the lunar surface. *Astronomochhesk. Zh.* **5**, 219–234.

Fischer, B. and Buhmann, J.M. (2003). Bagging for path-based clustering. *IEEE Trans. Pattern Anal. Mach. Intell.* **25**, no. 11, 1411–1415.

Fischler, M.A. and Bolles, R.C. (1981). Random sample consensus: a paradigm for model fitting with applications to image analysis and automated cartography. *Commun. ACM* **24**, no. 6, 381–395.

Fish, R.K., Ostendorf, M., Bernard, G.D. and Castanon, D.A. (2003). Multilevel classification of milling tool wear with confidence estimation. *IEEE Trans. Pattern Anal. Mach. Intell.* **25**, no. 1, 75–85.

Fitch, J.P., Coyle, E.J. and Gallagher, N.C. (1985). Root properties and convergence rates of median filters. *IEEE Trans. Acoust. Speech Signal Process.* **33**, 230–239.

Flynn, M.J. (1972). Some computer organizations and their effectiveness. *IEEE Trans. Comput.* **21**, 948–960.

Föglein, J. (1983). On edge gradient approximations. *Pattern Recogn. Lett.* **1**, 429–434.

Forgy, E.W. (1965). Cluster analysis of multivariate data: efficiency versus interpretability of classification. *Biometrics* **21**, 768–769.

Forsyth, D.A. and Ponce, J. (2003). *Computer Vision: A Modern Approach.* Pearson Education International, Upper Saddle River, NJ.

Forsyth, D.A., Mundy, J.L., Zisserman, A., Coelho, C., Heller, A. and Rothwell, C.A. (1991). Invariant descriptors for 3-D object recognition and pose. *IEEE Trans. Pattern Anal. Mach. Intell.* **13**, no. 10, 971–991.

Forsyth, R.S. (1982). *Pascal at Work and Play.* Chapman and Hall, London.

Foster, J.P., Nixon, M.S. and Prugel-Bennett, A. (2001). New area based metrics for automatic gait recognition. *Proc. British Machine Vision Assoc. Conf.*, pp. 233–242.

Fountain, T. (1987). *Processor Arrays: Architectures and Applications.* Academic Press, London.

Frankot, R.T. and Chellappa, R. (1990). Estimation of surface topography from SAR imagery using shape from shading techniques. *Artif. Intell.* **43**, 271–310.

Freeman, H. (1961). On the encoding of arbitrary geometric configurations. *IEEE Trans. Electron. Comput.* **10**, 260–268.

Freeman, H. (1974). Computer processing of line drawing images. *Comput. Surveys* **6**, 57–97.

Freeman, H. (1978). Shape description via the use of critical points. *Pattern Recogn.* **10**, 159–166.

Frei, W. and Chen, C.-C. (1977). Fast boundary detection: a generalization and a new algorithm. *IEEE Trans. Comput.* **26**, 988–998.

Freund, Y. and Schapire, R. (1996). Experiments with a new boosting algorithm. *Proc. 13th Int. Conf. on Machine Learning*, pp. 148–156.

Fu, K.-S. and Mui, J.K. (1981). A survey on image segmentation. *Pattern Recogn.* **13**, 3–16.

Gallagher, N.C. and Wise, G.L. (1981). A theoretical analysis of the properties of median filters. *IEEE Trans. Acoust. Speech Signal Process.* **29**, 1136–1141.

Garcia, C. and Delakis, M. (2002). A neural architecture for fast and robust face detection. *Proc. 16th Int. Conf. on Pattern Recogn., Québec, Canada (August 11–15)*, Vol. II, pp. 44–47.

Gårding, J. (1988). Properties of fractal intensity surfaces. *Pattern Recogn. Lett.* **8**, no. 5, 319–324.

Gavrila, D. (1999). The visual analysis of human movement: a survey. *Computer Vision Image Understanding* **73**, no. 1, 82–98.

Gavrila, D. (2000). Pedestrian detection from a moving vehicle. *Proc. European Conf. on Computer Vision*, Vernon, D. (ed.), *Dublin, Ireland* (*June*), pp. 37–49.

Gavrila, D.M. and Groen, F.C.A. (1992). 3D object recognition from 2D images using geometric hashing. *Pattern Recogn. Lett.* **13**, no. 4, 263–278.

Geiger, D., Liu, T.-L. and Kohn, R.V. (2003). Representation and self-similarity of shapes. *IEEE Trans. Pattern Anal. Mach. Intell.* **25**, no. 1, 86–99.

Gelsema, E.S. (ed.) (1994). *Special Issue on Genetic Algorithms, Pattern Recogn. Lett.* **16**, no. 8.

Geman, D. (1987). Stochastic model for boundary detection. *Image Vision Comput.* **5**, no. 2, 61–65.

Geman, S. and Geman, D. (1984). Stochastic relaxation, Gibbs distributions, and the Bayesian restoration of images. *IEEE Trans. Pattern Anal. Mach. Intell.* **6**, no. 6, 721–741.

Gerig, G. and Klein, F. (1986). Fast contour identification through efficient Hough transform and simplified interpretation strategy. *Proc. 8th Int. Conf. on Pattern Recogn., Paris* (*October 27–31*), pp. 498–500.

Ghosh, A. and Pal, S.K. (1992). Neural network, self-organisation and object extraction. *Pattern Recogn. Lett.* **13**, no. 5, 387–397.

Gibbons, A. (1985). *Algorithmic Graph Theory*. Cambridge University Press, Cambridge, UK.

Giblin, P.J. and Kimia, B.B. (2003). On the intrinsic reconstruction of shape from its symmetries. *IEEE Trans. Pattern Anal. Mach. Intell.* **25**, no. 7, 895–911.

Gibson, J.J. (1950). *The Perception of the Visual World*. Houghton Mifflin, Boston, MA.

Gil, J. and Kimmel, R. (2002). Efficient dilation, erosion, opening, and closing algorithms. *IEEE Trans. Pattern Anal. Mach. Intell.* **24**, no. 12, 1606–1617.

Godbole, S. and Amin, A. (1995). Mathematical morphology for edge and overlap detection for medical images. *Real-Time Imaging* **1**, no. 3, 191–201.

Goetcherian, V. (1980). From binary to grey tone image processing using fuzzy logic concepts. *Pattern Recogn.* **12**, 7–15.

Golightly, I. and Jones, D. (2003). Corner detection and matching for visual tracking during power line inspection. *Image Vision Comput.* **21**, no. 9, 827–840.

Golub, G.H. and van Loan, C.F. (1983). *Matrix Computations*. North Oxford, Oxford, UK.

Gong, S., McKenna, S. and Psarrou, A. (2000). *Dynamic Vision: From Images to Face Recognition*. Imperial College Press, London, UK.

Gonnet, G.H. (1984). *Handbook of Algorithms and Data Structures*. Addison-Wesley, London.

Gonzalez, R.C. and Wintz, P. (1987). *Digital Image Processing*. Addison-Wesley, Reading, MA (2nd edition).

Gonzalez, R.C. and Woods, R.E. (1992). *Digital Image Processing*. Addison-Wesley, Reading, MA.

Goulermas, J.Y. and Liatsis, P. (1998). Genetically fine-tuning the Hough transform feature space, for the detection of circular objects. In Davies, E.R. and Atiquzzaman, M. (eds.), *Special Issue on Projection-Based Transforms, Image Vision Comput.* **16**, nos. 9–10, 615–626.

Granlund, G.H. (1980). Description of texture using the general operator approach. *Proc. 5th Int. Conf. on Pattern Recogn., Miami Beach, Florida* (*December 1–4*), pp. 776–779.

Graves, M. and Batchelor, B.G. (eds.) (2003). *Machine Vision Techniques for Inspecting Natural Products*. Springer Verlag, London.

Graves, M., Batchelor, B.G. and Palmer, S. (1994). 3D X-ray inspection of food products. *Proc. SPIE Conf. on Applications of Digital Image Process.* **17**, Vol. **2298**, pp. 248–259.

Greenhill, D. and Davies, E.R. (1993). Texture analysis using neural networks and mode filters. *Proc. 4th British Machine Vision Assoc. Conf., Univ. of Surrey* (September 21–23), Vol. 2, pp. 509–518.

Greenhill, D. and Davies, E.R. (1994a). Relative effectiveness of neural networks for image noise suppression. In Gelsema, E.S. and Kanal, L.N. (eds.), *Pattern Recognition in Practice IV*, Elsevier Science B.V., pp. 367–378.

Greenhill, D. and Davies, E.R. (1994b). How do neural networks compare with standard filters for image noise suppression? *IEE Digest no. 1994/248, Colloquium on Applications of Neural Networks to Signal Processing, IEE* (*December 15*), pp. 3/1–4.

Greenhill, D. and Davies, E.R. (1995). A new approach to the determination of unbiased thresholds for image segmentation. *5th Int. IEE Conf. on Image Processing and Its Applications, Heriot-Watt University* (*July 3–6*), *IEE Conf. Publication* no. 410, pp. 519–523.

Gregory, R.L. (1971). *The Intelligent Eye*. Weidenfeld and Nicolson, London.

Gregory, R.L. (1972). *Eye and Brain*. Weidenfeld and Nicolson, London (2nd edition).

Grimson, W.E.L. and Huttenlocher, D.P. (1990). On the sensitivity of the Hough transform for object recognition. *IEEE Trans. Pattern Anal. Mach. Intell.* **12**, no. 3, 255–274.

Grimson, W.E.L. and Lozano-Perez, T. (1984). Model-based recognition and localisation from sparse range or tactile data. *Int. J. Robot. Res.* **3**, no. 3, 3–35.

Gruen, A. and Huang, T.S. (eds.) (2001). *Calibration and Orientation of Cameras in Computer Vision*. Springer-Verlag, Berlin, Heidelberg.

Guiducci, A. (1999). Parametric model of the perspective projection of a road with applications to lane keeping and 3d road reconstruction. *Computer Vision Image Understanding* **73**, 414–427.

Guru, D.S., Shekar, B.H. and Nagabhushan, P. (2004). A simple and robust line detection algorithm based on small eigenvalue analysis. *Pattern Recogn. Lett.* **25**, no. 1, 1–13.

Hall, E.L. (1979). *Computer Image Processing and Recognition*. Academic Press, New York.

Hall, E.L., Tio, J.B.K., McPherson, C.A. and Sadjadi, F.A. (1982). Measuring curved surfaces for robot vision. *IEEE Comput.* **15**, no. 12, 42–54.

Hampel, F.R., Ronchetti, E.M., Rousseeuw, P.J. and Stahel, W.A. (1986). *Robust Statistics: The Approach Based on Influence Functions*. Wiley, New York.

Hannah, I., Patel, D. and Davies, E.R. (1995). The use of variance and entropic thresholding methods for image segmentation. *Pattern Recognition* **28**, no. 8. 1135–1143.

Hansen, F.R. and Elliott, H. (1982). Image segmentation using simple Markov field models. *Comput. Graph. Image Process.* **20**, 101–132.

Haralick, R.M. (1979). Statistical and structural approaches to texture. *Proc. IEEE* **67**, no. 5, 786–804.

Haralick, R.M. (1980). Edge and region analysis for digital image data. *Comput. Graph. Image Process.* **12**, 60–73.

Haralick, R.M. (1984). Digital step edges from zero crossing of second directional derivatives. *IEEE Trans. Pattern Anal. Mach. Intell.* **6**, 58–68.

Haralick, R.M. (1989). Determining camera parameters from the perspective projection of a rectangle. *Pattern Recogn.* **22**, 225–230.

Haralick, R.M. (1992). Performance characterization in image analysis: Thinning, a case in point. *Pattern Recogn. Lett.* **13**, 5–12.

Haralick, R.M. and Chu, Y.H. (1984). Solving camera parameters from the perspective projection of a parameterized curve. *Pattern Recogn.* **17**, no. 6, 637–645.

Haralick, R.M. and Joo, H. (1988). 2D-3D pose estimation. *Proc. 9th Int. Conf. on Pattern Recogn., Rome, Italy (November 14–17)*, pp. 385–391.

Haralick, R.M. and Shapiro, L.G. (1985). Image segmentation techniques. *Comput. Vision Graph. Image Process.* **29**, 100–132.

Haralick, R.M., and Shapiro, L.G. (1992). *Computer and Robot Vision.* Volume I. Addison-Wesley, Reading, MA.

Haralick, R.M., and Shapiro, L.G. (1993). *Computer and Robot Vision.* Volume II. Addison-Wesley, Reading, MA.

Haralick, R.M., Chu, Y.H., Watson, L.T. and Shapiro, L.G. (1984). Matching wire frame objects from their two dimensional perspective projections. *Pattern Recogn.* **17**, no. 6, 607–619.

Haralick, R.M., Shanmugam, K. and Dinstein, I. (1973). Textural features for image classification. *IEEE Trans. Systems Man Cybern.* **3**, no. 6, 610–621.

Haralick, R.M., Sternberg, S.R. and Zhuang, X. (1987). Image analysis using mathematical morphology. *IEEE Trans. Pattern Anal. Mach. Intell.* **9**, no. 4, 532–550.

Haritaoglu, I., Harwood, D. and Davis, L.S. (2000). W^4: Real-time surveillance of people and their activities. In Special Section on *Video Surveillance. IEEE Trans. Pattern Anal. Mach. Intell.* **22**, no. 8, 809–830.

Harris, C. and Stephens, M. (1988). A combined corner and edge detector. *Proc. 4th Alvey Vision Conf.*, pp. 147–151.

Hart, P.E. (1968). The condensed nearest neighbour rule. *IEEE Trans. Inf. Theory* **14**, 515–516.

Hartley, R. and Zisserman, A. (2000). *Multiple View Geometry in Computer Vision.* Cambridge University Press, Cambridge, UK.

Hartley, R.I. (1992). Estimation of relative camera positions for uncalibrated cameras. *Proc. 2nd European Conf. on Computer Vision*: Sandini, G. (ed.). *Lecture Notes in Computer Science* **588**, Springer-Verlag, Berlin, Heidelberg, pp. 579–587.

Hartley, R.I. (1995). A linear method for reconstruction from lines and points. *Proc. Int. Conf. on Computer Vision*, pp. 882–887.

Hartley, R.I. (1997). In defense of the eight-point algorithm. *IEEE Trans. Pattern Anal. Mach. Intell.* **19**, no. 6, 580–593.

Harvey, N.R. and Marshall, S. (1994). Using genetic algorithms in the design of morphological filters. *IEE Colloquium on Genetic Algorithms in Image Processing and Vision, IEE (October 20), IEE Digest no.* 1994/193, pp. 6/1–5.

Harvey, N.R. and Marshall, S. (1995). Rank-order morphological filters: a new class of filters. *Proc. IEEE Workshop on Nonlinear Signal and Image Processing, Halkidiki, Greece (June)*, pp. 975–978.

Harwood, D., Subbarao, M. and Davis, L.S. (1985). Texture classification by local rank correlation. *Comput. Vision Graph. Image Process.* **32**, 404–411.

Hasler, D., Sbaiz, L., Süsstrunk, S. and Vetterli, M. (2003). Outlier modelling in image matching. *IEEE Trans. Pattern Anal. Mach. Intell.* **25**, no. 3, 301–315.

Haykin, S. (1999). *Neural Networks: A Comprehensive Introduction.* Upper Saddle River, Prentice-Hall, NJ (2nd edition).

Hebb, D.O. (1949). *The Organization of Behavior.* Wiley, New York.

Heijmans, H. (1991). Theoretical aspects of gray-level morphology. *IEEE Trans. Pattern Anal. Mach. Intell.* **13**, 568–582.

Heijmans, H. (1994). *Morphological Operators.* Academic Press, New York.

Heikkilä, J. (2000). Geometric camera calibration using circular control points. *IEEE Trans. Pattern Anal. Mach. Intell.* **22**, no. 10, 1066–1076.

Heikkonen, J. (1995). Recovering 3-D motion parameters from optical flow field using randomized Hough transform. *Pattern Recogn. Lett.* **16**, no. 9, 971–978.

Heinemann, P.H., Varghese, Z.A., Morrow, C.T., Sommer, H.J. III and Crassweller, R.M. (1995). Machine vision inspection of "Golden Delicious" apples. *Appl. Eng. Agric.* **11**, no. 6, 901–906.

Heinonen, P. and Neuvo, Y. (1987). FIR-median hybrid filters. *IEEE Trans. Acoust. Speech Signal Process.* **35**, 832–838.

Henderson, T.C. (1984). A note on discrete relaxation. *Comput. Vision Graph. Image Process.* **28**, 384–388.

Herault, L., Horaud, R., Veillon, F. and Niez, J.J. (1990). Symbolic image matching by simulated annealing. *Proc. British Machine Vision Assoc. Conf.*, pp. 319–324.

Hertz, J., Krogh, A. and Palmer, R.G. (1991). *Introduction to the Theory of Neural Computation*, Addison-Wesley, Reading, MA.

Hildreth, E.C. (1984). *Measurement of Visual Motion.* MIT Press, Cambridge, MA.

Hill, A. and Taylor, C.J. (1992). Model-based image interpretation using genetic algorithms. *Image Vision Comput.* **10**, no. 5, 295–300.

Hlaoui, A. and Wang, S. (2002). A new algorithm for inexact graph matching. *Proc. 16th Int. Conf. on Pattern Recogn., Québec, Canada (August 11–15)*, Vol. IV, pp. 180–183.

Ho, T.K., Hull, J.J. and Srihari, S.N. (1994). Decision combination in multiple classifier systems. *IEEE Trans. Pattern Anal. Mach. Intell.* **16**, no. 1, 66–75.

Hockney, R.W. and Jesshope, C.R. (1981). *Parallel Computers.* Adam Hilger Ltd., Bristol.

Hodgson, R.M., Bailey, D.G., Naylor, M.J., Ng, A.L.M. and McNeil, S.J. (1985). Properties, implementations, and applications of rank filters. *Image Vision Comput.* **3**, 4–14.

Hofmann, U., Rieder, A. and Dickmanns, E.D. (2003). Radar and vision data fusion for hybrid adaptive cruise control on highways. *Machine Vision and Applications* **14**, no. 1, 42–49.

Hogg, D. (1983). Model-based vision: a program to see a walking person. *Image Vision Comput.* **1**, no. 1, 5–20.

Holland, J.H. (1975). *Adaptation in Natural and Artificial Systems*, Univ. of Michigan Press, Ann Arbor, MI.

Hollinghurst, N. and Cipolla, R. (1994). Uncalibrated stereo hand-eye coordination. *Image Vision Comput.* **12**, no. 3, 187–192.

Horaud, R. (1987). New methods for matching 3-D objects with single perspective views. *IEEE Trans. Pattern Anal. Mach. Intell.* **9**, 401–412.

Horaud, R. and Brady, M. (1988). On the geometric interpretation of image contours. *Artif. Intell.* **37**, 333–353.

Horaud, R. and Sossa, H. (1995). Polyhedral object recognition by indexing. *Pattern Recogn.* **28**, no. 12, 1855–1870.

Horaud, R., Conio, B., Leboulleux, O. and Lacolle, B. (1989). An analytic solution for the perspective 4-point problem. *Comput. Vision Graph. Image Process.* **47**, 33–44.

Horn, B.K.P. (1975). Obtaining shape from shading information. In Winston, P.H. (ed.), *The Psychology of Computer Vision*. McGraw-Hill, New York, pp. 115–155.

Horn, B.K.P. (1977). Understanding image intensities. *Artif. Intell.* **8**, 201–231.

Horn, B.K.P. (1986). *Robot Vision*. MIT Press, Cambridge, MA.

Horn, B.K.P. and Brooks, M.J. (1986). The variational approach to shape from shading. *Comput. Vision Graph. Image Process.* **33**, 174–208.

Horn, B.K.P. and Brooks, M.J. (eds.) (1989). *Shape from Shading*. MIT Press, Cambridge, MA.

Horn, B.K.P. and Schunck, B.G. (1981). Determining optical flow. *Artif. Intell.* **17**, nos. 1–3, 185–203.

Horng, J.-H. (2003). An adaptive smoothing approach for fitting digital planar curves with line segments and circular arcs. *Pattern Recogn. Lett.* **24**, nos. 1–3, 565–577.

Hornik, K., Stinchcombe, M. and White, H. (1989). Multilayer feedforward networks are universal approximators. *Neural Networks* **2**, 359–366.

Horowitz, S.L. and Pavlidis, T. (1974). Picture segmentation by a directed split-and-merge procedure. *Proc. 2nd Int. Joint Conf. on Pattern Recogn.*, pp. 424–433.

Hough, P.V.C. (1962). Method and means for recognising complex patterns. *US Patent 3069654*.

Hsiao, J.Y. and Sawchuk, A.A. (1989). Supervised textured image segmentation using feature smoothing and probabilistic relaxation techniques. *IEEE Trans. Pattern Anal. Mach. Intell.* **11**, no. 12, 1279–1292.

Hsiao, J.Y. and Sawchuk, A.A. (1990). Unsupervised textured image segmentation using feature smoothing and probabilistic relaxation techniques. *Comput. Vision Graph. Image Process.* **48**, 1–21.

Hu, M.K. (1961). Pattern recognition by moment invariants. *Proc. IEEE* **49**, 1428.

Hu, M.K. (1962). Visual pattern recognition by moment invariants. *IRE Trans. Inf. Theory* **8**, 179–187.

Huang, C.T. and Mitchell, O.R. (1994). A Euclidean distance transform using greyscale morphology decomposition. *IEEE Trans. Pattern Anal. Mach. Intell.* **16**, no. 4, 443–448.

Huang, T.S. (ed.) (1983). *Image Sequence Processing and Dynamic Scene Analysis*. Springer-Verlag, New York.

Huang, T.S., Bruckstein, A.M., Holt, R.J. and Netravali, A.N. (1995). Uniqueness of 3D pose under weak perspective: a geometrical proof. *IEEE Trans. Pattern Anal. Mach. Intell.* **17**, no. 12, 1220–1221.

Huang, T.S., Yang, G.J. and Tang, G.Y. (1979). A fast two-dimensional median filtering algorithm. *IEEE Trans. Acoust. Speech Signal Process.* **27**, 13–18.

Hubel, D.H. (1995). *Eye, Brain and Vision*. Scientific American Library, New York.

Huber, P.J. (1964). Robust estimation of a location parameter. *Annals Math. Statist.* **35**, 73–101.

Huber, P.J. (1981). *Robust Statistics*. Wiley, New York.

Huber, P.J. (1985). Projection pursuit. *Annals Statist.* **13**, no. 2, 435–475.

Hueckel, M.F. (1971). An operator which locates edges in digitised pictures. *J. ACM* **18**, 113–125.

Hueckel, M.F. (1973). A local visual operator which recognises edges and lines. *J. ACM* **20**, 634–647.

Hufnagl, C. and Uhl, A. (2000). Algorithms for fractal image compression on massively parallel SIMD arrays. *Real-Time Imaging* **6**, no. 4, 267–281.

Hughes, G.F. (1968). On the mean accuracy of statistical pattern recognisers. *IEEE Trans. Inf. Theory* **14**, 55–63.

Hummel, R.A. and Zucker, S.W. (1983). On the foundations of relaxation labelling processes. *IEEE Trans. Pattern Anal. Mach. Intell.* **5**, 267–287.

Hush, D.R. and Horne, B.G. (1993). Progress in supervised neural networks. *IEEE Signal Processing Magazine* **10**, no. 1, 8–39.

Hussain, Z. (1988). A fast approximation to a convex hull. *Pattern Recogn. Lett.* **8**, 289–294.

Hussmann, S. and Ho, T.H. (2003). A high-speed subpixel edge detector implementation inside a FPGA. *Real-Time Imaging* **9**, no. 5, 361–368.

Huttenlocher, D.P., Klanderman, G.A. and Rucklidge, W.J. (1993). Comparing images using the Hausdorff distance. *IEEE Trans. Pattern Anal. Mach. Intell.* **15**, no. 9, 850–863.

Hwang, K. and Briggs, F.A. (1984). *Computer Architecture and Parallel Processing*. McGraw-Hill, New York.

Hwang, V.S., Davis, L.S. and Matsuyama, T. (1986). Hypothesis integration in image understanding systems. *Comput. Vision Graph. Image Process.* **36**, 321–371.

Iannino, A. and Shapiro, S.D. (1979). An iterative generalisation of the Sobel edge detection operator. *Proc. IEEE Comput. Soc. Conf. on Pattern Recogn. and Image Process., Chicago (May 31–June 2)*, pp. 130–137.

IBM (1986). Cluster estimation in Hough space for image analysis. *IBM Techn. Disclosure Bull.* **28**, 3667–3668.

IEE (1994). *IEE Colloquium on Genetic Algorithms in Image Processing and Vision*. IEE (October 20), *IEE Digest* no. 1994/193.

Ikeuchi, K. and Horn, B.K.P. (1981). Numerical shape from shading and occluding boundaries. *Artif. Intell.* **17**, 141–184.

Illingworth, J. and Kittler, J. (1987). The adaptive Hough transform. *IEEE Trans. Pattern Anal. Mach. Intell.* **9**, 690–698.

Illingworth, J. and Kittler, J. (1988). A survey of the Hough transform. *Comput. Vision Graph. Image Process.* **44**, 87–116.

Inmos Ltd (1987). *The Transputer Family 1987*. Inmos Ltd., Bristol.

Inmos Ltd (1988). *OCCAM 2 Reference Manual*. Prentice-Hall, London.

Ising, E. (1925), *Z. Physik* **31**, p. 253.

Ito, M. and Ishii, A. (1986). Three-view stereo analysis. *IEEE Trans. Pattern Anal. Mach. Intell.* **8**, no. 4, 524–532.

Jacinto, C.N., Arnaldo, J.A. and George, S.M. (2003). Using middle level features for robust shape tracking. *Pattern Recogn. Lett.* **24**, 295–307.

Jackway, P.T. and Deriche, M. (1996). Scale-space properties of the multiscale morphological dilation-erosion. *IEEE Trans. Pattern Anal. Mach. Intell.* **18**, no. 1, 38–51.

Jähne, B. and Haussecker, H. (eds.) (2000). *Computer Vision and Applications*. Academic Press, New York.

Jain, A.K. and Dubes, R.C. (1988). *Algorithms for Clustering Data*, Prentice-Hall, Englewood Cliffs, NJ.

Jain, A.K. and Farrokhnia, F. (1991). Unsupervised texture segmentation using Gabor filters. *Pattern Recogn.* **24**, 1167–1186.

Jain, A.K., Duin, R.P.W. and Mao, J. (2000). Statistical Pattern Recognition. *IEEE Trans. Pattern Anal. Mach. Intell.* **22**, no. 1, 4–37.

Jain, R. (1983). Direct computation of the focus of expansion. *IEEE Trans. Pattern Anal. Mach. Intell.* **5**, 58–63.

Jain, R.C. and Binford, T.O. (1991). Ignorance, myopia and naiveté in computer vision systems. *Computer Vision Graph. Image Process.: Image Understanding* **53**, 112–117.

Jang, J.-H. and Hong, K.-S. (1998). Detection of curvilinear structures using the Euclidean distance transform. *Proc. IAPR Workshop on Machine Vision Applications (MVA'98), Chiba, Japan*, pp. 102–105.

Jolion, J.-M. and Rosenfeld, A. (1989). Cluster detection in background noise. *Pattern Recogn.* **22**, no. 5, 603–607.

Jones, R. and Svalbe, I. (1994). Morphological filtering as template matching. *IEEE Trans. Pattern Anal. Mach. Intell.* **16**, no. 4, 438–443.

Juan, A. and Vidal, E. (1994). Fast *K*-means-like clustering in metric spaces. *Pattern Recogn. Lett.* **15**, no. 1, 19–25.

Kadyrov, A. and Petrou, M. (2001). The trace transform and its applications. *IEEE Trans. Pattern Anal. Mach. Intell.* **23**, 811–828.

Kadyrov, A. and Petrou, M. (2002). Affine parameter estimation from the trace transform. *Proc. 16th Int. Conf. on Pattern Recogn., Québec, Canada (August 11–15)*, Vol. II, pp. 798–801.

Kaizer, H. (1955). A quantification of textures on aerial photographs. MS thesis, Boston University.

Kamat-Sadekar, V. and Ganesan, S. (1998). Complete description of multiple line segments using the Hough transform. In Davies, E.R. and Atiquzzaman, M. (eds.), *Special Issue on Projection-Based Transforms. Image Vision Comput.* **16**, nos. 9–10, 597–614.

Kamel, M.S., Shen, H.C., Wong, A.K.C., Hong, T.M. and Campeanu, R.I. (1994). Face recognition using perspective invariant features. *Pattern Recogn. Lett.* **15**, no. 9, 877–883.

Kanesalingam, C., Smith, M.C.B., and Dodds, S.A. (1998). An efficient algorithm for environmental mapping and path planning for an autonomous mobile robot. *Proc. 29th Int. Symp. on Robotics, Birmingham*, pp. 133–136.

Kang, D.-J. and Jung, M.-H. (2003). Road lane segmentation using dynamic programming for active safety vehicles. *Pattern Recogn. Lett.* **24**, 3177–3185.

Kapur, J.N., Sahoo, P.K.and Wong, A.K.C. (1985). A new method for gray-level picture thresholding using the entropy of the histogram. *Comput. Vision Graph. Image Process.* **29**, 273–285.

Kasif, S., Kitchen, L. and Rosenfeld, A. (1983). A Hough transform technique for subgraph isomorphism. *Pattern Recogn. Lett.* **2**, 83–88.

Kass, M. and Witkin, A. (1987). Analyzing oriented patterns. *Comput. Vision Graph. Image Process.* **37**, no. 3, 362–385.

Kass, M., Witkin, A. and Terzopoulos, D. (1988). Snakes: active contour models. *Int. J. Comput. Vision* **1**, 321–331.

Kastrinaki, V., Zervakis, M. and Kalaitzakis, K. (2003). A survey of video processing techniques for traffic applications. *Image Vision Comput.* **21**, no. 4, 359–381.

Katatani, K. and Chou, T. (1989). Shape from texture: general principle. *Artif. Intell.* **38**, no. 1, 1–49.

Katz, A.J. and Thrift, P.R. (1994). Generating image filters for target recognition by genetic learning. *IEEE Trans. Pattern Anal. Mach. Intell.* **16**, no. 9, 906–910.

Keagy, P.M., Parvin, B. and Schatzki, T.F. (1995). Machine recognition of navel orange worm damage in x-ray images of pistachio nuts. *Optics in Agric., Forestry and Biological, SPIE* **2345**, 192–203.

Keagy, P.M., Parvin, B. and Schatzki, T.F. (1996). Machine recognition of navel orange worm damage in X-ray images of pistachio nuts. *Lebens.-Wiss. u.-Technol.* **29**, 140–145.

Keefe, P.D. (1992). A dedicated wheat grain image analyser. *Plant Varieties and Seeds* **5**, 27–33.

Kégl, B. and Krzyżak, A. (2002). Piecewise linear skeletonization using principal curves. *IEEE Trans. Pattern Anal. Mach. Intell.* **24**, no. 1, 59–74.

Kehtarnavaz, N. and Mohan, S. (1989). A framework for estimation of motion parameters from range images. *Comput. Vision Graph. Image Process.* **45**, 88–105.

Keller, J.M., Chen, S. and Crownover, R.M. (1989). Texture description and segmentation through fractal geometry. *Comput. Vision Graph. Image Process.* **45**, 150–166.

Kelley, R.B. and Gouin, P. (1984). Heuristic vision hole finder. In Pugh, A. (ed.), *Proc. 4th Int. Conf. on Robot Vision and Sensory Controls, London (October 9–11)*, IFS (Publications) Ltd., Bedford and North-Holland, Amsterdam, pp. 341–350.

Kender, J.R. (1981). Shape from texture. *Carnegie-Mellon University, Comput. Sci. Techn. Rep.* CMU-CS-81-102.

Kender, J.R. (1983). Shape from texture. *Carnegie-Mellon University, Techn. Report* CMU-CS-81-102.

Kenney, C.S., Manjunath, B.S., Zuliani, M., Hewer, G.A. and van Nevel, A. (2003). A condition number for point matching with application to registration and postregistration error estimation. *IEEE Trans. Pattern Anal. Mach. Intell.* **25**, no. 11, 1437–1454.

Kerbyson, D.J. and Atherton, T.J. (1995). Circle detection using Hough transform filters. *IEE Conference Publication* no. 410, 370–374.

Kesidis, A.L. and Papamarkos, N. (2000). On the grayscale inverse Hough transform. *Image Vision Comput.* **18**, no. 8, 607–618

Kessal, L., Abel, N. and Demigny, D. (2003). Real-time image processing with dynamically reconfigurable architecture. *Real-Time Imaging* **9**, no. 5, 297–313.

Kim, D.-S., Lee, W.-H. and Kweon, I.-S. (2004). Automatic edge detection using 3 × 3 ideal binary pixel patterns and fuzzy-based edge thresholding. *Pattern Recogn. Lett.* **25**, no. 1, 101–106.

Kim, D.Y., Kim, J.J., Meer, P., Mintz, D. and Rosenfeld, A. (1989). Robust computer vision: a least median of squares based approach. *Proc. DARPA Image Understanding Workshop, Palo Alto, CA (May 23–26)*, pp. 1117–1134.

Kim, H.-B. and Park, R.-H. (1990). Extraction of periodicity vectors from structural textures using projection information. *Pattern Recogn. Lett.* **11**, no. 9, 625–630.

Kim, M.J. and Han, J.H. (1998). Computation of a cross-section structure: a projection-based approach. In Davies, E.R. and Atiquzzaman, M. (eds.), *Special Issue on Projection-Based Transforms, Image Vision Comput.* **16**, nos. 9–10, 653–668.

Kimme, C., Ballard, D. and Sklansky, J. (1975). Finding circles by an array of accumulators. *Comm. ACM* **18**, 120–122.

Kimura, A. and Watanabe, T. (2002). An extension of the generalized Hough transform to realize affine-invariant two-dimensional (2D) shape detection. *Proc. 16th Int. Conf. on Pattern Recogn., Québec, Canada* (*August 11–15*), Vol. I, pp. 65–69.

King, T.G. and Tao, L.G. (1995). An incremental real-time pattern tracking algorithm for line-scan camera applications. *Mechatronics*, **4**, no. 5, 503–516.

Kirsch, R. A. (1971). Computer determination of the constituent structure of biological images. *Comput. Biomed. Res.* **4**, 315–328.

Kiryati, N. and Bruckstein, A.M. (1991). Antialiasing the Hough transform. *Computer Vision Graph. Image Process.: Graph. Models Image Process.* **53**, no. 3, 213–222.

Kitchen, L. and Rosenfeld, A. (1979). Discrete relaxation for matching relational structures. *IEEE Trans. Systems Man Cybern.* **9**, 869–874.

Kitchen, L. and Rosenfeld, A. (1982). Gray-level corner detection. *Pattern Recogn. Lett.* **1**, 95–102.

Kittler, J. (1983). On the accuracy of the Sobel edge detector. *Image Vision Comput.* **1**, 37–42.

Kittler, J., Hatef, M., Duin, R.P.W. and Matas, J. (1998). On combining classifiers. *IEEE Trans. Pattern Anal. Mach. Intell.* **20**, no. 3, 226–239.

Kittler, J., Illingworth, J. and Föglein, J. (1985). Threshold selection based on a simple image statistic. *Comput. Vision Graph. Image Process.* **30**, 125–147.

Kittler, J., Illingworth, J., Föglein, J. and Paler, K. (1984). An automatic thresholding algorithm and its performance. *Proc. 7th Int. Conf. on Pattern Recogn., Montreal* (*July 30–2 August*). pp. 287–289

Klassen, E., Srivistava, A., Mio, W. and Joshi, S.H. (2004). Analysis of planar shapes using geodesic paths on shape spaces. *IEEE Trans. Pattern Anal. Mach. Intell.* **26**, no. 3, 372–383.

Koch, C., Park, S., Ellis, T.J. and Georgiadis, A. (2001). Illumination technique for optical dynamic range compression and offset reduction. *Proc. British Machine Vision Assoc. Conf., Manchester, UK* (*September*), pp. 293–302.

Koenderink, J.J. and van Doorn, A.J. (1979). The internal representation of solid shape with respect to vision. *Biol. Cybern.* **32**, 211–216.

Koivo, A.J. and Kim, C.W. (1989). Robust image modelling for classification of surface defects on wood boards. *IEEE Trans. Systems Man Cybern.* **19**, no. 6, 1659–1666.

Koller, D., Weber, J., Huang, T., Malik, J., Ogasawara, G., Rao, B. and Russell, S. (1994). Towards robust automatic traffic scene analysis in real-time. *Proc. 12th Int. Conf. on Pattern Recogn., Jerusalem, Israel* (*October 9–13*), pp. 126–131.

Koller, Th.M., Gerig, G., Székely, G. and Dettwiler, D. (1995). Multiscale detection of curvilinear structures in 2-D and 3-D image data. *Proc. 5th Int. Conf. on Computer Vision* (*ICCV'95*), pp. 864–869.

Koplowitz, J. and Bruckstein, A.M. (1989). Design of perimeter estimators for digitized planar shapes. *IEEE Trans. Pattern Anal. Mach. Intell.* **11**, 611–622.

Kortenkamp, D., Bonasso, R.P. and Murphy, R. (eds.) (1998). *Artificial Intelligence and Mobile Robots*. AAAI Press/The MIT Press, Menlo Park, CA; Cambridge, MA; London, England.

Kuehnle, A. (1991). Symmetry-based recognition of vehicle rears. *Pattern Recogn. Lett.* **12**, 249–258.

Kulpa, Z. (1977). Area and perimeter measurement of blobs in discrete binary pictures. *Comput. Graph. Image Process.* **6**, 434–451.

Kumar, R. and Hanson, A.R. (1989). Robust estimation of camera location and orientation from noisy data having outliers. *Proc. Workshop on Interpretation of 3D Scenes, Austin, TX* (*November 27–29*), pp. 52–60.

Kung, H.T. (1980). The structure of parallel algorithms. *Adv. Comput.* **19**, 69–112.

Kuo, F.F. (1966). *Network Analysis and Synthesis*. Wiley, New York (2nd edition).

Kwok, P.C.K. (1989). *Customising thinning algorithms. Proceedings of the 3rd International Conference on Image Processing and Its Applications*, *Warwick* (*July 18–20*), *IEE Conf. Publ.* 307, 633–637.

Lacroix, V. (1988). A three-module strategy for edge detection. *IEEE Trans. Pattern Anal. Mach. Intell.* **10**, 803–810.

Lam, S.W.C. and Ip, H.H.S. (1994). Structural texture segmentation using irregular pyramid. *Pattern Recogn. Lett.* **15**, no. 7, 691–698.

Lamdan, Y. and Wolfson, H.J. (1988). Geometric hashing: a general and efficient model-based recognition scheme. *Proc. IEEE 2nd Int. Conf. on Computer Vision, Tampa, FL* (*December*), pp. 238–249.

Landgrebe, D.A. (1981). Analysis technology for land remote sensing. *Proc. IEEE* **69**, 628–642.

Lane, R.A., Thacker, N.A. and Seed, N.L. (1994). Stretch-correlation as a real-time alternative to feature-based stereo matching algorithms. *Image Vision Comput.* **12**, no. 4, 203–212.

Laurentini, A. (1994). The visual hull concept for silhouette-based image understanding. *IEEE Trans. Pattern Anal. Mach. Intell.* **16**, no. 2, 150–162.

Laws, K.I. (1979). Texture energy measures. Proc. Image Understanding Workshop, Nov., pp. 47–51.

Laws, K.I. (1980a). Rapid texture identification. *Proc. SPIE Conf. on Image Processing for Missile Guidance* **238**, San Diego, CA (July 28–August 1), pp. 376–380.

Laws, K.I. (1980b). Textured Image Segmentation. Ph.D. thesis, Univ. of Southern California, Los Angeles.

Leavers, V.F. (1992). *Shape Detection in Computer Vision Using the Hough Transform*. Springer-Verlag, Berlin.

Leavers, V.F. (1993). Which Hough transform? *Computer Vision Graph. Image Process.: Image Understanding* **58**, no. 2, 250–264.

Leavers, V.F. and Boyce, J.F. (1987). The Radon transform and its application to shape parametrization in machine vision. *Image Vision Comput.* **5**, 161–166.

Lebègue, X. and Aggarwal, J.K. (1993). Significant line segments for an indoor mobile robot. *IEEE Trans. Robotics and Automation* **9**, no. 6, 801–815.

Lee, M.-S., Medioni, G. and Mordohai, P. (2002). Inference of segmented overlapping surfaces from binocular stereo. *IEEE Trans. Pattern Anal. Mach. Intell.* **24**, no. 6, 824–837.

Lei, Y. and Wong, K.C. (1999). Ellipse detection based on symmetry. *Pattern Recogn. Lett.* **20**, no. 1, 41–47.

Lev, A., Zucker, S.W. and Rosenfeld, A. (1977). Iterative enhancement of noisy images. *IEEE Trans. Systems Man Cybern.* **7**, 435–442.

Levine, M.D. (1985). *Vision in Man and Machine*. McGraw-Hill, New York.

Li, H. and Lavin, M.A. (1986). Fast Hough transform based on bintree data structure. *Proc. Conf. Comput. Vision and Pattern Recogn., Miami Beach, Florida*, pp. 640–642.

Li, H., Lavin, M.A. and LeMaster, R.J. (1985). Fast Hough transform. *Proc. 3rd Workshop on Comput. Vision: Representation and Control, Bellair*, pp. 75–83.

Liao, K., Paulsen, M.R. and Reid, J.F. (1994). Real-time detection of colour and surface defects of maize kernels using machine vision. *J. Agric. Eng. Res.* **59**, 263–271.

Lin, C.C. and Chellappa, R. (1987). Classification of partial 2-D shapes using Fourier descriptors. *IEEE Trans. Pattern Anal. Mach. Intell.* **9**, 686–690.

Linsker, R. (1986). From basic network principles to neural architecture. *Proc. National Acad. Sciences, USA* **83**, 7508–7512, 8390–8394, 8779–8783.

Linsker, R. (1988). Self-organization in a perceptual network. *IEEE Computer* (March), 105–117.

Lippmann, R.P. (1987). An introduction to computing with neural nets. *IEEE Acoustics, Speech, Signal Process. Magazine* **4**, no. 2, 4–22.

Liu, L. and Sclaroff, S. (2004). Deformable model-guided region split and merge of image regions. *Image Vision Comput.* **22**, no. 4, 343–354.

Liu, M.L. and Wong, K.H. (1999). Pose estimation using four corresponding points. *Pattern Recogn. Lett.* **20**, no. 1, 69–74.

Lladós, J., Martí, E. and Villanueva, J.J. (2001). Symbol recognition by error-tolerant subgraph matching between region adjacency graphs. *IEEE Trans. Pattern Anal. Mach. Intell.* **23**, no. 10, 1137–1143.

Lockton, R. and Fitzgibbon, A. (2002). Real-time gesture recognition using deterministic boosting. *Proc. British Machine Vision Assoc. Conf., Cardiff, UK* (*September 2–5*), pp. 817–826.

Longuet-Higgins, H.C. (1981). A computer algorithm for reconstructing a scene from two projections. *Nature* **293**, 133–135.

Longuet-Higgins, H.C. (1984). The visual ambiguity of a moving plane. *Proc. R. Soc.* (*London*) **B233**, 165–175.

Longuet-Higgins, H.C. and Prazdny, K. (1980). The interpretation of a moving retinal image. *Proc. R. Soc.* (*London*) **B208**, 385–397.

Lowe, D. (ed.) (1994). *Special Issue on Applications of Artificial Neural Networks. IEE Proc. Vision Image Signal Process.* **141**, no. 4.

Lu, S. and Szeto, A. (1991). Improving edge measurements on noisy images by hierarchical neural networks. *Pattern Recogn. Lett.* **12**, 155–164.

Lüdtke, N., Luo, B., Hancock, E. and Wilson, R.C. (2002). Corner detection using a mixture model of edge orientation. *Proc. 16th Int. Conf. on Pattern Recogn., Québec, Canada* (*August 11–15*), Vol. II, pp. 574–577.

Lukac, R. (2003). Adaptive vector median filtering. *Pattern Recogn. Lett.* **24**, no. 12, 1889–1899.

Luo, B. and Hancock, E.R. (2001). Structural graph matching using the EM algorithm and singular value decomposition. *IEEE Trans. Pattern Anal. Mach. Intell.* **23**, no. 10, 1120–1136.

Luong, Q.-T. and Faugeras, O. (1997). Self-calibration of a moving camera from point correspondences and fundamental matrices. *Int. J. Computer Vision* **22**, no. 3, 261–289.

Lutton, E. and Martinez, P. (1994). A genetic algorithm for the detection of 2D geometric primitives in images. *Proc. 12th Int. Conf. on Pattern Recogn., Jerusalem* (*October 9–13*), Vol. 1, pp. 526–528.

Lutton, E., Maître, H. and Lopez-Krahe, J. (1994). Contribution to the determination of vanishing points using Hough transform. *IEEE Trans. Pattern Anal. Mach. Intell.* **16**, no. 4, 430–438.

Lyvers, E.R. and Mitchell, O.R. (1988). Precision edge contrast and orientation estimation. *IEEE Trans. Pattern Anal. Mach. Intell.* **10**, 927–937.

Ma, L., Tan, T., Wang, Y. and Zhang, D. (2003). Personal identification based on iris texture analysis. *IEEE Trans. Pattern Anal. Mach. Intell.* **25**, no. 12, 1519–1533.

Mackeown, W.P.J., Greenway, P., Thomas, B.T. and Wright, W.A. (1994). Contextual image labelling with a neural network. *IEE Proc. Vision Image Signal Process.* **141**, no. 4, 238–244.

MacQueen, J.B. (1967). Some methods for classification and analysis of multivariate observations. *Proc. 5th Berkeley Symp. on Math. Stat. and Prob.* **I**, pp. 281–297.

Magee, M.J., and Aggarwal, J.K. (1984). Determining vanishing points from perspective images. *Computer Vision Graph. Image. Process.* **26**, no. 2, 256–267.

Malik, J., Weber, J., Luong, Q.-T. and Koller, D. (1995). Smart cars and smart roads. *Proc. 6th British Machine Vision Assoc. Conf., Birmingham* (*September 11–14*), pp. 367–381.

Mallat, S.G. (1989). A theory of multiresolution image processing: the wavelet representation. *IEEE Trans. Pattern Anal. Mach. Intell.* **11**, no. 6, 674–693.

Mandelbrot, B.B. (1982). *The Fractal Geometry of Nature*. Freeman, San Francisco, CA.

Manthalkar, R., Biswas, P.K. and Chatterji, B.N. (2003). Rotation invariant texture classification using even symmetric Gabor filters. *Pattern Recogn. Lett.* **24**, no. 12, 2061–2068.

Maragos, P., Schafer, R.W. and Butt, M.A. (eds.) (1996). *Mathematical Morphology and Its Application to Image and Signal Processing*. Kluwer Academic Publications, Boston.

Marchant, J.A. (1996). Tracking of row structure in three crops using image analysis. *Comput. Electron. Agric.* **15**, 161–179.

Marchant, J.A. and Brivot, R. (1995). Real-time tracking of plant rows using a Hough transform. *Real-Time Imaging*, **1**, no. 5, 363–371.

Marchant, J.A. and Onyango, C.M. (1995). Fitting grey level point distribution models to animals in scenes. *Image Vision Comput.* **13**, no. 1, 3–12.

Marchant, J.A., Tillett, R.D. and Brivot, R. (1998). Real-time segmentation of plants and weeds. *Real-Time Imaging* **4**, 243–253.

Marino, F., Stella, E., Branca, A., Veneziani, N. and Distante, A. (2001). Specialized hardware for real-time navigation. *Real-Time Imaging* **7**, no. 1, 97–108.

Marr, D. (1976). Early processing of visual information. *Phil. Trans. R. Soc. (London)* **B275**, 483–524.

Marr, D. and Hildreth, E. (1980). Theory of edge detection. *Proc. R. Soc. (London)* **B207**, 187–217.

Marr, D. and Poggio, T. (1979). A computational theory of human stereo vision. *Proc. R. Soc. (London)* **B204**, 301–328.

Marshall, S. (2004). New direct design method for weighted order statistic filters. *IEE Proc. Vision Image Signal Processing* **151**, no. 1, 1–8.

Marshall, S., Harvey, N. and Shah, D. (eds.) (1998). *Proc. Noblesse Workshop on Non-linear Model Based Image Analysis, Glasgow (July 1–3)*. Springer-Verlag, London.

Marslin, R.F., Sullivan, G.D. and Baker, K.D. (1991). Kalman filters in constrained model based tracking. *Proc. 2nd British Machine Vision Assoc. Conf., Glasgow (September 23–26)*, pp. 371–374.

Matsuyama, T., Saburi, K. and Nagao, M. (1982). A structural analyzer for regularly arranged textures. *Comput. Graph. Image Process.* **18**, no. 3, 259–278.

Maybank, S.J. (1986). Algorithm for analysing optical flow based on the least squares method. *Image Vision Comput.* **4**, 38–42.

Maybank, S. (1992). *Theory of Reconstruction from Image Motion*, Springer-Verlag, Berlin, Heidelberg.

Maybank, S.J. (1996). Stochastic properties of the cross ratio. *Pattern Recogn. Lett.* **17**, no. 3, 211–217.

Maybank, S.J. and Faugeras, O. (1992). A theory of self-calibration of a moving camera. *Int. J. Computer Vision* **8**, no. 2, 123–151.

Maybank, S. and Tan, T. (eds.) (2004). *Special issue: Visual Surveillance. Image Vision Comput.* **22**, no. 7, 515–582.

Maybeck, P.S. (1979). *Stochastic Models, Estimation, and Control.* Volume 1. Academic Press, New York and London.

McFarlane, N.J.B. and Schofield, C.P. (1995). Segmentation and tracking of piglets in images. *Machine Vision Applications* **8**, no. 3, 187–193.

McGunnigle, G. and Chantler, M. (2003). Resolving handwriting from background printing using photometric stereo. *Pattern Recogn.* **36**, 1869–1879.

McIvor, A.M. (1988). Edge detection in dynamic vision. *Proc. 4th Alvey Vision Conf., Manchester (August 31–September 2)*, pp. 141–145.

Meer, P. and Georgescu, B. (2001). Edge detection with embedded confidence. *IEEE Trans. Pattern Anal. Mach. Intell.* **23**, no. 12, 1351–1365.

Meer, P., Mintz, D. and Rosenfeld, A. (1990). Least median of squares based robust analysis of image structure. *Proc. DARPA Image Understanding Workshop, Pittsburgh, Pennsylvania (Sept. 11–13)*, pp. 231–254.

Meer, P., Mintz, D., Rosenfeld, A. and Kim, D.Y. (1991). Robust regression methods for computer vision: a review. *Int. J. Comput. Vision* **6**, no. 1, 59–70.

Mémin, É. and Risset, T. (2001). VLSI design methodology for edge-preserving image reconstruction. *Real-Time Imaging* **7**, no. 1, 109–126.

Meribout, M., Nakanishi, M. and Ogura, T. (2002). Accurate and real-time image processing on a new PC-compatible board. *Real-Time Imaging* **8**, no. 1, 35–51.

Merlin, P.M. and Farber, D.J. (1975). A parallel mechanism for detecting curves in pictures. *IEEE Trans. Comput.* **28**, 96–98.

Mérõ, L. and Vassy, Z. (1975). A simplified and fast version of the Hueckel operator for finding optimal edges in pictures. *Proc. 4th Int. Joint Conf. on Artif. Intell., Tbilisi, Georgia, USSR*, pp. 650–655.

Michalewicz, Z. (1992). *Genetic Algorithms + Data Structures = Evolution Programs*. Springer-Verlag, Berlin, Heidelberg.

Mikic, I., Kogut, P.C.G. and Trivedi, M. (2000). Moving shadow and object detection in traffic scenes. *Proc. Int. Conf. on Pattern Recogn.* pp. 321–324.

Milgram, D. (1979). Region extraction using convergent evidence. *Comput. Graph. Image Process.* **11**, 1–12.

Minsky, M. and Papert, S. (1971). On some associative, parallel and analog computations. In Jacks, E.J. (ed.), *Associative Information Techniques*. Elsevier, New York, pp. 27–47.

Minsky, M.L. and Papert, S.A. (1969). *Perceptrons*. MIT Press, Cambridge, MA.

Mirmehdi, M. and Petrou, M. (2000). Segmentation of colour textures. *IEEE Trans. Pattern Anal. Mach. Intell.* **22**, no. 2, 142–159.

Mitra, P., Murthy, C.A. and Pal, S.K. (2002). Unsupervised feature selection using feature similarity. *IEEE Trans. Pattern Anal. Mach. Intell.* **24**, no. 3, 301–312.

Moganti, M., Ercal, F., Dagli, C.H. and Tsunekawa, S. (1996). Automatic PCB inspection algorithms: a survey. *Comput. Vision Image Understanding* **63**, no. 2, 287–313.

Mohr, R. and Wu, C. (eds.) (1998). Special Issue on *Geometric Modelling and Invariants for Computer Vision. Image Vision Comput.* **16**, no. 1.

Mokhtarian, F. and Bober, M. (2003). *Curvature Scale Space Representation: Theory, Applications and MPEG-7 Standardisation*. Kluwer Academic Publishers, Dordrecht.

Mokhtarian, F., Abbasi, S. and Kittler, J. (1996). Efficient and robust shape retrieval by shape content through curvature scale space. *Proc. 1st Int. Conf. Image Database and Multi-Search*, pp. 35–42.

Montiel, E., Aguado, A.S. and Nixon, M.S. (2001). Improving the Hough transform gathering process for affine transformations. *Pattern Recogn. Lett.* **22**, no. 9, 959–969.

Moore, G.A. (1968). Automatic scanning and computer processes for the quantitative analysis of micrographs and equivalent subjects. In Cheng, G.C., et al. (eds.), *Pictorial Pattern Recognition*. Thompson, Washington, DC, pp. 236–275.

Moravec, H.P. (1977). Towards automatic visual obstacle avoidance. *Proc. 5th Int. Joint. Conf. on Artificial Intelligence, Cambridge, MA (August 22–25)*, p. 584.

Moravec, H.P. (1980). Obstacle avoidance and navigation in the real world by a seeing robot rover. *Stanford Artif. Intell. Lab. Memo AIM-340*.

Môtus, L. and Rodd, M.G. (1994). *Timing Analysis of Real-Time Software*, Elsevier, Oxford, UK.

Mundy, J.L., and Zisserman, A. (eds.) (1992a). *Geometric Invariance in Computer Vision*, MIT Press, Cambridge, MA.

Mundy, J.L. and Zisserman, A. (1992b). Appendix—projective geometry for machine vision. In Mundy, J.L. and Zisserman, A. (eds.) (1992), op. cit., pp. 463–519.

Murray, D.W., Kashko, A. and Buxton, H. (1986). A parallel approach to the picture restoration algorithm of Geman and Geman on an SIMD machine. *Image Vision Comput.* **4**, no. 3, 133–142.

Nagao, M. and Matsuyama, T. (1979). Edge preserving smoothing. *Comput. Graph. Image Process.* **9**, 394–407.

Nagel, H.-H. (1983). Displacement vectors derived from second-order intensity variations in image sequences. *Comput. Vision Graph. Image Process.* **21**, 85–117.

Nagel, H.-H. (1986). Image sequences—ten (octal) years—from phenomenology towards a theoretical foundation. *Proc. 8th Int. Conf. on Pattern Recogn., Paris* (*October 27–31*), pp. 1174–1185.

Nagel, R.N. and Rosenfeld, A. (1972). Ordered search techniques in template matching. *Proc. IEEE* **60**, 242–244.

Nakagawa, Y. (1982). Automatic visual inspection of solder joints on printed circuit boards. *Proc. SPIE, Robot Vision* **336**, 121–127.

Nakagawa, Y. and Rosenfeld, A. (1979). Some experiments on variable thresholding. *Pattern Recogn.* **11**, 191–204.

Narendra, P.M. (1978). A separable median filter for image noise smoothing. *Proc. IEEE Comput. Soc. Conf. on Pattern Recogn. and Image Process., Chicago* (*May 31–June 2*), pp. 137–141.

Nesi, P. and Trucco, E. (eds.) (1999). *Special Issue on Real-Time Defect Detection. Real-Time Imaging* **5**, no. 1.

Newman, T.S. and Jain, A.K. (1995). A survey of automated visual inspection. *Comput. Vision Image Understanding* **61**, no. 2, 231–262.

Niblack, W. (1985). *An Introduction to Digital Image Processing*. Strandberg, Birkeroed, Denmark.

Nieminen, A., Heinonen, P. and Neuvo, Y. (1987). A new class of detail-preserving filters for image processing. *IEEE Trans. Pattern Anal. Mach. Intell.* **9**, 74–90.

Nightingale, C. and Hutchinson, R.A. (1990). Artificial neural nets and their applications to image processing. *British Telecom Technol. J.* **8**, no. 3, 81–93.

Nilsson, N.J. (1965). *Learning Machines—Foundations of Trainable Pattern-Classifying Systems*. McGraw-Hill, New York.

Nitzan, D., Brain, A.E. and Duda, R.O. (1977). The measurement and use of registered reflectance and range data in scene analysis. *Proc. IEEE* **65**, 206–220.

Nixon, M. (1985). Application of the Hough transform to correct for linear variation of background illumination in images. *Pattern Recogn. Lett.* **3**, 191–194.

Nixon, M.S. and Aguado, A.S. (2002). *Feature Extraction and Image Processing*, Newnes, London.

Noble, A., Hartley, R., Mundy, J. and Farley, J. (1994). X-ray metrology for quality assurance. *Proc. IEEE Int. Conf. Robotics and Automation*, Vol. 2, San Diego, CA (May), pp. 1113–1119.

Noble, J.A. (1988). Finding corners. *Image Vision Comput.* **6**, 121–128.

Noble, J.A. (1995). From inspection to process understanding and monitoring: a view on computer vision in manufacturing. *Image Vision Comput.* **13**, no. 3, 197–214.

North, D.O. (1943). An analysis of the factors which determine signal/noise discrimination in pulsed-carrier systems. *RCA Lab., Princeton, NJ, Rep. PTR-6C*; reprinted in *Proc. IEEE* **51**, 1016–1027 (1963).

Noyce, R.N. (1977). Microelectronics. *Scientific Amer.* **237** (September), 62–69.

O'Gorman, F. (1978). Edge detection using Walsh functions. *Artif. Intell.* **10**, 215–223.

O'Gorman, F. and Clowes, M.B. (1976). Finding picture edges through collinearity of feature points. *IEEE Trans. Comput.* **25**, 449–456.

Offen, R.J. (ed.) (1985). *VLSI Image Processing*. Collins, London.

Oflazer, K. (1983). Design and implementation of a single chip 1d median filter. *IEEE Trans. Acoust. Speech Signal Process.* **31**, 1164–1168.

Ogawa, H. and Taniguchi, K. (1979). Preprocessing for Chinese character recognition and global classification of handwritten Chinese characters. *Pattern Recogn.* **11**, 1–7.

Ohanian, P.P. and Dubes, R.C. (1992). Performance evaluation for four classes of textural features. *Pattern Recogn.* **25**, 819–833.

Ohta, Y., Maenobu, K. and Sakai, T. (1981). Obtaining surface orientation from texels under perspective projection. *Proc. 7th Int. Joint Conf. on Artif. Intell., Vancouver*, pp. 746–751.

Oja, E. (1982). A simplified neuron model as a principal component analyzer. *Int. J. Neural Systems* **1**, 61–68.

Oja, E. and Karhunen, J. (1993). Nonlinear PCA: algorithms and applications. *Report A18*, Helsinki Univ. of Technol., Espoo, Finland.

Ojala, T., Pietikäinen, M. and Mäenpää, T. (2002). Multiresolution gray-scale and rotation-invariant texture classification with local binary patterns. *IEEE Trans. Pattern Anal. Mach. Intell.* **24**, no. 7, 971–987.

Olague, G. and Hernández, B. (2002). Flexible model-based multi-corner detector for accurate measurements and recognition. *Proc. 16th Int. Conf. on Pattern Recogn., Québec, Canada (August 11–15)*, Vol. II, pp. 578–583.

Olson, C.F. (1998). Improving the generalized Hough transform through imperfect grouping. In Davies, E.R. and Atiquzzaman, M. (eds.), *Special Issue on Projection-Based Transforms, Image Vision Comput.* **16**, nos. 9–10, 627–634.

Olson, C.F. (1999). Constrained Hough transforms for curve detection. *Computer Vision Image Understanding* **73**, no. 3, 329–345.

Onyango, C.M. and Marchant, J.A. (1996). Modelling grey level surfaces using three-dimensional point distribution models. *Image Vision Comput.* **14**, 733–739.

Osteen, R.E. and Tou, J.T. (1973). A clique-detection algorithm based on neighbourhoods in graphs. *Int. J. Comput. Inf. Sci.* **2**, 257–268.

Otsu, N. (1979). A threshold selection method from gray-level histograms. *IEEE Trans. Systems Man Cybern.* **9**, 62–66.

Ouerhani, N. and Hügli, H. (2003). Real-time visual attention on a massively parallel SIMD architecture. *Real-Time Imaging* **9**, no. 3, 189–196.

Overington, I. and Greenway, P. (1987). Practical first-difference edge detection with subpixel accuracy. *Image Vision Comput.* **5**, 217–224.

Pal, N.R. and Pal, S.K. (1989). Object-background segmentation using new definitions of entropy. *IEE Proc. E* **136**, no. 4, 284–295.

Pal, S.K., Bhandari, D. and Kundu, M.K. (1994). Genetic algorithms for optimal image enhancement. *Pattern Recogn. Lett.* **15**, no. 3, 261–271.

Pal, S.K., King, R.A. and Hashim, A.A. (1983). Automatic grey level thresholding through index of fuzziness and entropy. *Pattern Recogn. Lett.* **1**, 141–146.

Paler, K. and Kittler, J. (1983). Greylevel edge thinning: a new method. *Pattern Recogn. Lett.* **1**, 409–416.

Paler, K., Foglein, J., Illingworth, J. and Kittler, J. (1984). Local ordered grey levels as an aid to corner detection. *Pattern Recogn.* **17**, 535–543.

Pan, X.-B., Brady, M., Bowman, A.K., Crowther, C. and Tomlin, R.S.O. (2004). Enhancement and feature extraction for images of incised and ink texts. *Image Vision Comput.* **22**, no. 6, 443–451.

Pan, X.D., Ellis, T.J. and Clarke, T.A. (1995). Robust tracking of circular features. *Proc. 6th British Machine Vision Assoc. Conf., Birmingham* (*September 11–14*), pp. 553–562.

Panda, D.P. and Rosenfeld, A. (1978). Image segmentation by pixel classification in (gray level, edge value) space. *IEEE Trans. Comput.* **27**, 875–879.

Parker, D.B. (1985). Learning-logic: casting the cortex of the human brain in silicon. *Technical Report TR-47*, Center for Comput. Res. in Economics and Management Sci., MIT, Cambridge, MA.

Parker, J.R. (1994). *Practical Computer Vision Using C.* Wiley, New York.

Patel, D. and Davies, E.R. (1995). Low contrast object detection using a MLP network designed by node creation. *Proc. IEEE Int. Conf. on Neural Networks, Perth* (*November 27–December 1*), Vol. 2, pp. 1155–1159.

Patel, D., Davies, E.R. and Hannah, I. (1995). Towards a breakthrough in the detection of contaminants in food products. *Sensor Review* **15**, no. 2, 27–28.

Patel, D., Davies, E.R. and Hannah, I. (1996). The use of convolution operators for detecting contaminants in food images. *Pattern Recognition* **29**, no. 6, 1019–1029.

Patel, D., Hannah, I. and Davies, E.R. (1994). Texture analysis for foreign object detection using a single layer neural network. *Proc. IEEE Int. Conf. on Neural Networks, Florida* (*June 28–July 2*), Vol. VII, pp. 4265–4268.

Pavlidis, T. (1968). Computer recognition of figures through decomposition. *Inf. Control* **12**, 526–537.

Pavlidis, T. (1977). *Structural Pattern Recognition.* Springer-Verlag, New York.

Pavlidis, T. (1978). A review of algorithms for shape analysis. *Comput. Graph. Image Process.* **7**, 243–258.

Pavlidis, T. (1980). Algorithms for shape analysis of contours and waveforms. *IEEE Trans. Pattern Anal. Mach. Intell.* **2**, 301–312.

Pavlidis, T. (1992). Why progress in machine vision is so slow. *Pattern Recogn. Lett.* **13**, 221–225.

Pearl, J. (1988). *Probabilistic Reasoning in Intelligent Systems: Networks of Plausible Inference.* Morgan Kaufmann, San Mateo, CA.

Pelillo, M. (1999). Replicator equations, maximal cliques and graph isomorphism. *Neural Computation* **11**, no. 8, 1933–1955.

Penman, D., Olsson, O. and Beach, D. (1992). Automatic X-ray inspection of canned products for foreign material. *Machine Vision Applications, Architectures and Systems Integration, SPIE,* **1823**, pp. 342–347.

Pentland, A.P. (1984). Fractal-based description of natural scenes. *IEEE Trans. Pattern Anal. Mach. Intell.* **6**, no. 6, 661–674.

Persoon, E. and Fu, K.-S. (1977). Shape discrimination using Fourier descriptors. *IEEE Trans. Systems Man Cybern.* **7**, 170–179.

Petrou, M. and Bosdogianni, P. (1999). *Image Processing: The Fundamentals.* Wiley, Chichester.

Petrou, M., and Kittler, J. (1988). On the optimal edge detector. *Proc. 4th Alvey Vision Conf., Manchester* (*August 31–September 2*), pp. 191–196.

Pfaltz, J.L. and Rosenfeld, A. (1967). Computer representation of planar regions by their skeletons. *Comm. ACM* **10**, 119–125.

Pham, D.T. and Bayro-Corrochano, E.J. (1992). Neural computing for noise filtering, edge detection and signature extraction. *J. Systems Eng.* **1**, 13–23.

Phong, B.-T. (1975). Illumination for computer-generated pictures. *Comm. ACM* **18**, 311–317.

Pickett, R.M. (1970). Visual analysis of texture in the detection and recognition of objects. In Lipkin, B.S. and Rosenfeld, A. (eds.), *Picture Processing and Psychopictorics*, Academic Press, New York, pp. 289–308.

Pietikäinen, M., Rosenfeld, A. and Davis, L.S. (1983). Experiments with texture classification using averages of local pattern matches. *IEEE Trans. Systems Man Cybern.* **13**, no. 3, 421–426.

Plummer, A.P.N. and Dale, F. (1984). *The Picture Processing Language Compiler Manual.* National Physical Laboratory, Teddington.

Pollard, S.B., Porrill, J., Mayhew, J.E.W. and Frisby, J.P. (1987). Matching geometrical descriptions in three-space. *Image Vision Comput.* **5**, no. 2, 73–78.

Postaire, J.G. and Touzani, A. (1989). Mode boundary detection by relaxation for cluster analysis. *Pattern Recogn.* **22**, no. 5, 477–489.

Pountain, D. and May, D. (1987). *A Tutorial Introduction to OCCAM Programming*. BSP Professional Books, Oxford.

Prati, A., Cucchiara, R., Mikić, I. and Trivedi, M. (2001). Analysis and detection of shadows in video streams: a comparative evaluation. *Proc. IEEE Int. Conf. Computer Vision and Pattern Recognition, Hawaii (December 11–13).*

Pratt, W.K. (2001). *Digital Image Processing*. Wiley-Interscience, New York (2nd edition).

Press, W.H., Teukolsky, S.A., Vetterling, W.T. and Flannery, B.P. (1992). *Numerical Recipes in C. The Art of Scientific Computing*. Cambridge Univ. Press, Cambridge (2nd edition).

Prewitt, J.M.S. (1970). Object enhancement and extraction. In Lipkin, B.S. and Rosenfeld, A. (eds.), *Picture Processing and Psychopictorics*. Academic Press, New York, pp. 75–149.

Prieto, M.S. and Allen, A.R. (2003). A similarity metric for edge images. *IEEE Trans. Pattern Anal. Mach. Intell.* **25**, no. 10, 1265–1273.

Princen, J., Illingworth, J. and Kittler, J. (1989a). A hierarchical approach to line extraction. *Proc. IEEE Comput. Vision and Pattern Recogn. Conf., San Diego*, pp. 92–97.

Princen, J., Illingworth, J. and Kittler, J. (1994). Hypothesis testing: a framework for analyzing and optimizing Hough transform performance. *IEEE Trans. Pattern Anal. Mach. Intell.* **16**, no. 4, 329–341.

Princen, J., Yuen, H.K., Illingworth, J. and Kittler, J. (1989b). Properties of the adaptive Hough transform. *Proc. 6th Scand. Conf. on Image Analysis, Oulu, Finland (June 19–22)*, pp. 613–620.

Pringle, K.K. (1969). Visual perception by a computer. In Grasselli, A. (ed.), *Automatic Interpretation and Classification of Images*. Academic Press, New York, pp. 277–284.

Pugh, A. (ed.) (1983). *Robot Vision.* IFS (Publications) Ltd., Bedford, UK/Springer-Verlag, Berlin.

Pun, C.-M. and Lee, M.-C. (2003). Log-polar wavelet energy signatures for rotation and scale invariant texture classification. *IEEE Trans. Pattern Anal. Mach. Intell.* **25**, no. 5, 590–603.

Pun, T. (1980). A new method for grey-level picture thresholding using the entropy of the histogram. *Signal Process.* **2**, 223–237.

Pun, T. (1981). Entropic thresholding, a new approach. *Comput. Graph. Image Process.* **16**, 210–239.

Rabah, H., Mathias, H., Weber, S., Mozef, E. and Tanougast, C. (2003). Linear array processors with multiple access modes memory for real-time image processing. *Real-Time Imaging* **9**, no. 3, 205–213.

Ramesh, N., Yoo, J.-H. and Sethi, I.K. (1995). Thresholding based on histogram approximation. *IEE Proc. Vision Image Signal Process.* **142**, no. 5, 271–279.

Rätsch, G., Mika, S., Schölkopf, B. and K.-R. Müller (2002). Constructing boosting algorithms from SVMs: an application to one-class classification. *IEEE Trans. Pattern Anal. Mach. Intell.* **24**, no. 9, 1184–1199.

Reed, T.R. and Du Buf, J.M.H. (1993). A review of recent texture segmentation and feature extraction techniques. *Computer Vision Graph. Image Process.: Image Understanding* **57**, 359–372.

Reeves, A.P. (1984). Parallel computer architectures for image processing. *Comput. Vision Graph. Image Process.* **25**, 68–88.

Reeves, A.P., Akey, M.L. and Mitchell, O.R. (1983). A moment-based two-dimensional edge operator. *Proc. IEEE Comput. Soc. Conf. on Comput. Vision and Pattern Recogn.* (*June 19–23*), pp. 312–317.

Ridgway, C. and Chambers, J. (1996). Detection of external and internal insect infestation in wheat by near-infrared reflectance spectroscopy. *J. Sci. Food Agric.* **71**, 251–264.

Ridgway, C., Davies, E.R., Chambers, J., Mason, D.R. and Bateman, M. (2002). Rapid machine vision method for the detection of insects and other particulate bio-contaminants of bulk grain in transit. *Biosystems Engineering* **83**, no. 1, 21–30.

Rindfleisch, T. (1966). Photometric method for lunar topography. *Photogrammetric Eng.* **32**, 262–276.

Ringer, M. and Lazenby, J. (2000). Modelling and tracking articulated motion from multiple camera views. *Proc. 11th British Machine Vision Assoc. Conf., Bristol, UK* (*September 11–14*), pp. 172–181.

Ripley, B.D. (1996). *Pattern Recognition and Neural Networks*. Cambridge Univ. Press, Cambridge, UK.

Rives, G., Dhome, M., Lapreste, J.T. and Richetin, M. (1985). Detection of patterns in images from piecewise linear contours. *Pattern Recogn. Lett.* **3**, 99–104.

Robert, L. (1996). Camera calibration without feature extraction. *Comput. Vision Image Understanding* **63**, no. 2, 314–325.

Roberts, L.G. (1965). Machine perception of three-dimensional solids. In Tippett, J. et al. (eds.), *Optical and Electro-optical Information Processing*. MIT Press, Cambridge, MA, pp. 159–197.

Robinson, G.S. (1977). Edge detection by compass gradient masks. *Comput. Graph. Image Process.* **6**, 492–501.

Robles-Kelly, A. and Hancock, E.R. (2002). A graph-spectral approach to correspondence matching. *Proc. 16th Int. Conf. on Pattern Recogn., Québec, Canada* (*August 11–15*), Vol. IV, pp. 176–179.

Rocket, P.I. (2003). Performance assessment of feature detection algorithms: a methodology and case study of corner detectors. *IEEE Trans. Image Process.* **12**, no. 12, 1668–1676.

Rogers, D.F. (1985). *Procedural Elements for Computer Graphics*. McGraw-Hill, New York.

Roh, Y.J., Park, W.S. and Cho, H. (2003). Correcting image distortion in the X-ray digital tomosynthesis system for PCB solder joint inspection. *Image Vision Comput.* **21**, no. 12, 1063–1075.

Rohl, J.S. and Barrett, H.J. (1980). *Programming via Pascal*. Cambridge University Press, Cambridge, UK.

Rosenblatt, F. (1958). The perceptron: a probabilistic model for information storage and organisation in the brain. *Psychol. Review* **65**, 386–408.

Rosenblatt, F. (1962). *Principles of Neurodynamics*. Spartan, New York.

Rosenfeld, A. (1969). *Picture Processing by Computer*. Academic Press, New York.

Rosenfeld, A. (1970). Connectivity in digital pictures. *J. ACM* **17**, 146–160.

Rosenfeld, A. (1979). *Picture Languages: Formal Models for Picture Recognition*. Academic Press, New York.

Rosenfeld, A. and Kak, A.C. (1981). *Digital Picture Processing*. Academic Press, New York (2nd edition).

Rosenfeld, A. and Pfaltz, J.L. (1966). Sequential operations in digital picture processing. *J. ACM* **13**, 471–494.

Rosenfeld, A. and Pfaltz, J.L. (1968). Distance functions on digital pictures. *Pattern Recogn.* **1**, 33–61

Rosenfeld, A. and Troy, E.B. (1970a). Visual texture analysis. *Computer Science Center, Univ. of Maryland Techn. Report TR-116.*

Rosenfeld, A. and Troy, E.B. (1970b). Visual texture analysis. Conf. Record for Symposium on Feature Extraction and Selection in Pattern Recogn. *IEEE Publication 70C-51C, Argonne, Ill.* (October), pp. 115–124.

Rosenfeld, A. and VanderBrug, G.J. (1977). Coarse-fine template matching. *IEEE Trans. Systems Man Cybern.* **7**, 104–107.

Rosenfeld, A., Hummel, R.A. and Zucker, S.W. (1976). Scene labelling by relaxation operations. *IEEE Trans. Systems Man Cybern.* **6**, 420–433.

Rosie, A.M. (1966). *Information and Communication Theory*. Blackie, London.

Rosin, P. (2000). Fitting superellipses. *IEEE Trans. Pattern Anal. Mach. Intell.* **22**, no. 7, 726–732.

Rosin, P. and Ellis, T.J. (1995). Image difference threshold strategies and shadow detection. *Proc. British Machine Vision Assoc. Conf.*, pp. 347–356.

Rosin, P.L. (1999). Measuring corner properties. *Computer Vision Image Understanding* **73**, no. 2, 291–307.

Rosin, P.L. and West, G.A.W. (1995). Curve segmentation and representation by superellipses. *IEE Proc. Vision Image Signal Process.* **142**, no. 5, 280–288.

Roth, G. and Levine, M.D. (1994). Geometric primitive extraction using a genetic algorithm. *IEEE Trans. Pattern Anal. Mach. Intell.* **16**, no. 9, 901–905.

Roth, G. and Whitehead, A. (2002). Some improvements on two autocalibration algorithms based on the fundamental matrix. *Proc. 16th Int. Conf. on Pattern Recogn., Québec, Canada* (*August 11–15*), Vol. II, pp. 312–315.

Rothwell, C.A. (1995). *Object Recognition through Invariant Indexing*. Oxford University Press, Oxford.

Rothwell, C.A., Zisserman, A., Forsyth, D.A. and Mundy, J.L. (1992a). Canonical frames for planar object recognition. *Proc. 2nd European Conf. on Computer Vision, Santa Margherita Ligure, Italy (May 19–22)*, pp. 757–772.

Rothwell, C.A., Zisserman, A., Marinos, C.I., Forsyth, D.A. and Mundy, J.L. (1992b). Relative motion and pose from arbitrary plane curves. *Image Vision Comput.* **10**, no. 4, 250–262.

Rousseeuw, P.J. (1984). Least median of squares regression. *J. Amer. Statist. Assoc.* **79**, no. 388, 871–880.

Rousseeuw, P.J. and Leroy, A.M. (1987). *Robust Regression and Outlier Detection*. Wiley, New York.

Rowley, H., Baluja, S. and Kanade, T. (1998). Neural network-based face detection. *IEEE Trans. Pattern Anal. Mach. Intell.* **20**, no. 1, 23–38.

Rumelhart, D.E., Hinton, G.E. and Williams, R.J. (1986). Learning internal representations by error propagation. In Rumelhart, D.E. and McClelland, J.L. (eds.), *Parallel Distributed Processing: Explorations in the Microstructure of Cognition*. MIT Press, Cambridge, MA, pp. 318–362.

Rumelhart, D.E., McClelland, J.L. and PDP Research Group (1987). *Parallel Distributed Processing, Explorations in the Microstructure of Cognition.*, MIT Press, Cambridge, MA.

Rummel, P., and Beutel, W. (1984). Workpiece recognition and inspection by a model-based scene analysis system. *Pattern Recogn.* **17**, 141–148.

Russ, J.C. (1998). *The Image Processing Handbook*. CRC Press, Boca Raton, FL (3rd edition).

Russell, S.J. and Norvig, P. (2003). *Artificial Intelligence: A Modern Approach*. Prentice Hall, Englewood cliffs, NJ (2nd edition).

Rutkowski, W. and Benton, R. (1984). Determination of object pose by fitting a model to sparce range data. *Interim Techn. Rep. Intelligent Task Automation Program, Honeywell, Inc.*, pp. 6–62–6–98.

Rutovitz, D. (1970). Centromere finding: some shape descriptors for small chromosome outlines. In Meltzer, B. and Michie, D. (eds.), *Machine Intelligence* **5**. Edinburgh University Press, Edinburgh, pp. 435–462.

Rutovitz, D. (1975). An algorithm for in-line generation of a convex cover. *Comput Graph. Image Process.* **4**, 74–78.

Sahoo, P.K., Soltani, S., Wong, A.K.C. and Chen, Y.C. (1988). A survey of thresholding techniques. *Comput. Vision Graph. Image Process.* **41**, 233–260.

Sakarya, U. and Erkmen, İ. (2003). An improved method of photometric stereo using local shape from shading. *Image Vision Comput.* **21**, no. 11, 941–954.

Sanchiz, J.M., Pla, F., Marchant, J.A. and Brivot, R. (1996). Structure from motion techniques applied to crop field mapping. *Image Vision Comput.* **14**, 353–363.

Sanfeliu, A. and Fu, K.S. (1983). A distance measure between attributed relational graphs for pattern recognition. *IEEE Trans. Syst. Man Cybern.* **13**, no. 3, 353–362.

Sanger, T.D. (1989). Optimal unsupervised learning in a single-layer linear feedforward neural network. *Neural Networks* **2**, 459–473.

Sangwine, S.J. and Horne, R.E.N. (eds.) (1998). *The Colour Image Processing Handbook*. Chapman and Hall, London.

Sanz, J.L.C. (ed.) (1988). *Special Issue on Industrial Machine Vision and Computer Vision Technology. IEEE Trans. Pattern Anal. Mach. Intell.* **10** (1).

Schaffalitsky, F. and Zisserman, A. (2000). Planar grouping for automatic detection of vanishing lines and points. *Image Vision Comput.* **18**, no. 9, 647–658.

Schildt, H. (1995). *C: The Complete Reference*. Osborne McGraw-Hill, Berkeley (3rd edition).

Schmid, C., Mohr, R. and Bauckhage, C. (2000). Evaluation of interest point detectors. *Int. J. Computer Vision* **37**, no. 2, 151–172.

Schneiderman, H., Nashman, M., Wavering, A.J. and Lumia, R. (1995). Vision-based robotic convoy driving. *Machine Vision and Applications* **8**, no. 6, 359–364.

Scott, G.L. (1988). *Local and Global Interpretation of Moving Images*. Pitman, London and Morgan Kaufmann, San Mateo, CA.

Sebe, N. and Lew, M.S. (2003). Comparing salient point detectors. *Pattern Recogn. Lett.* **24**, nos. 1–3, 89–96.

Sebe, N., Tian, Q., Loupias, E., Lew, M.S. and Huang, T.S. (2003). Evaluation of salient point techniques. *Image Vision Comput.* **21**, nos. 13–14, 1087–1095.

Seeger, U. and Seeger, R. (1994). Fast corner detection in grey-level images. *Pattern Recogn. Lett.* **15**, no. 7, 669–675.

Semple, J.G. and Kneebone, G.T. (1952). *Algebraic Projective Geometry*. Oxford University Press, Oxford, UK.

Ser, P.-K. and Siu, W.-C. (1995). Novel detection of conics using 2-D Hough planes. *IEE Proc. Vision Image Signal Process.* **142**, no. 5, 262–270.

Serra, J. (1982). *Image Analysis and Mathematical Morphology*. Academic Press, New York.

Serra, J. (1986). Introduction to mathematical morphology. *Comput. Vision Graph. Image Process.* **35**, 283–305.

Sewisy, A.A. and Leberl, F. (2001). Detection ellipses by finding lines of symmetry in the images via an hough transform applied to straight lines. *Image Vision Comput.* **19**, no. 12, 857–866.

Shah, M.A. and Jain, R. (1984). Detecting time-varying corners. *Comput. Vision Graph. Image Process.* **28**, 345–355.

Shaming, W.B. (1974). Digital image transform encoding. *RCA Corp. paper no.* PE-622.

Shapiro, L.G. and Haralick, R.M. (1985). A metric for comparing relational descriptions. *IEEE Trans. Pattern Anal. Mach. Intell.* **7**, no. 1, 90–94.

Shen, F. and Wang, H. (2002). Corner detection based on modified Hough transform. *Pattern Recogn. Lett.* **23**, no. 8, 1039–1049.

Shen, X. and Hogg, D. (1995). 3D shape recovery using a deformable model. *Image Vision Comput.* **13**, no. 5, 377–383.

Shirai, Y. (1972). Recognition of polyhedra with a range finder. *Pattern Recogn.* **4**, 243–250.

Shirai, Y. (1975). Analyzing intensity arrays using knowledge about scenes. In Winston, P.H. (ed.), *The Psychology of Computer Vision*. McGraw-Hill, New York, pp. 93–113.

Shirai, Y. (1987). *Three-dimensional Computer Vision*. Springer-Verlag, Berlin.

Shufelt, J.A. (1999). Performance evaluation and analysis of vanishing point detection techniques. *IEEE Trans. Pattern Anal. Mach. Intell.* **21**, no. 3, 282–288.

Shuster, R., Ansari, N. and Bani-Hashemi, A. (1993). Steering a robot with vanishing points. *IEEE Trans. Robotics and Automation* **9**, no. 4, 491–498.

Siebel, N.T. and Maybank, S.J. (2002). Fusion of multiple tracking algorithms for robust people tracking. In Heyden, A., Sparr, G., Nielsen, M. and Johansen, P. (eds.), *Proc. 7th European Conf. on Computer Vision (ECCV)*, Vol. IV, pp. 373–387.

Siedlecki, W. and Sklansky, J. (1989). A note on genetic algorithms for large-scale feature selection. *Pattern Recogn. Lett.* **10**, no. 5, 335–347.

Siegel, A.F. (1982). Robust regression using repeated medians. *Biometrika* **69**, no. 1, 242–244.

Silberberg, T.M., Davies, L. and Harwood, D. (1984). An iterative Hough procedure for three-dimensional object recognition. *Pattern Recogn.* **17**, 621–629.

Singh, R.K. and Ramakrishna, R.S. (1990). Shadows and texture in computer vision. *Pattern Recogn. Lett.* **11**, no. 2, 133–141.

Sjöberg, F. and Bergholm, F. (1988). Extraction of diffuse edges by edge focussing. *Pattern Recogn. Lett.* **7**, 181–190.

Skillicorn, D.B. (1988). A taxonomy for computer architectures. *IEEE Comput.* (*November*) **21**, 46–57.

Sklansky, J. (1970). Recognition of convex blobs. *Pattern Recogn.* **2**, 3–10.

Sklansky, J. (1978). On the Hough technique for curve detection. *IEEE Trans. Comput.* **27**, 923–926.

Sklansky, J., Cordella, L.P. and Levialdi, S. (1976). Parallel detection of concavities in cellular blobs. *IEEE Trans. Comput.* **25**, 187–196.

Slama, C.C. (ed.) (1980). *Manual of Photogrammetry*. Amer. Soc. of Photogrammetry, Falls Church, VA (4th edition).

Smith, S. and Brady, J. (1997). Susan—a new approach to low level image processing. *Int. J. Computer Vision* 23, no. 1, 45–78.

Snyder, W.E. (ed.) (1981). *Special Issue on Computer Analysis of Time-Varying Images. IEEE Comput.* (August) **14.**

Sobey, P.J.M. (1989). *The Automated Visual Inspection and Grading of Timber*. Ph.D. Thesis, University of Adelaide.

Soille, P. (2003). *Morphological Image Analysis: Principles and Applications*. Springer-Verlag, Heidelberg (2nd edition).

Song, J., Cai, M., Lyu, M. and Cai, S. (2002). A new approach for line recognition in large-size images using Hough transform. *Proc. 16th Int. Conf. on Pattern Recogn., Québec, Canada* (*August 11–15*), Vol. I, pp. 33–36.

Sonka, M., Hlavac, V. and Boyle, R. (1999). *Image Processing, Analysis and Machine Vision*. Brooks/Cole Publishing Company, Pacific Grove, CA (2nd edition).

Spann, M., Horne, C. and du Buf, J.M.H. (1989). The detection of thin structures in images. *Pattern Recogn. Lett.* **10**, no. 3, 175–179.

Spence, A., Robb, M., Timmins, M. and Chantler, M. (2003). Real-time per-pixel rendering of textiles for virtual textile catalogues. *Proc. INTEDEC 2003, Edinburgh* (September 22–24).

Spirkovska, L. and Reid, M.B. (1992). Robust position, scale, and rotation invariant object recognition using higher-order neural networks. *Pattern Recogn.* **25**, no. 9, 975–985.

Srinivas, M. and Patnaik, L.M. (1994). Genetic algorithms: a survey. *IEEE Computer* **27**, no. 6, 17–26.

Startchik, S., Milanese, R. and Pun, T. (1998). Projective and illumination invariant representation of disjoint shapes. In *Special Issue on Projection-Based Transforms, Image Vision Comput.* **16**, nos. 9–10, 713–723.

Steele, J.M. and Steiger, W.L. (1986). Algorithms and complexity for least median of squares regression. *Discrete Applied Math.* **14**, 93–100.

Stella, E., Lovergine, F.P., D'Orazio, T. and Distante, A. (1995). A visual tracking technique suitable for control of convoys. *Pattern Recogn. Lett.* **16**, no. 9, 925–932.

Stephens, R.S. (1991). Probabilistic approach to the Hough transform. *Image Vision Comput.* **9**, no. 1, 66–71.

Stevens, K. (1980). Surface perception from local analysis of texture and contour. *MIT Artif. Intell. Lab. Memo AI-TR-512.*

Stockman, G.C. and Agrawala, A.K. (1977). Equivalence of Hough curve detection to template matching. *Comm. ACM* **20**, 820–822.

Stone, J.V. (1990). Shape from texture: textural invariance and the problem of scale in perspective images of textured surfaces. *Proc. British Machine Vision Assoc. Conf., Oxford* (*September 24–27*), pp. 181–186.

Straforini, M., Coelho, C. and Campani, M. (1993). Extraction of vanishing points from images of indoor and outdoor scenes. *Image Vision Comput.* **11**, no. 2, 91–99.

Stroustrup, B. (1991) *The C++ progragmming language*. Addison-Wesley, Reading, MA, USA (2nd edition).

Sturm, P. (2000). A case against Kruppa's equations for camera self-calibration. *IEEE Trans. Pattern Anal. Mach. Intell.* **22**, no. 10, 1199–1204.

Sullivan, G.D. (1992). Visual interpretation of known objects in constrained scenes. *Phil. Trans. R. Soc.* (*London*) **B337**, 361–370.

Suzuki, K., Horiba, I. and Sugie, N. (2003). Neural edge enhancer for supervised edge enhancement from noisy images. *IEEE Trans. Pattern Anal. Mach. Intell.* **25**, no. 12, 1582–1596.

Tabandeh, A.S. and Fallside, F. (1986). Artificial intelligence techniques and concepts for the integration of robot vision and 3D solid modellers. *Proc. Int. Conf. on Intell. Autonomous Systems, Amsterdam* (*December 18–11*).

Tan, T.N. (1995). Structure, pose and motion of bilateral symmetric objects. *Proc. 6th British Machine Vision Assoc. Conf., Birmingham* (*September 11–14*), pp. 473–482.

Tan, T.N., Sullivan, G.D. and Baker, K.D. (1994). Recognizing objects on the ground-plane. *Image Vision Comput.* **12**, no. 3, 164–172.

Tang, Y.Y. and You, X. (2003). Skeletonization of ribbon-like shapes based on a new wavelet function. *IEEE Trans. Pattern Anal. Mach. Intell.* **25**, no. 9, 1118–1133.

Tao, W.-B., Tian, J.-W. and Liu, J. (2003). Image segmentation by three-level thresholding based on maximum fuzzy entropy and genetic algorithm. *Pattern Recogn. Lett.* **24**, no. 16, 3069–3078.

Taylor, J. and Coombes, S. (1993). Learning higher order correlations. *Neural Networks* **6**, 423–427

Theil, H. (1950). A rank-invariant method of linear and polynomial regression analysis (parts 1–3). *Nederlandsche Akad. Wetenschappen Proc.* **A53**, 386–392, 521–525 and 1397–1412.

Theodoridis, S. and Koutroumbas, K. (1999). *Pattern Recognition*. Academic Press, London.

Thomas, A.D.H. and Rodd, M.G. (1994). Knowledge-based inspection of electric lamp caps. *Engineering Applications of Artificial Intelligence* **7**, no. 1, 31–37.

Thomas, A.D.H., Rodd, M.G., Holt, J.D. and Neill, C.J. (1995). Real-time industrial visual inspection: a review. *Real-Time Imaging* **1**, no. 2, 139–158.

Tillett, R.D., Onyango, C.M. and Marchant, J.A. (1997). Using model-based image processing to track animal movements. *Comput. Electron. Agric.* **17**, 249–261.

Tissainayagam, P. and Suter, D. (2004). Assessing the performance of corner detectors for point feature tracking applications. *Image Vision Comput.* **22**, no. 8, 663–679.

Toennies, K., Behrens, F. and Aurnhammer, M. (2002). Feasibility of Hough-transform-based iris localisation for real-time application. *Proc. 16th Int. Conf. on Pattern Recogn., Québec, Canada* (*August 11–15*), Vol. II, pp. 1053–1056.

Torr, P.H.C. and Davidson, C. (2003). IMPSAC: synthesis of importance sampling and random sample consensus, *IEEE Trans. Pattern Anal. Mach. Intell.* **25**, no. 3, 354–364.

Torr, P.H.S. and Fitzgibbon, A.W. (2003). Invariant fitting of two view geometry *or* In defiance of the 8 point algorithm. *Proc. British Machine Vision Assoc. Conf., Norwich, UK* (September 9–11), pp. 83–92.

Torr, P.H.S. and Fitzgibbon, A.W. (2004). Invariant fitting of two view geometry. *IEEE Trans. Pattern Anal. Mach. Intell.* **26**, no. 5, 648–650.

Torreão, J.R.A. (2001). A Green's function approach to shape from shading. *Pattern Recogn.* **34**, no. 12, 2367–2382.

Torreão, J.R.A. (2003). Geometric-photometric approach to monocular shape estimation. *Image Vision Comput.* **21**, no. 12, 1045–1062.

Toulson, D.L. and Boyce, J.F. (1992). Segmentation of MR images using neural nets. *Image Vision Comput.* **10**, no. 5, 324–328.

Trucco, E. and Verri, A. (1998). *Introductory Techniques for 3-D Computer Vision.* Prentice-Hall, Upper Saddle River, NJ.

Tsai, D.-M. and Huang, T.-Y. (2003). Automated surface inspection for statistical textures. *Image Vision Comput.* **21**, no. 4, 307–323.

Tsai, D.-M., Hou, H.-T. and Su, H.-J. (1999). Boundary based corner detection using eigenvalues of covariance matrices. *Pattern Recogn. Lett.* **20**, no. 1, 31–40.

Tsai, F.C.D. (1996). A probabilistic approach to geometric hashing using line features. *Comput. Vision Image Understanding* **63**, no. 1, 182–195.

Tsai, R.Y. (1986). An efficient and accurate camera calibration technique for 3D machine vision. *Proc. Conf. on Comput. Vision Pattern Recogn., Miami, FL*, pp. 364–374.

Tsai, R.Y. and Huang, T.S. (1984). Uniqueness and estimation of three-dimensional motion parameters of rigid objects with curved surfaces. *IEEE Trans. Pattern Anal. Mach. Intell.* **6**, 13–27.

Tsang, P.W.M. (2003). Enhancement of a genetic algorithm for affine invariant planar object shape matching using the migrant principle. *IEE Proc. Image Vision Signal Process.* **150**, no. 2, 107–113.

Tsang, P.W.M. and Yu, Z. (2003). Genetic algorithm for model-based matching of projected images of three-dimensional objects. *IEE Proc. Image Vision Signal Process.* **150**, no. 6, 351–359.

Tsuji, S. and Matsumoto, F. (1978). Detection of ellipses by a modified Hough transform. *IEEE Trans. Comput.* **27**, 777–781.

Tsukune, H. and Goto, K. (1983). Extracting elliptical figures from an edge vector field. *Proc. IEEE Conf. on Computer Vision and Pattern Recogn., Washington, DC*, pp. 138–141.

Tuckey, C.O. and Armistead, W. (1953). *Coordinate Geometry*. Longmans, Green and Co. Ltd., London

Turin, G.L. (1960). An introduction to matched filters. *IRE Trans. Inf. Theory* **6**, 311–329.

Turney, J.L., Mudge, T.N. and Volz, R.A. (1985). Recognizing partially occluded parts. *IEEE Trans. Pattern Anal. Mach. Intell.* **7**, 410–421.

Tytelaars, T., Turina, A. and van Gool, L. (2003). Noncombinatorial detection of regular repetitions under perspective skew. *IEEE Trans. Pattern Anal. Mach. Intell.* **25**, no. 4, 418–432.

Tzionas, P. (2000). A cellular automaton processor for line and corner detection in gray-scale images. *Real-Time Imaging* **6**, no. 6, 461–470.

Uhr, L. (1978). Recognition cones and some test results; the imminent arrival of well-structured parallel-serial computers; positions, and positions on positions. In Hanson, A.R. and Riseman, E.M. (eds.), *Computer Vision Systems*. Academic Press, New York, pp. 363–377.

Uhr, L. (1980). Psychological motivation and underlying concepts. In Tanimoto, S. and Klinger, A. (eds.) *Structured Computer Vision–Machine Perception through Hierarchical Computation Structures*. Academic Press, London, pp. 1–30.

Ullman, S. (1979). *The Interpretation of Visual Motion*. MIT Press, Cambridge, MA.

Ullmann, J.R. (1969). Experiments with the *n*-tuple method of pattern recognition. *IEEE Trans. Comput.* **18**, 1135–1137.

Ullmann, J.R. (1973). *Pattern Recognition Techniques*. Butterworth, London.

Ullmann, J.R. (1974). Binarisation using associative addressing. *Pattern Recogn.* **6**, 127–135.

Ullmann, J.R. (1976). An algorithm for subgraph isomorphism. *J. ACM* **23**, 31–42.

Umeyama, S. (1988). An eigen decomposition approach to weighted graph matching problems. *IEEE Trans. Pattern Anal. Mach. Intell.* **10**, no. 5, 695–703.

Unger, S.H. (1958). A computer orientated towards spatial problems. *Proc. IRE* **46**, 1744–1750.

Unser, M. (1986). Local linear transforms for texture measurements. *Signal Process.* **11**, 61–79.

Unser, M. and Ade, F. (1984). Feature extraction and decision procedure for automated inspection of textured materials. *Pattern Recogn. Lett.* **2**, no. 3, 185–191.

Unser, M. and Eden, M. (1989). Multiresolution feature extraction and selection for texture segmentation. *IEEE Trans. Pattern Anal. Mach. Intell.* **11**, no. 7, 717–728.

Unser, M. and Eden, M. (1990). Nonlinear operators for improving texture segmentation based on features extracted by spatial filtering. *IEEE Trans. Systems Man Cybern.* **20**, no. 4, 804–815.

Urriza, I., Artigas, J.I., Barragan, L.A., Garcia, J.I. and Navarro, D. (2001). VLSI implementation of discrete wavelet transform for lossless compression of medical images. *Real-Time Imaging* **7**, no. 2, 203–217.

Vaillant, R., Monrocq, C. and Le Cun, Y. (1994). Original approach for the localisation of objects in images. *IEE Proc. Vision Image Signal Process.* **141**, no. 4, 245–250.

Van Gool, L., Dewaele, P. and Oosterlinck, A. (1985). Survey: Texture analysis anno 1983. *Comput. Vision Graph. Image Process.* **29**, 336–357.

van Digellen, J. (1951). Photometric investigations of the slopes and heights of the ranges of hills in the Maria of the moon. *Bull. Astron. Inst. Netherlands* **11**, 283–289.

van Dijck, H. and van der Heijden, F. (2003). Object recognition with stereo vision and geometric hashing. *Pattern Recogn. Lett.* **24**, nos. 1–3, 137–146.

Van Gool, L., Proesmans, M. and Zisserman, A. (1998). Planar homologies as a basis for grouping and recognition. In *Special Issue on Geometric Modelling and Invariants for Computer Vision. Image Vision Comput.* **16**, no. 1, 21–26.

van Wyk, M.A., Durrani, T.S. and van Wyk, B.J. (2002). A RKHS interpolator--based graph matching algorithm. *IEEE Trans. Pattern Anal. Mach. Intell.* **24**, no. 7, 988–995.

VanderBrug, G.J. and Rosenfeld, A. (1977). Two-stage template matching. *IEEE Trans. Comput.* **26**, 384–393.

Vapnik, V.N. (1998). *Statistical Learning Theory*. Wiley, New York.

Veenman, C.J., Reinders, M.J.T. and Backer, E. (2002). A maximum variance cluster algorithm. *IEEE Trans. Pattern Anal. Mach. Intell.* **24**, no. 9, 1273–1280.

Vega, I.R. and Sarkar, S. (2003). Statistical motion model based on the change of feature relationships: human gait-based recognition. *IEEE Trans. Pattern Anal. Mach. Intell.* **25**, no. 10, 1323–1328.

Vilnrotter, F.M., Nevatia, R. and Price, K.E. (1986). Structural analysis of natural textures. *IEEE Trans. Pattern Anal. Mach. Intell.* **8**, no. 1, 76–89.

Vistnes, R. (1989). Texture models and image measures for texture discrimination. *Int. J. Comput. Vision* **3**, 313–336.

von Helmholtz, H. (1925). *Treatise on Physiological Optics*, Volumes 1–3 (translated by Southall, J.P.C.). Dover, New York.

Voss, R. (1986). Random fractals: characterization and measurement. In Pynn, R. and Skjeltorp, A. (eds.), *Scaling Phenomena in Disordered Systems*, Plenum, New York.

Wagner, G.G. (1988). Combining X-ray imaging and machine vision. *Proc. SPIE* **850**, 43–53.

Waltz, D. (1975). Understanding line drawings of scenes with shadows. In Winston, P. H. (ed.), *The Psychology of Computer Vision*. McGraw-Hill, New York, pp. 19–91.

Wang, C., Sun, H., Yada, S. and Rosenfeld, A. (1983). Some experiments in relaxation image matching using corner features. *Pattern Recogn.* **16**, 167.

Wang, J.-G. and Sung, E. (2001). Gaze determination via images of irises. *Image Vision Comput.* **19**, no. 12, 891–911.

Wang, J.-G., Sung, E. and Venkateswarlu, R. (2003). Determining pose of a human face from a single monocular image. *Proc. British Machine Vision Assoc. Conf., Norwich, UK* (*September 9–11*), pp. 103–112.

Wang, L. and Bai, J. (2003). Threshold selection by clustering gray levels of boundary. *Pattern Recogn. Lett.* **24**, no. 12, 1983–1999.

Wang, S. and Haralick, R.M. (1984). Automatic multithreshold selection. *Comput. Vision Graph. Image Process.* **25**, 46–67.

Wang, S. and Siskind, J.M. (2003). Image segmentation with ratio cut. *IEEE Trans. Pattern Anal. Mach. Intell.* **25**, no. 6, 675–690.

Wang, T., Zhuang, X. and Xing, X. (1992). Robust segmentation of noisy images using a neural network model. *Image Vision Comput.* **10**, no. 4, 233–240.

Wang, T.W. and Evans, J.P.O. (2003). Stereoscopic dual-energy X-ray imaging for target materials identification. *IEE Proc. Vision, Image and Signal Process.* **150**, no. 2, 122–130.

Webb, A. (2002). *Statistical Pattern Recognition*. Wiley, Chichester, UK.

Weiman, C.F.R. (1976). Highly parallel digitised geometric transformations without matrix multiplication. *Proc. Int. Joint Conf. on Parallel Processing*, pp. 1–10.

Weimer, P.K. (1975). From camera tubes to solid-state sensors. *RCA Rev.* **36**, 385–405.

Werbos, P.J. (1974). *Beyond Regression: New Tools for Prediction and Analysis in the Behavioral Sciences*, Ph.D. Thesis, Harvard Univ., Cambridge, MA.

Wermser, D. and Liedtke, C.-E. (1982). Texture analysis using a model of the visual system. *Proc. 6th Int. Conf. on Pattern Recogn., Munich (October 19–22)*, pp. 1078–1081.

Wermser, D., Haussmann, G. and Liedtke, C.-E. (1984). Segmentation of blood smears by hierarchical thresholding. *Comput. Vision Graph. Image Process.* **25**, 151–168.

Weska, J.S. (1978). A survey of threshold selection techniques. *Comput. Graph. Image Process.* **7**, 259–265.

Weska, J.S. and Rosenfeld, A. (1976). An application of texture analysis to materials inspection. *Pattern Recogn.* **8**, 195–199.

Weska, J.S. and Rosenfeld, A. (1979). Histogram modification for threshold selection. *IEEE Trans. Systems Man Cybern.* **9**, 38–52.

Weska, J.S., Nagel, R.N. and Rosenfeld, A. (1974). A threshold selection technique. *IEEE Trans. Comput.* **23**, 1322–1326.

Weszka, J.S., Dyer, C.R. and Rosenfeld, A. (1976). A comparative study of texture measures for terrain classification. *IEEE Trans. Systems Man Cybern.* **6**, no. 4, 269–285.

Whelan, P.F. and Molloy, D. (2001). *Machine Vision Algorithms in Java*. Springer, New York.

White, J.M. and Rohrer, G.D. (1983). Image thresholding for optical character recognition and other applications requiring character image extraction. *IBM J. Res. Dev.* **27**, 400–411.

Wiehler, K., Heers, J., Schnörr, C., Stiehl, H.S. and Grigat, R.-R. (2001). A one-dimensional analogue VLSI implementation for nonlinear real-time signal preprocessing. *Real-Time Imaging* **7**, no. 1, 127–142.

Wiejak, J.S., Buxton, H. and Buxton, B.F. (1985). Convolution with separable masks for early image processing. *Comput. Vision Graph. Image Process.* **32**, 279–290.

Will, P.M. and Pennington, K.S. (1971). Grid coding: a preprocessing technique for robot and machine vision. *Artif. Intell.* **2**, 319–329.

Wilson, H.R. and Giese, S.C. (1977). Threshold visibility of frequency gradient patterns. *Vision Res.* **17**, 1177–1190.

Wilson, R.C. and Hancock, E.R. (1997). Structural matching by discrete relaxation. *IEEE Trans. Pattern Anal. Mach. Intell.* **19**, no. 6, 634–648.

Winston, P.H. (ed.) (1975). *The Psychology of Computer Vision*. McGraw-Hill, New York.

Witkin, A.P. (1981). Recovering surface shape and orientation from texture. *Artif. Intell.* **17**, 17–45.

Witkin, A.P. (1983). Scale-space filtering. *Proc. 4th Int. Joint. Conf. on Artif. Intell. Tbilisi, Georgi, USSR*, pp. 1019–1022.

Wolfson, H.J. (1991). Generalizing the generalized Hough transform. *Pattern Recogn. Lett.* **12**, no. 9, 565–573.

Wong, R.Y. and Hall, E.L. (1978). Scene matching with invariant moments. *Comput. Graph. Image Process.* **8**, 16–24.

Woodham, R.J. (1978). Reflectance map techniques for analysing surface defects in metal castings. *MIT Artif. Intell. Lab. Memo AI-TR-457.*

Woodham, R.J. (1980). Photometric method for determining surface orientation from multiple images. *Opt. Eng.* **19**, 139–144.

Woodham, R.J. (1981). Analysing images of curved surfaces. *Artif. Intell.* **17**, 117–140.

Woods, J.W. (1972). Two-dimensional discrete Markovian fields. *IEEE Trans. Inf. Theory* **18**, 232–240.

Worrall, A.D., Baker, K.D. and Sullivan, G.D. (1989). Model based perspective inversion. *Image Vision Comput.* **7**, 17–23.

Wu, A.Y., Hong, T.-H. and Rosenfeld, A. (1982). Threshold selection using quadtrees. *IEEE Trans. Pattern Anal. Mach. Intell.* **4**, 90–94.

Wu, H., Yoshikawa, G., Shioyama, T., Lao, S. and Kawade, M. (2002). Glasses frame detection with 3D Hough transform. *Proc. 16th Int. Conf. on Pattern Recogn., Québec, Canada* (*August 11–15*), Vol. II, pp. 346–349.

Wu, Z.-Q. and Rosenfeld, A. (1983). Filtered projections as an aid in corner detection. *Pattern Recogn.* **16**, 31–38.

Xie, Y. and Ji, Q. (2002). A new efficient ellipse detection method. *Proc. 16th Int. Conf. on Pattern Recogn., Québec, Canada* (*August 11–15*), Vol. II, pp. 957–960.

Xu, L. and Oja, E. (1993). Randomized Hough transform (RHT): basic mechanisms, algorithms, and computational complexities. *Computer Vision Graph. Image Process.: Image Understanding* **57**, no. 2, 131–154.

Yan, C., Sang, N. and Zhang, T. (2003). Local entropy-based transition region extraction and thresholding. *Pattern Recogn. Lett.* **24**, no. 16, 2935–2941.

Yang, G.J. and Huang, T.S. (1981). The effect of median filtering on edge location estimation. *Comput. Graph. Image Process.* **15**, 224–245.

Yang, J. and Li, X. (1995). Boundary detection using mathematical morphology. *Pattern Recogn. Lett.* **16**, no. 12, 1277–1286.

Yang, J.-D., Chen, Y.-S. and Hsu, W.-H. (1994). Adaptive thresholding algorithm and its hardware implementation. *Pattern Recogn. Lett.* **15**, no. 2, 141–150.

Yang, M.-T., Gandhi, T., Kasturi, R., Coraor, L., Camps, O. and Candless, J. (2002). Real-time implementation of obstacle detection algorithms on a Datacube MaxPCI architecture. *Real-Time Imaging* **8**, no. 2, 157–172.

Yazdi, H.R., and King, T.G. (1998). Application of "Vision in the Loop" for inspection of lace fabric. In Davies, E.R., and Ip, H.H.S., (eds.) (1998), pp. 317–332.

Yin, L., Astola, J. and Neuvo, Y. (1993). A new class of nonlinear filters—neural filters. *IEEE Trans. Signal Process.* **41**, no. 3, 1201–1222.

Yitzhaky, Y. and Peli, E. (2003). A method for objective edge detection evaluation and detector parameter selection. *IEEE Trans. Pattern Anal. Mach. Intell.* **25**, no. 8, 1027–1033.

Yoon, J.J., Koch, C. and Ellis, T.J. (2002). Shadowflash: an approach for shadow removal in an active illumination environment. *Proc. British Machine Vision Assoc. Conf., Cardiff, UK* (*September 2–5*), pp. 636–645.

Yuen, H.K., Illingworth, J. and Kittler, J. (1988). Ellipse detection using the Hough transform. *Proc. 4th Alvey Vision Conf., Manchester* (*August 31–September 2*), pp. 265–271.

Yuen, H.K., Princen, J., Illingworth, J. and Kittler, J. (1989). A comparative study of Hough transform methods for circle finding. *Proc. 5th Alvey Vision Conf., Reading* (*September 25–28*), pp. 169–174.

Yuille, A. and Poggio, T.A. (1986). Scaling theorems for zero crossings. *IEEE Trans. Pattern Anal. Mach. Intell.* **8**, 15–25.

Yuille, A.L., Kammen, D.M. and Cohen, D.S. (1989). Quadrature and the development of orientation selective cortical cells by Hebb rules. *Biol. Cybern.* **61**, 183–194.

Zahn, C.T. and Roskies, R.Z. (1972). Fourier descriptors for plane closed curves. *IEEE Trans. Comput.* **21**, 269–281.

Zayas, I.Y. and Flinn, P.W. (1998). Detection of insects in bulk wheat samples with machine vision. *Trans. Amer. Soc. Agric. Eng.* **41**, no. 3, 883–888.

Zayas, I.Y. and Steele, J.L. (1990). Image analysis applications for grain science. *Optics in Agric., SPIE* **1379**, 151–161.

Zhang, G. and Wei, Z. (2003). A position-distortion model of ellipse centre for perspective projection. *Measurement Sci. Technol.* **14**, 1420–1426.

Zhang, J. and Modestino, J.W. (1990). A model-fitting approach to cluster validation with application to stochastic model-based image segmentation. *IEEE Trans. Pattern Anal. Mach. Intell.* **12**, no. 10, 1009–1017.

Zhang, Z. (1995). Motion and structure of four points from one motion of a stereo rig with unknown extrinsic parameters. *IEEE Trans. Pattern Anal. Mach. Intell.* **17**, no. 12, 1222–1227.

Zheng, Z., Wang, H. and Teoh, E.K. (1999). Analysis of gray level corner detection. *Pattern Recogn. Lett.* **20**, no. 2, 149–162.

Zhou, H., Wallace, A.M. and Green, P.R. (2003). A multistage filtering technique to detect hazards on the ground plane. *Pattern Recogn. Lett.* **24**, 1453–1461.

Zhuang, X. and Haralick, R.M. (1986). Morphological structuring element decomposition. *Comput. Vision Graph. Image Process.* **35**, 370–382.

Zielke, T., Braukermann, M. and von Seelen, W. (1993). Intensity and edge-based symmetry detection with an application to car-following. *Comput. Vision Graph. Image Process.: Image Understanding.* **58**, no. 2, 177–190.

Zisserman, A., Marinos, C., Forsyth, D.A., Mundy, J.L. and Rothwell, C.A. (1990). Relative motion and pose from invariants. *Proc. 1st British Machine Vision Assoc. Conf., Oxford* (*September 24–27*), pp. 7–12.

Zucker, S.W. (1976a). Toward a model of texture. *Comput. Graph. Image Process.* **5**, 190–202.

Zucker, S.W. (1976b). Region growing: childhood and adolescence. *Comput. Graph. Image Process.* **5**, 382–399.

Zuniga, O.A. and Haralick, R.M. (1983). Corner detection using the facet model. *Proc. IEEE Comput. Vision Pattern Recogn. Conf.*, pp. 30–37.

Zuniga, O.A. and Haralick, R.M. (1987). Integrated directional derivative gradient operator. *IEEE Trans. Pattern Anal. Mach. Intell.* **17**, 508–517.

Author Index

Subject Index